D0933695

TIME MACHINES

Second Edition

TIME MACHINES

Time Travel in Physics, Metaphysics, and Science Fiction

Second Edition

Paul J. Nahin

Foreword by Kip S. Thorne

with 53 Illustrations

Springer

Paul J. Nahin
University of New Hampshire
Durham, N.H. 03824
USA

Library of Congress Cataloging-in-Publication Data
Nahin, Paul J.
    Time machines : time travel in physics, metaphysics, and science
fiction / Paul J. Nahin. — 2nd ed.
        p. cm.
    Includes bibliographical reference (p.   ) and index.
    ISBN 0-387-98571-9 (pbk. : alk. paper)
    1. Science fiction, American—History and criticism.   2. Science
fiction, English—History and criticism.   3. Time travel in
literature.   4. Metaphysics in literature.   5. Physics in
literature.   I. Title.
PS374.S35N34   1998
813′.0876209384—dc21   98-25145

Printed on acid-free paper.

Production managed by Lesley Poliner; manufacturing supervised by Joseph Quatela.
Typeset by Impressions Book and Journal Services, Inc., Madison, WI.
Printed and bound by R.R. Donnelley and Sons, Harrisonburg, VA.
Printed in the United States of America.

9 8 7 6 5 4 3 2 (Corrected second printing, 2001)

ISBN 0-387-98571-9   SPIN 10833772

Springer-Verlag   New York Berlin Heidelberg
A member of BertelsmannSpringer Science+Business Media GmbH

Like the original *Time Machines,*
this book is for my loving wife, Patricia Ann, whose journey
through space-time has not often wandered far from my own—
and who has for decades quietly tolerated huge, teetering piles
of books and papers in her otherwise immaculate home,
and
for my splendid cat Heaviside,
who kept me company on many a quiet night as I struggled
to understand time travel and who, like Fritz Leiber's
supercat Gummitch—IQ of 160 and planning to write a book
on spacetime—often gets a look in her eyes that says she knows
more about it than I ever will,
and, new for the second edition of *Time Machines,*
for my equally sweet, polydactyl cat Maxwell, who,
now that Heaviside has gotten sufficiently lazy and fat that she'd
rather go to bed early, has taken over the jobs of sniffing the
laptop screen while I write and now and then taking a swipe
at the printer with a huge paw.

There is only one kind of Science Fiction story that I dislike, and that is the so-called time-traveling. It doesn't seem logical to me. For example: supposing a man had a grudge against his grandfather, who is now dead. He could hop in his machine and go back to the year that his grandfather was a young man and murder him. And if he did this how could the revenger be born? I think the whole thing is the "bunk."

> —Letter to the Editor from a seventeen-year-old reader of *Astounding Stories* (December 1931)

"I'm not kidding you at all Phil," Barney insisted. "I *have* produced a workable Time Machine, and I am going to use it to go back and kill my grandfather."

> —opening line from "A Gun for Grandfather" (1987) by F.M. Busby

"The past—it's pretty damn solid, Phil. It's a little like a compost pile—fairly soft near the surface but packed hard further down, with all that Time piled on top of it."

> —Barney tells Phil about his astonishingly unique insight concerning the organic nature of spacetime

"Whether or not the laws of physics permit the existence of closed timelike curves (time machines) is one of the important problems in the research field of modern general relativity."

> —Chinese theoretical physicist Li-Xin Li (1996)

Calvin and Hobbes

by Bill Watterson

Time travel was once solely the province of science fiction writers. Serious scientists avoided it like the plague—except when writing fiction under pseudonyms or reading it in privacy.

How times have changed! One now finds scholarly analyses of time travel in serious scientific journals, written by eminent theoretical physicists such as John Friedman, Stephen Hawking, and Igor Novikov. Why the change? Because we physicists have realized that the nature of time is too important an issue to be left solely in the hands of science fiction writers. If we could understand time deeply, then that understanding would produce breakthroughs in our comprehension of the universe. If the fundamental laws of physics permit time travel, even just on subatomic scales, then our present understanding of quantum mechanics is flawed in ways that would explain how information gets lost down black holes—and ways that may have profoundly affected the birth of the universe.

Some science fiction writers are smart. So are some physicists. Smart physicists seek insight everywhere, including from clever science fiction writers who long ago began probing seriously the logical consequences that would ensue if the laws of physics permitted time travel.

For example, Igor Novikov's Principle of Self Consistency (his conjecture on how nature may enable the laws of physics to accommodate time travel) has its roots in science fiction. We physicists presumably could have arrived at this principle on our own, without the aid of science fiction, but our familiarity with science fiction's grandfather paradox in fact was a powerful motivation for the principle. Similarly, thought experiments based on science fiction gave rise to my own realization that if the laws of physics permit wormholes, then the relative motion of a wormhole's two mouths may convert it into a time machine.

Of course, this interplay of science fiction and serious physics has not been a one-way street. I fed the concept of a wormhole to Carl Sagan for use in his novel *Contact*, and from there wormholes made their way into "Star Trek" and human culture. Larry Niven, Bob Forward, and other science fiction writers have plumbed the minds of theoretical physicists (including my Ph.D. students at Caltech) in search of ideas for their novels.

The previous edition of Paul Nahin's book *Time Machines* was the first serious attempt I have seen to document the interplay between science and science fiction in the realm of time travel. It also was the most thorough compendium ever written on time travel in science fiction. Dr. Nahin's first edition was written soon after serious scientific research on time travel became respectable among physicists. In the intervening six years, such research has burgeoned. This second edition of Nahin's book has been revised accordingly. It now is not only the most comprehensive documentation of time travel in science fiction; it is also the most thorough review of serious scientific literature on the subject—a review that, remarkably, is scientifically accurate and at the same time largely accessible to a broad audience of non-specialists.

In browsing through this revised edition, I have been struck by the richness and complexity of the tapestry of ideas that Nahin presents. His interweaving of physics and fiction is done adeptly and keeps the book flowing.

As a theoretical physicist, I keep wanting Dr. Nahin to boil the complexity down into a few key ideas and principles, from which all the rest follows. Were I writing such a book, I would do so. But then, inevitably, the book would become not an overview of all the interwoven time travel themes in fiction and science, but instead an overview of my own personal conjectures about how nature really fits them together—my speculations about what the laws of physics really regard as important and true and what those laws deem secondary or false. Because we physicists are still at a rather early stage in our struggle to understand the nature of time and the attitudes of fundamental physical law toward time travel, it is quite appropriate (albeit overwhelming) for Nahin to weave the full tapestry even-handedly. Like a good journalist, Nahin simply reports what he sees in the physics and science fiction literature, commenting lucidly and often pointedly on the interconnections, contradictions, and controversies but leaving it to his readers to form their own final judgments.

It may well be that by the time the fifth edition of this book is published, we physicists will finally have mastered the laws of quantum gravity (which control the nature of time) and will have learned their implications for time travel. *I think* we will have discovered that there *is* a universal "chronology protection" mechanism that forbids time travel on macroscopic, human scales but that time travel is essential to the fabric of spacetime on subatomic, "Planck-length" scales and thereby has a profound influence on the fundamental laws of physics. However, until the day I am proved right (or wrong), Nahin's book, with its complex tapestry of ideas and possibilities, may well remain the most readable and thorough treatise on time travel in science and science fiction.

Kip S. Thorne
The Feynman Professor of Theoretical Physics
California Institute of Technology
April 1998

WHAT'S NEW IN THE SECOND EDITION

> My [works] contain twenty thousand matters worthy of attention . . . gathered from some two thousand books . . . yet I cannot doubt that I too have omitted much, for I am but a man. . . .
> —from Pliny the Elder's *Historia Naturalis* (A.D. 77)

The second edition of *Time Machines* is a corrected, revised, and expanded edition of the original *Time Machines* published in 1993. The basic structure of the book is unchanged, but now that I've upgraded my word processing software to include a spell-checker, there are a lot fewer misspelled words. I have also streamlined my citation notation to the published literature so as to be far less harsh on the eye. In addition, new material includes (1) discussions of additional stories and movies that I missed or that have appeared since 1993 (but see the "Bibliographic Adieu"); (2) discussions of the physics literature that has appeared since 1993; (3) several new illustrations; (4) two new Tech Notes, one on the connection between time and gravity and one on the cosmic string time machine invented by Gott at Princeton; (5) a greatly expanded Tech Note on wormholes and wormhole time machines, because this particular form of time machine continues to receive intense study; (6) a glossary of important terms, an addition prompted by many of the numerous e-mail suggestions I received from all over the world on how to improve *Time Machines*. Finally, the biggest change of all has been the acquisition of the book by Springer-Verlag/New York from the American Institute of Physics, where it is now being published under a dual Springer/AIP Press Logo. I was most fortunate that not only my book but also my AIP editor, Maria Taylor, went to Springer. I would be lost without Maria.

In addition to my remarks in the original Acknowledgments, I wish first to thank the well-known authority on the spacetime physics of wormholes, Matt Visser at Washington University, St. Louis. Professor Visser reviewed the entire manuscript for Springer-Verlag and provided me with a long, detailed, no-nonsense commentary. Some of his remarks were highly critical, but all were written in the spirit of constructive help and I am most grateful. It should go without saying (but I'll say it anyway!) that any remaining errors of fact or expressions of (perhaps) outrageous personal opinion are totally my doing. Professor Visser gave me my chance, and if I have missed or ignored any of his warnings, it isn't *his* fault. I also want to thank my University of New Hampshire freshman general education honors seminar class in the fall semester of 1993. Those students used *Time Machines* as the primary text, and many of their insights and suggestions have been incorporated in this edition of *Time Machines*. My production editor at Springer, Lesley Poliner, and my copyeditor Connie Day have both been wonderful to work with, and the physical creation of this book owes much to their considerable efforts. Finally, as I have noted, I received communications from all around the planet concerning the first edition of *Time Machines*. Almost without exception, people wrote to tell me of their pleasure in at last finding a book on time travel that was readable without requiring a Ph.D. in mathematical physics, and yet was not one of those embarrassingly silly books that appear from time to time claiming the government really does have a time machine and—like UFOs and space aliens—is covering it all up.

But even more gratifying, nearly all of my correspondents *did* have a suggestion or two (or three) on how I could have done better ("You missed the following wonderful story. . . ," or "You overlooked the following physics paper. . . ," or "You really should get a dictionary. . . ," etc., etc.) And, of course, nearly all those suggestions were right! So, to all who wrote to me, my thanks, and I want you to know that I have indeed carefully considered your comments. To my chagrin, my sharp-eyed readers were so numerous that I simply cannot list everybody who wrote. But their correspondence did not go unheeded—I read every e-mail and letter, and I almost always learned something new. There were also three issues that were repeatedly raised in that e-mail that I would like to respond to here.

1. *Do I personally believe in time travel to the past?* I believe in physics, so my scientific positions are subject to change as we learn more physics. Indeed, that's precisely why time travel is interesting even to physicists who can't stand the idea (such as Hawking and Visser). Right now known physics doesn't forbid time travel, and such physicists want to discover the new physics that they believe will finally forbid time machines.

2. *Why do I emphasize the principle of self-consistency in the book and write without enthusiasm about splitting universes and multiple time tracks?* Because those are my personal views. I could, of course, be wrong. (But if you want to see "authority" cited, then take a look at note 46 for Chapter 4.) I feel I have done my scholarly duty by devoting space to those ideas; it isn't necessary that I remain neutral. If *you* like splitting universes, however, that's okay with me. You might even be right!

3. *When I criticize a particular science fiction story as being illogical, does that mean I think the author stupid and all of his work incompetent?* No, it just means I think that *particular* story is scientifically flawed. That doesn't mean it isn't a good story. Good stories don't have to be true, but good physics does. The mark of an unsophisticated reader is a confusion of what an author's story character says with what the author believes in real life. Some very good physicists have written what I consider scientifically flawed stories, but I'm willing to bet a good meal that those stories were written to entertain, not to teach anybody physics. The second word in *science fiction* is important!

Now, with all that said and mindful of Pliny's ancient words, I look forward to receiving even more mail about *Time Machines,* Second edition. You can write to me at paul.nahin@unh.edu.

Paul J. Nahin
Durham, New Hampshire
April 1998

> Time travel is the thinking person's UFO, an improbability
> that nevertheless resonates with mysterious and sometimes
> marvelous possibilities.
> —Richard Schickel, from his film review in *Time*
> (December 4, 1989) of *Back to the Future, Part II*

As a youngster, at about age ten, I discovered science fiction stories.
I read them, *devoured* them (to the great distress of my teachers), because the mind-expanding ideas of that tiny, often unfairly maligned
corner of literature excited my imagination. The stories of alien in-
telligences, parallel universes, voyages to other worlds, robots, the
fourth dimension, paranormal powers, and intergalactic wars—and
of course *all the gadgets*—were wonderful. BEMs (bug-eyed monsters) and telepathic mutants, beings that often populated the same
stories (particularly those about radiation-blasted, post-nuclear-war
societies), were fun too. Still, it was the gadgetry of science fiction
that most fascinated my pre-teenage mind. I loved reading about
faster-than-light hyperspace rocket ships, invisibility and tractor
beams, supercomputer brains, Dirac radios, matter transmitters, instantaneous language translators, disintegrator ray guns, and time
machines. Ah, those time machines. How I loved time travel stories!

I read comic books a lot then, too. When Superman solved a
problem by traveling into the past by flying faster than light or
when Walt Disney's nutty inventory Gyro Gearloose invented a

time machine to deliver a letter that just had to get there yesterday—well, that was pretty neat. I was a science-fiction-contraption junky. Time machines offered a special thrill that made them king of all the gadgets, however, and any story with a time machine had me hooked. And if an author made brain-boggling logical paradoxes a critical part of his tale, then my happiness on Earth was complete. If there were causal loops, and characters running into themselves, and poor unfortunates unwittingly changing the past (or even killing their youthful grandfathers so that they caused themselves never to have existed)—for that I'd risk even the terrible wrath of my parents, who wanted me asleep in bed by ten, not reading a book by flashlight under the covers.

Mom and Dad subscribed to the view expressed by Vladimir Voinovich's time traveler in the 1986 novel *Moscow 2042,* who declared, "Science fiction . . . is not literature, but tomfoolery like the electronic games that induce mass idiocy." And there *is* a lot of nonsense in science fiction, sure, but there's a lot of nonsense in every genre—the point is to be discriminating.[1] Even in those long-ago days of the early 1950s, I sensed that there is much that is very good, indeed, in science fiction.

So science fiction was what I liked, but I also liked science and mathematics, and unlike some of my high school friends who also read science fiction but who liked football and cars better than algebra, I went to college and majored in electrical engineering. And now I teach electrical engineering and rarely read science fiction for recreation. Still, even as I prepare my lectures, sit in department meetings, and sleep through faculty senate debates on how outrageous the new university parking fee is, I find that my mind is still in love with time travel and time machines. I am now forty years and three engineering degrees removed from that technically naive teenage boy of the early 1950s, but the fascination with time travel (or "chronomotion") is as strong as ever. What I wouldn't give to be a chronoviator!

I still remember the surprise and joy I felt when I learned that Kurt Gödel, among the twentieth century's greatest mathematicians, had in 1949 actually published solutions to Einstein's general relativity gravitational field equations that appear to suggest the possibility of time travel to the past. It has been known since 1905 that Einstein's special theory of relativity allows travel into the future, but to return had been thought impossible—impossible both physically and philosophically. After all, what sense could there be to the classic paradox of a time traveler going back in time to kill his grandfather as a baby? How then could the time traveler come to be born? But Gödel's mathematics seemed to be saying that time travel to the past does make some sort of sense! Could it be, I recall wondering in amazement, that what I had reluctantly come to admit could be only a fantastic dream *might really be possible?*

Today there are individual scientists all over the world seriously studying the physics and the philosophical implications of time travel. The two principal players are Kip Thorne at the California Institute of Technology and Igor Novikov at the Academy of Sciences Astro Space Centre of the P.N. Lebedev Physical Institute in Moscow. These two men have also developed groups of advanced graduate students who work on various aspects of time machines and time travel physics for their dissertation topics. It is to be expected that as these students receive their degrees and move on to their own teaching and research positions, the level of scientific activity in this exciting new area of physics will continue to grow worldwide.

Realistically, however, Gödel's astonishing theoretical discovery and the subsequent work of Thorne and Novikov notwithstanding, I suspect many physicists would still agree with words published over fifty years ago in the British magazine *Tales of Wonder* (I.O. Evans, "Can We Conquer Time?" Summer 1940): "Of all the fantastic ideas that belong to science fiction, the most remarkable—and, perhaps, the most fascinating—is that of time travel. . . . Indeed, so fantastic a notion does it seem, and so many apparently obvious absurdities and bewildering paradoxes does it present, that some of the most imaginative students of science refuse to consider it as a practicable proposition." Still, not all physicists view the time-travel-paradox arguments as convincing. Provocative, yes, of course, but many are not yet prepared to write "signed, sealed, and delivered" at the end.

The theological side of this book may strike some readers as a bit odd for a topic usually treated with heavy doses of thick mathematics, but I have learned that theology is a necessary dimension to any informed discussion on time travel. A literary linkage between time travel and theology has, in fact, existed for a long time. As pointed out in Paul Alkon's *Origins of Futuristic Fiction,* "The first time-traveler in English literature is a guardian angel who returns with state documents from 1998 to the year 1728 in Samuel Madden's *Memoirs of the Twentieth Century*" (published in 1733, more than two-and-a-half centuries ago). Madden was an Irish–Anglican clergyman whose book was satire rather than science fiction, but its time-traveling aspect was a first. As Alkon also points out, "Madden [was] the first to write a narrative that purports to be a document from the future. He deserves recognition as the first to toy with the rich idea of time-travel in the form of an artifact sent backward from the future to be discovered in the present."

Much has happened in the area of time travel studies since Madden's time. I therefore set myself two goals as I wrote this book. First, I attempted to find and read every time-travel story ever anthologized in the English language [and many others in their original (and only) magazine appearances]. I have also examined the popular-culture treatments of time travel in the comics and in the movies. From these sources I extracted the central fictional speculations on what time travel might be like and how the logical time travel paradoxes were treated. This book demonstrates that the fascination with rational time travel is not a modern idea, but rather one that can be traced back to the mid-eighteenth century. One conclusion that is evident from this book is that even the very early science fiction writers tackled conceptual issues now taken quite seriously by philosophers and physicists. Science fiction writers have been enormously inventive with the concept of time travel (some of the ideas appearing now in the philosophical literature appeared decades ago in science fiction), and I have attempted to weave their more startling ideas into the text. As Robert Silverberg wrote in the "Introduction" to his 1977 anthology *Trips in Time,* "The only workable time-machine ever invented is the science-fiction story." And where else but in a time-travel story would Einstein's gravitational field equations appear as the playtime chant of a bright child of the future (in Fritz Leiber's "Nice Girl with 5 Husbands")?

As for my second goal, I have discussed, using prose only, the physical theory of time travel, including Einstein's special and general theories of relativity. For those who also appreciate a more mathematical approach, however, nine Tech Notes appear as appendices, but I strongly emphasize that it is *not* necessary to understand calculus (or even algebra) to read this book. Even the Tech Notes themselves, which contain all the mathematics in this book, can be usefully read simply for their historical content.

What all this means is that one does not have to be either a science fiction fan or a mathematical whiz to read this book. The only requirement is to have enough curiosity and imagination to be fascinated by one of the oldest of fantasies—a fantasy that at least some of the best physicists in the world are now coming to think may not be so fantastic as to be impossible. This book does not advocate one view or another on time travel, although I do of course have my personal opinions that I am not shy in expressing. However, when I do offer a personal opinion, I have tried to be careful to state it as such.

The book contains a selection of illustrations concerned with time travel and the fourth dimension from early science fiction magazines and contemporary pop-culture art, and also a number of line illustrations in support of the technical discussions. The published literature cited includes over 2100 science fiction stories and papers from the professional journals in physics, mathematics, electrical engineering, philosophy, and theology. I believe this careful citation of sources, and the extensive bibliography, will make the book of interest and value to scholars as well as to science fiction fans (and other fiction fans who enjoy stories with intellectual content that are "good reads").

The book is structured with seven major components.

1. Chapter 1, an overview of the entire book, touches on a wide assortment of topics, including the history of time travel in fiction, philosophical skepticism, the nature of time, black holes, time-travel paradoxes and their historical development, Gödel's time travel universe, and Frank Tipler's astonishing discovery of "how to make" a time machine.

2. Chapter 2 introduces the fundamental concept of four-dimensional spacetime, hyperspaces, the connection of each to time travel, and the uses (good *and* bad) by science fiction writers of these concepts in time travel stories.

3. Chapter 3 addresses in depth the nature of time; the distinction between past and future; and the uses of time, reverse time, and multiple-time dimensions in works of science fiction.

4. Chapter 4, which in my opinion is the most important chapter in the book, is devoted entirely to the time travel paradoxes and how they have long been discussed in science fiction and are now also beginning to be considered in prominent physics journals.

5. An Epilogue offers a somewhat poetic wrap-up of the book.

6. Nine technical appendices, using no mathematics beyond differential and integral calculus, explain the relativity of simultaneity, motion-induced time dilation, four-dimensional spacetime, the Lorentz transformations, Minkowski diagrams, light cones, metrics, intervals, faster-than-light speeds, time travel in Gödel's and Tipler's works, and the wormhole time machine analyses in the work of Thorne and Novikov.

7. The Bibliography represents the most extensive citation guide published to date on the fictional and scientific literature of time travel.

The Bibliography plays such a central role in this book that some elaboration on it is appropriate. The references, in chronological order for each author (starting with the most recent publications), are from the vast time travel literature in the fields of mathematics, physics, metaphysics, philosophy, theology, and (very critically) science fiction.

The early science fiction time travel stories nearly always first appeared in paperback monthly magazines that are now almost impossible to find. These "pulps," printed on inexpensive wood-pulp acid-based paper, are literally burning up, slowly but relentlessly, as they oxidize, turn yellow, and crumble in library archives. As Raymond Chandler once wrote, "Pulp paper never dreamed of posterity." I therefore decided to reference the stories, whenever possible, through the anthology collections that have reprinted and preserved the classics of this subgenre. As a last resort, I have cited the magazines themselves if they are available on microfilm. I make no claims for an exhaustive bibliography. The stories, books, and papers listed are the ones I found particularly useful, but if I have overlooked an item you feel should have been included, I would very much appreciate hearing about it.

The bibliography is strictly alphabetical; thus De Camp is in the *d*'s and Von Hoerner is in the *v*'s. In the text, when referring to an item in the bibliography, I have used the form of author (year), as in Busby (1987). If the same author has multiple items in the same year, the year is followed by a lower-case letter, as in Weingard (1979b). Stories in anthologies that include multiple works referenced in this book are cited in the form author (anthology code), as in Padgett (SFAD). The anthology codes are listed at the start of the bibliography. There are two exceptions to this. One is the multivolume anthology *The Great Science Fiction Stories* currently being published by DAW Books. This excellent series is appearing at the rate of two volumes each year, each volume treating just one year (Volume 1, issued in 1979, contains stories from 1939). Stories that appear in the DAW series are listed, along with the appropriate volume number, in the bibliography proper rather than with the anthologies. (In several cases, a story can be found in more than one anthology, but I have listed only the one I happened to use first.) The other exception occurs for single-author anthologies; here I just list all the story titles, one after the other, in the book's entry in the bibliography proper under the author's name.

Now and then I refer to adventures in time and space experienced by various superheros of the comics. My source for most of these bits of nearly lost popular-culture history is the marvelous, multivolume, scholarly work-of-love by Michael L. Fleisher, *The Encyclopedia of Comic Book Heros*. A research bibliography that I found particularly detailed and useful (available in the Reference Department of most university libraries), even if a bit dated now, is *Index to Science Fiction Anthologies and Collections* (Contento). Invaluable for any student of time travel in science fiction is Donald B. Day's *Index to the Science-Fiction Magazines 1926–1950*. More recent citations can be found in *The M.I.T. Science Fiction Society's Index to the S-F Magazines, 1951–1965*. For access to the novel-length literature from all over the world and all time periods, no serious student of fictional time travel could begin without reading the encyclopedic *Anatomy of Wonder: A Critical Guide to Science Fiction* (Barron). The most recent source, with three thousand (!) synopses of stories from the early pulps, is the extraordinarily useful *Science-Fiction: The early years* (Bleiler).

This book is for all those who agree, as I do, with Professor David Park, who wrote in his book *The Image of Eternity,* "Is time travel in principle (never mind the difficulties) a possibility? It has received some thought in the past and deserves some more." As my book shows, it is such thought by *physicists* that will lead to a better understanding of time travel and to discovering whether time travel is consistent with the known laws of physics. Readers of this book will, I believe, find that this process of discovery is no mere boring, scholarly drudgery but rather an exciting, ongoing adventure. As Professor Adolph Baker wrote in his book *Modern Physics and Antiphysics,* "Physics is engaged neither in the development of time machines nor in the fabrication of bombs. But it is the business of physicists to make flights of fancy that carry them far beyond the boundaries imposed by current technology." Even the most conservative critics of time machines and time travel will surely admit the truth of this, and I have worked hard to make this book a flight *well* beyond known boundaries. This is, I believe, a book for the adventurous in spirit.

Paul J. Nahin
Durham, New Hampshire
October 1992

A number of individuals helped me during the seven years I worked on this book. John Pokoski, Chair of the Electrical and Computer Engineering Department at the University of New Hampshire (UNH), was always supportive of what must have seemed to be a project very much unlike what my colleagues were going. The UNH Library makes up for its occasionally incomplete holdings with some superb staff—in particular, Karen Fagerberg and Susanne Gerred Seymour of Inter-library loan, who almost never failed me in my quest for obscure books and microfilm reels of ancient science fiction magazines. I am especially indebted to Texas A&M, Dartmouth College, the University of California at Riverside, the Claremont Colleges, the California State Universities at Northridge and Fullerton, Mount Holyoke College, the New York City Public Library, and the University of Delaware (with special thanks to Tom Muth of that institution's microfilm unit) for giving me access to their archives. David Deutsch of the Oxford University Mathematical Institute kindly provided me with a draft preprint of his research into the quantum mechanics of time travel. John Archibald Wheeler, Professor Emeritus of Physics at Princeton University, wrote me a long and friendly letter on his memory of Gödel's views on time travel to the past (and he included his own opinion of the concept as well). Matt Visser at Washington University, and Frank Tipler at Tulane, replied to my questions about time travel with thoughtful answers. Kip Thorne, professor of physics at Caltech, graciously sent me several preprints of the work he and his students had completed on closed timelike curves, as did Igor Novikov of the Nordisk Institut for Teoretisk Fysik/Copenhagen and the P.N. Lebedev Physical Institute in Moscow. (Igor and Kip showed far more decency and manners in their prompt replies to my endless electronic and postal mail queries than I did in literally deluging them with questions.) Hans Moravec, of the Robotic Institute at Carnegie Mellon University, took time from his sabbatical at Thinking Machines Corporation to send me a copy of his unpublished research on the computational implications of backward time travel. Edwin Taylor, past editor of *The American Journal of Physics,* and Gregory Benford, professor of physics at the University of California at Irvine, who is well known to science fiction readers around the world, have been supportive of my efforts since they first read the entire book in typescript. Nan Collins and Alice Greenleaf of the UNH College of Engineering and Physical Sciences Word Processing Center typed all the mathematics in the Technical Notes as well as the seemingly endless revisions of the Bibliography. The final editing and writing for this book were done while I was on sabbatical leave at Harvey Mudd College in Claremont, California, during the fall semester of 1991, and I gratefully thank the College for giving me the financial and administrative support required for that work. Special thanks, in particular, to Professor John Molinder, chairman of the Engineering Department at Harvey Mudd, who originally suggested that a semester in southern California was just what I needed after two winters in Virginia and fifteen more in New Hampshire. He was right. To teach again at Harvey Mudd, where I began my teaching career more than twenty years ago, was like a wonderful trip back in time.

As a final comment, I would like to say to all readers that although I have done my best to write a scholarly book, I really do not know whether I have succeeded. I hope, of course, that everything I have had to say on the pages that follow is clear, correct, and if not always brilliantly witty then at least minimally interesting—but self-doubts do remain. This book was very difficult for me to write; it was not easy for me to hold in my mind all the seemingly countless and conflicting opinions, theories, and philosophical speculations concerning time travel. There were more than a few times when I despaired and wondered if I had gotten myself into a job simply too big for me. The experience was a bit like walking from Boston to Los Angeles, backwards on my knees. As a Harvard professor stated many years ago, upon his retirement: "Is scholarship pleasure or pain? If someone cuts your leg off, you know it's pain, but with scholarship you never know." When I read that, I knew exactly what he meant!

Paul J. Nahin
Durham, New Hampshire
October 1992

CONTENTS

The three pioneers behind time travel as we think of it today. The scientific pioneers were Albert Einstein (1879–1955) and Kurt Gödel (1906–1978), good personal friends who are shown here in 1954 at the Institute for Advanced Study in Princeton, New Jersey, in a photo taken by Richard Arens. It was Einstein's 1916 general theory of relativity ("theory of gravity") that Gödel used as the basis for a 1949 paper that was the first to show that the general theory does not forbid time travel into the past. The literary pioneer of time travel was, of course, Herbert George Wells (1866–1946), who is shown here as a college freshman cut-up around 1885. The photograph was taken as

a prank by an unknown friend while Wells was a student in a biology course given by Thomas Huxley, at the Normal School of Science in South Kensington (a branch of the University of London). A far too thin and impoverished Wells was then still a teenager, and *The Time Machine* lay a distant ten years in the future.

Einstein/Gödel photograph courtesy of the American Institute of Physics Emilio Segré Visual Archives of the AIP Niels Bohr Library. Wells photograph courtesy of the Rare Books and Special Collections Department of the Library of the University of Illinois at Urbana-Champaign.

An Overview of Time Travel

Woodn't it be grate to go bacc in tyme and korrect your mistakes? Wouldn't it be great to go back in time and correct your mistakes?

—motto of *Time Twisters* comics

What if you knew the course of history . . . before it happens? Like Kennedy's assassination . . . the World Series . . . the stock market? Most of us would die for a chance to replay.

—from the paperback cover blurb of Ken Grimwood's novel *Replay*

Take me back—up the hill—to my grave. But first: Wait! One more look. Good-by, world. Good-by, Grover's Corners . . . Mama and Papa. Good-by to clocks ticking. . . .

—the despairing words of Emily Webb, after she returns from the dead to relive her twelfth birthday but finds only unhappiness from knowing the future (from Thorton Wilder's *Our Town*)

The Mystery of Time Travel

To travel in time.

Could there possibly be a more exciting, more romantic, more wonderful adventure than that? I don't think so, and in this opening section and in the next, I want to show you in some detail just how fascinating many writers (and their readers) have found the concept of time travel. And this was *long* before physicists discovered time travel lurking in Einstein's general theory of relativity.

Before the arrival of humans on the surface of the Moon in 1969, the only other fantastic or cosmic voyage that could compare with time travel was traveling through space. During the seventeenth and eighteenth centuries, in fact, such voyages were the center of a genre of fiction (now called *science* fiction) called the "imaginary voyage" or "extraordinary voyage." Marjorie Hope Nicolson's 1948 book *Voyages to the Moon* carefully documents just how popular that form of literature was (it still is). Since 1969 the first such voyages have become history, of course, and time travel has replaced space travel as the modern "imaginary voyage."

I am willing to wager that, given a random collection of middle-aged adults, at least three-quarters of them would respond enthusiastically when asked whether time travel interests them. Make that a group of children over the age of five, and the vote would certainly be unanimous. As specific examples for why I make this claim, consider the popularity enjoyed by Edith Nesbit's description of a trip into the past in her 1907 juvenile novel *The Story of the Amulet,* the modern juvenile novels of Madeleine L'Engle, the 1960s television cartoon series "Mr. Peabody's Improbable History,"[1] and the 1981 film-fantasy romp through history, *Time Bandits.* For youngsters who are very young, even of pre-kindergarten age, there is *Professor Noah's Spaceship* (Wildsmith), in which Professor Noah takes endangered animals through a time warp back into the past (to just after the Flood), where they can once again lead happy lives. Nicely done versions of the classic time travel tales "The Biography Project" (Gold) and "Over the River & Through the Woods" (Simak) are part of *The Bank Street Book of Science Fiction* targeted for older pre-teenagers. One of the more interesting Danny Dunn gadget stores uses a time machine (Williams and Abrashkin), as does *The Trolley to Yesterday* (Bellairs); and for older, more sophisticated children, there are such books as *The Green Futures of Tycho* (Sleator) and *Alistair's Time Machine* (Sadler).

And, of course, for children of all ages (including the author), there is "Doctor Who" and his time-traveling police telephone booth (Haining). This British TV series, still seen on both sides of the Atlantic, had a somewhat less campy American counterpart in the 1966–1967 series "Time Tunnel." That show, about two time travelers caught in a secret government time machine experiment gone wrong, took itself far more seriously than does "Doctor Who." Still, other than the quality of the special effects, there was nothing new in these shows—I recall watching, as a youngster in the early 1950s, time travel episodes on the television pioneers "Space Patrol" and "Captain Z-Ro."

This fascination with time travel has actually been "scientifically" documented. In an intriguing 1976 study by Cottle, several hundred men and women were asked to consider the possibility of spending an hour, a day, and a year back in both their personal past (since birth) and their historical past (before birth). They were further told that it would cost them $10,000 to purchase such time travel services. Their responses indicated that 10% would be willing to spend this much money for an hour into the historical past, 22% for a day and 36% for a year. As might be expected, the numbers rose as the cost dropped, and if such trips were free, the interest was almost universal. Less scientific was the poll conducted in 1988 by the editors of *Seventeen* magazine. Published in the March issue, "The Best of Times" opened with the provocative question "Given a trip in a time machine, where would you get off?" The answers ranged from Troy 1200 B.C. to Victorian England to the "so cool Fifties." For the young women who make up the readership of the magazine, their responses show they view the past as a romantic "place."

Rod Serling used his fascination with time as the inspiration for many of the episodes on the enormously popular television series "The Twilight Zone." During its five-year run (1959–1964), it often presented stories that dealt in some manner with time—and specifically with time travel. Several of the most popular episodes on the two "Star Trek" series also involved time travel. Such interest is easy to understand. To travel in time, as William Blake wrote in "Auguries of Innocence," would be to

> Hold infinity in the palm of your hand
> and eternity in an hour.

As a modern writer (Paul) expressed it, "Time travel [is] the ultimate fantasy, the scientific addition to the human quest for immortality." And a philosopher—Smith (1985)—has correctly observed that "the popular appeal of time travel . . . is no doubt due to a nostalgia for the past, which is almost an omnipresent aspect of the human condition."

Some writers have recognized that time travel fantasy has a strong appeal for adults as well as for children. For example, recall the old saying "Time is money," and then ask yourself whether it wouldn't be wonderful if you could save time just like money. This adult fantasy idea has been used many times, as in *Tourmalin's Time Cheques* (Anstey), "The Time Exchange" (Knight), and "Time Bank" (Bilenkin). The German writer Wolfgang Jeschke's best-known book, for another adult example, is *Der Zeiter* (*The Time Person*), a collection of time travel stories for grownups; the central story, which can be found in English translation, has the texture of a fairy tale. In this story, two princes of the far future build a time machine, and then the younger one sends his brother (the heir to the throne) 11,000 years into the past, to A.D. 1619, and thereafter fears his brother's return. The brothers Grimm would surely have written a similar tale if the time machine concept had been popular in the early nineteenth century. And the means of time travel in the Norwegian Johan Wessel's 1781 comedy–fantasy *Anno 7603* actually *is* a fairy. Maurice Maeterlinck's 1908 play *The Blue Bird* also enlists a fairy to start two children on a search through time to find the bluebird of happiness (the moral is that it isn't in either the past or the future but is always in the present).

Ray Bradbury, a writer more of fantasy than of science fiction (although most readers probably associate him with the latter), is fond of imagining how time machines could aid famous yet tragic writers that he admires (*any* writer's fantasy). For example, in "The Kilimanjaro Device" a time machine is used to transport Ernest Hemingway away from the time of his suicide back to happier days. And in "Forever the Earth," a time machine temporarily transports Thomas Wolfe out of his death bed in 1938 into the far future of 2257 for one last writing assignment. Performing a task that Bradbury would surely applaud are the time travelers in "The Fort Moxie Branch" (McDevitt), who roam the past in search of the lost masterpieces of writers, writers unknown as well as the famous (including Hemingway and Wolfe). A similar time travel fantasy (which mentions several time travel paradoxes but mostly gets them wrong) is "Two Guys from the Future" (Bisson), in which the future saves works of art that would otherwise be lost to fire, war, and the like.

Some of the best modern science fiction stories have played with the fantasy appeal of time travel by having gifts arrive by accident from the future, as in "Of Time and Third Avenue" (Bester), "Mimsy Were the Borogoves" (Padgett), "Something for Nothing" (Sheckley), "Thing of Beauty" (Knight), "The Little Black Bag" (Kornbluth), and "Child's Play" (Tenn). More often than not, the fairy tale moral of these stories is that such unearned gifts generally bring grief. A related story is "Service Call" (Dick), in which a *repair man* from the future mistakenly tries to fix a marvelous invention before it has been invented; the gadget turns out, in fact, to be a terrible thing indeed. The common fairy tale theme of "The Three Wishes," in which the recipient ends up using the final wish to undo the unforeseen consequences of the first two, is in fact the precursor to all modern change-the-past stories. The editor's introduction to Harness (FCW) expands on this idea, telling us (with reference to another age-old form of adult fantasy) that "time travel stories about the Civil War have one thing in common with pornography; they serve to titillate an impulse [in the case of time travel stories, the impulse to change history] and to frustrate it."

For Better or For Worse®   **by Lynn Johnston**

The one fantasy that is common to all decision-making beings.

The editor of the science fiction pulp magazine *Thrilling Wonder Stories* used the powerful emotional hook of changing the past in a 1950 passage hyping a time travel story coming in the next issue: "What's the biggest mistake you ever made? Don't worry about it. You may have pulled some awful boners in your time, but there's a sure-fire remedy for them all. It's simple. Just look up at that old time-clock on the wall—and turn it back to the moment just preceding your terrible blunder. Then make your corrections—and set your time-clock back to the present. You may be starting a new chain of error, but why fret? You can go back in time again. . . ."

A character in *The Bird of Time* (Effinger) captured the fantasy appeal of time travel with a simple statement: "The past . . . is the home of romance." More scholarly is Wachhorst (1984): "The time-travel [film] romance is an attempt to reenchant the world, to regain a sense of belongingness, to reinstate the magical, autocentric Universe of the child and the primitive." On a less poetic level, time travel and the movies and stories about it fascinate us because they turn our everyday world view upside down and inside out. Such movies and stories make us *think*. It is not surprising, I think, that the top film of 1985 in terms of both excellence and popularity (as determined in a *Boxoffice* magazine poll) was *Back to the Future*. And for sheer excitement, it is impossible to outdo the two time travelers in "The World of the Red Sun" (Simak); as they activate their time machine, one yells to the other, "Kiss 1935 good-bye!"

Still, perhaps not everybody is interested in time travel, as one adventurer from A.D. 2156 (who is stranded 79,062,156 years in the past) learns to his shock and horror in "When Time Was New" (Young). Offering to trade his advanced knowledge for rescue by aliens who happen to be visiting Earth at the time, he is told, "Mr. Carpenter, if we had wanted time travel, we would have devised it long ago. Time travel is the pursuit of fools. The pattern of the past is set, and cannot be changed. . . . And as for the future, who but an imbecile would want to know what tomorrow will bring?" And in "Out of the Past" (Eshbach), one scientist chastises a colleague with these harsh words: "If you don't stop this senseless theorizing upon something that's an obvious impossibility, you'll find yourself working alone! Your ridiculous ideas sound like the ravings of a madman. Anyone with average intelligence realizes that the mere thought of traveling through time is absurd."

I will, of course, not follow *that* line of argument here, nor will this book be sympathetic with the attitude of Koestler (1971), who, finding much of science fiction boring, declared (somewhat irrelevantly), "Time machines are no escape from the human condition." That writer, I think, told us more about his own lack of imagination than he provided insight into time travel. To go forward in time, you see, would be to discover what wonders the future will bring, or perhaps to witness (as did H. G. Wells' time traveller) the final stages of the decay of the human race.

That last experience might be quite an unpleasant sight, too. "The Choice" (Hilton-Young), for example, tells the story of a time traveler who is given the choice of whether or not to remember the future after returning home—once back, all he can recall is that he decided to forget. The classic "Twilight" (Campbell) has a similar theme, but in that tale we are told the grim details of the fate that awaits humanity (a time traveler from A.D. 3059 goes seven million years into the future to find that machines, not humans, rule the world). Even more humiliating for men is the future in "An Adventure in Time" (Flagg), in which a time traveler from 1950 finds that the world a thousand years later is run by "mechanicals" (robots) and women—who keep men around mostly for breeding purposes! In "John Sze's Future" (Pierce), a nuclear physicist is temporarily displaced by accident to A.D. 2178 and finds that the very name of his profession has become an obscenity in a world run by psychologists (since the "atomic blowup of 1987"). Time travelers to a war-torn future in "Judson's Annihilator" (Wyndham) find that it is disease bombs and airborne bacilli that will do the world in, and explorers in "The Time Cave" (Sheldon) step through a "time fault, a kind of warp that took us into the future" and find the skeleton of a man who hunted (will hunt) with bow and arrow and who worshipped religious icons resembling the mushroom clouds of nuclear explosions. A twist on this appears in "Now and Then" (Rosenbaum), in which a time traveler thinks (basing his judgment on the primitive men he encounters) that he has gone back to 100,000 B.C. On his return to the present, he learns that he was actually just fourteen years in the *future*.

Other stories, too, have used time travel to the future as a literary device to warn against what the end of the world might be like. The cautionary story "Breakfast at Twilight" (Dick), for example, describes how a family (and their home) are suddenly transported seven years into the future, into a two-year-old atomic war in a world that is a horror of destruction and death. They (and we) are told this happened because of a robot missile attack: "The concentrated energy must have tipped some unstable time fault . . . a *time quake*. . . . The release of energy . . . sucked your house into the future." A second missile attack later sends them all back to their correct time, and the neighbors think the house was ruined by the explosion of a faulty water heater! (The father doesn't try to explain to them what really happened—he knows the reaction he'd get to a tale like that.) The story closes with his understanding of what the future holds: "And when [the war] really came, when the five years were up, there'd be no escape. No going back, tipping back into the past, away from it. When it came for them all, it would have them for eternity."

In "The Failed Men" (Aldiss) the theme is more up-beat. When the inhabitants of the 3157th century find that their descendants many hundreds of millions of years later are in a degenerative, collapsing state, their humanitarian response is to organize a relief agency, the "Intertemporal Red Cross"! The response in "Stalking the Sun" (Eklund), to a similar discovery about the people in the future a half-million years hence, however, is a shocker. Flipping Aldiss' idea on its head, Eklund has the men of the far future receive no pity; rather, they are hunted like game animals by time travelers from the past. Yet another twist, one that combines the ideas of Aldiss and Eklund, enlivens "The Man Who Liked Lions" (Daley), in which a time traveler from an advanced *prehistoric* civilization (that long ago died out) kills modern humans. In his own time he once hunted the ancestral apes of humans, and he is bitter that their descendants, not his, have inherited the earth.

A future world of post-nuclear ruin is the dramatic setting for the 1964 movie *The Time Travellers* and also for the 1968 film *Planet of the Apes,* which has an astronaut plunge through a time warp into a distant future where English-speaking apes are the dominant species. An interesting variation on "travel to the future" is given in the 1989 film *The Navigator,* in which individuals of fourteenth-century England, who are on a religious quest to save their village from an outbreak of the Black Death, are temporally displaced, through a time warp in a cave, into our world of the twentieth century.

The resulting shock when 1348 meets 1988 is profound. The same idea was used in a more frivolous way in the 1995 French film *The Visitors,* in which a twelfth-century knight and his squire arrive in modern times with the aid of black magic. All of these examples, I assert, convincingly refute the words of Godfrey-Smith (1980), who rashly declared that "forward [time] travel may be dismissed as boring."

The use of time machines to see the end of the world, or at least the end of humankind, is a common theme in science fiction; see, for example, *The House on the Borderland* (Hodgson), "Alas, All Thinking!" (Bates), "Terror Out of Time" (Williams), "Twilight" (Campbell), "Wanderers of Time" (Wyndham), and "When We Went to See the End of the World" (Silverberg). So seductive is this theme that Edmond Hamilton once even made do ("The Man Who Evolved") without a time machine, instead using cosmic rays to speed up evolution to allow his hero to see what lies at the end of the road for humankind. In a twist on this theme, the same author *did* invoke a time machine in "In the World at Dusk"; an end-of-the-world scientist tries to use the past to repopulate the dead Earth of the future. A similar idea was used in the 1958 film *Terror from the Year 5000,* in which a female time traveler arrives in the present on a search for a healthy man to rejuvenate the battered genes of a radioactive future. This same idea was repeated in print in the novel *Millennium* (Varley) and in "The Comedian" (Sullivan).

A sobering variation on the future-getting-something-from-the-past theme occurs in "The Figure" (Grendon). Appearing in 1947, this story reflects the concern, even in those early days of the atomic age, about the long-range implications of radioactive contamination from nuclear weapons. After being told of massive insect invasions following the atomic bomb blasts in New Mexico, Hiroshima, and Nagasaki, a government research team reaches an unknown time in the future by using a "grab," a gadget that can "warp space-time curvature so that anything 'Here-Then' would be something 'Here-Now.'" The object retrieved by the team is a three-foot-high silver statue, done in great detail and with impressive skill. It is an obviously intelligent and majestic figure—the figure of a beetle!

As long ago as 1856 (considerably before the fundamental idea of a "time machine" in the spirit of H. G. Wells was conceived), people were fascinated by the idea of seeing the future. In that year an article appeared in *Harper's* (Anonymous) that speculated on what A.D. 3000 might be like. It wasn't very good prophecy, falling far short of even 1920! (The description of that far future is given by an anonymous narrator whose presence out of his own time is never explained.) Jules Verne, too, tried his hand in 1889 at prophecy (and had no better luck), with a look a thousand years ahead in "In the Year 2889." He did better with his long-lost novel *Paris in the Twentieth Century,* written in 1863 and discovered in 1989.

If we look at languages other than English, we can find even earlier tales that at least weakly attempt to give some sort of physical explanation or mechanism for time travel (though not by machine). In the 1824 Russian story "Plausible Fantasies or a Journey in the 29th Century" (Bulgarin), for example, the hero is swept overboard while sailing and awakens a thousand years later (in the year A.D. 2824) wrapped in the preserving herb Radix Vitalis ("Root of Life"). This may not be terribly convincing, but it is better than having it all turn out to be a dream à la Dickens' *Christmas Carol,* with its trips to Christmas Past and Future. However, a fellow countryman of Bulgarin regressed a decade later from even this limp attempt at explaining movement in time; in "The Year 4338" (Odoevski), he used hypnotic trances to transport his narrator to any country and period of time. In this way the author was able to speculate about the future, but not through time travel as physicists think of it today.

Mesmerism was also used by John Macnie, a University of North Dakota professor, in his 1883 novel *The Diothas.* This work (reprinted in 1890 as *A Far Look Forward*) has its hero visit the ninety-sixth century (and later return to his own time by falling over a waterfall). It's pretty poor reading today, but at the end there *are* a few brief, tantalizing words on the idea of a traveler in time being his own ancestor, on the possibility of his reading an old newspaper while in the future and thus learning details of his life as it will be after his return to the past, and on an infinity of parallel worlds. All these are puzzles that later would fascinate generations of physicists, philosophers, and science fiction writers. Time travel by dreaming (for example, Wells' *The Dream*) was once commonly used as a literary device in stories whose real purpose was social commentary. (The use of periodic dying and resurrection in Edwin Lester Arnold's nineteenth-century novel *The Wonderful Adventures of Phra the Phoenician* is, however, beyond the scope of this book!) A somewhat similar idea, "sleeping into the future," can be found in *Looking Backward* (Bellamy); *A.D. 2000* (Fuller), whose hero sleeps with the aid of the "wonder-gas" ozone!; *When the Sleeper Wakes* (Wells); and "Pausodyne" (Allen)—all first published before the turn of the century. Fuller, in particular, failed hilariously in predicting the future. In his novel the science of A.D. 2000 has radio *telegraphy* (Morse code, only) but doesn't know how it works!

Early twentieth-century writers for the pulp science fiction magazines also used sleeping to get a story character into the future, and they didn't prove to be any better at "dreaming up" prophecy than was the anonymous writer for

Harper's, Verne, or Fuller. In "A Visitor from the Twentieth Century" (Donitz), for example, we read of an architect who falls asleep after working hard all day designing a "new" New York City for a contest. When he wakes up he is in the last days of the twenty-first century and receives (along with the readers of *Amazing Stories*) a tour of the far future. Technology has advanced so fast, he is told, that to fly from San Francisco to New York takes "only" ten hours! (When this story was published in 1928, of course, it took *days* to fly across the United States.) We learn at the end (with no astonishment) that it had all simply been a dream.

Sleeping into the future is a literary technique of ancient origin. As we learn from Stern (1936), "about 600 A.D. Gregory of Tours told the story of the Seven Sleepers of Ephesus who slept for three hundred and seventy-two years. . . . And it was Frederick II's Boniface who in one day lived forty years. . . . Let us not forget, too, that Mohammed had 90,000 conversations with God while he washed his hands. So the mind vanquishes time." And, of course, Sleeping Beauty was comatose for a century in her ancient fairy tale. Harry Geduld points out in his Introduction to Wells (1987) that this popular idea can also be found in more modern "olden" times, such as in Mercier's *L'An Deux Mille Quatre Cent Quarante* (1771), in which an eighteenth-century sleeper wakes up in the twenty-fifth century. The sleeping-to-the-future device was later resurrected in the famous "Rip Van Winkle" tale by Washington Irving in 1819 and, again, in 1836 in Mary Griffith's "Three Hundred Years Hence," and these were followed in 1887 by W. H. Hudson's *A Crystal Age,* in which a botany explorer falls off the edge of a ravine, knocks himself out, and wakes up an undisclosed but very long time later.

This hoary device has been used in modern times, too, to explain how Buck Rogers, born in the last days of the nineteenth century, could still be around in the year A.D. 2432. (He was buried in a mine cave-in but was kept alive in suspended animation for 500 years by the release of a mysterious radioactive gas.) In "The Man Who Awoke" (Manning) we have a sleeper who snoozes 3000 years into the future—presumably measured from 1933, the year this tale appeared in *Wonder Stories.* Four years later, the even more outrageous story of an ancient Greek warrior–sailor, stranded in Central America, was told in "Past, Present, and Future" (Schachner). The warrior used volcanic gases as a drug to sleep thousands of years; while he slept he was worshipped as the Mayan god Quetzal. And in "The Fourth Dynasty" (Winterbotham) we have an embalmer(!) who perfects a fluid that gives a human body "a hardness of diamond, which could withstand even the erosive action of wind and water." This magical stuff allows the embalmer to be resurrected in the 1500th millennium.

By 1940 the sleeping-into-the-future technique was so well known that lazy authors hardly bothered to explain it to their readers. In that year's January issue of *Fantastic Adventures,* for example, Don Wilcox's story "The Robot Peril" tells of characters who arrive in the year 2089, and the *how* of how they do it is quick and to the point: "All three had emerged from a century and a half of temporary death from a pit of absolute zero, into which they had been hurled by a mad experimenter." Even a decade later, at least one writer made an even bigger fool of himself with a story of this same sort called "The Long Dawn" (Loomis); it appeared in a 1950 issue of *Super Science Stories* and was simply packed with scientific nonsense. In it, for example, we read of a prehistoric nuclear physicist (from the Jurassic period!) who arrives in modern times with his two telepathic pets (a pterodactyl and a tyrannosaur, I kid you not) after a drug-induced hundred-million-year sleep in a time vault. One wonders just what the editor who bought it was drinking!

The far superior 1958 story "Two Dooms" (Kornbluth) has a physicist working on the Manhattan Project fall asleep and wake up 150 years later—to find the Germans and the Japanese in America as occupying forces. He returns to the present in an equally mysterious fashion. Woody Allen used this same idea in his 1973 comedy film *Sleeper.* Victor Rousseau's 1917 dystopian novel, *The Messiah of the Cylinder,* in which the hero travels a century into the future via suspended animation through freezing (using a vacuum cylinder equipped with a hundred-year clock) is just a bit more sophisticated than Allen's film. Leo Szilard, one-time collaborator with Einstein on the design for a refrigerator, used the freezing idea of Rousseau to let his hero in "The Mark Gable Foundation" travel ninety years into the future, where all sorts of odd developments have taken hold. Note, however, that Boussenard's novel *10,000 Years in a Block of Ice* had already used freezing before the end of the nineteenth century to span a period more than a hundred-fold longer, which in turn was pre-dated by William Clark Russell's 1887 novel *The Frozen Pirate.* The "corpsicles" in Larry Niven's 1976 novel *A World Out of Time* are a more modern invocation of freezing into the future.

None of the characters in these stories is really a time traveler, of course, but rather each is a person "out of time." (If you aren't quite sure about the "of course," I'll be coming back to this point soon to discuss what I think is a convincing *philosophical* argument.) And finally, in his classic 1956 novel *The Door Into Summer,* Robert Heinlein combined the cold-storage method of reaching the future with a true time machine that allowed his hero to return to the present (the future's past, of course). Freezing might seem to be pretty much an exhausted cliché by now, but Hollywood rediscovered it in the 1990s, releasing in a six-year interval five films based on freezing into the future: *Late for Dinner, Encino Man, Forever Young, Demolition Man,* and *Austin Powers.*

If freezing and sleeping aren't mechanisms for *real* time travel, then what is? Professor H. Bruce Franklin (1966) has stated that "when one says time travel what one really means is an extraordinary dislocation of someone's consciousness in time." By this criterion, for example, it might reasonably be argued that the doomed Peyton Farquhar, captured Confederate sympathizer, time-traveled, because in the brief moments it took for him to drop with a noose around his neck in Ambrose Bierce's "An Occurrence at Owl Creek Bridge," he lived (in his mind) through a day. His "trip" through time ended, however, with a broken neck beneath the bridge. It seems to me, then, that Franklin's definition of time travel is far too broad, and I will take a narrower position on time travel in this book. In addition to excluding hanging, I will also take the position that smoking marijuana or taking amphetamines and/or LSD to achieve a non-linear, hallucinogenic experience of time will not count as time travel, even though I must admit some distinguished writers have used such approaches (consider the 1822 *Confessions of an English Opium-Eater* by Thomas De Quincey, in which we read, "I sometimes seemed to have lived for 70 to 100 years in a night.")

Soon after De Quincey, Tennyson used drugs to freeze time in his "The Lotus Eaters," written over the period 1833–1842. The setting in that work is a place "in which it seemed always afternoon," similar to the land where Odysseus nearly lost his crew to the same narcotic plant. H. G. Wells himself was not above using such a method, as shown by his tale "The New Accelerator." In that story we learn of Professor Gibberne's discovery of a drug that makes the personal time of its taker run thousands of times faster than usual; the whole world appears to be a frozen instant to anyone who drinks the "New Accelerator." The "frozen instant" theme was also used in "The Einstein Slugger" (Wellman) to speed up a boxer *without* drugs, and more recently it appeared in "The Six Fingers of Time" (Lafferty). There is also mention by Wells of a "Retarder," a drug for achieving the opposite temporal result.

An example from the horror story genre of the use of drugs to time-travel is "The Ancestor" (Lovecraft and Derleth), in which a medical doctor states, "I have found that a combination of drugs and music, taken at a time when the body is half-starved . . . makes it possible to cast back in time," but his experiments go terribly wrong (in a horror story, one can only say "*of course* they go wrong"!) The same idea, with a slightly less ghastly ending, is Du Maurier's *The House on the Strand.*

In *An Age* (Aldiss) we find such drug-induced time travel aptly renamed *mind travel;* Aldiss requires his travelers to leave a jar of blood behind in the present as a homing signal for successful mental return. This sounds (to me, at least) more like "time travel for drugged vampires" than real science! Although there is still much talk of drugs, mind travel is somewhat less outrageous in *The Gap in the Curtain* (Buchan), in which several characters are able to read a newspaper a year in the future.

The appearance in 1927 of Dunne's *An Experiment with Time,* on mind travel to the future via precognitive dreaming, attracted a fair amount of attention—see, for example, Besterman (1933). John William Dunne (1875–1949) was a military man, not a mystic, and he attempted a rational explanation for his experiences of dreaming his way through time. J. B. Priestley (1964) described Dunne as an "old regular-officer type crossed with a mathematician and engineer," and H. G. Wells himself thought highly of Dunne's ideas on time. However, Dunne's explanation of mind travel involves an infinite regress of times—a philosophical horror—and his time travel by dreaming is of no more interest for us in this book than is hanging or drugs. This, then, is the last I'll have to say on any of those forms of "time travel."

Machineless Time Travel Without Dreams or Drugs

In "The Hero Equation" (Arthur) we find mind travel without drugs or hanging or dreams; rather, a college professor discovers an equation that, if you just *think* about it, results in sending your personality back in time. The same idea received a powerful, vastly more serious treatment in "The Primal Solution" (Norden), which describes a Jewish survivor of Nazi Germany who mind-travels back to the Vienna of 1913 to kill the then twenty-four-year-old Adolf Hitler. More "literary" is the novel *Breakthrough* (Grimwood), which describes a woman who has her epileptic seizures controlled via electrodes implanted in her brain; when one particular electrode is energized, she mind-travels into the body of a woman in Victorian London.

In a popular series by "Kelvin Kent" (one of many pen names of Henry Kuttner), which appeared in *Thrilling Wonder Stories* during the late 1930s and 1940s, the central character also visits the past with the aid of similar gadgetry, but its inventor is careful to explain (in "World's Pharaoh") that ". . . it isn't a time machine. There's no such thing. . . . You didn't travel in time. Your mind merely took possession of the brain and body [of a person in that time]." "Quantum Leap," on television from 1989 to 1993, also combined machines with mind travel. The show details the time travel experiments of its genius–hero (IQ of 267, holder of six doctorates, speaker of seven modern and four dead languages, and expert in the martial arts), which begin with the "quantum accelerator machine." But a glitch in the hardware results in only his *mind* being inserted into the bodies of past people.

Achieving mind travel, also with the aid of a machine (but with no need for a host mind in the past), is "When the Graves Were Opened" (Burks). First published in a 1925 issue of *Weird Tales* magazine, this is the horrifying story of a time traveler who *watches* (how that is done, without a body and its eyes, isn't explained) the Crucifixion and the events that follow (from which comes the title—see *Matthew* 27:52). Perhaps the cleverest use of mind travel is in "Clap Hands and Sing" (Card). In that tale the inventor of mind travel uses it, as an old man, to travel back to *his own mind* in the past so that he can make love to the young girl he passed-by in his youth. At the end of the story we learn that she, too, has traveled back to *her* mind, on the same day, to make love to him! Most recently this approach to time travel fiction can be found in Robert Silverberg's *Letters from Atlantis*.

Perhaps it is all merely a matter of personal taste when reading fiction, but mind travel is not what *I* mean by time travel. And simply beyond the pale are "time portals in the Bermuda Triangle that look like green fog" (as in the awful 1976 made-for-TV movie *Lost in Time*); using wings (!) to fly through time, as in *Blake's Progress* (Nelson); and what I call the Shirley MacLaine method of time travel—the "channeling" offered by the New Age Time Travel Agency in Gore Vidal's irreverent (but awfully funny) spoof of the gospel stories, *Live from Golgotha*.

We will also not be interested in this book in such stories as *Monsters From Out of Time* (Long), which talks of "time shifts" caused by "prehistoric monsters still surviving deep within the Earth, changed through long ages of evolutionary adaptation into animals with the ability to generate billions of electron volts." Why such creatures (unbelievable in their own right) would disturb "the very keystone of matter itself" and the "spacetime continuum as well" is never explained. Equally unconvincing is "Millennium" (Jackson), a story written in the early days of the atomic age; a man travels into the far future merely because he happens to have a piece of uranium in his pocket. Marginally better is the mention in "Flight Through Super-Time" (Smith) that "movement in the Time dimension could be controlled, accelerated or reversed, by the action of some special force," but this story reduced itself to babble-talk in the very next sentence with the admission that its genius–hero "succeeded in isolating the theoretic Time force, though without learning its ultimate nature and origin." The reader begins to suspect a joke when told that the heart of the "time device" resembles a large hourglass!

Transporting one's self into the past either by means of psi-powers, as in "Psi-Man" (Dick), or by means of sheer willpower, as in *The Time Stream* (Bell), *Time and Again* and *Time After Time* (Finney), "The Ambiguities of Yesterday" (Eklund), and *Bid Time Return* (Matheson), are also out. The latter approach, in particular, is quite often found in time travel stories written for young women. As observed by Friend (1982), "the heroines are abruptly shot into the past through their own personal upheaval during a time of extreme emotional pain and embarrassment." These stories, such as "The Earlier Service" (Irwin), which tells of a sensitive young girl who suddenly finds herself about to be sacrificed during a Black Mass nearly six hundred years in the past, depend far more on romanticism than on rationality and science, but this does not mean they cannot generate tremendous emotional appeal. (In particular, Matheson's novel was made into the beautiful 1980 film *Somewhere in Time.*)

In *Kindred* (Butler) a modern black woman's travels back to the antebellum South (where she lives a parallel, episodic life as a slave) are initiated by feelings of illness and waves of nausea. In a similar fashion, *A Traveler in Time* (Uttley) tells us of a young girl who finds herself slipping back and forth across more than three centuries, usually while daydreaming, but in one instance after falling down a flight of stairs; and in *The Victorian Chaise Longue* (Laski) all it takes for one young woman to travel back to the Victorian England of 1864 is to fall asleep on an antique couch. These are all well-written stories, but they aren't *science*-based.

This sort of machineless time travel is not always restricted to young women, I hasten to add: When the undeniably macho Batman and Robin made their first trip through time (a 1944 visit to ancient Rome), it was via hypnosis. Only in later adventures did they make use of the "time-box" invented by Professor Carter Nichols, the Gotham City scientist regarded as "the world's foremost authority on time travel." In *The Devil on the Road* (Westall) a tough-talking English biker visits the witch-burning England of three hundred years ago—with the aid of his "time-cat"! In "Time Reversal" (Sauvy) the emotional trauma of almost certain death in an accident temporarily transports a tough young male mountain climber back in time sixty years. (Trauma-induced time travel made an early appearance in the movies, almost certainly because of its potential for violent dramatic effect; in the 1925 silent film *The Road to Yesterday*—based on the 1907 New York stage play of the same name—five people are hurled into the distant past when they are involved in a spectacular train crash.) And *The Devil in Crystal* (Marlow) gives us the fifty-year-old Mr. James Tidburnholme, who suddenly finds himself transported from the war-torn London of 1943 to the self-satisfied London of 1922. What it doesn't give us is any explanation of *how* this happened (except for a few references to the devil).

Also out for us in this book are such literary devices as a knock on the head by a crowbar á la Mark Twain's *A Connecticut Yankee in King Arthur's Court.* (This same idea was used by Chauncey Thomas, for travel to the year 4827, in his 1891 novel *The Crystal Button.*) As old and now somewhat stale as this idea is, it still reappears from time to time, as in "The Bones of Charlemagne" (Pei), in which a linguistics professor travels back to ninth-century France after receiving a bump on the head in an auto accident; in "Odd" (Wyndam), in which a man in 1906 is knocked into 1958 by being run over by a tram; and in *The Christ Commission* (Mandino), in which a best-selling author is sent back to the time of Christ by being punched by a drunk in a bar. Public Television used this device in the 1990 movie *Brother Future,* in which a streetwise black Detroit teenager is hit by a car—and learns about slavery when he wakes up in 1822. Interesting, yes, but it isn't physics. Out, too, in this book, is the device of simply getting lost in a strange place, as is the time traveler lost on the coastal moors of France in "The Demoiselle d'Ys" (Chambers), who then meets and falls in love with a young lady while temporarily three centuries in the past. This was a good read in the late nineteenth century, but today it is derivative fiction (and of course such stories *always* lack science!).

More interesting—that is, more rational—at least superficially, is the approach used in L. Sprague de Camp's classic 1944 science fiction novel *Lest Darkness Fall,* written by his own admission as a modern *Connecticut Yankee.* In this book the "granddaddy of all lightning flashes" hurls the archaeologist-hero back to the Rome of A.D. 535. The how of it is unexplained, but at least the implied electrical mechanism avoids the supernatural and the fantastic. This particular mechanism has since become quite popular (nearly all modern time travel movies have lots of sparks and arcs!), and a version of it is used in the 1980 movie *The Final Countdown,* in which a fully armed nuclear aircraft carrier is sent back through a violent electrical storm to the time just before the Japanese attack on Pearl Harbor. The dramatic tension in this film is the obvious question of "Should—indeed can—the past be changed?" Similarly, recall how the "flux capacitor" in the time machine of the original *Back to the Future* film gets recharged in 1955.

An electrical disturbance is also the mechanism for the time travel in "The Incredible Slingshot Bombs" (Williams); the intense fields produced by a high-voltage transmission line open a "time portal" to the future, into a factory making miniature atomic bombs. When the local village fool stumbles through the portal, he mistakes the bombs for slingshot pebbles and brings them home—chaos soon follows. And in "Uncommon Castaway" (Bond) a World War II submarine is sent back to Biblical times when its electrical systems short-circuit during a German depth-charge attack. The sub returns to the present by yet another electrical effect (it is nearly hit by a lightning bolt while in the past).

Finally, it should go almost without saying that time travel by the trivial means of moving across time zones is without scientific interest in this book. This is a common experience today (travelers can leave London on the Concorde and arrive in New York "before" they left), but one can actually trace the idea as far back as 1841, to Poe's short story "Three Sundays in a Week." A long expansion of Poe's idea was printed in one of the early science fiction magazines (Verrill), but the readers' responses to it were uniformly negative. And correctly so—it simply isn't *time travel*. An earlier variation on Poe's idea, however, the 1904 novel *The Panchronicon* (Mackaye), *is* worthy of special notice. An Edwardian literary time machine with style, the Panchronicon swings on a rope tether around a steel post erected at the North Pole. By "cutting the meridians" faster than the sun does, it travels through space *and* time, from 1898 New Hampshire to the London of three centuries earlier. Unlike Verrill's silly tale, Mackaye's novel specifically treats such time travel paradoxes as changing the past, meeting yourself, and causal loops. These are ideas that do not appear even in Wells' intellectually (vastly) superior *The Time Machine*—ideas that will be discussed all through this book. *The Panchronicon* also specifically denies the incorrect assertion common in early science fiction that a backwards-moving time traveler would grow younger.

A story that illustrates, by counter-example, my unhappy view of machine-less time travel is the novel *Mastodonia* (Simak), which uses a godlike creature marooned on Earth (because of a spaceship crash centuries earlier) to make "time tunnels." Even as he wrote this mystical hocus-pocus, Simak seems to have had the feeling that perhaps it was pretty weak stuff, and he tried to explain his position through words spoken by one of his characters. Finding disbelief in those who are told about his machineless method of time traveling, the character complains, "The whole trouble was that I couldn't tell them about some machine—a time-travel machine. If I could have told them we'd developed a machine, they'd have been more able to believe me. We place so much trust in machines; they are magic to us. If I could have outlined some ridiculous theory and spouted some equations at them, they would have been impressed." I believe Simak has this all wrong. We trust in machines not because they are magic but for precisely the opposite reason. They are not magic, but rather *rational*. And to dismiss mathematics so brusquely is to admit that some nonnatural—some supernatural—influence is at work. The supernatural is precisely what this book is *not* about. This book is about *physics*.

Time Travel by Machine

In this book we are interested in physical time travel *by machines* that manipulate matter and energy in a *finite* region of space (I'll elaborate on the significance of the requirement *finite* as we get further into the book). In addition, the machine must have a *rational explanation*. For example, much like the aircraft carrier in *The Final Countdown* film, another Navy vessel in *The Ship That Sailed the Time Stream* (Edmondson) travels into the past via a "rational" (at least for a literary work!) explanation—the ship's still, made from an oddly shaped copper coil in a vacuum jar, proves to be a time machine when powered by lightning bolts during storms. For this book, such a rational explanation is found in Einstein's general theory of relativity, his theory of gravity. (His *special* theory of relativity applies in those situations where there is no gravity.)

Until Einstein, the theory of gravity used by scientists was Newton's—a theory that, although it is amazingly accurate for any situation encountered on Earth, does have observable errors in certain astronomical applications. In addition, Newton's theory is a descriptive one; it makes possible the calculation of gravity effects without offering any explanation for gravity itself. Einstein's theory not only gives the right answers, even in those cases where Newton's theory doesn't, but it also explains gravity. It does this by treating the world as a four-dimensional structure in which all four dimensions (three of space and one of time) are in a certain sense on equal footing. The resulting Einsteinian description of the world is that of a unified *spacetime,* whereas Newton's theory keeps space and time separate and distinct.

From the first (1905) it has been known that Einstein's special theory allows of time travel into the future. To return, however, to travel into the past, had been thought impossible. Yet since 1949 it has been known that the general theory, which so far has passed every experimental test it has been subjected to, does allow time travel to the past under certain conditions. It is this availability of a *theory* that separates time travel speculations from the fantasy speculations with which it is often unjustly lumped—speculations that *are* in the province of quacks, such as ESP, astrology, and mind over matter (spoon bending).

In his general theory Einstein showed how spacetime can be either flat (in the no-gravity, special-relativity case) or curved (with gravity), and he did this not by verbal, philosophical handwaving but by writing mathematical equations—the famous gravitational field differential tensor equations. These complicated equations are very difficult to solve in most cases. But in certain special cases they have been solved. Those solutions show how matter and energy, and spacetime, interact. As the popular shorthand phrase puts it, "Curved spacetime tells matter how to move, and matter tells spacetime how to curve."

At microscopic, local levels, general relativity has causality built in, but on larger scales the situation can be a good deal more complicated. The phrase *large scale* actually covers a lot of ground, inasmuch as everything from the radius of an elementary particle to the radius of the universe—a ratio of 10^{41} in linear size—is fair game for general relativity. In the large scale, in fact, curved spacetime can (according to some solutions to the field equations) lead to violations of causality—that is, to time travel to the past.

In 1949 the mathematician Kurt Gödel (1906–1978) found one such solution to the field equations that describes the movement of mass-energy not only through space but *backward in time* along what are called *closed timelike lines* (or CTLs), a special case of *world lines* in spacetime (see Tech Note 4). These world lines are such that if a human traveled on one, always at a speed less than that of light, he would see everything around him happening in normal causal order from moment to moment (for example, the second hand on his watch would tick clockwise into the future), but eventually the world line would close back on itself and the traveler would find himself in his own past. That is what the physics and the mathematics in Gödel's solution imply. That is what I mean by saying there is a *scientific, rational* basis for discussing time travel.

It is particularly important to note that travel along one of the closed timelike world lines discovered by Gödel requires a *machine,* some kind of accelerating rocket ship. This particular machine does *not,* however, generate CTLs where none existed before (CTL *creation* is what a so-called *strong* time machine does) but rather simply makes use of the CTLs that are inherent in Gödel's spacetime; a Gödelian rocket ship then, is an example of a *weak* time machine.

The dual requirement of a machine with a rational explanation eliminates the previously discussed time travel stories, as well as stories like Manly Wade Wellman's 1940 classic novella *Twice in Time.* In that story a "time projector" throws a time traveler to his destination in spacetime and does not itself go with him. For the return trip, a second projector is needed. Its gruesome feature is the need to have a good supply of organic material at the destination, out of which to reassemble the traveler's tissues. As the hero (about to be tossed back from 1938 to the Florence, Italy, of 1470, during the days of the Medici) instructs a friend, "Tomorrow, at this time, have a fresh veal carcass, or a fat pig, brought here. That's for me to materialize myself back." This is more a ghoul's view of time travel than a physicist's! For the purposes of this book, Wellman's machine is less interesting, although it is described in some detail, than the time machine in *The World Below* (Wright), because Wright attempts to explain his machine rationally by making a fleeting reference to Einstein's curved space.

Years later, Brian Aldiss' timeslip novel *Frankenstein Unbound,* an historical fantasy in which the hero travels back from the year 2020 to 1816, where he meets Mary Wollstoncraft Godwin and Frankenstein's monster, at least attempts to give a rational (though machineless) explanation for it all; a nuclear war has supposedly unhinged spacetime. The novel was made into an interesting 1990 movie. Aldiss has repeated the same kind of idea in *Dracula Unbound,* except that there *is* a time machine in this novel.

It is interesting to note that even at the early date of 1930, Wright was not the only writer of fiction to understand that rational backward time travel would somehow be intimately linked to Einstein's work. In that same year, for example, Edmond Hamilton, one of the pioneering writers for science fiction magazines, published a quite sophisticated story, "The Man Who Saw the Future," in which Einstein and other scholars of relativity are repeatedly mentioned. In that story a young man is hauled before the Inquisitor Extraordinary of the King of France to explain his mysterious disappearance and subsequent reappearance in an open field, amid thunderclaps and in plain sight of many onlookers. As the story unfolds, we learn that the young man was transported five centuries into the future, from A.D. 1444 to 1944, by scientists working in twentieth-century Paris. The thunderclaps were produced by spacetime "rotations," as the atmospheres of 1944 and 1444 were reversed. The Inquisition finds this tale preposterous, of course, and the first time traveler is burned at the stake as a sorcerer.

A few years later a spacetime explanation was provided for a curious mind-travel-into-the-future tale, "Terror Out of Time" (Williamson). There we learn (from the story's stereotypical gadget-nutty professor) that he has discovered "a [magnetic current] that enables me to warp space-time along the time dimension. . . . It enables me to bring the brain of my subject, in the present, into very intimate contact with the brain of some other human being, perhaps a million years distant in time, or a hundred million! They are brought together, so to speak, by bending the time dimension." Then, we are told, thoughts can be transferred between the two brains by a process called neuroinduction. This is, at least, an attempt at a rational spacetime explanation for mind travel.

Early science fiction writer Philip F. Nolan (whose 1928 tale "Armageddon 2419 A.D." in *Amazing Stories* inspired the Buck Rogers character) used the idea of curved spacetime, as well, in "The Time Jumpers." In that story the then standard fictional character of the brilliant, lone inventor has cornered the world market in something called dobinium. He did this because dobinium is the power source for his time-car and, as he explains, "when induced to activity through the bombardment of rays from the cosmic generator, its [the dobinium's] emanations formed the basis of the complex reactions of pure and corpuscular energy by which I was able to cut the *curvature of space-time* [my emphasis] and hurl a material object backward along the time coordinates." Well, whatever all that might mean, slightly less sure about relativity as the basis for rational time travel is the opening of the 1938 story "The Loot of Time" (Simak). There we read that time travel was invented by a scientist "kicked out of Oxford for saying Einstein's relativity theory was all haywire." By the end, however, we are told that in the fifty-sixth century, time machines *do* work on the "warping of world lines principle."

Even as early as 1937 the New York stage was sophisticated enough that its audiences could understand the time travel play *The Star-Wagon* (Anderson), which used a machine based on the fourth dimension, not fantasy. Indeed, the *New York Times* actually found time travel by machine rather old hat, referring in a review of the play (September 30, 1937) to "the hackneyed theme of the time machine." Some days later (October 10) the paper gave the play an even unhappier assessment, saying that "the theories of time, which seem to be breaking out like a rash in many quarters today, are worth more searching study than the *Star-Wagon* discloses." (The word *rash* apparently referred to J. B. Priestley's time play *Time and the Conways,* which was then getting a far more appreciative reception on the London stage.) Well, despite the sophisticated sniffing of the *New York Times,* the romantic plays of Priestley and Anderson are interesting to us because they are rational.

And finally, a story written for youngsters, "The Little Monster" (Anderson), bluntly states a rational explanation for time travel. When a young boy asks his physicist-uncle how his time machine works, he is told, "Come back when you know tensor calculus and I'll explain to you about n-dimensional forces and the warping of world lines." *That's* the attitude I want to encourage in the readers of this book.

H. G. Wells—Why His Time Machine Won't Work

It was H. G. Wells who in 1895 introduced readers to the first time machine in the modern sense, in his story about a Victorian scientist who masters the fourth dimension and makes a gadget that takes him to the time of the last days of Earth. There were many imitators in the years that followed. In the 1929 novel *The Time Journey of Dr. Barton* (Hodgson), for example, a scientist moves through spacetime, 2000 years into the future, in a Wellsian-type machine. But such early machine stories always had a limited readership.

Furthermore, because most Victorians thought the notion of time travel was simply outrageous, such stories generally excited a skeptical response. One of the initial reviews of *The Time Machine,* by the literary editor of the *Spectator,* called the time machine concept "hocus-pocus" and the tale itself a "fanciful and lively dream," and the *Daily Chronicle's* reviewer used the word *bizarre* (both of these reviews are reprinted in Parrinder's book). Indeed, such reactions continued for decades. Nearly twenty years later, a professor in the Columbia University School of Journalism (Walter Pitkin) took his turn at criticizing Wellsian time travel. Initially, Pitkin just hinted at the specific target of his general displeasure about time travel, saying only that it was a frivolous example drawn from contemporary fiction. Later he revealed the target to be "one of the wildest flights of literary fancy which that specialist in wild flights, H. G. Wells, has indulged in . . . his amusing skit, *The Time Machine.*"

The first significant use of machine-based time travel in popular culture was in the comic strip "Alley Oop." The earlier strip "Brick Bradford" had used a time machine, too—the "Time-Top"—but its influence fell far short of that of "Oop." When "Brick Bradford" appeared in movie theaters in 1947 as a Saturday afternoon serial, the "Time-Top" was there for a trip back to the eighteenth century, but it had already lost its chance to be the first movie time machine. That distinction belongs to the "Time-Ball" in the 1941 British film *Time Flies.*

It was "Alley Oop" that really brought time travel by machine into the public consciousness. Oop was an ape-man (with a Brooklyn accent and a pet dinosaur named Dinny) who originally lived in the very ancient past, and at first, all the action took place in the past. The strip, which was launched in 1934, was popular for several years, but by 1939 it was losing its readership. That's when its creator, V. T. Hamlin, got the idea of freeing Oop from his own remote time and letting him roam through all the ages. Enter the Wellsian-type time machine invented by Dr. Wonmug (who, of course, is actually Einstein: *Ein stein* = "*One mug*"!).

After Oop, other time machines began to appear regularly in the comics. Wonder Woman, for example, often traveled in time. Wonder Woman's trips in time were not accomplished by wishing or dreaming or drugs, but rather by machine—the wonderful "space transformation machine." Superman, too, has often journeyed both forward and backward in time. Not infrequently he uses machines, or something called "time rays," but unlike Wonder Woman and Batman, Superman also has the ability to make *himself* into a time machine. He does this by virtue of his superpower to move faster than light. In 1949, for example, Superman returned to his destroyed home planet of Krypton *before* its destruction. An explanatory text tells us, "Time seems to stand still as Superman races at cosmic velocity, passing the speed of light . . . faster . . . until Superman breaks through the time barrier." Some of the science here is correct: Moving at a speed greater than that of light automatically allows backward time travel (but see Tech Note 7 for why physicists today don't believe Superman could even have *reached* the speed of light, much less punched through it).

Wells, too, *almost* had time travel by machine right. But not quite right. As Tech Note 4 explains, a real time machine must move in space (like Superman) as well as in time. All the theoretical models for time travel discussed in this book (Tipler cylinders, black holes, Gödel rockets, cosmic strings, spacetime wormholes, and superluminal spacetime warp drives) require spatial displacement. Wells' machine, however, did not move; it always remained in the Time Traveller's laboratory (or at least on the spot where the laboratory would have been) unless he pushed it about *after* a trip. Such Wellsian-type time machines are common in science fiction (I even used one in my "Newton's Gift"), but in fact they really just won't do. They raise a number of troublesome problems, at least one of which is fatal. To take the worst first, such a machine would run into itself!

Consider: There sits my time machine as I prepare for the first time journey ever, a trip back to the late-Mesozoic era to hunt dinosaur. I load my Continental .600 with Nitro Express cartridges the size of bananas, make sure my blood-proof boots are laced up tight, kiss my wife good-bye, and climb in. I pull the lever. Now, Wellsian-type time machines don't jump over time but rather travel through time. (See the Time Traveller's own description of how things looked to him, a description faithfully and spectacularly reproduced in the 1960 film.) Therefore, the time machine will instantly collide with itself at the micromoment *before* I pull the lever! The resulting destruction obviously introduces a nice paradox: Given that this happens before I pull the lever, how did I manage to pull it? [Many of the early science fiction writers were not totally oblivious to collision problems. For example, in order to avoid materializing inside of an object in the distant past or future, it was common to combine the time machine with an airplane, as in "The Time Valve" (Breuer), "Via the Time Accelerator" (Bridge), and "The Time Cheaters" (Binder), although why this avoids the problem of colliding with the atmosphere is never explained. One writer, Lafleur, thought a Wellsian time traveler would get "a severe case of the 'bends' " if his body materialized in air! Wells' own Time Traveller worried about the collision problem, as well.] Of course, one might argue that Wells' machine does actually move, because it is attached to the Earth, which is certainly moving, but it is not clear why this should result in the time machine arriving in the temporal past of the Earth, rather than in some past region of space (almost surely a vacuum.)

The general problem of where things are for time travelers has been nicely illustrated by the physicist Gregory Benford in his novel *Timescape*. In that story the world of 1998 is on the verge of total ecological collapse, and an attempt is made to change the past by aiming a backward-in-time message via faster-than-light tachyons (see Tech Note 7) at the pivotal year 1963. When the principal scientist involved in this effort is explaining the process to a potential financial backer, he is asked, "Hold on. Aim for *what*? Where *is* 1963?" The scientist replies, "Quite far away, as it works out. Since 1963, the Earth's been going round the Sun, while the Sun itself is revolving around the hub of the galaxy, and so on. Add that up and you find 1963 is pretty distant."

An understanding of the question "Where is everything?" actually goes quite a bit further back in fiction. For example, after looking through a TV-like gadget to view the past, one character in the 1941 story "Dead End" (Jameson) complains, "You said you'd find Captain Kidd's treasure, but all I can see is fog and static." He is told that's because "It's too far back—1698 or thereabouts. The Earth was billions of miles from here then, and there are too many cosmic rays between." The "cosmic rays" are presumably the cause of the interference.

But let's suppose we ignore this concern about where things are for a time traveler, as do most science fiction stories. Still another problem with a true Wellsian-type time machine is that because it travels *through* time, the machine must *always* appear to be located in the same place. For example, to travel from Ford's Theater today to Ford's Theater on the evening of Good Friday, April 14, 1865, in a misguided attempt to save Lincoln from Booth's bullet (see Chapter Four for an explanation of *why* it would be futile to try this), a Wellsian-type time machine would have to occupy every instant of the intervening century and more. For observers outside the machine, the machine would appear to have been sitting in the same place all those years. There is an amusing illustration of a failure to understand this point by the scriptwriters of the 1989 film

Time Trackers, who have time travelers hide their Wellsian machine from accidental discovery by "parking" it five seconds in the future! One modern story that gets this point right is "The Very Slow Time Machine" (Watson).

Wells, himself, was well aware of the loss of story-telling credibility that such immobility entails, and he tried to explain the problem away (for the details of his explanation, see note 1 of Chapter Four). An immediate implication of this immobility is that if you are being chased by an angry mob somewhere in time (perhaps because you unwittingly violated a sensitive social taboo), then hopping into your Wellsian-type time machine isn't going to help, because the machine just sits there. The mob could simply take its deliberate time in first building a roaring fire and then pushing the machine (and you) into it. As Cook (1982) — and what an appropriate author on this point! — puts it so nicely, "You might as well try to escape danger by taking a nap."

As usual, this idea had appeared in science fiction long before the scholars thought of it. For example, in the 1940 tale "Murder in the Time World" (Jameson), a criminal sends the bodies of his victims into the far future with a Wellsian-type time machine that automatically unloads its ghastly cargos. Then he himself uses it to escape into the future. The police, however, having learned of his foul deeds, simply build a cage around the location of the machine and arrest him when he appears twenty-three years later!

To argue about time travel in terms of a Wellsian-type time machine, whether for or against, is to build your house of cards in the middle of a storm; doing so has as little meaning as discussing evolution in terms of unicorns. Still, one can still find such flawed analyses — e.g., Dummett (NOT) — in the philosophical literature.

Traveling to the Future

In fact, it *is* theoretically possible to travel by machine as far as you wish into the future and to see it for yourself, a conclusion solidly endorsed by the solidly accepted special theory of relativity. As the physicists Deser and Jackiw (both severe critics of time travel) have written (1992), "After 1900, special relativity made scientific discussion of time machines possible." What they are referring to is the fact that, by traveling in a rocket ship fast enough (but never, unlike Superman, faster than the speed of light) and far enough, one could leave Earth, loop out on a vast journey perhaps halfway across the universe, and then return hundreds, thousands, even millions of years in the future. You could do this, in fact, with the apparent passage of "personal time" (as measured by your wrist watch or the beating of your heart) as brief as you'd like.

Physicists call personal time "proper time." This astonishing statement (see Tech Notes 2, 5, and 6) put a lot of Victorian-era trained scientists utterly into shock.

Some modern philosophers, however, are still not quite sure about this particular way of traveling into the future and have confused it with simply being asleep or being frozen. After first mentioning special relativity's time dilation via high-speed motion, for example, one philosopher—Mellor (1981)—goes on to declare, "Rip van Winkle was a time traveller . . . and so in its way is every hibernating animal. All in all, real forward time travel is . . . really only an overly grand description of processes slowing down or stopping." Because of this Mellor calls forward time travel a "trifling topic"—agreeing with him is Smith (1985), who says "forward time travel I regard to be . . . not an especially remarkable phenomenon"—and he has taken this position in apparent ignorance of proper time, despite his own words acknowledging special relativity. Ten years later, Swartz (1991) made another philosophical argument for freezing being a legitimate way to time-travel into the future. The Victorian shock over the relativity of time has obviously not yet totally vanished.

The distinction between the proper (local) time of a time traveler on a rocket ship and the time of those who are not fellow time travelers (such as the stay-at-homes back on Earth) is the scientific argument for eliminating freezing (for example) as a means of *true* time travel into the future. That is, suppose that when you deposit a friend in his deep freezer, you compare his wrist watch with yours and find they agree. Years later, when you thaw your friend out, you will find that the two watches still agree (both are powered and maintained at a constant temperature, let us say, by 100-year nuclear batteries!) But if another friend's watch agreed with yours as he was climbing into a rocket ship to begin the space journey described in Tech Note 6, then upon his return to Earth years later (as measured by *you*), you would find his watch is far *behind* yours. The recorded passage of time has been the same for you and your frozen friend, but not for you and your space-traveling friend. This conclusion was, for many turn-of- the-century scientists, simply too much of a discontinuity from all they held dear about the nature of reality.

An excellent discussion of the distinction between the proper time of a time traveler and the time of his friends who "stay home" has been given by Gallois (1994). Gallois begins his analysis by asking what appears to be a question with an obvious answer. Suppose, he says, it is 1998 and you suddenly wake up in a hospital and are told that you have been in a coma for the past two weeks. You are also told that you were in an auto accident two weeks ago, that you suffered temporary neural damage, and that the eventual reversal of such damage always, at some time within four weeks after the damage occurs, causes a day of excruciating pain *if you are conscious at the time.* Would you prefer the day of damage reversal to be in the past two weeks (when you were in the coma) or in the next two weeks? The answer seems obvious. After all, if the day that damage reversal occurs has already happened, then you simply slept through it and missed the pain. To prefer the day of pain to be in the future (when you will be awake) seems absurd. Now let's add time travel to the equation.

All is as before, but now you immediately leave the hospital upon regaining consciousness to take a time trip back to 1892, where you will stay for two weeks. Again, it seems clear that you would prefer to have had the day of pain in the past two weeks (in 1998), not in the next two weeks (in 1892). Note that the *next* is a reference to your proper time, because whereas 1892 is the global past, it is *your* personal future. Thus your preference would be to have the day of pain in the *recent* personal past of 1998, not the *distant* past (globally) of 1892. Now, let's put another time travel twist to this story.

All is as before in the original tale, except now you are told in the hospital that the auto accident happened just after you made a time trip to 2092: As you walked out of your time machine in 2092, you were hit by a car. The two weeks that you were in the coma were in 2092, before you were judged fit enough (although still unconscious) to make the time journey back to 1998. When would you now prefer to have the day of pain? Clearly, as always, in your *personal past,* which is the global future. Time *is* different for those who time travel and those who don't!

It did not take long for the early science fiction magazine writers to pick up on the concept of time distortion induced by high speed; it appeared as early as 1930 in "The FitzGerald Contraction" (Breuer). And two years later, "The Time Express" (Schachner) used the so-called time dilation effect to travel into the future *without* a rocket ship, exploiting instead mysterious rays that vibrated a time traveler's body at nearly the speed of light (don't try this at home!) We are not told, however, how the fictional time travelers in these tales managed to return to their own time.

As a brief aside, Schachner's story is also interesting because it pioneered the use of time travel for touring purposes. The story opens with a description of an ad for a time machine company that is constructed in neon lights twenty-feet high:

> HOOK'S TOURS THROUGH TIME. PERSONALLY
> CONDUCTED. ALL-EXPENSE TOURS INTO THE FU-
> TURE. VISIT YOUR GREAT GRANDCHILDREN IN

THE TRADITIONAL HOOK COMFORT. THE LAST
WORD IN TOURS.

Far more sophisticated is "Out Around Rigel" (Wilson), wherein a space traveler returns from a high-speed trip out to the blue supergiant star Rigel in the constellation Orion. The 900 or so light years of the round trip had taken just six months of ship or personal time (proper time), but a thousand years of back-home time. The traveler returns to Earth to find all he left behind long dead and reduced to dust: "Sometimes I waken from a dream in which they are all so near . . . all my old companions . . . and for a moment I cannot realize how far away they are. Beyond years and years." Another tale that achieves a similar emotional effect, this time via a Wellsian-type time machine rather than by Wilson's spaceship, is *The Year of the Quiet Sun* (Tucker). There a time traveler is trapped in post-nuclear war times, where there is no energy to power his machine. As the story ends, he finds the woman he had loved and left behind in the past. She is now the elderly widow of another man, having married his rival for her hand because the time traveler never returned.

A trip into the future does not *have* to be serious or sad. For example, "Far Centaurus" (Van Vogt) tells the story of a spaceship crew that sets off for Alpha Centauri, more than four light years distant. They survive the trip, which requires 500 years of personal time, with the aid of a preserving drug (shades of Radix Vitalis!). Long before they arrive at the star, however, the secret of faster-than-light travel is discovered and they arrive at their destination only to find a *human* reception committee! As shown in Tech Note 7, knowledge of such speedy space travel is equivalent to knowing the secret of time travel, so the shocked crew is sent back in time, to Earth, to just one year after they originally left—and they listen to their own early radio communications arriving from deep space.

Science fiction writers have thought of several clever uses for forward time travel. Lester del Rey, for example, explored the interesting idea of time travel to the future for military gain, to get advanced weaponry, in his "Unto Him That Hath"; the humiliating, ironic result is that the gadgets brought back from the future are so advanced that nobody can figure out how they work. A chilling version of this idea appeared more recently in the undated Number 17 issue of *Time Twisters* comics. A medieval sorcerer, forced to fight a knight, reaches into the future for the "ultimate" weapon. (It isn't clear why, with such an ability, the sorcerer needs any help.) He obtains a World War II submachine gun but unfortunately sights along the barrel in the wrong direction as he *pushes* the trigger. And "Quit Zoomin' Those Hands Through the Air"

(Finney) describes the failed use of the Wright brothers' first airplane to win the Civil War (it was temporarily borrowed from the Smithsonian Institution by a time-traveling Union Major-cum-Harvard College professor from 1864.)

The idea of using future knowledge to prevail in the past actually appeared quite early in science fiction, as in the 1931 story "The Man from 2071" (Wright). That tale is told from the viewpoint of a person in the far future who is visited by a man from the long-ago time of A.D. 2071. This visitor is, in fact, the inventor of the first time machine, who wants to learn the secret of atomic energy so that he can use it to rule the past. (Remember, it wasn't until 1938 that Otto Hahn split the uranium nucleus, and in 1931 it probably *did* seem that discovery of the secret of atomic energy could still be centuries away, even more distant than the development of time travel.) This story is particularly interesting because the time traveler is told he can't possibly succeed in his evil goal: "Remember that all the history of your time is written. . . . It is in the books of Earth's history, with which every child of this age, into which you have thrust yourself, is familiar. And those histories do not record the domination of the world by yourself. . . . You are struggling . . . against a fate that has been sealed for centuries!"

Later in this book you'll see that, in fact, the foregoing fictional argument actually explains why most philosophers and physicists believe that all stories of changing the past are simply fantasy. For example, the Cambridge philosopher D. H. Mellor (almost certainly unaware of Wright's tale) expressed the common view that science fiction is mostly nonsense, and probably necessarily so when it comes to time travel stories. He wrote (near the beginning of his 1981 book *Real Time*) that "inconsistency is a commonplace in science fiction, especially about time travel. [Inconsistency] is always threatening to happen in time travel fiction, where it is usually avoided by time travellers being very careful, or by invoking different possible universes coming in and out of existence along some further dimension of time. I propose to ignore this sort of nonsense: no one could seriously defend the possibility of time travel by such means." Indeed not, although we'll see as we go along that Mellor is correct in asserting that story writers have resorted to such devices. A principle of self-consistency, however, *is* generally assumed by physicists who are doing serious time travel analyses, which renders Mellor's concern about "inconsistency" moot.

Finally, let me end this section with what may be the most cynical (but perhaps the most realistic) use of time travel to the future, *The Year of the Quiet Sun* (Tucker). In that novel the first time machine is used only to serve the tawdry political needs of a failed president. Three time travelers are sent on various forward penetrations of time to see who opposes him, what their tactics are, whether he is reelected, and so on. Who can doubt that politicians of all parties would be tempted to use a time machine in this way?

Traveling to the Past

The real adventure in time travel would be to go backward in time: to visit the past. I believe this longing for the past—even if only subconsciously, as in Jack Finney's classic "I'm Scared"—accounts for the sweet pleasure most people get from experiencing almost any recreation of times gone by, as shown by the steady market in old-time radio tapes and the popularity of nostalgia movies like Woody Allen's *Radio Days*[2] and of weekly television shows about a romanticized past, such as "Happy Days," "The Waltons," and "The Little House on the Prairie." The weekly theme of the 1979 TV show "Time Express" involved going back in time to "do things right."

Nostalgia for the past is nothing new. The Victorians were nostalgic, too, as evidenced, for example, in the writings of William Morris. His 1890 *News from Nowhere* was a mirror of socialist utopian perfection (as viewed in a dream set sometime in the twenty-first century), which reflects Morris' idealized image of the past. Even though all the events described occur in the future, there is much talk of the past, and his descriptions of places are generally as he remembered them from his childhood.

Morris' idea of the perfect future found its inspiration in nostalgia, not in wonder-gadgets-to-come. Indeed, he wrote *News* as a rebuttal to Bellamy's *Looking Backward* (published two years earlier), which found its salvation-to-come in a highly regimented, machine-oriented society. Morris' was a popular view but, perhaps ominously, Bellamy's work is the better known today. When *News from Nowhere* first appeared in serial form in *The Commonweal*, it enjoyed the same sort of enthusiastic reception as Garrison Keillor's nostalgic "News from Lake Wobegon" (a segment of the radio show "The Prairie Home Companion"). *News from Nowhere* was not Morris' first excursion into literary longing for the past. In his earlier *The Earthly Paradise* (1868) he wrote,

> Come again,
> Come back, past years! Why will ye pass in vain?

And in his *A Dream of John Ball* (serialized in *The Commonweal* in 1886–1887), Morris' hero finds the Middle Ages a nice place to wake up after having fallen asleep in the nineteenth century—a claim that anyone who has studied the Middle Ages may find a bit hard to believe.

As the promotional text on the videotape package of the 1986 movie *Peggy Sue Got Married* says, "to do it again" is "the golden opportunity almost everyone has longed for at least once." Writing less romantically, one philosopher— Sorensen (1987)—declared that a "major source of interest in the time travel question is our general fascination with the exotic and the child-like frustration we sometimes feel at being confined to the present. We wish that the benefits of moving through space could be supplemented with the benefits which would accrue from movements through time."

Robert Silverberg, a science fiction writer who has used the time travel theme often and effectively, expressed Sorensen's wish clearly when he wrote his tale "Ms. Found in an Abandoned Time Machine": "Suppose you had a machine that would enable you to fix everything that's wrong in the world. . . . The machine can do anything . . . it gives you a way of slipping backward and forward in time. . . . Call this machine whatever you want. Call it Everybody's Fantasy Actualizer. Call it a Time Machine Mark Nine." Silverberg gives a masterful demonstration of what he means by that in "Many Mansions," a tale of the year 2006 when time machines actually exist. Even so, the characters use their imaginations to explore their fantasy worlds and wishes—wishes that could (if they *really* wanted them) be realized with a real time machine. Time travel fiction is, of course, the ultimate in escapist literature!

Less happy with time travel to the past was Jack Feathersmith in "Blind Alley" (Jameson), who made a pact with the devil in order to be sent back a half-century to the "good old days" of 1902. Like many others who have made similar arrangements, Mr. Feathersmith overlooked a few subtle loopholes in the contract.[3] Equally unenthusiastic is the character in "Hobson's Choice" (Bester) who delights in pointing out all the bad points of living in the past. Tell him *when* in time, and he quickly ticks off the disadvantages. To live in ancient Greece would let you rub shoulders with Aristotle, sure, but you already know what he said and you'd pretty soon regret the lack of modern plumbing. The years of the American Revolution might let you exchange greetings with George Washington, but you'd also have to put up with cholera in Philadelphia, malaria in New York, and the fact that if you needed an operation there would be no anesthesia anywhere.

The Victorian Age appeals to twentieth-century romantics, but before you go back, you'd better have your eyes and teeth checked and make sure you're not poor. And you absolutely had better not be a member of a religious minority (or of any minority at all). The time-traveling historian in *Doomsday Book* (Willis) has her appendix prophylactically removed and is also advised to have her nose cauterized against all the awful stinks of her destination, the fourteenth century. The medical concern, in particular, serves as a dramatic source of conflict in *The Victorian Chaise Longue* (Laski), as the time traveler (who suffers from tuberculosis) falls asleep in the present and wakes up in 1864—where she finds her illness is untreatable. Similar medical concerns are pondered by characters in *Breakthrough* (Grimwood) and *Chorale* (Malzberg), and in "The Hertford Manuscript" (Cowper) we learn the fate of Wells' Time Traveller—he journeys back to the seventeenth century where his time machine malfunctions. Trapped in the year 1665, he perishes in the Great Bubonic Plague then sweeping London. More down-to-Earth are the dietary concerns of the time traveler in *Timeshare* (Dann) who likes to visit the 1940s and 1950s; he tries to avoid fatty, cholesterol-laden foods and to eat lots of fruit, but he often finds that difficult to do.

On the other hand, poor health isn't *all* there is with backward time travel. The time traveler from the future in "Time Track" (Sprague) makes a very good living in the past by winning bets on yet-to-happen events whose outcomes he knows. Time travel to the past in *Time Tunnel* (Leinster) allows a Parisian curio shop, in the present, to offer remarkably authentic looking newspapers from 1804 whose only "flaw" is that they appear to be fresh off the press—which of course they are! And the failed professor in "April in Paris" (Le Guin) finds the Paris of 1482 infinitely better than the Paris of 1961. Moving in the backward direction, too, is the time-traveling anthropologist in the novel *Mists of Dawn* (Oliver), who solves the mystery of why the museum skeletons of ancient man, dug up in the nineteenth century, are not present in any past more remote than 25,000 B.C. (see also Oliver's "Transfusion"). Traveling back even further is the high school history teacher in a 1932 Jack Williamson novella, *The Moon Era.* In that story an antigravity spaceship that "moves along the fourth dimension" turns out to be a time machine. The hero uses it to visit the Moon of the ancient past, where he has many fantastic adventures.

A popular fictional application of time travel is to use the past as a hiding place; perhaps Silverberg had Clifford Simak's story "Project Mastodon" in mind when he wrote of his "Fantasy Actualizer." In that story Simak visualized the past as a sanctuary for those wishing to escape from the war worries of modern times. Simak liked this theme enough to repeat it in his *Mastodonia* and in the famous short story "Over the River & Through the Woods." Silverberg himself devoted an entire novel, *The Time Hoppers,* to the same theme. In that work a woman who has just lost her husband to this form of escape comes to realize that "her husband had put thumb to nose and disappeared on a one-way journey to the past. "Good-by, Beth, good-by, kids, good-by, lousy twenty-fifth century," he might have said, as he vanished down the time tunnel. The coward couldn't face responsibility."

A Beatles song nicely expresses this lure of the past—"I need a place to hide, that's why I believe in yesterday"—and in the words of the involuntary time traveler in *The Devil in Crystal* (Marlow), "Put back the Universe and give me yesterday!" Ray Bradbury's sad story "The Fox and the Forest" also makes effective use of this idea, with a bomb designer and his wife fleeing from the evils of 2155 back to the Mexico of 1938. Jack Finney was influenced by that story to write the similar "Such Interesting Neighbors," and he graciously has one of his characters allude to Bradbury's tale. J. B. Priestley repeated the theme by having a time traveler flee into the past from an alien invasion of the far future in "Mr. Strenberry's Tale," as does the entire population of Earth in Simak's *Our Children's Children*. James Gunn's "The Reason Is with Us" gives an excellent tutorial on how a time traveler from the future could hide from an oppressive State in the twentieth century. In a reversal of this, the cruel tyrant in "Flight from Tomorrow" (Piper) flees ten thousand years into the past to escape from those he once oppressed. And the occasionally incoherent 1989 film *The Time Guardian* has humanity fleeing the forty-first century to the twentieth, in an attempt to evade an army of killer robots.

In addition to using time traveling to find peace, Simak also expounded on the opposite theme in "Project Mastodon": time travel to the past for military purposes. That is, one could move troops into the past while on home ground, then travel in ordinary fashion to the proto-territory of the enemy, and finally return to the present to appear in dramatic and astonishing fashion in the surprised foe's midst (an idea briefly mentioned again in Simak's later novel *Mastodonia*). Simak therefore offered up the ultimate fantasy of time travel for both doves and hawks in the same story.

Simak's idea was developed in greater detail in "Time Column" (Jameson), which observed that such a tactic would be the ultimate Trojan Horse. The past is also used in an innovative way for military gain in "Not To Be Opened—" (Young). In that story the Earth of a thousand years in the future is ruled by a non-human dictator, and the oppressed masses are unable to arm themselves for revolt. But back into the past travels an agent to arrange for the construction of weapons, which are then stockpiled in hidden caverns—from which the weapons can be dug up for use ten centuries later. The past is used in that tale as a sanctuary and repository from which to make war in the future.

More conventional are two "what-if" Civil War stories, "The Long Drum Roll" (Turtledove) and "The Chronicle of the 656th" (Byram). In Turtledove's

One use for a time machine that science fiction has missed!

DILBERT reprinted by permission of United Feature Syndicate, Inc.

story, twenty-first-century assault rifles end up in the hands of Confederate troops (this story was later expanded to novel length in *The Guns of the South*), and in Byram's work a World War II regimental combat team is transported 80 years backward in time. More gruesome is the idea in "Brooklyn Project" (Tenn), which posits that Newton's third law of motion (for every action there is an equal but opposite reaction) holds for time travel. This "time recoil" is said to be so violent that no human could survive it. As an enthusiastic government hack explains to reporters at a press conference, "Do you realize what we could do to an enemy by virtue of that property alone? Sending an adequate mass . . . into the past while it is adjacent to a hostile nation would force that nation into the future—all of it simultaneously—a future from which it would return populated only with corpses!"

An indirect military application of backward time travel is mentioned in Simak's "Project Mastodon"—the mining of early uranium deposits before they have had time to reduce themselves to lead via radioactive decay. In the same spirit, the main character in "The Wings of a Bat" (Ash) is employed as a time traveler for the Mining and Processing Branch of Cretaceous Minerals, Inc., and "Wildcat" (Anderson) has an oil-drilling crew in the Jurassic period. This idea of using the perishable resources of the past has been pursued by others, too; *The Last Day of Creation* (Jeschke), for example, details a diabolical military plot to extract Middle Eastern oil five million years in the past, and "Death of a Dinosaur" (Moskowitz) has a food distributor selling frozen dinosaur steaks (on this particular idea, see also "A Statue for Father" by Isaac Asimov).

The most direct use of the past's unique resource, *itself*, is seen in "History in Reverse" (Laurence). After purchasing the motion picture rights to H. G. Wells' *Outline of History,* the head of a movie studio uses a time machine to send his ace cameraman into the past to get live action footage. Prehistoric animals, the last ice age, Cheops building his pyramid, the destruction of Pompeii by the eruption of Vesuvius, the Battle of Hastings, Columbus, all the *originals* of these historical events appear in the final film. A more sophisticated treatment of the same idea can be found in the 1947 story "E for Effort" (Sherred). Years later the idea was developed even further in the very funny novel *The Technicolor Time Machine* (Harrison), in which a movie director uses the eleventh century as a realistic setting for a picture. But this isn't necessarily always the result: The videograph gadget in "An Eye for History" (Derleth) is a failure because the films it produces of the past don't look "authentic enough" to Hollywood's moguls!

Who Else Might Be Interested in Time Travel

You don't have to be a dinosaur-steak fan or a movie producer to be fascinated with time machines; criminals and lovers are likely to be interested too. As the science fiction writer Larry Niven (1971) put it, "If one could travel in time, what wish could not be answered? All the treasures of the past would fall to one man with a submachine gun. Cleopatra and Helen of Troy might share his bed, if bribed with a trunkful of modern cosmetics." Or, as the tragically flawed inventor in Anthony Boucher's "Elsewhen" dreamed, before using time travel to commit the perfect murder, "The Great Harrison Partridge would have untold wealth. He could pension off his sister Agatha and never have to see her again. He would have untold prestige and glamour, despite his fat and baldness, and the beautiful and aloof Faith Preston would fall into his arms like a ripe plum." (Partridge's dreams are shattered, however, because he overlooks a detail about time travel, one that he wouldn't have missed if he could have read this book!)

Instead of viewing the past as an aid to crime, some authors have seen it as a possible dumping ground for criminals—a convenient place to get them out of the way—as in "My Object All Sublime" (Anderson), "In the Upper Cretaceous with the Summerfire Brigade" (Watson), and "Hawksbill Station" (Silverberg).

After all, there can be no breakout from the prison of the past—at least not without a time machine. As explained in "Vincent Van Gogh" (Gansovsky) on what might happen to criminal recidivists in a world that has time travel. "If you cannot live among people, then off to the reptiles—one hundred or one hundred twenty million years before the present. There you wouldn't freeze in a tropical preglacial climate, and you could nourish yourself on plants. But there is no one to talk with, boredom, and in the end you offer yourself up as an afternoon snack to a tyrannosaurus." And using mind travel for a particularly innovative form of criminal punishment is "Just Like Old Times" (Sawyer), in which the consciousness of a serial killer in 2042 is transmitted by the Chronotransference Institute back to the Cretaceous era—into the brain of a tyrannosaurus.

On the other side of the law-and-justice coin, a police officer in "The Future Is Ours" (Dentinger) travels to the future to learn new crime-fighting techniques, only to find that in the thirtieth century it is socially acceptable to be a crook, so there are no police. And in "The Face in the Photo" (Finney), criminals escape into the past with the help of a physics professor.

Museum curators would seem obvious clients for time machine companies, as would collectors of extinct species who are working for zoos. Indeed, in the first case we find just such a business, "Time Researchers," in *The Lincoln Hunters* (Tucker). T-R, whose corporate motto is "We Sift the Sands of Time," hires itself out as a futuristic version of Indiana Jones, recoverers of lost historical artifacts (for instance, it makes original sound recordings of one of Lincoln's unreported speeches) for customers who can pay the substantial charges [see also "Pushkin's Photograph" (Bitov)]. In "Man of Distinction" (Shaara), Genealogy, Inc uses its "time scanner" to provide its clients with a list of distinguished predecessors—its corporate motto is "An Ancestor for Everybody." And in *The Cross-Time Engineer* (Frankowski) we read of the Historical Corps, whose time travel agents are "writing the definitive history of mankind." Using the same idea, the 1991 film comedy *The Spirit of '76* has time-traveling historians from 2176 visiting the past in an attempt to reconstruct the lost records of the founding of America.

Another historical use for time travel to the past, one of the more unusual I have come across, was suggested in a philosophical article by Graves and Roper (1965), which considers a hoary question: "If everything in the Universe doubled in size overnight while we slept, could we tell what had happened when we woke up next morning?" The usual answer to this puzzle, called by philosophers the Universal nocturnal expansion, is *no* [but see Nerlich (1991) for why that is true only in Euclidean space, which is not the space of the actual universe]. Graves and Roper, however, suggest that even in Euclidean space, "all" we need do is take a yardstick back to yesterday and compare it with itself! A clever idea, yes, but as so often happens, this scholarly concept was introduced even earlier by a science fiction writer; see "Success Story" (Sycamore).

Several other uses for the past are discussed in *Mastodonia* (Simak). After going into the time travel business, for example, *Time Associates* receives a request from a United States senator who wants to send the disadvantaged of today back into the remote past, where they could have a fresh start with a virgin Earth. Yet another use comes from a religious fringe group that wants to purchase exclusive rights to the time of Jesus—not to visit, but to prevent *anyone* from visiting. The group fears that such visitors would "learn the truth," which might contradict the very legends that form the heritage of Christianity [this same concern is expressed in *The Fury Out of Time* (Biggle)]. And, in what may be the most ingenious idea of all, *Time Associates* itself does not directly do business in the present, but rather 150,000 years in the past in a "new" country called Mastodonia. You see, the corporate lawyer has determined that legally this means the company is thus a *foreign* company doing business outside the United States, so it is not liable for taxes to the IRS!

The tourist trade, in general, is a booming business in science fiction time travel, with dinosaur hunting at the top of the list: "Poor Little Warrior!" (Aldiss), "A Sound of Thunder" (Bradbury), and *Mastodonia* (Simak). For those who like absolutely nothing left to the imagination, the bloody dinosaur hunts described in David Drake's "Time Safari" and "Boundary Layer" are quite graphic. Far more cerebral are the stories in L. Sprague de Camp's short-story collection *Rivers of Time*. Starting with the original tale, the classic "A Gun for Dinosaur," these stories continue the adventures of Reginald Rivers, his partner the Raja, and their time-safari-for-hire dinosaur hunts. The earliest fictional use of time travel to the past for hunting—that I know of—was not for dinosaur hunting, however. In "The Loot of Time" (Simak), first published in 1938, a time machine inventor raises money for continued research by transporting hunters back 70,000 years to the Old Stone Age to shoot saber-toothed tigers, woolly mammoths, and cave bear.

Historical tours to great events of the past are described in *Up the Line* (Silverberg), "When We Went to See the End of the World" (Silverberg), "Vintage Season" (Kuttner and Moore)—the basis for the 1991 movie *Grand Tour,* which has a terribly muddled change-the-past ending not in the original story—and "Let's Go to Golgotha!" (Kilworth). Even the mundane may one day be the target for time-traveling tours. In "The Tourist Trade" (Tucker), for example, curious crowds from the very far future show up, nightly, inside the home of a twentieth-century family, much as tourists today visit Stonehenge. Flipping this idea on its head is "Closing the Timelid" (Card), a story in which the people four centuries hence have become so jaded that even trips back in time to watch Michelangelo paint the ceiling of the Sistine Chapel, or Handel write *Messiah,* have become boring.

As a symbol of the prehistoric past, perhaps nothing equals the dinosaur in power and vulnerability. After ruling Earth for hundreds of millions of years, about sixty-five million years ago they simply vanished. Did they die out because of an asteroid impact—or were they hunted to extinction? Asimov's "Day of the Hunters" (an expansion of his "Big Game") has a twist on who the hunters were, as does "The King and the Dollmaker" (Jeschke). In "The Twentieth Voyage of Ijon Tichy" (Lem) there is an ingenious time travel explanation for the death of the dinosaurs that has nothing to do with hunting. Rather, the first experiments in time travel run amok and inadvertently terrorize the far past. Besides creating Meteor Crater in Arizona and causing the ice ages (to say nothing of the canals of Mars!), a tremendous burst of "time-machine radiation" also kills off the dinosaurs.

In "Dinosaurs" (Landis) the dinosaurs are wiped out when a Soviet missile attack is diverted into the past. And at the other end of the spectrum, "Boundary Layer" (Drake) tells us the dinosaurs ended with a whimper when a captured tyrannosaur, temporarily brought into the present, returns to the Cretaceous era infected with ornithosis from bird droppings (a modern disease, to which the dinosaurs would have had no immunity). Clearly, time travel (in the minds of many science fiction writers) has not been kind to dinosaurs! Even more grandiose than killing dinosaurs, however, is the concept in "T" (Aldiss), in which an entire planet is destroyed by a time machine miscalculation in the past, thus creating the asteroid belt between Mars and Jupiter.

The obvious advantages to be gained by knowing the future, or knowing the history of things to come, receive an interesting presentation in "What We Learned from This Morning's Newspaper" (Silverberg). An entire block of homeowners wakes up on Monday, November 22, to find the *New York Times* for Wednesday, December 1, on their doorsteps. No explanation for this is given, except that perhaps it is some sort of fluke of the fourth dimension. The happy suburbanites do the obvious, and use the week's advance notice to make money on the stock market. And the developers of a time machine in "History in Reverse" (Laurence) use their knowledge of the stock market's past behavior, and a small test model, to go back a few days into the past to make enough money to complete a bigger machine. On the other hand, "Snulbug" (Boucher) takes the view that knowing the future provides no advantage at all; for example, if you try to change the future, you will get stuck in a temporal loop until you decide to give up the attempt. (As we'll see in Chapter Four, this idea of *repeating* closed time loops is not correct.)

In "Forever Is Not So Long" (Reeds) knowledge of the future proves to be a curse. On the night of the first human trip in time, the traveler says good-bye to his fiance and departs ten years into the future. It is England 1931 that night, and the future still looks bright and inviting. When he steps out into 1941, however, he of course finds devastation and death. Indeed, he learns that he was (will be) blown to bits at Dunkirk and that his wife, "now" a widow, lost an arm in a German air attack. So back he goes to 1931, where the words of greeting from his wife-and-widow-to-be are hollow with irony: "We'll be the happiest people in the world, Steve. The happiest, gayest, most in love people in the world. And we'll go on being that, Steve—forever."

Agreeing with Reeds on this gloomy theme is the Russian writer Dmitri Bilenkin. In his "The Ban" a physicist tries to discourage research into "left-spiraling photons," particles that can carry information into the past. He does this because he is convinced that knowledge of the future will cause only unhappiness. And in another of Bilenkin's stories, "The Inexorable Finger of Fate," a man tunes in a news broadcast from tomorrow; the result is only grief. (A reversal of this idea of transporting information from the future to the past is in the non-time-machine story "Stolen Centuries" (Kline). In that tale a man plans to travel into the future via drug-induced suspended animation. He reveals his reason for wanting to do this when he tells another character, "Do you realize that you are the last man I am going to see for three centuries! . . . In three hundred years I'll be alive . . . the hero of the hour—with [valuable] historical knowledge that will have been long forgotten.")

Egomaniacs would find time machines useful, too. In "Ghost Lecturer" (Watson), for example, the inventor of the "Roseberry Field" uses it to "yank past geniuses out of time, supposedly to honor them so they would know their lives had been worthwhile in the eyes of the future. But then he would go on to tell them—oh so kindly—where they had gone wrong or fallen short of the mark. And how much more we knew nowadays. 'You almost got it right, boy! You were on the right track, and no mistake. Bravo! *But. . . .*" Watson makes the interesting observation that one can easily imagine playing this pathetic game of Whig history with scientists, but what could even the most talented modern do to upstage a Mozart or a Shakespeare? (A similar idea is the basis for *Timescoop* (Brunner), which describes the ultimate family reunion—not only of those from the present, but also of long dead ancestors plucked from the past!)

And finally, the very funny "From the Annals of the Onomastic Society" (Watson) tells us of the future a quarter of a million years hence in which marriage patterns have resulted in the complete disappearance of diversity in surnames. *Everybody's* last name in the future is Chang. Accordingly, back to our present day travels one particular Mr. Chang, to collect the most precious resource of the past: a complete listing of the world's surnames.

Some Problems

The seductiveness of the past has never been written about better than by Jack Finney, who practically made writing fiction on this theme his specialty; see in particular his "The Third Level." The past isn't necessarily any better than the present, however, as is made clear in the humorous story "Status Quondam" (with the perfect generic Latin title, one suitable for any story about time travel to the past) by P. Schuyler Miller. Tired of the mid-twentieth century, that tale's hero travels back to the Greece of the fifth century B.C., only to be grateful in the end to escape literally with *just* his skin. (His initial attempt to return to the present falls a bit short, with his appearance in the nude in 1895, in the midst of a group of Victorians touring "modern" Athens.)

Another science fiction writer has, in fact, made the interesting observation that "in the nude" may actually be the most reasonable way for a time traveler to make his appearance. As we are told in "Barrier" (Boucher), nakedness is the one costume common to all ages, and we are asked, "Which would astonish you more, a naked man, or an Elizabethan courtier in full apparel?" With that question in mind, consider the time traveler from A.D. 5050 in "Time Tourist" (Murphy) who journeyed back to the twentieth century wearing historically authenticated clothing—a blue serge Hoover coat, a Sinatra silk butterfly tie, shiny red and yellow marching band pants, an opera hat, and ballet shoes. We are not surprised to learn that he created quite a stir. The time travelers in the 1984 movie *The Terminator* and its 1991 sequel *Terminator 2,* however, follow Boucher's suggestion and arrive from the future naked. And in *The Year of the Quiet Sun* (Tucker), it is said to be very expensive to transport mass through time; that novel's travelers travel nude simply to save money.

A less amusing aspect of backward time travel is illustrated by the visit to the past experienced by the protagonist of J. B. Priestley's story "Look After the Strange Girl," who suddenly and without explanation finds himself a half-century in the past, at a 1902 English upper-crust party. It is the year after Queen Victoria's death, and the merrymakers still share the common nineteenth-century conviction that God smiles with particular brightness on

England. Our time hopper, however, knows of the approaching horrors of the Great War and even the dates of the deaths of some of the people present. In lonely words (similar to those of the space traveler in Wilson's "Out Around Rigel") he realizes that "distance in time [is] apparently harder to bear than distance in space . . . back in his own time he would have felt less desolate, he was certain, if he had suddenly found himself wandering on South Cape, Tasmania, half the globe away from home. Was home, then, more in time than in space?"

The unstated horror of a trip backward in time is, of course, that it would bring the dead past, filled with all its dead occupants, alive again, literally resurrected from dank and moldering graves. The top of Mount Everest, the bottom of the Marianas Trench, the sands of Mars—none of these exotic places can even be mentioned in the same breath with the *past*. I recall, with a shiver, the opening words of a story that perfectly captures the mystery, the fascination, and, yes, the sheer terror that the past holds for us. See if you react as I do to the first line in Reginald Bretnor's "The Past and Its Dead People": "When Dr. Flitter came into the room, it seemed as though the past and its dead people came in with him, clinging to him like stale surgery smells, like the cold sweat of ancient autopsies."

Emily Webb's words in the last quote that opens this chapter are perhaps a warning that we should not ignore. In *The Devil in Crystal* (Marlow), for example, a time traveler anticipates a meeting with a long-dead lover as he "shivered with a renewal of horror. . . . She ought to be grateful to him for having raised her from the dead, even briefly." This horror of the "dead alive again" is graphically detailed in *The Victorian Chaise Longue* (Laski). Once the central character realizes that her mind has traveled backward in time, almost a century, into the body of another, the terror begins: "This body I am in, it must have rotted filthily, this pillowcase must be a tatter of rag, the coverlet corrupt with moth, crisp and sticky with matted moths' eggs, falling away into dirty crumbling scraps. It's all dead and rotten—these hands, all this body stinking, rotten, dead. She shuddered, and knew she was shuddering in a body long ago dead. Her flesh crawled away, and it was flesh that had turned green and liquescent and at last become damp dust with the damp crumbling coffin wood."

The same ghoulish feeling is expressed by one time traveler (who has fled the oppression of a twenty-fourth century tyrant) to another in *Time Echo* (Lionel), as they ponder their situation five hundred years in the past: "The main point seems to be how we are going to get back to our time. I thought when I first came here I would be able to escape permanently, but living in the past is like living with the dead." When one of the time-traveling characters in *Voices*

After Midnight (Peck) dances at a nineteenth-century masquerade party, he, too, suddenly sees his partner as she will be in his time: "[She] was lowering her mask. I opened my mouth to scream . . . It was a skull turned brown by time and stained by the graveyard years. I looked around me, and everybody in the fading ballroom was like her . . . skulls stared at me because all of these people were dead, long dead. And I had no business being here." And the time travelers in "Ghost Ship" (Simons) call all they encounter in the past *ghosts* (the ship in the title is the particular temporal target of the story, the *Titanic*.)

Even the "mere" communication across time can result in horror, and no writer has better captured the essence of that experience than have Ambrose Bierce and H. P. Lovecraft. Bierce's "John Bartine's Watch" is the story of a Tory traitor who was executed by the American rebels, only to reappear in a ghastly manner, across a hundred years, in the body of his great-grandson. Lovecraft's weird, macabre tale "The Shadow Out of Time" tells about Professor N. Wingate Peaslee of Miskatonic University and of his astonishing discovery that his mind has been among grotesque creatures that inhabited Earth 150 million years ago. These stories by Bierce and Lovecraft are time travel horrors in the marvelously disgusting (I use these particular words in admiration, by the way!) tradition of Poe. When Wells' brilliant work *The Time Machine* appeared in serial form in 1895 in the *New Review,* the *Review of Reviews* called the author "a man of genius" but then ended its notice—see Parrinder (1972)—with "he has an imagination as gruesome as that of Poe." Indeed, Poe himself experimented with time travel via exchanges of personal identity [see his 1844 "A Tale of the Ragged Mountains" in Franklin (1966)], but as I have already indicated, that is not a time travel mechanism of scientific interest in this book.

This potential for literary horror almost certainly explains why more than one writer has speculated that Jack the Ripper (whose identity is still unknown) may have been (and perhaps still is) a time traveler—see "A Toy for Juliette" (Bloch) and "The Prowler in the City at the Edge of the World" (Ellison). A time traveler, by definition, can strike at literally any time, and to imagine Jack suddenly materializing behind you while you walk on a dark, foggy London street *is* a bit unnerving.[4] The 1979 movie *Time After Time* played with this idea: Jack fled to modern-day San Francisco after stealing a time machine, with H. G. Wells himself in hot pursuit.

Another concern is that a trip into the past would require careful planning and not just a little crafty caution. As the opening line of L. P. Hartley's 1954 novel *The Go-Between* says, "The past is a foreign country: they do things differently there." And to be different may possibly invite unpleasant attention, as the time traveler in *Berkeley Square* (Baldeston) learns to his unhappy surprise. By revealing the knowledge of a 1928 man to people he meets in 1784, he first draws nervous glances and then arouses outright hostility and fear. The time traveler in *The Devil in Crystal* (Marlow) knows the danger he runs right from the start. As he thinks to himself while in the midst of friends, "Will they notice anything funny about me? They'd tear me to pieces if they found out! Alien, as no one else could be alien. They couldn't stand it if they knew! The only thing would be to kill me." Certainly people notice something funny about the old man who can literally walk through time in "The Ghost" (Van Vogt). His ability to tell them their futures boosts both their curiosity and their level of nervousness whenever he appears on the scene.

And finally, as I've mentioned before, to travel in time is to run the risk of great emotional trauma. I know of no stories that better capture the pathos a trip back in time might have, or the plight of being trapped out of one's normal time in the foreign land of yesterday, than "The Man Who Came Early" (Anderson), "The Doctor" (Thomas), and "The Ugly Little Boy" (Asimov). Anderson's sad tale is of a modern soldier suddenly hurled a thousand years into the past, as was the archeologist in *Lest Darkness Fall* (De Camp), by a powerful electrical storm. Even his high-technology weapons cannot save the soldier from a brutal end in the violent tenth-century world of warrior Iceland. Thomas' story tells of the tragic fate of a medical doctor, the lone survivor of the first time travel expedition gone terribly wrong. Trapped half-a-million years in the past, he is doomed to live out his life among primitive cave dwellers. And Asimov's novella, perhaps one of the most moving of all time travel stories, tells us of a nurse who refuses to let a little Neanderthal boy be sent back to the past alone. She has come to love him as her own, and when the scientists who plucked him out of time grow weary of him, she goes back with him—to unknowable, almost surely fatal dangers.

Backward in Time—Can It Really Be Done?

The question I want to address before going any further is the obvious one. After all the preceding discussion of literary time travel, is *physical* time travel to the past possible? It is one thing to say that special relativity permits travel to the future, but it is quite another to ask whether one could return (and the present is, of course, the future's past). As we study this question throughout the rest of this book, you will find that as Professor Monte Cook (1982) put it, "travel through time is an odd business" and, in particular, that "travel to the past" is indeed when "time travel is at its oddest."

Science fiction writers, ironically, are generally not enthusiastic boosters of what Polish writer Stanislaw Lem calls "chronomotion" (or, alternatively, "chronokinesis," a word invented by science fiction writer and editor Anthony Boucher). Isaac Asimov unequivocally rejected backward time travel, even though a glance at the Bibliography of this book shows that he loved to write stories using the idea. For example, he wrote (1984b) that "The dead give-away that true time travel is flatly impossible arises from the well-known 'para-doxes' it entails. . . . So complex and hopeless are the paradoxes . . . so whole-sale is the annihilation of any reasonable concept of causality, that the easiest way out of the irrational chaos that results is to suppose that true time travel [Asimov's term for travel at will via a machine under human control both for-ward and backward in time—what is today called a *strong time machine*] is, and forever will be, impossible."

Asimov was consistent in his negative attitude toward time travel. In an essay prompted by the movie *Peggy Sue Got Married,* he wrote (1986a): "In many science-fiction stories, the trip into the past is by way of some futuristic machine that can take you through time at will. . . . That, however, is totally impossible on theoretical grounds. It can't and won't be done." And later, in the Introduction to Damon Knight's time travel story "Anachron," Asimov stated, "To my way of thinking it is precisely because time travel involves such fascinating paradoxes that we can conclude, even in the absence of other evi-dence, that time travel is impossible."

Asimov's fellow science fiction writer, Arthur C. Clarke, is more cautious and hedges just a tiny bit, first declaring (1972) that "I feel certain (well, prac-tically certain!) that time travel is impossible" and later (1985) that "I do not take time travel very seriously; nor, I think, does anyone else." Certainly Basil Davenport didn't; writing in his Introduction to Olaf Stapledon's impressively massive (if often ponderous) "history of the universe," Davenport observed (1953) "There appears to be more objective evidence for apparitions—ghosts—. . . than for the possibility of time travel."

Some early science fiction fans shared Davenport's view; in the January 1933 issue of *Astounding Stories,* one letter to the editor declared, "No one can travel forward or backward in time. . . . I believe that it will be possible, me-chanically at some time, to communicate with the spirits of the departed, but this other—I can't see it!" And if a science fiction believer in the possibility of time travel (rather than simply a *story writer* using the idea) did surface, others were quick to pounce. For example, when P. Schuyler Miller (an early teenage fan who later authored at least two time travel classics) wrote a letter to the ed-itor at *Astounding Stories* (June 1931) stating that "there is nothing in physics

. . . to prevent yourself from going into the past . . . and shaking hands with yourself or killing yourself," he got this in reply (*Astounding Stories,* December 1933): "P. S. Miller once wrote that time traveling is not incompatible with any laws of physics. . . .' he don't know from nothin.'" This intemperate letter did, however, offer up an interesting technical objection: "Time traveling is impossible because it is contrary to the laws of conservation of mass and energy," a concern I will pursue (with the modern response) in Chapters Two and Four of this book.

Taking the contrary position and arguing for the rationality of time travel is the eminent philosopher David Lewis, who began a famous essay (1972) with the lines "Time travel, I maintain, is possible. The paradoxes of time travel are oddities, not impossibilities. They prove only this much, which few would have doubted: that a possible world where time travel took place would be a most strange world, different in fundamental ways from the world we think is ours." Robert Forward, until 1987 a senior scientist on the Director's staff at the Hughes Research Laboratories (and the author of several imaginative science fiction novels) thinks similarly. Writing as a long-time researcher in experimental general relativity, he declared (1988) with admirable optimism, "Some of us now living may . . . wonder [after time machines have actually been invented] about those ancient philosophers of the twentieth-century who worried so much about those 'time-machine paradoxes.'"

There are many besides Asimov, Davenport, and Clarke, however, who think Lewis and Forward are wrong. The "hard science" science fiction writer Larry Niven, for example, is so convinced of the impossibility of time travel that he has elevated its denial to a metaphysical law (1971):

> Niven's law: If the universe of discourse permits the possibility of time travel, and of changing the past, then no time machine will be invented in that universe.

A weakened form of Niven's law is the idea that if a universe does allow the creation of a time machine, then its use will change the past to *undo* the invention! As I write, for example, the wormhole time machine (see Tech Note 9) is the one being talked about most in a serious way, and one commentator (C. J. S. Clarke) has invoked a sort of Niven-type argument against it (1990). After conceding the remote possibility that a stable wormhole might somehow be formed, Clarke asks what might happen if one really tried to use it as a time machine. His answer: "The most likely outcome [is] that the back-reaction of the physical fields on the space-time would destroy the time machine."

Clarke's view was actually anticipated in fiction, in the story "Time Bomb" (Zahn), with its vivid descriptions of quantum field back-reactions on all attempts to build a theoretically possible time machine. Still, I think Niven himself would probably not accept even the slight possibility of time travel. He simply says, "No time machines! Ever! Period!" (as with Asimov, however, the Bibliography testifies to Niven's fascination with the concept). David Gerrold dedicated his now-classic time travel novel *The Man Who Folded Himself* to Niven: "A good friend who believes that time travel is impossible. He's probably right." And just as blunt as Niven is the well-known physicist Milton Rothman, who flatly declared (1988) that "no one will ever build a time-travel machine."

An interesting variation on Niven's law is an anthropocentric twist that says time travel paradoxes are actually acceptable as long as there are no conscious, intelligent entities in the universe to be aware of them—see Gribbin and Rees (1989). That is, a paradox isn't really a paradox unless there is somebody around to be bothered by it! This strikes me, personally, as a conceited (and totally unacceptable) argument, but some science fiction writers have actually used it. For example, in the novel *Dead Morn* (Anthony and Fuentes) a time traveler from 2413 visits the revolutionary Cuba of the late 1950s and early 1960s in an attempt to change history (and thus patch up his own rather dismal future). While in the past, he not only discovers the presence of a "paradox shield" that prevents him from doing things like killing his own ancestors but also finds that "the shield had to act in a natural or coincidental fashion, lest it create paradox itself by overt manifestations." Such a shield puts a crimp on the time traveler's actions, of course, but not for long, because when you are writing fiction you can make up your own rules. So the authors have their time traveler overwhelm the shield by *embarrassing* it—"He could fight the shield by forcing it to be obvious." Apparently, according to this line of argument, when caught with its fingers in the cookie jar, spacetime physics simply blushes and withdraws its objections!

Another science fiction writer, Murray Leinster, had one of the characters in "Interference" actually offer up the following curious "proof" of the impossibility of time travel: "There was a guy proved that if you could travel in time you'd have to pass through all the time in-between where you started from and where you went, and passing through means being there, so you'd have to spread out through all the time in-between. And if you were spread out over a coupla hundred years it would be the same thing as being spread out over a coupla hundred miles. . . . Not healthy."

One well known philosopher, J. J. C. Smart, has offered an even more succinct (if no more convincing) impossibility proof (1963): "Motion is rate of change of *space* [my emphasis] with respect to time, and so we cannot have motion through time." Philosophers seem much more prone to this sort of "proof by grammar" than are physicists, and I personally am unpersuaded by Professor Smart's argument. But there is no denying that he speaks for many. One author, to demonstrate how utterly implausible was Ernest Lawrence's plan to build a one-hundred-million-volt cyclotron just before the start of the World War II, wrote that "Lawrence's project for a hundred million volts was no more practicable than a time machine."[5] For Professor Smart, and those of his persuasion, time traveling is even more unlikely than (as declared by Robert Louis Stevenson) the "welding ice and iron."

The Problem of Paradoxes

What might happen if a time traveler changed the past? This concern is described in *The Anubis Gates* (Powers). When the time traveler in that novel (a work shot through more with magic than with physics) finds himself stranded in the London of 1810, he takes solace with "I could invent things—the light bulb, the internal combustion engine, . . . flush toilets. . . ." But then he thinks better of doing that: "no, better not to do anything to change the course of recorded history—any such tampering might cancel the trip I got here by, or even the circumstances under which my mother and father met. I'll have to be careful."

The above is really just a recent treatment of the change-the-past paradox that was already well established even in early science fiction. In a story originally published in 1936, "Tryst in Time" (Moore), we can find the paradox explicitly stated, along with a possible solution that is similar to the kind of explanations that have appeared more recently in the philosophical literature; see, for example, Fulmer (1980):

> "Suppose you landed in your own past?," queried Eric.
> Dow smiled.
> "The eternal question," he said. "The inevitable objection to the very idea of time travel. Well, you never did, did you? You know it never happened!"

Reasoning the same way was Batman in 1944, during his first comic strip trip in time back to ancient Rome. Once there, Batman reveals his secret identity without concern because, as he correctly explains, history shows that his identity is not known to the "present" (the future of ancient Rome, of course) as knowledge from the past. Thus, revealing his identity in the past cannot possibly pose a danger.

Martin Gardner (1974), however, who for years wrote the acclaimed "Mathematical Games" column in *Scientific American,* disagrees with Batman and thinks that time travel will be, at best, enormously difficult because of the paradoxes that philosopher David Lewis so boldly accepted in the previous section. The classic change-the-past paradox is, of course, the so-called grandfather paradox, which poses the question of what happens if an assassin goes back in time and murders his grandfather before his (the time-traveling murderer's) own father is born. If his father is never born, then neither can be the assassin—so how can he go back in the first place to murder his grandfather!? (In an amusing observation, Hollinger [1987] notes that "this is never posed as a 'Father Paradox.' It is as if [science fiction] is evading the Oedipal aspects implicit in its favorite model of temporal paradox." Soon afterward, however, Caltech physicist Kip Thorne began to refer to the paradox of a time traveler killing his own *mother*, so now time travel *is* sexually unbiased when it comes to murder in the past.)

One story, "Status Quondam" (Miller), pushes the grandfather paradox to its logical limit to show the supposed dangers of combat for a time traveler in the past. Having traveled to Greece in the fifth century B.C., the protagonist suddenly realizes (with just a little exaggeration): "Ninety-five generations back you'd have more grandfathers than there are people on Earth, or stars in the Galaxy! You're kin to everyone. . . . You as much as take a poke at anyone, and the odds are you won't even get to be a twinkle in your daddy's eye." An earlier (1933) story, "Ancestral Voices" (Schachner), illustrated this point in a graphic way—in A.D. 452 a time traveler shoots and kills one of Attila's Huns (who would have been his great-grandfather many times removed); the result is that 50,000 of the Hun's descendants vanish. So dramatic did Schachner's readers find this that the following year, in "The Time Imposter," he repeated the idea.

An even vaster result of such tampering with the past is in "The Mosaic" (Ryan). In that story the Moslem defeat by Christians in 732 "originally" is a Moslem victory. Centuries later the first time traveler (a Moslem) accidently changes the victory into a defeat and, at the crucial moment, vanishes "with all the suddenness of a bursting bubble. And with him into nothingness, across the gulf of Time" went all the history after 732, having been instantly changed to "our" world's history that records the ancient victory of Cross over Crescent. In Chapter Four we will see why this particular view, while entertainingly provocative, is illogical.

The idea of killing ancestors in the past is used to great effect in the 1984 film *The Terminator.* In that film a time-traveling robot-killer from the future of 2029 appears in the Los Angeles of the present to murder the woman who will (already has?) give(n) birth to a son who has (will have?) enemies in the

future. In the 1985 movie *Back to the Future,* a comic spoof of this same potential sexual paradox, Marty McFly uses the plutonium-powered "flux-
capacitor" to travel back to the 1955 high school days of his parents. When he
accidently almost prevents his parents from marrying (the *required* reason for
his own existence—remember, it's the 1950s!), Marty is horrified to see himself fading out of an old picture of the family that he carries in his wallet.

As one philosopher has observed—Horwich (1987)—it is, of course, not
necessary to kill your grandfather, or even more remote ancestors (as in
Schachner's stories), to cause a paradox. Using the Marty McFly approach you
could commit (or at least *try* to commit) what Horwich calls "autofanticide,"
the killing of yourself as an infant. This idea was actually anticipated in fiction
by the Russian author of "Vanya" (Grigoriev), and an even earlier version of
the idea appears in the 1940 pulp science fiction tale "Perfect Murder" (Gold).
One science fiction solution to the paradoxes of killing your ancestors or your
younger self is that somehow the murder fails to occur: The gun jams, the
blade snaps, the poison is old, a gust of wind blows the flaming arrow off-
target, the murderer faints just before she can do the foul deed, *whatever.* No
matter how many times the murderer tries and no matter how clever her
scheme, she fails. Such an explanation was offered by C. L. Moore in her 1936
story "Tryst in Time."

This particular problem of the unchangeability of past events is of special
interest to theologians, because it is directly related to the question of free will
versus fatalism—that is, are humans the creators of the future, or are they mere
fated puppets of destiny? Is a time traveler to the past unable to alter events because that was the *only* way they *could* happen? The Bible offers us no definitive
help on this issue. In his *Guidance to the Duties of the Heart,* the eleventh-
century Jewish philosopher Bahya ibn Paquda lists several scriptural texts in
support of predestination and yet offers another list in support of free will; for
instance, compare Psalm 127 with Job 34:11. Bahya aptly presents his lists in
the form of a dialogue between the (rational) mind and the (emotional) soul.
In this dialogue the mind attempts to ease the soul of its concern with the "ills
of the body," one of which is the conflict between free will and fatalism.

That same ill lurks in the 1950 story "Typewriter from the Future"
(Worth), which offers a solution: "The answer is quite simple. When the man
goes back in time and kills his grandfather, and returns to his own time again,
he finds to his surprise that he made a mistake. It was not his grandfather at all!
And no matter how many times he goes back and kills his grandfather . . . he
always finds he made a mistake." A twist on this idea is that a recurrent, psychological failure of will is the critical factor, as in "Time Enough" (Knight).

That moving story has a thirty-year-old man continually trying to relive and reshape the painful boyhood incident that warped his emotional development. He fails again and again, but of course there is always "time enough" to try once more.

Or perhaps the time traveler never gets a second chance. In one of Robert Silverberg's stories, "The Assassin," the doomed hero journeys back to 1865 to save Lincoln from Booth, but his "time-distorter" is quickly taken away from him by suspicious guards. Its internal workings tick and they think he is an assassin with a bomb. They destroy it, haul him away to his fate, and Lincoln goes on to meet his.

L. Sprague de Camp's famous story "A Gun for Dinosaur" puts forth the terrifying suggestion that nature itself will take any corrective action required to avoid paradoxes—or "chronoclasms," as they are called in "The Chronoclasm" (Wyndham). De Camp's story has two big-game hunter-guides using "Professor Prochuska's time machine at Washington University" (built with the aid of a "cool thirty million"-dollar grant from the Rockefeller Foundation) to run a safari-for-hire business that takes hunters back to the late Mesozoic era. When a disgruntled client tries to go back to the day *before* a previous trip to shoot the guides (who had displeased him, or rather *would* do so the next day), we learn just how nasty De Camp thinks Mother Nature might be in order to avoid a paradox. (The guides had, of course, *not been* shot on the safari, so they *could not be* shot): "The instant James started [to ambush the guides] the space-time forces snapped him forward to the present to prevent a paradox. And the violence of the passage practically tore him to bits [making his body look] as if every bone in it had been pulverized and every blood vessel burst, so it was hardly more than a slimy mass of pink protoplasm."

Ray Bradbury's classic 1952 story "A Sound of Thunder" also uses the "dinosaur hunters in the past" idea, but it *does* allow for changing the past. (This story was the inspiration for a wonderfully funny episode of the 1990s prime-time television cartoon series "The Simpsons," in which the moronic Homer Simpson accidently converts his toaster into a time machine. While in the past, he kills off the dinosaurs by sneezing on them. This scenario is *not* a paradox, however, but rather an "explanation" for the known historical record, a form of explanation I'll take up later in the book.) In Bradbury's tale, a client fails to follow the instructions of his guide to do *nothing* in the past except shoot a dinosaur that is about to die for "other reasons" anyway—he accidently kills a butterfly. This results in enormous changes in history, as indicated by the "before" and "after" versions of the time machine company's ad:

before
TIME SAFARI, INC.
SAFARIS TO ANY YEAR IN THE PAST.
YOU NAME THE ANIMAL.
WE TAKE YOU THERE.
YOU SHOOT IT.

after
TYME SEFARI INC.
SEFARIS TU ANY YEER EN THE PAST.
YU NAIM THE ANIMALL.
WE TAEK YU THAIR.
YOU SHOOT ITT.

Bradbury describes the death of the butterfly as having started the knocking "down [of] a line of small dominoes and then big dominoes and then gigantic dominoes, all down the years across Time." This is, of course, a somewhat unconvincing argument. After all, previous dinosaurs, when shot, must have fallen to the ground and flattened a lot of butterflies! With such threats for every decision, no matter how seemingly innocent, hanging over the head of a time traveler, it would take a brave soul to do much more, while in the past, than just stand still and breathe.

Indeed, the novel *Krono* (Harness) introduces the curious idea of prospective time travelers to the past being required first to file Historical Impact Statements with the proper authorities. Not all would receive permission. In "The Rescuer" (Porges), for example, we have the story of a time traveler who takes a rifle and five thousand rounds of explosive bullets back to Golgotha. His intention—to be history's first Rambo by picking off any Roman soldier who gets within a hundred yards of Jesus! As outrageous as this concept is (but who among those now reading this won't admit to at least a momentary thrill at the idea, and perhaps even a secret willingness to do it themselves, if they could), it isn't the story's peak. That comes when the reader is reminded that it was Christ's *desire* to die on the Cross, that he *had* to die for our sins; to prevent that from happening would change all of history for the last two thousand years. What, then, should the time traveler's colleagues do when they discover his plan? Should they stop him or not? What might happen if they do interfere?

A twist on the idea of a time traveler saving Jesus appears in "Un Brilliant Sujet" (Rigaut), in which the time traveler commits many disturbed acts in the past, including murdering the infant Jesus with an injection of potassium cyanide. David Gerrold has used changing the past with an equally chilling effect in *The Man Who Folded Himself*. A time traveler in that novel experiments with "making things different." In his words (and the last word is particularly horrible), "Once I created a world where Jesus Christ never existed. He went out into the desert to fast and never came back. The twentieth century I returned to was—different. Alien."

This is an old idea. Mark Twain had Satan give a good lecture on it in an early draft (written before the turn of the century) of his last novel, *No. 44, The Mysterious Stranger:* "If at any time—say in boyhood—Columbus had skipped the triflingest little link in the chain of acts projected and made inevitable by his first childish act, it would have changed his whole subsequent life, and he would have become a priest and died obscure in an Italian village, and America would not have been discovered for two centuries afterward. I know this. To skip any one of the billion acts in Columbus' chain would have wholly changed his life. I have examined his billions of possible careers, and in only one of them occurs the discovery of America."

A far more horrifying tale is "Brooklyn Project" (Tenn), by Penn State English professor Philip Klass, who writes under the pen name of William Tenn. In that classic story the first major experiment in time travel results in horrendous changes, with humans as observers at the start but in the end altered into things with "purple blobs" on "slime-washed forms." The irony of the story derives from the reader's knowledge of what has happened, while the once-human characters "see" nothing happen.

Taking the opposite approach in tampering with the past is "Thus We Frustrate Charlemagne" (Lafferty). Here the tampering is intentional, done for the specific purpose of observing the changes in the present that are caused by a single, precise alteration centuries earlier. In this story the past is treated as something like a lump of Play-Doh. A similar approach to time travel is taken in *Lord Kelvin's Machine* (Blaylock). As Herbert (1988) poetically declares, "Disembarking in the past without warning, uninvited guests from the future would be free to change crucial events, the effect of these changes rippling through time to modify the present drastically. The ability to willfully change the past amounts to a virtual omnipotence over human history. No mere temporal tourists, these doughty travelers in time. Omniscience, omnipresence, and omnipotence are the traditional attributes of divinity. The power to execute U-turns on time's one-way street would make time travelers nothing less than gods." (You should realize that this is *not* the prevailing philosophical *or physical* view of the powers of time travelers!)

Herbert's view also seems, at first, to be expressed in Robert Frost's famous poem "The Road Not Taken." It opens with "Two roads diverged in a yellow wood,/ And sorry I could not travel both." Then later come the lines "And both that morning equally lay/ In leaves no step had trodden black./ Oh, I kept the first for another day!" Could these words be interpreted to mean one could later return and "do things" differently? The ending of the poem, however, makes it clear (I *think*) that Frost was consciously thinking of the *crucial* (unchangeable) nature of decisions: "Two roads diverged in a wood, and I—/ I took the one less traveled by,/ And that has made all the difference."

From that I think Frost would not have agreed with the presentation in "All Our Yesterdays" (Farrell), a tale of the far future in which watching the past is a leisure-time activity for the upper classes. One of these idlers shows his girl-friend first the execution of a criminal a thousand years in the past (in 1949, the year this story was published) and then the actual crime itself, which demonstrates that the condemned man was really innocent. She then pleads with the "watcher" to change the past, to stop the execution. He at first resists, saying, "My dear girl, don't be absurd! To alter the objective past would be like kicking out the bottom block of a tower. We are built on the past. . . . If [the executed man lived, some of his] descendants would be alive today. And who knows what alterations they would have made?" She persists, however, and he finally yields—with disastrous results. With the past changed, the watcher is no longer a privileged upper-class idler but rather is a beggar on Mars pan-handling for the rocket-fare back to Earth!

The idea that the past might be changed and then consciously changed back again if things don't improve is, interestingly, *not* a common one in fiction. But it does occur now and then; the earliest example that I know of is in the 1937 stage play *The Star-Wagon* by Maxwell Anderson. There a proverbial "self-taught" inventor, when told by his wife of thirty-five years that perhaps their marriage has been a mistake, travels back to 1902 to give them both a chance to make a new future. When once again we see them in 1937, matters are much worse with their new mates. So, back again to 1902 goes the time traveler, to put things *back* to the way they were.

More recently, the "change the past until you get it right" idea appeared in the novel *Changing the Past* (Berger)—the dust jacket of his book cleverly shows an eraser against a blank background. The classic film based on what-would-happen-if-you-could-change-the-past (and then change it back if you don't like what happens) is, of course, the 1946 *It's a Wonderful Life*. More recently, the same theme was the basis for the 1990 movie *Mr. Destiny*. Both of these pictures are fantasy, not science fiction, in as much as they use supernatural forces—angels!—to "explain" the changes.

The Fictional Origins of "Change the Past"

In a January 1963 personal letter to the editor of *The Magazine of Fantasy and Science Fiction,* Robert Heinlein wrote (GG): "Mark Twain invented the time-travel story; six years later H. G. Wells perfected it *and its paradoxes* [my emphasis]. Between them they left little for latecomers to do." How a man as widely read as Heinlein, who is the author of two of the best time travel short stories ever written ("By His Bootstraps" and "All You Zombies—"), could write such an erroneous statement is a mystery to me. *A Connecticut Yankee in King Arthur's Court* and *The Time Machine,* certainly both works of genius, are *not* pioneers in paradox. And Heinlein's own contributions are proof enough that there was a lot left to do with time travel well after 1900.

The very first story to be written that even hints at the particular time travel paradox of changing the past seems to be by the Unitarian minister Edward Everett Hale (1822–1909), best known today as the author of the 1863 story "The Man Without a Country." Hale wrote "Hands Off" in 1881 and published it anonymously in *Harper's New Monthly Magazine* with the express purpose of stirring up some theological debate (which it didn't), and he had no idea that he would come to be recognized by literary scholars as a pioneer in the yet-to-be invented genre of science fiction.

Hale's story opens with the mysterious words "I was in another stage of existence. I was free from the limits of Time, and in new relations to space." These words are spoken by an unnamed narrator who seems to have just died and who finds himself, in his new "form," observing "some twenty or thirty thousand solar systems" while in the company of "a Mentor so loving and patient." Under the guidance of this Mentor, in an attempt to "improve" history, the narrator alters the Biblical account of Joseph and his imprisonment in Egypt. At first, subsequent history *is* better, but then humanity sinks into irreversible depravity. In the end the narrator watches the last handful of humans kill each other at a particularly symbolic place for the Christian world: "The last of these human brutes all lay stark dead on the one side and on the other side of the grim rock of Calvary!" There would be no Crucifixion and Resurrection for the salvation of humankind, which naturally much disturbs the narrator. But the Mentor calms him, saying "Do not be disturbed, you have done nothing." It has, you see, been just an experimental world, an alternate Universe, and the narrator has learned the lesson of "Hands Off."

Hale's story, of course, is a better Sunday sermon than it is a change-the-past time travel story, and the device of experimenting on a not-really-real Earth is disappointing from a modern science fiction point of view. But Hale's story almost certainly did have an immediate if indirect impact. There is no absolute proof, but with its appearance in a national magazine, it seems likely that Hale's story was read by Edward Page Mitchell (1852–1927), an editor on the daily New York newspaper the *Sun*. Just six months after Hale's story appeared, the *Sun*, in its issue of September 18, 1881, printed Mitchell's "The Clock That Went Backward." That tale, published anonymously, was both the first to use a machine for time travel *and* the first to incorporate a time travel paradox as its central idea. The story predates Wells' *Time Machine* by fourteen years, and Wells' novel did not include a paradoxical element. [Wells' failure to use paradox in his famous novel surprises most modern readers, and in fact one of the very first reviewers criticized him in detail on this lapse. See the 1895 review from *Pall Mall Magazine* by Israel Zangwill in Parrinder (1972)].

The mechanism of Mitchell's time machine, an eight-foot-high, sixteenth-century Dutch clock, is quite simplistic, even bordering on fantasy. It is simply stated that if the clock runs backward, then it travels backward in time—a rather disappointing explanation. The same idea was used eight years later (still six years before Wells) by Lewis Carroll in his *Sylvie and Bruno,* which featured the "Outlandish Watch." This marvelous timepiece "has the peculiar property that, instead of *its* going with the *time,* the *time* goes with *it*." That is, the time its hands are set to *is* the time. So attractive is this idea that it is still occasionally used; examples include "The Time Piece" (Wilhelm). Like Mitchell, Carroll did not provide a scientifically plausible reason for the remarkable time travel property of his clock. The story is typical Carroll nonsense, and, indeed, because his story was for children, Carroll *wanted* to avoid anything plausible!

It was Wells who took the final, crucial step of presenting a science-based rationale for the workings of a time machine. This was the original contribution that Wells introduced in his *The Time Machine*. Ever since these stories of Hale, Mitchell, and Wells, then, the question of changing, or trying to change, the past has been one of the fundamental puzzles of time travel via a scientifically explained machine. The fascination of this puzzle is difficult to exaggerate.

Wells, in particular, was not oblivious to the possibility of paradox in time travel, but his failure to use it seems to indicate that he simply did not know how to respond to such puzzles. In the opening of *The Time Machine,* for example, during the dinner party at which the Time Traveller tries to convince his friends of the possibility of a time machine, one of them observes that "It would be remarkably convenient for the historian. One might travel back and verify the accepted account of the Battle of Hastings, for instance." To that another guest replies, "Don't you think you would attract attention? Our ancestors had no great tolerance for anachronisms." The Time Traveller has no reply to that, but after Wells there were those who had a great deal to say about time paradoxes.

An old 1960s television science fiction program, "The Outer Limits," for example, did an interesting production of a story about changing the past (and the future) called "The Man Who Was Never Born" (available today on videotape). A horribly deformed man from two hundred years in the future travels backward in time to kill the scientist whose research will lead to the medical disaster that mutilates humankind (hence the time traveler's terrible physical appearance). He arrives rather too far back in time, however, and finds his target has not yet been born. He locates the mother to be, though, and prevents her marriage by the simple act of bringing her back with him to his world of the future. Thus is the birth of her baby prevented, and this indeed changes history. But there is an unexpected twist—the title refers not only to the unborn scientist but also to the time traveler himself, because he was never born in the "new" future, either. He vanishes, leaving a very confused young lady to try and sort things out.

Another author made the point that perhaps the past might be changed, but what happens after the change may be rather unpredictable. As his cleverly titled story "Demotion" (Locke) opens, Mars has been settled for a century and a half, since five years after the end of the six-month World War III in 1960. (World War II had ended after a Lieutenant Charles Leslie had bombed Berchtesgaden and killed Hitler in 1943—this is clearly not "our history.") Faced with an apparently unresolvable military crisis with the settlers on Mars, the "Temporal Lab" concludes that the planet was opened up for colonization too soon. The initial rocket landing on Mars was headed by the then General Charles Leslie, so the Temporal Lab reaches back in time to disable the rocket and thus abort the landing and delay colonization.

Alas, the crisis is not undone. Things *are* a bit different back in the Temporal Lab, yes, but nobody notices because history after 1965, and memories, have been revised. The problem with Mars is now thought to be solvable by not letting World War III end so quickly. This will be accomplished by canceling the event that resulted in the war's termination—the use of the first atomic bomb by none other than Colonel Charles Leslie (apparently history before 1965 has been altered, too). So once again, Leslie will be denied. This is done, but still the Mars crisis persists. Additional changes in the Temporal Lab have also occurred, of course, but of course nobody notices.

The new fix is to let World War II go on for a while more, and so the Berchtesgaden bombing is undone. Hitler lives and the Mars crisis at last ends. The story closes with Charles Leslie (who has had the high points of his entire military career blotted out, although of course nobody knows this because these milestones didn't happen!) leaving the Air Force after World War II, getting a Ph.D. in mathematics, and becoming a professor. (Presumably this is yet another demotion.) Professor Leslie's area of research is on how to change the past, so the end result has been a future (with a history different from ours) changing the past to end up with our history via the manipulation of the life of a single individual, who turns out to be the inventor of the theory that allows such changes to be made. The next time you have trouble going to sleep, try mulling that over for a while!

The irony of a time traveler unwittingly changing the past (but not knowing it because his memory is also altered), while the reader of course does know, is attractive to writers, and the idea periodically recurs in fiction. See, for example, "Don't Look Back" (Gribbin), which tells the "real" stories behind the early (as history now records them) deaths of the musicians Buddy Holly and John Lennon. A much longer short story, which consequently has more room to develop carefully the idea of time travel changing the past, is the brilliant "Vincent Van Gogh" by the modern Russian writer Sever Gansovsky. This is the story of a time traveler who attempts to make money by acquiring some of Van Gogh's paintings from the great artist himself. The traveler finds, however, that his presence in the past always causes unforeseen disturbances, and he learns first-hand why society has passed "The Universal Law for the Preservation of the Past," forbidding the mucking up of history by time travelers.

Ways to Avoid Paradoxes

The various forms of changing-the-past paradoxes, and of what seems to be a lack of free will, are of course prominent among the reasons why philosophers and theologians have been so attracted to the question of the possibility (or impossibility) of time travel. If a time traveler could visit the past, and were able to change events, then paradoxes like killing your grandfather would result. On the other hand, if the time traveler could not change events, then *why* not? Is free will simply an illusion? And what if we could see the future? Could we then change it? Does that question even make any sense? That is, does it mean anything to talk about changing events that have not yet happened?

Interesting variations on *physical* backward time travel in science fiction, which neatly avoid the paradoxes caused by interacting with and affecting the past (we'll see later that such interactions are not *sufficient* for paradox to occur), use gadgets that enable story characters simply to hear or see the remote past without introducing paradox-producing changes—a concern specifically mentioned, for example, in the mind-transference story "Balsamo's Mirror" (De Camp). This is, in fact, such an extraordinarily popular idea among science fiction writers that it has actually been mentioned in the technical literature. An example of this can be found in "Time Shards" (Benford); Benford (who is a physicist) writes in an afterword that his story about what we might hear in a sound recording accidently made on a clay pot in the year A.D. 1280 was motivated by an actual letter, on just that possibility, to the world's preeminent electrical engineering journal—see Woodbridge (1969).

In fact, both Benford's and Woodbridge's ideas were anticipated years before in the story of "The Sound-Sweep" (Ballard).[6] Using a gadget called the sono-vac, the sound-sweep removes extraneous and discordant noises such as coughing, crying, and the mumbling of prayers that have recorded themselves in the walls of churches. The residues of previous performances, and of the shuffling disturbances of crowds, get similar treatment in theaters. This story, in turn, was preceded by *decades* by Gardner Hunting's novel *The Vicarion,* in which has a gadget of that name reproduces scenes and sounds from the past using impressions indelibly imprinted on the ether (this device "makes vicarious experiences easy"). Hunting tried to show how such an instrument would violently disrupt society, an idea reinforced in Isaac Asimov's "The Dead Past."

Science fiction is simply bursting with gadgets for simply *seeing* the past, a limitation that avoids the paradoxes of changing the past. For example, there is the Palaeoscope in "The Time Valve" (Breuer), which is more constrained than the Vicarion; it has to be in actual physical contact with the object whose history is to be viewed. And E. T. Bell (who for many years was a professor of mathematics at Caltech) created "television in time" in his novel *Before the Dawn.* With this gadget one can see all the history of any object that has ever been exposed to light, no matter how long ago. Its descriptions of ancient flash floods a thousand feet high, and of the death agonies of decapitated dinosaurs, still make absolutely wonderful, horrifying reading. A similar gadget called Tempevision enlivens *Traveller in Time* (Mitchell). In "Through the Time-Radio" (Coblentz) Eskimos of the forty-first century use a device—which is much like the Vicarion, as is the obscurascope in "Matter is Conserved" (Palmer)—to watch the destruction of New York City in the twentieth century. The posthumously published, incomplete story of "The Dark Tower" by C. S. Lewis, written before 1940, tells the story of the eerie chronoscope that can see through time. And there is a gadget for filming the past in "E for Effort" (Sherred).

Some stories have gadgets that can do much more than just see the past. The inventions in these stories allow for *affecting* the past, not just looking at

This illustration from "The Time Eliminator" (*Amazing Stories,* December 1926)
shows the inventor demonstrating his gadget to his future wife and father-in-law. Able
to look back in time, the gadget's screen is displaying scenes from the older man's
courtship, decades in the past.

Illustration for "The Time Eliminator" by Frank R. Paul, © 1926 by Experimenter Publishing
Co.; reprinted by permission of the Ackerman Science Fiction Agency, 2495 Glendower Ave.,
Hollywood, CA 90027 for the Estate.

it. In "Private Eye" (Padgett), for example, there is a device that can both hear and see the past, and in "Brooklyn Project" (Tenn) there is the chronar for filming the past. And Horace Gold, the editor of *Galaxy Science Fiction* magazine, invented the Biotime Camera that can film the past in "The Biography Project." [This last contrivance was preceded by the similar "history-scanning machine" in "Forgotten Past" (Morrison)]. The time viewer in "One Time in Alexandria" (Franson) enables its archeologist-user to learn the horrifying answer to how the ancient Library at Alexandria was *really* burned to the ground. John Wyndham conceived of the ultimate in voyeuristic time-peeking in "Operation Peep."

Reversing the time direction from the past and toward the future was the Prognosticator in "The Push of a Finger" (Bester), a great computer that can extrapolate from the initial conditions of the present into the future with such accuracy that it produces conversations that will take place a thousand years hence! Equally (if not more) impressive is John W. Campbell's probability wave tube in "Elimination" (also called, à la C. S. Lewis, the "chronoscope"); it can not only see all of the past but also display all possible futures. Eric Frank Russell's psychophone in "Mechanical Mice," which looks something like an old crystal radio set, lets its inventor copy inventions from the future. Similar to this is the scanner in "What You Need" (Padgett), which provides its inventor with a livelihood selling people what they will need to survive an imminent crisis, such as a pair of scissors to cut away a scarf suddenly caught in a printing press. More utilitarian is the chrono-camera, an investigative tool used by the police in *Time Bomb* (Tucker) to film what happened (*will* happen) at a crime scene *before* the crime occurs. More ambitious is "The Evil Eye" (Gillespie) and its camera that produces instant pictures of how whatever it photographs will appear twenty years in the future. "Johnny Cartwright's Camera" (Bond) does the same thing, but it has a reach of only one day into the future, whereas "A Most Unusual Camera" (Serling) can peek just five minutes ahead. Lloyd Biggle, Jr.'s Chronus is a TV-display gadget that shows the unalterable future to police detectives assigned to the DFC, a "Department of Future Crime." In "Guilty as Charged" (Porges) an unnamed device that can look into the future reveals what a criminal trial of A.D. 2183 will be like. And a character in "Controlled Experiment," a 1963 episode of "The Outer Limits," uses a Martian temporal condenser to watch a murder that will occur. In a spooky analysis that treats the effect that such future timeward-looking instruments as these would have on society if they were in widespread use, Damon Knight's story of the Ozo is a cautionary tale in the best tradition of "Be careful what you wish for—you might get it."

Perhaps if she had read any of these "time-viewing" gadget stories, Virginia Woolf would have reconsidered some wishful words (1976) she wrote shortly before her death in 1941: "Is it not possible—I often wonder—that things we have felt with great intensity have an existence independent of our minds; are in fact still in existence? And if so, will it not be possible, in time, that some device will be invented by which we can tap them? . . . Instead of remembering here a scene and there a sound, I shall fit a plug into the wall; and listen in to the past. I shall turn up August 1890."

As an alternative to simply watching the past, another fictional way to squirm out of the change-the-past/free-will quandary was conceived by John R. Pierce, who wrote in "Mr. Kinkaid's Pasts," published decades ago, "There is no unique past! The uncertainty principle of Heisenberg, which philosophers use to assure us that the world is not a predestined machine, without room for free will, leading to one unique future, just as decisively contradicts the idea of a unique past . . . there is an infinity of pasts which are consistent with all the evidences in our present Universe, and any of these pasts is as much the real past as any other." (This seemingly bizarre idea has since found a scientific home in the many-worlds interpretation of quantum mechanics, which we will discuss at length in Chapter Four.)

In other words, neither does the state of the world now determine the state tomorrow nor was the state today determined by the state yesterday. According to this reasoning, no paradox can result from tampering with the past because the state of the past doesn't matter anyway. This view reminds me of an assertion, written before Pierce's, by the Russian religious philosopher Nikolai Berdyaev (1874–1948):[7] "The past which we so much admire has never really existed—it is but a creation of our own imagination which cleanses it from all evil and ugliness. The past we so love belongs to eternity and never existed in bygone ages: it is merely a composite of our present; there was another present in the past as it actually was, a present with all its owns evils and shadows."

Unlike Pierce's story character, who argues that one can tweak the past without limit and without paradox, the approach of Fritz Leiber is probably the classic among those stories that assert it is very difficult to change the past. In "Try and Change the Past", a sequel to his 1958 novel *The Big Time* about the Change War (a war fought by time travelers who attempt to change the past), Leiber wrote, "Change one event in the past and you get a brand new future? Erase the conquests of Alexander by nudging a Neolithic pebble? Extirpate America by pulling up a shoot of Sumerian grain? Brother, that isn't the way it works at all! The space-time continuum's built of stubborn stuff and change is anything but a chain-reaction. Change the past and you start a wave of changes moving futureward, but it damps out mighty fast. Haven't you ever heard of temporal reluctance, or the Law of the Conservation of Reality?"

Or, as Leiber wrote in the original novel itself, which poetically describes the "Change Winds" as a ripple effect of a change in the past that blows into the future "thousands of times faster than time moves," "Most of us [begin as participants in the Change War] with the false metaphysic that the slightest change in the past—a grain of dust misplaced—will transform the whole future. It is a long time before we accept with our minds as well as our intellects the law of the Conservation of Reality: that when the past is changed, the future changes barely enough to adjust, barely enough to admit the new data. The Change Winds meet maximum resistance always." L. Sprague de Camp put forth the same philosophy in his 1941 classic novel *Lest Darkness Fall:* "History is a four-dimensional web. It's a tough web. . . . If a man did slip back [in time], it would take a terrible lot of work to distort it. Like a fly in a spider web that fills a room."

Much later, Brian Aldiss' novel *An Age* went beyond Leiber and De Camp with its denial that any change at all could result from the actions of time travelers: "They were walled off completely from the reality all round them. They were spectres, unable to alter by the slightest degree the humblest appurtenances of this world [of the past]—unable to kick the smallest pebble out of the way—unless . . . by haunting it they altered the charisma of the place."

The immutability of the past, with its consequent immunity to paradox, is not a new idea or even one unique to science fiction. We find it, for example, in "Roads of Destiny," one of O. Henry's most celebrated stories from the turn of the century. In this tale a young poet is faced with deciding among three possible choices. The reader follows him down all three and finds that ultimately it makes no difference—each results in his death by a single bullet from the same pistol. This fatalistic philosophy is also the theme in "When the Bough Breaks" (Padgett). In that story, time travelers from the future inform a man in the present that they are going to make some changes but that nothing serious will happen as a result. "The past," he asks, "you mean it's plastic?" To that the reply is "Well, it affects the future. You can't alter the past without altering the future, too. But things tend to drift back. There is a temporal norm, a general level. In the original time sector [we were not here]. Now that's changed. So the future will be changed. But not tremendously." An eerie Russian novel, *The Strange Life of Ivan Osokin* (Ouspensky), about a young man sent twelve years back in time by a magician, expresses a harder version of Padgett's "temporal norm." As the magician explains to his client, "If you go back . . . you will do the same things again and a repetition of all that happened before is inevitable. You will not escape from the wheel; everything will go on as before."

More successful at changing the past, although he comes to regret it, is the time traveler in "The Histronaut" (Seabury). Seabury, a professor of political science, wrote his tale as a parody (and a splendid one it is) of the arms race between the Soviet Union and the United States—only now the ultimate weapon is not the Bomb but rather the (time) Machine. As Seabury tells the reader, the secret of time travel may be discovered by physicists, but its use as a weapon will be decided by historians. (Physicists, however, might take consolation from the reaction of the time-traveling historian in Ward Moore's *Bring the Jubilee* when he meets the inventor of the novel's time machine: "I looked at her with respect. Anyone, I thought, can read a few books and set himself up as a historian: to be a physicist means genuine learning.") Seabury tells his readers that "at the Desert Springs Conference of Historiographical Manipulation, attended by a carefully selected group of Harvard and Berkeley historians, the matter was broached as calmly and fully as men can broach the fantastic. The selective manipulation of history dwarfed even the decision to use nuclear weapons in the Second World War: tampering with history was dangerous precisely because of the inability . . . to judge the infinite ramifications of even the slightest change." Despite this inability, the histronaut is given an assignment involving somewhat more than just a slight change—he travels back to 1917 and assassinates Lenin! When he returns, he finds 1968 Washington, DC, under curfew by order of the *German* Governor-General.

Equally successful in changing the world is a researcher in time travel in the 1956 tale "Aristotle and the Gun" (De Camp). After deciding that what is wrong with the world can be traced back to the scientific method getting off to a late start, the time traveler thinks he can correct matters by visiting 340 B.C. and educating Aristotle on the proper scientific attitude. (Aristotle believed that observing the world was inferior to pure thinking about how things *ought* to work.) This the time traveler does, with utterly disastrous consequences. He returns to the present to find a scientifically retarded world that makes him a slave. In his cell he writes on a wall the bitter lesson he has learned too late: "Leave Well Enough Alone."

De Camp may well have written his story as a twist on Asimov's classic 1949 story "The Red Queen's Race," which puts forth a position now found in many modern philosophical papers. An idealistic physics professor, convinced that the world's political problems are the result of the comparative newness of scientific thought and tradition, tries to change the past (and thus the present) by sending a Greek translation of a modern chemistry text back two thousand years to the Hellenic days of Leucippus, Lucretius, and Democritus. He dies in the attempt but succeeds in the transmission. When a government investigator—called in because the professor drained an entire nuclear reactor to power his time machine!—discovers that the transmission takes a day to travel back a hundred years, he fears "our" world will vanish in twenty days, to be replaced by a "new" one. In the end, however, he learns you can't change the past. As one of the professor's colleagues tells the investigator, "While you are right that any change in the course of past events, however trifling, would have incalculable consequences . . . I must point out that you are nevertheless wrong in your final conclusion. *Because THIS is the world in which the Greek chemistry text WAS sent back.*"

It is amusing to note that a few years later a time machine, sounding much like Asimov's, appeared in the philosophical literature. Gale (1964) called such a gadget "fantastic, yet nevertheless, logically conceivable," but he still could not bring himself to call it a time machine. His name for a device that runs backward through time at a rate of a hundred years per minute is the "History of Philosophy Machine"! This issue, of time travel to the past itself requiring time (a point I must admit I fail to understand), was hinted at again nearly half a century after Asimov's story, in "The Plot to Save Hitler" (Thompson). This tale, which agonizes at length over the dangers of changing the past, has a man travel back to 1903 to kill the then fourteen-year-old Adolf Hitler. A Chronolab agent is therefore sent in hot pursuit to stop the would-be assassin. It's a *hot* pursuit because the agent was quickly "dispatched into the past by scientists who feared the unknowable consequences of delay." But why such concern? What matters is not when the pursuing agent leaves the present to begin his hunt, but rather the time in which he arrives in the past, in time (so to speak) to foil the potential killer. This is an old mistake in science fiction—"A Star Above It" (Oliver) made it decades ago, and it was glaringly awful *then.*

In their stories Leiber, Asimov, and O. Henry have free will falling to defeat in the face of the overpowering dominance of predestination. There are, however, far more subtle problems with time travel than the apparent conflict between free will and predestination. There are other paradoxes that need to be addressed, and we will have *much* more to say about them soon. Unfortunately, most physicists who have not thought very hard about time travel simply ignore these paradoxes, except to invoke them to "prove" that time travel is impossible. In this way they are much like Zeno, who used his famous paradox to "prove" that motion is impossible—a proof shown to be false by the very act of moving one's hand to write down the paradox! The actual paradox really has nothing at all to do with motion, of course, as Zeno himself knew, but was instead the puzzle of how apparently logical reasoning could result in such an obviously incorrect conclusion. (Zeno's argument, by the way, is used in an entertaining way in the time travel tale "Dark Interlude" by Mack Reynolds and Fredric Brown.)

To prove by paradox is a risky business, and all that such lines of reasoning may show is a lack of imagination and insight. Physicist/philosopher David Malament of the University of Chicago forcefully rebutted the dangers of arguing by paradox (1984):

> [One] view is that time travel . . . is simply absurd and leads to logical contradictions. You know how the argument goes. If time travel were possible, one could go backward in time and undo the past. One could bring it about that both conditions P and not-P obtain at some point in spacetime. For example, I could go back and kill my earlier infant self making it impossible for that earlier self ever to grow up to be me. *I simply want to remark that arguments of this type have never seemed convincing to me. . . . The problem with these arguments is that they simply do not establish what they are supposed to* [my emphasis]. To be sure, if I could go back and kill my infant self, some sort of contradiction would arise. But the only conclusion to draw from this is that if I tried to go back and kill my infant self then, for some reason, I would fail. Perhaps I would trip at the last minute. The usual arguments do not establish that time travel is impossible but only that if it *were* possible, certain actions could not be performed.

This last point still rankles, of course, because it pulls in the issue of free will versus fatalism. But that is a different issue from time travel.

Still, the grandfather paradox is commonly put forth by those who deny the rationality of even thinking about time travel but who are seemingly unaware of the fatal logic in their position. For example, here is Rothman (1988): "the time-travel paradox, a classic in science fiction, forces us to consider what would happen if you could travel back in time so as to prevent the mating between your father and your mother. In that event you could not be born. But if you were not born you could not travel back in time to prevent that fateful meeting. But then you would be born, etc., etc. Do you exist, or don't you? The paradox can not be untangled."

Is Rothman correct?

In fact, he is not. Rothman *is* correct when he says the grandfather paradox is a classic. We find a version of it, for example, in the 1930 novel *The World Below* (Wright), in which a character called the Professor tells us that "It is obviously impossible to project anything into the past, which is fixed irrevocably.

Otherwise there would be no finality, and the confusion would be intolerable. . . . [F]or instance, upon reading of a long-past murder, I could project myself into the past, and intervene to save the victim. In such an event the murder would both have occurred, and been prevented: which is absurd." But of course Malament's argument, and similar ones found in Lewis (1976) and Horwich (1987), *do* untangle the illogic of the Professor's paradox.

There are other, far more subtle time travel paradoxes than the now moot one put forth by Rothman. Modern philosophers mention grandfather-type paradoxes only to refute them à la Malament, and even the generally low-level and humorous Blumenthal *et al.* (1988) has the grandfather paradox right. The more interesting paradoxes will be discussed at length in Chapter Four.

Where Are All the Time Travelers?

Arthur C. Clarke (1985) discusses a much more interesting objection to time travel than the "kill-yourself-in-the-past" or other grandfather-type paradoxes. He writes, "The most convincing argument against time travel is the remarkable scarcity of time travelers. However unpleasant our age may appear to the future, surely one would expect scholars and students to visit us, if such a thing were possible at all. Though they might try to disguise themselves, accidents would be bound to happen—just as they would if we went back to Imperial Rome with cameras and tape recorders concealed under our nylon togas. Time traveling could never be kept secret for very long." As a skeptic in "Time's Arrow" (McDevitt) explains Clarke's problem to a would-be time machine inventor, "If it [time travel] *could* be done, someone will eventually learn how. If that happens, history would be littered with tourists. They'd be *everywhere*. They'd be on the *Santa Maria,* they'd be at Appomattox with Polaroids, they'd be waiting outside the tomb, for God's sake, on Easter morning."

From the moment after the first time machine was constructed, through all the rest of civilization, there would be numerous historians, to say nothing of weekend sightseers, who would want to visit every important historical event in recorded history. They might each come from a different time in the future, but all would arrive at destinations crowded with temporal colleagues—crowds for which there is no historical evidence!

Philosophers are well aware of Clarke's objection, and indeed the objection considerably predates Clarke, having been mentioned, for example, by Farley (1950). As another example, Fulmer (1980) writes, "Actually I know of only one argument against the possibility of time travel that seems to carry any weight at all. This is the fact that it does not appear ever to have happened.

That is, it might be argued that there will be no time trips from 1985 to 1975 because we were here in 1975 and saw no time travelers. But this argument is far from conclusive." Professor Fulmer then mentions some possible explanations, but it is his last—one contrary to his position—that I find the most technically intriguing: "Finally, and even less interesting to philosophers, there might be some pettifogging physical limitation on time travel: perhaps the energy expenditure varies as the fourth power of the time traversed, making only very short trips feasible, and its discovery lies too far in the future for its effects to have yet been felt."

Isaac Asimov used a variation of Fulmer's energy-limitation idea in "Button, Button," the story of Otto Schlemmelmayer's machine that can retrieve objects from the past—but only if they don't weigh too much. The problem is one of energy, the relationship being an inverse exponential; all the power in the universe could bring forward maybe two grams. Robert Heinlein uses a similar idea in *The Door into Summer.* As one character in that novel explains, "Now if there was some way to photograph the Crucifixion . . . but there isn't. Not possible . . . there isn't that much power on the globe. There's an inverse-square law tied up in [time travel]."

Clarke, always intellectually honest and burdened with no ideological positions to defend, presents some other possible rebuttals from science fiction to his own objection. As Clarke writes, "Some science-fiction writers have tried to get round this difficulty [about where all the time travelers are] by suggesting that Time is a spiral; though we may not be able to move along it, we can perhaps hop from coil to coil, visiting so many millions of years apart that there is no danger of embarrassing collisions between cultures. Big game hunters from the future may have wiped out the dinosaurs, but the age of *Homo sapiens* may lie in a blind region which they cannot reach."

This particular idea of time as a spiral was quite popular in early science fiction. Typical is the 1931 tale "A Flight into Time" (Wilson), in which the time traveler suddenly finds himself not in 1933 but in 2189. His situation is "explained" to him thus: "[The] time stream is curved helically in some higher dimension. In your case, a still further distortion brought two points of the coil into contact, and a sort of short circuit threw you into the higher curve." In "The Sands of Time" (Miller), published in 1937, we find the same spiral-time concept; with a sixty-million-year pitch to the time helix, there is no danger of a grandfather paradox. The next year Isaac Asimov also used the idea in his first attempt at professional writing, and this despite his life-long unhappiness with the concept of time travel. Unlike Miller's success, however, Asimov's suggestively titled "Cosmic Corkscrew" was rejected. It was never published and is now lost. [Though perhaps not for long. See "Cosmic Corkscrew" (Burstein) for the tale of a time traveler from not too far in our future who journeys back to 1938 to retrieve a copy of Asimov's just-finished, soon-to-be-lost story.] Spiral time is also the central scientific theme in the play *I Have Been Here Before* by J. B. Priestley.

The spiral time mentioned by Clarke, Wilson, and Miller is a close cousin to circular time. Circular time is the theme in "Wanderer of Time" (Fearn), in which the final thoughts of the time traveler (who has just been executed in an electric chair) are "They had done all this before somewhere—would do it again—endlessly, so long as Time itself should exist. Death—transition—rebirth—evolution—back again to the age of the amoeba—upward to man—the laboratory—the electric chair—Eternal. Immutable!"

More successful in his experiment with circular time than is Fearn's character is the time traveler in "Flight to Forever" (Anderson), who finds, after a trip one hundred years into the future, that he can't get all the way back because the required energy rises exponentially with increasing penetration into the past (recall Professor Fulmer's suggestion). Still, it is very cheap in energy to go forward in time in this story, and so he does that, in search of help from the future's advanced technology. He never finds what he needs, however, and so goes forward right into the collapse of the universe and through a new Big Crunch that forms an identical new cycle of time. He thereby returns home just before he left. Again we have, as in Fearn's story, an eternal recycling of identical, circular time.[8] The experience is so terrifying that our hero decides to suppress what he has learned, and maybe *that's* why there are no apparent time travelers!

There are, in fact, theoretical models of spacetime that do lead to time travel to the past by the indirect method of traveling in the opposite direction, as, for example, in Weingard (1979b). But as Professor Weingard writes, this "is not quite what we are after. True, one can travel back in time *in the sense* of being able to travel to one's past. But this is done, not by really traveling back in time but by traversing the whole history of the Universe."

Stephen Hawking (1992) has invoked the "missing time travelers" argument as experimental evidence for his theoretical studies in support of his version of "Niven's Law," the so-called Chronology Protection Conjecture. Hawking denies the possibility of time travel to the past because "we have not been invaded by hordes of tourists from the future." It is amusing to note that the writer of a spoof on the use of time travel as the basis for financial investing in the past also invoked the same argument—two years *before* Hawking (does Hawking read *Barron's*?) As we read in Queenan (1990), ". . . if time travel was possible, someone from the future would eventually either discover a time tunnel or build a time machine and come visit us."

Not all physicists find the lack (so far) of mass visitations from the future to be as compelling as Hawking does in demanding rejection of time travel to the past. For example, in Kriele (1990b) we read, "The assumption of chronology is a serious drawback because chronology violation can not be ruled out by physical reasoning. We are only able to conduct local [i.e., small-scale] experiments, but causality violation in general relativity is a global [i.e., large-scale] effect, and so the lack of experience cannot give evidence of its absence. So chronology violation can only be discussed on philosophical grounds, grounds which have often enough turned out to be nothing but prejudice." Then, after observing that all such philosophical arguments, if non-trivial, are based on the free-will issue, Professor Kriele goes on to suggest that Einstein's general relativity will eventually prove to be the limiting case of an even more general theory of gravity, in which " 'free will' is something like a second-order effect and therefore it is possible that the classical limit space-time of our world contains closed timelike curves [implying the possibility of time travel to the past] though we enjoy the comfort of free will."

Still, Hawking's question about missing time-traveling tourists *is* a puzzle, no doubt about it, one that echoes Enrico Fermi's half-century-old question about alien intelligent life in the universe—if "they" exist, *where are they?* The common point to the stories "Absolutely No Paradox" (Del Rey) and "Invasion" (Podolny), which attempt to answer that question about time travelers, is that they are some*when* else—i.e., that time travel is possible only into the future. There are no time travelers from the future in our *now* because such travelers can journey only into the future, and time travel wasn't invented in our past. A "proof" of the impossibility of backward time travel has even been offered by an economist(!), who claims that the fact we observe positive interest rates "is proof that time travelers do not and can not exist." That is, Reinganum (1986) argues that time travelers from the future, if they were actually in the past, could, by virtue of their advanced knowledge of things to come, make financial killings so numerous and extensive as to drive interest rates to zero. Interest rates are not zero, however, and thus time travelers are not among us, or so concludes this "proof." Other even less serious economic arguments for why there are no time travelers from the future in our present are advanced in Queenan (1990).

More interesting than arguing about greedy time travelers is the speculation that backward time travel *is* possible but is extraordinarily dangerous. If so, it seems reasonable that it would be difficult to get anyone to do it willingly. In

Hot Wireless Sets (Compton), we learn of volunteer Roses Varco, the local village fool, who does time-travel because he does not really understand what is about to happen to him. Another disturbing variation on the danger issue was put forth in "The Man from When" (Plachta); there will be only *one* time traveler to the past; his first and last experiment will destroy the Earth—and he will be from eighteen minutes in the future. The suggestion of danger in time travel is actually an old concern. In the 1930 story "The Time Ray of Jandra" (Palmer), for example, in what was probably intended to be a horrifying scene but that today reads more like a stupendous and silly joke, a time traveler moves into the future by means of a "time-ray." Unfortunately, the ray works differently on the various chemical elements, and it doesn't at all on either hydrogen or oxygen. Thus the time traveler—or at least *much* of him—and his machine do vanish into the future, but left behind are "several gallons of water spilled on the floor"!

The most imaginative and disturbing answer to the question "Where are all the time travelers?" that I have come across in fiction is offered in "Time Payment" (Shaara). In that story backward time travel is possible, and, in fact, one of the inventors of the first time machine has just returned from 1938. Still, despite this success, the inventors are puzzled by "The Problem": "But if *we* have time traveled, then obviously men of the future have time traveled. They will be able—*are* able to come back. [So] where are they?" They finally conclude that there can only be two possible answers. Either there is nobody in the future, or time travel is so dangerous (Is that why the future might be empty—humanity misused time travel and killed itself off?) that all who invent it will suppress it. And that is what they decide they must do. In an afterword, Shaara writes that the original ending to his story was that there are no time travelers from the future because there are no humans in the future—in the story's present, the sun is already bloating as the prelude to going nova! The other ending was printed at the behest of the editor of *The Magazine of Fantasy and Science Fiction,* which published the story in 1954. Shaara calls the new ending "blah," but I find the scientists' intentionally suppressing the secret of time travel because it would be the ultimate weapon to be a particularly noble, if perhaps unrealistic, act. It also adds to the list of rebuttals to Hawking's chronology protection conjecture.

A Russian writer answered Clarke's puzzle by hypothesizing, in "The Founding of Civilization" (Yarov), that "a cardinal rule of [time travel] categorically forbids [time travelers] to stop at any point in time." One might wonder what the point of time travel would then be, but Yarov's story introduces the wonderful sport of "TM racing": seeing who can go back in time the farthest in the least time. Imagining time machines to be something like sports cars—"Once the machines were warmed up it was difficult to idle them, to restrain them from bucking backward into time before the race began"— Yarov talks about these ghostly apparitions whisking across the ages causing various reactions in those who happen to see the passage of a chronoviator. The superstitious shrink back in horror, while courageous philosophers write of "atmospheric effects"! But under no circumstances does a time traveler stop—until one does, briefly, after suffering an accident in 33,000 B.C., with astonishing repercussions. The ghostly appearance of time travelers "on the move," so to speak, is given a similar treatment in "Retroactive" (Shaw), a story flawed by seeming not to know what to say about paradoxes.

An interesting answer to the question "Where are all the time travelers?" is that they *are* here but (contrary to Clarke's view) have somehow managed to avoid giving themselves away. In "The Fox and the Forest" (Bradbury) the risk of a visitor from the future being too talkative was avoided by each time traveler having a psychological block installed to prevent any knowledge of the future, or of the mechanism of time travel, from being revealed to the past. The same idea was used by Lester del Rey in his spooky 1942 tale "My Name Is Legion," about the ideal fate for the then-still-living Adolf Hitler. Norman Knight's "Short-Circuit Probability" reverses the idea; the traveler in that tale uses hypnotic deception on those around him.

On the other hand, the private investigator in "Unborn Tomorrow" (Reynolds), hired to find a time traveler, somewhere, anytime, finds hundreds of them—at the annual *Oktoberfest* in Munich, a beer celebration that "makes the New Orleans Mardi Gras look like a quilting party." Time travelers like to go there, you see, because everybody gets so drunk there is no danger of giving yourself away. The psychologists of 3046 in "Unthinking Cap" (Pierce), however, who bring people of the past forward in time to study, go far beyond mere alcohol with a particularly diabolic means for ensuring the silence of the time travelers after they are returned to the past. Just before being returned, they are given a "forgetting machine"—in the form of a close-fitting black plastic cap—that purges unhappy memories. At first, this is a wonderful release from hurtful memories, but inevitably, the time travelers eventually wash away *all* their memories (including those of their trips into the future) while playing with the gadget.

There Will Be Time (Anderson) offers an inventive twist on secrecy and deception, with a hero who travels back to the Crucifixion, the one historical event he assumes *all* time travelers will wish to visit. He goes there to *advertise* himself as a time traveler via his appearance, such as his modern haircut. He is therefore thrilled when approached by a stranger who quite directly asks (in Latin, the one language immune to the passage of vast periods of time), "Es tu peregrinator temporis?" ("Are you a time traveller?"). In "102 H-Bombs" (Disch), time travelers from 3650 are *themselves* ignorant of their origin; they are children of the future whose fertilized ova were transported back to the twenty-first century and implanted in surrogate mothers.

And finally, one shouldn't overlook the possibility that a time traveler might find the judgments of his audience of little account. Consider, for example, the visitor from the year 2999 in Robert Silverberg's brilliant 1970 novel *Vornan-19* (reprinted under the title *The Masks of Time*). When the time traveler arrives in the world of 1999, he is initially greeted with some skepticism. When asked why anyone should believe his claim to be from the future, he replies, "Why, feel free to believe none of it. I'm sure it makes no difference to me."

Skepticism and Time Travelers

A thought-provoking possibility for explaining the discouraging scarcity of certified time travelers is the central thesis of a fascinating paper in the philosophical literature. Sorensen (1987) argues that nobody would believe a time traveler even if he willingly confessed and revealed his knowledge of the future or even gave the details of his time machine. Indeed, Sorensen makes the rather astonishing assertion that even the time traveler himself would have doubts! This shocking suggestion deserves some elaboration, especially because Sorensen invokes a philosophical authority to buttress his position, the patron saint of skeptics, the Scot David Hume (1711–1776). An important caveat, explicitly stated in Sorensen's paper, is crucial to keep in mind: "The key question will not be 'Is time travel possible?' We shall instead ask whether it is possible to justify a belief in a report of time travel." This gets to the real heart of Clarke's puzzle.

Much of the resistance to the idea of time travel lies in sheer skepticism. For many, time travel (to the past, in particular) is simply too much out of the ordinary to be taken seriously. For many, time travel would literally be miraculous. Hume's great work, *An Enquiry Concerning Human Understanding*,[9] contains a section on how a rational person should react to a claim that a miracle has occurred. Hume proclaimed that a miracle *by definition* violates scientific law and that, because such scientific laws are rooted in "firm and unalterable experience," any violation of one or more of these scientific laws immediately provides a refutation for the report of a miracle. In Hume's own words,

> Nothing is esteemed a miracle, if it ever happened in the common course of nature. It is no miracle that a man, seemingly in good health, should die on a sudden; because such a kind of death, though more unusual than any other, has yet been frequently observed to happen. But it is a miracle, that a dead man should come to life; because that has never been observed in any age or country. . . . When anyone tells me, that he saw a dead man restored to life, I immediately consider with myself, whether it be more probable, that this person should either deceive or be deceived, or that the fact, which he relates, should really have happened. I weigh the one miracle against the other; and according to the superiority, which I discover, I pronounce my decision, *and always reject the greater miracle* [my emphasis].

It is a strict interpretation of Hume that Sorensen adopts in claiming that a time traveler would have no success (among rational persons, anyway) with tall tales of "different places." As Sorensen explains, "Clearly the time traveler cannot persuade a reasonable person by baldly asserting 'I am a time traveler.' The improbability of his claim places a heavy burden of proof on him. But perhaps he could shoulder the burden by means of artifacts, predictions, and demonstrations." Sorensen dismisses all of these possibilities, however, by reminding us of the slightly sleazy history of parapsychology and ESP, which run counter to known scientific laws, but which have still duped "many a respected scientist." Any artifact, prediction, or demonstration of time travel, argues Sorensen, is more likely to be the result of deception and fraud than of actual time travel: "Should the time traveler take observers for a spin in his time machine, the skeptics will have us compare their adventures with seances." The rational reaction to such a spin around the centuries, according to Sorensen's presentation, would be like that of a magician who can not figure out how a colleague has just done his newest act: "Nice trick! How did you do it?"

For many, such skeptical reactions to time travelers seem dogmatic in the extreme—the response of one with no imagination, no spirit, and a head full of cement. Humean skepticism requires, it seems, the rejection of anything and everything that is profoundly surprising, leaving the world a place of utter predictability and boredom. As Robert Sheckley put it in "Something for Nothing," "When the miraculous occurs, only dull, workaday mentalities are unable to accept it." Sorensen answers this harsh criticism as follows: "Humeans respond by distinguishing between surprises. Most surprises in science do not violate accepted scientific laws. The strange wildlife in Australia was not excluded by biology, X-rays were not precluded by physics." Sorensen does well, however, to avoid mentioning such profound surprises as, for example, the spectrum of black-body radiation and, later, the photoelectric effect, which were not in the domain of known classical science at the beginning of the twentieth century. Those puzzling, surprising, *totally mystifying* effects required new science—the discovery of the quantum concept by Max Planck. A strict Victorian-age Humean, as described by Sorensen, would have wrongly rejected the experimental reports of all quantum phenomena and would also (perhaps just as wrongly) have rejected all reports of time travel.

Not all modern philosophers subscribe to the strict Humean definition (as described by Sorensen) that a miracle requires a violation of one or more of the scientific laws of nature. Ahern (1977) defines a miracle as any event that "can be explained *only* [my emphasis] by reference to the intervention of a supernatural force." Time travel, by that interpretation, is not a miracle because general relativity, not God, is all that is required. Indeed, Professor Ahern asserts that his definition is actually the true Humean one.

C. S. Lewis, late professor of Medieval and Renaissance Literature at Cambridge University, absolutely rejected Hume's view on how a rational person should react to certain surprising events. Lewis, one of the most thoughtful modern writers on Christian theology, had no patience with skeptics (or, as he called them, materialists), such as Sorensen describes. Professor Lewis graphically illustrates (1986) the dug-in position of such skeptics as follows: "If the end of the world appeared in all the literal trappings of the Apocalypse; if the modern materialist saw with his own eyes the heavens rolled up and the great white throne appearing, if he had the sensation of being himself hurled into the Lake of Fire, he would continue forever, in that lake itself, to regard his experience as an illusion and to find the explanation of it in psychoanalysis, or cerebral pathology." If the end of the world would receive such a skeptical response, then a mere time traveler would surely have no hope at all of being believed.

Lewis would certainly have rejected Sorensen's most astonishing assertion: "So far I have concentrated on the time travel question from the perspective of the time traveler's audience. What about the time traveler himself? Can he at least know he is a time traveler?" Sorensen argues that a time traveler, if authentic, should be able to convince us, and that if he can't (and he *cannot* if we are true Humean skeptics), then the traveler must entertain doubts, too! No matter, says Sorensen, that the time traveler has memories of his adventures, and no matter that he knows in his heart that he speaks the truth. Using words that echo Lewis' sarcasm, Sorensen quickly dismisses the importance of the time traveler's self-knowledge, declaring such memories to be merely the symptoms of some deep psychosis and the traveler's introspective sincerity to be a product of gross self-deception.

In Hume's defense, it should be clearly understood that he was not arguing for disbelief in absolutely anything surprising, but rather for rational analysis. Historically, the context of Hume's times was that of what he took to be non-rational arguments for a belief in God. That is, as Professor Peter Heath puts it in a wonderfully entertaining as well as scholarly essay,[10] Hume was "an exposer of bad arguments in rational theology." For Hume, second-hand (or even more remote) tales of the return of a man from the dead—the claim that literally kept Christianity alive after Christ's execution—were suspect. As Heath explains, "Hume . . . makes no attempt to deny the supposed facts; he simply argues that they are consistent with other explanations and other analogies of a less ambitious kind. There is no right to attribute to the causes of such phenomena abilities more extensive than are needed to produce the observed effects."

Sorensen specifically mentions the traditional Humean response to astonishing reports when he cites earlier writers on time travel in the philosophical literature. In one of those analyses, for example, Putnam (1962) argues for the reasonableness of a rational belief in time travel ("I have been amused and irritated by the spate of articles proving that time travel is a 'conceptual impossibility'") by claiming that there is a *mathematically consistent* explanation for such a belief. (What Putnam meant by this is explained in detail in Tech Note 4, in the form of Minkowski spacetime diagrams.) Ten years after Putnam wrote, however, came Weingard's (1972a) Humean-style rebuttal to Putnam. Weingard showed how to explain all of the time travel phenomenon that Putnam describes *without* invoking time travel. This isn't to say that Weingard doesn't require some pretty astonishing gadgets, and other things, himself—they include matter transmitters ("beam me up, Scotty!") and anti-matter humans—but he doesn't need a time machine. A resurrected Hume would surely applaud Weingard's analysis, although he would almost certainly doubt his own fresh existence.

Not everybody would be happy with Weingard's gadgets for avoiding time travel, however. Those gadgets are, like a time machine, incredible, and as Professor Challenger said in Arthur Conan Doyle's "The Disintegration Machine," "You cannot explain one incredible thing by quoting another incredible thing." An interesting science fiction exposition of Professor Challenger's Humean philosophy occurs when a copy of the *New York Times* shows up a week early in "What We Learned from This Morning's Newspaper" (Silverberg). It seems the only explanation is either that the paper really is from he future or that it is a hoax. The first-person narrator of the tale provides us with his reason for believing the former: "I don't find either notion easy to believe but I can accept the fourth-dimensional hocus-pocus more readily than I can the idea of a hoax. For one thing unless you've had a team the size of the *Times'* own staff working on [a hoax] it would take months to prepare it." For those who prefer classical authority to that of science fiction, I like Aristotle's observation that "Plausible impossibilities should be preferred to unconvincing possibilities." We have, at least, a theory for time travel—general relativity—but nothing at all to support Weingard's matter transmitters and anti-matter people.

What would Arthur Clarke think of skepticism toward those who claim to have a time machine? It was with his thoughts about the difficulty time travelers would have in maintaining low profiles that we began the previous section, so we might wonder what Clarke would think of Humean skepticism as it is related to time travel. It would be best to ask Clarke personally, of course, but my guess is that he would have little patience with such incredulity. The surprise of being confronted by a time traveler would soon turn to awe and pleasure IF—and I emphasize the IF—Clarke were taken for a spin in the stranger's machine. He would surely quote his own famous "Clarke's Third Law" (1972) to explain the wonder of it all: "Any sufficiently advanced technology is indistinguishable from magic."

Or perhaps he would recall the opening line of S. Fowler Wright's 1930 novel *The World Below:* "Applied science is always incredible to the vulgar mind." In a similar vein, Robert Heinlein's character Lazarus Long once asserted that "One man's 'magic' is another man's engineering. 'Supernatural' is a null word." Or, as the story "The Four-Dimensional Roller-Press" (Olsen) put it in 1927, "These things [four-dimensional objects] sound like miracles; but, after all, what are miracles but phenomena which, on account of our ignorance, we cannot explain?"

As a matter of fact, even Hume could be convinced of quite strange matters, and I think Sorensen does interpret the philosopher a little too narrowly. In his essay concerning Hume's position on holding a belief in God,[10] Professor Heath wonders whether there is "empirical evidence [imaginable] which would persuade any reasonable mind of the real existence of an infinite God." Heath answers his own question as follows: "If the stars and galaxies were to shift overnight in the firmament, rearranging themselves so as to spell out, in various languages, such slogans as I AM THAT I AM, or GOD IS LOVE— well, the fastidious might consider that it was all very vulgar, but would anyone lose much time in admitting that this settled the matter? . . . Confronted with such a demonstration, the hard-line Humean [but not Hume, himself, I think] could continue, of course, to argue that, for all its colossal scale, the performance is still finite, and so cannot be evidence of more than the finite, though immense power that is needed to achieve it."

Heath concludes with what I think is the perfect rebuttal to anyone who would refuse to admit to time travel, even after taking a quick trip backward a few tens of millions of years to the late-Mesozoic era to hunt *Tyrannosaurus rex,* and even after seeing the instant photographs of the dead monster with the skeptic's own foot on its head, or of his boots dripping a bloody puddle of unholy size on the floor of the time machine. Writing about the Humean-unconvinced, even when faced with a rearranged firmament, Professor Heath notes, "But this now seems a cavil, designed only to prove that even omnipotence is powerless against the extremer forms of skeptical intransigence." Where God would fail to convince, a simple time traveler could hardly hope to do better!

In fiction from long ago, it is easy to find examples of how out of the ordinary, or even unthinkable, time travel was viewed (or not viewed) by most people. The puzzling 1875 story "Who Is Russell?" (Eggleston) tells of a man who suddenly, without explanation, appears in the midst of a Union military camp during the American Civil War. This man quickly displays strange lapses in his background, as well as knowledge of many different things well beyond anything that could be called common. The details of the story are not important here, but if it were published in a modern science fiction magazine, the stranger in this tale would almost surely be identified quickly in most readers' minds as a time traveler. In 1875, however, the author's narrator found his punch line in "his firm conviction that the quiet, gentle, well-behaved, modest gentleman, so singularly gifted . . . is, in plain terms, the devil!" Time travel certainly never entered the author's thoughts, or if it did, he lost his nerve at the idea of using it in his pre-Wells story.

Modern science fiction writers have often used skepticism as a means of building conflict and tension in their time travel stories. The skeptical reception, for example, is extreme for the time traveler in "The Oldest Soldier" (Leiber). In that story a soldier-in-time, who has fought in wars from the ancient past to a billion years in the future, finds that nobody believes him when he speaks openly of his temporal adventures while on a visit to a bar in the present. Everybody merely thinks it is all a hilarious gag. Often the skeptical reaction is less benign. As the inventor of the time viewer explains in "E for Effort" (Sherred), "I've watched scribes indite the books that burnt at Alexandria; who would buy, or who would believe me, if I copied one? What would happen if I went over to the Library and told them to rewrite their histories? How many would fight to tie a rope around my neck if they knew I'd watched them steal and murder and take a bath? What sort of a padded cell would I get if I showed up with a photograph of Washington, or Caesar? Or Christ?" The padded cell was indeed the fate of the time traveler in "The Ambassador from the 21st Century" (Shay), who journeyed from A.D. 2007 back to 1952 to warn of a future war—he was committed to a mental institution to receive help for his "illusion."

It would be more convincing, of course, if a scientific explanation could be offered for the claimed time travel. The time-traveling tourist stranded in the past in "No Future In It" (Haldeman) is used to a skeptical reaction because he can provide his questioners no explanation for his situation. When asked how time machines work, he can only reply, "How the hell should I know? I'm just a tourist. It has something to do with chronons. Temporal Uncertainty Principle. Conservation of coincidence. I'm no engineer." Somewhat more successful (perhaps) is the time traveler in "The Garden" (Gor). Born in the year 2003, he turns up in 1975 and tries to convince an interrogator of his origin. And apparently he does; the time traveler later tells a new friend in the past of 1975, "What amazed me . . . was that he really believed me in the end." But the friend doesn't buy that, replying "He did? I think he just pretended. A scientist isn't likely to believe a thing that is against all logic."

If his reception committee is a crowd of conservative, cautious Humeans, a time traveler is doomed. In "Spectator Sport" (MacDonald), for example, we learn of the awful fate of a time traveler who, on arriving in the future, falls into the hands of skeptics. It is similar to the fate the time travelers from 2030 are warned about in "The Time Hoaxers" (Bolton). As one of them relates in this early (1931) tale of time travel skepticism (the editorial introduction called it "a curious study of psychology"), "Our wisest men advised against [this trip to the past]. They said we could hope to be received only as impostors and fakirs, that . . . we would find only twentieth-century barbarians, suspicious, ill-tempered, likely to do us bodily harm." Similarly, when the renegade time-traveling villain in the film *Time Trackers* is confronted in the medieval past, he simply laughs at a threat to reveal his true identity. Go ahead, he says in effect, the only thing your talk of time machines from the future will accomplish is for people to think *you* are crazy!

Manly Wade Wellman attempted to avoid the problem of a time traveler needing to convince strangers of his true identity by using the interesting idea of the time traveler convincing his own earlier self. In a story with the appropriate title "Who Else Could I Count On?" Wellman has John (the older) travel back forty years to meet John (the younger), to enlist the aid of his younger self in alerting the world to the coming war. He succeeds in doing this, but the closing words "Lord have Mercy!" of John (the younger) do have the ring of astonishment to them. This story is a greatly condensed version of ". . . backward, O Time!", in which the final words of shock, uttered once the younger version realizes his visitor is himself, are "Good grief!" Indeed. It is amusing to note that in C. S. Lewis' eerie, unfinished story "The Dark Tower," which tells the tale of the chronoscope—an invention that "does to time what the telescope does to space"—the persistent skeptic in the story is a Scot, surely created by Professor Lewis in the image of Hume.

Einstein, Gödel, and the Past

The eleventh-century Persian poet-philosopher Omar Khayyam was blunt in his evaluation of the likelihood of reliving the past. As he so beautifully wrote in one of the quatrains of the *Rubaiyat,*

> The Moving Finger writes; and, having writ,
> Moves on: nor all your Piety nor Wit
> Shall lure it back to cancel half a Line,
> Nor all your Tears wash out a Word of it.

Quite a bit later the English poet Thomas Heywood, in his 1607 play *A Woman Killed with Kindness,* had one of his characters express a similar thought:

> O God, O God, that it were possible
> To undo things done, to call back yesterday;
> That Time could turn up his swift sandy glass
> To untell the days, and to redeem these hours.
> Or that the Sun
> Could, rising from the west, draw his coach backward,
> Take from the account of Time so many minutes,
> Till he had all these seasons called again,
> . . .
> But O! I talk of things impossible,
> And cast beyond the moon . . .

Wyn Wachhorst, a modern scholar of popular culture, has an odd way of rejecting time travel. Transporting Wells' Victorian Time Traveller to the wrong century and invoking Einstein, he writes (1984), "We are indebted to H. G. Wells not only for the notion of voluntary time travel but also for the image by which we conceive it: a sunny, Edwardian [sic] gentleman perched on an ornate steam-age contraption that moves through time in much the same manner that a streetcar moves across town. This spatialized view of time, along with its Newtonian catechism, has increasingly gone the way of bowler hats and high button shoes in the new world of Einstein and quantum mechanics."

Wachhorst's final comment is ironic because it is Einstein's field equations that actually provide the basis for the modern theory of time travel.

Well, you might say, such deniers of time travel are just storytellers, philosophers, poets, and magazine essayists. What do *they* know about the possibility of time travel? It is what physicists and mathematicians think that is important because, after all, if a time machine is ever built, it will be as a result of new understandings at a profoundly deeper of level of mathematical physics than we have today. And, curiously enough, there *are* physicists and mathematicians who do not discount the possibility of time travel.

For example, Kurt Gödel, described in an obituary notice as one of the greatest mathematical logicians of all time,[11] published (1949b) a solution to Einstein's field equations for general relativity—a solution applicable to a rotating universe composed of a perfect fluid at constant pressure. In this universe, called a Gödelian universe, there exist closed timelike world lines in spacetime.[12] As Gödel put it, such world lines imply that it is "theoretically possible in [such a universe] to travel into the past, or otherwise influence the past." These world lines are the possible paths of space travelers who always move into the local future, but who nevertheless eventually arrive back in their own past. In fact, such *natural* closed timelike world lines pass through *every point* in Gödel's spacetime; that is, time travel in Gödel's universe is *not* the result of a machine manipulating mass and energy on a local scale. In Gödel's spacetime, time travel is a natural phenomenon!

It is an astonishing fact that one of the great twentieth-century physicists, Hermann Weyl, a colleague of both Einstein and Gödel at the Institute for Advanced Study in Princeton, wrote (1952) the following anticipatory passage *three decades before* Gödel:

> It is possible to experience events now that will in part be an effect of my future resolves and actions. Moreover, it is not impossible for a world-line (in particular, that of my body), although it has a time-like direction at every point, to return to the neighborhood of a point which it has already once passed through. The result would be a spectral image of the world more fearful than anything the weird fantasy of E. T. A. Hoffmann [an early nineteenth-century German writer of the eccentric] has ever conjured up. In actual fact the very considerable fluctuations of the [components of the metric tensor, discussed in Tech Note 4] that would be necessary to produce this effect do not occur in the region of the world in which we live. . . . *Although paradoxes of this kind appear, nowhere do we find any real contradiction to the facts directly presented to us in experience* [my emphasis].

It would be thirty years after Weyl wrote these amazing words before Gödel finally presented his rotating universe that showed just how those "considerable fluctuations" might actually occur.

In the pivotal year 1949, in an invited essay, Gödel specifically mentioned (1949a) the paradoxical aspect of his time travel result: "By making a round trip on a rocket ship in a sufficiently wide course, it is possible in these [rotating] worlds to travel into any region of the past, present, and future, and back again, exactly as it is possible in other worlds to travel to distant parts of space. This state of affairs *seems* [my emphasis] to imply an absurdity. For it enables one, e.g., to travel into the near past of those places where he has himself lived. There he would find a person who would be himself at some earlier period of his life. Now he could do something to this person which, by his memory, he knows has not happened to him."

Gödel defended his statements about the paradoxes of a time traveler meeting himself in the past with what I think is an astonishingly unconvincing argument (particularly so for a logician) based primarily on *engineering* limitations: "This and similar contradictions, however, in order to prove the impossibility of the worlds under consideration, presuppose the actual feasibility of the journey into one's own past. But the velocities which would be necessary in order to complete the voyage in a reasonable time are far beyond everything that can be expected ever to become a practical possibility.[13] Therefore it cannot be excluded *a priori,* on the ground of the argument given, that the space-time structure of the real world is of the type described." That is, Gödel was trying to head off critics of his rotating-universe model who might point to the time travel result as proof that the model had to be flawed.

In a reply to Gödel, Einstein wrote (1949) "Kurt Gödel's essay constitutes, in my opinion, an important contribution to the general theory of relativity, especially to the analysis of the concept of time. The problem here involved disturbed me already at the time of the building up of the general theory of relativity, without my having succeeded in clarifying it. . . . the distinction 'earlier-later' is abandoned for world-points which lie far apart in a cosmological sense, and those paradoxes, regarding the *direction* of the causal connection, arise, of which Mr. Gödel has spoken. . . . It will be interesting to weigh whether these are not to be excluded on physical grounds."

Two science fiction stories, both published in 1947 and two years before Gödel's paper, treat time travelers doubling back on their own world lines. In "Pete Can Fix It" (Jones), the meet-yourself concept is not essential to the story and, in fact, seems to have been tacked on at the end. I mention this because it

"Miss! Oh, Miss! For God's sake, stop!"

In a chapter on time travel in Watzlawick (1976) the author says that this 1957 *New Yorker* cartoon combines time and space travel. If true, that would be remarkable indeed. It had been only eight years since Gödel showed how backward time travel might be done via a rocket ship, and could even the readers of *The New Yorker* be so sophisticated as to chuckle over a joke based on a paper in *Reviews of Modern Physics*? At first glance it does seem to be a visit to the Garden of Eden, in an attempt to change the past. But of course it isn't; the creatures aren't human (look closely at their heads). Time travel was surely not the artist's intention.

Drawing by Whitney Darrow, Jr.; © 1957 The New Yorker Magazine, Inc.

was published by the premier science fiction editor of the day (John W. Campbell at *Astounding Science Fiction*), who could (and, in my opinion, should) have tightened up the tale by deleting the time loop. What makes this worth commenting on is the sharp contrast between Jones' story and the other tale, "Castaway" (Chandler). Chandler's appeared in *Weird Tales,* a magazine outlet for fantastic supernatural and horror stories. And yet "Castaway" does a far superior job of presenting the oddness of a loop in time than does Jones' "science fictional" story. "Castaway" precisely projects the mysterious nature of a closed loop in time, a mystery that clearly fascinated Gödel.

The story opens with a man on a ship spotting the signal fire of a castaway on a Pacific island, as well as the tiny, distant figure of a man waving and jumping about. While sailing in to help, the ship hits a mine left over from the war, and the would-be rescuer becomes a castaway, too. After swimming to the island, he can find no trace of who built the fire, although there are footprints about it. Exploring the island, he finds the remains of a crashed interstellar spaceship(!), powered by a drive unit based on "temporal precession," an idea borrowed from the theory of gyroscopes. The man, curious, turns the drive on and thus sends himself backward in time one day. He then spots a ship on the horizon, builds a signal fire, waves and jumps about, recognizes the approaching ship *as his own,* . . . and so the loop nearly but not quite closes. The man apparently rushes off into the jungle, terror-stricken at the thought of meeting himself. Gödel would have loved it—and who knows, maybe he *did* read it!

Gödel's analysis was later attacked (and also defended), but always on the basis of how to interpret his bizarre solution, not on whether he had made a mathematical error. For example, North (1965) writes of Gödel's solution: "This property [of time travel] must be judged an absurdity by anyone committed to the ordinary modes of speech." And in Chari (1960) we read that Gödel's solution is a "bizarre conception" and a "mere mathematical curiosity," and the previously mentioned obituary notice itself states that a "novel feature of the Gödel Universe is that it contains closed timelike world-lines. . . . The unpalatable consequences [of time travel to the past] represent not so much time travel as a breakdown of causality. For this reason Gödel's solution cannot be taken too seriously as a model for the real Universe."

Other physicists have criticized the critics, however, as in Pfarr (1981):

> What strikes us as rather precarious is the exclusion of Gödel's space-time . . . only because of closed timelike world lines. . . . Rather than rule out such solutions by means of reference to logical paradoxes . . . it should be investigated

> whether physical reasons can be presented which require an
> exclusion. [Note that this was Einstein's position, too.] Dur-
> ing the past 30 years [now, as I write *Time Machines 2,* nearly
> 50] since Gödel's discovery his solution could not be ex-
> cluded by cosmological arguments alone. It seems that the
> "physical grounds" for excluding this solution which Ein-
> stein mentioned in his reply to Gödel have to be looked for
> beyond the theory of gravity.

Ozsvath and Schucking (1962) thought that one possible explanation for excluding the time travel in Gödel's solution, based on "physical grounds," could be developed using the observation that Gödel's universe is infinite. Perhaps, they argued, a rotating, *finite* universe would be free of time travel. The primary concern of those authors was that Gödel's rotating universe fails to incorporate Mach's principle, the claim that the inertia of any object is determined by the distribution of all the rest of the mass in the universe. Einstein himself originally believed that the general theory satisfied this principle, and until Gödel's counterexample in 1949, all known solutions to the gravitational field equations were, in fact, in agreement with Mach's claim.

Ozsvath and Schucking, speculating that Mach's principle might be retained in a finite rotating universe, thus decided to search for a new counterexample—a rotating solution that is finite and does not allow time travel to the past and yet still violates Mach's principle. In this quest, they succeeded. Their desire to find such a solution was particularly understandable; Gödel had tantalizingly claimed (1952) that in 1950 he had found such models himself. Unfortunately, Gödel gave no details, but Ozsvath and Schucking mention in their 1962 paper that they had received a personal communication from Gödel that one of the temporally benign models he had had in mind was, in fact, their model. In the 1962 paper you can find an outline sketch of this model, and in a later paper (1969) Ozsvath and Schucking published a much more detailed analysis. And thus was Mach's principle finally banished from the general theory [see Rindler (1994) for more on general relativity and Mach's principle, and see Bondi and Samuel (1997) for a rebuttal].

Theoretical time travel could not so easily be dismissed, however, for some years later other time travel solutions were discovered, such as Som and Raychaudhuri (1968). Banerjee and Baneji (1968), and De (1969). The conclusion in the last paper was that a general relativistic situation violating causality *could* be formulated. That is, time travel to the past seems to be no mere anomaly of Gödel's particular solution, but rather is built right into the basic gravitational field equations of general relativity.

Quantum Mechanics, Black Holes, Singularities, and Time Travel

A fundamental objection, based on general relativity, to Gödel's ideas and to the very idea of time travel to the past is that, in a very deep sense, general relativity is *known* to be incomplete. That is, it is incompatible with quantum mechanics, which is the physics of the very, *very* small—the physics of objects on the order of just a single big molecule. In quantum mechanics, the discrete nature of the atomic world appears in such phenomena as the photoelectric effect, in which light acts like individual particles rather than as continuous waves. General relativity works beautifully on a cosmological scale, but like Maxwell's theory of electromagnetism, and unlike quantum mechanics, it utterly fails when applied deep in the interior of the atom. Quantum theory, however, seems to work *everywhere*. As physicist Nick Herbert writes in his excellent book *Faster Than Light,* "As far as we can tell, there is no experiment that quantum theory does not explain, at least in principle. . . . Though physicists have steered quantum theory into regions far distant from the atomic realm where it was born, there is no sign that it is ever going to break down."[14]

As we noted earlier in this chapter, one of the central concepts in relativity is the *world line,* which is the complete story of a particle in spacetime. A world line assigns a definite position to the particle at each instant of time. This is a classical, pre-quantum concept, however, and today physicists use the probabilistic ideas of quantum mechanics to describe the position and momentum of a particle, once they get down to the atomic scale of matter. Quantum theory is a discrete theory in which the values of physical entities vary discontinuously (in "quantum jumps"), whereas in classical theories the values of physical entities are continuous. The difference between the two types of theories is something like the difference between sand and water. Mixing the two theories—the classically smooth general relativity and the discrete quantum mechanics—to get something called *quantum gravity* is the Holy Grail of physicists today, and nobody has more than an obscure idea how to do it. As Kar (1992) writes, in a paper that attempts to identify the essential features that a quantum gravity theory should exhibit (such as locality), "it has proved difficult to conceive, even in outline, of *any* theory which combines the fundamental principles of general relativity and quantum field theory. . . ."

Just one of the more curious results of the fusing of quantum mechanics with general relativity may be *quantum time;* see Kraghand and Carazza (1994). That is, in quantum gravity the smallest increment of time that has physical meaning—sometimes called the *chronon,* a term first used in a 1935 non-time-travel story, "The Ideal," by Stanley G. Weinbaum—may have a non-zero value. As we'll see later, much of the present controversy over the possibility of time machines hinges on what is called the quantum gravity cut-off. This is a cut-off of destructive spacetime stresses that tend to grow toward infinity whenever a time machine attempts to form. This process goes under the general name of the *back-reaction* and is conceptually similar to a stretched rubber band growing ever more taut as it is stretched, an effect that resists more stretching (and, of course, if stretched too far it breaks).

The so-called cut-off of those stresses, at some *finite* value, is supposed to occur when the terminal phase of the growth would take place in less than the minimum possible time interval. The cut-off happens because nothing can actually occur in less than the minimum time. The debate is over just what the minimum duration is and over whether the cut-off would occur before the stresses could reach *finite* values high enough to destroy the putative time machine anyway. That is still all very speculative, of course, and the search for a way to connect general relativity and quantum mechanics continues. This is all related to time travel via a fantastic sequence of discoveries made during the last sixty years in relativistic physics.

General relativity predicts that a sufficiently massive star—greater than about four times the mass of the Sun—will, when its fuel is nearly exhausted and its nuclear fires are beginning to fade, experience a truly spectacular death called total gravitational collapse. When its fuel-starved, weakened radiation pressure is no longer able to keep an aged star inflated against the collapsing force of its own gravity, the star will implode and crush itself into what is called a *black hole,* a term coined in 1967 by the Princeton physicist John Wheeler in an address before the American Association for the Advancement of Science. A black hole is a compact object with a gravitational field so strong that even light cannot escape—hence it is black—and whose center is a singularity in spacetime. The singularity is a place where physicists believe spacetime either is terribly weird or no longer even exists. Cataclysmic views of the collapse of matter are actually quite old. In Lucretius' first-century B.C. *The Nature of the Universe,* for example, we find the following "wonderfully dreadful" imagery on what it would be like if matter itself "gave way": "The ground will fall away from our feet, its particles dissolved amid the mingled wreckage of heaven and earth. The whole world will vanish into the abyss, and in the twinkling of an eye no remnant will be left but empty space and invisible atoms. At whatever point you allow matter to fall short, this will be the gateway to perdition." These words were actually prompted more by earthquakes than by black holes, but they apply equally well to both phenomena.

As theoretical physicist Paul Davies has dramatically written (1981), "once gravity runs out of control, spacetime smashes itself out of existence at a singularity," or, to quote Stephen Hawking (1976), "A singularity is a place where the classical concepts of space and time break down as do all the known laws of physics." Best of all, I think, is the anonymous observation that a singularity is "where God is dividing by zero." One view of a singularity is that it is a place in spacetime that has infinite density and a gravitational field that is infinitely strong.[15] The curvature of spacetime at this sort of singularity, sometimes called a *crushing* singularity, is also infinite. That is the sort of singularity that physicists believe to be at the center of non-rotating black holes. Historically, however, the occurrence of infinities in physical theories has been the signal that the theories have simply been extended too far.

Perhaps, then the infinities mean that singularities occur only in unrealistic, idealistic applications of the theory; perhaps it is only *perfectly* spherical collapsing stars that can wind up as black hole singularities. The Russian astrophysicists I. M. Khalatnikov and E. M. Lifshiftz tried, in the 1960s, to establish exactly that view, but they were forced to abandon it when Stephen Hawking and Roger Penrose showed in 1970 that singularities are unavoidable in the general theory. This result has worried many. Even Einstein's own one-time assistant, Peter Bergman, once declared that "a theory that involves singularities and involves them unavoidably, moreover, carries within itself the seeds of its own destruction." In the case of a realistic, crushing singularity, perhaps all it means is that once the collapsing star has fallen into a region even smaller than an electron, Einstein's general relativity is no longer valid and fails. This would be similar to the behavior of Newton's theory, which fails at speeds comparable to that of light. Einstein, himself, was quite clear on this point. In his book *The Meaning of Relativity,* he wrote (concerning the use of relativity to study the origin of the universe as a "big bang"), "For large densities of field and matter, the field equations [of general relativity] and even the field variables which enter into them will have no real significance. One may not therefore assume the validity of the equations for very high density of field and of matter, and one may not conclude that the 'beginning of the expansion' must mean a singularity in the mathematical sense."

Another way to characterize a singularity is to say that it is where world lines end. This certainly includes the curvature singularities we have just discussed, but there is another possibility that does not have infinite spacetime curvature — a spacetime containing such singularities is called *geodesically incomplete* (see the end of Tech Note 5). The distinction between the two kinds of singularities, crushing and incomplete, is of great importance in time machine analyses, as we will see in the next section. Both kinds of singularities do signal, of course, that something unusual is afoot, perhaps even the failure of general relativity. If physicists stay away from a singularity during their mathematical discussions, however, then the conclusions of general relativity are certainly correct.

So what *does* the theory predict? Some of the amazing answers to that question came from a series of theoretical investigations undertaken by the American physicist J. Robert Oppenheimer and his students in the late 1930s—answers so astonishing that nobody *really* believed them to be other than mathematical (not physical) results. Einstein himself shared that view, and in a now nearly forgotten 1939 paper, he stated his belief that singularities "do not exist in physical reality . . . for the reason that matter cannot be concentrated arbitrarily." To start, general relativity theory says that around the singularity of a black hole, at a distance directly proportional to the mass of the collapsed object, the formation of a so-called *spacetime event horizon* should occur. The event horizon is a surface in spacetime through which anything can fall into the hole but through which nothing, not even photons of light, can escape. The singularity at the black hole center is therefore not visible (that is, not "naked") to a remote observer. The singularity is said to be "clothed" by the horizon, and for an observer outside the horizon of any black hole, the only observable properties of the hole are its mass (via its gravitational effects), its angular momentum (rate of spin), and its electric charge.

There are, in fact, several fundamentally different types of black holes. If the collapsed star forms a nonrotating, spherically symmetric, uncharged object, then the result is called a *Schwarzschild* black hole after the German astronomer Karl Schwarzschild, who found the first exact solutions to Einstein's general relativity equations just months after Einstein published them. (Even *Einstein* hadn't yet solved them, and he apparently thought they were too complicated to be solved. When he saw Schwarzschild's result, Einstein wrote to say, "I had not expected that the exact solution to the problem could be formulated. Your analytical treatment of the problem appears to me splendid.") The radius of the spacetime event horizon is called, in this case, the *Schwarzschild radius*.[16] Soon after that, the Finn Gunnar Nordström and the German Heinrich Reissner independently found the solution for a non-rotating, *charged* black hole (where *charge* refers either to electrical or magnetic charge, although no magnetic monopoles—the magnetic analog to positive and negative electrical charge—have yet been observed).

More interesting, however, are the theoretical time travel properties of *rotating* black holes.[17] These *Kerr–Newman holes* are named after the New Zealand mathematician Roy Kerr, who first solved (1963) the general relativity gravitational field equations for the spacetime region exterior to the horizon of a spinning uncharged hole, and Ezra Newman, a University of Pittsburgh physicist who two years later extended Kerr's solution to rotating, *charged* black holes. The Kerr–Newman solution is intriguing because it implies that the interiors of such rotating holes are portals into other spacetime regions that are otherwise inaccessible from our universe. In these holes the central singularity is a *ring*—a geometrical shape with a hole in it through which matter may be able to pass without destruction. Further, some of those regions may be past (or future) versions of "our" universe. That is, such black hole interiors may contain ring-portals that are "doors into time machines." An excellent discussion of the details of such black hole portals is given by Kaufman (1977).

This is all very speculative, of course, and the eminent physicist John Wheeler, for one, is not convinced that there is any sense at all to the notion of black hole time travel. As he related in a poignant story (1981): "I received one day a long distance telephone call from a distinguished Washington lawyer. 'My wife and I have lost our only child, our twelve-year-old son. Without him nothing has any meaning for us. We're willing to run any risk, pay any price, do whatever it takes to be transported back in time to his company. We have heard that black holes exist and that time goes backward in the neighborhood of a black hole. Is that true?' I had to tell him, 'I'm so sorry; no.' "

Wheeler's objection to black hole time travel is based on the quantum field fluctuations of gravity fields,[18] which are related to the uncertainties inherent in our knowledge of the values of physical entities (this is ignoring the formidable engineering problems posed by the lawyer's request). Such fluctuations, vanishingly small in systems of everyday size, increase dramatically at very tiny distances on the order of twenty orders of magnitude smaller than the nucleus of an atom. In the microscopic region of spacetime that the matter forming a black hole is falling into, these fluctuations might conceivably result in effects that preclude the formation of the end-stage singularity. There does seem to be increasing astronomical evidence for black holes, but the matter is still open as I write (1998) and perhaps such things do not actually exist; or even if they do exist (which *is* my personal opinion), maybe they contain no singularities.

When the theory of quantum gravity is at last developed, we will probably learn whether black holes really are potential time machines—although at least one physicist, Misner (1969a), has already asserted that the coming of quantum gravity will *not* prevent the formation of singularities.[19] And another, Soleng (1989), takes a pessimistic position with regard to using singularities for time travel even if they do occur. He begins by saying "Einstein's theory

. . . runs repeatedly into singularity problems. These problems may, however, be swept under the rug by saying that the extreme conditions of the very early Universe or those inside black holes, are beyond the region of validity of the theory. In these cases quantum effects will dominate." But he remains highly skeptical of the correctness of trying to sweep time travel problems under the rug: "The causality problems of the Gödel model . . . cannot be avoided [by appealing to quantum effects]. Rather they point at a serious flaw in our understanding of space-time. Can we accept a theory that allows such paradoxical solutions, or is it possible [to eliminate] the bizarre possibility of closed timelike curves?"

Another physicist, Brandon Carter, has been equally skeptical of black hole time travel. He concluded (1968), in a discussion of the Kerr black hole solution, "When the charge or angular momentum [is sufficiently large] causality violation is of the most flagrant possible kind in that it is possible to connect any event to any other by a future-directed timelike line." That is, backward time travel is allowed, even though the traveler is always moving at less than the speed of light into his local future (see Tech Note 4). Carter did not like this conclusion at all, declaring such a result to be "pathological" and calling such a timelike path "vicious." Indeed, a spacetime like Gödel's, in which a closed timelike path exists at *every* point, is termed by Carter as *totally vicious*. So disturbed was Carter about the possibility of time travel that he felt "the breakdown in general relativity may be [so severe that] the whole theory may have to be abandoned." Several years later Simpson and Penrose (1973) agreed, writing "the presence of closed timelike curves near the [ring singularity of a Kerr black hole] would seem to argue against any too close relation [to] physical reality." It is a paradoxical fact, however, that the 1970 Hawking and Roger Penrose theorem states that singularities *must* occur if certain assumptions are made, one of which is that backward time travel is impossible. Singularities, the entities that so horrify Carter and Soleng precisely because they seem to *allow* time travel, were declared to be inevitable if time travel could *not* occur!

What should we think about the foregoing objections? Even before general relativity, Pitkin (1914) wrote a stinging criticism of the concept of Wellsian time travel, and yet he began his analysis with a cautionary note that should be kept in mind today: "May it not be that our inability to leap into the fiftieth century, A.D., seems impossible to us, merely because of certain prejudices we entertain or certain facts and tricks of which we are still hopelessly ignorant? Assuredly, this is not a foolish query. Its answer, whatever that may be, carries immeasurable consequences for metaphysics." The assumption of global causality has recently been weakened in the Hawking and Penrose theorem, in fact—see Kriele (1990a)—so it seems that singularities and time travel to the past might be able to coexist. And even more recently, Cvetic *et al.* (1993) found singularity-free solutions to the gravitational equations that still allow time travel to the past. In their words, they have found "causally non-trivial exact solutions to Einstein's equations without space-time singularities . . . in which [the] physics of closed timelike curves can be studied." That is, they found that there *are* solutions to the field equations that allow time travel into the past "without the difficulties that the curvature [that is, crushing] singularity presents for black holes." Still, time travel to the past is so strange that many physicists have long tried to prove it impossible.

In one early attempt to discover a physical reason for rejecting the possibility of time machines that require rotation for their operation, Charlton (1978)—who refers to the "displeasing nature of time machines"—suggests that such machines will radiate away enough of their angular momentum to damp the rotation below the rate required to generate closed timelike lines in spacetime. Since that early speculation, the search for physical mechanisms that can prevent the creation of *any* form of time machine, rotating or otherwise, has grown ever more sophisticated and abstract. Some physicists are so appalled at the potential they see for paradoxes if a time machine could exist that they believe there simply *must* be such mechanisms. This view is illustrated by Visser (1994a), who, in some frustration over the failure to find such mechanisms, simply declares that "it [Hawking's Chronology Protection Conjecture] should be taken to be an axiom, rather than attempting to prove it by calculation."

More recently, however, other physicists have come to recognize the danger in drawing conclusions from what may only *seem* to be paradoxical. For example, after proving a theorem that begins by hypothesizing the formation of a naked singularity via gravitational collapse, Clarke and de Felice (1982) showed that the end result could be chronology violation throughout all of spacetime and thus that "this means an observer could receive information from his own future." Shaken but not defeated, these two physicists then ended their paper with the bold words "Unless a suitable alternative is available it seems we have to face the hard but also challenging perspective of accepting these extreme consequences of general relativity as a real possibility in nature." And as a second example of not losing your nerve, after a brief description of how very *un*restrictive general relativity is in imposing constraints on the geometry of spacetime, Yurtsever (1990) observes that "it is neither suggested nor warranted by the theory to discard any entire class of space-times as 'unphysical,' regardless of how strange and counterintuitive their properties may be. Attitudes that lead to such selective, ad hoc dismissals of space-time phenomena may be misleading and counterproductive." Yurtsever reminds his readers that it was not so long ago that physicists rejected the event horizons of black holes as "unphysical," and yet today's physicists actively search for black holes and believe that they have good evidence that such things do indeed exist.

Agreeing with Yurtsever's philosophy is another physicist, Politzer (1994), who wrote, "While the phenomena [due to a time machine] may offend our sensibilities, to dismiss their possibility on that basis may be premature." Similarly, the prediction by the general theory of gravitational radiation—literally ripples in spacetime—was once viewed with great skepticism. Today, however, most theoreticians believe that such radiation exists and that its detection from astronomical sources is merely a matter of time and improved instrumentation. Still, not everybody agrees; certainly not Visser (1994b), who calls the possibility of time machines "extremely disturbing" and an "unfortunate happenstance" and opens his analysis with a philosophical stance, declaring, "If [certain conditions exist] then the battle against time travel is already lost, the spacetime is diseased, and *it should be dropped from consideration* [my emphasis]." Event horizons and gravitational radiation as "crazy ideas" have now been replaced by spacetimes with acausal behavior. Perhaps, warns Yurtsever, present-day physicists might do well to remember the hasty judgements of the past.

Tipler's Time Machine

The time travel property of rotating black holes makes them a favorite of science fiction writers—witness *Re-entry* (Preuss) and *The Forever War* (Haldeman). There is talk of a "micro-black-hole cluster," too, in "Houston, Houston, Do You Read?" (Tiptree), in which a spaceship is sent three centuries into the future. The major difficulty in all of this, of course, is how to get one's hands on a black hole! Is there any other form of "time machine" that is consistent with the general theory of relativity? Yes, there is.

In 1974 a young physics graduate student at the University of Maryland, Frank Tipler, caused a bit of a stir when he published what seemed to be quite specific construction details for a time machine. Indeed, the final sentence in his paper couldn't be clearer: "In short, general relativity suggests that if we construct a sufficiently large rotating cylinder, we create a time machine." Nobody had ever before made such a statement in a respectable physics journal,[20] and best of all, there were no apparent singularities involved (as with black holes). However, a close look at Tipler's analysis does turn up some difficulties.

What Tipler had actually done was to show that if one had an *infinitely long, very dense* cylinder rotating with a surface speed of at least half the speed of light (the rotation speed is such that the centrifugal forces are balanced by gravitational attraction), then this allowed the formation of closed timelike lines connecting events in spacetime. This means that by moving around the surface of such a fantastic cylinder, one could travel through time into the past—but not to earlier than the time of the creation of the cylinder.[21] Tipler's cylinder would also enable a time traveler to return to her original time, to go "back to the future." Tech Note 8 shows a simple illustration—based on a similar one in Tipler's dissertation—that demonstrates how the cylinder works as a time machine. No one, in fact, disputes this. It *is* true. On paper.

But Tipler did *not* prove that this time travel property holds for cylinders of even very long but finite length, which are the only kind we could actually build from a finite amount of matter; he merely suggested that such might be the case. This suggestion seems reasonable, because if the time traveler orbits at the midpoint of the cylinder, near the surface, then the gravitational end-effects of sufficiently remote ends of the cylinder might be negligible. Similar mathematical approximations are routinely made, for example, when calculating the electrical effects of charged cylinders of finite length. But as Thorne (1970) warns, "Extrapolation from cylindrical symmetry to reality is very dangerous, since spacetime is not even asymptotically flat [see Tech Note 4 for what that means] around an infinite cylinder." The issue of whether a spinning, finite-length cylinder can create closed timelike lines is still open. As Bonner (1980) puts it, "in some respects an infinite cylinder may be a model for a long finite one, and the possibility cannot be dismissed that a time machine might be associated with a long, but finite rotating system." In Gribbin (1983), for example, we find the estimate that a 10-to-1 ratio of cylinder length to radius may be enough for Tipler's cylinder to be "infinite."

There is, however, another potential problem besides the length of a cylinder. There is a strong likelihood that a Tipler protocylinder would collapse under its own internal gravitational pressure before it could be made nearly long enough to be even "approximately infinite." That is, such a finite-length cylinder might actually crush itself along its long axis into a pancake-shaped blob, something like what happens to a long cylinder of jello stood on-end. An ordinary can of jellied cranberry sauce will also sometimes display this curious behavior.

The required rotational speed causes a problem, too. We are not talking about cylinders the diameter of a pencil or even of a large water pipe. Recall that for a given surface speed, the larger the diameter, the less the centrifugal acceleration at the surface. It is easy to calculate that even a huge cylinder 10 kilometers in radius—and so by Gribbin's estimate at least 100 kilometers in length—would have, with a surface speed of half the speed of light, a surface acceleration *two hundred billion* times the acceleration of Earth's surface gravity. No known form of ordinary matter could spin that fast and not explosively disintegrate. But then, Tipler cylinders would not be ordinary in any sense of the word. Tipler has estimated that the required density for a time machine cylinder would be 40 to 80 orders of magnitude above that of nuclear matter.[22] Made from such superdense stuff, a finite cylinder would typically be as massive as the Sun but many trillions of times smaller. Showing no lack of imagination, Tipler has himself suggested (1977) the possibility of speeding up the rotation of an existing star as an alternative approach to that of actually tying to build a cylinder.[23] Of course, this would be a project for a far-future society with a very advanced technology.

Tipler's scientific pessimism about time travel via one of his cylinders is shown by the words he used to open his 1977 paper: "[A]ny attempt to evolve [a time machine] from [normal] matter will cause singularities to form in space-time. Thus, if by the word 'manufacture' we mean 'construct using only ordinary materials *everywhere*,' then the theorems of this paper will conclusively demonstrate that a [time machine] cannot be manufactured." Much less pessimistic is Ori (1993), who makes *two* pointed observations. First, Tipler's theorems apply only to singularities of the incomplete kind, not to the more convincingly fatal crushing (or curvature) type. Second, Ori makes an even more telling observation—one I think *so* telling that I'll quote him at length.

> The standard interpretation of Tipler's theorems is to say that the appearance of a singularity in a given [spacetime] model indicates that this model is unrealistic and cannot be physically realized. Even for future-generation engineers it will probably be impossible to use "singular matter" for the construction of their time machine. However, the theory of black holes provides an obvious counterexample to this interpretation. For, by applying this interpretation to the black hole singularity theorems one could conclude that black holes can never form. This analogy makes it clear that Tipler's theorems can bear a very different interpretation, namely, that the construction of a time machine is perhaps possible, but that the causality violation will then inevitably lead to the formation of a singularity. This possible interpretation suggests that one should discard a time-machine model due to a singularity only if this singularity appears sufficiently early [in the process of "making" the time machine] that it can causally interfere with the occurrence of causality violation.

Even less concerned about singularities are Headrick and Gott (1994), who wrote, "it would seem that a successful attempt to manufacture [a time machine] within a finite region of space will be accompanied by the creation of a singularity. . . . This does not immediately imply, however, that with a sufficiently advanced technology one could not make a time machine. *There is no reason to suspect spacetime singularities could not in principle be created through deliberate human action* [my emphasis]." These views, of course, were welcome news for science fiction writers, who had been using Tipler cylinders almost since Tipler first wrote of them.

Indeed, science fiction writer Poul Anderson used Tipler cylinders (he called them "T-machines") in his 1978 novel *The Avatar*. He describes such cylinders as scattered about the universe by ancient, altruistic aliens called "the Others," to be used by any with the wits to decipher how. Anderson recognized the obvious problems with Tipler cylinders and had one of his characters say of T-machines, "I have no doubt whatsoever that here is the product of a technology further advanced from ours than ours is from the Stone Age." Indeed, in his 1977 paper Tipler had written that whatever these cylinders might be made of, it could only be called "unknown material." This requirement for supermatter seems to be a feature of time machines in general—as discussed in Tech Note 9, for example, the wormhole time machine requires what physicists call "exotic" conditions. So, alas, Tipler cylinders are out, at least for a while, as practical time machines—but even "no time machine, ever" Larry Niven liked the idea of them well enough to lift Tipler's title for a short time travel story, "Rotating Cylinders and the Possibility of Global Causality Violation."[24] The significance of Tipler's result, and of Gödel's, too, is that they hold out at least a little hope for the physical possibility of time travel to the past.[25] The fact that their particular mechanisms for achieving it are not possible (in the engineering sense) is irrelevant. Other, methods yet unthought of—what Forward (1988) calls "future magic"—could someday perhaps be engineering possibilities.

There is, however, still the major objection to time travel to consider, the one that bothered Einstein about Gödel's solution, and it is the one that is the hardest to wave away. It is the problem of causality violation. This has already been briefly discussed in this chapter, but the issue warrants much more attention and a chapter all its own. That is Chapter Four. But first we need to take a closer look at time itself, the "stuff" or "thing" or . . . ? that we are interested in traveling "through" or "around" or "across" or . . . ?

2

On the Nature of Time, Spacetime, and the Fourth Dimension

I do not believe that there are any longer any *philosophical* problems about Time; there is only the physical problem of determining the exact physical geometry of the four-dimensional continuum that we inhabit.

—Professor Hilary Putnam (1967)

What, then, is time? I know well enough what it is, provided that nobody asks me; but if I am asked what it is and try to explain, I am baffled.

—Saint Augustine, *Confessions*

The Fourth dimension is just a hypothetical math concept. Or else it's time, or something. Just a lot of sci-fi crud.

—Laura's reaction in *The Boy Who Reversed Himself* (Sleator) after an oddly behaving new boy in school says he has been in the fourth dimension

Science fantasy enthusiasts know that "time is the fourth dimension" and that Einstein's theories are the hopeful but as yet unrealized basis for all kinds of marvelous possibilities, from time machines to weird spacetime warps in the fabric of the Universe which enable the gifted to pass through discontinuities into the exotic world of the *n*th dimension. Einstein's theories do *not* in fact suggest these things.

—a more than slightly exasperated comment by two physicists in their book on relativity (Sears and Brehme, 1968)

What Is Time?

Christian clerics had identified time as something unusual long before science fiction writers and their time travel stories. We can, in fact, trace their interest back at least fifteen centuries to Saint Augustine. Certainly the seventeenth-century Spanish Jesuit Juan Eusebius Nieremberg caught the spirit of wonder that time holds for the devout when he wrote, in his *Of Temperance and Patience,* that *"Time* is a sacred thing; it flows from Heaven. . . . It is an emanation from that place, where eternity springs. . . . It is a *clue* cast down from Heaven to guide us. . . . It hath some assimilation to Divinity."

Going outside of Christianity, we can find other equally strong reactions to the mystery of time. From Plutarch's *Platonic Questions* we learn that when the question of time's nature was put to Pythagoras, he simply uttered the mystical "time is the soul of this world." The *Laws of Manu* of Hinduism, the *Torah* of Judaism, the *Koran* of Islam, and the revealed truths of Gautama Buddha are all full of references to time. It is, in fact, to the pagan gods of Greek mythology that we owe *our* "modern" image of Chronos, or Father Time.

Not just the Greeks made time a god. In the *Bhagavad Gita* (*Song of the Lord*), the central religious-romantic epic of Hinduism that predates Christ by five centuries, one of the characters reveals his divine nature and declares his power thus: "Know that I am Time, that makes the worlds to perish, when ripe, and bring on them destruction." And in the even older Egyptian *Book of the Dead,* which dates back over three thousand years, the newly deceased was thought literally to become one with time itself. The merging of time and the resurrection of the body after death is shown in the line "I am Yesterday, Today and Tomorrow, and I have the power to be born a second time."

In contrast to these poetical views, one anonymous wit pushed his brain to the limit and found that the best he could come up with as an answer to "What is time?" was that "Time is just one damn thing after another." This may or may not have been the same person who declared that "Time is what keeps everything from happening at once." The introduction to Farley (1950) attributes this last insight to the early science fiction writer Ray Cummings, and the words do appear in his 1921 story "The Time Professor." Cummings repeated them a few years later in his novel *The Man Who Mastered Time,* and he certainly thought the phrase to be his; in an April 1931 letter to the editor of *Astounding Stories* he repeated them yet again, in a way that implies he was not quoting someone else. He used the same concept, with more romantic imagery, in his 1946 novel *The Shadow Girl:* "This same Space; the spread of this lawn . . . what would it be in another hundred years? Or a thousand? This little space, from the Beginning to the End so crowded with events and only Time to hold them apart!"

A modern philosopher—see Horwich (1987)—has expressed Cummings' sentiment in more scholarly fashion, in words that echo Saint Augustine's: "Time is generally thought to be one of the more mysterious ingredients of the Universe." However, perhaps the most pragmatic approach to the meaning of time is the one expressed by the English essayist Charles Lamb in a letter he wrote in 1810: "Nothing puzzles me more than time and space and yet nothing puzzles me less, for I never think about them."

Long before 1940 the science fiction connection between the fourth dimension and time was common; see, for example, the 1927 story "The Machine Man of Ardathia" (Flagg). Indeed, by 1939 Robert Heinlein had the central character in his first published science fiction story, "Life-Line," assert that "you have been told that time is a fourth dimension. . . . It has been said so many times that it has ceased to have any meaning. It is simply a cliche that windbags use to impress fools." To illustrate Heinlein's point, a typical story of the pre-1940 period is "An Adventure in Futurity" (Smith) from a 1931 issue of *Wonder Stories;* it is the tale of a man from 1930 who befriends a visitor from A.D. 15,000 and so receives an invitation to visit the future. The machine that accomplishes this visit is briefly (and somewhat confusingly) described as making "possible a journey in that fourth-dimensional space known as time."

The author of the bizarre little 1937 tale "Down the Dimensions" (Bond) got his story twist with a play on the idea of time as the fourth dimension. In that story a professor invents a gadget that will carry him downward in spatial dimension, from the everyday three-dimensional space to a two-dimensional plane, and finally to a one-dimensional line. But after completing the last "dimensional collapse" to a line, he finds that time is the *first*—not the fourth—dimension.

What, then, *can* we say about time? Despite the bold words in the first quotation at the beginning of this chapter, from the distinguished twentieth-century Harvard professor Hilary Putnam, I suspect that most people would tend to agree more with the second quotation, from the distinguished early-fifth-century Christian theologian who earlier had been a distinguished late-fourth-century sinner. The passage of fifteen hundred years has, in fact, done little to clarify the meaning of time, and most of us might actually agree most of all with Laura in the third quotation.

Ray Bradbury wrote a beautifully poetic passage about the mystery of time in "Night Meeting," one of the splendid substories in his episodic masterpiece

The Martian Chronicles. A man of A.D. 2002, who is one of the modern inhabitants of Mars, somehow meets the ghostly image of a long-dead Martian one cold August night. The conditions are just right for such a cross-time encounter. As the man thinks to himself, "There was a smell of Time in the air tonight. He smiled and turned the fancy in his mind. There was a thought. What did Time smell like? Like dust and clocks and people. And if you wondered what Time sounded like it sounded like water running in a dark cave and voices crying and dirt dropping down on hollow box lids, and rain. And, going further, what did Time *look* like? Time looked like snow dropping silently into a black room or it looked like a silent film in an ancient theater, one hundred billion faces falling like those New Year balloons, down and down into nothing. That was how Time smelled and looked and sounded. And tonight . . . tonight you could almost *touch* Time."

Lovely words, yes, but they don't really tell us what time *is*. Perhaps Einstein can tell us. In the *New York Times* of December 3, 1919, we find him quoted as follows: "Till now it was believed that time and space existed by themselves, even if there was nothing—no Sun, no Earth, no stars—while now we know that time and space are not the vessel for the Universe, but could not exist at all if there were no contents, namely, no Sun, no Earth, and other celestial bodies." Less than two years later Einstein stated this view again (*New York Times,* April 4, 1921): "Up to this time the conceptions of time and space have been such that if everything in the Universe were taken away, if there were nothing left, there would still be left to man time and space." Einstein went on to deny this view of reality, saying that, according to his general theory of relativity, time and space would *cease to exist* if the universe were empty. This has the ring of one of Einstein's favorite philosophers, Spinoza, who declared in his *Principles of Cartesian Philosophy* that "there was no Time or Duration before Creation." In a correspondence with Samuel Clarke—Newton's friend who translated Newton's *Optiks* into Latin—Leibniz (who began the correspondence in late 1715) expressed similar ideas: "Instants, consider'd without the things, are nothing at all; . . . they consist only in the successive order of things."

The pragmatic scientist would certainly agree with Leibniz. After all, what could it even mean to talk of time unless you can measure it? And what you use to measure time is a clock—some kind of changing configuration of matter such as spinning gears, ticking pendulums, and rotating dial pointers. Mere unchanging matter, alone, is not sufficient to measure time, because a still clock measures nothing. *Changing* matter seems to be required. Yet, not surprisingly, not everybody agrees. The counterview, the view that time has nothing to do with change, was expressed in an interesting manner by a science fiction fan in a letter to the editor of *Wonder Stories* (January 1931): "Just one thing, you have these time-traveling yarns, good stuff to read all right, but bunk, you know; because if there's no such thing as time, which there isn't, only change, how can one travel in . . . something that doesn't exist. To our planet which goes around the Sun there is simply a turning and warming of one side and then the other, i.e., years, days, minutes, etc., is something purely artificial, invented by man to tell him when to do certain things, work and stop work. . . ."

Going even beyond the ideas of Einstein, Spinoza, Leibniz, and our science fiction fan, at least one metaphysician—Taylor (1987)—feels that time would have no meaning, even in a massive universe, without the additional presence of conscious, rational beings. This sounds very much like an echo of the French philosopher Henri Bergson, who in 1888 mysteriously declared (1910) that time is "nothing but the ghost of space haunting the reflective consciousness." A few years before Taylor, however, a fellow philosopher—McCall (1976)—had argued for exactly the opposite view, that temporal passage is independent of the existence of conscious beings.

All this divergence of opinion perhaps explains why even a lightweight movie such as Mel Brooks' 1987 *Spaceballs* can get a laugh from a time joke. Even kids know that the characters, when talking about time, haven't the slightest idea of *what* they are talking about. The movie, which is a spoof on such classic films as *Star Wars, The Wizard of Oz,* and *Raiders of the Lost Ark,* quickly reaches a point of crisis. To find out what to do next, the evil Lord Helmet and his chief henchman decide on a novel approach: They will look at an instant videotape of their own movie! (Instant videos are available *before* the movie is finished.) Perplexed at watching on a television screen everything that he is doing as he does it (the screen correctly shows an infinite regression of television screens, each being watched by a Lord Helmet), Lord Helmet initiates the following rapid-fire exchange. It is, of course, a clever take-off on Abbott and Costello's "Who's on First?"

> "What the hell am I looking at? When does this happen in the movie?"
> "Now! You're looking at now, sir. Everything that happens now, is happening, now."
> "What happened to then?"
> "We're past that."
> "When?"
> "Just now. We're at now, now."
> "Go back to then."

"When?"

"Now."

"Now?"

"Now."

"I can't."

"Why?"

"We missed it."

"When?"

"Just now." [The henchman then sets the video to rewind.]

"When will then be now?"

"Soon."

We may laugh at this, even dismiss it as mere movie madness, but could any of us *really* do much better if, like Saint Augustine, we were backed into a corner and asked to explain time? Somehow, I think even Professor Putnam would find it difficult to know where to begin. He might even become as confused as the time traveler in the 1968 French film *Je t'aime, Je t'aime,* whose oscillations in time from present to past and back again leave him so befuddled that he decides he'd rather be dead.

Speculations on the Reality of Time

The mystery of time was well captured by R. H. Hutton, the literary editor of the *Spectator,* when he wrote—see Parrinder (1972)—in his 1895 review of Wells' *Time Machine* that "the story is based on that rather favorite speculation of modern metaphysicians which supposes *time* to be at once the most important of the conditions of organic evolution, and the most misleading of subjective illusions . . . and yet Time is so purely subjective a mode of thought, that a man of searching intellect is supposed to be able to devise the means of traveling in time as well as in space, and visiting, so as to be contemporary with, any age of the world, past or future, so as to become as it were a true 'pilgrim of eternity.' "

Novelist Israel Zangwill wrote a similar but much more analytic review of Wells' novel for the *Pall Mall Magazine* (also included in Parrinder). Zangwill was the only Victorian reviewer to attempt a scientific analysis of time travel. Although he thought Wells' effort was a "brilliant little romance," Zangwill also thought the time machine—"much like the magic carpet of *The Arabian Nights*"—was simply "an amusing fantasy." Zangwill continued in his review with what was even then a common idea about a way one might actually be able, at least in principle, to look backward in time; one could travel far out into space by going faster than light and then watch the light from the past as it catches up. (Note carefully that that was written in 1895, ten years before Einstein and special relativity's apparent limit on possible speeds.) In this way, Zangwill wrote, one could watch "the Whole Past of the Earth still playing itself out."

Indeed, even before Zangwill, the well-known French astronomer Camille Flammarion had made this dramatic idea a centerpiece of his 1887 novel *Lumen*. That book, a best-seller in Europe even before its appearance in English, describes how a man just dead (in 1864) instantly finds his spirit on the star Capella, where he is able to watch the light then arriving from the Earth of 1793. Specifically, he watches the French Revolution play itself out and sees himself as a child. Flammerion may have, in fact, been inspired to write his novel by an essay written several years earlier (in 1883) by the British physicist J. H. Poynting. Poynting's essay, which opens with the statement that it was, in turn, inspired by an anonymous pamphlet published "thirty or forty years ago" on the same topic, specifically mentions watching historical events from Capella.

By the beginning of the twentieth century the idea of watching the past by outrunning light had drifted down into the juvenile literature, as in Jean Delaire's 1904 novel *Around a Distant Star,* in which a young man builds a spaceship that can travel at two thousand times the speed of light. With it he and a friend travel to an Earth-like planet nineteen hundred light-years distant and use a super-telescope to watch the crucifixion (and then the resurrection) of Jesus. Early magazine science fiction also found the idea of looking backward in time with delayed light to be a romantic one, as in the murder-romance "The Time Reflector" (England). And in "Faster Than Light" (Sharp) a scientist loses his wife to a rival who kidnaps her and then escapes in a faster-than-light rocket ship headed for parts unknown. After years of searching for them with his own brilliant invention of the ampliscope (several quantum leaps beyond the telescope), the scientist finally locates the couple, skipping from planet to planet light-years distant. His only pleasure, then, is to use his own faster-than-light craft to outrun the images of his lost love and watch them over and over. But eventually he comes to realize the ultimate futility of it all. As the last line of the story says, "It would be senseless, I knew, chasing on and on after yesterdays." The 1938 story "Time on My Hands" (Weisinger) uses the same idea in a very short story that specifically cites Flammarion.

A different way to look backward in time is found in a curious idea—that of looking *forward* in time, an idea that assumes time is a "closed loop." Plato (circa 400 B.C.), for example, thought of time as certainly having a beginning, but his conception did not have time extending off into the indefinite future as does the modern, everyday view. Rather, Plato visualized time as curving back on itself—as *circular* in nature. This was a reasonable reflection of what Plato could see everywhere in nature, with the seemingly endless repetition of the seasons, the regular ebb and surge of the tides (the old English word *tid* was a unit of time), the unvarying alternation of night and day, and the rotation of the planets in the sky. Whatever might be observed today would, it seemed obvious, happen again in the future. Several fictional examples of circular time were discussed in Chapter One; another is "Night Broadcast" (Hobana), in which a television signal from the past is picked up by a gadget that probes the future—"by going far enough into the future one comes upon what we call the past." James Joyce's novel *Finnegans Wake,* which opens in mid-sentence and ends with the first part of the same sentence, is another exercise in cyclical time. This view of time has a powerful, ancient visual symbol, the Worm Ouroborous, or World Snake, that eats its own tail endlessly.

Plato's most famous student, Aristotle, was a keen observer of physical fact, and for him, time was *motion* (in a world in which nothing moved, there would be no time), and he expressed this view in his famous metaphor "Time is the moving image of eternity." For Aristotle, time and change were inseparably intertwined. For Aristotle the world had existed for eternity, and the circularity of time was a central and powerful image; using his vivid illustration, it is equally true in circular time that we live both before and after the Trojan War.

In the West it was the Christian theological doctrine of unique historical events that gave rise to linear time in the minds of the common folk. The creation of the world and of Adam and Eve, Noah and the cataclysmic Flood, the Resurrection—these were all events that occurred in sequence, *once*. None would happen again, so for Christianity, circular time just would not do. In addition, it has been argued that the major spiritual content of Christianity, and a significant reason for its popular support in the face of brutally harsh Roman suppression, is that it brought the *expectation of change* into the static world of ancient times. It was, in fact, in ancient religious teachings that our modern belief in linear time had its origin, a belief that most people today (including the most hardened agnostic physicist) find to be as natural as Plato and Aristotle found circular time.

Just to show how one can find support for almost any view in the same religious dogma, *Ecclesiastes* 1:9 would seem to be a claim not for linear time but for circular time: "The thing that hath been, it is that which shall be; and that which is done is that which shall be done; and there is no new thing under the Sun." Interestingly, the little-known American poet Joseph Stickney thought circular and linear time might be one and the same. Because of his death from a brain tumor at the early age of thirty, little of his work has survived, but in one fragment called "The Soul of Time" he used the mathematician's view of a straight line as the limiting case of the circumference of a circle with infinite radius:

> Time's a circumference
> Whereof the segment of our station seems
> A long straight line from nothing into naught.

Circular time, with its closed topology, was favorably presented in Stephen Hawking's famous book *A Brief History of Time.* In it he concludes that there is no need for God because in circular time there is no first event and hence no need for a First Cause. Vigorous philosophical rebuttals were quick to come, of course; see Craig (1990) and Le Poidevin (1991). Gott and Li (1998) have also argued for no first cause, suggesting that, via time travel, the universe could be its *own* cause.

Even though linear time was the norm after Christ, there were still enough questions about time to perplex the deepest of thinkers, and the next two thousand years resulted in plenty of thinking. Discourses on time by such philosophers as Descartes, Spinoza, Hobbes (who in the seventeenth century associated the points of a straight line with the instants of time), Kant, Nietzsche, and Hegel can be found by the yard in any decent university library. Nearly all (if not indeed all) of these presentations have metaphysical, even theological, underpinnings.

For example, Descartes is generally believed to have argued for a discontinuous, atomistic nature of time (recall the *chronon* from Chapter One). This is the modern view, because in his *Meditations* (1641), in particular in the third meditation on God's existence, he apparently argues that God must continually recreate the world at each *separate* moment of its existence. That is, the world is continually recreated in a discontinuous succession of *individual* acts by God—but see Arthur (1988) for a refutation. Then, with Newton's discussion of *absolute time,* which is the belief that time is the same everywhere in the universe (for more on this, see Tech Note 1), there was for the first time a *physicist* writing about time. But despite Newton's genius, the mystery of time remained a mystery.

In 1905 Einstein's name appeared among the contributors to the study of time, and at last something besides metaphysical speculation on the subject was added to the body of human thought. Einstein's paper introduced the idea of *relative time,* which is the belief that the passage of time is not the same

Calvin and Hobbes   by Bill Watterson

CALVIN AND HOBBES © 1994 Watterson. Dist. by UNIVERSAL PRESS SYN-DICATE. Reprinted with permission. All rights reserved.

everywhere but rather depends on local conditions (for more on this, see Tech Note 2). In retrospect, Einstein's paper seems to be the perfect reply to the comment by Isaac Barrow—Newton's teacher and the first Lucasian professor of mathematics at Cambridge, the chair held today by Stephen Hawking—that "because *Mathematicians* frequently make use of Time, they ought to have a distinct idea of the meaning of the Word, otherwise they are Quacks."

Then, just three years after Einstein, along came a second astonishing paper by the Cambridge philosopher John Ellis McTaggart. This paper (1908) claimed to prove that whatever time might be thought to be (even by Einstein), it really wasn't that because time wasn't even real! The method of the paper is to deny the reality of time via an infinite-regress argument that one philosopher, Mink (1960), has called the *pons asinorum* of the riddle of time. That is, the argument is so difficult to follow that it is the "bridge of asses" over which most people balk at crossing. As McTaggart's opening sentence freely admits, "It doubtless seems highly paradoxical to assert that Time is unreal, and that all statements which involve its reality are erroneous."

McTaggart began his analysis by observing that there are two separate and distinct ways of talking about events in time. Following his symbolism, one can say that events are either future, present, or past, (the so-called A-series), or one can say that events are temporally ordered by each being later than some other events, earlier than others, and simultaneous with still others (the so-called B-series). He then continued by asserting that time requires change and followed that with the observation that the A-series (but not the B-series) incorporates such change. That is, if event X is earlier than event Y, then X is *always* earlier than Y and thus there is no change in this (or in any other) example of a B-series. As a specific example, let Y be the birth of a child, and let

X be the birth of its mother. In contrast, if X is first in the future, then is in the present, and finally is in the past, then we have an example of change (and hence of *time*) in an A-series—for example, let X be the next time you blink.

With this rather pedestrian start, McTaggart then pulled the rabbit out of the hat. It makes no sense, he argued, to talk of the "future," "present," and "past" of an event because these terms are mutually exclusive. That is, no two of these predicates can apply at once, and yet, paradoxically, every event possesses all three and thus we have a contradiction. It therefore, concludes McTaggart, makes no sense to talk of future, present, or past. And because it makes no sense to talk of them, they do not exist, and so there can be no A-series and hence no change, and thus there can be no time. McTaggart of course realized just how astounding all that would appear, and in fact he played devil's advocate (D.A.) in his paper by trying to anticipate the various objections people could raise. Of course, he always managed to refute the D.A. at every turn.

The predicates of future, present, and past are really not incompatible for any event, the D.A. says some will claim, because the *real* predicates we should use are "*was* future," "*is* present," and "*will be* past," and these *can* be possessed all at once by any event. Nice try, counters McTaggart, but that will not solve the problem. By allowing such modified predicates, we must actually allow for all nine possibilities, some of which are *still* incompatible. That is, the "was," "is," and "will be" could each be potentially attached to "future," "present," and "past"; and, for example, "was past" is incompatible with "will be future."

Oh, counter-counters the D.A., we can eliminate that concern by allowing even more complex modified predicates to arrive at a third level of structure, such as "is going to have been past" and "was going to be future," and those *are* compatible. But then McTaggart bats that argument away, too, by displaying new incompatibilities, as well as by showing that the process of ever-increasing predicate complexity is a vicious infinite regress that drags along the seeds of its own doom at every step.[1] There is no escape from incompatibility, he says, and hence we were fated to fail at the very first step, and so there is no time.

Well! What can one do when presented with such an argument—one that seems to claim philosophers can wrest free the secrets of nature by pondering the historical accidents of English syntax? As David Hume once said, "Nothing is more usual than for philosophers to encroach on the province of grammarians, and to engage in disputes of words, while they imagine they are handling controversies of the deepest importance and concern." One modern philosopher—Christensen (1974)—seems to agree, at least in the case of McTaggart's "proof," and he has been pretty blunt with his evaluation: "McTaggart's famous argument for the unreality of time is so completely outrageous that it should long ago have been interred in decent obscurity. And indeed it would have been, were it not for the fact that so many philosophers are not sure that it has ever really been given a proper burial, and so from time to time someone digs it up all over again in order to pronounce it *really* dead. These periodic autopsies reveal that something more remains to be said." That is certainly true, in as much as McTaggart's disarmingly innocent argument has caused disagreement and furrowed brows among philosophers for decades.

It is, in fact, easy to find examples of the continuing debate over McTaggart's analysis. At least one philosopher—Smith (1986a)—has actually argued that McTaggart did not really understand his own proof, though his conclusion about the unreality of time is correct. Smith was then quickly refuted in turn by Oaklander (1987). And Currie (1992) shows how McTaggart's ideas have found their way into modern philosophical debates on the meaning of time in the cinema, particularly in the analysis of *anachrony,* the telling of a story out of normal time sequence, such as occurs in time travel movies. My own reaction is that although McTaggart's "proof" may mean something to grammarians, is simply has no significance when it comes to the business of physics.

Other sorts of metaphysical proofs for the unreality of time have been offered besides McTaggart's. For example, it has been argued that time is unreal, at least in a world empty of consciousness, because the concepts of past, present, and future could not possibly have any meaning unless events could be remembered, experienced, and anticipated. Or, for a second example, some have held time to be unreal, at least in a deterministic world (as some incorrectly argue *fatalistic* four-dimensional spacetime to be), because any event whose occurrence followed from present conditions, and from physical laws, would exist *now.* This position, which seems to say that everything should happen at once, I fail to understand sufficiently to be bothered by it—but see Gale (1963) for a rebuttal if it bothers you.

Debates between those who believe in the common-sense idea that present, past, and future are attributes of events (the "tensers") and those who deny it (the four-dimensional spacetime, block-universe "detensers") continue to rage across the pages of philosophy journals. In Weingard (1977) we even find one philosopher who sees merit to *both* positions.

Less than a month before his death, Einstein revealed his feelings about the meaning of present, past, and future. In a letter written on March 21, 1955 to the children of his dearest friend, Michele Besso, who had just died, Einstein wrote—with full knowledge that his own illness would be his last[2]—"And now he has preceded me briefly in bidding farewell to this strange world. This signifies nothing. For us believing physicists, the distinction between past, present, and future is only an illusion, even if a stubborn one." Later in this chapter I'll return to these curious words and speculate on what Einstein may have meant by them.

Has the Past Been for Ever?

Our modern concept of linear time as a straight line extending from the dim past through the present and disappearing into the misty future gives rise immediately to twin questions: "Did time have a beginning?" and "Will time ever end?" As Stearns (1950) puts it, "Endings and beginnings are rooted in the very conception of time itself." I'll limit myself for now to the question of whether the past is finite or infinite in duration and will take up the possibility of time having an ending later.

Early Biblical scholars, of course, believed that the answer was *finite* (the world came into being because of a First Cause, God's creation of everything), and they expended vast quantities of energy (and, need I say it, time itself) on calculating the date of creation. Martin Luther, for example, argued for 4000 B.C. as roughly when everything, including time, began. Johannes Kepler adjusted this by a notch, to 4004 B.C., and later the Calvinist James Ussher, Archbishop of Armagh and Primate of All Ireland, tweaked it again. His date is the most impressive of all, at least in detail: the first day of the world was 4003 years, seventy days, and six hours before the midnight that started the first day of the Christian era. Six days after that first day of the world, Adam was made, and as a final dash of specificity, this last date was declared to be Friday, October 28!

Ironically, then, although Christian theology may be given credit for introducing linear time, it certainly did not provide very much of it. The beginning of time was just six thousand years or so ago, and of course the End—in the form of the Battle of Armageddon—has been awaited (with varying degrees of eagerness) for the last thousand years. The famous, ancient question that this view of history naturally prompts—"What was God doing before he created the world?"—has the equally famous, ancient answer "Creating Hell for those who ask that question." But see Leftow (1991) for how a modern philosopher has put a curious twist on that answer.

The discovery in the seventeenth century of geological time cast a certain amount of skepticism on those early calculations. With the discovery that the very Earth itself could be decoded for its history, the lure of trying to decode a book of admittedly finite age declined for most people (although it cannot be denied that modern Creationists still find such a task to have its rewards). Geological time was discovered to be a *chasm* of time extending backward for billions of years, a duration that is really incomprehensible for the human brain. It has become fashionable for geologists to refer to such enormous durations with the apt term *deep time,* a subtle play on the metaphor of the "ocean of time."

In his "The Future," the nineteenth-century English poet Matthew Arnold used a related watery image, the flow of time as a river (an image discussed in more detail in the next chapter), to express the brief duration of an individual life. For each of us, the past is simply myth and the future but speculation:

> Vainly does each, as he glides,
> Fable and dream
> Of the lands which the river of Time
> Had left ere he woke on its breast,
> Or shall reach when his eyes have been closed.
> . . .
> Only the thoughts,
> Raised by the objects he passes, are his.

It is nothing less than humbling to historians who pause to think on how little of the past is known—that is, recorded. As the ever-available anonymous wit once put it, "History is a damn dim candle over a damn dark abyss." H. G. Wells, in his 1944 doctoral thesis at the University of London, wrote,[3] "A thousand years is a huge succession of yesterdays beyond our clear apprehension." Even as enormous as the age of the Earth is, it is not infinite. But of course our planet *is* very old, and the universe itself is many billions of years older. Is the age of the universe also the duration of the past? Or is the past itself actually *infinite?*

An implicit assumption of the infinity of the past (and of the future, too[4]) can be found in Book Three of Lucretius' science poem *De Rerum Natura (On the Nature of Things),* where, just before the birth of Christ, Lucretius argues for the irrationality of fearing death: "The bygone antiquity of everlasting time before our birth was nothing to us. Nature holds this up to us as a mirror of the time yet to come after our death. Is there anything in this that looks appalling, anything that means an aspect of gloom? Is it not more untroubled than any sleep?"

Whitrow (1978) traces the origins of rational analyses of the duration of the past back as far as the sixth century A.D. The argument presented then by the Christian philosopher Joannes Philoponus of Alexandria (who is otherwise known as John the Grammarian) is simply that the world could *not* have been for ever because that implies that an infinity of successive acts could have taken place, which (according to Philoponus) is impossible. A variation on this is the claim that if the past were infinite in extent, then everything would have happened by now! Infinity was just too big for the ancient mind. (Zeno's hoary pre-Christian paradoxes, as is well known today, are based on subtle errors in the use of infinity.) Even as late as the twelfth century, the debate among Christian theologians was not about the possibility of an infinite past but rather about whether the Biblical "six days of Creation" actually had taken place simultaneously. For many in this debate, the past was clearly finite in duration—see Gross (1985). Not all Christians accepted that conclusion, however, and the following century saw St. Thomas Aquinas (a follower of Aristotle) arguing for the opposite view of an infinite past.

Thomas' contemporary, St. Bonaventure, however, argued for a *finite* past, and it is with Bonaventure that we start to see some mathematical sophistication—see Sweeney (1974) and Baldner (1989). Bonaventure argued that in a world infinitely old, the Sun would have made an infinite number of its annual trips around the ecliptic. But for each such trip the Moon would have made twelve monthly trips around the earth, and so this second infinity would be twelve times as great as the first one, and how could that be? Infinity is infinity, and how can something be twelve times bigger than infinity? This argument doesn't have any strength today because of Cantor's work on the concept of infinity, but it is still clever.

Agonized, convoluted theological analyses of God, infinity, and eternity continued long after Aquinas and Bonaventure. Two examples should capture their flavor. Consider first this one, on the supposed immortality of the soul. If $A = B$, then $2A = 2B$. Next, let A = "half alive" and B = "half dead," where $A = B$ in the same sense that a glass that is half full is also half empty. Then, $2A$ = "fully alive" and $2B$ = "fully dead." Thus to be dead is to be alive, and so the soul is immortal. Outrageous? Yes, indeed, but the "reasoning" does have a certain charm! A second example, from the seventeenth century, is a particularly instructive illustration of how intertwined are the concepts of world creation, divine eternity, infinity, and the finitude (or not) of time.

After publication of the English political philosopher Thomas Hobbes' *Leviathan* in 1651, with its arguments against the power of the Church and for civil power (with some criticism tossed in, as well, for universities), Seth Ward counterattacked. Ward, who was both a minister (later a bishop) in the Anglican Church and Savilian Professor of Astronomy at Oxford, was greatly offended by the secular nature of *Leviathan*. Even before *Leviathan*, in fact, Ward certainly would not have liked Hobbes' earlier denial of the existence of immaterial substances (such as souls). Ward's 1652 book *A Philosophical Essay*

Towards Eviction of the Being and Attributes of God, the Immortality of the Souls of Men, the Truth and Authority of Scripture, was the first of a two-punch reply to Hobbes. The second came in 1654 with the appearance of Ward's *Vindiciae academiarum.* In both of these works Ward attempted to undermine Hobbes' credibility by attacking his mathematical ability. (Hobbes had long been fascinated by, and was considered an expert on, the ancient problem of "squaring the circle"—a task that has been known to be impossible only since 1882.) In his *Essay* Ward also attempted to defend the view that the world has a finite age—that is, it had a specific moment of creation, presumably by God. In an opening note, in fact, Ward cited Hobbes' rejection of immaterial substances as the motivation for his penning of *Essay.*

To support his view of a finite age for the world, Ward invoked infinity in an interesting way. He argued that nothing is permanent, certainly not humans. Each is created; one can imagine tracing a chain of creation events backward in time through successive generations. Now, there are only two separate and distinct possibilities to where this chain could lead to in the past. First, it could terminate, after a finite number of generations, at a *first* generation, i.e., with the "creation" of the first human. If that is the case, then, said Ward (in effect), "case closed!" If that is not the case, however, then the chain of successive generations never terminates, i.e., the chain is infinitely long. But that, argued Ward, is nonsense—how could anything infinitely long have an *end* (in the present)?

Why Ward thought this an unanswerable paradox is hard to understand; after all, one can imagine a line in some coordinate system *beginning* at the origin and yet still being infinitely long; an example is the positive *x*-axis. This counter-example was not put forth by Hobbes in his own defense but rather was offered by one of Ward's own colleagues at Oxford, John Wallis, the Savilian Professor of Geometry. As for Hobbes, he was little bothered by Ward's argument. As he pointed out, surely with a smile on his face, Ward was in danger of impaling himself as a theologian on his own sword: Ward's argument "proved" the finite age not only of the world but of *everything,* including God.

Similar problems with infinity lay behind Kant's rejection of an infinite past. It is interesting to note that Kant, somewhat paradoxically, thought an infinite *future* a possibility. Why did Kant think time could be infinite in one direction but not in the other? Bennett (1971) tells us that Kant "failed to make himself clear," and that seems to be true because Kant's argument was that the duration of the future is less problematical than that of the past because it is only the past that influences the present. The best I can do in "explaining" this is to speculate that if the present depends on an *infinite* past, then perhaps Kant thought that much influence was too much for the present to handle! In any case, Kant's view fails if we consider the possibility of backward time travel—the possibility that the future could also influence the present.

The question of the duration of time is still the source of lively philosophical debate. Consider, for example, Whitrow (1978), which is an echo of Philoponos' claim that an infinite past is an impossibility because that implies that an actual infinity of events would already have happened (which is, "of course," absurd). To add scientific support to this assumed absurdity, Whitrow cites the prediction from general relativity of a singularity in spacetime at some *finite* past time, i.e., the prediction that time—and everything else—had its beginning in the now famous Big Bang.

There is, however, a philosopher for every point of view. Whitrow's analysis provoked a flurry of replies, with Popper (1978) and Bell (1979) disagreeing. Then Craig (1979) replied to support Whitrow and refute Popper, but Craig himself was refuted by Small (1986). That didn't put an end to the debate, however. The following year saw another paper—Smith (1987)—support the logical possibility of an infinite past; it was met in turn by a mathematical rebuttal in Eells (1988). And, indeed, that same year Smith (1988) stated that he believed only in the *logical* possibility of an infinite past and that, in fact, he really believed that the universe is of finite age and that it originated in an uncaused (no God required) Big Bang. Indeed, Smith *had* previously argued (1985a) for a finite past. Smith has been rebutted (are you surprised?) by Smith and Weingard (1990), who argue that Smith has overlooked the complications of quantum effects under non-singular conditions. Smith has, of course, replied (1991, 1994). An even more pointed rejection of Smith's arguments came from the theologian Wm. Lane Craig (1993).

An interesting parallel debate on the same issue took place during these same years between Craig and the Canadian philosopher Julian Wolfe—see Wolfe (1971), Craig (1978, 1980, 1981), and Wolfe (1985). When will debate on the age of the past end? Not until the end of the (infinite?) future, I'd wager!

There was one reply to Whitrow, however, that addressed the scientific view of the beginning of time that Professor Whitrow had raised. That paper, Weingard (1979c), pointed out that although general relativity and its predicted spacetime singularity in the distant past may indeed allow for a finite past, that still does not totally close the door to the possibility that the Big Bang was a continuation in time from a previous contraction phase of the universe, and so on, *ad infinitum*. To quote T. S. Eliot ("Little Gidding"):

What we call the beginning is often the end
And to make an end is to make a beginning.
The end is where we start from.

Even without entertaining such an oscillating, accordion-like universe, it is possible to have a universe that originated in a single Big Bang a finite time in the past but that has *no first instant!* This astonishing statement shocks most at first encounter, but it is simply the cosmological version of a well-known mathematical result. The instant $t = 0$ is *not* actually part of spacetime, because the Big Bang was quite literally a singular event for which the laws of space-time physics fail. Thus all instants of time in spacetime are greater than zero—but there is no smallest number greater than zero. If you name a number, no matter how small, I can name one smaller, such as one-half of yours. Of course, if there really is merit to the idea of the chronon, well . . .

In an ingenious observation that seems to have been missed by most philosophers, E. A. Milne, a professor of mathematics at Oxford, suggested in his 1948 book *Kinematic Relativity* that with general relativity it is conceivable to have both a Big Bang and an infinite past. [More recently, Misner (1969b) has made a similar observation.] Pointing out that to talk meaningfully of time implies a clock to measure it by, Professors Milne and Misner looked for a Universal Clock that would be far more durable than our heartbeats, Big Ben, the rotation of the Earth about the Sun, the finest watch, or anything else that exists only transiently. They suggested the expansion rate of the universe itself as the ideal clock. As we go back in time to the Big Bang, the expansion rate rises to infinity, and as Misner puts it, "We see the Universe ticking away . . . quite actively. *The Universe is meaningfully infinitely old because infinitely many things have happened since the beginning* [Misner's emphasis]."

In this view cosmic time is taken as proportional to the negative of the log-arithm of the normalized volume V of the universe (that is, $V = 1$ represents infinite volume, so time "stops" at the end of the expansion). Thus, because V goes to zero as we go backward in time, time runs ever faster as we go farther into the past. This view puts the Big Bang (with $V = 0$) infinitely long ago, so it has the virtue of sidestepping the issue of what happened before the creation of the universe and of time—see also York (1972). Thus the answer to the question that is this section's title is *yes,* the past is infinite when time is mea-sured by the largest clock imaginable, the universe itself.

The debate over the length of the past continues as hot and heavy today as it was in medieval times. For example, in his editorial of August 10, 1989 ("Down with the Big Bang") the editor of *Nature,* John Maddox, declared the standard explosive model of the universe to be "philosophically unacceptable," because "the implication is that there was one instant at which time literally began and so, by extension, an instant before which there was no time." For Maddox, this meant that the Big Bang "is an *effect* [my emphasis] whose *cause* [my emphasis] cannot be identified or even discussed." The usual (non-time-travel) use of the words *cause* and *effect* is that the cause happens first and then the effect occurs, but if the Big Bang (the effect) is the origin of time, then how (asked Maddox) could there be a cause of the Big Bang *before* that beginning? (For creationists the answer is of course obvious—God. Creationists also of course avoid the question of God's cause, merely saying he needs no cause. They also avoid explaining why one couldn't simply say that of the Big Bang itself.)

In raising his objection, Maddox was in good philosophical company; Aristotle had long ago (in his *Physics*) declared an instant in time with no predecessor to be an absurdity. Simply being on Aristotle's side is, however, no longer so impressive as it once was. A low-key reply to Maddox, based on Misner's idea, quickly came from Lévy-Leblond (1989, 1990): Because the Big Bang happened "infinitely" long ago, the very idea of an instant before that is simply meaningless. More vigorous was Adolf Grünbaum's reply (1990). Maddox, said Grünbaum, had created his own confusion by asking the wrong question. The right question, for Grünbaum, is "Did the universe have a beginning?" and Maddox's "What *caused* the beginning?" is a question not for physics but rather for theological metaphysics. Maddox himself had introduced religion into the debate with the claim that creationists love the Big Bang because it seems to endorse science by "imagination"; Maddox thereby stained the Big Bang model with his unfair (I think) juxtapositioning of it with the pseudo-science of creationism.

Grünbaum had actually been thinking about the beginning of the universe long before Maddox's editorial. In June 1988 he gave a talk on his ideas in Moscow and then wrote them all up for publication. His paper (1989) appeared just a month after the *Nature* editorial. Of course, as befits a topic as big as the Big Bang, Grünbaum's arguments provoked still another reply; see Narlikar (1990). And then see Grünbaum (1993)!

Time and Clocks

Looking for clocks to measure time is clearly of fundamental importance. Going to the other extreme for the ultimate clock, from the universe clock of Milne and Misner down to the microscopic, was a mathematician—Ellis (1974)—who offered "a radical and far-reaching declaration; the variation of an elementary particle's time *is* the variation of its radius; time is size and size is time; old is big and young is small, or vice versa; time is but a grand illusion." No one, to my knowledge, has pursued this curious suggestion, which sounds like something Humpty Dumpty might have said to Alice. Still, as in the universe clock, it is *change* that is central in Ellis' particle clock. The same relationship impressed the Elizabethan poet Edmund Spenser, who wrote in his *Faerie Queene* of the end of time:

> . . . when no more *Change* shall be,
> But stedfast rest of all things firmely stayd
> Upon the pillours of Eternity,
> That is contrayr to *Mutabilitie.*

The Milne/Misner clock is an odd clock by all ordinary standards; the universe is simply not the sort of thing we usually associate with being a clock. Clocks are supposed to run at a fixed, uniform rate, whereas the Milne/Misner cosmic clock runs faster as we go backward in time and slower as we go forward. Of course, to say one clock is not running at a uniform rate implies that we have a second clock that *is* running at a uniform rate that we can use as a standard of comparison. And how do we know that this, the master clock, is indeed running at a uniform rate? Do we need a third, super-master clock (and then a fourth, and . . .)? Are we again facing a McTaggart-like infinite-regress horror of clocks? The answers to these questions are almost paradoxical. Special relativity tells us that time does *not* always pass at the same rate, and yet the same theory tells us that just one clock will do because all clocks are related in their timekeeping.

No one would question that our lives are intimately entangled with time and hence with the clocks we use to measure it. As we read in Mendilow (1952), life in modern times is ruled by the clock: "Achievement is estimated in terms of the length of time taken to accomplish our purposes, for time is money, and in a changing Universe we have no time to waste or lose. While factories turn out thousands of new appliances to save time, the entertainment industry spends millions on amusements to kill time. Life seems to be resolving itself into a feverish scramble for the last drinks before the inexorable barmaid calls out her fatal 'Time, gentlemen, time' and shuts up shop for good." Or, in the wonderfully morbid words of Philip Van Doren Stern in his editorial introduction to the 1947 fantasy anthology *Travelers in Time,* "The clock is a dreadful instrument, the impartial ruler of the brief span of consciousness which lies between the warm darkness of the womb and the cold everlasting night of the grave."

We might think this slavery to clocks is a fairly new development, but consider the following passage from "The First Voyage" (to Lilliput) of Captain

Lemuel Gulliver, written more than two hundred fifty years ago.[5] Upon searching Gulliver, the Lilliputians made an inventory of their findings, which included his watch, or, as they called it, "a wonderful kind of Engine." The inventory continues: "And we conjecture it [the watch] is either some unknown Animal, or the God that he worships: But we are more inclined to the latter Opinion, because he assured us . . . that he seldom did any Thing without consulting it. He called it his Oracle, and said it pointed out the Time for every Action of his life."

Gulliver's mechanical watch has an ancient ancestry. The very first clocks to measure the passage of time—as opposed to just counting off the days by the alternation of night and day, and the years by the cycling of the seasons—were the shadow clocks of Egypt, dating back at least to 1500 to 2000 B.C. and almost certainly much earlier than that. Progress was steady if sporadic, and by Newton's time, seagoing chronometers were precise enough to make worldwide navigation possible. By the nineteenth century the balance-spring-and-wheel clock proved to be astonishingly accurate (an error of just one second per year was achievable), and further significant advances had to await twentieth-century technology. The invention of the electric-quartz clock increased accuracy to one second of error in decades, and today's atomic-gas clocks have reached the nearly incomprehensible accuracy of one second of error in a hundred centuries.

For our purposes in this book, however, the clock of most interest is one that has never actually been built, the *photon clock*. To visualize this clock, imagine two parallel mirrors with a "pendulum of light,"—a photon—reflecting endlessly back and forth with each pair of bounces being a single tick-tock of the clock. The idea of a photon clock has been around in physics for well over half a century. This clock is important because, as shown in Tech Note 2, it allows us to derive (using nothing more than simple algebra and elementary algebra) the central result of special relativity: the conclusion that time is not Newton's absolute time (see Tech Note 1) but in fact is Einstein's relative time, which depends on the spatial state of the observer relative to the clock. This mixing together of space and time was one of the greatest intellectual insights in all of human history, but a large number of other astonishing ideas have been derived from relativity. We will begin to examine them in the next two sections.

Hyperspace and Wormholes

The idea of a fourth dimension to space is viewed by many as simple-minded nonsense,[6] as Laura stated in the third quotation at the opening of this chapter. And in his 1897 Presidential Address to the American Mathematical Society,

Simon Newcomb declared, "The introduction of what is now very generally called hyper-space, especially space of more than three dimensions, into mathematics has proved a stumbling block to more than one able philosopher." Einstein (1961) later stated the issue more bluntly: "The non-mathematician is seized by a mysterious shuddering when he hears of "four-dimensional" things, by a feeling not unlike that awakened by thoughts of the occult."

To see just how right Einstein was with his observation, consider the reaction one Egyptian philosopher had (in 1929) to Einstein's own writings, as reported by Ziadat (1994): "We have no doubt in our mind that nobody can understand it (the fourth dimension), including Einstein himself. The incomprehensibility of these assumptions [of general relativity] is due to their nature. They deal with the fourth dimension . . . and the reality of time and space. They can only be described by a mathematician's hypothesis or by religious faith." This reaction is easy to understand—after all, anybody can "see" that there are exactly three spatial dimensions, and that is that!

Ford Madox Ford's failed 1901 novel *The Inheritors,* the tale of an insidious hyperspace invasion of our world, illustrates Einstein's assertion with an example from the time before science fiction magazines. When the novel's narrator is bluntly told by an invader that she is from the fourth dimension (an idea inspired by Ford's appreciation of how much success his acquaintance H. G. Wells had enjoyed with it), he recoils from the claim with the words "If you expect me to believe that you inhabit a mathematical monstrosity, you are mistaken. You are, really." And who can blame the skeptical narrator. How can there be *four* spatial dimensions?

For science fiction writers, of course, the fourth dimension (and hyperspace, in general) is a major concept. Writing in *Analog,* today's premier "hard science" fiction magazine, the physicist John Cramer (1985b) has nicely summed up what is so fascinating about the idea of an extra dimension or two, or perhaps even more, from a fictional point of view. "Are there hidden dimensions not accessible to us, dimensions in which we could go adventuring, dimensions within which malevolent hyperdimensional aliens may be lurking, ready to pierce our flimsy paper-thin three-space bodies with their terrible hyper-sharp claws?" The early science fiction magazines encouraged this lurid imagery. Witness the editorial blurb opening the 1939 other-dimensional monster story "Into Another Dimension" (Duclos), which stated that "it was a strange world in which Lester and Florence found themselves. A world of sudden death and strange science, ruled by inhuman beasts."

Just what *is* hyperspace? It is a space of higher dimension than the one we obviously seem to live in. As you'll see later in this chapter, our universe appears to be a four-dimensional (three spatial and one temporal) world called *spacetime*. This four-dimensional world can, at least mathematically, be thought of as the boundary or surface of a five-dimensional hyperspace. This is analogous to the way a one-dimensional closed plane curve bounds its two-dimensional interior and to the way the two-dimensional space of the surface of a sphere bounds the three-dimensional space of the sphere itself. For the inhabitants of *Sphereland* (Burger), then, an example of what they would consider hyperspace (excluding time) is the interior and exterior of the sphere that the surface they live on bounds. This interesting idea appeared quite early in science fiction. For example, in the remarkably sophisticated "The Gostak and the Doshes" (Breuer), from a 1930 issue of *Amazing Stories,* an eccentric scientist exclaims, "A mathematical physicist lives in vast spaces . . . where space unrolls along a fourth dimension on a surface distended from a fifth."

There are some interesting geometrical implications to hyperspace. For example, for beings in sphereland there are *two* ways to travel from one pole to the other: the usual way, *on* the surface of the sphere, and the hyperspace way, which takes them *through* the sphere along the polar diameter. In imagery generated by thinking of the sphere as an apple, and of the hyperspace path as a tunnel through the apple, it has become popular to call all such shortcuts through any hyperspace of any dimension *wormholes;* see, for example, Morris and Thorne (1988).

The general theory of relativity predicts the existence of such wormholes in spacetime, and in fact they were first "discovered" theoretically in the mathematics of relativity as early as 1916, by the Viennese physicist Ludwig Flamm. Later analyses were done by Einstein himself (1935a). Cohen (RAG) discusses wormholes as a possible model for pulsars, as opposed to the more usual model of pulsars as rotating neutron stars. Ori (1991b) has presented theoretical analyses suggesting that the interior of a charged black hole may be the entrance to a wormhole. All of these various solutions to the field equations are often generically called "Einstein–Rosen bridges" in the physics literature, and recently the term has appeared in fiction, too—see *Einstein's Bridge* (Cramer). (Nathan Rosen was Einstein's co-author in the 1935 paper.)

The term *wormhole* was coined in the 1950s by John Wheeler, who used wormholes to show how (1962a) electric charge could be thought of as lines of force trapped in the changing topology of a multiply connected *empty* space—see also Brill (MWM) and Misner and Wheeler (1957). (Indeed, Wheeler has claimed that the observation of what we call electricity is experimental proof that space is *not* connected simply.) More poetically, I have even seen wormholes referred to as spacetime subways! For much more about wormholes and how they are related to time travel, see Tech Note 9.

The use of hyperspace portals for explaining some observed physical phenomenon actually appeared in the scientific literature long before Wheeler's electricity example. In his 1928 book *Astronomy and Cosmogony,* for example, the British theoretician Sir James Jeans devoted a chapter to what were then called nebulae, the island-universes we now call galaxies. At the end of his discussion on the arms of spiral galaxies, Jeans offered the following speculation: "Each failure to explain the spiral arms makes it more and more difficult to resist a suspicion that spiral nebulae are the seat of types of forces entirely unknown to us, forces which may possibly express novel and unsuspected *metric properties of space* [my emphasis]. The type of conjecture which presents itself, somewhat insistently, is that the centers of the nebulae are of the nature of 'singular points,' at which matter is poured into our universe from some other, and entirely extraneous, spacial dimension, so that, to a denizen of our universe, they appear as points at which matter is being continually created." This, in everything but name, is a wormhole.

What, then, would hyperspace be like? It is immediately obvious in the case of sphereland that the hyperspace or wormhole path can always be shorter than the surface path, although in "FTA" (Martin) we find a clever tale based on precisely the opposite view. Even with the "shorter path" view, getting around in hyperspace may not be a trivial task. "The Mapmakers" (Pohl), for example, tells the story of how one of the first spaceships to explore hyperspace gets lost. As Pohl puts it, the trouble with hyperspace travel is that "You go in at one point, you rocket around until you think it's time to come out, and there you are. Where is 'there'? Why, that's the surprise that's in store for you, because you never know until you get there. And sometimes not even then." Pohl's idea plays a central role, too, in the 1957 novel *Tunnel in the Sky* (Heinlein), in which a "hyperspace gate" is discovered by accident during failed time travel experiments.

Isaac Asimov's "Take a Match" asks a similar question and arrives at the same answer: "When you took the Jump . . . how sure were you *where* you would emerge? The timing and quantity of the energy input might be as tightly controlled as you liked . . . but the uncertainty principle reigned supreme and there was always the chance, even the inevitability of a random miss . . . a paper-thin miss might be a thousand light-years." And in "The Trouble with Hyperspace" (Sharkey), instantaneous travel through hyperspace is related to backward time travel—just *how* that occurs is explained in Tech Note 9—and the serious causality problems thereby created.

"Get into that vibrator!" snarled Bentley. "Get in, I say!"

This illustration of a "super science" gadget accompanied a 1939 story by Maurice Duclos in *Fantastic Adventures*. The gadget operated by vibrating an object faster than light, whereupon the Lorentz–FitzGerald contraction formula predicts an *imaginary* size for the object—which really means (so we are told) that the object has entered "another plane of existence." The inventor (the fellow with the gun) is inviting his grim-faced assistant to give the gadget a try. (The original caption reads "Get into that vibrator! Get in, I say!")

Illustration for "Into Another Dimension" by Kenneth J. Reeve, © 1939 by Ziff-Davis Publishing Co.; reprinted by arrangement with Forrest J. Ackerman, Holding Agent, 2495 Glendower Ave., Hollywood, CA 90027.

A common way to visualize hyperspace shortcuts is to imagine the beginning and the end of a journey as points A and B in the two-dimensional surface of a piece of paper. Then imagine that the paper is folded so as to position A over B, perhaps with A almost touching B. The distance from A to B *through hyperspace* (the three-dimensional space in which the folding operation was performed) can clearly be much less than is the distance through "normal" space, (the distance covered by a trip that always remains in the paper). Indeed, this is the specific example used in the 1948 tale "The Möbius Trail" (Smith) to explain the instantaneous "space-warp" (wormhole) teleportation device invented by the story's hero. Such imagery actually appeared quite early in science fiction, as in the 1928 story "The Space Bender" (Rementer). There, a gadget is used to "bend space" so that Earth and Venus touch! In that tale the fourth dimension is purely spatial, and there is no talk of time machines or time travel; it is a space travel story.

The idea of a folded spacetime allowing time travel was quick to enter science fiction, however, as in the 1930 story "In 20,000 A.D." (Schachner and Zagat). The "machine" in that story is a stand of trees(!) that forms a time portal through which the narrator (one Thomas Jenkins) walks 18,000 years into the future. In an editorial footnote (a device commonly used in early pulp fiction to inject scientific verisimilitude), we are told that "Jenkins had evidently fallen into a warp in space. The [stand of trees] was a pucker—a fault, we might say, borrowing a geologic term—in the curvature of space. Through this warp he had been thrown clear out of our three dimensions into a fourth. There he slid *in time* over the other side of the ridge or pucker, into the same spot in the three-dimensional world, but into a different era in time. Notice that he had not traveled an inch in space; all his journeying had been purely in time."

That is a terribly flawed explanation, with its talk of space rather than spacetime and of the time traveler somehow leaving space entirely during his trip, but some authors eventually learned to do better. For example, folded spacetime as a mechanism for time travel is used in a cautionary tale "Pete Can Fix It" (Jones) on the potential horrors of the atomic bomb. In that story, originally published two years after the atomic bombings of Japan, the world fifteen years hence experiences a terrible atomic war. As the time traveler in the tale explains, "During the unprecedented release of atomic energy that arose during the simultaneous bombings of our cities, something happened to the very continuum in which we exist. . . . A crook, a twist, a fold—explain it how you will, I accidently stumbled upon an electronic circuit that would create a field that would enable passage from one folded section [of spacetime] to the adjacent section. The fold proved to be about fifteen years in length. . . ."

The idea of hyperspace folding has broken free from science fiction and can now be found in modern stories in other genres. For example, in the Stephen King story "Mrs. Todd's Shortcut" a woman keeps finding ever shorter ways to drive from Castle Rock, Maine, to Bangor. As the crow flies it is 79 miles, but she gets the journey down to 67 miles and later 31.6 miles. When doubted, she replies: "Fold the map and see how many miles it is then . . . it can be a little less than a straight line if you fold it a little, or it can be a lot less if you fold it a lot." The doubter remains unconvinced: "You can fold a map on paper, but you can't fold *land*." For our purposes here—the creation of wormholes in spacetime—we have to imagine much more: the folding of four-dimensional spacetime through a five-dimensional hyperspace. The folding imagery has even appeared in the movies; spacetime folding is demonstrated with a piece of paper as an explanation for the faster-than-light spaceship in *Event Horizon* (perhaps the worst film of 1997).

Despite the final quote that opened this chapter, it is not just science fiction fans who take the concept of hyperspace seriously. For example, in Whiston (1974) we find a mathematician writing that "most science fiction addicts are familiar with the notion of 'hyperspace,' a higher dimensional space-time bounded by Space-Time through which, in the far distant future, interstellar voyages shortcut the (otherwise unsurmountable) distances between the stars. The purpose of this article is to demonstrate that any . . . relativistic space-time model is the boundary of some . . . five-dimensional hyperspace." That, of course, is just what Breuer's magazine character said—in 1930!

The idea of dimensional gateways between a plurality of worlds is an old one in science fiction, and even earlier. For example, the great H. G. Wells himself used the portal idea in 1895 (the same year *The Time Machine* appeared in book form) in his *The Wonderful Visit*. That novel, not science fiction by any stretch of definition, concerns the strange adventures of an angel who, in some unexplained manner, literally flies into our world where he is shot in the wing by a Vicar's gun! There is simply a brief bit of speculation about the fourth dimension: "There may be any number of three dimensional Universes packed side by side."

With just a bit more plausibility, mirrors have long been used as spatial portals (probably because their reflective behavior is amazing to most people). Like his friend Lewis Carroll, George Macdonald had the central character in his 1895 fantasy novel *Lilith* enter another world through a mirror. Unlike Carroll's *Through the Looking Glass*, however, Macdonald's work was specific about the nature of the new place; in *Lilith* it is hyperspace, "the region of the seven dimensions."

More conventional is the "hole in space" leading to a different world in the 1930 story "The Lizard-Men of Buh-lo" (Flagg), which seems very much to be a wormhole (it is not-very-convincingly "explained" as caused by a gadget that rotates at super high speed). And Jack Williamson, for another example, used the portal idea in his 1931 story "Through the Purple Cloud," which somewhat implausibly posits that the fabric of spacetime might somehow be torn asunder by the puny energy of a mere chemical reaction. He repeated the idea in the 1933 story "In the Scarlet Star," which serves up a gadget that lets a man step through a dimensional portal into another world that is hinted to be our world's ancient past.

More recently, the 1959 "Triple-Time Try" (Collins) has a geologist bouncing around in the remote past, from Silurian to Carboniferous to Cretaceous to Miocene times, because a big meteor strike on the California coast had released "enough mass-conversion-energy to breach the space-time continuum;" somehow, three alternative realities are temporarily linked by the breach. And in Friedman (1988) we even find a physicist with a sense of humor, who suggests that C. S. Lewis' wardrobe, in his children's book *The Lion, the Witch, and the Wardrobe,* is actually a wormhole connecting our world and that of Narnia! *Stonewords* (Conrad), a change-the-past story, uses a similar idea—a wormhole staircase in an old house—to connect the world of 1870 to the present.

An interesting fictional discussion of hyperspace unfolds in "Avoidance Situation" (McConnell). The author, who is an academic psychologist, has put himself into this story of a starship captain who explains to the crew psychologist how he feels about hyperspace (or *subspace,* as it is called in the story): "God forsaken. That's just what it is. Completely black, completely empty. It frightens me every time we make the jump through it . . . it frightens me because—well, because a man seems to get lost out there. In normal space there are always stars around, no matter how distant they may be, and you feel that you've got direction and location. In subspace, all you've got is nothing—and one hell of a lot of that. It's incredible when you stop to think about it. An area—an opening as big as the whole of our Universe, big enough to pack every galaxy we've ever seen in it . . . and not a single atom of matter in it . . . until we came barging in to use it as a shortcut across our own Universe."

The vastness of hyperspace got a more down-to-Earth treatment from the early science fiction writer Bob Olsen, who wrote the following verses in 1934, in the introduction to his "The Four Dimensional Auto-Parker."

> I read a yarn the other day—
> A crazy concept, I must say.

It states that objects have extension
In what is called the "Fourth Dimension."

In hyperspace one could, no doubt,
Make tennis balls turn inside out;
And from a nut remove the kernel
And not disturb the shell external.
A crook could pilfer bonds and stocks,
Then laugh at prison bars and locks;
One step in this direction queer,
And presto! He would disappear!

Let's hope, in planning new inventions,
They'll give us cars with four dimensions.
When searching for a parking place
We sure could use some hyperspace!

The next two sections in this book elaborate on the various properties of hyperspace mentioned in Olsen's amusing doggerel. By 1966, however, the year Larry Niven's "Neutron Star" was published, he could use the following dramatic opening sentence with complete confidence that his readers would know what he talking about: "The *Skydiver* dropped out of hyperspace an even million miles above the neutron star." It's hard to imagine how one could do much better than that in creating exciting images in a reader's head!

Monsters in Hyperspace

As Professor Cramer implied in the previous section, speculation about aliens in hyperspace has long been popular. The 1932 story "The Einstein See-Saw" (Breuer) limits itself to mere beasts with "rows of teeth that came together with a snap," ripping the hero's trousers. In "The 32nd of May" (Ernst), however, things are more serious. The narrator, after visiting his friends the Bartons, gets up to leave just as the clock is about to strike midnight. As the clock rings out with the eleventh note, he passes "between two mirrors, each facing each other at an angle allowing both my face and my back to be seen by me"— and then he stumbles.

He stumbles, in fact, into a strange, alien place where he is confronted by a two-dimensional creature that "watched me with callous interest out of its inhuman eye." After observing a lengthy battle between this first creature and a second, over which will have the right to kill him, the narrator manages in the nick of time to stumble back into his friends' living room. As he does so, he hears the clock strike the twelfth note of midnight, and one of his hosts exclaims, "How funny! You know, for just a fraction of a second after you tripped, I couldn't see you! I guess that means another trip to the oculist." But the narrator knows better, as he tells the reader. "I passed between the mirrors in the Bartons' oddly angled living room. I fell—into another world, or plane, or dimension, or whatever you wish to call it, where unimaginable creatures seemed to fight. . . . It would seem that there are powers in untried combinations of angles undreamed of by man—and that perhaps geometry is a bridge between worlds. And it would seem that by chance the mirrors formed an angle that transported me instantly from one plane to another. But your guess is as good as mine." As this tale shows, the connection between hyperspace and geometry, and the possibility of there being a difference in the rates of time flow in two different spaces, were common ideas in even early science fiction.

The violence that fictional hyperspace creatures show toward one another could also be directed against humans. As the editorial lead-in to the 1931 "Hell's Dimension" (Curry) luridly announces, "Professor Lambert deliberately ventures into a Vibrational Dimension to join his fiancee in its magnetic torture-fields." (There is a hint at the possible distortion of time in hyperspace in this story, too; when eventually rescued by a genius-colleague, the professor and his lady learn that they haven't been gone for just a few hours as they had thought, but rather for days.) A less violent but equally hectic tale is "The Captured Cross-Section" (Breuer), in which a mathematician loses *his* fiancee to a fourth-dimensional being.[7] It is really his own fault, because after building a gadget that can rotate the fourth dimension into three-dimensional space, he accidently traps one of the higher-dimensional beings in it. There is a brief struggle, the young woman screams (wouldn't you?), and then she vanishes. The mathematician is quick to understand what has happened: "There is only one possible conclusion—the struggles of the fourth-dimensional creature swept her out into hyperspace."

Aliens from the fourth dimension entering *our* world are presented as even more threatening in "The Monster from Nowhere" (Bond). This is the story of an explorer who manages to catch such a creature while on expedition to the Maratan Plateau in Upper Peru. We and the narrator soon learn more, from the explorer who has brought the "thing" back to civilization, when the fellow delivers a mini-tutorial on how all that we see in our world is a three-dimensional cross section of the higher-dimensional monster. At last it sinks into the narrator's head just what he is looking at: "This time I got it. I gasped: 'Then you think that *thing* in the work-shed is a cross-section of a creature from the . . .' Yes, Len. From the Fourth Dimension!" The "thing" eventually escapes and returns to its world, but not before killing one man and taking another with it back into the fourth dimension.

The early readers of magazine fiction who found the idea of hyperspace monsters intriguing must have thought Clifford Simak's 1932 "Hellhounds of the Cosmos" the ultimate story. After first telling us of his odd theory of cosmic evolution—it is downward rather than upward—the professor in the tale reveals the precise nature of the strange creatures attacking Earth. They are creatures whose ancestors were higher-dimensional beings that have degenerated down to the mere fourth dimension. They view Earthlings, who are of course only three-dimensional, as "fodder, something to be eaten as we eat vegetables and cereals."

The situation becomes so precarious that the professor finally cries out in despair, "We are facing an invasion of fourth dimensional creatures. . . . We are being attacked by life which is one dimension above us in evolution. We are fighting, I tell you, a tribe of hellhounds out of the cosmos. They are unthinkably above us in the matter of intelligence. There is a chasm of knowledge between us so wide and so deep that it staggers the imagination." (As I type Simak's hysterical prose, I can't help but recall, from the editor's introduction to the anthology FSFS, that so much in early magazine fiction was "science that was claptrap and fiction that was graceless.")

Hollywood has made little use of the idea of invasions from other dimensions, and then only as a joke. An example is the insipid 1984 film *The Adventures of Buckaroo Bonzai,* in which the hero—so talented he moonlights both as a brain surgeon and as a rock singer—battles invaders from the eighth dimension. More recently, the 1991 movie *Xtro II* details a top-secret, underground government experiment that transports scientists across the dimensions to a parallel world. The film quickly degenerates into a third-rate copy of *Alien* (which it shamelessly robs), with an "other world" monster eating people in various air shafts.

Such films are hollow copies of the dimensional adventures of heros in another form of popular culture, the comics. Wonder Woman, for example, was in the fourth dimension in 1944 battling a villain with the curious name of Anton Unreal, and in 1958 she was in the X-dimension (to which she traveled with the aid of Professor Alpha's gadget, the "X-dimension machine"). In 1940 Batman and Robin were in the fourth dimension, too (if only in a dream), and in 1965 Superman's Metropolis was invaded by "It," a "whirling behemoth" thrust out of its world by a vicious tornado that had torn a hole in something called the "dimensional barrier." Superman, of course, rose to the occasion and thrust "It" back to where it belonged (wherever that was).

"Beings from other dimensions" is an idea that some believe had its genesis well before the science fiction writers got hold of it. Bailey (1972), for example,

From Nowhere

A giant maw appeared in mid air, and swooped downward. "Look out!" yelled Berch, yanking at his brother's arm

This illustration from the original magazine appearance of Nelson Bond's "Monster from Nowhere" shows the monster. For some reason, the three-dimensional cross section of the fourth-dimensional monster looks amazingly like an "ordinary" monster.

Illustration for "Monster from Nowhere" by Jay Jackson, © 1939 by Ziff-Davis Publishing Co.; reprinted by arrangement with Forrest J. Ackerman, Holding Agent, 2495 Glendower Ave., Hollywood, CA 90027.

asserts that a visitor from some realm of non-Euclidean spacetime is described in Guy de Maupassant's 1887 story "The Horla," but I fail to find de Maupassant's fictional speculation for the origin of his creature to be quite that specific. Bailey also claims that Ambrose Bierce's 1893 story "The Damned Thing" presents another visitor from a dimension beyond the third, and again I don't see it in my reading of the work. That story seems to me to be more a precursor to Wells' *Invisible Man*, although in other stories, Bierce certainly did express an interest in higher-dimensional spaces.

Predating the stories of Bierce and de Maupassant is the tale "What Was It? A Mystery" by Fitz-James O'Brien, a writer some critics believe to be the equal of Poe and Lovecraft in the horror genre. That story by the Irish-born American, who died in the Civil War, first appeared in the March 1859 issue of *Harper's New Monthly Magazine,* and it is a first-person narrative of an incident "so awful and inexplicable in its character that my reason fairly reels at the bare memory of the occurrence." After retiring for the evening in a house reputed to be haunted, the narrator suddenly feels a "Thing" drop upon his chest in the dark. He is forced into a "struggle of awful intensity" for his very life. Overcoming it at last, he incapacitates the Thing and lights the gas burner to take a look—only to find that his attacker, like the Damned Thing, is invisible. Not to be stopped by that, the narrator then chloroforms the Thing and has a plaster cast made of it. The result shows a small, heavily muscled creature that looks like a "ghoul, capable of feeding on human flesh." Unable to determine what else it might eat, the narrator watches the creature soon starve to death and then secretly buries it. Was it perhaps from a dimension beyond ours?

I don't think so. The feature that these three particular nineteenth-century creatures share is their malignancy, not their other-dimensionality. O'Brien's Thing eats human flesh, the Horla drives one man mad, and Bierce's Damned Thing kills in a most gruesome manner. In my opinion, commentators who see hyperspace in nineteenth-century tales of mysterious creatures are simply seeing too much. It was just too early for any writer's readers, much less the writers themselves, to have made such a connection.

However, as Simak's "hellhounds" demonstrates, the science fiction of the early twentieth century is different from earlier works when it comes to hyperspace monsters. Over time, writers have become more sophisticated than Simak was, and we find stories of hyperspace aliens who don't *always* win[8] and aren't always murderous; an example is "The Captured Cross-Section" (Breuer). In fact, many modern stories of the fourth dimension don't offer any aliens at all. In Ray Bradbury's "The Shape of Things," for example, we have just the opposite: a fourth-dimension story based not on horror, hellhounds, or savage hyperbeasts but rather on love and compassion. In that tale, despite (indeed, because of) the latest in high-tech hospital gadgetry, a baby is born in hyperspace. As the doctor explains to the understandably stunned father with a smooth bit of pseudo-scientific obfuscation, "The child was somehow affected by the birth pressure. There was a dimensional distructure caused by the simultaneous short-circuitings and malfunctionings of the new birth-mechs and the hypnosis machines. Well, anyway, your baby was born into . . . another dimension."

The baby, named Py, is healthy but odd-appearing to his parents; he looks like a small blue pyramid (hence his name) with three eyes and six appendages. Fearing for his emotional development and learning that the doctors cannot bring Py into "normal" three-dimensional space, it is decided that the parents will join their child in four-dimensional space. The romantic appeal of escape into the fourth dimension with the aid of helpful hyperspacians will probably never fade: You can find the same idea in "Tangents" (Bear), a story that first appeared in a 1985 issue of the slick, mass-market magazine *Omni*.

Not only writers for science fiction paperback magazines (the "pulps") were fascinated by other-dimensional worlds. In 1930, for example, *Amazing Detective Tales* published a short, still scary story of the perfect murder. In "Murder in the Fourth-Dimension," written by fantasy and horror writer Clark Ashton Smith, the fourth dimension proves to be the perfect place for disposing of the body until the murderer realizes he has made a fatal mistake — made all the more awful when it proves not to be quite so fatal after all. The physical conception of the fourth dimension is rather primitive in this tale, which speaks of "the theory that other worlds or dimensions may co-exist in the same space with ours by reason of a different molecular structure and vibrational rate, rendering them intangible for us." The modern view, of course, is that our three dimensions are just three of the four (or five or more) dimensions that constitute hyperspace, just as Flatland's two dimensions are two of our three dimensions. "Molecular vibration rate" has nothing to do with anything. As a second example not drawn from pulp, in the comics one of Superman's more interesting adversaries is Mr. Mxyzptlk (pronounced *mix-yez-pittle-ick*), a being with seemingly magical powers from the Land of Zrfff in the fifth dimension. His powers aren't really magic, however, but "merely" the result of his two extra dimensions.

Space as the Fourth Dimension

The idea in fiction of the fourth dimension as a space dimension, although its history stretches well back into the nineteenth century, is itself predated by much academic speculation and commentary. Aristotle, writing in 350 B.C., declared in his essay "On the Heavens" that "the three dimensions are all that there are." And that was just the beginning observation, leading up to a virtual explosion of activity over twenty-one hundred years later, in the nineteenth century. Indeed, one scholar on the history of the concept of hyperspace, Bork (1964), found that by 1911 there were at least 1800 papers on n-dimensional geometry, three-fourths of them written before 1900.

In 1873, for example, we find an essay in *Nature* that refers to well-known mathematicians who even earlier had shown that they had an inner assurance of the reality of transcendental space.[9] Just five years later the eminent mathematical physicist Peter Tait, in a comment tossed out in casual passing, tells us that "Prof. Klein, of Munich, some time ago showed, *as is well known* [my emphasis], that knots cannot exist in four dimensions."[10] Tait went on to indicate the basis, for some, for believing in a fourth spatial dimension; it offered one way to explain otherwise inexplicable occurrences, such as ghosts, the reading of sealed letters, and rope tricks: "It is some time since [a friend] told me *his* jocular mode of arguing from Klein's discovery—that all the secrets of the spiritualistic 'rope-trick' could be at once explained by supposing that *inside* the mysterious cabinet (in which the tambourines and the musical boxes fly about) space was of four dimensions—so that the well-corded performers were at once loosened from their bonds on entering it!"

The American philosopher Charles Sanders Peirce (1839–1914) was an early advocate for the four-dimensionality of space. Just what Peirce thought the nature of the fourth dimension to be is somewhat unclear, but the context suggests that he took it to be spatial. He thought three-dimensional space to be "perverse" because of the existence of incongruous counterparts (such as left- and right-handed gloves), and this was apparently strong evidence for him that space could not be three-dimensional. Of course, incongruous counterparts exist in all n-dimensional spaces, but Peirce preserved the special purity of the fourth dimension by suggesting—see Dipert (1978)—that all physical objects, although capable of motion in the fourth direction, could themselves have no extent in that direction.

But is it really possible that there could be four spatial dimensions? After all, we experience only three independent directions. In Ouspensky (1981) we read "By an *independent direction* we mean . . . a line lying at right angles to another line. Our geometry . . . knows *only three* such lines which lie simultaneously at right angles to one another and are not parallel in relation to each other. Why are there only three and not ten or fifteen? This we do not know." Indeed, from Beichler (1988) we learn that in an 1888 talk to the Philosophical Society of Washington, Simon Newcomb (see Note 22 for more on Newcomb) dismissed the view that space must necessarily be three-dimensional as an "old metaphysical superstition." Yet, despite Newcomb's open-mindedness, it has been shown that in the framework of classical physics there are several powerful reasons for why there must be *exactly* three spatial dimensions.

It might be thought that the three-dimensionality of space is "obvious" because we use three numbers to locate a position in space. In a 1938 radio address, for example, the Polish physicist Leopold Infeld (at that time a colleague of Einstein's at the Institute of Advanced Study) used an elaborate discussion of coordinate counting to give a popular explanation of the four-dimensionality of spacetime. Counting coordinates to determine dimensionality is flawed, however; in 1878, Georg Cantor showed how to establish a (discontinuous) one-to-one mapping between the points of a one-dimensional space (e.g., a line segment) and the points of a two-dimensional space (e.g., a square). An even greater bombshell followed in 1890, with Giuseppe Peano's discovery of a continuous one-dimensional curve that passes through *every* point in a square. That is, Peano showed how a one-dimensional curve could *completely fill* a two-dimensional space. Do the points reside in a two-dimensional or a one-dimensional world? Coordinate counting, at least here, seems to fail in determining dimensionality. The beginning of a scientific explanation for the dimensionality of space appears in Kant, who believed the *three* dimensions of space and Newton's inverse square law for gravity are intertwined, but he offered nothing beyond philosophical speculation to support his conviction.

The origin of Kant's view is quite old. Historians are of the general opinion that the second-century A.D. Greek astronomer, Ptolemy, argued in an essay for the impossibility of more than three spatial dimensions, but that essay did not survive the fall of the Roman Empire, so we don't know what arguments he presented in reaching that conclusion. Centuries before Ptolemy, in fact, the ancient Greeks had already begun to suspect that there was something special about the third dimension. They knew of the infinity of regular two-dimensional polygons, and they knew that only five regular polyhedrons are possible in three dimensions (the so-called Platonic solids). Rather than prompting a search for physical reasons, however, these early observations resulted only in philosophical speculations and mysticism. It wasn't until much later that physics began to appear in discussions on the dimensionality of space.

Beginning with the work of Paul Ehrenfest in 1917, we can find the idea that the Poisson–Laplace equation, a second-order partial differential equation that describes the potential functions for both Newtonian gravity and electrostatics, does not allow for stable planetary or electronic orbits in any space with dimensionality greater than three. Further, the distortionless, reverberation-free propagation of both electromagnetic and sound waves is possible only in spaces of dimensions one and three. These conclusions have been shown to hold even when we go beyond classical physics to general relativity and quantum mechanics. On an even more abstract level, *if* we believe in the existence of a unified field theory of gravitation and electromagnetism—see Misner and Wheeler (1957) and Wheeler (1962)—and *if* we believe that the Maxwell and the Einstein equations correctly describe the electromagnetic field and spacetime, respectively, *then* only in a four-dimensional world do these equations mathematically determine their fields with equal strength. Because time is one of these dimensions, we are again back to three dimensions for space. Mathematical proofs of all of these statements can be found in the literature.[11]

Using a slightly different approach, some have advanced a biological-topological argument for why space could not have fewer than three dimensions. In all of our common experience, complex intelligent life is always found to occur as an aggregate of a vast number of elementary cells interconnected via electrical nerve fibers. Each such cell is connected to several others, not all immediate neighbors, by these fibers. If space had only one or two dimensions, then such highly interconnected nets of cells would be impossible, because the overlapping nerve fibers would have to intersect, which would result in their mutually short-circuiting one another.

Some nineteenth-century academics, either ignorant of or willing to ignore all the above, associated the fourth dimension with the luminiferous ether. This mysterious stuff, by analogy with all known wave phenomena, was thought to be necessary to give light something in which to move, particularly in a vacuum. For example, Karl Pearson of London's University College attempted a typical Victorian explanation of various optical and chemical phenomena via a mechanical model of something he called an "ether squirt." This was motivated by Lord Kelvin's observation that under certain conditions, two sources of incompressible liquid would attract each other with an inverse-square-law force. This seemed so suggestive of gravitational and electromagnetic behavior that it appeared that such sources (and their analogous sinks) might allow a mechanical explanation of those phenomena.

Pearson thought of the luminiferous ether as being Kelvin's liquid (pulsating atoms would be its sources and sinks), and he imagined the ether flowing into and out of our three-dimensional world through atomic portals. Where, then, did the ether come from, and where did it go? Pearson cautiously hinted at the fourth dimension: "From whence the squirt comes into three-dimensional space it is impossible to say; the theory limits our possibility of knowledge of the physical Universe to the existence of the squirt. It may be an argument for the existence of a space of higher dimensions than our own, but of that we can know nothing." A few years later came a half-joking response,[12] which called the whole business of ether squirts "A Holiday Dream," and which said that it compels "us to the supposition of a fourth dimension, which belongs to the domain of nightmares, not of dreams, and we try to shake ourselves from the idea." Well, nightmarish or not, as Beichler (1988) shows, such scientific speculations on the nature of the ether and its possible explanation in terms of a four-dimensional hyperspace were not at all uncommon during the last two decades of the nineteenth century.

There are, in fact, two particular nineteenth-century individuals who are most closely identified with bringing the fourth dimension out of academia and into public consciousness: the mathematician Charles Howard Hinton (1853–1907) and Herbert George Wells (1866–1946). Hinton was no angle-trisecting crank, having earned an M.A. at Oxford, an appointment in the mathematics department at Princeton, and then another at the University of Minnesota.[13] Later, with the help of the eminent astronomer Simon Newcomb, he obtained a position at the Naval Observatory in Washington, DC, and was on the staff of the United States Patent Office at the time of his sudden death. Hinton was a man to be taken seriously.

Hinton's first published essay, "What Is the Fourth Dimension?," appeared in 1880 and then in book form in 1884 as part of his *Scientific Romances*. That book received a generally favorable review in *Nature*.[14] The four-dimensional-space essay, itself, almost certainly had impact. At one point he wrote, "We might then suppose that the matter we know extending in three dimensions has also a small thickness in the fourth dimension," an idea that was used a few years later by the well-known mathematician W. W. Rouse Ball (1891) in an attempt to explain gravity. Ball specifically cited Hinton as having intellectual priority, but he also claimed that he was unaware of Hinton's work until after his own was completed. Hinton was extremely inventive, and he also proposed four-dimensional-space models for static electricity.

Henderson's massive work (1983) on the influence on art and literature of the idea of a fourth spatial dimension, gives a complete summary of that influence on nineteenth-century literary works, in particular. H. G. Wells is, of course, the best known of the authors, but many well-known writers working outside the field of science fiction also used the concept, including Dostoevski in *The Brothers Karamazov* (1880) and Oscar Wilde in "The Canterville Ghost"

(1891). Wilde's use of the fourth dimension in "The Canterville Ghost" is brief, but it does show the fascination this idea held for supernaturalists; the ghost, at one point, makes a quick retreat by disappearing through the wainscoting, "hastily adopting the Fourth Dimension of Space as a means of escape."

The identification of the fourth dimension with the spirit world can actually be traced as far back as the mid-seventeenth century, to the philosopher-poet and Cambridge Platonist Henry More. Two centuries later the idea of two parallel worlds (ours and another that is inhabited by the spirits of the dead and that shares a common temporal dimension with us but is displaced in space) was used to great effect in fiction by Elizabeth Phelps in her internationally best-selling novel *The Gates Ajar* (1868). Phelps wrote to offer ease from the terrible emotional pain suffered by the legions who had lost loved ones in the American Civil War but had received little comfort from traditional nineteenth-century religions. Before the end of the nineteenth century at least two "non-fictional" religious books appeared that interpreted hyperspace as the dwelling place of God himself—Alfred Taylor Schofield's *Another World* (1888), which declared that space to be of four spatial dimensions, and Arthur Willink's *The World of the Unseen* (1893), which took the even bolder leap into a hyperspace with an *infinity* of spatial dimensions.

Some years after Phelps but still before the end of the century, Scottish writer Robert Barr wrote "The Hour Glass," in which a ghost lays claim to his lost timepiece, an hourglass, across nearly two centuries; there is the strong hint in this story of a connection between the time dimension and the supernatural. The use of higher dimensions in supernatural fiction was continued by modern writer Algernon Blackwood, who liked the concept enough to use it more than once. Some of his stories explicitly mention nineteenth- and early twentieth-century mathematicians, such as Gauss, Lobachevski, Einstein, Minkowski, Bolyai, and Hinton, who all worked with extensions of Euclid's geometry to beyond the third dimension. Running through Blackwood's stories (1949) are repeated references to a new direction at right angles to the three known ones. Other writers have also connected ghosts and the fourth dimension. For example, Ray Cummings "explained" ghosts in his "Into the Fourth Dimension," a 1926 story published in *Science & Invention* (Hugo Gernsback's magazine that was a precursor to the science fiction pulps soon to appear). Goulart (1975) wrote a modern story that has the inventor of the first time machine as a "renowned ghost detective and occult investigator," and Cartur (SSFT) links ghosts with spatial dimensions beyond the third.

All sorts of fourth-dimension tricks must have been common knowledge even before the turn of the century: One can find them integrated into stories appearing in popular, general-readership magazines of the day. For example, "The Conversion of the Professor" (Griffith), with the subtitle "A Tale of the Fourth Dimension," is essentially a love story in which a curmudgeonly math professor is convinced by a meeting with a four-dimensional version of himself that he should not block his daughter's marriage plans. The dramatic revelation comes when he sees two rings interlocked without either being broken, a feat impossible in the third dimension but not in the fourth.

Popular fascination with space having a fourth dimension reached a peak with an essay contest run by *Scientific American* in 1900. The top prize of $500 for the best explanation of four-dimensional space attracted 245 entries from all over the world; the very best of them have been preserved by Manning (1960), first published in 1910. It wasn't long, of course, before these ideas found their way into science fiction. "The Fifth-Dimension Catapult" (Leinster) was typical of those early stories. As its title implies, the story is about the fifth dimension[15]—its hero is Tommy Reames, a playboy genius who in his spare time writes such papers as "On the Mass and Inertia of the Tesseract" and "Additions to Herglotz's Mechanics of Continua." The dialogue reads at times like Raymond Chandler. Informed that the inventor of the catapult is marooned in the fifth dimension, Tommy "pulled out a cigarette case and lit a cigarette and said sardonically, 'The fifth dimension? That seems rather extreme. Most of us get along very well with three dimensions. Four seems luxurious. Why pick on the fifth?'" There is a beautiful girl, and there are gangsters too (but no Philip Marlowe), in this potboiler.

Another early use of space as four-dimensional is in the 1928 "Four Dimensional Transit" (Olsen). An awkward rewrite of Verne's *Around the World in Eighty Days,* Olsen's work has a professor and his crew fly into hyperspace and around the world and to the moon and back, in less than a day. They do this with a plane equipped with a four-dimensional rudder! A more interesting illustration of the fourth dimension as a spatial dimension is given in "The Vanishing Man" (Hughes), the tragic 1926 story of a math professor who learns how to move into hyperspace and back. A colleague catches him at it and, once over his astonishment, asks what is behind it all. The professor replies, "My assumption is that the fourth dimension is just another dimension—no more different in kind from length, say, than length is from breadth and thickness, but perpendicular to all three. Now suppose that a being in two dimensions—a flat creature, like the moving shadows of a cinematograph—were suddenly to grasp the concept of a third dimension [as in Edwin Abbott's 1880 classic fantasy *Flatland*] and so step out of the picture. He might move only an inch, but he would vanish completely from the sight of the world."

The professor has learned how to step out of 3-space and into 4-space, but when asked to explain *how,* all he can say is "How can I explain? It's just the *other* direction. It's *there*!" His colleague can't see it but nonetheless is quick to grasp the practical implications: "This is power! Think of it! A step, and you are invisible! No prison cells can hold you,[16] for there is a side to you on which they are as open as a wedding ring! No ring is secure from you: you can put your hand *round the corner* and draw out what you like. And, of course, if you looked back on the Universe you had left, you would see us in sections, open to you! You could place a stone or a tablet of poison right in the very bowels of your enemies!"

Early science fiction was, in fact, literally overrun with nutty professors whose experiments with the fourth dimension go amiss. Some of those stories were interesting, even instructive, but often they were simply silly, like the 1934 tale "Scandal in the 4th Dimension" (Long) and "The Worlds of If" (Weinbaum), in which an off-the-wall professor uses "polarized light, polarized not in the horizontal or vertical planes, but in the direction of the fourth dimension" to peer into worlds that are "time parallel" to ours. And in "The Dangerous Dimension" (Hubbard), originally published in 1938, we find a confused professor who discovers the secret of *negative* dimensions—with it he can instantaneously teleport to *anywhere.*

In "Dr. Fuddles' Fingers" (Bond), however, we meet another, far more interesting professor, whose right hand has been modified through an accident to exist in four-dimensional hyperspace. To finance his research, he uses his "talent" to become the perfect pickpocket, able to reach into any wallet no matter how well secured. He also can, indeed, reach right into the very bowels of his fellow man. And he *does.* When he demonstrates his hand to the policeman who has arrested him for being a thief, the astonished officer chokes on a lemon drop. Dr. Fuddles, of course, removes the drop from the poor fellow's windpipe with ease.

Some of the best science fiction stories are concerned simply with space itself as four-dimensional rather than with any creatures, monsters, or ghosts that may exist there. A classic example is "No-Sided Professor" (Gardner), which, in addition to teaching its readers that a *Möbius band* is a single-sided surface with a single edge,[17] comes complete with technical footnotes—just like a journal article. In Gardner's wonderfully imaginative story we learn of a fantastic discovery by Professor Stanislaw Slapenarski that goes beyond Möbius. Slapenarski has found a *no*-sided surface! When he demonstrates it, by folding an oddly cut piece of paper that promptly vanishes with "a loud pop," he is openly challenged by an skeptical colleague. Enraged, the Professor

An experiment in hyperspace goes astray in this illustration from Bob Olsen's "Four Dimensional Surgery" (*Amazing Stories,* February 1928). The young man is pulling on "hyper-Forceps" in an attempt to retrieve a surgeon who has fallen out of 3-space (along with his patient, a professor of non-Euclidean geometry, who suffers from gall-stones!)

knocks his critic out and, powerful man that we are told he is ("He was built like a professional wrestler"), folds the hapless fellow up into a no-sided professor, and the critic departs in an explosion. A stunned witness to this asks, "Can he . . . be brought back?" "I do not know, I do not know," Slapenarski wailed. "I have only begun the study of the surfaces—only just begun. I have no way of knowing where he is. Undoubtedly it is one of the higher dimensions, probably one of the odd-numbered ones. God knows which one.'"

It is amusing to note that if Bond's Dr. Fuddles had turned his right hand over in the fourth dimension, then he would have had two left hands (see Note 16 again). Curiosity about the relationship of space and handedness can be traced back at least to Kant, and for a technical spacetime discussion of Kant's view, see Earman (1971). Professor Earman points out, for example, that Kant erred in his claim that left-handed and right-handed objects cannot alternately occupy the same space. For Kant, it was "obvious" that there simply is no non-deforming, continuous transport that makes a left hand into a right hand. Kant was wrong, however, and the fact that such a transport does exist can be demonstrated in two dimensions by sliding a two-dimensional hand around a Möbius strip, just as described in Gardner's tale. That, of course, effectively flips the hand over in three-dimensional space.

To turn a three-dimensional right hand into a left hand we would have to flip it over in a four-dimensional space. That is a trick Kant (understandably!) missed. Hollywood, for some reason, seems not to have been fascinated by these possibilities for wonderful visual effects. The only film I know of with space as the fourth dimension is the 1959 *4D Man,* in which a man is shown walking through walls. This dramatic effect is as inaccurate as it would be for a flatland film maker to make "3D Man" showing the hero moving *through* a closed curve, instead of the correct image of the hero suddenly vanishing on one side (as he leaves flatland along the direction of the third dimension) and then suddenly reappearing on the other side.

Two modern fictional classics of the spatial fourth dimension are "—And He Built a Crooked House" (Heinlein) and "A Subway Named Möbius" (Deutsch). In Heinlein's story an architect learns the danger of building on top of an earthquake fault in Los Angeles. He builds a house that is a tesseract—a four-dimensional cube—or at least a house that looks the way such an object would appear in ordinary 3-space. *If* the house were actually in 4-space, then it would have wonderful properties, as the architect explains to his initially somewhat reluctant clients: "That's the grand feature about a tesseract house, complete outside exposure for every room, yet every wall serves two rooms and an eight-room house requires only a one-room foundation. It's revolutionary." It certainly is, especially when an earthquake pushes it over the edge of stability—with the architect and his clients inside.[18] The architect finally figures out what has happened: "This house, while perfectly stable in three dimensions, was not stable in four dimensions. I had built a house in the shape of an unfolded tesseract; something happened to it, some jar or side thrust, and it collapsed into its normal shape—it folded up. . . . From a four-dimensional standpoint this house was like a plane balanced on an edge. One little push and it fell over, collapsed along its natural joints into a stable four-dimensional figure."

In Deutsch's story the Boston subway (the author was a Harvard professor) suddenly extends itself into hyperspace after its connectivity becomes so high—indeed, it becomes beyond calculation—that a train could travel from any one station to any other station in the whole system. Trains, in fact, can go to even more places than that, and they begin to disappear by wandering off into the fourth dimension. The management needs the services of a topologist, fast, and fortunately "The best in the world is at Tech [presumably this is a reference to M.I.T.]," but unfortunately he is also on a missing train. This story of the convoluted Boston subway system was originally published in a 1950 issue of *Astounding Science Fiction* magazine at the suggestion of Isaac Asimov. Asimov later wrote that he suspected the story was the inspiration for the Kingston Trio's popular song in the late 1950s, "Charlie and the MTA," about "the man who rode forever 'neath the streets of Boston / He's the man, who never returned."

Time as the Fourth Dimension

The idea of time, rather than space, as the fourth dimension is much more current these days. In a subtle little joke, for example, the young couple in "When the Bough Breaks" (Padgett), who are visited by time travelers from five hundred years in the future, live in Apartment 4-D. As with the spatial interpretation, the time interpretation is an old one. In fact, Bork (1964) has traced the idea back to the late eighteenth century, finding references to the idea in pre-1800 works of the French mathematical physicists d'Alembert and Lagrange. In fact, Meyerson (1985) even quotes from a 1751 passage written by d'Alembert that appears to indicate that it is some unknown person to whom the credit is really due: "I have said [that it is] not possible to imagine more than three dimensions. A clever acquaintance of mine believes, however, that duration could be regarded as a fourth dimension and that the product of time and solidity would be in some way a product of four dimensions; that idea can be contested, but it seems to me that it has some merit, if only that of novelty."

Still, it wasn't until a curious letter appeared in *Nature* during 1885 that the concept of time as the fourth dimension was mentioned seriously in an English-language scientific journal. The author, mysteriously signing himself only as "S.," began by writing, "What is the fourth dimension? . . . I [propose] to consider Time as a fourth dimension. . . . Since this fourth dimension cannot be introduced into space, as commonly understood, we require a new kind of space for its existence, which we may call time-space." Who was this prophetic writer? Nobody knows, but Bork speculates that it was an acquaintance of H. G. Wells.[19]

The idea of time as the fourth dimension entered the popular mind around 1894–1895 with the publication of the first of Wells' so-called "scientific romances," *The Time Machine*. Then, after that pioneering use of time as the fourth dimension, science fiction writers quickly adopted the idea as the basis for one of their most popular subgenres. Murray Leinster, for example, one of the first with a four-dimensional space story, was just as quick to capitalize on time as the fourth dimension; his "The Fourth Dimensional Demonstrator" is typical. Leinster's very first published story, "The Runaway Skyscraper," in fact, interprets the fourth dimension as temporal in nature. First appearing in 1919 in *Argosy* magazine, it is the incredible tale of a Manhattan skyscraper (and its 2000 occupants) sent backward in time several thousand years because its foundation slips—in an unexplained way—along the fourth dimension. The scientific sophistication of the story is primitive, just one of the many logical flaws being a vivid description of the involuntary time travelers living forward-in-time lives even as their wrist watches run backward. Indeed, when Hugo Gernsback reprinted the tale in one of the early issues of *Amazing Stories*, a reader complained about this very point. Gernsback felt compelled to defend the story in his November 1926 editorial ("Plausibility in Scientifiction"), but he could muster only a weak rebuttal based on an author's right to "poetic license."

A little more than a decade after Leinster's story, time travel and the fourth dimension were linked in an almost casual manner by Vita Sackville-West, an early writer of feminist novels and non-fiction. In her "An Unborn Visitant" a woman in the Edwardian England of 1908 is visited by her not-yet conceived daughter, a time-traveling flapper from 1932. The clash of their two viewpoints of the world is amusing. The explanation for this odd encounter is simply a brief mention of Einstein, and then "I can't stop to tell you about Einstein now. . . . For the moment I can only tell you that I'm living in the fourth dimension. . . ."

More technical is the discussion in "Doorway of Vanishing Men" (McGivern), in which a clerk transforms the main entrance to a department store into a time machine by building a tesseract. The claim made is that the fourth dimension of the four-dimensional cube/doorway is time. (That tale appeared in the July issue of *Fantastic Adventures*, five months after Heinlein's tesseract house appeared in the February issue of *Astounding Science Fiction*. It is interesting to speculate whether McGivern was inspired to write his tesseract-time story by reading Heinlein's tesseract-space treatment.)

Some stories have discussed both space and time as the fourth dimension. A major example is "Star, Bright" (Clifton). The enigmatic title is soon explained. Star is a little girl whose mental abilities are so far beyond those of a genius that she invents a new category for herself—she is a "Bright." At age three she discovers the Möbius strip and astonishes her father (who is a mere genius or, as Star calls him, a "Tween"—between a "Bright" and the ordinary "Stupids") by using a crayon to show its one-sidedness (see Note 17). By age six Star has discovered how to use the Möbius strip to make a Klein bottle and then a "twisted cube," i.e., a tesseract à la Heinlein's earthquake house,[20] and then, as Star explains to her father, she manipulates "the twisted cube all together the same way you did Klein's bottle. Now if you do that big enough, all around you, so you're sort of half twisted in the middle, then you can [teleport] yourself anywhere you want to go." Star later discovers that spacetime is just a Möbius strip—an idea later played with in the technical literature, as in Weingard (1977)—and she learns how to time-travel up and down the strip. The story ends with Star getting off the strip in a new present, and with her frantic father trying to catch up with his lost child.

Another story that invokes a fourth dimension involving both space and time is "The Maladjusted Classroom" (Nearing). Here we encounter what Professor Ransom (of the Mathematics Faculty at an unnamed university) calls "A three dimensional Möbius strip that twists through the fourth dimension—it's called a Klein bottle." Ransom, you see, has accidently made such a thing from a bicycle tire (!) and he ends up sending a colleague on a wild ride through it. Indeed, when the poor fellow emerges from hyperspace, he finds himself both miles distant *and* an hour backward in time. This is explained, in a blink, with only the words "Fourth dimension. Time factor. *You* know. . . ."

Much more serious in substance, and certainly in consequence, is "Technical Error" (Clarke). In that tale an electrical engineer named Nelson is caught in the middle of an enormous electromagnetic field surge produced by a short circuit in a power plant. As a physicist explains to the shocked board of directors of the utility, "It now appears that the unheard-of current, amounting to millions of amperes . . . must have produced a certain extension into four dimensions. . . . I have been making some calculations and have been able to satisfy myself that a 'hyperspace' about ten feet on a side was, in fact, generated: a matter of some ten thousand quartic—not cubic!—feet. Nelson was occupying that space. The sudden collapse of the field [when the overload beakers finally broke the circuit] caused the rotation of that space."

Being rotated through 4-space has inverted the unlucky Nelson (see Note 16), and to bring him back to normal he must be flipped again. The physicist brushes aside a question about the fourth dimension as time, asserting that the only issue is one of space. Poor Nelson is, therefore, again subjected to a stupendous power overload—only now he disappears! Too late, the physicist realizes that the fourth dimension is both space and time and that Nelson has been spatially flipped *and* temporally displaced into the future. (To understand the particularly monstrous fate of Nelson, just ask yourself what the result would be if he should materialize *inside* matter sometime in the future!)

In a similar fashion, when one of the characters in "Yesterday Was Monday" (Sturgeon) becomes displaced in time, he asks for an explanation from a higher-dimensional being that appears on the scene: " 'Just where is Tuesday?' he asked. 'Over there [and when the being extends its hand, the hand disappears].' 'Do that again.' 'What? Oh—Point toward Tuesday? Certainly.' " The being explains the physics of the situation to the astonished time traveler thus: "It is a direction like any other direction. You know yourself there are four directions—forward, sideward, upward, and—*that* way! . . . It is the fourth dimension—it is duration." In the same way, in "The Middle of the Week After Next" (Leinster) a mad inventor discovers how to make a substance whose atoms resist being pushed by "pushing back at right angles to all of the other directions." That is, to push on this stuff is to risk being pushed "off into the fourth dimension [which we are told is time] . . . into the middle of the week after next."

H. G. Wells on Space and Time

It was H. G. Wells who pioneered time travel as we think of it in this book (but see the qualifying remarks about stationary versus moving time machines in Chapter One and in Tech Note 4). We are therefore indulging no mere idle curiosity when we look carefully at what this literary genius thought of space and time. His *The Time Machine* has never been out of print (something most books more than a century old cannot claim), and it is now recognized as one of the modern classics of the English language. But it didn't come without some worry. Wells, who was at first uncertain just how to present his revolutionary new work, wrote to the editor at the *New Review* for an opinion on the opening chapters. Back came a letter (dated September 1894) properly declaring of *The Time*

Machine that "It is so full of invention & the invention is so wonderful . . . it must certainly make your reputation." No one who reads it today can doubt that Wells' editor was correct.

The novella opens with "The Time Traveller" expounding on a recondite matter to a group of his friends.[21] As he asserts, "There is no difference between Time and any of the three dimensions of Space except that our consciousness moves along it." When asked to say more about the fourth dimension, he replies, "It is simply this. That Space, as our mathematicians have it, is spoken of as having three dimensions, which one may call Length, Breadth, and Thickness, and it is always definable by reference to three planes, each at right angles to the others. But some philosophical people have been asking why *three* dimensions particularly—why not another direction at right angles to the other three?—and have even tried to construct a Four-Dimensional geometry. Professor Simon Newcomb was expounding this to the New York Mathematical Society only a month or so ago."[22]

Wells was, I think it is important to realize, not primarily motivated to write *The Time Machine* by an interest in either the fourth dimension or time travel. Rather, as argued in Philmus (1969), he was attempting to refute the nearly suffocating, unjustified (to his mind), smug optimism of the Victorian age. (He wasn't alone in this attitude—indeed, his contemporary Grant Allen wrote in *his* 1895 time travel novel *British Barbarians* of a visitor from the twenty-fifth century who is appalled at the hypocrisy of Victorian customs and taboos.) And so, on his journey into the future to the year A.D. 802,701, the Time Traveller discovers the awful decay of humanity in the cannibalistic subjugation of the Eloi by the Morlocks, the end result of class warfare between the working class (Morlocks) and the idle, parasitic upper class (Eloi).

Wells' pessimistic attitude concerning the future is reflected in his nonfictional writing before 1895, too. For example, in 1891 Wells wrote,[23] "There is a good deal to be found in the work of biologists quite inharmonious with such phrases as 'the progress of the ages,' and the 'march of the mind'. . . . There is no . . . guarantee in scientific knowledge of man's permanence or permanent ascendancy . . . so far as any scientist can tell us, it may be that . . . Nature is, in unsuspected obscurity, equipping some now humble creature with wider possibilities of appetite, endurance, or destruction, to rise in the fullness of time and sweep *homo* away into the darkness from which his Universe arose. The Coming Beast must certainly be reckoned in any anticipatory calculations regarding the Coming Man."

Two years later, and while still polishing the prose of *The Time Machine*, Wells was if possible even more depressing:[24] "The life that has schemed and struggled and committed itself, the life that has played and lost, comes at last to the pitiless judgement of time, and is slowly and remorselessly annihilated. This is the saddest chapter of biological science . . . the tragedy of Extinction . . . the most terrible thing that man can conceive as happening to man [is] the Earth desert through a pestilence, and then two men, and then one man, looking extinction in the face."

Still, Wells was not always gloomy about the future. On January 24, 1902, he delivered an invited lecture to the Royal Institution[25] called "The Discovery of the Future," an invitation that demonstrates his highly visible and admired position in the elitist world of Victorian/Edwardian British science. He did, indeed, talk of such possible calamities as pestilence, cometary impact (it had been only five years since publication of his short story "The Star" in which just such an event is graphically described), atmospheric poisoning, and the extinction of the Sun, but his final sentence was prophetic as well as poetic: "All this world is heavy with the promise of greater things, and a day will come, one day in the unending succession of days, when beings, beings who are now latent in our thoughts and hidden in our loins, shall stand upon this Earth as one stands upon a footstool, and shall laugh and reach out their hands amidst the stars."

It was in that address that Wells made it clear what his answer would be to the question briefly touched on in Chapter One—How sensitive is the future to events in the past?—an issue *not* treated in *The Time Machine:* "I must confess I believe that if by some juggling with space and time Julius Caesar, Napoleon, Edward IV, William the Conqueror, Lord Roebery and Robert Burns had all been changed at birth, it would not have produced any serious dislocation of the course of destiny. I believe that these great men of ours are no more than . . . the pen-nibs Fate has used for her writing, the diamonds upon the drill that pierces through the rock." Wells, in this same address, came quite close to asserting that there will never be a real time machine: "The portion of the past that is brightest and most real to each of us is the individual past, the personal memory. The portion of the future that *must remain darkest and least accessible* [my emphasis] is the individual future."

What irony! The father of the time machine seems to have had no faith in his own conception. But, no matter, *The Time Machine* was an immediate and enormous success; it was reviewed even in *Nature,*[26] a journal not usually given to commenting on works of fiction. That review read, "Ingeniously arguing that time may be regarded as the fourth dimension . . . the author of this admirably-told story has conceived the idea of a machine that shall convey the traveler either backward or forward in time. Apart from its merits as a clever piece of imagination, the story is well worth the attention of the scientific reader . . . from first to last the narrative never lapses into dullness."

That's a nice review, of course, but Wells never thought he was actually writing *science*. In 1934, in fact, in the preface to a collection of his novel-length scientific romances (*Seven Famous Novels,* Knopf), including *The Time Machine,* Wells made his mind-set abundantly clear: "These stories of mine collected here do not pretend to deal with possible things; they are exercises of the imagination. . . . They are all fantasies; they do not aim to project a serious possibility; they aim indeed only at the same amount of conviction as one gets in a good gripping dream." Wells then went on to write that all attempts before at writing fantastic stories depended on magic. But not his works. "It occurred to me that instead of the usual interview with the devil or a magician, an ingenious use of scientific patter might with advantage be substituted."

Some of Wells' literary contemporaries utterly failed to grasp this point. For example, the English essayist and novelist M. P. Shiel once criticized Wells' assertion in *War of the Worlds* that there are no bacteria on Mars (thus explaining why it is earthly germs that finally do the invaders in), rightfully pointing out that such a dearth of bacteria would leave unexplained how anything once dead on Mars could decay, a necessary condition for there to be room for the next generation. Though correct on this point, Shiel incorrectly thought that no time traveler could journey past the time of her birth or that of her death, so he isn't a perfect example of scientific rectitude, either.

And Wells' own good friend, the English novelist Arnold Bennett, took him to task in an 1897 review of *The Invisible Man,* opining that "Mr. Wells seems actually to have overlooked a scientific point. If the man was invisible his eyelids must have been transparent, and his eyes, without their natural shield, must speedily have become useless from simple irritation." Wells quickly responded with a letter to Bennett: "You raise the point of the transparent eyelids in your review, but there is another difficulty behind that which really makes the whole story impossible. I believe it to be insurmountable. Any alteration of the refractive index of the eye lenses would make vision impossible. . . . And for vision it is also necessary that there should be . . . an opaque cornea and iris. *On these lines you would get a very effective short story but nothing more* [my emphasis]."

Wells clearly understood that he wasn't writing science at all (and with the last of his novel-length "scientific romances," *The First Men in the Moon,* done by 1901, he never again returned to that form) but rather was writing about the human condition, both past and present, and he was perfectly willing to make an "error" for the sake of a story. Indeed, as he wrote in his 1934 preface about *The Time Machine,* the fourth dimension is simply a "magic trick for a glimpse of the future." And for most readers, for over a century now, Wells' first novel has delivered that magic.

Of course, not everybody is taken by Wells' story. Such persons might agree with the response of the Devil in "Enonch Soames" (Beerbohm) to the question "*The Time Machine* is a delightful book, don't you think? So entirely original!": "It is one thing to write about an impossible machine; it is quite another thing to be a Supernatural Power." Agreeing with the Devil was Lafleur (1940), a philosopher who dredged up the grandfather paradox and wrote, "The fallacies of Wells are obvious enough . . . a man could go back into past time and change the course of history, even to the extent of bringing it about that he would never be born, and hence never take his backward trip into the past!"

More imaginative souls than the Devil and Lafleur *were* caught up by Wells, however. As reported in Ramsaye (1926), the British motion picture pioneer Robert Paul read *The Time Machine* and was so swept up by its theatrical possibilities that he immediately wrote to Wells. The two men met, and then a short time later Paul received a British patent on a mechanical gadget to simulate the sensation of an actual journey through time. Paul's machine was the forerunner of similar so-called motion simulators, at least one of which (at the Luxor Hotel in Las Vegas) purports to be a time machine experience.

It is a curious observation that, after explicitly introducing time as the fourth dimension, Wells returned to the alternative idea of the fourth dimension as spatial. For example, in "The Plattner Story" an incompetent chemistry teacher is literally blasted into hyperspace by an experiment gone wrong. And in "Davidson's Eyes" a man can apparently see through a "kink in space" to the other side of the planet, an effect induced while "stooping between the poles" of a "big electro-magnet" which gave "some extraordinary twist to his retinal elements" (an idea that Wells then has his narrator amusingly dismiss with an airy "seems mere nonsense to me").

In his 1897 novel *The Invisible Man* Wells also hints at the same idea, however, when he has the central character explain the secret of his discovery: "I found a general principle . . . a formula, a geometrical expression involving four dimensions." An interesting film that connects the two ideas of time travel and invisibility is the 1984 *The Philadelphia Experiment.* It tells the story of a 1943 Navy experiment that uses an electronic "invisibility cloak" in the spirit of the Klingon gadget in "Star Trek," but the experiment goes wrong and sends an entire destroyer forward in time into the 1980s. (This film has developed a cult following of people who believe it depicts actual events!) The supposed link between invisibility and time has been advanced more recently, more poetically, by Coates (1987). There we read that "the thinning out of matter at high speed (says Wells) renders the Time Traveller invisible and invulnerable [see Note 1 for Chapter Four]. . . . So long as he moves . . . he can slip through matter with the ease of a ghost (ghosts being, after all, the prototypical time travelers); he is himself the laser beam of time. Hence Wells' Time Traveller and his Invisible Man are related figures; for as long as he travels, the Time Traveller *is* the invisible man (for we cannot see time)."

Spacetime and the Fourth Dimension

The poet Henry Van Dyke wrote in his 1904 "The Sun-Dial at Wells College" that

> The shadow by my finger cast
> Divides the future from the past:
> Before it, sleeps the unborn hour,
> In darkness, and beyond thy power:
> Behind its unreturning line,
> The vanished hour, no longer thine:
> One hour alone is in thy hands,—
> The NOW on which the shadow stands.

The very next year Einstein's theory of special relativity appeared, and three years after that Minkowski's spacetime interpretation of special relativity. Van Dyke's beautiful poetry was dealt a mighty blow by those developments in mathematical physics, and in the rest of this section we'll see how that came to pass.

The modern view of reality, that the past and present and future are joined together into a four-dimensional entity called *spacetime,* can be attributed to the work of Hermann Minkowski (1864–1909), Einstein's mathematics professor when he was a student in Zurich. Minkowski gave spacetime to the world during a famous address at the 80th Assembly of German Natural Scientists and Physicians meeting in Cologne on September 21, 1908. Entitled "Space and Time," his remarks were electrifying then and still are today. He began dramatically:[27] "Gentlemen! The views of space and time which I wish to lay before you have sprung from the soil of experimental physics, and therein lies their strength. They are radical." Then came the famous line, quoted in so many freshman physics texts and philosophy papers, concerning spacetime: "Henceforth space by itself, and time by itself, are doomed to fade away into mere shadows, and only a kind of union of the two will preserve independence."

Minkowski explained what spacetime is in these words to his audience:

> A point of space at a point of time . . . I will call a *world-point*.
> The multiplicity of all thinkable *x, y, t* systems of values we
> will christen the *world*. With this most valiant piece of chalk I
> might project upon the blackboard four world axes. . . . Not
> to leave a yawning void anywhere, we will imagine that
> everywhere and everywhen there is something perceptible.
> To avoid saying "matter" or "electricity" I will use for this
> something the word "substance." We fix our attention on the
> substantial point which is at the world-point *x, y, z, t,* and
> imagine that we are able to recognize this substantial point at
> any other time. Let the variations *dx, dy, dz,* of the space co-
> ordinates of this substantial point correspond to a time ele-
> ment *dt.* Then we obtain, as an image, so to speak, of the
> everlasting career of the substantial point, a curve in the
> world, a *world-line*. . . . The whole Universe is seen to resolve
> itself into similar world-lines, and I would fain anticipate my-
> self by saying that in my opinion physical laws might find
> their most perfect expression as relations between these
> world lines. . . . *Thus also three-dimensional geometry becomes a
> chapter in four-dimensional physics* [my emphasis].

With those words Minkowski gave mathematical expression to the philosophical exposition of Wells' Time Traveller. But not everybody understood
Minkowski. In a little-known yet quite erudite essay, published after the first
experimental verification of general relativity (the bending of starlight passing
through the Sun's gravitational field), an anonymous author presented an optical analogy to help those who thought relativity simply "a mathematical
joke." Signing himself only as "W. G.," he included the following passage:[28]

> Some thirty or more years ago [it was forty] a little *jeu d' es-
> prit* was written by Dr. Edwin Abbott entitled "Flatland." . . .
> Dr. Abbott pictures intelligent beings whose whole experi-
> ence is confined to a plane, or other space of two dimensions,
> who have no faculties by which they can become conscious
> of anything outside that space and no means of moving off
> the surface on which they live. He then asks the reader, who
> has consciousness of the third dimension, to imagine a
> sphere descending upon the plane of Flatland and passing
> through it. How will the inhabitants regard this phenome-
> non? They will not see the approaching sphere and will have
> no conception of its solidity. They will only be conscious of
> the circle in which it cuts their plane. This circle, at first a
> point, will gradually increase in diameter, driving the inhabi-
> tants of Flatland outward from its circumference, and this
> will go on until half the sphere has passed through the plane,
> when the circle will gradually contract to a point and then
> vanish, leaving the Flatlanders in undisturbed possession of

their country. . . . Their experience will be that of a circular obstacle gradually expanding or growing, and then contracting, and they will attribute to *growth in time* what the external observer in three dimensions assigns to a movement in the third dimension. Transfer this analogy to a movement of the fourth dimension through three-dimensional space. Assume the past and future of the Universe to be all depicted in four-dimensional space, and visible to any being who has consciousness of the fourth dimension. If there is motion of our three-dimensional space relative to the fourth dimension, all the changes we experience and assign to the flow of time will be due simply to this movement, *the whole of the future as well as the past existing in the fourth dimension* [my emphasis]."

W. G.'s words are a clear and unequivocal statement of the so-called *block universe* concept of four-dimensional spacetime, of reality as a once-and-forever entity. Of course, one can find the block universe concept in the writings of the ancients, too. Consider, for example, the fifth-century B.C. Greek philosopher Parmenides on reality: "It is uncreated and indestructible; for it is complete, immovable, and without end. Nor was it ever, nor will it be; for now it *is*, all at once, a continuous *one*." And in Thomas Aquinas' *Compendium Theologiae,* written in the thirteenth century, we find "We may fancy that God knows the flight of time in His eternity, in the way that a person standing on top of a watchtower embraces in a single glance a whole caravan of passing travelers." This is the block universe idea, too, but whereas for Parmenides it was metaphysics and for Aquinas it was theology, for Einstein and Minkowski it was physics.

The block universe concept may explain the enigmatic statement made by Einstein at the death of Michele Besso (it is quoted near the beginning of this chapter). As Horwitz, Arshansky, and Elitzur (1988) so nicely put it,

It seems that Einstein's view of the life of an individual was as follows. If the difference between past, present, and future is an illusion, i.e., the four-dimensional spacetime is a "block Universe" without motion or change, then each individual is a collection of myriad of selves, distributed along his history, each occurrence *persisting on the world line, experiencing indefinitely the particular event of that moment* [my emphasis]. Each of these momentary persons, according to our experience would possess memory of the previous ones, and would therefore believe himself identical with them; yet they would all exist separately, as single pictures in a film. Placing the past, present and future on the same footing this way, destroys the notion of the unity of the self, rendering it a mere illusion as well.

It appears by his words that Einstein was indeed in agreement with the block universe concept and that he was attempting to give Besso's family some reason to believe that Michele still lives "somewhen." There is, however, the additional (and to some minds, perhaps, the rather awful) logical implication that if Michele is still living then there are also other Micheles still dying, a ghoulish sentiment Einstein surely did not mean to convey.

Not everybody believes that this view of spacetime was Einstein's, however. Karl Popper, an Austrian philosopher of science, wrote twenty-eight years after the scientist's death that "Einstein was a strict determinist when I first visited him in 1950: he believed in a 4-dimensional Block-Universe. But he gave this up."[29] Shortly before he wrote those words, however, Popper must have learned something new to convince himself of his final comment, because just three years earlier he had declared Einstein to be a determinist in his Foreword to the anthology CPP. Popper presents no evidence to bulwark his claim of Einstein's philosophical conversion, and it would seem that the Besso letter still offers the best insight into his actual view of spacetime shortly before his death.

Use of the term *block universe* is generally thought to have originated with the Oxford philosopher Francis Herbert Bradley (1846–1924) who, in his 1883 *Principles of Logic,* wrote, "We seem to think that we sit in a boat, and are carried down the stream of time, and that on the bank there is a row of houses with numbers on the doors. And we get out of the boat, and knock at the door of number 19, and, re-entering the boat, then suddenly find ourselves opposite 20, and having then done the same, we go on to 21. And, all this while, the firm fixed row of the past and future stretches in a *block* [my emphasis] behind us, and before us." The house numbers would seem to be Bradley's way of referring to the centuries. Note that he wrote these words twelve years before *The Time Machine* and that they preceded Minkowski by a quarter-century.

But this origin of *block universe* may not be so clear-cut as I have made it appear. Bradley, who was frequently criticized by the Harvard psychologist William James (James argued for free will and indeterminism, concepts disallowed in a block universe), may have been mocked on the idea by James during an address to the students of the Harvard Divinity School in March 1884 ("The Dilemma of Determinism"). At that time, a year after Bradley's book was published, James spoke of a deterministic world as being a "solid" or "iron block." (See Note 29 for why I think James was really arguing against fatalism.) However, writing the year *before* Bradley's book, in the April 1882 issue of *Mind,* James wrote (with obvious disdain) of "the universe of Hegel—the *absolute block* [my emphasis] whose parts have no loose play," as having "the oxygen of possibility all suffocated out of its lungs" and as being a universe in which "there can be neither good nor bad, but [only] one dead level of mere fate." So, perhaps, the chain of evolution of the term *block universe* is actually from Hegel to James and then, finally, to Bradley.

The contribution of *mathematical* spacetime to physics has a much clearer origin: It derives from Minkowski, not Bradley or Einstein (who often gets credit for it even though he did not use the concept in special relativity three years earlier, in 1905). Eventually Einstein did come to appreciate the power and conceptual beauty of four-dimensional spacetime, however, and it came to play a central role in his ideas about gravity. Indeed, in Einstein's general theory of relativity, gravity *is* curved spacetime. The starting point for general relativity was Minkowski's creation, however—as Rindler (1977) says, it is Minkowski who truly deserves the title "father of the fourth dimension." The spacetime diagrams that are the basic conceptual tool for discussing time travel (see Tech Note 4) are often called *Minkowski diagrams.*

Of course, it is true that Newton's physics also talks about an analytical (as opposed to a merely philosophical) space and time long before Minkowski and Einstein, but Newtonian spacetime is something very different from Minkowski's self-described "radical" view; see Stein (1967), Earman and Friedman (1973), and Berger (1974). In the Newtonian view there is a universal time, a *cosmic* time, which is the same time for everyone, everywhere in the universe. At every instant a cosmic simultaneity exists—see Hawking (1968) for a precise definition of cosmic time, and how it requires that time travel to the past be impossible. Newton's space is Euclidean; that is, parallel lines never meet, through any point exterior to a line exactly one parallel line can be constructed, all triangles (no matter how large) have an interior angle sum of 180°, and so on. For Newton, space and time were absolutely and uniquely separable. They were, as philosophers are inclined to say, "distinct individuals." Minkowski changed all that. For Minkowski, space and time are only relatively separable, and the separation is different for observers in relative motion. For Newton, space and time are the *background* in which physical processes in the world evolve. For Minkowski, spacetime *is* the world.

Taking the Minkowskian view of the primacy of spacetime as the ultimate building-block stuff of reality has been Princeton professor of physics John Wheeler, who wrote—see Misner and Wheeler (1957)—"There is nothing in the world except empty curved space. Matter, charge, electromagnetism . . . are only manifestations of the bending of space. *Physics is Geometry.*" This idea has been picked up by at least one writer of fiction; in the 1987 novel *Moscow 2042* (Voinovich) a time traveler sounds a lot like Wheeler when he says, "Anyone with even a nodding acquaintance with the theory of relativity knows that nothing is a variety of something and so you can always make a little something out of nothing."

Wheeler developed his ideas (which many physicists find, to use Minkowski's word, radical) in detail in his 1962 book *Geometrodynamics;* see also Wheeler (1962b). That same year saw a hint in his work, described in Baierlein, Sharp, and Wheeler (1962), that time itself might, like spacetime, find explanation in geometry. Since then Wheeler's ideas have continued to evolve, and more recently it is not geometry but something else, called *pregeometry,* that for him is the basic building-block stuff of the world. An extended discussion of the evolution in Wheeler's ideas can be found in Grünbaum (1973).

In Williams (1951b), a famous philosophical paper by an advocate of the block universe interpretation of spacetime, states, "I . . . defend the view of the world . . . which treats the totality of being, of facts, or of events as spread out eternally in the dimension of time as well as the dimensions of space. Future events and past events are by no means present events, but in a clear and important sense they do exist, now and forever, as rounded and definite articles in the world's furniture." In an even more famous paper (1951a) Professor Williams makes clear his belief that the passage of time is a myth; he poetically declared "the total of world history is a spatio-temporal volume, of somewhat uncertain magnitude, chockablock with things and events."

Williams did, indeed, embrace four-dimensional spacetime, and this is demonstrated by the following incredible passage, perhaps his best-remembered words: "It is then conceivable, though doubtless physically impossible, that one four-dimensional area of the time part of the manifold be slewed around at right angles to the rest, so that the time order of that area, as composed by its interior lines of strain and structure, run parallel with a spatial order in its environment. It is conceivable, indeed, that a single whole human life should lie thwartise of the manifold, with its belly plump in time, its birth at the east and its death in the west, and its conscious stream running alongside somebody's garden path."

Good Lord!

Now, I am willing to admit that Professor Williams probably wrote that wonderful passage mostly for effect, but I ask you—what, if anything, does it *mean?* It is marvelous to read and yet it remains (for me) mysterious. It should come as no surprise that Professor Williams originally presented his papers to the Metaphysical Society of America, rather than to the American Physical Society. But this passage was perhaps not without impact in areas far removed from metaphysics; some years later, for example, there appeared a science fiction story, "The Rubber Bend" (Wolfe), that reads as though it had been inspired by Williams. A scientist discovers how to bend his perception of the four dimensions so as to view verticality as duration and duration as verticality. Thus, he is in October while sitting, but when he stands up he is in November! (As bizarre as this may seem, such coordinate interchanges actually *do* occur in the theory of time machines. See, for instance, Tech Note 8 on the operation of Tipler's cylinder.)

I should tell you now that, despite the enthusiastic embrace of the block universe by Williams and others (including Einstein), there are those who have been harsh in their criticism of Minkowski's spacetime. The major philosophical problem with the block universe interpretation of four-dimensional spacetime is that it looks like fatalism disguised as physics. It seems to be a mathematician's proof of a denial of free will dressed up in geometry. For example, the block universe is savaged by Geach (1968) for being fatalistic, but then Geach's remarks provoked ridicule from Smart (1972), who feels that Geach's interpretation of free will as an ability to change the future is either trivial or absurd. Indeed, Geach's argument *is* inconsistent. On the one hand, he does not like the block universe because of its implication that the future already exists, but on the other hand, if the future is not there, then what sense can be made of being able to change it? How do you *change* something that does not exist?

Reichenbach (1956) tells a charming story that vividly demonstrates the compelling need many humans have to deny a fatalistic world.

> In a moving picture version of *Romeo and Juliet,* the dramatic scene was shown in which Juliet, seemingly dead, is lying in the tomb, and Romeo, believing she is dead, raises a cup containing poison. At this moment an outcry from the audience was heard: "Don't do it!" We laugh at the person who . . . forgets that the time flow of a movie is unreal, is merely the unwinding of a pattern imprinted on a strip of film. Are we more intelligent than this man when we believe that the time flow of our actual life is different? Is the present more than our cognizance of a predetermined pattern of events unfolding itself like an unwinding film?

Most people in the Western world would answer *yes* to Reichenbach's question. Most such people find Omar Khayyam's *Rubaiyat* to be a beautiful poem, but still they reject its fatalistic message: "And the first Morning of Creation wrote, What the Last Dawn of Reckoning shall read." Indeed, William James quoted these very words in his 1884 address to the students of the Harvard Divinity School when he argued against fatalism and the block universe.

Equally unhappy with the block universe was Herbert Dingle, who declared in his last book before his death (see Tech Note 5 for more about Dingle and his book) that "it is to Minkowski that we owe the idea of a 'spacetime' as an objective reality—which is perhaps the chief agent in the transformation of the whole subject [of relativity] from the ground of intelligible physics into the heaven (or hell) of metaphysics, where it has become, instead of an object of intelligent inquiry, an idol to be blindly worshipped. . . . Reduced to its essence, Minkowski's [work] is a piece of pure mathematics—as such, extremely elegant and admirable, but insofar as it purports to contribute to physics, as it does, calamitous." And in his 1965 British Academy lecture, P. T. Geach (1968) attacked the Minkowskian view as "very popular with philosophers who try to understand physics and physicists who try to do philosophy." Geach clearly believes that both groups have failed. Finally, P. F. Strawson, in an introduction to Geach's essay, put in his two cents by calling the four-dimensional, spacetime view of reality nothing but "fanciful philosophical theorizing."

Besides fatalism, another reason for the stinging words of these critics is that in Minkowski's spacetime it seems that events don't *happen*—they just *are;* there seems to be no temporal process of *becoming* in Minkowski's spacetime. Everything is already there, and as what we perceive as the passing of time occurs, we become conscious of ever more of Minkowski's "world-points," or *events,* that lie on our individual world lines. Hermann Weyl (1885–1955), a German mathematical physicist who in his last years was a colleague of Einstein and Gödel at the Institute for Advanced Study in Princeton, expressed that very interpretation in words that have become famous,[30] words that sound very much like those of Wells' Time Traveller: "The objective world simply *is,* it does not *happen.* Only to the gaze of my consciousness, crawling upward along the life line of my body [Minkowski's world-line], does a section of the world [spacetime] come to life as a fleeting image in space which continuously changes in time [creating what we call the *now* or the *present*]."

A modern story that copies Wells' title, "The Time Machine" (Jones), was clearly influenced by this, but rather than using metal and glass and ivory, it has the human brain and memory as the time machine—"The time machine operates on an organic, electrochemical basis." Weyl's view of the block universe is adopted in this tale, because after telling us that the universe is a four-dimensional ring, the author then says, "Time is merely the attention [of an observer] constantly and involuntarily operating on a different part of the ring, giving the impression of movement and animation to what is in fact a static object."

Weyl was skillful at finding poetic ways to express the world-line view of reality, but not everybody is convinced by the poetry because it seems to deny the common-sense idea of time "flowing," and of temporal passage; it effectively says that time is mind-dependent, a mere illusion. The philosophers J. J. C. Smart and Max Black had no sympathy with Weyl on that issue. Observing that four-dimensional spacetime is *timeless,* as indeed Weyl's own words claim, Professor Smart wrote (1955b), "Within the Minkowski representation we must not talk of our four-dimensional entities changing or not changing. This sort of mistake is often made in expositions on relativity. We often read of light signals being transmitted from one point of Minkowski space to another. This is liable to lead to metaphysical error, such as that of consciousness crawling up world-lines."

Black was even harsher in his objection to Weyl. As he wrote in no uncertain terms,[31] "this picture of a 'block Universe,' composed of a timeless web of 'world-lines' in a four-dimensional space, however strongly suggested by the theory of relativity, is a piece of gratuitous metaphysics." Mundle (1967), another philosopher, has gone beyond mere name calling to put forth a fascinating speculation on what Weyl's description of four-dimensional spacetime might really imply. In a generally critical reply to Smart, Mundle declares, "If the physical world is conceived as a 4-D manifold, it is, as Smart acknowledges, logically impossible for a physical thing, a 4-D solid, to move or otherwise change. . . . Since Smart is conceiving the physical world and its contents as changeless [it must be] our states of consciousness which change as we become successively aware of adjacent cross-sections of the 4-D manifold [that is aware of Weyl's "crawling"]. But this makes sense only if we, the observers, are *not* in Space-Time."

That is, our conscious minds somehow must exist on a level beyond anything physics can tell us about. This is, of course, a radical view that does not enjoy much support among scientists. Mundle makes it clear, in fact, that he is of a different mind from Smart when it comes to science; he wrote that "Smart seems to have been influenced mainly by Physics" and that "we must not be blinded by science." As Sherlock Holmes would no doubt reply to that, "What a singular statement."

Still, no matter how odd Mundle's suggestion is, a close variant of it was used in a science fiction story even as Mundle wrote. In *The Technicolor Time Machine* (Harrison) time travel takes place in the "extratemporal continuum." And in Isaac Asimov's famous time travel novel *The End of Eternity,* the time police (called Eternals) oversee the endless centuries from a place outside of time called Eternity. This idea has no foundation in physics, however, and it is Minkowski's block universe idea that will play a central role in this book.

We can actually find the block universe in fiction *before* Minkowski. In the 1875 tale "The True Story of Bernard Poland's Prophecy" (Eggleston), for example, we read of a man who sees his own death in the American Civil War, years in the future. Here Bernard speaks to his unnamed friend, the narrator.

> "Do you know," said Bernard, presently, "I sometimes think prophecy isn't so strange a thing. . . . I really see no reason why any earnest man may not be able to foresee the future, now and then. . . ."
>
> "There is reason enough to my mind," I replied, "in the fact that future events do not exist, as yet, and we can not know that which is not, though we may shrewdly guess it sometimes. . . ."
>
> "Your argument is good, but your premises are bad, I think," replied my friend, . . . his great, sad eyes looking solemnly into mine.
>
> "How so?" I asked.
>
> "Why, I doubt the truth of your assumption, that future events do not exist as yet. . . . Past and future are only divisions of time, and do not belong at all to eternity. . . . To us it must be past or future with reference to other occurrences. But is there, in reality, any such thing as a past or a future? If there is an eternity, it is and always has been and always must be. But time is a mere delusion. . . . To a being thus in eternity, all things are, and must be present. *All things that have been, or shall be, are* [my emphasis]."

Also consider Wells' Time Traveller's speech to his friends at the fateful dinner party that opens *The Time Machine*. "There is no difference between Time and any of the three dimensions of Space except that our consciousness moves along it . . . here is a portrait of a man at eight years old, another at fifteen, another at seventeen, another at twenty-three, and so on. All these are evidently sections, as it were, Three-Dimensional representations of his Four-Dimensional being, *which is a fixed and unalterable thing* [my emphasis]." Remember, those words were written in 1895, thirteen years before Minkowski and his world-lines, and of course decades before Weyl's famous words quoted earlier in this section. Wells' passage made a considerable impression on at least one well-known physicist of the time, who references it in his early book on relativity.[32] And in another book on relativity published the same year,[33] we find the same interpretation of Minkowski's spacetime as a block universe: "With Minkowski, space and time become particular aspects of a single four-dimensional continuum. . . . [A]ll motional phenomena . . . become timeless phenomena in four-dimensional space. The whole history of a physical system is laid out as a changeless whole."

The block universe concept appeared very early in magazine science fiction stories. In the 1927 "The Machine Man of Ardathia" (Flagg), for example, a time traveler from the future and a man in the present (who is the narrator) have the following exchange:

> "I have just been five years into your future."
> "My future!" I exclaimed. "How can that be when I have not lived it yet?"
> "But of course you have lived it."
> I stared, bewildered.
> "Could I visit my past if you had not lived your future?"

And in the 1929 novel *The Man Who Mastered Time* (Cummings), a time traveler who is about to set out on his first journey to the world 28,200 years in the future (where he has already observed a beautiful young girl via a mechanism that lets him see through time) tells his friends, "You say that girl *will be* living in the future. I say she *is* living in the future. She is living just as you and I are living—right here in this exact spot we call New York—within a few hundred yards of this room. She is separated from us, not by space, but only by time."

More gruesome are the fiendish thoughts of the stereotypical mad scientist of pulp fiction in the 1932 "The Time Conqueror" (Eschbach), who wants to float a disembodied brain in a vat. He believes that such a brain, cut off from the ordinary five senses, will develop a compensatory sixth sense that will be aware of the "hyper-world" of spacetime. He clearly has the block universe in mind when, at one point, he crazily cackles, "It will see the past and future, lying together and existing simultaneously!"

"Life-Line" (Heinlein), published in 1939, uses world lines as its central scientific concept. The story draws an analogy between a world line and a telephone cable, as follows. The beginning and end points in spacetime for the world line (birth and death) are associated with breaks (faults) in the telephone cable. Now, by sending a signal up and down the cable and measuring the time delay until the arrival of the echo produced by such discontinuities, a technician can both detect and locate the faults. In the same manner, Heinlein's story gadget sends a signal of unspecified nature up and down a world line and thus locates the birth and death "discontinuities." Knowledge of the latter date, in particular, causes financial stress among life insurance companies, and an examination of that tension (not weird physics) is the fictional point of the story. Less scientific and more mystical is "Damnation Morning" (Leiber), where we learn of an eternal time-war in which the soldiers are recruited from the newly dead, who are resurrected by being cut out of their life-line [world line] and given the freedom of the fourth dimension (whatever *that* means).

The block universe idea that past and present coexist got dramatic treatment in the 1937 tale "Temporary Warp" (Long), the story of a high school teacher who invents a "spacetime warp" theory and who is deceived by an evil industrialist into implementing it in the form of a gun. The weapon produces incredible effects when it is tested; for example, an allosaurus appears, which we are told is "a carnivorous dinosaur of the Jurassic Age, the most frightful engine of destruction that ever walked the Earth!" By the story's end the teacher explains what has happened to a crowd of breathless newspaper reporters.

> Spacetime was warped slightly. . . . The Einsteinian spacetime continuum buckled. . . . Because it was superficial, only a little of the past, a little of the future broke through. The folds of the warp distorted spacetime evanescently, erratically skirting the vast gulf where the past lies buried and lightly tapping the vast stores of the future. It is a truism of modern speculative physics that the past and the future exist simultaneously and coextensively in higher dimensions of space. De Sitter has speculated as to the possibility of seeing an event before it happens. It is quite possible, gentlemen. Events of the far future already exist in spacetime."

That "explains" the dinosaur. In the teacher's words, "You tell me that two men saw an incredible beast. . . . They swear it looked like a dinosaur. I think it was a dinosaur, gentlemen. It broke through when the warp tapped the past."

The static view of the block universe was also invoked by one science fiction fan who wrote in support of time travel, after another fan had cited a resulting failure of mass/energy conservation in an attempt to disprove the notion of time travel. The exchange began with a letter to the editor at *Astounding Stories* (November 1937), written in response to the story "The Time Bender" (Saari): "Let us say that there is, at a certain time, 'x' amount of matter in the Universe, and 'e' amount of energy. Then if a man of 'a' mass travels backward in time to this particular instant aforementioned, the total amount of matter is thus 'x' plus 'a', while if no other such mass changing occurrences take place, the amount of matter in that future is 'x' minus 'a'. Only a corresponding loss and gain respectively in the amount of energy could explain this conservation of energy, advocates [of time travel] say what they may. But you can't rob or add energy to a Universe nilly-willy! Or perhaps time doesn't enter in on the matter. Perhaps you can add matter in a Universe provided you take it away on some future date." That fan's concern obviously made an impression on science fiction writers: The necessity for conservation of energy is stated in many of the time travel stories that appeared after the publication of his letter. Examples include *Lest Darkness Fall* (De Camp), "Time Waits for Winthrop" (Tenn), and *The Time Hoppers* (Silverberg).

A reply was received by the magazine in a letter (January 1938) from another science fiction fan: "[A recent letter] implies that the idea of time travel is incompatible with the law of conservation of mass and energy. I believe [the] reasoning is wrong [and that the] difficulty lies primarily in the assumption that a body moved in time is transported into a different Universe. According to Einstein, time and the three normal dimensions are so related as to form a continuous, inseparable medium we call the spacetime continuum. Time is in no way independent of the other components of our Universe. Hence a fixed mass [a time traveler and his machine] moved in time is by no means lost from the Universe, the action being analogous to a shift along any other dimension." The block, or frozen, universe of Minkowski is clearly reflected in those words.[34]

The block universe concept rapidly made an impression on popular culture outside of science fiction, as well. For example, in a 1928 New York stage play, "Berkeley Square" (Balderston), which the author said was suggested by Henry James' unfinished (and, in my opinion, unreadable) novel *The Sense of the Past,* the action alternately takes place in the years 1784 and 1928. To explain how that can be, the character who is doing the time traveling says to another,

> "Suppose you are in a boat, sailing down a winding stream. You watch the banks as they pass you. You went by a grove of maple trees, upstream. But you can't see them now, so you saw them in the *past,* didn't you? You're watching a field of clover now; it's before your eyes at this moment, in the *present.* But you don't know yet what's around the bend in the stream there ahead of you; there may be wonderful things, but you can't see them until you get around the bend, in the *future,* can you?" Then, after this prologue about the stream of time, comes the block universe idea: "Now remember, *you're* in the boat. But *I'm* up in the sky above you; in a plane. I'm looking down on it all. I can see *all at once* the trees you saw upstream, the field of clover that you see now, and what's waiting for you, around the bend ahead! *All at once!* So the past, present, and future of the man in the boat are all *one* to the man in the plane."

And finally, the theological conclusion: "Doesn't that show how all Time must really be one? Real Time—real Time is nothing but an idea in the mind of God!" This powerful play was made into a 1933 movie of the same name, and it appeared again in 1951 as the film *I'll Never Forget You.*

Decades later we find the same idea used in *Doctor Brodie's Report* (Borges), which tells the strange story of the discovery of the long-lost first-person manuscript of David Brodie, D.D. (found "among the pages of one of the volumes of Lane's *Arabian Night's Entertainments,* London, 1839"). That manuscript describes the strange ape-men discovered by Brodie, who calls them "Yahoos." Among their many unusual characteristics is the ability of Yahoo witch doctors to see the future. As the Brodie manuscript puts it, "Hundreds of times I have borne witness to this curious gift, and I have also reflected upon it at length. Knowing the past, present, and future already exist, detail upon detail, in God's prophetic memory, in his Eternity, what baffles me is that men, while they can look indefinitely backward, are not allowed to look one whit forward."

Real men may not be able to look forward in time, but in the comics there are no such restrictions. Mr. Mxyztplk, for example, in one of his misadventures with Superman in 1954, begins selling the *Daily Mpftrz* in competition with the *Daily Planet.* Unlike a traditional newspaper that reports what has happened, the *Daily Mpftrz* prints what will happen. As Mr. Mxyztplk says, "You see, as a resident of the fifth dimension, I can get all the news I want from the *fourth* dimension!" The science editor at the *Daily Planet* explains the meaning of that to his boss, Perry White: "That's right Mr. White, . . . many physicists consider *time* the fourth dimension . . . so if Mr. Mxyzptlk can travel from the fifth dimension to our three-dimensional world, he most likely *is* able to see the future!" This leaves unanswered the question of why he continues to challenge Superman when he knows he will be defeated—he *always* is!

Spacetime, Omniscience, and Free Will

Some old theology on God's omniscience, as discussed in Aquinas' *Summa Theologiae,* is seemingly lent support by Minkowski's spacetime: "Now although contingent events come into actual existence successively, God does not, as we do, know them in their actual existence successively, but all at once; because his knowledge is measured by eternity, as is also his existence; and eternity which exists as a simultaneous whole, takes in the whole of time. . . . Hence all that takes place in time is eternally present to God." Somewhat paradoxically, however, Aquinas did make a distinction between past and future. In that same work he declares that "God can cause an angel not to exist in the future, even if he cannot cause it not to exist while it exists, or not to have existed when it already has." For Aquinas, then, whereas the past is rigid and unchangeable, the future is plastic, which is *not* the block universe view of spacetime.

As pointed out in Craig (1985), this does not mean Aquinas thought God had to view all events simultaneous with one another. (The story "The Weed of Time" (Spinrad) graphically describes what a nightmare that could be!) Rather, using an interesting analogy, Craig says that Aquinas could have thought of the relationship between God and events as being similar to that between the center of a circle and all the points on the circumference. That is, each point on the circumference has its own identity, coming before and/or after any other point, but the center is related to each and every point on the circumference in precisely the same way. The center, then, is "eternity" and the circumference is the temporal series ("one thing after another") of reality. Saying that God is *eternal* is thus very different from saying he is *everlasting*. The first means he is outside of time, whereas the second means he is a temporal entity but has neither beginning nor end—for more on this second interpretation, see Wolterstorff (1995). Craig supports the first interpretation, using Aquinas' words from *Summa Contra Gentiles:* "The divine intellect, therefore, sees in the whole of its eternity, as being present to it, whatever takes place through the whole course of time. And yet what takes place in a certain part of time was not always existent. It remains, therefore, that God has a knowledge of these things that according to the march of time do not yet exist."

The issue of God's eternity and his relationship to spacetime is presently a hot topic among theologians with a scientific inclination. Practically every issue of the learned journal *Religious Studies,* for example, carries an article on the subject, often invoking relativity theory to support some argument. The Bible itself, amusingly, is a confusing guide in such matters. For example, consider the Old Testament story of King Ahab (*First Kings* 21). Ahab, King of Sumeria, coveted Naboth's vineyard, but Naboth would not sell. The King retreated, but his wife Jezebel arranged for Naboth's downfall and judicial murder and thus caused the arrival of all his property into her husband's hands. This angered God, who commanded Elijah to prophesy disaster on Ahab's house. Ahab responded with sackcloth, and at that God shifted the disaster to the house of Ahab's son—which certainly raises concerns about a benevolent God, but that's another story.

The point here is that God, declared to be omniscient, seems to have been *surprised* at Ahab's penitence! God is aware of everything in this tale, but only as it happens; that is, God's knowledge is subject to growth. This Hebrew concept of God as a participant in history is at odds with the contemporary Christian conception of divine knowledge of all that has been, all that is, and all that will be. And before leaving this topic, I should mention that Biblical support for the modern conception of divine eternality also exists, of course. It includes *Malachi* 3:61 ("For I am the Lord, I change not"), *John* 8:58 ("Before Abraham was, I am"), and *James* 1:17 ("the Father . . . with whom is no variableness").

In the *New Review* serialization of *The Time Machine*, in a passage that does not appear in the now-classic version of the story, the Time Traveller explains his view of the connection between omniscience and the block universe to his dinner guests:

> "I'm sorry to drag in predestination and free-will, but I'm afraid those ideas will have to help. . . . Suppose you knew fully the position and properties of every particle of matter, of everything existing in the Universe at any particular moment of time: suppose, that is, that you were omniscient. Well, that knowledge would involve the knowledge of the condition of things at the previous moment, and at the moment before that, and so on. If you knew and perceived the present perfectly, you would perceive therein the whole of the past. If you understood all natural laws the present would be a complete and vivid record of the past. Similarly, if you grasped the whole of the present, knew all its tendencies and laws, you would see clearly all the future. To an omniscient observer there would be no forgotten past—no piece of time as it were that had dropped out of existence—and no blank future of things yet to be revealed. . . . [P]resent and past and future would be without meaning to such an observer. . . . He would see, as it were, a Rigid Universe filling space and time. . . . If 'past' meant anything, it would mean looking in a certain direction, while 'future' meant looking the opposite way."

Wells' "Rigid Universe" certainly sounds like the block universe, and he seems to have believed that it held important implications for the concept of free will.

With Einstein's discovery of the relativity of simultaneity (see Tech Note 1), however, there seems to be a problem with how there can be divine knowledge of any sort in a relativistic, four-dimensional spacetime. In some frames of reference event A is observed before event B, whereas in other frames the temporal order is reversed (this point is discussed at the end of this section). If God is to be actively involved in human affairs, then what is *his* frame of reference? Does God have a special frame in which he is immune to the relativity of simultaneity and in which he imposes an absolute order on the sequence of becoming of events? Does it make any sense to say God enjoys what might be called "divine immediacy"? What should we think, in fact, of a God who follows rules different from those that govern all he is supposed to have made? Theologians have argued these questions for decades—see Wilcox (1961), Ford (1968), Fitzgerald (1972), Hasker (1985, 1987) and Reichenbach (1987)—whereas physicists (including Einstein) have been unaware, unimpressed, or just plain uninterested.

But that, perhaps, is a mistake. Consider, for example, an interesting exchange between Hasker (who believes free will and divine foreknowledge are not compatible) and Reichenbach (who thinks Hasker has made a serious error in blurring the distinction between changing and affecting the past, a distinction treated at length in Chapter Four). Reichenbach (1987) presents some of his comments in terms of a time traveler to the past:

> Consider the following. Parsons (P) has invented a special machine which allows him to go back in time. He enters the machine in 1986 and finds himself in the presence of or, perhaps better, observing, Quigly (Q) in 1876. P is an authority on Q, and knows immediately the situation Q is in. Not only that, but he remembers reading about the particular decision or act which Q made in that situation. Thus one might argue that from P's perspective what Q decides is as if already done. It is not already done, since P is standing there waiting for Q to do it. He has gone back in time. Yet from P's perspective, which is of one come back from the future, it is as if already done, since he knows what Q does decide. Since P strongly believes in the unalterability of the past, it is not within Q's power to do something other than what Q in fact does in that situation. From Q's perspective his decision is not already made nor is the action taken, so that it is in his power at that time to do either x or y. From his perspective, that he will do x rather than y is indeterminate; it is not yet done, though at the same time he can grant that P knows what he will do because for him it is as if he has already done it.

To that Hasker (1987) replied as though he had at last heard one too many "philosophers' stories". "It should be abundantly clear by now that the fact that such stories are in some way imaginable and intuitively graspable says nothing about their logical coherence." Given the present interest among physicists in the concept of time travel, however, I think Haskers wouldn't write that today.

One possible answer to at least some of the concerns just cited can perhaps be found in a paper written by a philosopher and two mathematicians—Bennett *et al.* (1949)—that describes a five-dimensional unified field theory in which the fifth dimension is initially given the provocative label of the "eternity" axis. But then the authors lost their nerve and quickly elected to rename it "anti-time." A theological interpretation of that theory is given in Stromberg (1961). It is curious to note that science fiction anticipated that terminology by decades; in the 1932 "The Time Conqueror" (Eshbach) one character says, "Beyond the fourth there is a fifth dimension. . . . Eternity, I think you would call it. It is the line, the direction perpendicular to time."

The idea of supernatural beings existing outside of mortal time is an old one in theology, and it can also be found in the secular literature long before science fiction got hold of it. For example, in the first act of Lord Byron's 1821 poem *Cain,* the fallen angel Lucifer tells Cain and his wife that

> With us acts are exempt from time, and we
> Can crowd eternity into an hour,
> Or stretch an hour into eternity.
> We breathe not by a mortal measurement,
> But that's a myst'ry.

Black and Smart would certainly be unhappy with Lucifer's claim, and Weyl himself might have been rather surprised at how far afield his views have been taken by others. In another philosophical rebuttal to Weyl's block universe views, Dobbs (1969) wrote, "While philosophers may be forgiven intellectual extravagances of this kind, I think it is a pity when they receive encouragement from theoretical physicists, even if these share a common mystical experience." Physicists, no doubt, will be flattered to learn that they have such a profound influence on philosophers—excluding, of course, Mundle of the previous section!

The idea of tampering with the future (which "already exists" in a block universe) receives an interesting treatment in "What We Learned from This Morning's Newspaper" (Silverberg). There, an event in the present that occurs "before it should" (a heart patient learns that her obituary notice will be in next week's *New York Times* when that paper arrives "early") so upsets the future that spacetime is destroyed! That story takes the view that the future is delicate. As one character puts it to the sister of the lady who is soon to die, "The future mustn't be changed. . . . For us the events of . . . the future are as permanent as any event in the past. We don't dare play around with changing the future, not when it's already signed, sealed and delivered in that newspaper. For all we know the future's like a house of cards. If we pull one card out, say your sister's life, we might bring the whole house tumbling down. You've got to accept the decree of fate. . . . You've got to."

Disagreeing with that plea is the time traveler in the 1941 story "The Man Who Saw Through Time" (Raphael), who, while visiting the future ten years hence, learns he will be arrested, tried, and executed for the murder of a colleague and friend. To avoid that, to change the future, he returns to the present and successfully arranges for his own immediate death. The 1932 novel *The Gap in the Curtain* (Buchan), on the other hand, agrees with the block universe view, arguing that the future is nothing less than petrified. See also "Final Audit" (Disch).

Before Minkowski, the debates over fatalism and free will had been the exclusive province of philosophers, theologians, and lawyers (if a person has no control over his or her actions, then can we morally and ethically punish that person if those actions happen to be criminal?) After Minkowski and his spacetime, the physicists (at least a few of them) joined the debates. According to Williams (1951b), the major motivation behind these debates is "the age-old dread that God's foreknowledge of our destiny can in itself impose the destiny upon us." The implication is, of course, that God is "outside of time" and so can take in the entire Minkowskian block universe at a glance (hence his foreknowledge). For an argument in support of this view, see Lackey (1974).

The relativistic view of the universe as a timeless four-dimensional spacetime seems to many to provide scientific, mathematical support for the conclusion that not only is the past fixed, but so is the future. Does that mean the future is what it will be—and if so, then why bother agonizing over the many apparent decisions each of us faces every day? If the answer is yes, and the future will be what it will be, then Christian theologians are left with the puzzling task of explaining what could possibly be meant by the Biblical exhortation (*Deuteronomy* 30:19) "I call Heaven and Earth to record this day against you, that I have set before you life and death, blessing and cursing; therefore *choose* [my emphasis] life, that both thou and thy seed may live."

This issue has bothered philosophers for a very long time. The so-called Master Argument (the name reflects its supposed invulnerability to rebuttal), for example, comes down to us from its origins in ancient, pre-Christian times, in the *Discourses* of the first-century A.D. Roman Stoic philosopher Epictetus. That argument can be summarized as follows—see Purtill (1974, 1975):

1. The future follows from the past;
2. The past is unchangeable;
3. What follows from the unchangeable is unchangeable;

Therefore,

4. The future is unchangeable.

This certainly does seem to be fatalistic, in effect arguing that all events in a block universe spacetime are recorded in a "Book of Destiny." As the title character in Shakespeare's *Macbeth* says, expressing the view that we are all mere players following a script (Act 5, Scene 5),

> Life's but a walking shadow, a poor player
> That struts and frets his hour upon the stage
> And then is heard no more: it is a tale
> Told by an idiot, full of sound and fury,
> Signifying nothing.

Since ancient times many great works of literature have adopted that view, recounting tales of the foretold fates of men, such as Sophocles' *Oedipus*. It is, in a block universe, as though our conscious experience of the world is no different from that of Reichenbach's watcher of a projected film image.

That view is the central issue in Boethius' influential *De Consolatione Philosophiae* (circa A.D. 500), which was written during a year of imprisonment before his execution for treason; perhaps he wondered during that year if his fate could have been anything different. Certainly he might have taken some consolation in fatalism, but in fact he tried to argue that God's vision of *all* temporal reality does not limit the freedom to act. According to Boethius, "The expression 'God is ever' denotes a single Present, summing up His continual presence in all the past, in all the present . . . and in all the future." That is, God sees in one timeless and eternal moment all that has been and will be freely chosen.

Kurt Vonnegut called this ability "chrono-synclastic infundibulated" vision in *The Sirens of Titan,* a novel that was meant to be a parody of God's omniscience. In his novel *Slaughterhouse-Five,* Vonnegut tries to see a happy side to omniscience with his creation of the alien inhabitants of the planet Tralfamadore, who can see in four dimensions: "When a Tralfamadorian sees a corpse, all he thinks is that the dead person is in a bad condition at that particular moment, but that the same person is just fine in plenty of other moments." That sort of prose lacks Einstein's polish and Weyl's poetry—Blackford (1985), for example, calls Vonnegut's expression of the block universe "a variety of facile mysticism"—but it does actually seem to reflect Einstein's block universe view of reality in the Besso letter.

Boethius, however, would not I think have appreciated Vonnegut's wit, and neither would the British scientist Oliver Lodge (1850–1940), who in one of his innumerable popular essays, an essay on time and the fourth dimension (1920), wrote, "Is the future all settled beforehand, and only waiting to be 'pushed through' into our three-dimensional ken? Is there no element of contingency? No free will? I am talking geometry, not theology."

Chaucer prepared a translation of *Consolatione,* and he was obviously inspired by it when he wrote his very long poem on the nature of love (*Troilus and Criseyde*) more than six hundred years ago (Book IV.140).

> Some say "If God sees everything before
> It happens—and deceived He cannot be—
> Then everything must happen, though you swore
> The contrary, for He has seen it, He."
> And so I say, if from eternity
> God has foreknowledge of our thought and deed,
> We've no free choice, whatever books we read.

But Chaucer's poetry misstates Boethius' philosophy when Troilus declares that divine foreknowledge is incompatible with free will. Modern, purely philosophical rebuttals to Chaucer's position can be found in Mavrodes (1984) and Plantinga (1986), which argue that God's omniscience (a fundamental teaching in the theistic religions of Christianity, Judaism, and Islam) *is* compatible with free will (also a fundamental belief in those same religions). Both of these scholarly papers, I should tell you, depend much more on the nuances of grammar than most physicists would like.

A more recent paper in the philosophical literature makes the connection between spacetime physics and free will explicit. As Craig (1988) puts it, "For philosophers in either field, philosophy of science and philosophy of religion are too often viewed as mutually irrelevant. . . . This is unfortunate, because sometimes the problems can be quite parallel and a consistent resolution is required. One especially intriguing case in point concerns, in philosophy of science, the possibility of . . . time travel and, in philosophy of religion, the relationship between divine foreknowledge and human freedom."

Craig could also have included science fiction writers in his group of people interested in both spacetime physics and free will. In "Turn Backward, O Time" (Kubilius), for example, a man in the twenty-fifth century is about to travel back into the past to escape criminal prosecution. He is asked where he'd like to go, and he replies, "I do not understand the paradoxes—what if I choose to build gravity-deflectors in Ancient Rome?" When he is told (correctly) that that can't happen because it didn't happen, he persists: "But if I can choose any period, it means that I can alter history at will—which presumes that the present can also be changed." Then, at last, we get the explicit answer that bothers nearly everyone: "The real answer is that in the final analysis your decision to choose a certain time period is already made, and the things you will do [in the time traveler's proper time] are already determined. Free will is an illusion; it is synonymous with incomplete perception." That same idea also appears in "Beep"(Blish); when one character says, "What you are saying is that the future is fixed, and that you can read it, in every essential detail," the response is "Quite right . . . both those things are true."

Those story characters have, of course, run into the ancient dilemma made famous by Aristotle in his *De Interpretatione,* where he asked a question now famous among philosophers, "Will there be a sea fight tomorrow?" Aristotle began his equally famous answer by first posing the following premise: If a statement about some future event is, eventually, shown to be true (or false), then that statement was true (or false) from the moment it was made. Consider, then, the following two assertions: (A) "It is true that there will be a sea fight tomorrow" and (B) "It is true that there will not be a sea fight tomorrow." Surely, argued Aristotle, (A) and (B) cannot both be true, but equally surely, one of them must be true. Suppose it is (A) that is true. Then there is nothing that can be done to prevent the sea fight and so the future is fated. Suppose, however, it is (B) that is true. Then there is nothing that can be done to cause the sea fight and so the future is fated. The conclusion is the same no matter which assertion is the true one; the future is fated.

As might be expected, those who like the fatalistic block universe like this conclusion. But Aristotle himself was not one of those—he disliked his own conclusion so much that he struggled to find a way around it. On the other hand, philosophers such as Professor Williams, who believe in a fatalistic universe, reject Aristotle's rejection of his own logic! Williams (1951b) goes so far as to call Aristotle's final reasoning "a tissue of error" and "swaggeringly invalid." Possibly so, but the philosophical debates over Aristotle's problem have not ceased to this day; see Linsky and Williams (1954), Anscombe (1956), Williams (1978) and Rudavsky (1983, 1988) for more arguments and counter-arguments galore on the merits and faults of Aristotle's reasoning.

The debate over omniscience and its sometimes surprising implications took on a fresh twist in 1960 with a new puzzle, this time from a physicist. That year William Newcomb posed what has since become famous as "Newcomb's problem." (Ten years later, Newcomb co-authored an equally famous paper on the impossibility of sending information into the past, a topic treated at length in Chapter Four). Newcomb's problem, still unresolved to everyone's satisfaction, is easy to state.

You are presented with two boxes; you may keep the contents of either Box A or Box B alone, or you may keep the contents of both boxes. Box A is transparent, and you can see that it contains $1000. Box B is opaque, but you are told that it either is empty or contains $1,000,000, depending on whether an omniscient being has foreseen that you will select both boxes or that you will select Box B alone, respectively. The being's omniscience is temporally long-range; that is, it placed its money into the boxes a *year* ago. What should you do?

There are two ways to reason. The first school of thought says to take both boxes, because then you get the sure-fire $1000 in Box A and also whatever is in Box B. After all, whatever is in Box B has been there for a year, and it would be foolish not to take it *now*. The second school of thought disagrees, saying that the first school is ignoring the given factor of omniscience. If you decide to take both boxes, then the omniscient being—having foreseen your decision—will have put nothing into Box B. Therefore, you should ignore the sure-fire $1000 in Box A and take "just" the sure-fire $1,000,000 in Box B that the omniscient being placed in it (a year ago) because it knew that is what you would do now! What is one to make of these contradictory arguments?

The first school is, of course, appalled at the willingness of the second school to give up free will. I personally fall into the second school because this is a problem of logic, and as the problem is stated, the omniscience is a *given;* I simply fail to see what puzzles. The difficulty, for me, isn't that of giving up free will but rather the assumption of omniscience. But that is a *given,* so. . . . For many this is too simplistic an answer, however, and for them the problem continues to puzzle. See Brams (1983), who describes the omniscient being as simply a *superior* but not supernatural being, and Ahern (1979), who frankly takes it as God. Most recently, Schmidt (1998) has presented what I find the most incredible discussion of Newcomb's Paradox in the scholarly literature. To "explain" omniscience, he imagines a population of super-small "dwarfs" made of "tinion" (matter, yet to be discovered, that is much smaller than the "gigantions" *we* are made of)! These dwarfs then use classical physics to predict human decisions to be made in the future. Though fun to read, this is simply a philosopher's fairy tale—a literary technique about which I will have much more to say (none of it good) in the next section.

Does the Future Already Exist? Is the Past Still Around?

Since 1951, and the appearance of Williams' papers, philosophers have become steeped in relativity theory, and their more recent papers have exhibited greater sophistication. For example, Capek (1965) responds directly to Williams' myth-of-passage (of time) paper, calling it "an interesting piece of science fiction"; in his paper Capek likened the block universe interpretation of Minkowski's spacetime to a giant refrigerator. In a play on Williams' title, "The Myth of Passage," Capek turned the tables and called the concept Williams espoused the "myth of frozen passage."

Capek felt that Williams was unaware of the nuances of special relativity; Capek, in fact, appears to be the first philosopher to have appreciated the fact that the temporal ordering of potentially causal events in both the future and the past of one observer is invariant for any other observer, whatever that second observer's motion may be. It is only in *elsewhere*, that region of spacetime that is not causally linked to an observer's *now*, that temporal ordering can be reversed as seen by some other observer—assuming that speeds are limited by the speed of light. For an analytical discussion of this issue, see Tech Note 4.

Williams responded to Capek in a paper printed immediately after Capek's in the same book. Williams' work was a rather poetic essay, but it failed to address the substantive technical points raised by Capek. Certainly it failed to change Capek's mind; ten years later he wrote (1975) that Donald Williams "is clearly unaware of the basic difference between classical and relativistic space-time." After Capek, the philosophers who wrote on spacetime did at least refer to that difference. For example, Le Poidevin (1990) makes the point that the proper task for philosophers is not to try to show that time has any particular topology (such as circular versus linear) by logical necessity, but rather to explore the consequences of its empirical, contingent properties.

Some years after Williams' poetic essays, his philosophical style was harshly dismissed with these words from a member of the "modern school" of philosophical thought, John Earman (1970b)—they are the opening passage of a critical "guide to spacetime for philosophers": "Space-time is the basic spatiotemporal entity. Many philosophers have mouthed this truth, but few have swallowed it, and very few have digested it. . . . [A]n appreciation of this truth is crucial to what is commonly referred to as the philosophy of space and time. . . . [I]n large measure the lack of progress in this area can be traced to the fact that philosophers have not taken seriously the corollary that talk about space and time is really talk about the spatial and temporal aspects of space-time." I take that to mean that Earman was telling philosophers they had better learn some physics! For more on the same theme, see also Earman (1970a).

What provoked Earman to write this was his perception that philosophers were not talking science when they wrote of space and time but rather were in the business of telling each other irrelevant stories and myths. The curious philosophical approach of "telling tales" reached its peak in the early and mid-1960s. The story of how this story telling came about is interesting, and one can easily find its influence in much of the philosophical literature on time travel. Space-time story telling seems to have started with a paper by the Oxford philosopher Anthony Quinton (1962), who argued that although there can be multiple, disjointed spaces (science fiction's "parallel worlds"), there can only be a single time. The issue here is not the validity of that assertion but rather Quinton's technique for arriving at it: myth construction. Though it was popular among philosophers of the "older" school, myth construction usually strikes those trained in the technical sciences as perhaps interesting (even physicists, after all, can enjoy a good fairy tale now and then), but certainly quaint and nearly always totally beside the point.

In his paper Quinton tells a fairy tale about how he thinks someone can live continuously in time and yet, via dreaming, be in two different spatial worlds—when awake in one world, the person is asleep in the other. He argues that this multispatial myth is plausible but that a search for an analogous multitemporal myth is doomed from the start. Another philosophical paper, Swinburne (1965b), rebutted Quinton by using a counter-myth about the "warring tribes of Okku and Bokku." That second story, the details of which are not important for us here, started a vigorous debate in the literature; see Skillen (1965), Swinburne (1965a), and Hollis (1967b), all of which presented their readers with even more stories.[35]

It was this continual spinning of hypothetical tales that caused Earman to write in his 1970b paper that "the procedure for arriving at answers to these questions [about space and time] adopted by Quinton and most of the other authors is, to say the least, a curious one: a story is told about a mythical land—usually called something like the land of the Okkus-Bokkus—and then we are asked what we would say if confronted by experiences like those of the Okkus-Bokkusians. As often happens with such a question, people have said all sorts of things, not all of which are interesting or enlightening."

Smart (1967) was even less gentle in his rejection of the philosophers' fairy tale approach to spacetime physics: "Swinburne, Skillen, and Hollis follow Quinton in inviting us to say what we should think in certain strange circumstances which they describe within common-sense language [as opposed to scientific terminology]. I must say that if I found myself in the circumstances which they describe (for example Swinburne's savage tribe disappearing when a medicine man waves his magic wand [thus "the land of Okkus-Bokkus" is an outrageous pun!]) I just would not know what to think. Probably I should simply conclude that I had gone mad, or at any rate madder than I normally am. . . . It looks as though these writers are inviting us to consider what we should say if we knew no science."

It is in Smart's essay, in fact, that we at last find the modern view of time and space, the issue originally raised by Quinton. Smart notes that the issues of space and time lose whatever philosophical mysteries they may seem to have when one thinks in terms of relativistic spacetime. As Smart correctly explains, "The reason why there could be two totally disparate space-times is simply the quite obvious one that two totally disparate four-dimensional spaces can exist within a suitable five-dimensional space. There is no difficulty in mathematical conceivability here. Now let one of these four-spaces be our own space-time world, and let the other four-space be more or less similar, in accordance with whatever story you wish to tell about it."

At about the same time that Smart's article appeared, one last apparent attempt at myth making was published by Hollis (1967a), this time about two men—named Peabody and Snooks—and a tribe of two-dimensional, live tiddly-winks! That would seem clearly to be an example of the reprehensible sorts of doings that so irritated Smart and Earman, but in fact Hollis was actually demonstrating the essential worthlessness of philosophical stories (so Smart erred in lumping Hollis in with the fairy tale philosophers). As Hollis wrote, "Whenever a human being produces an argument which opens 'Suppose I had 23 senses . . .,' 'Suppose I were God . . .,' 'Suppose I experienced objects extended in four spatial dimensions . . .,' we can protest that the argument is worthless. For in supposing that he has transcended our human point of view, he has also transcended the limits of our understanding." As Hollis amusingly concluded his paper, such opening sentences are the signatures of myths from "The Philosopher's Fairy Book." He is prepared to accept the failure of his message to convince many of his colleagues, however, and he says he is still waiting for the paper that begins "Twice upon a time in another space no distance in any direction from here . . ."!

Despite that concern, there *were* others besides Earman, Smart, and Hollis who understood that telling fairy tales just would not do. For example, one of the quotations that opens this chapter comes from Putnam (1967), which, like Williams' paper, argues for the block universe view of spacetime—but is written in the scientific language of Einstein, not the poetic prose of Williams. Putnam asks us to imagine two observers, one stationary (him) and the other moving at a very high speed (you). At some particular instant the moving observer (you) passes very close by (arbitrarily close) the stationary one (him). At this point of "closeness," the two of you share a common event in spacetime, so both of you experience the same present or *now*. But, argues Putnam, his future light cone and yours are tilted relative to each other (please note that you absolutely *must* read Tech Note 4 to understand this sentence). Thus, says Putnam, as he pulls the rabbit out of the hat, "It is well known that, as a consequence of Special Relativity, there are events [in spacetime] which lie 'in the future' according to *my* coordinate system and which lie in the present of . . . *your* coordinate system."

Because all of the events in the present of the moving observer (you) are clearly real, then (says Putnam) the future of the stationary observer, which contains at least some of those same events, is at least partially real, too. That is a pretty dramatic conclusion, no doubt about it, but Putnam is simply wrong on this. In Tech Note 4 it is shown that if an event is in the future of one observer, then it is in the future (not the now) of the moving observer as well. In an astonishing coincidence, Rietdijk (1966) appeared just a few months before Putnam's paper, and that was followed a decade later by Rietdijk (1976), in which much of the same theme and many of the same conclusions as Putnam's appear.[36]

Putnam's argument was quickly refuted by Harris (1968), which demonstrated that, despite superficial appearances, Putnam's reasoning is *not* consistent with special relativity. Specifically, Harris showed that distant simultaneity (here *distant* refers to the separation of two events) can be determined only after the fact—perhaps long after—and that the moving observer will not know the reality of all the events simultaneous with the "local now" he shares for an instant with the stationary observer *until* those events are in his (and the stationary observer's) past. Reitdijk apparently anticipated this rejoinder (and perhaps this is why Harris failed to cite Reitdijk as well as Putnam) when he wrote that "only an extreme positivism—'that which can not yet be observed does not yet exist'—can possibly withstand the conclusion" of a fatalistic block universe spacetime.

On that point, in fact, my personal sympathies lie with Rietdijk. Being unknown is not equivalent to being undetermined. On the other hand, Capek (1975) makes the cogent observation that if something is unobservable, then, as eventually came to be understood in the historical cases of the ether, caloric, and phlogiston, there is nothing to be gained by postulating its existence. On the *other* hand, Sklar (1974) has objected to Capek's position by claiming that to say an event does not exist because it is not happening at our time is as absurd as saying that an event does not exist because it is not happening at our location.

Thus Putnam felt a bullet whiz by while Reitdijk escaped—but not for long. A second refutation of Putnam appeared in Stein (1968), which leveled a verbal barrage at Putnam, but this time there was one for Reitdijk, too. The rebuttal was along the same lines as Harris'; Stein stated that Putnam and Reitdijk had misapplied special relativity. Unlike Harris, however, Stein was at times absolutely scathing: "Putnam . . . has used . . . a collection of principles of philosophical interpretation derived from dubious authority, subjected to no adequate critical examination, and in point of fact demonstrably inappropriate to the scientific context in which they are employed. . . . [T]hese defects seem to me to merit comment precisely as symptoms of a *prevalent* laxness in philosophical discussion, which is very much to be deprecated. Technical mistakes and errors of detail can be corrected . . . but the lowering of critical standards in philosophical discourse itself precludes understanding and is the death of philosophy."

Well, that seems a pretty harsh dismissal of poor Professor Putnam! Would no one defend him? (Putnam himself seems never to have replied.) For a year there was silence, and then a paper by Fitzgerald (1969) appeared, written in a tutorial, non-confrontational style that must have brought forth a sigh of relief from the beleaguered Putnam. Fitzgerald presented a summary of the "traditionally competing philosophical views about the ontological status of the future" and then followed that with analyses of how each such view transforms after reformulation in relativistic terms. Both Putnam and Stein are cited, but no mention is made of the extreme disagreement between Putnam and his critics. Fitzgerald then adds a little mathematics by invoking the Lorentz transformation (see Tech Note 3), but of the distant simultaneity issue *he does not say a word*.[37] The precise technical points central to the dispute between Putnam and his critics are ignored in Fitzgerald's paper.

A few years later a defense of substance by Weingard (1972b) finally appeared—at least it was sort of a defense. Weingard opened with "Hilary Putnam concludes that all events in special relativistic spacetime, whether past, present, or future, are equally real. . . . Although I believe this conclusion is correct, I think Putnam's argument is not." That is, Weingard rejected Putnam's specific arguments (and thus ducked Stein's specific rebuttals) and yet retained the view of spacetime as a block universe in which everything already is—the view expressed by Weyl and the view that Einstein also appears to have held when writing the Besso letter of condolence. Like all critics since, Weingard faulted Putnam for a failure to understand what distant simultaneity does and does not mean. He then used the arbitrary conventionality about the two-way versus one-way velocity of light to arrive at a conclusion that agrees with Putnam.[38]

In Fitzgerald (1985) we at last see that the idea that the relative simultaneity of events for an observer exists only outside the light cone of that observer, in *elsewhere*, has at last become part of the philosophical literature as written by philosophers. (Others also knew what was going on, such as Earman, but they were rare.) Indeed, that very point was explicitly stated even earlier by Godfrey-Smith (1979), who also mentions Reitdijk's error in reasoning.

The debate over the reality of the future continues to this day. In Zemach (1977), for example, we see a failure to understand Harris, Stein, and Capek because we find the author concluding with "Hence, it is nonsense to say that 'the past' or 'the future' [is] not fully real; each point-event in the Universe is past with respect to some . . . and future with respect to others; all are equally real." For the purposes of time travel, it is, of course, mandatory to accept the reality of past and future, an idea that Capek (1983) emphatically denied when he wrote that "fortunately, such fantasies have not the slightest basis in the physics of relativity." It is not clear why Capek used such a judgmental word as *fortunately*, as though something awful would otherwise occur, but in any case not all philosophers refuse to accept the equal reality of past and future. In Smart (1981), for example, we find a philosopher who doesn't refuse but who also admits that "it is very hard to convince those who are not professional philosophers that the future is real." As an example of how absurd he finds that to be, Smart asks us to "conceive of a soldier in the twenty-first century . . . cold, miserable and suffering from dysentery, and being told that some twentieth-century philosophers and non-philosophers had held that the future was unreal. He might have some choice things to say."

Putting aside the relevance (or the lack of it) of Smart's imaginary exercise, it is interesting to note that his scenario had been used in science fiction decades earlier, in "Soldier" (Ellison). In that story, originally published in 1957 (and available today on videotape as an episode from television's "The Twilight Zone"), a soldier from the far future is hurled back to our time by the beam energy flux of enemy weapons fire. The soldier tells tales of the utter horror of future warfare (and Harlan Ellison really outdoes himself with monstrous, graphic images of horror in his descriptions of telepathic "brain burners" and other terrors), and that generates a backlash to war. But even as peace engulfs the world, there are those who wonder whether the future *can* be changed—perhaps it *is* rigidly frozen into a Minkowskian/Capek refrigerator universe, into a Wellsian Rigid universe. Perhaps today's peace movement, led by the soldier of tomorrow, is precisely what leads to the future at war.[39]

The idea of a fixed future is also illustrated in the 1979 film *Time After Time*, in which H. G. Wells himself pursues Jack the Ripper (who has stolen the time machine—in this movie it is *Wells* who is the Time Traveller!) across time. When Wells reads a newspaper in modern-day San Francisco, he learns of the deaths of two of Jack's latest victims. Traveling back to a few days before the murders, Wells then sets out to prevent the killings. In both attempts, however, and in agreement with the block universe, he fails.

We find a softer version of the "bringing the fixed future into existence" idea in the mainstream novel *Woman on the Edge of Time* (Piercy), the story of a poor woman who is institutionalized in a mental hospital. This powerful novel, which on the surface tells the tale of a woman in mental contact with the world of 2137, is really a metaphor about the dichotomy between those in power (the male doctors who perform surgical experiments on her brain) and those who aren't. When she realizes the utopian world of the future she has experienced has no such dichotomy, she tries to help bring about that very future by destroying the powers that are suppressing her in the present: She murders her doctors by putting poison in their coffee!

And as a final fictional example related to the reality of the future in the block universe, consider "Stalking the Sun" (Eklund). In that story two time travelers from the past, hunting men a half-million years in the future for sport, argue this very issue. When one of them wonders about the morality of killing future men, he quickly rationalizes it, saying, "This time, it isn't a real time. It's the future, but the future does not truly exist. To kill here is not real." His companion isn't so easily convinced, however. He replies, "The timewarp was discovered twenty-five years ago. In that time, this time [the future] has not changed. My opinion is only my opinion. But I think this is it. The future. The only future. I think we're moving toward this no matter what we do. This is it."

Arguing the opposite view is "Nobody Here But Us Shadows" (Lundwall), a story in which one can go "Upside" (into the future) and examine alternatives ("probability lines"). One only looks, however, and never interacts. Indeed, the time machine is locked from the outside to prevent the traveler/observer from exiting. This is done because on one of the first such trips the observer *did* take something back, a girl. Years later, when an observer journeys upside along that same probability line, he finds it has utterly vanished. That once-possible future was destroyed by the removal of what was then only a possibility (the girl) and by her injection into the present as an out-of-time fact.

An eerie twist on that last story idea appears in "A Few Minutes" (Janifer), which describes a time machine that lets one explore what would have been the future if different choices had been made in the unchangeable past. Even more flexible is *Lightning* (Koontz), in which a time traveler can visit his own future and then make earlier (but still in his own future) changes to "correct" events he doesn't like. Koontz makes some fairly slippery arguments concerning time travel. For example, his time traveler can change his own future but not his own past, and it doesn't matter that for someone "already" living in the future the past *is* changed. But hey—this is fiction, not physics!

As for the question of the persistence of the past, I find particularly evocative a passage from Grant Allen's Introduction to his 1895 sleeping-into-the-future novel, *British Barbarians:* "I am writing in my study on the heather-clad hill-top. When I raise my eye from my sheet of foolscap, it falls upon miles and miles of broad open moorland. My window looks out upon unsullied nature. Everything around is fresh and pure and wholesome. . . . But away below in the valley, as night draws on, a lurid glare reddens the north-eastern horizon. It marks the spot where the great wen of London heaves and festers." It *is* tempting to imagine Allen somehow still there in his study in 1895, and of heaving and festering late-Victorian London, too, with H. G. Wells himself in the middle of it, still reading the first rave reviews of *The Time Machine.*

Even if events in all their infinitely infinite spatial and temporal web—what Gold (MDT) calls the "world map"—are really laid out in a four-dimensional block universe, there still remains the great mystery of why we see them unfold in the particular sequence that we do. Why not in reverse order? Why, indeed, do we see what we call *time* run from what we call the past to what we call the future, and, indeed, what do we really mean by *past* and *future?* As we'll see in the next chapter, these are not easy questions, and nearly everybody who has thought about them believes we are not yet even close to knowing the answers.

On that gloomy note, it is appropriate to end with a few more words from Saint Augustine's *Confessions,* words that followed the second quotation at the opening of this chapter:

> I confess to you, Lord, that I still do not know what time is. Yet I confess too that I do know that I am saying this in time, that I have been talking about time for a long time, and that this long time would not be a long time if it were not for the fact that time has been passing all the while. How can I know this, when I do not know what time is? Is it that I do know what time is, but do not know how to put what I know into words? I am in a sorry state, for I do not even know what I do not know!

Amen.

3

The Arrows of Time

On a microscopic level there is *no preferred* direction for time. The equations of motion don't give a damn whether time moves forward or backward.

—Eisenberg (AO)

If space is "looking-glassed" the world continues to make sense; but looking-glassed time has an inherent absurdity which turns the world-drama into the most nonsensical farce.

—Eddington (1929)

Now we are also in a position to shed some light on a problem being frequently discussed nowadays: the possibility that the direction of the flow of time is reversed. This subject used to be considered fit for science fiction only . . .

—Zwart (1972)

Of all the problems which lie on the borderline of philosophy and science, perhaps none has caused more spilled ink, more controversy and more emotion than the problem of the direction of time. . . . [T]he main problem with "the problem of the direction of time" is to figure out exactly what the problem is or is supposed to be!

—Earman (1974)

Talk of the flow of time or the advance of consciousness is a dangerous metaphor that must not be taken literally.

—Smart (1954)

The Language of Time Travel

For the phrases *flow of time* and *direction of time* to have any objective meaning at all, it must be somehow possible to identify a difference between past events and future ones. The special moment at which that distinction occurs is known as the *now* or the *present,* and as events make the transition associated with that distinctive difference between past and future, we say that the now moves or flows. Philosophers—and physicists, too, who after all are human beings with human senses like everybody else—call this common feeling that we all have, of the passage of time, the *psychological arrow of time.*

In principle, so it would seem, we can achieve perfect knowledge of what has happened but only imperfect prediction of what might happen. This seems to be at least a start at being able to tell past from future. And, in fact, the nature of the distinction between the two intervals of time seems obvious: We can remember past events, but not future ones! As philosophers have so nicely put it, events in the past have formed *traces,* such as skulls, footprints in the sand, fossilized skeletons, surgical scars, photographs, taped recordings, carved stones, and the like, whereas future events appear not to have formed traces. But is that necessarily so? Is it impossible for future events to create traces? The common-sense answer is *yes,* certainly, there must be a temporal asymmetry in trace formation because of cause and effect; that is, traces are the effects of prior causes. That line of reasoning leads quickly to the fundamental issue of causation.

Part of the problem we have with backward time travel and cause and effect is our language. The distinct and separate concepts of the temporal ordering of events, and of causality, have become merged in everyday thought. It is considered obvious to modern minds that if event A causes event B, then A must happen first. Most people find it virtually inconceivable that it could be otherwise. There is, however, at least one historical example of a similar merging of concepts that is parallel to our modern mixing of time order and causality—an example that shows how an issue could seem obvious and natural to the minds of the period and yet today seems confused, odd, peculiar, even laughable.

I quote from Csonka (1969), who presented this example in a paper on advanced (i.e., inverted causality) effects:

> Ancient Egypt was an essentially one-dimensional country strung out along the Nile, which flows from south to north. The winds were conveniently arranged to be predominantly northerly. To go north, a traveler could let his boat drift, while with a sail he could move south against the slow current. For this reason, in the writing of the ancient Egyptians,

"go downstream (north)" was represented by a boat without sails, and "go upstream (south)" by a boat with sails. The words (and concepts) of north-south and up-downstream became merged. Since the Nile and its tributaries were the only rivers known to the ancient Egyptians, this caused no difficulties until they reached the Euphrates, which happened to flow from north to south. The resulting confusion in the ancient Egyptian mind is recorded for us to read today in their reference to "that inverted water which goes downstream (north) in going upstream (south)."

An interesting example of how we exhibit a similar confusion concerning time was commented on by Francis Bacon as long ago as 1605, in his *The Advancement of Learning*. He pointed out that although each generation thinks of itself as living through days of youth (and thinks of the past as "the good old days"), it is we who are really the "ancients of the world" (to paraphrase Tennyson), the most distantly removed from the start of the world of all who have lived. We make the error of reckoning time *backward from our present,* rather than from the instant of the beginning.

Often we can work our way free of the difficulties we create with language, but only through common agreement. For example, the chairman of the board calls a meeting to order with mixed tenses by declaring, "The meeting *will* take place *now*" and then saying, at the end, "We will meet again *next* month, *same* time." We all know what these sentences mean, but only by our cultural heritage and not by the process of applying logic. The language problem causes trouble for both fictional time travelers and the physicists/philosophers who study the possibility of time machines and time travel. So—reader beware!

Does Time Have a Direction?

The answer seems obvious. *Of course* time has a direction; everybody knows it flows from past to future. There is a curious language problem here, however, because we also like to say that the present recedes into the past, which implies a "flow" in the opposite direction, from future to past. Given this uncomfortable situation of snarled syntax, and returning to the question of the previous section, can we at least distinguish past from future, whichever way time might flow? That at least would be a start, but even this simplified question is not so simple to answer. The central debate here is between what is called *objective time* (the idea that time really does flow) and *mind-dependent time* (the belief that time's flow is simply an illusion, an artifact of our incomplete perception of reality).

The idea of time flowing is a popular one, and it repeatedly appears in the time travel literature as the "river of time" or the "ocean of time." The deep psychological appeal of this sort of "water language" has, not surprisingly, attracted the attention of philosophers. We can find one of the earliest expressions of the concept in the *Meditations* of the second-century A.D. Roman emperor and Stoic philosopher Marcus Aurelius, who wrote, "Time is like a river made up of events which happen, and a violent stream; for as soon as a thing has been, it is carried away, and another comes in its place, and this will be carried away, too."

A most interesting essay on why such metaphors often seem so intuitively appropriate is offered by Smart (1949). Smart points out that the seductiveness of these metaphors is sufficiently great, in fact, that we often find them in scientific writing as well. For example, Newton wrote specifically of time flowing, as discussed in more detail in Tech Note 1. As for why such metaphors have a powerful grip on our imaginations, I think we need look no further back than to Kant. As he wrote in *Critique of Pure Reason,* "Time is nothing but the form of inner sense, that is, of the intuition of ourselves and of our inner state. . . . [B]ecause this inner intuition yields no shape, we endeavor to make up for this want by analogies." And what better than a rushing stream of water to represent our feeling of time rushing by?

Still, no matter how intuitive such metaphors may be, they can still easily befuddle us as well. To quote Smart, "Time a river! A queer sort of river that. Of what sort of liquid does it consist? Is time a liquid? A very peculiar liquid indeed!" A classic paper by Donald Williams (1951a), discussed in the previous chapter with respect to the block universe, rejects the liquid view of water. In the course of his discussion, Williams presents a truly staggering collection of entertaining examples of 'time as metaphor,' of which I repeat here only a few. Time flies, goes, flows, marches, rolls. The evolution of our lives is like "a moving picture film, unwinding from the dark reel of the future, projected briefly on the screen of the present, and rewound into the dark can of the past." For some, time is a snowball with the past in the center, and ever new presents accrete around it (presumably as the snowball rolls down the hill of history)!

Charles Nordmann opened and closed his 1925 book *The Tyranny of Time* with the following gloomy but too true summary of the overwhelming sense we all have of the inexorable, one-way "flow" of time. (The ellipses in what follows denote over 200 pages!) "Nothing can equal the bitter sweetness of dreaming on the banks of Time, that impalpable and fatal river strewn with dead leaves, our wistful hours carried down stream like rudderless wrecks. . . . In the eternal wave which rocks us, carries us along, and soon swallows us up, there is no rock to which we can fasten our frail barques; the very buoys we put out to measure our course are only floating mirages; and on the mysterious foundation of things our anchors slide along and fail to bite." A young person sees time, from Nordmann's perspective, as an ocean on which golden mornings arrive like waves from the future, whereas for an older person, liquid time is a nightmare flood, a swollen black torrent sweeping him first into the yawning abyss of the past and, ultimately and finally, into the eternal silence of the dark grave.

The same depressing sentiment is captured in the famous opening words to the nineteenth-century English poet Austin Dobson's "The Paradox of Time." That poem turns the metaphor of moving time upside down, in a way that actually predates Dobson; Dobson freely admitted that his opening was inspired by a nearly identical couplet attributed to the sixteenth-century French poet Pierre de Ronsard:

> Time goes, you say? Ah, no!
> Alas, Time stays, *we* go.

The metaphor of time as a flowing river was ready-made for the early science fiction storytellers, such as Caltech math professor Eric Temple Bell. His *The Time Stream* began to appear in December 1931 as a serial in the science fiction magazine *Wonder Stories,* and Bell, writing as John Taine, made great use in his novel of the idea of time as a flowing stream, a stream in which one could swim into either the future or the past. There is strong evidence that Bell wrote the book in July of 1921 but was unable to find a publisher for a decade, so odd was the premise. By the time of its publication, others had gotten into print first.

The 1930 story "The Time Annihilator" (Manley and Thode), for example, played with the erosive nature of time in a dramatic way. As two time travelers speed into the future to rescue a friend, one of them describes the scene for us: "We huddled together in the whirling time-girdling machine, cutting through the years as a ship's prow breasts surging waves. I could not help but think of the years as waves, beating in endless succession on the sands of eternity. They wore all away before them with pitiless attrition. Time seemed to eat all with dragon jaws."

Manley and Thode's watery image of time was taken a step further three years later in "Wanderers of Time" (Wyndham), a story in which a large number of adventurers from all across time find themselves stranded at precisely the same place. One of them offers his theory of what is behind this remarkable coincidence: They all have faulty time machines, like faulty boats, and all have hit the same snag on the "river of time." As he explains, "You may turn boats adrift on a river at many points, and they will all collect together at the same serious obstacle whether they have traveled a hundred or two miles. We are now at some period where the straight flow of time has been checked—perhaps it is even turning back on itself. . . . [W]e have struck some barrier and been thrown up like so much jetsam." More recently, in a story of time travel via a storm (as in de Camp's *Lest Darkness Fall* and in *The Final Count-down* movie), a giant water wave literally does the deed in "Of Time and Kathy Benedict" (Nolan): "There was no longer any doubt in her mind: the wave . . . had carried her backward eighty years, through a sea of time."

The "flow" of time does have its critics, of course. The philosopher Max Black argues (1959b) that questions about the direction of time are meaningless because there can be no direction to something that (he asserts) does not flow. His reasoning is that if time does flow, then he ought to be entitled to ask *how fast* it flows. That requires, in turn, a metatime or supertime for measuring the flow rate of "ordinary" time. But because supertime must flow too, we would then need a super-supertime, and so off we trip into what would appear to be the black hole of a McTaggert-like infinite regress.[1] As we noted in Chapter One, Dunne (1958) tried to explain precognition in terms of an infinite regress of times, an attempt that attracted nothing but polite rejection from philosophers; see Broad (1935), for example. Not so polite, in fact, was Zeilicovii (1986), who bluntly stated the general position of most philosophers on infinite regresses of times, opining that "the very idea of super (or hyper)-time is indeed repulsive in its redundancy and its aroma of dilettante physics."

A hierarchy of hypertimes has not bothered other analysts, however, and an entire subfield of specialty among philosophers (and some physicists, too) in time analysis is that of what is called multidimensional time. Indeed, Webb (1960) sarcastically rejected the infinite regress complaint as a valid objection—he called it "a crushing and unanswerable position" but actually meant just the opposite—and stated that it is not at all clear (at least, not to him) why supertime must flow. After all, Webb said, we measure the flow of a river with respect to its banks without requiring that the banks themselves flow. That actually strikes me as being a point worth debating, but I have not been able to find any mention of it in the later philosophical literature. The idea of multiple time dimensions is particularly attractive for one sort of time travel (I'll take it up at the end of this chapter) that enjoys far more popularity among science fiction writers and philosophers than it does with physicists.

Black's objection to talk of time "flowing" was based, at least in part, on the fact that there are uses of the word *direction* that are not directly tied to something flowing. For example, consider the statement "He is facing in the direction of north." Black argued that that is mere pointing and is not at all the same as moving north. He then dismissed the possibility of there being any meaning to the direction of time, writing that making an analogy of time "with a sign-post or an index finger is too far-fetched to be worth considering." That claim is, of course, an affirmation of the myth-of-passage view mentioned in Chapter Two that was made famous in Williams (1951a).

Not just philosophers have objected to the idea of time flowing. In his novel *October the First Is Too Late*, which deals with a world in which different parts of Earth simultaneously experience widely different eras of the past, the British cosmologist Fred Hoyle calls the river of time a "grotesque and absurd illusion," and a "bogus idea." Another more recent fictional work that agrees with Hoyle's non-moving image of time appears in *Roadmarks* (Zelazny). In that novel we read of "the Road" along which characters can travel but which itself does not move; exits from the Road lead to the various centuries. (This imagery was no doubt motivated by the central role of the automobile in our modern civilization.) *Roadmarks* is a clever work, with many allusions to the paradoxes of time travel, but the author's explanation of the Road's origin as having been built by *dragons* (!) greatly undermines its interest for physicists.

Cause and Effect

The philosophical literature is full of discussions about potential causal relationships between events. One of the more famous of these discussions illustrates that cause and effect can be pretty slippery concepts. Kim (1974) asks what at first appears to be an almost trivial question: Did Socrates' death cause the widowhood of Xanthippe? The quick and easy answer is "Of *course*—she was his wife and it was his death that *causes* us to say she was then a widow. What could be more obvious?" Kim, however, provides some interesting commentary that might make you reconsider, or at least to become aware of how very different are the questions that concern philosophers and physicists.

Suppose we agree that there are two events to be considered, Socrates ceasing to live and Xanthippe becoming a widow. Those events occurred at different places (in prison, and wherever Xanthippe happened to be, which was almost certainly *not* the prison). Kim argues that "the two events occur with absolute simultaneity . . . [and so] we would have to accept this case as one in which causal action is propagated instantaneously through spatial space." This is an interesting argument, but perhaps Tech Note 1 on the *relativity* of distant simultaneity weakens Kim's assertion. And for the conclusion which he draws from that assertion, just *what* is propagating instantly? If it isn't mass-energy (as "widowhood" would appear not to be!) then special relativity isn't bothered. And finally, as Kim himself admits, is Xanthippe becoming a widow *really* an event distinct from the event of Socrates dying, or is it just another way of saying the same thing, that Socrates died?

The central puzzle of time travel to the past is its apparent denial of causality—that is, its denial of the belief that we live in a world where every effect has a cause and that the cause always happens first. *First* we flip the switch and *then* the kitchen light comes on. It is *never* the other way around. So deeply embedded is the temporal ordering of cause and effect in our feelings about how the world—and all the rest of the cosmos—works that the philosopher J. L. Mackie calls causation the "cement of the universe." Without causality, says Mackie, everything would come unglued and fall apart. For example, when electrical engineers design electronic systems that they intend actually to construct (as opposed to doing mere "paper designs"), they insist that the design be a *causal* one. By that they mean the system must have no output before an input is applied; that is, the system must not be able to anticipate (foresee) the future, the application of an input. Now that might all seem self-evident, but there are some subtle problems to consider.

For example, it has become almost a cliché to say that nothing can go faster than light; that is what physicists mean by *relativistic causality*. No cause can produce an effect at a distant location sooner than the time lapse required for a light pulse to make the trip. Classical mechanics, however, the science of Newton's laws that engineers use all the time, is *not* relativistically causal. Push the left end of a rigid rod, for example, and the right end moves *instantly*. Most of the time the lack of this form of causality causes no problems, but the fact remains that the mechanics all engineers learn first in school is flawed on a fundamental level. A rigid rod is an impossibility in Einstein's mechanics.

Indeed, I occasionally daydream about how, after a discussion of causality, a traditional engineering professor would respond if challenged on this issue by a bright student. Causality might not look so obvious, after all, if such a student stuck her hand up in class and said "Professor, you've told us that everything that happens in nature is due to a cause. That what we see happening all around us, as the world unfolds, is the domino-process of cause-effect-cause-effect, and so on into the future. But suppose, Professor, that at some instant, somehow, every particle in the world suddenly reversed its velocity vector. Wouldn't that mean, given the time-reversible nature of the classical equations of motion, that the world would then run backward in time along the same path it had followed up until the instant of reversal? Wouldn't that mean that what was effect is now cause and that what was cause is now effect? And if cause and effect can change roles like that . . . well, Professor, what do our words *mean?*"

An amusing, and instructive, cartoon illustration of the idea of reversing all the velocity vectors in a system appeared on the cover of the November 1953 issue of *Physics Today*. That issue contains an article on the 1949 nuclear magnetic resonance experiments by E. L. Hahn, which in a certain sense dealt with just such reversed systems. In that illustration a group of runners on a circular race track begin at the starting line in a coherent state, i.e., all lined up together. Then, as they run around the track at various speeds, they gradually spread out into what appears to be an incoherent state. But that incoherence is an illusion because if, at a prearranged instant (signaled in the cartoon by a pistol shot), they all turn around and run in reverse, they will all arrive back at the starting line *together*. The initial coherence of the runners was actually never lost, despite the superficial appearance of disorder, and the coherent state could be recovered at any time by a reversal of velocity vectors.

An almost magical application of velocity vector reversal is used in the technique called optical phase conjugation, discussed in Giuliano (1981). That is a process to "time-reverse" the severe distortion that beams of light can suffer during atmospheric propagation; for example, by effectively reversing the velocity vectors of photons, one can remove the turbulence blurring induced in satellite pictures made from light moving *upward* from the Earth's surface. This actually works, so perhaps my hypothetical student has asked a good question.

Let me immediately short-circuit one possible answer our beleaguered professor might give in desperation, a response based on the idea that somehow the equations of physics are not all time-reversible. In fact, it was discovered decades ago that in certain very rare, fundamental particle decay processes (involving the neutral K-mesons, or *kaons*) there is the hint that perhaps nature can indeed distinguish between the two directions of time. In particular, kaons violate what is called CP-symmetry, and so the so-called TCP theorem says T-symmetry must also fail (see Note 11 for Chapter Two). So important was that discovery—see Christenson *et al.* (1964)—that the principal investigators, James Cronin and Val Fitch, received the 1980 Nobel Prize in physics for their work. Later, Casella (1968, 1969) reported on the direct observation of the failure of T-symmetry, thereby providing experimental evidence in support of the TCP theorem. In an astonishing example of literary prescience, the use of K-mesons in a machine for affecting the past, and therefore the present, is mentioned in the story "Target One" (Pohl). Published in 1955, that tale appeared years before the peculiar, T-symmetry-violating decays were first observed!

Could K-mesons account for the physical processes that we see evolve in time in one direction but not the other? As Hurley (1986) puts it so nicely, "The decay of the neutral K meson is not time-reversal invariant; perhaps it is this ubiquitous meson which is responsible for the cream diffusing uniformly throughout our coffee in the morning. Possibly, but again this conjecture cannot account for the computer models [of diffusion processes that, like cream in coffee, also display a bias for one temporal direction over the other] which have no neutral K mesons." Still, the tiny chink that K-mesons appear to have made in the once-solid rock of time direction indistinguishability is an active area of research and speculation.

In general, of course, even with such a chink, the fact that the classical laws appear to be insensitive to a direction of time, whereas the real world—which seems in no way dependent on the arcane properties of kaons—seems distinctly asymmetric, is a puzzle of the first rank. As Earman (1969) wrote, "The Universe seems asymmetric with respect to past and future in a very deep and non-accidental way, and yet all the laws of nature are purely time symmetric. So where can the asymmetry come from?" Lots has been written on that question; Hurley (1986) says, "There are few paradoxes which have been resolved so often as the time-asymmetry paradox." In other words, the question is still an open one that continues to be debated.

For example, Hutchinson (1993) discusses some curious mathematical examples he interprets as meaning, in the context of classical mechanics, that there are physical systems that are temporally irreversible *in principle*. A reply by Savitt (1994), however, argues that Hutchinson has, at most, shown only that classical mechanics is perhaps not deterministic. And that, Savitt argues (convincingly, I believe), is not equivalent to showing a failure of time reversibility. There is, in fact, powerful experimental evidence that, with the rare exceptions of kaons, the classical laws of physics (including general relativity and quantum mechanics), *are* time-reversible.

Perhaps the most compelling of such evidence comes from the *reciprocity theorem* that electrical engineers routinely use when designing radio antennas. The theorem is easy to state. Suppose two electrical engineers, Bob in Boston and Lois in Los Angeles, send radio signals to each other. Bob sends his messages by exciting his antenna with a time-varying electrical current, which thus launches electromagnetic radiation into space. Lois' distant antenna intercepts some of this radiation, which then creates a (very tiny) signal current in itself.

The reciprocity theorem then says the following: Suppose Bob makes a tape recording of his excitation signal and mails it to Lois. Lois then plays Bob's tape back as the excitation to *her* antenna so as to create the same current that Bob did in his; then the signal current induced in Bob's antenna as it intercepts her launched radiation will be the very same (very tiny) signal that Lois measured in her antenna as a result of Bob's transmission. This result is completely independent of the details of the two antennas, which can be totally different in design, and it is independent of the details of the propagation path between Boston and Los Angeles as long as those details don't change with time. The reciprocity theorem is true—it can be *measured* to be true as accurately as one wishes to perform this experiment—because of the reversibility of physics right down to the electronic level. The answer to the professor's problem of explaining the reversal of causality to his student has not yet been found in the known laws of physics.

Now, to make things even more interesting, consider the problem of mutual or *simultaneous* causation. This can lead quickly to several interesting questions. For example, when a ball sits motionless in a depression on a pillow, is the ball motionless because it sits in a depression, or is there a depression because the ball sits motionless? When two leaning dominoes, A and B, hold each other, is A nearly upright because of B or is it B that is nearly upright because of A? When two children bob up and down on a see-saw, whose motion is the cause and whose is the effect? There are other puzzles, too, that involve mutual causation.

For example, causation is usually thought to be transitive: If A causes B, and if B causes C, then A causes C. But if A and B are mutually causative, then "A causes B" coupled with "B causes A" leads to "A causes A" (and "B causes B"). That is, mutual causation seems to imply *self*-causation! Except for those theologians who like this kind of conclusion (it lets them answer the question "Who made God?" with "He made Himself"), hardly anyone likes self-causation. But how do we avoid the conclusion that perhaps the mutual causation of two leaning dominoes represents an experimental proof that God could have made himself? This is certainly outrageous stuff, but don't you wonder how our poor professor would respond if asked?

That last example is actually a far more esoteric one than we need to illustrate how our ordinary, everyday concept of cause and effect can be turned inside out by going only a little bit beyond the routine. Consider, for example, the problem of the data processing of recorded time signals, such as information written onto magnetic tape or floppy disk. Typical applications that produce such recordings include the strata-probing seismic echoes from dynamite explosions set by oil exploration geologists; arms control compliance-monitoring stations that listen for the acoustic rumbles of both earthquakes and underground nuclear test—and then try to tell one from the other; and the gathering by various military intelligence agencies of turbine shaft/propeller noise signatures emitted by different kinds of submarines. In each of those situations, the raw information is recorded and later processed with a certain degree of unhurried calm and leisure. That pool of oil, after all, has been underground for several hundred million years, and waiting a few more days or weeks for a computer analysis of the explosion echo isn't going to make much difference.

Such after-the-fact processing of recorded data is said to be done "off-line, in non-real time." When we play the tape or disk back in the lab, however, we can do all sorts of neat things, like speed up the playback (make time "run fast") or slow it down (make time "run slow") or even play it *backwards* (make time "run in reverse or backward"). For various technical reasons, generically called spectrum shifting, such tricks are often quite useful. Now, the way we retrieve magnetically recorded information is, of course, to run it through a playback machine with a "read head" that senses the magnetic flux variations. The electrical signal produced by the read-head is just like the original signal and in fact we can pretend we don't know it is really coming off a tape or disk but rather believe that it is the original signal. For the new high-quality digitally recorded music tapes and disks, in fact, it is virtually impossible to distinguish the original from the playback.

Now, suppose we construct our playback machine with *two* read-heads, the new head sensing the recording slightly before the old head does. The two heads produce the same electrical signal, of course, but the signal from the new head is *ahead* in time compared to the signal from the old head. The new head is "seeing the future" of the old head! We can use these two signals, the old head representing "now" time and the new head representing "future" time, to build real systems that are *not* causal. The causality violation occurs in non-real time, of course, not *our* time, but no matter; some absolutely astonishing signal processing can be achieved this way. The universe is about fifteen billion years old, and pretending that time has shifted a few milliseconds or so doesn't seem to be doing too much violence to reality.

Two heads are often used on radio call-in talk shows to catch inappropriate remarks from intemperate callers and prevent them from being broadcast. A short time-delay is introduced by first recording remarks "live" on tape, and then a few seconds "upstream," a read head regenerates the remarks for broadcast. A five-second delay is generally sufficient, so what is heard on a radio receiver *now* actually occurred five seconds in the *past*. A caller can get terribly confused if he does not turn his own receiver off, because one ear hears the present on the telephone while the other ear listens to the past over the radio. Two intriguing fictional treatments that explore just how confusing "living out of sync" might be are "Man in His Time" (Aldiss) and "The Man Who Saw Too Late" (Binder). The first is the story of an astronaut who returns from a trip to Mars and finds he is 3.3077 minutes ahead of everybody else; the second tells us what it might be like to have a three-minute delay in just your vision. The 1956 British film *Timeslip* develops this same idea, with an atomic scientist's perception advanced seven seconds into the future as the result of an accidental radiation exposure. The three fictional characters in these tales all become horribly confused and disoriented.

The mixing of the concepts of temporal ordering (that of before and after) and causality is also a source of potential confusion. Despite my previous words, it is not necessarily obvious that it is proper to reason that if A causes B, then A happens first. That was Immanuel Kant's view, but David Hume thought the reasoning should go the other way: If A happens before B (and if A and B are causally linked) then A is the cause and B is the effect. It is all too easy to fall into the trap of the Egyptian sailors and think that one interpretation implies the other.

Backward Causation

All of the previous material has fueled countless arguments about what is called *backward, reverse,* or even *retro* causation. What is generally meant by *forward* causation is, of course, that any event that occurs at time t is caused by events that all occurred at an earlier time. Backward causation says that at least one of the causing events occurs *after* time t—backward causation is clearly a close relative of time travel. Indeed, Brown (1992) uses the terms *time traveler* and *retro-causal engineer* interchangeably. The topic, understandably, is at the root of many hot philosophical debates, though not everyone thinks those debates are illuminating. For example, Earman (1976) writes, "Causation as a topic of philosophical discussion refuses to die. Each year, books and articles on causation continue to pour forth. Of course, all this activity may simply be a symptom of the necrophilia that infects so much of philosophy." Why does Earman take such a harsh position? He offers, as one reason, his disdain for the common philosophical proof of the impossibility of backward causation: By definition, a cause is always before its effect. Yes, that's the entire "proof." One can, of course win *any* argument by defining the answer to be what it is that you wish to believe.

More interesting, and certainly more pertinent to time travel, is the argument that if backward causation were possible, then one could change the past, but that cannot be done because the past is dead and gone and thus unchangeable. That does seem to be a pretty solid argument against backward causation—it appears, for example, in Mellor (1987)—but Earman rebuts it by pointing out that the very same logic could also be applied to the future, and thereby the usual uncontested forward causation would also denied. That is, one could argue that whatever the future will be, *will be* (literally "by definition"), so one cannot change the future. A similar argument appeared even earlier in Smart (1958), where we find "suppose that someone says 'I can change the future. I can do *this* or I can do *that*.' Well, then, suppose that he does *that*. Has he changed the future? No, because doing *that* was the future."

The reversal by backward time travel of the "usual" causal order of events is beautifully illustrated in the romantic love story "The Dandelion Girl" (Young). In that gentle tale a man on vacation by himself, without his wife along, meets a young lady—and they fall in love. The man loves his wife, too, though, and he realizes even as the young lady leaves him for the last time, never to return, that it is all for the best. But she really hasn't gone that far away from him, as we soon discover. She is a time traveler from the future, and after leaving him she goes even further back in time, back an additional twenty years. She does this because she has learned that he met his wife twenty years earlier, and so she goes back to be *that* woman! Thus the usual causal order of the two events "a long marriage" and the "pre-marriage courtship" has been reversed (if we accept the fact that the man doesn't remember what his wife looked like when they married).

Actually, our everyday uses of cause and effect are not nearly so straightforward as one might think, even when they are under far less stress than backward causation and time travel inflict. Consider, for example, the endless problems that are easy to imagine in the legal world. If a man falls off the roof of a ten-story building and is electrocuted as he plunges through power lines while still twenty feet above ground, was gravity or electricity the cause of death? Or was it both? If an athlete breaks a leg during a sporting event and later dies under anesthesia at the hospital, what was the cause of death—the accident on the playing field or the accident in the operating room? If a man with a paper-thin skull dies after being accidently struck on the head by another who, in turn, was bumped into the first one by a third individual, who (or what) was the cause of death? As demonstrated by these examples and in Mackie (1992), one clearly does not have to discuss time travel to get into serious trouble about cause and effect. But *with* time travel, and the resultant backward causation, things can become even more perplexing. For example, we normally think it foolish to prepare, now, for an event that has already happened, but the prudent time traveler about to visit an ice age would be wise to pack a fur coat before getting into his time machine!

Some philosophers have presented interesting but somewhat odd (in my opinion) arguments for why they believe in backward causation. Consider, for example, the light bulb example in Nerlich (1979), who credits Taylor (1964) for his inspiration. At some time in the past a light is on. Then, later, at *now,* we turn it off. Nerlich asks us to think of this as an event *now* that causes the light to *have been* on. Clearly the "have been" refers to a time before *now,* so we have (claims Nerlich) an example of doing something *now* that makes an earlier thing have happened. If that leaves you a bit stunned, then try an example you can find in Mavrodes (1984). Imagine that you have every intention of going to Toronto next week. Then, later, you change your mind and stay home. Your later act (changing your mind) prevented God from *earlier* foreknowing that you would go to Toronto! Mavrodes fails to say why he doesn't think an omniscient God would have known you were going to change your mind. Neither of these examples gets to what is really meant by backwards causation via time travel to the past. You can find more discussions of the same sort in Freddoso (1982) and Fischer (1984), but I doubt that many physicists would be much impressed by any of it.

Forrest (1985) provides, I think, a good start at explaining why so many other philosophers (and not just a few physicists) have adopted the "common sense" position of rejecting backward causation. As he writes, "Part of the answer, no doubt, is a confusion between affecting and altering [the past—a topic treated at length in the next chapter]. We cannot alter the past. But then we cannot alter the future either, although we can affect it. However, I take the common-sense rejection of backward causation to be, for the most part, quasi-empirical. It is based on a thought experiment. Think how you would set about affecting the past. By building a time-machine, perhaps? But how would you build one? We have no idea how to start. Yet, by contrast, we can work out how to affect the future . . . we just move our bodies." But, as Forrest also argues, if we accept that we can't *change* the past (which means there is no way we could actually observe backward causation), then there still exists the possibility that past events were as they were because of events in the future.

Are there actual phenomena that justify a belief in the possibility of effect before cause in real time (not just in tape recorder time)? The only example I know of, and a controversial one at that—but see also Rietdijk (1978, 1981, 1987) on claimed retroactive effects in quantum mechanics—is a theoretical result from Dirac's formulation of electrodynamics. Classical theory models electric charges as point objects of zero size, which causes problems when one tries to calculate certain details, such as the total field energy of a single electron: The answers come out as infinity. In an attempt to find more reasonable—that is, finite—answers to the calculations, Dirac returned to the idea (originated by Lorentz) of viewing charges as extended objects in space, while retaining the validity of Maxwell's equations right down to a point. To calculate how such objects will behave mechanically, however, one has to include what are called the self-interaction forces, such as the force that one side of an electron exerts on the other side.

When it is all worked through, one arrives at Dirac's third-order differential equation of motion, which involves a force term proportional not to the first time derivative of the velocity (which is the acceleration), but rather to the second derivative. This force is proportional to the first derivative of the acceleration, and is a quantity of direct interest mostly to the designers of automobile suspensions, who call it the *jerk*. There is no other force in physics, at least not in Newtonian physics, that shows that sort of dependence, and there are some curious consequences. For example, in Dirac's theory an electron experiencing no external force can still continually accelerate, exhibiting what is called the "runaway solution."

Dirac (1938) showed how the runaway solution can be eliminated by picking a particular value for what until then was an arbitrary constant of integration. But that trick causes, in turn, a second problem called pre-acceleration. That is, if a charged particle is subjected to an external disturbance (Dirac considered a passing pulse of electromagnetic radiation), then the charge will start to move *before* the pulse reaches the electron's position. Now *that* does seem to be a pretty clear example of backward causation. The time interval during which the so-called *pre*-acceleration occurs is very brief, on the order of the time it takes light to travel across the width of the spatially extended charge (about 10^{-24} second for an electron), but no matter. The apparent crack in the door of causality may be slight, but that is enough to satisfy some philosophical analysts seeking scientific support for backward causation. The technical details of Dirac's theory can be found in Plass (1961) and Davies (1977), as well as in Dirac's beautifully written original 1938 paper.

Not everybody likes the apparent failure of causality in Dirac's theory, of course. Davies, for one, is clearly uneasy about it all, calling pre-acceleration "unpleasant" acausal behavior. On the other hand, however, one can find believers, such as Earman (1974). Earman's paper provoked a strong reply from Grünbaum (1976) who argued that the whole business is a non-problem.

Grünbaum makes, in fact, a very interesting point. He argues that because Dirac's equation is non-Newtonian, we have no reason for coupling force and acceleration together as a cause-and-effect pair. In Newtonian mechanics we do use that particular coupling, yet we do not think of force and velocity as such a cause-and-effect pair because there is an integration operation involved in getting from one to other. Similarly, in Dirac's theory we have an integration operation separating force and acceleration.

The exchanges over these matters became quite heated, indeed personal. For example, in his argument for the seeming backward-causation result from Dirac's theory, Earman wrote, "I believe that backward causation is a conceptual possibility and that the question of whether backward causation exists in nature is a question which must be settled not by armchair philosophers but by natural philosophers." Earman had some critical words specifically for Adolf Grünbaum's rejection of pre-acceleration as a case for backward causation. Grünbaum (1976) struck back at Earman by ending his paper with "Earman's uncritical depiction of Dirac's preacceleration as an example of retrocausation is a case of an armchair philosopher misinterpreting what a natural philosopher has wrought." The battle of words continued with Earman (1976) and the replies of Grünbaum and Janis (1977, 1978). See also Nissim-Sabat (1979).

One curious aspect to that debate is that many of the modern commentators seem not to have paid much attention to what Dirac himself had to say about pre-acceleration. As a Nobel laureate, it hardly seems likely that he would let such a result pass unnoticed, and indeed, his paper contains the following physical explanation: "It would appear here that we have a contradiction with elementary ideas of causality. The electron seems to know about the pulse before it arrives, and to get up an acceleration. . . . The behavior of our electron can be interpreted in a natural way, however, if we suppose the electron to have a finite size. There is then no need for the pulse to reach the center of the electron before it starts to accelerate. It starts to accelerate . . . as soon as the pulse meets its outside. Mathematically, the electron has no sharp boundary."

A recent paper by McKeon and Ord (1992) hints at a fascinating connection between travel backward in time and Dirac's relativistically correct, quantum mechanical description of the electron. They show that in flat, two-dimensional spacetime (see Tech Note 4), the assumption of time travel to the past leads in a natural way to Dirac's equation. If, on the other hand, time travel only into the future is assumed, then additional assumptions are required to derive Dirac's equation. This connection between the Dirac theory and time travel to the past makes some philosophers and physicists nervous, but it didn't seem to bother Dirac. In fact, he went on in his paper to show how the pre-acceleration implies the possibility of building a device for sending a faster-than-light signal backward in time. Science fiction writers were, of course, quick to grasp that idea (whereas most physicists and philosophers were nonplussed), and such gadgets were dubbed "Dirac radios"; for instance, see "Beep" (Blish).

One of the more disturbing aspects of backward causation is that it seems to allow for the possibility of *causal loops* and for the potential breaking of such loops, a prime ingredient in many of the very best time travel stories. For example, suppose there is a gadget such that if I push its control button now, then today's lecture notes will have appeared in the gadget's output tray *yesterday*. Indeed, yesterday I found today's notes there, and, in fact, I am about to go to class to deliver that lecture. A mighty good one it is, too, so I think I'll send it back to yesterday in just a few minutes with the help of the gadget. But I haven't yet pushed the button. What if I now decide to let the entire day pass without pushing the button? Why did the notes appear so I could use them today? Modern philosophers call this possibility of breaking a causal loop a *bilking paradox*. In Chapter Four, we'll see how such paradoxes have appeared in the physics and philosophical literature since the 1940s.

In science fiction, by contrast, such paradoxes have been discussed since long before World War II. In a letter to the editor at *Astounding Stories* (June 1932), for example, a fan clearly stated his objection to time travel with the aid of a bilking paradox. He suggested the following experiment: Immediately publish an open offer to the inventor of time travel (who will be born, presumably, at some future date) to travel back to one week before the offer is published. But of course (argued the fan), we'd have a pretty problem if we then decided not to publish the offer after the inventor showed up! As that fan wrote, "Paradoxical? I'll say so, if time travel is possible." That fan didn't know of what seems to be a generic limitation on time machines, however: that one can't travel back to a date before the date of the time machine's creation (recall Tipler's rotating cylinder from Chapter One). Thus that fan's particular bilking paradox actually has no weight. A similar bilking paradox had actually appeared earlier in the mainstream, non–science fiction literature the year before, in the 1931 novel *Many Dimensions* (Williams).

For another fictional example of a bilking paradox, consider "The Time Cheaters" (Binder), the story of time travelers who, just before they begin a trip into the future, see Earth invaded by Martians (there is an amusing reference in this 1940 tale to Orson Welles' famous 1938 radio-drama hoax of just that "event," based on H. G. Wells' *War of the Worlds*). At first the invaders are unbeatable, but then the defending military forces of Earth suddenly and mysteriously acquire a fantastically powerful new weapon. It is not long before the time travelers realize where it came from—they themselves will go into the far future and then return with it to what is now their own past (when the weapon first appeared). But then they wonder what might happen if they don't go, if instead they "cheat time." After all, they reason, why bother now to hunt for the weapon when the invasion has already been defeated? We are told that this potential bilking paradox is a "sinister conception, crawling evilly within their brains, like an unanswerable enigma."

Some philosophers, and practically all physicists, agree with that last assessment about bilking paradoxes, and so they believe there is simply nothing more to say. That is, puzzles like the one in "The Time Cheaters" show that causal loops (and backward causation) are impossible. Many feel this way about time loops and backward causation because, as is well known, time travel to the past can create all sorts of paradoxes. But such paradoxes are often offensive only to human, culturally biased intuitions about how "things ought to work," and not to the laws of physics, which are indifferent to a reversal in the direction of time—which of course underlies what time travel is all about. As the great American chemist G. N. Lewis expressed (1930) it, "Our common idea of time is notably unidirectional, *but this is largely due to the phenomena of consciousness and memory* [my emphasis]."[2] Lewis' work caught the eye of the editor at *Astounding Stories,* who summed it up for his readers in a half-page essay ("Two-Way Time," September 1931) that contained dramatic words hinting at backward causation: "A new theory of time . . . reveals the possibility that events now occurring are among the factors that decided Caesar nearly 2,000 years ago to cross the Rubicon."

Lewis' willingness to accept causality violations is not a popular view today. For example, in Visser (1990) we read, "It is fair to say that most conservative physicists have very serious reservations about the admissibility and reality of causality-violating processes. Causality violation (i.e., the existence of a 'time machine') is such an extreme violation of our understanding of the cosmos that it behooves us to be as conservative as possible about introducing such unpleasant effects into our models." Visser then declares closed timelike loops to be verboten because "the existence of closed timelike loops leads to such unpleasant situations as meeting oneself five minutes ago." Visser sums up his philosophical position nicely with "any theory that is 'just a little bit causality violating' is 'just a little bit inconsistent.' "

Agreeing with Visser is at least one philosopher who believes that the "association of causality with a particular temporal direction is not merely a matter of the way we speak of causes, but has a genuine basis in the way things happen" and that there is indeed an asymmetry with respect to past and future that is bound up with our concept of intentional action. But Dummett (1964) goes even further than that when he continues with the claim that being an agent of cause is not a necessary condition for seeing the asymmetry. Being an observer is enough; even an immobile yet intelligent tree (!) could detect the difference between past and future. Alas, there seems no immediate hope for putting that particularly fascinating idea to an experimental test.

Others, however, are not so sure about such matters as are Visser and Dummett; see Yurtsever (1990), and also Weir (1988), who presents an explanation for causal loops in terms of an oscillating universe—see Schmidt (1966) for more on that—and circular time. For our purposes in this book, we will adopt the position of Australian philosopher Huw Price (1984). Despite the present lack of any experimental evidence for acausal phenomena, Price has convincingly argued that it is the job of philosophers to ensure that physicists do not ignore major and promising avenues of thought in a mistaken belief that ideas concerning backward causation are nothing but philosophical traps for the foolish. In taking that position, Price is simply echoing the words of Bertand Russell, who long ago, in his 1912 Presidential Address to the Aristotelian Society ("On the Notion of Cause"), declared, "The law of causality, I believe, like much that passes muster among philosophers, is a relic of a bygone age, surviving, like the monarchy, only because it is erroneously supposed to do no harm." And I do think Nerlich (1979) was correct when he ended his paper: "[T]he concept of cause is powerless to solve the problems posed by the concept of time. The fundamental laws of physics present our most careful, best established and most sophisticated understanding of time. Notoriously, nothing in these laws endorses the idea of the flow of time nor of the direction which is basic to our conception of it. Nor are these laws causal (in the sense of singling out causes) even when they are deterministic. The concept of cause is not a fundamental one and cannot illuminate the darker corners in our understanding of the fundamental concept of time."

What Does "Now" Mean?

In Gale (1963) there is an overview analysis of various reality-of-time arguments other than McTaggart's, whose rejection of the reality of time we discussed in Chapter Two. Gale also presents an amusing gastronomical interpretation of time when discussing examples of the sort of "time-talk" one often finds in the philosophical literature: "New slices of salami are continually being cut from a nonexistent chunk of salami called the future. *The* present is the slice on top of the pile. The past are the pieces beneath this, and even though they are not present they still continue to exist in the same way that the top slice of salami does. This is a version of the River-of-Time metaphor, since it treats becoming as a type of motion; and it faces humiliation before the embarrassing question of how fast the pile of salami slices grows."

It also runs into trouble with the psychological aspect of time that makes the now seem not like a zero-duration instant, but rather as having, itself, some extent. For example, if you watch the second hand of a ticking clock, you *see it move,* and if we hear someone knock at the door, we hold each entire knock from start to finish in our *now.* These are phenomena that have led to the concept of the so-called *specious present* with non-zero duration; that is, the top slice of time-salami is not infinitely thin at all but rather has some thickness, a concept usually attributed to the American psychologist William James. So—just how thick *is* the slice of salami called the *now?*

Advocates of the doctrine of process theism, the concept that God is a temporal entity that participates in becoming, have actually tried to calculate the duration of the now for God. That might seem like a resurrection of the "How-many-angels-can-dance-on-the-head-of-a-pin?" debates from a thousand years ago, but the analysis does have a certain undeniable charm to it. The idea is that God, through his divine immediacy, his cosmic omniscience, his whatever you care to call it, is aware of what is happening at every instant with every mentality in the universe. As Baker (1972) says, "God must be able to prehend the satisfaction of every actual entity of the temporal process. God's omniscience requires this."

The theologian John Cobb (1965) has used that requirement for omniscience as the basis for calculating what is essentially the speed of thought in God's mind! As Cobb writes, perhaps with just a bit of obscurity, "We may ask how many occasions of experience would occur for God in a second. The answer is that it must be a very large number, incredibly large to our limited imaginations. The number of successive electronic occasions in a second staggers the imagination. God's self-actualizations must be at least equally numerous if he is to function separately in relation to each individual in this series. Since the electronic occasions are presumably not in phase with each other or with other types of actual occasions, still further complications are involved."

To demonstrate how to calculate what he calls the "temporal extension of the divine experience," God's "now" if you will, Baker provides several examples in his paper. I'll repeat one of them that is particularly interesting. Suppose first, he says, that because every entity must be "creatively related to God," then we may conclude that "God's life must be synchronized with the lives of every actual entity." This assumption of divine synchronization avoids what Cobb described above as the "further complications" due to a lack of phasing. Next, suppose that the durations of the nows of whatever life-forms we find in the universe may be different, but not very different, from our own approximately one-tenth of a second (see Tech Note 1). For the sake of calculation, let's suppose that there are just 12 minds in the entire universe, with only slightly different rational durations (in units of seconds) for each individual "now":

1/9	1/10	1/11
2/19	2/21	2/23
3/28	3/31	3/34
4/37	4/41	4/45

If we were to write these fractions with a common denominator, then we would have to determine the least common denominator, a number easily calculated to be nearly five trillion; more precisely, each second God must experience 4,842,179,260,380 nows. And that's for just 12 minds! Imagine the trillions upon trillions of minds that there may actually be in the universe, utilizing a continuum of now-durations—the number of God's experiences per second must be beyond finite expression.

Analyses like that make many think that the salami analogy of time, at least when applied to God, is actually baloney. And as for the concept of a duration for the "now," the English mathematician A. A. Robb expressed a dissenting view to the notion of a specious present with its non-zero duration when he wrote, somewhat enigmatically, that "the present properly speaking does not extend beyond itself." For physicists there is nothing in physical theory that marks the present moment as unique, and therefore nothing that reflects an actual "flow" of time, nothing that models the events of "now" becoming part of the past and the events of the future becoming "now"—recall Weyl's view (as well as Black's sharp response to it) from the previous chapter. For everyone, however, even physicists, there is the powerful psychological sense that time *does* flow. [And here we have yet another "reason" why some philosophers reject backward causation,—for instance, see Waterlow (1974): It is in the "wrong" temporal direction for a continuous becoming.] Despite that, the relativistic, four-dimensional block universe view of spacetime that many physicists so dearly love clearly seems to have no room for an objective theory of time flow.

All events in the block universe simply have coordinates in spacetime, and there is nothing corresponding to "have been" (past), "are" (present), or "will be" (future). There is no "moving now" in the block universe except for its subjective presence in our conscious minds. All we can say from physics is that events are ordered in an earlier/later sequence, and even that relatively weak condition is true only for causally related events. As shown in Tech Note 4, non-causally related events *can* have different temporal orderings for different observers in relatively moving reference frames. For such observers, then, "earlier/later" has no more meaning than does "past/future," and indeed both pairs of words become empty of meaning. (But just to show how philosophers can accommodate nearly anything, see Brier (1972) for how backward causation can be cast in a manner that avoids making reference to the temporal sequence of cause and effect, contrary to Waterlow!)

Science fiction has, not surprisingly, made good use of the idea of relativity of language. In the 1958 story "Two Dooms" (Kornbluth), for example, we learn why one character thinks there is some truth to the speculation that Hopi children of the American Southwest understand Einstein's relativity theory as soon as they can talk: "The Hopi language—and thought—had no tenses and therefore no concept of time-as-an-entity; it had nothing like the Indo-European speech's subjects and predicates, therefore no built-in metaphysics of cause and effect. In the Hopi language and mind all things were frozen together into one great relationship, a crystalline structure of space-time events that simply were because they were."

Kornbluth was clearly influenced in writing that passage by the work of the amateur American ethnolinguist Benjamin Lee Whorf (1956). A fascinating essay on the differences in the conceptions of time between Indo-European languages and that of the Hopi, based on the writings of Whorf, is Cox (1950). In particular, Cox imagines how physics (and general relativity) might have developed in a society using a tenseless language. For example, Cox speculates that the lack of the concept of relative simultaneity might have resulted in the Lorentz spacetime transformation equations (see Tech Notes 1 and 3) as an automatic assumption of Hopian mathematics! A year after Kornbluth's story, however, at least one philosopher began to have doubts about Whorf's claims—see Black (1959a)—and more recent scholarship in Malotki (1983) has conclusively shown that Whorf's assertion that the Hopi language is tenseless is simply incorrect.

Modern physics takes the view that time is simply a *parameter,* a *label,* just as the tick marks along the axes of a graph denote different values of a spatial parameter. When we draw an *x-y* graph, we do not think of either *x* or *y* as moving, and similarly a physicist would argue that when we draw a spacetime diagram, we should not think of time as moving. As mentioned before, physicists call our feeling of moving time *psychological time* because the equations of physics provide no physical interpretation for a moving "now." That is not to say that psychological time does not have great fascination and significance in human affairs, but just that the physicist's view is that it has no role in physics. Psychological time simply has no place in a universe devoid of something like the physical processes in a brain that give rise to what we call consciousness.

The philosopher Max Black has stated in no uncertain terms (see Note 31 for Chapter Two) how he feels about the profound difference between psychological time and the way physicists think of time: "So wide is the gap between the common-sense notion of time and the physicist's t that clarity would be fostered if physicists were to imitate the psychologist's practice of talking about g rather than about intelligence by referring to their own concept as t rather than as 'time.' It is not shocking to be told that t may have a unique origin like the absolute value of the temperature scale, or may have several dimensions, or even that it may 'run backward'; it is only when such aphorisms are transformed into corresponding statements about time that paradox emerges and philosophical hackles rise."

The one view of the block universe that Black accepts, that there is no time flow except in the minds of conscious beings (including, I suppose, Dummett's tree), has received strong advocacy from the philosopher Adolf Grünbaum (1963). Indeed, I'd hazard that most physicists and a majority of philosophers agree with Grünbaum, who declared, "I believe the issue of determinism vs. indeterminism is *totally irrelevant* to whether becoming is a significant attribute of the time of physical nature independent of human consciousness." My only quibble with that is, as I've argued before, that the block universe should be associated not with determinism but with *fatalism*.

But not everybody agrees with Black and Grünbaum. Indeed, Grünbaum himself quotes Hans Reichenbach, who took the opposite stance (saying that time *does* flow) with equal vigor. Writing in 1925, Reichenbach asked, "What does 'now' mean? Plato lived before me, and Napoleon VII will live after me. But which one of these three lives *now*? I understandably have a clear feeling that *I* live now. But does this assertion have an objective significance beyond my subjective experience?" Reichenbach went on to answer that question in the affirmative and to deduce that the block universe is an incomplete representation of reality: "In the condition of the world, a cross-section called the present is distinguished; the 'now' has objective significance. *Even when no human is alive any longer, there is a 'now'* [my emphasis]. . . . In the four-dimensional picture of the world, such as used by the theory of relativity, there is no such distinguished cross-section. But this is due only to the fact that an essential content is omitted from this picture."

And what is Reichenbach's missing essential content? Feeling that the block universe is unacceptably fatalistic—in his words of ridicule, "the morrow has already occurred today in the same sense as yesterday"—he found his answer in the antithesis of determinism, the probabilistic theory of quantum mechanics. Laplacian physics argues that given total information about the state of the world *now,* one could calculate perfectly the future or the past; one could both predict and retrodict. In contrast, quantum mechanics distinguishes past from future in a fundamental way. Quantum mechanics does not deny that in principle we can know the past with exquisite accuracy, because each and every event leaves traces, evidence that is available to all with the means to find and decode it. This is called the *archivalist* view of the past.

But quantum mechanics also takes as a postulate that there is an irreducible uncertainty to the future. The instant that this uncertainty is crystallized into fact was taken by Reichenbach to be the very definition of what we mean by "now." The ever-increasing record of the past, in turn, defines the movement of the "now." Reichenbach believed that with these observations, he had at last captured the moving now in mathematical theory, that he had elevated the present from psychology to physics, and that he had shown that the flow of time is independent of the need for a conscious mind.

Grünbaum's own associate, however, Baker (1975) has written a powerful philosophical analysis of the time-flow issue that comes down solidly in support of the opposite position, the view that the moving now is only in our minds and is not an intrinsic attribute of reality. (Others, such as Smith 1985c, have since written also to refute Grünbaum.) Baker's most interesting point is that a mind-independent flow of time is incompatible with the relativity of simultaneity because it implies a universal now, a concept that Einstein showed is an illusion. [But see Peacock (1992) for a provocative essay by an author willing to stick his neck out in critique of Einstein!] Baker's objection also applies to the thesis of *probabilism,* which says that at any instant there are many alternative futures, but just one past. This instant is simply a synonym for the moving now, however, and so supporters of probabilism, such as Maxwell (1985), conclude that special relativity must be deficient because of its denial of cosmic simultaneity. That, of course, is not the majority view among physicists—see Dieks (1988), Maxwell (1988, 1993), and Stein (1991).

Early science fiction stories are full of theories about the nature of "now," and the vast majority of them have no basis in scientific thought. Some of them *are* ingenious, however, and even though they are largely the pet ideas of the authors (and no one else's), perhaps they resulted in some of the young readers of the science fiction magazines of the 1930s and 1940s thinking about deeper matters than did "Buck Rogers," "The Lone Ranger," or "Terry and the Pirates." For example, according to the 1934 story "Today's Yesterday" (Ray), time is a wave, and the moving now we experience is carried on a crest of that wave, just as a piece of wood is carried along on the crest of a water wave. There are time waves both ahead and behind the crest we happen to be on, of course (so we are told), and each such crest carries a different now for a different reality—hence the curious title.

More recently, in the story "The Trouble with the Past" (Eisenstein) about object duplication via time travel (discussed in more length in Chapter Four), nine (!) inadvertent copies of the same person from the year 2314 meet in 1870 to try to figure out what is going on. Part of their discussion is the following analysis of the "present".

> "Gentlemen, I think I understand," said the first James Thomas.
> Eight faces turned toward him, and he felt as though he were looking into multiple mirrors.
> "We hold that time is a single instant—the instant of the Present—which travels through Duration—do we not?"
> Eight heads nodded.
> "We assume that time passes in a manner analogous to the stringing of an infinite number of beads. Each bead is the instant of Now when it is last on the chain. Beads are continually being added, and each one is only the Now until another is placed after it."
> "Yes, that is my theory," said another James Thomas. "It can also be likened to the process of knitting. No matter how many stitches are knitted, there is only one last stitch, only one Now."

Einstein, too, was greatly bothered by the place of the "now" in time, perhaps even more than were James Thomas and his "friends." In an autobiographical essay, the philosopher Rudolf Carnap recalled one of his conversations with Einstein in the early 1950s, at the Institute for Advanced Study in Princeton:[3] "Once Einstein said that the problem of the Now worried him seriously. He explained that the experience of the Now means something special for man, something essentially different from the past and the future. That this experience cannot be grasped by science seemed to him a matter of painful but inevitable resignation. . . . Einstein thought . . . that there is something essential about the Now which is just outside of the realm of science." For a counterview, however, see Nor (1992), who claims to have found a geometrical explanation for the flow of time, and which Nor believes gives an objective, mathematical reality to the moving present.

Irreversibility

Well, no matter whether time actually flows or not, most of us still believe we have had a past and hope we will have a future. Each of us thinks we can easily tell one from the other, too. We have, in fact, many not so subtle indications from our everyday lives of the obvious direction to time. Nearly all of these indications have the common theme of *irreversible change*. As the British mathematician J.J. Sylvester (1869) put it, "The whirligig of time brings about its revenges." Ovid, who died when Christ was a teenager, said the same in his *Metamorphoses* with the famous words "Time, the devourer of all things." The image of time as devourer of all that is mortal was brilliantly presented by James Barrie in his *Peter Pan,* with the crocodile who had swallowed a ticking clock chasing Captain Hook all about Neverland.

As grim as that sentiment is, people seem to want to find new ways of expressing the same thought. In his infinitely sad time play *Time and the Conways,* J. B. Priestly wrote, "There's a great devil in the Universe, and we call it Time." That certainly appears to be true. No one yet has escaped the biological decay processes of time, whether it be the evolution of a loved pet from kitten to large window cat to death or the appearance of aches and pains in our own bodies that once seemed immune to them. Inanimate objects are not immune to this aspect of time, either. Logs and cigarettes burn in the stove and ashtray, but they never unburn. Our cars rust but never derust. An explosion has *never* been seen to reverse itself, to form a dynamite stick or a bomb casing out of a collapsing fireball.

Our world seems, indeed, literally to be built on an irreversible movement toward chaos, death, and decay. Rubens in the seventeenth century, and Goya in the nineteenth, painted well-known works showing that movement with the use of truly horrific symbolism. Those paintings, both of which are known today as "Saturn Eating One of His Children," depict Saturn, the Roman planet-god often associated with the Greek Chronos, or Father Time, doing literally what the title says. The words that open the best-known poem of the seventeenth-century Englishman Robert Herrick ("To the Virgins, to make much of Time") make the same point, in an only slightly less brutal way:

> Gather ye rosebuds while ye may,
> Old Time is still a-flying;
> And this same flower that smiles to-day,
> To-morrow will be dying.

That observation, of time as destroyer, runs all through the works of Shakespeare, the consummate interpreter of human experience. One hardly knows where to begin to select a quote on time from his writings, but perhaps a particularly good example is from the poem *The Rape of Lucrece,* in which time is called the "eater of youth," and time's function is declared to be

> To fill with worm-holes stately monuments,
> To feed oblivion with decay of things,
> To blot old books and alter their contents,
> . . .
> And waste huge stones with little water-drops.

And time shows no favorites. In *Cymbeline* we are reminded that

> Golden lads and girls all must,
> As chimney-sweepers, come to dust.

Two centuries or so later, Lewis Carroll repeated the message. In *Through the Looking-Glass* Alice tells Humpty Dumpty "one can't help growing older." And speaking of Humpty Dumpty, his famous fall provides a dramatic example of a one-way evolution from past to future; he wasn't convinced that Alice was correct, but once he had splattered, then

> All the King's horses and all the King's men
> Couldn't put Humpty Dumpty together again.

While we are on the subject of Mr. Dumpty, it is also worthwhile to note that nobody has ever figured out how to unscramble an egg. Why is that? One philosopher—Margenau (1954)—speculated that it is because of such "irreversible organic phenomena" taking place in our brains that our flow of consciousness is always in the same direction.

More subtle than the undignified undoing of a prideful egg is the phenomenon of memory, which seems trivial only because most people have not thought very carefully about it. We remember the past while remembering nothing about the future. We might, in fact, be tempted to use the phenomenon of memory to answer the question of how to tell past from future: Anything you can remember *is* the past. But that is a circular definition, a point discussed by Smart (1954), who observed that to ask why memory is always of the past "is as foolish as to ask why uncles are always male, never female." In *Through the Looking-Glass* the White Queen tells Alice that "it's a poor sort of memory that only works backward," but except for the claims of clairvoyants, it seems that is the only sort of memory any of us has. Why is that so? (Of course, that would *not* be the case for a time traveler while in the past. His personal past, which of course he *would* remember, would be the future for the world around him.)

For physicists, the question of the direction of time is one of profound mystery. There seems, in fact, to be no fundamental reason why time should *not* be able to go from future to past—but then what would "future" and "past" *mean?*—even though no one has ever observed time to do so. All the laws of classical physics, including general relativity, and quantum mechanics, too (except for kaons, of course), involve time in such a way that they ignore its sign. In other words, replacing t with $-t$ results in a perfectly valid description of something that could actually happen. But not all such possibilities are observed really to occur. Why not?

In an unpublished paper written in 1949, while doing the work that would bring him a share of the 1965 Nobel prize in physics, the young genius Richard Feynman wrote—see Schweber (1986)—"The relation of time in physics to that of gross experience has suffered many changes in the history of physics. The obvious difference of past and future does not appear in physical time for microscopic events. . . . Einstein discovered that the present is not the same for all people [see Tech Note 1 on the relativity of simultaneity]. . . . [I]t may prove useful in physics to consider events in all of time at once and to imagine that we at each instant are only aware of those that lie behind us. The complete relation of this concept of physical time to the time of experience and causality is a physical problem which has not been worked out in detail. It may be that more problems and difficulties are produced than are solved by such a point of view."

Feynman did not elaborate on what he meant by the "problems and difficulties" with that point of view (which is clearly that of the block universe), but surely he had the logical paradoxes of time travel high on his list. Certainly time travel was on the minds of others during that period. As the Yale philosopher Henry Margenau (1954) wrote in a tutorial on Feynman's work, "The theory of quantum electrodynamics developed by Feynman incorporates reversals in the course of time and thereby cherishes, in the minds of many, an *age-old phantasy* [my emphasis] of more than scientific appeal."

Because the individual classical equations of microscopic physics are time-reversible, the distinction between past and future for individual particles disappears. The equations are said to be symmetric with respect to time; the algebraic sign of t is irrelevant in the classical laws. It must be understood, however, that there is a crucial distinction to be made here. When a physicist says *time reversal,* she is talking about a system evolving *backward* in *forward* time—that is, all the individual particle velocity vectors are instantly reversed at once. This is distinct from the time-reversed worlds of philosophers and science fiction writers (which are discussed in the next section), in which time itself "runs backward." The physicist's point of view is expressed clearly in Lewis (1930): "Every equation and every explanation used in physics must be compatible with the symmetry of time. Thus we can no longer regard effect as subsequent to cause. If we think of the present as pushed into existence by the past, we must in precisely the same sense think of it as pulled into existence by the future." More than three decades later the two mathematicians Penrose and Percival (1962) presented similar ideas: "In classical dynamics, the past completely determines the present, and therefore, by symmetry, the future also completely determines the present."

Besides the physics, there is also an interesting theological connection to time reversal. As Mehlberg (BIPT) puts it "If all natural laws are time reversal invariant and no irreversible processes occur in the physical Universe then there is no inherent, intrinsically meaningful difference between past and future. . . . If this is actually the case, then all mankind's major religions which preach a creation of the Universe (by a supernatural agency) and imply, accordingly, a differentiation between the past and the future . . . would have to make appropriate readjustments."

There are, as you might suspect, dissenters to the view that the classical laws of physics are necessarily time-reversible. Dirac himself wrote (1949) that "I do not believe there is any need for physical laws to be invariant under time and space reflections, although all the exact laws of nature so far known do have this invariance." Dirac did not, unfortunately, elaborate on just why he felt that way, but with the later discovery of kaons his position is seen to have been "ahead of its time"! And in fact, the actual macroscopic world does appear to be decidedly asymmetric. The puzzle of why that is so has generated a vast scientific literature, including entire books on the physics of time reversibility, such as Sachs (1987), Davies (1977), and Zeh (1989). Some philosophers, such as Grünbaum and Reichenbach, find an explanation for the asymmetry in the statistical irreversibility of complex processes. Others, like Earman, disagree because such asymmetry can also be found in at least one non-statistical case, as Finkelstein (1958) and Davies (ET) remark in their analyses of the one-way nature of the event horizon of a black hole.

Worlds in Reverse

Philosophers and writers of speculative fiction were the first to wonder what things might be like in a world where the time asymmetry is reversed—in a world where time 'runs backward.' Indeed, fascination with the idea of time reversal actually dates back thousands of years, long before science fiction: It can be found in Plato's dialogue *Statesman,* written (most probably) fifteen years before Plato's death in 347 B.C. At one point Plato offers an extended description of the world suddenly running backward in time in the ancient past. After one character is told that at that remote time "all mortal beings halted on their way to assuming the looks of old age, and each began to grow backward," he asks, "But how did living creatures come into being, Sir? How did they produce their offspring?" The answer is shocking: "Clearly . . . it was not of the order of nature in that era to beget children by intercourse. . . . [I]t is only to be expected that along with the reversal of the old men's course of life and their return to childhood, a new race of men should arise, too—a new race formed from men dead and long laid in Earth. . . . Such resurrection of the dead was in keeping with the cosmic change, all creation being now turned in the reverse direction." There is also a hint of such a world in the Old Testament, a story that is told in both *Second Kings* 20 and *Isaiah* 38. There we are told that when afflicted by a fatal illness, King Hezekiah of Judah pleaded for his life, a request that God granted. When given his choice of having God move the shadow on a sundial either ten degrees forward or ten degrees backward as the sign of this gift, Hezekiah naturally enough elected to see time run backward.

The reversed-time world is an important philosophical concept. Before the turn of the century, for example, Francis Bradley (one of the early originators of the block universe, you will recall), thought about reversed-time worlds and concluded that they would be quite odd:[4] "Let us suppose . . . that there are beings whose lives run opposite to our own. . . . [I]f in any way *I* could experience *their* world, I should fail to understand it. Death would come before birth, the blow would follow the wound, and all must seem to be irrational." More sympathetic with the possibility of reversed-time worlds was W. R. Inge (1921). In his November 1920 Presidential Address to the Aristotelian Society, in the Conference Hall of the scholarly University of London Club, he felt safe enough to make time-reversed existence his topic. Still, even though he was more flexible on the possibility of such a world than was Bradley, Inge did feel it necessary to conclude with "I have not discovered Mr. Wells' Time Machine."

Thirty years later the philosopher J. N. Findlay took Bradley's position of supporting a skeptical attitude toward the possibility of time-reversed worlds. Writing in a book review,[5] Findlay declared, "The reversed world in question wouldn't merely strike us as queer, but definitely crazy: it would be a world where what is wildly and intrinsically *improbable* was always occurring. It would, in fact, be much more startling than the original asymmetry that led us to think of it." The question of backward-running time so fascinated Findlay that some years later he posed it as a problem for the readership of the journal

Analysis to consider. In Findlay and McGechie (1956) he presents both his own negative view of time-reversed worlds and that of the best response (which came from McGechie) received to his posed problem.

Findlay showed admirable open-mindedness by awarding the title of "best" to an argument that refutes his own position. Findlay, however, remained unconvinced about the concept of reversed-time worlds, stating that "I continue to feel that a total reversal of my experiences is a terrifying possibility."[6] And anybody who has watched a movie film or videotape played backward would almost certainly agree with Findlay; see "This Side Up" (Banks) for a story about the temporal confusion that might be caused by projecting a film in the wrong direction of time.

Despite Findlay's "terror," however, the prevailing view today is that to the inhabitants of a time-reversed (or what is sometimes called a *counterclock*) world, *nothing would look odd!* That is a fairly new idea, and even the relatively recent philosophical literature sometimes, as in Dummett (1964), betrays a misunderstanding of how a time-reversed world would appear to its occupants. The modern idea that a time-reversed world would appear normal to someone living in it seems to have been explicitly stated first by Smart (1954), but I do not find the arguments presented there persuasive. Agreeing with me on that point is Whitrow (1980), who calls Smart's reasoning "fallacious or, at best, trivial."

More compelling analyses appear in Schumacher (1964), Narlikar (1965), and Stannard (1966). Narlikar and Stannard introduce the concept of a universe made of matter in which every particle interaction occurs with a reversed time sense. For example, a neutron does not decay into a proton, an electron, and an antineutrino (see Note 11 of Chapter Two), but rather neutrons are *created* by the collision of those other particles. As Stannard points out, the spacetime diagrams (see Tech Note 4) of such interactions are merely those of "our world" with the time axis reversed; he calls such matter interactions *faustian* (from Goethe's play *Faust,* in which the normal flow of time is routinely violated) and hypothesizes that interactions between normal (our) and Faustian matter cannot occur. Indeed, matter with a reversed time sense is thought by many to be antimatter in our world,[7] and such interactions would be spectacular! (That particular objection to time travel is raised several times in Robert Silverberg's novel *Vornan-19.*) Stannard's hypothesis is a sort of censorship principle, and if true, it eliminates Bradley's concern about living in a reversed world.

The philosopher J. R. Lucas has argued that even if beings from two such oppositely directed worlds could meet, they still could not communicate.[8] "If two beings are to regard each other as communicators, they must both have the same direction of time. It is a logical as well as a causal prerequisite." Now, this matter is well worth some effort to understand, because it is intimately tied to time travel. At first blush, Lucas' words seem almost self-evident, and after a little thought, they seem absolutely irrefutable. The philosopher Murray MacBeath, however, takes exception.

MacBeath (1983) opens with a story to demonstrate that the persuasive power of Lucas' position is only superficial. In that story of Jim and Midge, Jim is one of us, whereas Midge is a Faustian time-antagonist—what Rothman (1987) calls a "retro-friend." In his analyses MacBeath uses capitalized words and symbols for the time-reversed Midge, and lower case for Jim, as shown in Figure 3.1. As MacBeath explains, "While Jim and Midge are together a face-to-face conversation is hardly likely to get off the ground. To make this clear let us say that they are together from t_0 until t_{10} on Jim's time-scale, and from T_0 until T_{10} on Midge's TIME-scale; t_0 is then the same temporal instant as T_{10} and, in general, $t_n = T_{10-n}$. If Jim at . . . t_2 asks Midge a question, and Midge hears the question at T_8, she will answer at T_9, and Jim will hear his question answered at t_1, before he asked it. What is more, if Jim is inexpert at interpreting backward sounds, and at t_4 asks Midge to repeat her answer, Midge will hear that request at T_6, BEFORE she has heard the original question; and her puzzled reply at T_7 will again be heard by Jim at t_3, before he has uttered the request."

Certainly that is a mess in time, and Lucas seems to be on safe ground with his denial of the possibility of communication between Jim and Midge. MacBeath, however, shows how to refute all of Lucas' arguments if Midge and Jim are allowed to be clever about how they send their messages back and forth—that is, if we give up some of our usual ideas of what a conversation is like. MacBeath's analyses are far too lengthy and detailed to present here, but this simplified diagram, a version of one he offers himself, should enable us to follow the logic of his approach.

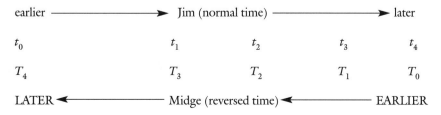

Figure 3.1 Opposite time directions in counterclock worlds.

We imagine that Jim and Midge will not actually talk and so will not have to decipher backward-spoken language. Rather, they will exchange messages via handwritten notes or computer-generated text displayed on monitor screens. We can, in fact, imagine that Jim and Midge are separated by a window that is proof against all penetration but light.[9] Now, at t_0 Jim brings a computer to the window. He programs it to wait for four days, until t_4, and then to display the following message on its screen: "This message is from Jim, who experiences time in the sense opposite to yours. Please study the following questions and display your answers on a computer screen three days from now." Jim's message ends with the list of questions.

Because that all took place at t_4, Midge sees Jim's message and questions at what we will now call T_0. As requested, she brings her computer to the window, enters the answers to Jim's questions, and programs the machine to display them (after a three-day delay) on its screen. Thus, at T_3, which is Jim's t_1, Jim sees Midge's computer screen light up with "Hi, Jim. This is Midge. The answers to your questions are at the end of this message. Now, I've got some questions for *you*. Please display the answers two days from now." Midge's message ends with answers to Jim's questions and her list of questions.

Jim sees Midge's message at t_1, enters the answers to her questions and sets his machine to answer after a two-day delay. At t_3, which is Midge's T_1—and by now you see how the process goes. It's cumbersome, sure, but it works. Or at least it does if everybody follows the rules. What if they don't? MacBeath provides other, increasingly complicated analyses that treat some of the more subtle problems that can be imagined to occur in this method of exchanging messages. I will mention just two of them, which have direct analogs with what we normally think of as "time travel."[10]

For the first problem, consider Jim's initial message, created at t_0 to be sent at t_4. He receives Midge's answer as described above at t_1, *before* his message is displayed through the window. So, what happens if at t_2 Jim cancels the message and it is *not* displayed? He has already gotten Midge's reply, but how can that happen if he does not send his message? This is, of course, a bilking paradox, and we will discuss it in Chapter Four when we look at paradoxes in general.

A second curious problem is the possibility of creating a causal message loop. For example, let's say that at t_0 Jim suddenly decides to send a message through the window. (His reason for this sudden urge will be explained in the next few lines.) He thinks all night about what to send and, at t_1, finally settles on the following: "Greetings to the people on the other side of the window. This message comes from Jim, who hopes you will reply." Midge immediately sees Jim's message through the window (at her time T_3) and so is suddenly caught up with the desire to respond. She thinks all night about what to send and, hoping to be witty, she finally decides (at time T_4) on the following echo to Jim's message: "Greetings to the people on the other side of the window. This message comes from Midge, who hopes you will reply." Jim immediately sees this through the window (at what is his time t_0), and so now we know why he decided to send his original message!

And Jim *will* send his original message—he has to because Midge replied to it (forget about the bilking paradox issue, for now). Causal loops are very strange things, wherein events happen in the present because of events in the future that, in turn, are caused by those present events when they are "later" in the past. Unlike the grandfather paradox, which most philosophers believe is now "solved," causal loops have yet to be given a satisfactory explanation. MacBeath does not like causal loops, admitting that although he believes them to be conceptually possible, he also finds causal loops to have "so queer a smell" that he prefers to avoid thinking about them. Later, in Chapter Four, *we* will think about them and, not surprisingly, find that some of the most puzzling (and entertaining) time travel stories are based on causal loops.[11] In another of MacBeath's papers (1982) he calls such tales "loopy stories"!

It should now be clear that the Faustian existence of a counterclock world would indeed seem quite strange to us, because its inhabitants would remember what we call the future. But, of course, our world would seem just as strange to them, because what we remember, our past, is *their* future. As Stannard points out, "This rather odd situation, it will be recalled, has a close parallel in the special theory of relativity where two observers in relative motion are equally convinced that it is the other's clock that is running slow [see Tech Note 3]." This rather bizarre (to us) feature of reverse time has of course intrigued writers of fiction. When Merlyn the magician makes his first appearance in T. H. White's 1939 masterpiece *The Once and Future King*, for example, he explains how he knows the futures of others: "Ordinary people are born forward in Time, if you understand what I mean, and nearly everything in the world goes forward too. . . . But I unfortunately was born at the wrong end of time, and I have to live backwards from its front, while surrounded by a lot of people living forwards from behind. Some people call it having second sight."

What may have put the idea in White's mind for his time-reversed magician is only speculation today, but perhaps it was something he might have read a decade before, in a fellow Englishman's writing. I am referring to Eddington's 1929 book *The Nature of the Physical World*, where one finds this passage: "In

'The Plattner Story' H. G. Wells relates how a man strayed into the fourth dimension and returned with left and right interchanged. . . . In itself the change is so trivial that even Mr. Wells cannot weave a romance out of it. [But see Note 16 for Chapter Two for how science fiction writer Arthur C. Clarke and physicist George Gamow *did* make nice little tales out of this idea, one that Eddington all too quickly discarded.] But if the man had come back with past and future interchanged, then indeed the situation would have been lively."

Whether or not those words influenced English fantasy, they certainly had some effect on American science fiction. In his 1979 memoir *The Way the Future Was,* pulp editor Frederik Pohl writes that Eddington's book (which Pohl incorrectly attributes to Sir James Jeans) had given him the idea for a story using the reversed-time twist. But before Pohl could publish it, an even better tale (says Pohl) arrived from Malcolm Jameson. Jameson, too, had read Eddington's book, and the result was the novella-length "Quicksands of Youthwardness," which Pohl published in *Astonishing Stories* as a three-part serial during 1941–1942. Sad to say, however, Jameson's tale is both pretty awful and devoid of *any* connection with Eddington's suggestion. One can only wonder might have been in the story that Pohl says *he* discarded—did *it* have the future remembered? Unfortunately, Pohl says he can't remember!

Remembering the future does occur in "This Way to the Regress" (Knight), where everybody knows what will happen (as they live backward) by reading "prediction books." What distinguishes that story from many others on the same theme is an interesting conversation a student has with a philosophy professor about how things would be if time went the other way, as in *our* world.

> "How can we tell? The reverse sequence of causation may be just as valid as the one we are experiencing. Cause and effect are arbitrary, after all."
> "But it sounds pretty far-fetched."
> "It's hard for us to imagine, just because we're not used to it. It's only a matter of viewpoint. Water would run downhill and so on. Energy would flow the other way—from total concentration to total dispersion. Why not?"

The student is unconvinced, however, and when he tries to visualize such a peculiar world (*our* world, don't forget!) it gives him a half-pleasant shudder." Imagine, he thinks in wonder, never knowing the date of your own death.

In a review of Reichenbach (1956),[12] Hilary Putnam restates the problems of a time-reversed world in the form of a provocative question: "How do you know that one man's future isn't another man's past?" He begins by making the interesting observation that for us to be able just to observe a backward-running universe, we would have to provide our own normal radiation source, because the counterclock stars in such a universe *suck* radiation rather than suppling it.[13] (This imagery conjures up the "back-rays" in David Lindsay's 1920 novel *Voyage to Arcturus,* which are described as "Light which goes back to its source," an explanation "explained" by asking, "Unless light pulled, as well as pushed, how would flowers contrive to twist their heads round after the sun?") Putnam concludes his comments about reversed time with a cautious warning: "It is difficult to talk about such extremely weird situations without deviating from ordinary idiomatic usage of English. But this difficulty should not be mistaken for a proof that these situations could not arise." This challenge is no doubt why so many writers outside of science fiction and fantasy have tackled the question of what it would be like if time ran backward. For example, the mid-twentieth-century German-American composer Paul Hindemith was intrigued by the possibility of a musical palindrome, and he showed that the idea of time reversal could be treated outside of both prose and physics. His one-act musical skit *Hin und Zuruck* (*Here and Back*) has an angel appear at midpoint to reverse time, and then the very music itself reverses.

In fiction, the narrator (a "professor of astronomy and higher mathematics") of Edward Bellamy's 1886 story "The Blindman's World" suddenly finds himself on Mars. (An aside: This literary device was therefore used long before Edgar Rice Burroughs used it in 1912 to put *his* hero, John Carter, on the red planet!) There he encounters beings who know the future up to their deaths, and whose memories of the past are "scarcely more than a rudimentary faculty." The entire tale is in the form of a conversation between the professor and one such being, who argues (quite persuasively) for the virtues of his 'backward' existence compared to that of earthlings (the Martian name for Earth is the story's title). Bellamy realized that his story implies a fatalistic block universe. "No one could have foresight . . . without realizing that the future is as incapable of being changed as the past," he wrote.

We find the first (rather weak) attempt at an explanation of reversed time in the Outlandish Watch of Lewis Carroll's *Sylvie and Bruno;* because of a "reversal-peg," some things in that story go in reverse—for instance, people walk backwards—but other things still go forward, such as human speech. Mark Twain, in a dramatic use of reversed time in Chapter 32 of his last, posthumously published novel *No. 44, The Mysterious Stranger*—see Gibson (1969)—decided he would have to go outside rationality for reversed time. In that work a supernatural being called "No. 44" reverses the world's time, and the narrator tells us that "everywhere weary people were re-chattering previous conversations backward . . . where there was war, yesterday's battles were being refought, wrong-end first; the previously killed were getting killed again . . . we saw Henry I gathering together his split skull. . . ."

That dramatic recital is logically flawed, however, because other descriptions of the reversed-time world have people "scared and praying . . . gazing in mute anguish" at the Sun moving through the sky in the wrong direction. In a truly time-reversed world, of course, no one would perform such acts unless they had previously performed "reversed fright and praying" at the Sun moving in the right direction. A similar error is made in "The Man Who Never Lived" (Wandrei). The time travel mechanism in that story, originally published in 1934 by *Astounding Stories,* is the mystical hocus-pocus of "mental monism," which sends a philosophy professor back to "before the beginning of time." The professor somehow manages to send a description of what he sees, as he moves backward through time, to a friend (the narrator) in the present. But some of these descriptions are of actions in forward time: "There are shells bursting all around and men die by thousands in gouts of flesh and blasts of flame."

Such mainstream writers of the first half of the twentieth century as F. Scott Fitzgerald and the Cuban writer Alejo Carpentier also applied their considerable imaginations to the idea of backward-running time. Fitzgerald's famous 1922 story "The Curious Case of Benjamin Button" is a bit strange in its unique way, with the backward-living Benjamin Button appearing not from the grave but from his mother's womb! (How he fit seems an obvious physical flaw, to say nothing of the logical error.) Carpentier's beautiful story "Journey to the Seed," which tells of a dying man suddenly reversing his time sense, is not so blatantly erroneous as is Fitzgerald's, but it suffers from yet another logical flaw also present in the Fitzgerald story: Why are these individuals who are living backward in a forward-running world the only ones doing so, and why can they both understand, and be understood by, all those around them?

Science fiction writer Nelson Bond tried his hand at this sort of fantasy version of reversed time. His 1941 story "The Fountain," about an old man who finds Ponce de Leon's Fountain of Youth, explores the problems of a man who is continually forgetting as he lives backward in forward time; that is to say, he is younger in year $x + 1$ than he was in year x. Many readers will recognize that this story idea was partly anticipated in Nathaniel Hawthorne's 1837 short story "Dr. Heidegger's Experiment," but the continual loss of memory is Bond's original touch. Even more curious is *The Man Who Lived Backward* (Ross), a 1950 novel of a man who lives backward in a different way from Bond's character. Born in 1940, the man wakes up each morning, lives a normal forward day, goes to bed at night, and wakes up in yesterday! This process continues until he dies in 1865, in a futile attempt to save Lincoln (whose fate he has known from reading history books.) There is no explanation for the character's astonishing condition, but the problems such a life would pose are cleverly worked out.

To find science fiction writers speculating on reversed time is, of course, really not surprising. The 1938 reversed-time story "The Man Who Lived Backwards" (Hall), for example, gives an outstanding treatment of the logical nuances of reversed time in which people talk backward, and there is a marvelous bathroom scene of a man un-washing his hands. The tale tells of a young physics teacher who is "twisted into a reversed Time Stream" by an electrical discharge. As he lives backward in time, he observes everybody about him appearing to run in reverse, but even more puzzling is that they have developed a "dreadful, granite-like hardness." We soon learn why:

> For a while he could not understand the impenetrable hardness of external objects which he had experienced; it seemed they ought rather to be of intangible transiency, much as a dream, since he was re-viewing the Past. But a moment's thought gave him the logical answer. The Past is definite, shaped, unalterable, as nothing else in Creation is. Therefore, to argue that he could move or alter any object here [the past] was to argue that he could change the whole history of the world or cosmos. Everything he saw about him had happened, and could not be changed in any way. On the other hand, he was fluid, movable, alterable, since *his* future still lay before him, even if it had been reversed; he was the intruder, the anomaly. In any clash between himself and the Past, the Past would prove irresistible every time.

This is, I believe, a unique presentation of the unchangeability of the past. Why Hall's young teacher could displace the molecules of the Past's air, however, is left unanswered.

Another early brush at the idea of time-reversed worlds appears in Wilson Tucker's 1955 novel *Time Bomb*. There we find the intriguing idea of political assassination by time bomb with the bombs actually *time traveling* to their targets. A policeman begins to suspect what is happening when it becomes evident that one of the explosions was actually an implosion: "The time bomb . . . had been going in and had carried the force of the blast with it. Inward. Into the past. He frowned at that. A backward explosion? An explosion which ran counter to the normal flow of time, to the normal method of living? . . . How would an explosion appear to a man if the blast happened in the opposite manner? If it began exploding now, in this moment, but continued backward instead of forward? Would it be an implosion?"

Philip K. Dick's 1967 novel *Counter-Clock World* is perhaps the definitive reversed-time horror treatment.[14] Dick's world was once our world, but then, as in Plato's tale, time suddenly begins to run backward. People still alive reverse their direction of aging (but still think, walk, and talk in forward time), and dead, buried people come alive again and emerge from graveyards as the "Sacrament of Miraculous Rebirth" is intoned by a priest; all live their way back to the womb, just as in Plato's tale 1600 years earlier.[15] Such imagery is powerful stuff, but the novel is terribly flawed on a logical level.

Dick says nothing, for example, about what happens in the case of cremated people whose ashes were scattered and nothing about missionaries who were devoured by cannibals. And he certainly is *not* consistent in his presentation. He delights in describing characters disgorging food as they "eat" but avoids obvious (if indelicate) speculations on what happens at the other end. Of course Dick's handling of those two processes, which are called "increting" and "exgesting" in "This Way to the Regress" (Knight), is understandable from the point of view of finding a publisher, and perhaps we should actually be grateful to Dick for his restraint! Indeed, Putnam (1962) has described a human living backward in time as "not a *person* at all, but a human *body* going through a rather nauseating succession of physical states." However, to the logical mind Dick's silence on these issues is a disturbing omission.[16]

Dick has men paste whisker stubble on their faces each morning and slowly absorb it, but I looked in vain for the obvious bathroom scenes of people unbrushing their teeth, with toothpaste appearing on the brush and then slipping smoothly back up into the tube. Apparently, although it is never so stated, only biological processes are time-reversed. Dick also loves playing word games, such as making "You're a horse's mouth" and "You're full of food" epithets of verbal abuse. "Food" is the new universal curse. This is fun for a while (before it gets repetitious), and it is amusing for a while to have people say "hello" as they part and "good-bye" as they meet, (but it doesn't make much sense for characters who are living their ever-younger lives in forward time to say such things.) The same logical flaw appears as well in the de-aging, reversed-time stories "Victims of Time" (Rao) and "The Man Who Never Grew Old" (Leiber).

One thing Dick's often illogical novel does do, with great effect, is to discuss the theological side of reversed time. For example, after a once-dead, now-resurrected character admits to having no memory of much of anything beyond the grave, he is told, "I guess that disproves God and the Afterlife." To that the modern Lazarus replies, logically enough, "No more so than the absence of pre-uterine memories disproves Buddhism." Another science fiction writer has also used reversed time to explore a religious theme quite thoughtfully. In "The Very Slow Time Machine" (Watson) we read of the VSTM that suddenly appears one day in 1985 in the National Physical Laboratory, to the understandable astonishment of all present. It soon becomes apparent that the VSTM is actually traveling backward through time, and its arrival was really its departure. This confusing state of affairs is gradually explained as the story unfolds, and so far as I can tell, all of Watson's descriptions make logical sense *if* one accepts the possibility of time travel—admittedly a pretty big if!

Some of MacBeath's ideas on backward-in-time communication (recall Midge and Jim) are anticipated in Watson's story; the time traveler sealed inside the VSTM communicates with our world via appropriately timed, handwritten messages displayed through a window just as Jim and Midge communicated. The story idea is that a young man in the year 2020 wants to appear in the year 2055 as the Messiah. To go forward those 35 years requires—according to Watson's assumed "laws of time travel"—a preliminary trip backward of the same duration. That initial part of the journey is something like the slow arming of a crossbow, done in reverse at the rate of minus one second of external time per second of internal machine time. Hence, the name VSTM.

The forward half of the trip, on the other hand, is done instantaneously, in the same way that the arrow from the crossbow is launched to its target. The lonely, backward part of the trip drives the would-be holy man crazy, however, but in reverse fashion so that at his first appearance in the laboratory (which is really the instant of departure, so the time traveler has been in the machine, alone, for 35 years), he appears insane to the external observers. Thereafter they gradually see him become less crazy (not to belabor this, but remember that the time traveler is moving *backward* through time during the first part of his journey). As this amazing tale ends, the narrator wonders about the appearance of the VSTM in 2055: "What then, when God rises from the grave of time, *insane?*"

Another tale that is equally careful with its logic is "The Chronokinesis of Jonathan Hull" (Boucher). There we are given a rational explanation for all that happens, an explanation that involves the ill-fated experimenter Jonathan Hull, who develops a machine to achieve a reversal of the time sense. There are none of Dick's dead bodies coming alive and clawing free from dank graves here, and the rest of the world not only walks backward to Hull's eyes (as it should) but talks backward, too. This story is, in fact, an extremely ingenious examination of the real physical effects of reversed time, including the details of how to survive in a world that is going in the opposite time direction from oneself.[17]

Reversed time appears to be simply too bizarre for the movies. The only example I know of from that medium is the Czechoslovakian film *Happy End* from thirty years ago. That movie opens with a close-up of a head in a coffin. As the camera pulls back, we see that there is no body. It is the end (or the beginning) of a man who has just been (or is about to be) guillotined, and the rest of the film shows us why in reverse. The makers of this picture certainly didn't take any of it very seriously, using Mack-Sennet-era music throughout as the score. When *Time* magazine took notice (June 28, 1968) of the film, a witty reviewer thought it simply too long: ". . . the whole conceit might have made a delightful short. Much too hour an is it of minutes 73 but."

The Philosophy and Physics of Reversed Time

In the 1950s, Brown University philosophy professor Richard Taylor thought he had uncovered a non-controversial interpretation of the notion "moving forth and back in time." He first published his ideas in academic journals (1955, 1959) but later collected those somewhat difficult papers together in a section of his book *Metaphysics*. In that book Taylor is careful to distance himself from any suspicion of actually endorsing time travel. Without quite coming out and just saying that, he wrote that his analysis avoids "inconsistent accounts of the kind familiar to the literature of science fiction" and that it is important not to "lapse into imaginative but incoherent fables of science fiction . . . a temptation almost irresistible to most persons unaccustomed to thinking philosophically." Such words are, I think, a perfect example of the pot calling the kettle black! Read on, to see why I say this.

Taylor's reasoning is an excellent illustration of the "grammarian approach" to time. The writing is scholarly, the logic seductive, the *physics* sorely wanting. He starts by arguing that the idea of moving forth and back in space is certainly noncontroversial and also that, given any statement about such movement, we can write an analogous statement about movement in time by making appropriate spatial-temporal interchanges. Consider, then, this statement where O denotes some object or event:

> O is at $place_1$ at $time_1$, and also at $place_1$ at $time_2$; it endures from $time_1$ through $time_2$ but is *then* (i.e., at some time within that temporal interval) at places other than $place_1$.

That statement describes situations of spatial movement that we all see happen all the time. Next, consider the transformed statement that is created by swapping all the spatial and temporal terms in the foregoing statement:

> O is at time$_1$ at place$_1$, and also at time$_1$ at place$_2$; it extends from place$_1$ through place$_2$ but is *there* (i.e., at some place within that spatial interval) at times other than time$_1$.

Taylor asserts that this new statement also describes situations that are perfectly possible. For example, he asks us to imagine two towns—place$_1$ and place$_2$,—that simultaneously experience at time$_1$ an earthquake (which is O). This earthquake also occurs at every place between the two towns, but at one of the in-between places it occurs at a time other than time$_1$. Therefore, says Taylor, the earthquake can be said to move forth and back in time, simply because he arrived at the second statement by swapping all spatial and temporal references in the first statement!

I find it incredible that anyone other than a philosopher could possibly imagine that this bizarre argument has captured what is meant by the words "moving forth and back in time." And not even all philosophers liked it. Otten (1977) demonstrated the basic error by constructing a specific example of a simple physical situation that satisfies Taylor's first statement (which Taylor claims describes moving forth and back in space) but clearly does *not* involve moving forth and back in space. Otten's clever counter-example is simply a long train, with one end *parked* in one of the towns and the other end *parked* in the other town. This train (O) completely satisfies Taylor's first statement—and yet the train is *not* moving; it has nothing to do with spatial movement. Thus if the first statement is flawed, then certainly so is the second statement dealing with time, because it would be derived from the first one.

Otten ends his analysis with the observation that the only way he can imagine to talk coherently about moving forth and back in time *is*, unlike Taylor, to use "the quite controversial notion of time travel." Not all philosophers, however, are convinced that Taylor was merely pushing words around. Swartz (1991), for example, admiringly says of Taylor's 1955 paper that it was a "revelatory" experience for him. And a similar grammarian approach to traveling backward in time, that of swapping temporal and spatial terms, occurs in Mayo (1961)—but see the rebuttal, by a philosopher, in Dretske (1962). Most modern philosophers, however, when addressing the issue of moving about in time, have been more willing and able to put physics, not grammar games, into their analyses.

To be fair, we must acknowledge that bizarre misconceptions about backward time travel have not been limited to philosophers. Consider these words written by a professor of physics:

> Let us consider time travel in the manner of H. G. Wells. Suppose that I were really to travel in time back to my fifth birthday. Here are some children sitting around a table. I am five years old and know nothing of the time machine in my future. *If I really go back, then all traces of the intervening years, inside and outside me, are gone* [my emphasis; how this follows escapes me, and no explanation is given]. There is nothing remarkable about the birthday party. It is indistinguishable from the original one; in fact it *is* the original one. There are no consequences to time travel. A statement that time travel can, or can not, or does, or does not take place is unverifiable and therefore, in my logic as a physicist, meaningless. What is usually called time travel should be called lack of time travel; Wells's picture is that I take my present mind back to past events. This I take to be a fiction.

In that amazing passage, Park (1971) has described a view of backward time travel that, in fact, has no connection at all with physics. Particularly ironic, indeed, is that years before he wrote these astonishing words, the proper physics of time travel had already appeared in the *philosophical* journals.

In support of the logical possibility of time travel to the past, for example, Putnam (1962) asks us to imagine the spacetime diagram of one Oscar Smith who, in Figure 3.2, is at spatial location A next to his time machine. At time t_0 Oscar has not yet gotten into his time machine. A little later, at time t_1, we suddenly see not only Oscar at A but also *two more* Oscars who have appeared (apparently out of thin air) moving away from spatial position B! Between t_1 and t_2 we see the original Oscar at A and the two spontaneously created Oscars at B (for a total of three Oscars, labeled in the figure as Oscar$_1$, Oscar$_2$, and Oscar$_3$) move forward in time—but one of the new Oscars (Oscar$_2$) lives a decidedly odd existence in that his life seems to be running in reverse!

Eventually, at time t_2, the original Oscar$_1$ and the weird, reverse Oscar$_2$, merge and seemingly annihilate one another, vanishing into thin air to leave only a single Oscar (Oscar$_3$) for all time after t_2. Putnam argues that, although strange, what has just been described is still sensible and that, indeed, the very fact that we can draw the spacetime diagram of Figure 3.2 supports the case for backward time travel. He claims this because although the spacetime diagram does show time increasing upward for all three Oscars (that is the time direction for an external observer), there is actually no "spontaneous creation" or "mutual annihilation" and all is sensible *if* Oscar$_2$ is understood actually to be a time traveler into the past with *his* time direction thus pointed opposite to that of the "other two" Oscars. There is, of course, just *one* Oscar!

Weingard (1972a) takes exception to Putnam's suggestion but offers what is in my opinion an even less plausible mechanism for Putnam's kinked space-

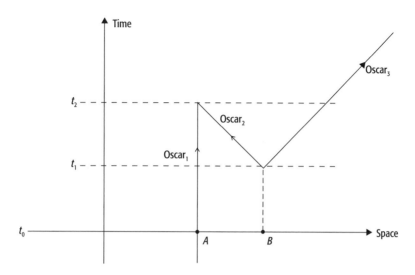

Figure 3.2 A time traveler and his world line.

time diagram than is Putnam's time travel. Weingard advocates, instead, an explanation based on matter transmitters and anti-matter, the latter an idea he credits to Feynman (who actually got it from John Wheeler). Indeed, in Feynman (1949a) we find the famous suggestion that a positron that appears to us to be moving forward in time is actually an electron traveling backward through time (see Note 7 again). Logically, of course, that greatly weakens Weingard's view, because it puts him in the position of using anti-matter (explained in terms of backward time travel) to argue against backward time travel! But let's ignore that concern, give Weingard the benefit of the doubt, and see how anti-matter and backward time travel are connected.

Feynman asks us first to imagine the process shown in Figure 3.3. Gamma ray *A* spontaneously creates an electron-positron pair, with electron$_2$ moving off to some distant region while the positron soon meets with electron$_1$, resulting in mutual annihilation and the production of gamma ray *B*. This description involves three particles, and each segment of the kinked line is a distinct particle. But Feynman said there is another way to look at this, a way that involves just *one* particle. According to Feynman, the kinked line in the figure is the world line of a single *electron;* the middle segment that we called a positron is just the electron traveling backward in time, and so we must reverse the arrow on it (indicating the opposite of the direction shown in Figure 3.3).

There are two central questions at this point. First, why is a positron (with a positive electric charge) moving forward in time mathematically (and physically) equivalent to a negatively charged electron moving backward in time?

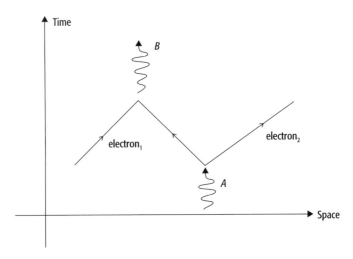

Figure 3.3 Anti-matter through backward time travel.

The answer is that the reversal in charge sign, which results from the reversal of the electron's time direction, follows from the TCP theorem that is discussed in Note 11 for Chapter Two. And second, what *causes* the electron suddenly to move backward in time? Picturesquely, the electron is recoiling from the emitted energy burst of gamma ray *B*. You can understand this as follows. Just as momentum and space are complementary variables—momentum conservation is the consequence of the indifference of the laws of physics to space direction—and just as a particle can reverse its direction of motion in space if it loses enough momentum, so too can a particle reverse its direction of motion in time if it loses enough energy. That follows because energy and time are another pair of complementary variables: Conservation of energy is the consequence of the indifference of the laws of physics to time direction.

Similarly, the absorption of the energy of gamma ray *A* by the electron that is recoiling backward in time causes a second recoil, giving the world line of what we originally called electron$_2$. This reinterpretation of a kinked spacetime diagram was described as follows in Feynman's famous words: "It is as though a bombardier flying low over a road suddenly sees three roads and it is only when two of them come together and disappear again that he realizes that he has simply passed over a long switchback in a single road."

In a latter paper (1979b), Weingard presented another line of attack against Putnam's interpretation of spacetime diagrams as lending support to time travel. There Weingard observed that the presence of the time-reversed Oscar$_2$ shows that the "world of the Oscars" is not *temporally orientable*. A temporally orientable spacetime is one in which *every* point in it agrees with its local neighbors on the directions of past and future—a condition clearly *not* satisfied for the case of Oscar$_2$! As Weingard pointed out, two of the best-known time travel spacetimes in general relativity, the Gödel universe and the Kerr (spinning) black hole, *are* temporally orientable, so the ambiguity of Oscar$_2$ (as to whether he is traveling backward in time, as opposed to living forward "in reverse") does not occur. That is, Weingard *agrees* with Putnam's acceptance of the conceivability of time travel to the past, but not with his use of Feynman's concept of anti-matter as time-traveling matter.

Still, even early on, Feynman's idea was greeted with great enthusiasm by many physicists. The Japanese physicist Yoichiro Nambu, for example, was almost poetic when he opened a highly technical paper (1950) as follows:

> The space-time approach to quantum electro-dynamics, as has been developed by Feynman, seems to offer a very attractive and useful idea to this domain of physics. His ingenious method is indeed attractive, not only because of its intuitive procedure which enables one to picture to oneself the complicated interactions of elementary particles, its ease and relativistic correctness with which one can calculate the necessary matrix elements or transition probabilities, but also because of its way of thinking which seems somewhat strange at first look and resists our minds that are accustomed to causal laws. According to the new standpoint, one looks upon the world in its four-dimensional entirety. A phenomenon that will come into play in this theatre is now laid out beforehand in full detail from immemorial past to ultimate future and one investigates the whole of it at a glance. The time itself loses sense as the indicator of the development of phenomena; there are particles which flow down as well as up the stream of time; the eventual creation and annihilation of pairs that may occur now and then, is no creation nor annihilation, but only a change of directions of moving particles, from past to future, or from future to past; a virtual pair, which, according to the ordinary view, is foredoomed to exist only for a limited interval of time, may also be regarded as a single particle that is circulating round a closed orbit in the four-dimensional theatre; a real particle is then a particle whose orbit is not closed but reaches to infinity. . . .

Of course, Feynman's radical idea has its critics, too, particularly among philosophers. Smith (1985), for example, wonders what it would be like for an electron to travel backwards in time. Here is his answer to this question.

> If the whole world is but one of God's films, then time reversal is the "film of the world" played in reverse. Consider an electron e_1. At time t_0 it is at (x_0, y_0, z_0). At time t_1 it is at (x_1, y_1, z_1).

If the direction of time for the electron was reversed, then the electron would be observed on the "film of the world" to travel back along the same path as it did before, i.e., back to (x_0, y_0, z_0). If God stopped the "film of the world" and examined the charge of e_1, then He would find that it was *negative,* not *positive.* Hence the electron travelling backwards in time is simply that: an *electron* traveling backwards in time, it is not a positron. Time reversal does not result in a reversal of charge. Thus, the Stückelberg–Feynman position is incorrect. . . ."

Well, so much for the mathematical integrity of the TCP theorem (see Note 11 for Chapter Two again). The self-confidence required to flick the theorem aside, simply on the basis of what one imagines God would "see" if He stopped the world, is pretty breathtaking. Indeed, one would be forced to admit admiration if only for its daring, if there weren't an obvious reply: What matters is not what God would "see" if he stopped the world (assuming, of course, that that sequence of words has any meaning), but rather what observers stuck in spacetime with an arrow of time *opposite* that of the electron's arrow would see. When God *stops* the film of the world, then presumably neither He nor the electron *has* an arrow of time!

Smart, for another example, who you will recall from the start of this chapter rejected as nonsense the ideas that time has a direction and is flowing, found Feynman's interpretation equally hard to swallow. He wrote (1958) that "One at once smells a category mistake here. After all, does an electron normally move *forward* in time? Clearly not. (If it did, how fast would it move? How many seconds per second?) Both 'forward in time' and 'backward in time' are equally nonsense." (See Note 7 again.) Perhaps the most startling puzzle of Feynman's interpretation is that it seems to say that the same electron can be at two different places in space at the same time (a point also raised by Smart). Of that, Reichenbach (1956) said, "The concept of physical reality is shaken to its very foundation."

Two physicists—Graves and Roper (1965)—answered Reichenbach's worry by arguing that the "same time" is that of some observer but that for a clock traveling with the electron and recording the local or *proper* time (see Tech Note 5), time is always increasing; that is, when the electron is in different places, the proper time is also different. It is interesting to note that Reichenbach (1956), published years before Graves and Roper wrote, had already rejected their explanation based on the use of different clocks. He felt that the confusion of temporal ordering of events along the kinked world line as a function of which clock you used, observer's or electron's, was "the most serious blow the concept of time has ever received in physics."

In a reply to Graves and Roper, Earman (1967a) sharply rejected their point of view. Earman particularly disliked Graves and Roper's attempt to argue about backward time travel in terms of bent-back world lines, which have to reverse direction sharply in order to stay within the requirement of special relativity that nothing of substance participating in a causal event chain can move faster than light (see Tech Note 4). The significance of special relativity in discussions of backward time travel is weak, however, and the thesis of this book is accurately contained in Earman's closing words, "General relativity seems to offer more hope than special relativity to the proponent of time travel." Indeed, Tech Note 8 discusses how the world line of a time traveler to the past can always satisfy special relativity's causality constraint.

Entropy as Time's Arrow

Whereas the reversed-time fictional worlds of Dick, Fitzgerald, and Carpentier are of course metaphors on the human condition,[18] Boucher's "The Chronokinesis of Jonathan Hull" is clearly trying to appeal to the analytical, logical mind of the scientist. After a little mood setting with some scientific-sounding babble words on how the method involves "the rotation of a temporomagnetic field against the natural time stream," Boucher has one of his characters face up to the real puzzle of it all: "How can a man live backward? You might as well ask the Universe to run in reverse entropy." That cogent question brings us, in fact, to the first scientific explanation developed to explain the observed asymmetric nature of time.

It was the Englishman A. S. Eddington who gave the picturesque name the "arrow of time" to the observed asymmetric nature of time's direction from past to future. He was also one of the popularizers of an explanation (1929) for the arrow, using the famous second law of thermodynamics. The second law of thermodynamics states that a measure of the internal randomness or disorder—what is the called the *entropy*—of any closed system (that is, one free of external influences) continually evolves toward that of maximum disorder, toward the condition called *thermodynamic equilibrium*. Indeed, so striking is this increase in entropy S with time in a macroscopically large system that the increase in entropy has come to be thought of as actually *defining* the direction of time. Eddington, however, was not the originator of the entropy concept. The history of entropy can be traced back to before the turn of the century, to the great Austrian scientist Ludwig Boltzmann (1844–1906) and his famous H-theorem of 1872. The quantity H in that theorem is directly related to the now more familiar entropy.[19]

The steady increase in entropy is a phenomenon often observed in the everyday world. A drop of ink in a glass of water spreads out in an expanding cloud, a cloud we never see collapse backward into an ink drop. A long rod of metal, initially hotter at one end than the other, evolves toward a constant temperature along its entire length. We never see a uniformly warm rod spontaneously begin to cool at one end and grow hot at the other. A hot bath grows cold—nobody has ever seen a bath at room temperature suddenly, all by itself, begin to heat up and then boil in the middle of the tub while the edges freeze. In all of these cases, the end (future) state represents greater internal randomness or disorder than does the beginning (past) state.

The first formal entropy model for the direction of time was put forth in a 1907 paper by the Austrian physicist Paul Ehrenfest (1880–1933), who was a friend of Einstein's, and his Russian-born wife Tatyana (1876–1964), who was a skilled mathematician and her husband's occasional collaborator. In their paper the Ehrenfests developed the idea, one of the mainstays of physics, of the so-called *entropic clock*. This clock, a statistical model based on the then new probability mathematics of Markov chains—after the Russian mathematician A. A. Markov (1856–1922)—describes how gases diffuse, and it is both a simple and a powerful concept. The Ehrenfest model is illustrated in Figure 3.4.

Formally, as derived by Boltzmann in 1877, the entropy of a system in a given state is denoted by S, a quantity proportional to W, which in turn is the

---

Figure 3.4 This illustration is based on the discussion in Wheeler (1979), which in turn is based on Cocke (1967). Imagine two urns, I and II, each containing n balls. Initially, at time $t = 0$, all of the balls in Urn I are black and all of the balls in Urn II are white. Then, at time $t = 1$ (in arbitrary units), a ball is selected at random from each urn and (instantaneously) placed in the other urn. This select-and-transfer procedure is repeated at times $t = 2, 3. \ldots$ At any given time each urn always contains n balls, but only at $t = 0$ are the colors of the balls in a given urn necessarily identical. The phrase "*selected at random*" means (for example) that the probability of selecting a black ball from an urn containing b black balls is b/n. At any given time, we completely describe the state of both urns by specifying the number of black balls in Urn I (or the number of white balls in Urn II, etc.). It is easy to write a computer simulation of this physical process, and the two plots in this figure (created by a MATLAB simulation written by the author) show how the fraction of black balls in Urn I evolves toward 0.5 as time increases. Both plots are for $n = 100$ balls, but they were created using different random-number sequences to simulate the random ball selections. The important observations are that (1) the evolution of the state of the two-run system is toward 50% black balls in Urn I for both simulations (and would be for "almost all" sequences of random numbers) and (2) the evolution is *not* monotonically decreasing from 100% black balls to 50% black balls but rather has never-ending fluctuations about 50% that may, in fact, be rather large in both amplitude and duration.

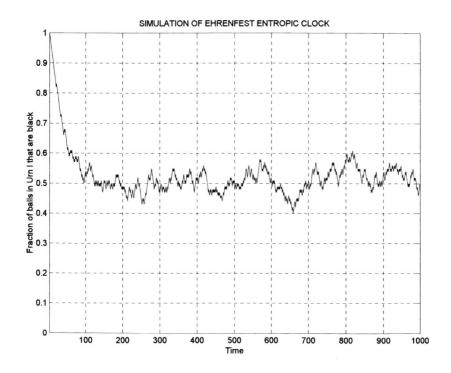

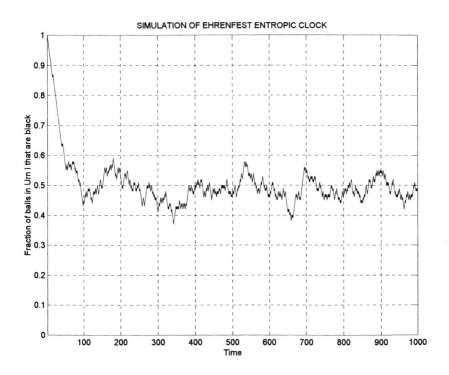

number of different possible ways the state can occur as a result of all possible variations of the system's internal, microscopic structure. The calculation of W in practical, everyday systems can be quite complicated, but in various idealized systems it is straightforward. Consider, for example, a vacuum cylinder with a thin membrane dividing the interior into halves. Suppose that we insert (to be specific) six molecules into the left half of the cylinder. If we define the microscopic state of the system to be the number of molecules in the left half, then initially $W = 1$ because there is just *one* way to put all six molecules on the left side. This represents the state of *minimum* entropy, the state of maximum order that is most distant from thermodynamic equilibrium. If we now puncture the membrane, then the molecules, once confined to the left side, are free to move about the entire cylinder. At any instant of time we can imagine counting the number of molecules on the left side—suppose that at some particular instant we count five, with one molecule having moved to the right side. Then $W = 6$, because there are six ways to pick the molecule that has moved from left to right, and so the entropy of the system has increased.

We think of the thermodynamic equilibrium state as being the state with equal numbers of molecules in both halves of the cylinder, and that state has the *maximum entropy*, which is associated with $W = 20$. With such a small number of molecules, it is not clear that W (and so S) will *inexorably* increase with time; perhaps, after one of the six molecules has gone to the right, it simply returns to the left side before any of its companions have joined it on the right side. Such an event is called a *reversal*, and it happens with some non-zero probability. But the more molecules in the cylinder (instead of six, make the number a million million million—still a small amount of gas in our everyday world, hardly enough to fill a sewing thimble), the more likely it becomes that the value of S *will* monotonically increase with time.

That is, low entropy was in the past, and high entropy will be in the future. The increase in entropy defines a direction to time, so entropy has come to be called the *thermodynamic arrow of time*. More accurately, it should be called the probabilistic or statistical arrow of time. These statements are all accepted today as certainly being true—as far as they go. But there is clearly a puzzle here, too. The puzzle is that the motion of each of the molecules, individually, is time-reversible, whereas the statistical or average behavior of the many is not. Now, the use of averages over large collections implies the loss of detailed information about the individual molecules. Thus what we have is the puzzle of how it can be that by *reducing* our knowledge of a system through statistical averaging we then find it displaying a *new* property (asymmetric time evolution) that we did not see before. And if that question isn't troublesome enough, we also have two more puzzles called the "reversibility" and the "recurrence" paradoxes.

The reversibility paradox is actually the issue raised earlier in this chapter: The classical equations of physics work just as well with time running in either direction, so why *don't* things actually go backward? This question, originally raised by Lord Kelvin in 1874, was brought to Boltzmann's attention in 1876 by the German physical chemist Johann Loschmidt (1821–1895), one of Boltzmann's professors at the University of Vienna. Boltzmann's answer to this paradox was that it *is* imaginable that a world could run backward if initial conditions were suitable. For example, if all the velocity vectors of every particle in an equilibrium state were reversed, then the system *would* unwind backward in time toward its original nonequilibrium condition. That is, a system in thermodynamic equilibrium, the state of highest entropy, could evolve toward one of low entropy. Boltzmann even suggested that such might be the case for regions in our own universe—that there might be beings in a world somewhere "out there" who experience time running counter to our earthly experience. He said that in 1877, and it is a remarkable statement for a conservative nineteenth-century professor, nearly a century before Dick's *Counter-Clock World*! But, argued Boltzmann, from most given states there are vastly more ways for entropy to increase than there are for it to decrease, and that is why we see what we see, a continuous *increase* in entropy.[20]

The recurrence paradox is something entirely different; it is based on a result established in 1890 by the great French mathematician Henri Poincaré (1854–1912) in his paper "On the Three-Body Problem and the Equations of Dynamics." Motivated by the question of the stability of the motion of three masses governed by Newton's laws of mechanics, Poincaré showed that starting from almost any initial state, any fixed volume system with a finite amount of energy and a finite number of degrees of freedom will return *infinitely often and with arbitrarily little deviation* to almost every previous state. If you wait long enough, implies Poincaré's astonishing theorem, Pearl Harbor will happen again—and again, and again, and. . . . In 1896 the German mathematician Ernst Zermelo (1871–1953) used this result, which philosophers call the eternal return, to claim that there could be no truly irreversible processes and thereby cast doubt on the idea that entropy always inexorably increases.

Even for very small systems, however, such as a mere handful of molecules, the recurrence time is extremely large, and this was, in essence, Boltzmann's reply to Zermelo's concern. For example, if the gas-filled cylinder has just one hundred molecules (not six) and if transitions from one side of the cylinder to the other side take place at the rate of one million per second, then Blatt

(1956) has calculated the recurrence time to be something like thirty million billion years! And for the universe itself, the recurrence time is simply incomprehensible. Mathematicians call 1 followed by a hundred zeros a *googol,* and the recurrence time in years for the universe has been estimated to be 1 followed by a googol of zeros (a so-called *googolplex* of years).[21]

An expanding universe (such as the one we appear to live in) would also seem to violate the Poincaré condition of fixed volume. Eddington (1935) put this as follows in one of his 1934 Messenger Lectures at Cornell University: "In an expanding space any particular congruence becomes more and more improbable. The expansion of the Universe creates new possibilities of distribution faster than the atoms can work through them, and there is no longer any likelihood of a particular distribution being repeated. If we continue shuffling a pack of cards we are bound sometime to bring them into their standard order—but not if the conditions are that every morning one more card is added to the pack."

Further, Tipler (1980) shows that if one considers general relativity rather than simply classical dynamics, then the recurrence theorem is simply no longer true. Tipler concluded that "in general relativity, singularities intervene to prevent recurrence. General relativistic Universes are thought to begin and end in singularities of infinite spacetime curvature, and these singularities force time in general relativity to be linear rather than cyclic." To arrive at this conclusion, however, Tipler assumes that gravity is always attractive and that spacetime satisfies a special condition (the so-called *Cauchy condition,* discussed in Tech Note 9) that avoids such bizarre situations as backward causation. But the first assumption is violated in wormhole time machines (see Tech Note 9 again), and the second is *by definition* violated in any general relativistic spacetime that supports time travel, such as Gödel's spacetime.

An interesting connection among time-reversed worlds in which entropy is the arrow of time, anti-matter, singularities, and faster-than-light tachyons is made in Gott (1974). Gott (the inventor of the cosmic string time machine—see Tech Note 10) discusses how, if quantum effects should *prevent* the Big Bang from being a singularity, then it appears possible that our observed universe with "normal" time direction and matter may be joined to two other universes, one containing tachyons and the other being a time-reversed, anti-matter universe. That would seem to be a wonderful idea for science fiction writers, but I haven't seen it yet used in a story.

Before going any further with entropy as "time's arrow," I should mention that it is possible to discuss entropy changes *without* introducing the concept of a direction to time. This was done by the great quantum physicist Erwin

Schrödinger (1950). You can most easily understand his argument by imagining a movie film of *two* isolated systems, A and B, each undergoing internal changes. To avoid any circular reasoning based on *assuming* a direction to time, let's further imagine that we *don't know* the temporal direction of the film. That is, the frames of the film are numbered consecutively as 1,2,3, . . . , but we don't know whether frame 1 is the first frame or the last frame. From the information recorded on each frame (such as thermometers measuring the temperatures of A and B) we can compute the entropy of each system, frame by frame. For example, let's write S_{A_i} = entropy of system A in frame i, and S_{B_j} = entropy of system B in frame j. Examination of many such movie films shows that the entropies of isolated systems "almost always" evolve in parallel—that is, in the same manner (as shown in Figure 3.4, for example). Thus

$$(S_{A_i} - S_{A_j})(S_{B_i} - S_{B_j}) \geq 0$$

This follows because, although we don't know the temporal order of frames i and j, which means we don't know whether the two factors in the inequality are positive or negative, we *do* know that they almost always have the same sign *independent* of whether entropy increases or decreases with increasing or decreasing time. Hence the product of the two entropy changes is non-negative no matter what the "arrow of time" may be.

It didn't take long for science fiction writers to incorporate entropy as time's arrow into time travel. In a 1930 tale there is the brief statement that entropy is behind the operation of "The Time Valve" (Breuer). And in the 1937 "Temporary Warp" (Long) the inventor of a "warp gun" explains, "The stupendous distortion of the warp may actually bring about a sort of kink in spacetime, and result in a reversal of entropy"—and when the gun is fired, a woman who is hit by the warp ages seventy years in seconds (which is, of course, exactly the opposite of what we would expect from a "reversal of entropy"). Later, in 1938, *Amazing Stories* published "Time for Sale" (Farley), the story of a college student about to flunk his senior physics course. The examination is scheduled for the following day, but he needs a week and a half of study time. To his rescue comes ENTROPY, INC., a company that sells time by placing its clients inside a "time-cabinet" in which the increase in local entropy is greatly accelerated. To someone looking through a window at the interior of the time-cabinet, the occupants would appear as characters in a speeded-up movie. Referring to Eddington by name, the author tells us that "entropy is what makes time irreversible—is what gives us the feeling of the flow of time." More recently "ARM" (Niven) repeated this idea in a far more sophisticated story about a so-called "time compressor" in which time runs

500 times faster than normal time, whereas in "Half-Past Eternity" (Mac-Donald) the speed-up factor is 36,000!

To travel into the future using an entropy cabinet does have some sense to it. "Simply" *slow* the increase of entropy inside the cabinet relative to the rate of increase outside. And indeed, Farley's story has this twist, too—the hero enters the modified time cabinet and then, after just a few hours have past *for him,* emerges to find that years have gone by on the outside. And that is just fine with him, as it turns out, because the daughter of the married woman he originally loved has had time to grow up and . . . surely you can finish the romantic story line yourself!

Also citing Eddington is a 1941 entropy tale that offers a curious twist; when the central character in "The Man Who Lived Next Week" (O'Brien) arrives in the future, his *clothing* has changed (but it returns to its original state when he returns to his original time). The return part of that journey, a trip to the *past,* suggests a curious problem, however, because an entropy time-cabinet makes sense only for *local* control of time. For example, travel into the global *past* via the local manipulation of entropy is not possible, because it would require the reversal and decrease of entropy for the entire external world. (Even that fantastic idea *was* once used, in the 1946 story "The Bacular Clock" (Bond), to run the entire world backward to undo a terrible train accident.)

In the melodramatic 1942 story "Prisoner of Time" (Cross), one of early science fiction's stereotypical mad scientists uses the entropy interpretation of the arrow of time for revenge. There we read of Bryce Field, "a master-scientist, a demon, cruel, ruthless," who is rejected in love by the stupendously beautiful Lucy Grantham. Her lack of enthusiasm is perhaps understandable; he is described as having "a lean-jawed, sunken-eyed" appearance, along with "lank, untidy hair sprawled across his massive forehead." As Lucy tells him at one point, "I could never love you; you are too clever, too brilliantly scientific." After hearing that, it is no surprise that before we are more than a page or two into this perhaps unintentionally hilarious tale, we learn that Bryce has Lucy strapped to a steel table in an underground laboratory-in-a-cave. Then he tells her of her fate: "You are going on a long journey, my dear. So long a journey that even I, master-scientist, do not know when it will end. A journey into the future—alone! . . . You, Lucy, shall be the victim of entropy! . . . I have discovered how to make a [globe] of non-time. Entropy will be halted. . . . You will be plunged into an eternal 'now.'" And so the mad Doctor Field throws the switch on the wall of his "instrument-littered" cave on July 17, 1941, and Lucy remains "suspended" in time until the outside world reaches the date of August 9, 2450. That is the day she is at last dug up from the cave by "big and muscular" engineer Clem Bradley and his "square-jawed" sidekick Buck Cardew, who use a "warp in spacetime" to release Lucy from her "globe of non-time."

Poul Anderson used entropy in a similar but vastly more "scientific" way in his "Time Heals," originally published in 1949 by *Astounding Science Fiction* magazine. There he describes how a scientist has discovered "a field in which entropy was held level." As Anderson explains, "An object in such a field could not experience any time flow—for it, time would not exist"; this is because time flow is a *change* in entropy, and the 'change' of a level (or constant) field is zero. Anderson speculates in his story about how such a field could have fantastic home uses ("Imagine cooking a chicken dinner, putting it in the field, and taking it out piping hot whenever needed, maybe twenty years hence!"). But its real use in the story is as a stasis generator for preserving fatally ill people until medical science has learned how to cure their diseases. This is, then, a high-tech method of suspended animation, of true time travel into the future that is different from simply freezing (a clock in such a field would not age or measure the passage of proper time). The gadget that does all this is called, somewhat sinisterly, the "Crypt." Anderson also tells us that the Crypt makes a great bomb shelter, too, because "not even an atom bomb could penetrate a stasis field." The reason for that is intriguing: "The field requires a finite time in which to collapse—only there is no time in it." The interior of the Crypt is, quite literally, a frozen block of time more rigid and unyielding than the strongest steel.

Somewhat later came Arthur C. Clarke's 1959 story about the tragic end of a geologist fifty million years in the past. This aptly named tale, "Time's Arrow," uses the idea of entropy to justify a time machine. Entropy has also clearly fascinated Robert Silverberg. For example, when a newspaper from the future (an idea used in the 1997 television series "Early Edition") appears on people's doorsteps in "What We Learned from This Morning's Newspaper," the initial astonishment is replaced by puzzlement as the papers rapidly disintegrate. That is the result, we are told, of "entropic creep." The explanation continues, informing us that it is sort of like a strain in a geological fault (Silverberg lives in Oakland, California, now and then the site of large to huge earthquakes, and it isn't surprising that he uses this particularly imagery): "Entropy you know is the natural tendency of everything in nature to come apart at the seams as time goes along. These newspapers must be subject to unusually strong entropic strains because of their anomalous position out of their proper place in time."

And in Silverberg's "In Entropy's Jaws" we see the randomness that underlies entropy running through the entire story. The central character is a telepathic "Communicator" who is "burned-out" by an information overload while mentally linking two clients. He thus becomes "unstuck in time"—like Vonnegut's Billy Pilgrim in *Slaughter-House Five,* like the time traveler in the film *Je t'aime, Je t'aime,* and like Chris McAllister in the novel *Weapon Shops of Isher* (Van Vogt), he starts oscillating, wildly and uncontrollably, back and forth between past and future. In fact, as the title implies, he is literally masticated by the teeth of time. The play on entropy and the job of Communicator is clear as well, because entropy plays a central role in the mathematics of information theory.[22] Silverberg is fascinated by this image of swinging forward and backward through time, and he repeated it in his appropriately titled novel *Project Pendulum.*

Other Arrows of Time

Despite all of the previous discussion, we must still admit that it is not clear whether the evolution of a system from past to future is always accompanied by an irreversible increase in entropy—that is, by an inexorable increase in some measure of the system's "disorder." It can be calculated statistically that entropy is *very likely* to increase monotonically in systems of macroscopic size, but that is not the same as certainty. Eddington, therefore, was wrong when he dramatically announced (1929), "The law that entropy always increases holds, I think, the supreme position among the laws of nature. If someone points out to you that your pet theory of the Universe is in disagreement with Maxwell's equations—then so much the worse for Maxwell's equations . . . but if your theory is found to be against the second law of thermodynamics, I can give you no hope; there is nothing for it but to collapse in deepest humiliation."

Contrary to Eddington, my position is that the second law is *not* on the same level with, for example, the fundamental conservation laws, which *never* fail in classical physics. Also contrary to Eddington, Maxwell's equations are on a higher, not a lower, rung of the ladder of "fundamentality" (if I may coin a word) than is the second law. In classical physics, Maxwell's equations *never* fail. The steady increase of entropy, on the other hand, *can* and *does* fail; there can be fluctuations in the thermodynamic evolution of a system so as to have, at least for a time, a decrease in entropy (take another look at Figure 3.4). All we can say for sure is that for macroscopically sized systems, even small fluctuations in entropy are most unlikely. To quote no less an authority than the combined genius of Gilbert and Sullivan (from their opera *H. M. S. Pinafore*), we can say of the possibility of failure in the supposed inexorable increase of entropy: "What, never?/No, never!/What, never?/Well, hardly ever."

Still, for most physicists entropy is just too useful a concept to give up even though it does not *always* increase with increasing time for an isolated system. One interesting way to argue for the irreversibility of a system whose basic micro-laws are reversible is given by Morrison (ET). Morrison finds the unavoidable perturbing effects of outside influences on a system to be sufficient always to preclude the possibility of an actual velocity-reversal of the micro-components of the system. He finds, in fact, that it takes almost nothing to disturb a system to the point where it will never unwind in reverse. As he explains, "One may estimate that a gravitational force exerted by a falling apple a kilometer away over an arc of ten centimeters is ample to mix up the trajectory of a mole of normal gas, in a time of milliseconds!" Because of this, Morrison concludes that the entropic arrow of time would be independent of the expansion/contraction state of the universe (see Note 15 and the following discussion on the so-called *cosmological* arrow of time). But Morrison's logic is merely begging the question, because his supposed explanation for time asymmetry has the asymmetry built into it: He assumes that any perturbing influence will always *increase* entropy, which is precisely what he is trying to explain.

The idea that the universe began in some sort of Big Bang process fifteen billion years or so ago is the generally accepted view today. The puzzle of that event, one that has been described as literally being a 'fireball' explosion, is that it must have been *fantastically* hot. This means that at the beginning (of everything) there was complete thermodynamic disorder, which from the earlier discussion means *maximum* entropy. Thus we immediately have a huge paradox. How can the entropy of the universe be continuously increasing if it was as large as possible right from the start? This long-standing puzzle was addressed in Davies (1983b), which discusses the so-called *inflationary universe model* as the solution (see also Tech Note 9).

The inflationary model has a very high expansion rate for the early universe, much higher than the rate in the standard hot Big Bang model. In the standard model, the entropy puzzle occurs because of the ability of all particle processes to readjust rapidly to the ever-changing state of the universe; the so-called relaxation times of all particle processes were short compared to the expansion rate of the universe. That means the actual entropy of the universe would, indeed, be the maximum possible at every instant (and so, of course, we have the entropy paradox). In the inflationary model, however, the expansion rate of the early universe was temporarily so high that the relaxation times of particle processes were very *long* compared to the expansion rate. This means that the maximum *possible* entropy of the universe, at every instant, would greatly exceed the *actual* entropy. As Davies put it, an "entropy gap" was created during the inflationary phase of the early universe. The result was the thermodynamic arrow of time that we observe today, as the universe tries to "catch up" and reduce its entropy deficiency. As one might expect for such an imaginative, speculative suggestion, not all have agreed. In Page (1983, 1984), for example, we find rebuttals that assert that the inflationary model of the early universe really *assumes*—as does Morrison's idea—the thermodynamic arrow of time, rather than explains it. Davies was not convinced by Page, however, and he replied in turn (1984). The entropy paradox (and the origin of the thermodynamic arrow of time) remains, I believe, one of the central outstanding problems in modern cosmology.

Well, whatever you may think of Morrison's and Davies' approaches, there are also serious philosophical problems with invoking entropy as the explanation of time. For example, events in the past leave traces, artifacts taken to be ordered states—or at least more ordered than their immediate surroundings. The classic example is a footprint in the sand, which is clearly a highly organized structure compared to the surrounding sandy beach. This is the trace of a *past* event; such a trace was all the evidence, for example, that Robinson Crusoe needed to conclude that another human had walked that way. But now consider Earman's famous counter-example (1974), that of a bombed city. Certainly there are traces aplenty of the past bombing, and in fact one has to be careful not to trip over or fall into them! The puzzle, of course, is in trying to argue that random bomb craters, strewn rubble, and crushed buildings somehow constitute a more organized state than did the original city and the surrounding unbombed areas.

For a second example, consider the situation described in Denbigh (1989), of a cloud of non-colliding particles all initially moving toward each other. At first the radius of the smallest sphere that contains the cloud decreases with time, but eventually, as the particles move past one another, the radius will grow without bound. Indeed, that inexorable increase of the radius could be taken as defining the direction of time that points toward the future. But in what sense is the disorder of the particle cloud increasing? After all, as the cloud expands, it looks the same at all times; only its scale (radius) changes. One might reply that this is an example of an open or unbounded system, whereas the entropic gas clock is defined as a closed, bounded system. Let's admit that, but the question still stands: What has entropy to do with our expanding-into-the-future cloud? Perhaps nothing.[23] Perhaps what is needed is a new arrow of time.

So far we have looked in some detail at two of the so-called arrows of time: the subjective, psychological feeling we have of time "flowing," which has no explanation in physics, and the thermodynamic, statistical quantity of entropy. A third arrow that I have hinted at is the so-called *cosmological arrow* of the expansion of the universe. This is an arrow not nearly so obvious as the first two. Only in modern times (since the 1920s), as a result of the work of the American astronomer Edwin Hubble, has science become aware that the universe *is* expanding. An interesting speculation about the thermodynamic and cosmological arrows, that has been made many times, is that if the cosmological arrow should ever reverse—that is, if the universe should ever begin to contract—then the thermodynamic arrow would also reverse, e.g., see Gold (1962). The reasoning is that the thermodynamic arrow *follows* the cosmological arrow in an *expanding* universe because that universe can always swallow up ever more electromagnetic radiation as it is produced by any physical process. Why shouldn't the thermodynamic arrow continue to follow the direction of the cosmological arrow even during a contraction?

The usual objection to that suggestion is simple enough. If the direction of time did reverse, then we would see (so goes this argument) all sorts of odd events that would require enormously improbable physics, such as a shattered glass mirror reassembling itself. The error in that objection is subtle but equally simple. *It presupposes the retarded causality of our expanding universe.* In a contracting universe with a reversed thermodynamic arrow of time, however, there would be *advanced causality,* and thus there would be nothing at all improbable about such things as self-assembling mirrors. An interesting analysis of the relationship between the thermodynamic and cosmological arrows of time is given by Schulman (1973).

More recently, Davies and Twamley (1993) observed that "The mere reversal of the cosmological expansion will not of itself serve to reverse the direction of thermodynamic and electrodynamic processes, any more than the compression phase of a piston-and-cylinder cycle in a heat engine serves to reduce the entropy of the confined gas." Citing Cocke (1967)—the inspiration for Figure 3.4, you'll recall—as offering a way around the question, Davies and Twamley then mention Hawking's interest in the relationships among the various temporal arrows. At one time, Hawking thought (1985) he had discovered a connection between the thermodynamic and cosmological arrows, but, at least partially because of arguments presented by Page (1985), Hawking has since abandoned that claim. For an analysis of the reasoning that led Hawking to his original claim, and for for an account of why it is incorrect, see Hawking, Laflamme, and Lyons (1993). Hawking, in fact, has labeled (1994) his original claim "my greatest mistake in science," and in that paper has quite openly (and most entertainingly!) discussed his interest in the arrows of time. Indeed, it was to be the subject of his doctoral dissertation, but "I . . . needed something more definite, and less airy fairy than the arrow of time, for my PhD, and I therefore switched to singularities and black holes. They were a lot easier."

Yet another arrow of time is the *electromagnetic arrow*, which refers to the fact that electromagnetic radiation is always observed to propagate into the future, never into the past. This is a mysterious fact, because Maxwell's equations for the electromagnetic field, like all the other laws of physics, have no intrinsic time sense. And yet another arrow was proposed, in anticipation of Morrison's falling apple, when Penrose (1979) suggested that gravity might be the basis for a temporal arrow. The motivation for his suggestion is the observation that a fluid mass under the influence of a non-uniform gravity field is subjected to tidal forces that cause *increasing* disturbances with *increasing* time. One concern about this so-called gravitational arrow, a concern that Penrose himself discusses, is that he has built in the very time asymmetry he is trying to explain. He did this by specifically forbidding the existence of white holes (the time-reversed version of black holes).

Despite that concern, or perhaps because of it, Bonner (1988) attempted to correlate Penrose's arrow with the electromagnetic arrow for a particular case in which the physical details can be calculated. Alas, for Bonner's special case of the collapse of a spherically symmetric, hot, radiating ball of gas (an idealized model for a star), the two arrows point in opposite directions! The gravitational arrow of time seems to have faded from view in recent years, but we will discuss the electromagnetic arrow in some detail in Chapter Four.

The search for an arrow of time lurking somewhere in the laws of physics has continued unabated, and technical and philosophical discussions, respectively, of how that search has been extended to the domain of quantum gravity can be found in Moffat (1993) and Liu (1993). As in nearly everything connected with quantum gravity, however, results so far are all extraordinarily tenuous, speculative, and debatable. As Liu writes, "Much investigation, both in physics of quantum gravity and in philosophy of reversibility, has to be carried out before one could have a clear sense of what exactly quantum gravity is trying to reveal about the secret of time."

Multidimensional Time

Could there be such a thing as more than one direction to time's arrow at each instant? At first this seems an absurd idea, something akin to a man jumping onto his horse and riding off in all directions at once. But philosophers, and to some extent physicists, have started to take a serious look at the concept of multidimensional time. As with so many other of the radical concepts associated with time travel, though, science fiction writers were dealing with multi-dimensional time long before it became a respectable topic in learned philosophical and physical journals.

In "Elsewhen" (Heinlein), for example, a story originally published in the September 1941 issue of *Astounding Science Fiction,* we find a professor asking his redundantly named class in speculative metaphysics "Why shouldn't time be a fifth, as well as a fourth, dimension?" In response to a generally skeptical reception to that, the professor goes on to say, "I believe in the existence of a two-dimensional time scheme. . . . Ordinarily, most people think of time as a track they run on from their births to their deaths. . . . Think of this time track we follow over the *surface of time* as a winding road [it is the imagery of a surface that gives the professor *two* time dimensions]. . . . Once in a while another road crosses at right angles. Neither its past nor its future has any connection whatsoever with the world we know." (This story has a particularly amusing scene in which one of the professor's students accidently "jumps time tracks" and enters a new track with his time arrow pointing backward.)

The year before, the same magazine had published another tale, "Bombardment in Reverse" (Knight), that went well beyond a mere two time dimensions. We are told in that story of two countries on an alien planet at war in the distant future. The war is a stalemate until one side begins to fire a gun at its foe from just two miles from its target in the heart of enemy territory— and from the middle of next week! The gun's shells are true "time bombs." This is not mere ordinary time travel along one's time track, however, but a multidimensional effect. Using a photograph of the gun in actual operation to support his astonishing discovery, an agent for the side being shelled reports to his superior that "the gun and its crew are existing along another time axis at right angles to the direction of our "normal time," so that from our point of view they are existing perpetually in the same instant." That explains why the gun crew can (will?) operate without interference in next week's future, as they are in their adversary's time only for the instant that the two time tracks intersect. Indeed, the spy used the same trick to obtain his undetected photograph: "I secured the photograph by orienting myself along still another time axis at right angles to that of the gun, and approached it as an instantaneous, invisible entity." By the story's end both sides are using and counter-using this technique, evading each other "to and fro along an ever increasing complexity of mutually perpendicular time axes." In fact, the final count exceeds 75 time axes, making Heinlein's two-dimensional time look rather skimpy by comparison.

Well, of course, 75 time directions *is* science fiction, and physicists are not so enamored of multidimensional time as are science fiction writers.[24] For example, Eddington long ago wrote that he found the idea of any region of spacetime with two-dimensional time to "defy imagination."[25] More recently, Dorling (1970) has looked at the idea more analytically. Dorling showed that the extremal property of timelike geodesics (see Tech Note 4) would fail for multidimensional time, which he then associates with the stability of matter. In addition, he makes a connection between multidimensional time and a failure of causality. And most recently, Isham (NP) has written on the possibility that a viable theory of quantum gravity might support the idea of multiple time dimensions. That might occur if the topology of spacetime, at dimensions smaller than the Planck length, changes from the smooth, essentially flat appearance it displays in the macroscopic to what is called the "quantum foam." (see Tech Note 9). Of what *more* than one time dimension might mean, however, Isham echoed Eddington by writing, "Physics in a spacetime of . . . two timelike dimensions would be very weird indeed. . . ." That is probably a fair statement of how most physicists presently think of multidimensional time.

One modern philosopher agrees—see Zeilicovici (1989)—calling the idea that there could be more than one dimension to time a "rather wild possibility" and then a "fairy-tale." But not all philosophers think this way, and many are in fact fascinated by the possibilities of multidimensional time. Why their interest in something so different from anything we actually experience? Where does the motivation come from? Of what *use* is multidimensional time? I think the answer is that it offers a theoretical model that supports those philosophers who argue that it makes sense to say there is meaning to the view that the past can be changed.

In a certain trivial sense, of course, the past is always changing. For each of us the past is the set of all events that have happened, arranged in a before/after temporal order,[26] and this set is continually increasing, i.e., changing. That is not, however, what most people mean by a changeable past. What *is* meant is that there may be some kind of change in the temporal ordering of events, or that an event that once was (or wasn't) a member of the set of past events no longer is (or is now) a member. Two-dimensional time offers a way to make sense of such possibilities, which one-dimensional time simply cannot do. To see how that works, let's follow the presentation in Meiland (1974), a paper that forcefully argues that it does make sense to talk about altering the past.[27]

Meiland is aware that some might find his model *ad hoc,* even "incredibly weird" (in his own words), but he justifies his efforts by taking a refreshingly enlightened, non-Humean view of what he thinks is the proper response to meeting purported time travelers: "If strange machines containing people in futuristic garments and speaking strange tongues (or perhaps using ESP instead of speech) were to appear and were to claim to be from the future, we might very well begin to search for a theory of time that allows their claim to be true." In Figure 3.5 you can see how Meiland has tried to do just that. The dashed diagonal line, marked with the points t_1, t_2, . . . represents our usual one-dimensional image of time. The horizontal lines P_1t_1, P_2t_2, . . . (which we can simply call P_1, P_2, . . ., for short) are the pasts for the present instants t_1, t_2, ... That is, P_1 is the past with respect to the present t_1, P_2 is the past with respect to the present t_2, and so on. The dashed vertical lines allow us to locate any moment in the past. For example, the intersection point A of P_4 with Pt_1 is the location of t_1 in the past with respect to the present t_4.

With this model, Meiland then analyzes in detail several interesting special cases. Suppose that t_1 and t_2 are one year apart and that there are similar time separations between all adjacent, marked present moments on that diagonal. Let us further suppose that a time traveler at t_4 journeys backward three years to t_1; then he will arrive at point A in the diagram of Figure 3.5. Assume he stays in the past two years—then his temporal locations lie along the dashed diagonal line $ABC;$ that is, at B he is three years in the past of t_5, and at C he is

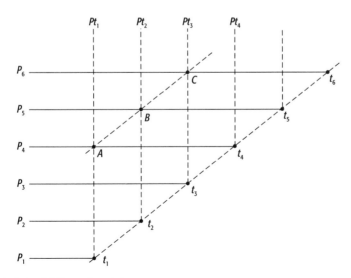

Figure 3.5 Multidimensional time.

three years in the past of t_6. From the diagram, then, we can imagine the time traveler saying, as he climbs into his time machine at t_4, "One year from now I'll be two years from now." That rather astonishing statement makes sense when we take both uses of *now* to be t_4 and observe that B (one year from A) is two years in the past with respect to t_4. Williams (1951a), a critic of time travel, uses what he claims to be the absurdity of such statements to support his rejection of time travel. One of Meiland's avowed reasons for developing his two-dimensional model of time was, in fact, to be able to reply to Williams.

As a final example of his time model, Meiland briefly discusses how it does away with the grandfather paradox. He does not "explain" the paradox but rather argues that the paradox simply does not exist. Take the extreme case of that paradox, that of time travel suicide: At t_4 the time traveler journeys back three years to A, kills his younger self, and then travels forward three years to mere moments after t_4. On the solid diagonal time line, the time traveler is seen to disappear at t_4 and then to reappear almost immediately. In the past a man vanishes at A. The one-dimensional paradox is avoided, however, because the traveler visited not t_1 but rather t_1's location at A in the past of t_4; the traveler's existence along the solid diagonal time line is uninterrupted except during his very brief trip into the past.

Meiland's time model is undeniably fascinating, but in fact it has no theoretical justification at all, as far as I know.[28] It is also simply not necessary to assume two-dimensional time to explain Meiland's "strange machines containing people in futuristic garments" from the future; it is possible to do so with one-dimensional time in four-dimensional spacetime. The grandfather and suicide paradoxes, too, are understandable without two-dimensional time. In the next chapter, we'll see how all this can be done.

4

Time Travel Paradoxes and (Some of) Their Explanations

Time travel is so dangerous it makes H-bombs seem like perfectly safe gifts for children and imbeciles. I mean, what's the worst that can happen with a nuclear weapon? A few million people die: trivial. With time travel we can destroy the whole Universe, or so the theory goes.

—*Millennium* (Varley)

He felt the intellectual desperation of any honest philosopher. He knew that he had about as much chance of understanding such problems as a collie has of understanding how dog food gets into cans.

—a time traveler, perplexed by paradoxes,
in "By His Bootstraps" (Heinlein)

"There's a lot we don't know about time travel. How do you expect logic to hold when paradoxes hold, too." "Does that mean you don't know?" "Yes."

—excerpt from a conversation between two paradox-puzzled
time travelers in "Bird in the Hand" (Niven)

What *was* this time traveling? A man couldn't cover himself with dust by rolling in a paradox, could he?

—the incredulous Editor,
astonished at the disheveled appearance of Wells' Time
Traveller upon his return from A.D. 802,701 and beyond

Paradoxes

A few years ago Quentin Smith, a philosopher who believes in a finite length to the past, wrote an analysis (1985b) to refute the logical arguments by Kant for *his* belief in an infinite past. That paper has nothing to do with the paradoxes of time travel, but in the course of presenting his reasoning Smith wrote the following passage:

> Why does the sun arise in the morning and not at some other time? Why do the hands of a properly functioning clock point to 12:00 at noon and midnight and not at other times? Why does the death of a person occur at a later time than his birth? The answer in all these cases is: Because by the very nature of these events they could not occur at other times. It belongs to the very nature of the sun's rising that it occur in the morning and not in the afternoon or evening. It belongs to the very nature of the hands of a properly functioning clock to point at 12:00 at noon and midnight and not at other times. And it belongs to the very nature of death to occur at a time later than a person's birth.

But what of a time traveler born in 1940 who, in 1998, enters his time machine, pushes a few buttons, and then boldly steps out into the Cretaceous period seventy million years earlier—and is promptly eaten for lunch by a passing *Tyrannosaurus rex*? Perhaps Smith himself would say that there is no contradiction between that and his third claim, because in the time traveler's *proper* time his death does indeed come after his birth. For many, however, for a man to die before his mother is born is a paradox plain and simple, say what you will about proper time.

The opening quote, from John Varley's imaginative novel *Millennium* (which was made into a movie in 1989), is typical of one common reaction to the threat of time travel paradoxes. This view assumes that nature has no stomach for paradoxes and that if one should be forced upon her, then the universe would be torn apart. Varley calls this the *cosmic disgust theory* and expresses it in the form of a petulant note to the offending time traveler:

> If you're going to play games like that,
> I'll take my marbles and go home.
> Signed,
God.

Varley was not the first to suggest such a grim view. In the 1967 novel *The Day After Tomorrow* (Klein) we read of the time engineers who manipulate history: "The time engineers knew that an error on their part would mean either a possible catastrophic change in the future of [their home-base planet] . . . or a cataclysm capable of destroying a least part of the galaxy and of spreading like a wave to the ends of the universe." Another story that adopts this view, and that also pre-dates Varley, is "Rotating Cylinders and the Possibility of Global Causality Violation" (Niven), which is about a paradox on the verge of occurring through the use of a Tipler-cylinder time machine. Rather than letting the paradox occur, the universe "decides" to avoid the problem by eliminating the perpetrators of the attempted paradox via a local nova! In a weaker form we saw the same response, nature preemptively protecting herself against time travel paradoxes, in De Camp's "A Gun for Dinosaur" discussed in Chapter One.

Such extreme, catastrophic visions of nature snuffing out paradoxes before they actually occur can be found in many other stories, as well. In *Black in Time* (Jakes), for example, so much paradoxical tampering with the past is done that finally the hero "heard a sound forevermore to be carried in his mind and recalled in nightmares: Time creaked. It was a great, cosmic creak, accompanied by a rush of wind for a trillionth of a second." On the other hand, time travel paradoxes do not always have to be gloomy, as shown in the following (hypothetical?) exchange between father and son in the (perhaps?) not too distant future.

> "Hey, Dad. Can I borrow the time machine tonight?"
> "Sure, son. Just be sure you have it back before you leave."

Fredric Brown was a master of the special category of science fiction story called the "short-short," in which everything happens in 500 words or less. As you might expect, the oddities of time travel were natural attractions for a quirky talent such as Brown's. For example, in "Experiment" the inventor of the first time machine demonstrates it to two colleagues by sending a brass cube five minutes into the future. After being placed in the machine, the cube vanishes and then five minutes later reappears. No paradoxes there—it is a trip into the *past*, as we'll soon learn, that has the potential for deadly repercussions.[1]

The inventor next declares that at three o'clock he will again place the cube into the time machine. Until then he will hold the cube in his hand. Thus, at five minutes before three the cube will vanish from his hand and immediately appear in the time machine (because five minutes after that, he says, at three o'clock, he will send it five minutes back into the past).[2] And indeed, at five minutes before three the cube *does* simultaneously vanish from his hand and appear in the time machine. Then, slightly before three, as the three men stand pondering what has happened, one of the observers asks what will happen if the inventor does *not* send the cube back at three o'clock? "Wouldn't there be a paradox of some sort involved?" he wonders. Curious, the inventor tries it (this is a bilking paradox, as discussed in Chapter Three), and the universe promptly vanishes in compliance with Varley's cosmic disgust theory.

The word *paradox* means different things to different people. Let me give you just two examples of paradox, one from theology and one from mathematics. In both Christian and Jewish theology, God is supposed to be omnipotent. Now, when he made humans, he either did or did not give them free will. If he did, then it follows that he cannot control the acts of humans, which means he is *not* omnipotent. And if he did not give humans free will, then the only way God can escape being responsible for the evil humans do is to suppose that he didn't create humans with free will because he couldn't. Again, then, God ends up being denied omnipotence. Thus we have the so-called Paradox of Omnipotence: No matter what God did concerning free will, he cannot be omnipotent, but that is in conflict with an all-powerful God.

For my second example, consider the Banach–Tarski paradox, which dates from the early 1920s. This is a result in pure mathematics that asserts that it is possible to cut a solid sphere (say, an orange) into a finite number of pieces and to then rearrange them into another solid sphere that is larger than the Earth! To a physicist, of course, who wonders about such things as conservation of mass-energy, this makes no sense, but for mathematicians who concern themselves only with the topics of measurability and the axiom of choice, the physicists' objection is dismissed as the grumblings of mere "practical folk." To the physicists, of course, the mathematicians appear to have rediscovered a modern version of the medieval debate on how many angels can dance on the head of a pin. And yet the mathematicians have *proved* the theorem. Paradox!

Time travel, of course, is full of paradoxes. A paradox, according to the usual dictionary definition, is something that appears to contain contradictory or incompatible parts, thus reducing the whole to seeming nonsense. And yet, truth is also evident in the whole. The history of science and mathematics has left a long trail of paradoxes, and those that involve time travel are merely among the most recent. Not all of the puzzles of time travel involve physics or logic, either. For example, Dwyer (1978) observes, "Doubtless time travel will raise a host of legal difficulties, e.g., should a time traveler who punches his younger self (or vice versa) be charged with assault? Should the time traveler who murders someone and then flees into the past for sanctuary be tried in the past for his crime committed in the future? If he marries in the past can he be tried for bigamy even though his other wife will not be born for almost 5,000 years? Etc., etc. I leave such questions for lawyers and writers of ethics text-books to solve." Interestingly, the ethical issues raised by time travel have, since Dwyer wrote, become a significant point of discussion in popular enter-tainment. Television's "Quantum Leap," for example, based essentially all four years of its episodes on the ethics of tampering with the past—see Wiggins (1993). And in the 1994 film *Timecop* a member of the "time police" (individ-uals charged with the mission of thwarting all who would change the past) is tempted by the possibility of preventing the murder of his own wife ten years before.

In the 1947 story "Time Twister" (Flagg and Wright) we see a hint of an-ticipation of the sexual paradox that is winked at in the original *Back to the Fu-ture* film. In that tale we find the following exchange between the inventor of a time machine and his helper:

> "You mean to say," he questioned incredulously, "that I could go back a hundred years?"
> "If you had the proper machine in which to travel, yes."
> "But that'd take me back to before I was born."
> The Professor smiled tolerantly.
> "Look at this diagram, Hank. This line is the time contin-uum. It incorporates space, too. [The authors didn't actually print a diagram with the story, but surely the Professor is using a Minkowski spacetime diagram, as discussed in Tech Note 4.] This dot is you. It doesn't matter when you were born, or when you will die. You exist right now, that's the fact. Traveling into the past or future wouldn't make you grow any younger or older. Such a thought is naive. Let me demonstrate the mechanics of it for you. If . . . we calculate with non-Euclidean mathematics . . ."
> "It don't sound reasonable," the farmhand objected. "If I went back—"
> "I know," interjected the Professor, "if you went back you might meet your own father as a young man and you'd be older than he, or maybe he and your mother would be kids going to school."
> "Haw, haw! That'd be funny, that would."

One way science fiction writers have of responding to the puzzles of time travel paradoxes is just to give up and to concede that the logical puzzles are overwhelming. In the 1941 story "Dead End" (Jameson), for example, the in-ventor of the Chronoscope (a gadget that can view the past) explains, "There is no time travel machine. Such a thing is a logical impossibility, treated seri-ously only by half-cracked writers of fantasy. Such a machine would lead at once into a hopeless paradox." Three decades later, in his introduction to Vance (1973), Robert Silverberg wrote, "We believe, of course, that time travel is a logical impossibility." Silverberg wrote that for the same reason that a history graduate student, a character in *Time of the Fox* (Costello), blurts out upon learning that the Columbia University physics department is doing experiments in time travel, "I'm no physicist, but even I know the logical difficulties with time travel. It's open season on coherent history, with goofy paradoxes. . . . Lots of fun for stories, but absolutely crackers as a real possibility."

Paradoxes offend common sense. They can also irritate. The specialists who study time travel paradoxes in *Chorale* (Malzberg), for example, make their colleagues in the Department of Reconstruction (of the past) uncomfortable with their constant worrying about altering history: "The paradoxologists were a stuffy bunch, and no one liked them very much." But are there really paradoxes at all? Or is it true, as the extraordinary boy-prodigy in "Vanya" (Grigoriev) who invented a time machine exclaimed when his teacher asserted that some questions could never be answered because "Nature is full of paradoxes": "Ah, Professor, what nonsense! Nature is harmonious; it is we who bring the paradoxes into it." Saying the same are Abramowicz and Lasota (1986), in a paper on the circular orbits of photons around black holes: "There are no paradoxes in physics, but only in our attempts to understand physical ideas by using inadequate reasoning or false intuition."

As the time traveler in the 1931 story "Via the Time Accelerator" (Bridge) coolly declared to a friend after an astonishing adventure in the year A.D. 1,001,930, "Paradoxical? My dear fellow, the Einstein Theory is full of apparent paradoxes, yet to him who understands it there is no inconsistency whatever. Give me another cigarette, will you, Frank?" Equally unconcerned is the character in the very funny 1941 tale "The Best-Laid Scheme" (De Camp) who, at the end, tells his friend, "My dear Collingwood, don't drive yourself crazy trying to resolve the paradoxes of time travel. The [time machines] are gone. . . . Have a drink." Somewhat more concerned about time travel paradoxes, however, was the time traveler in the 1940 story "The Time Cheaters" (Binder), who told his partner (just before their first trip into the future) that "I'm not sure any more about getting back. There're some unpredictable terms in the time-travel equation—paradoxes. Maybe we *won't* get back."

The concern expressed by Binder's time traveler was not shared by *Rip Hunter—Time Master*, a comic book series published in the early 1960s. Rip Hunter, inventor of the "time sphere," was the leader of a "famous foursome" of time travelers who operated out of a secret mountain laboratory. *Rip Hunter* was unconcerned with the real puzzles of time travel, and paradoxes played no role in any of the stories. Time travel was simply a device to get the characters into a new story setting for each issue. The formula for the stories was to send Rip and his pals into the past to film history for museums and photo-archives; there they would suffer some accident that would lead to a crisis. For example, in one 1964 tale, while Rip and his fellow travelers are filming the interior of Nazi Germany, the time sphere (which could fly) is shot down by anti-aircraft fire. Before that particular story was over, the temporarily stranded time travelers had arranged a meeting between Hitler and Napoleon (!), with no one expressing the slightest concern over the paradoxes of disturbing history.

Early Science Fiction Speculations on Time Travel Paradoxes

The late 1930s and the 1940s are generally thought of as the golden age of science fiction magazines. Before that came (of course) the pre-golden age, the first decade of the "scientifiction pulps," which is generally dated from the appearance in April 1926 of Hugo Gernsback's *Amazing Stories.* Gernsback's earlier publications, *Science & Invention* and *Radio News,* had printed science fiction from time to time, as had many of the "ten-cent family magazines" since the 1890s. Frank Munsey's *The Argosy,* which began in 1896, was the first all-fiction pulp, and his *The All-Story Magazine,* begun in 1905, was another all-fiction adventure pulp. Both magazines had often published the story form called the scientific romance (this term was used as early as 1888 by C. A. Hinton for his essays on the fourth dimension; recall the discussion of Hinton's work from Chapter Two), but they carried other sorts of stories, too.

There were many other adventure and "weird story" pulps, such as *The Popular Magazine* (1911), *The Cavalier* (1908), and *The Thrill Book* (1919), but *Amazing Stories* was the first pulp to be devoted totally to science fiction. With its motto of "Extravagant Fiction Today—Cold Fact Tomorrow," and with the illustration on the contents page showing Jules Verne bursting from his grave in a heroic pose made famous years later by Superman, there could be no doubt as to what kind of fiction the reader would find under the dramatic, multicolored cover art. And not all of what readers found in *Amazing Stories* could be dismissed with a condescending smile. For example, "The Tissue-Culture King" by the well-known English biologist Sir Julian Huxley was reprinted in the August 1927 issue, the year after it first appeared in the *Cornhill Magazine* (in England) and in *The Yale Review* (in America). This cautionary tale about the possible evil uses of biology was no mere adventure or western story transplanted into outer space, as many of the pulp offerings admittedly were. It was decades ahead of its time, and indeed its message was prophetic.

It is interesting to note that some very early, popular fiction, in mass-audience magazines of the nineteenth century, came very close to presenting their readers with "time travel" long *before* Wells' *Time Machine*. Consider Edward Bellamy's "The Old Folks' Party," which appeared in the March 1876 issue of *Scribner's Monthly*. In this very interesting story a group of teenagers, who belong to a weekly discussion club, agree that at their next meeting they will come dressed and behaving as they believe they will be dressing and behaving fifty years in the future. Also attending will be the grandmother of one of the young ladies. The meeting of the "old folks" takes place, and it invokes such powerful feelings of mortality that at last one of the young men can stand it no more: "Suddenly Henry sprang to his feet and, with the strained, uncertain voice of one waking himself from a nightmare, cried:—'Thank God, thank God, it is only a dream,' and tore off the wig, letting the brown hair fall about his forehead. Instantly all followed his example. . . ." The young people then begin to laugh with relief at once again being young, until they notice the grandmother is crying. Her granddaughter instantly knows what is wrong and says, "Oh, grandma, we can't take you back with us." That is perhaps as close to time travel as one could get, short of fantasy, in pre-Wellsian fiction.

It is in *Amazing Stories* that we find the first *non*-fictional speculations about time travel by machine in a pulp magazine. Gernsback started those speculations by reprinting Wells' *Time Machine,* which in turn sparked a fair number of readers' letters that were published in the magazine's "Discussions" section. Typical is the following comment from a letter in the July 1927 issue: "In the 'Time Machine' I found something amiss. How could one travel to the future in a machine when the beings of the future have not yet materialized?" (See Tech Note 6 for an answer to that reader's question.) More interesting was the letter from the reader in Cleveland, Ohio, who wrote in the same issue,

> How about this 'Time Machine'? Let's suppose our inventor starts a "Time voyage" backward to about A.D. 1900, at which time he was a schoolboy. . . . [H]is watch ticks forward although the clock on the laboratory wall goes backward. Now we are in June 1900, and he stops the machine, gets out and attends the graduating exercises of the class of 1900 of which he was a member. Will there be another "he" on the stage? Of course, because he *did* graduate in 1900. . . . Should he go up and shake hands with this "alter ego"? Will there be two physically distinct but characteristically identical persons? Alas! No! He can't go up and shake hands with himself because . . . this voyage back through time only duplicates actual past conditions and in 1900 this strange "other he" did *not* appear suddenly in quaint ultra-new fashions and

congratulate the graduate. How could they both be wearing the same watch they got from Aunt Lucy on their seventh birthday, the same watch in two different places at the same time. Boy! Page Einstein! No, he cannot be there because he wasn't there in 1900 (except in the person of the graduate). . . . The journey backward must cease on the year of his birth. If he could pass that year it would certainly be an effect going before a cause. . . . Suppose for instance in the graduating exercise above, the inventor should decide to shoot his former self. . . . [H]e couldn't do it because if he did the inventor would have been cut off before he began to invent and he would never have gotten around to making the voyage, thus rendering it impossible for him to be there taking a shot at himself, so that as a matter of fact he *would* be there and *could* take a shot—help, help, I'm on a vicious circle merry-go-round. . . . Now as to trips into the future, I could probably think up some humorous adventures wherein [the inventor] digs up his own skeleton and finds by the process of actual examination that he must expect to have his leg amputated because the skeleton presents positive proof that this was done.

All of the ingenious puzzles in this letter (signed only with the initials "T.J.D.") intrigued Gernsback, and no doubt it was no coincidence that the same issue featured a new, original time machine story "The Lost Continent" (White). This is a tale of a scientist who transports an entire ship at sea 14,000 years back in time and causes it to hover over lost Atlantis! That story provoked a sharp letter from a reader who claimed its logic had a fatal flaw—the story's author indicated that the Atlantians observed the time travelers when, "of course" (asserted the reader), the time travelers must actually have been invisible. The reader explained his reasoning as follows, beginning by defining A as one of the Atlantians.

Now A lived his life, thousands of years ago, and died. All right, now let us pass on in time 14,000 years. Now, back we come in time when A is again living his life. Lo and behold, this time A sees before he dies a strange phenomenon in the sky! He sees the shipload of people observing him. And yet these people are necessarily observing him during his one and only lifetime, wherein he certainly did not, could not, have observed them."

Gernsback printed this letter in his September 1927 editorial "The Mystery of Time" and concluded by saying, "I do . . . agree . . . that the inhabitants of Atlantis would probably not have seen the . . . travelers in time." Other readers felt this same way, because after Gernsback published yet another time machine tale in 1927—"The Machine Man of Ardathia" (Flagg)—the same invisibility argument again appeared in the "Discussions" column of the magazine.

Two years later an amateurishly written tale appeared, "Paradox" (Cloukey), in which a man travels in time from 1928 to 2930 with the aid of an "astounding machine based on advanced electro-physics and the non-Euclidean theory of hyperspace." The purpose of that story was twofold: to present several of the classic paradoxes of time travel and then to make the claim that although the simple minds of twentieth-century people cannot understand the explanations of the paradoxes (perhaps this is why the author offers none!), the paradoxes are all trivial to the scientists of the thirtieth century. The author, who was ignorant of how to plot a story as well as incapable of writing realistic dialogue, still managed vastly to entertain the readers of *Amazing Stories Quarterly* with the sheer mystery of the paradoxes. Letters poured in to the magazine from young fans, demanding more time travel fiction.

In the December 1929 issue of *Science Wonder Stories,* Gernsback published Henry F. Kirkham's story "The Time Oscillator." (By that time Gernsback had lost control of *Amazing,* and *Science Wonder* was part of his comeback as a publisher of pulp "scientifiction.") That story plays with the question of the role of time travelers in the past. Could they actually participate in events ("mix into the affairs of the period," in Gernsback's words) or would they just be unseen observers? This question, obviously inspired by the earlier discussion in *Amazing Stories,* intrigued Gernsback as much as it did his readers, and along with Kirkham's story he printed a challenge entitled "The Question of Time-Traveling":

> In presenting this story to our readers, we do so with an idea of bringing on a discussion as to time traveling in general. The question in brief is as follows: Can a time traveler, going back in time—whether ten years or ten million years—partake in the life of that time and mingle in with its people; or must he remain suspended in his own time-dimension, a spectator who merely looks on but is powerless to do more? Interesting problems would seem to arise, of which only one need be mentioned: Suppose I can travel back into time, let me say 200 years; and I visit the homestead of my great great great grandfather, and am able to take part in the life of his time. I am thus enabled to shoot him, while he is still a young man and as yet unmarried. From this it will be noted that I could have prevented my own birth; because the line of propagation would have ceased right there. Consequently, it would seem that the idea of time traveling into a past where the time traveler can freely participate in activities of a former age, becomes an absurdity. The editor wishes to receive let-

ters from our readers on this point; the best of which will be published in a special section.

(This question was clearly "in the air"; that same month Edward L. Re-menter's "The Time Deflector" appeared in *Amazing Stories,* which also ad-dressed this particular puzzle of time travel.)

Gernsback's challenge did not go unnoticed, and over the next year or so he published a large number of reader responses in the magazine's letters column, "The Reader Speaks." Indeed, a few months later, in his introduction to "An Adventure in Time" (Flagg), Gernsback wrote that ever since the publication of Kirkham's tale, "there has been a great controversy among our readers as to the possibility of time flying and the conditions under which it may be done." Most of those letters, and the ones that followed, are interesting but not par-ticularly profound—with one exception, a letter written by a fourteen-year-old. That letter, which appeared in the February 1931 issue, may well have served as inspiration for several of the classic time travel tales published during the next twenty years:

> Some time ago you asked us (the readers) what our opinions on time traveling were. Although a bit late, I am now going to voice four opinions. . . .
>
> (1) Now, in the first place if time traveling were a possi-bility there would be no need for some scientist getting a headache trying to invent an instrument or "Time-Machine" to "go back and kill grandpa" (in answer to the age-old argu-ment of preventing your birth by killing your grandparents I would say: "who the heck would want to kill his grandpa or grandma!") I figure it out thusly: A man takes a time ma-chine and travels into the future from where he sends it (under automatic control) to the past so that he may find it and travel into the future and send it back to himself again. Hence the time machine was never invented, but!—from whence did the time machine *come?*
>
> (2) Another impossibility that might result could be: A man travels a few years into the future and sees himself killed in some unpleasant manner,—so—after returning to his cor-rect time he commits suicide in order to avert death in the more terrible way which he was destined to. Therefore how could he have seen himself killed in an entirely different man-ner than really was the case?
>
> (3) Another thing that might corrupt the laws of nature would be to: Travel into the future; find out how some inge-nious invention of the time worked; return to your right time; build a machine, or what ever it may be, similar to the one you had recently learned the workings of; and use it until the time you saw it arrive, and then if your past self saw it as

you did, he would take it and claim it to be an invention of his (your) own, as you did. Then—who really *did* invent the consarn thing?

(4) Here's the last knock on time traveling: What if a man were to travel back a few years and marry his mother, there by resulting in his being his own "father"?

Jim H. Nicholson
40 Lunado Way
San Francisco, Cal.

Gernsback's reply that immediately followed this letter was favorable, opening with "Young Mr. Nicholson does present some of the more humorous [?] aspects of time traveling. Logically we are compelled to admit that he is right—that if people could go back into the past or into the future and partake of the life in those periods, they could disturb the normal course of events." Gernsback also apparently still liked the "invisibility of time travelers" view, too; he had only a few months earlier again published such a tale, "The Time Ray of Jandra" (Palmer).

Nicholson's letter *is* ingenious, and it anticipated the central ideas of many science fiction tales. For example, his item (2) is a precise plot outline of the 1941 story "The Man Who Saw Through Time" (Raphael), which was mentioned in Chapter Two, and a version of item (4) was used, decades later, in Robert Heinlein's famous short story "All You Zombies—." However, in the next several sections you will also see why, contrary to Gernsback's view, Nicholson's comments are *not* logical. As a final comment on Nicholson, his imaginative interest in science fiction served him well in later life. He eventually became a co-founder and president of American International Films, a company that made such "classics" as *X (The Man with the X-Ray Eyes)* and *Attack of the Crab Monsters* (and, outside of the science fiction genre, *Beach Blanket Bingo*)! American International also made the classic 1964 film *The Time Travelers,* one of the more intelligent science fiction time travel movies. Nicholson died in 1972.

Two Basic Time Travel Paradoxes

Williams (1951a) makes the claim that a time traveler about to set out on a trip a century into the past is also about to utter a contradiction when he says, "Five minutes from now, I'll be a hundred years from now." How can he be *both* times from now? Smart (1963) calls that a "neat argument against the possibility of time travel," and even Horwich (1987), who ultimately rejects Williams' "paradox," begins his discussion of it with the title "Is 'time travel' an oxymoron?" In Chapter Three we saw how such Williams-type statements can make sense with multidimensional time—but what if we limit ourselves to the one temporal dimension we actually know? Williams uses his paradox to deny the rationality, in particular, of the closing, haunting words to Wells' *Time Machine,* when the narrator speculates about the fate of the Time Traveller: "He may even now—if I may use that phrase—be wandering on some plesiosaurus-haunted Oolitic coral reef, or beside the lonely saline seas of the Triassic Age." Williams would have declared equally nonsensical the bold claim of the time traveler in "Time's Arrow" (McDevitt) who tells a friend that "it *is* possible to reverse the arrow of time in the macroworld. Tonight you and I will have dinner in the nineteenth century."

Horwich correctly rebuts Williams by observing that if the five minutes and the hundred years are measured in different reference systems, then the time traveler's assertion can make perfect sense. Indeed, Smart raises that possibility as well in his paper. And, indeed, Smart even admits that all that is needed for travel into the future is a high-speed rocket and, with it, it *is* possible to be a century from now in five years *if* the century is in Earth time and the five years are in rocket time. But as for trips to the past, Smart says they are not possible because "fast rockets will not enable us to experience past ages." Those words lose much (if not all) their strength because of the time-travel-into-the-past results of Gödel, however (which Smart might assert to be irrelevant because our universe seems not to be rotating), and from other more recent work (see Note 12 for Chapter One).

Smart's form of argument against time travel dies hard. Years later another version of it appeared once more in Christensen (1976), a paper written by yet another philosopher who, like Williams and Smart, rejects time travel.

> Consider a sample statement asserting the occurrence of time travel: "I stepped into the time machine *and then* I saw Caesar being stabbed." We may re-word this to read "My stepping into the time machine was earlier than Caesar's being stabbed"; but because it is also true that Caesar's stabbing was earlier than my time-machine entry, we have a flat contradiction—unless the first "earlier than" involves something other than ordinary time.

Christensen claims that only the equally flawed concept of meta-time ("whatever *that* might mean," he writes) could possibly save the day, but the same reply can be given to him as to Smart. He has failed to distinguish between the proper time of the time traveler and the time of non–time travelers. He continues to make this same error, in fact, repeating his erroneous argument in a book published seventeen years later—see Christensen (1993).

A misunderstanding of the different time rates of time travelers and of non–time travelers is not limited to philosophers. For example, Ijon Tichy's encounter with the doomed time traveler (who ages and dies as he travels into the future) in *Memoirs of a Space Traveler* (Lem) is based on that same mistake of confusing proper time and cosmic time. The reverse error occurs in the 1949 story "I Died Tomorrow" (Worth), when a time traveler to the future is killed in the year 4000 but returns to life when a colleague brings his body back to the present. Similar to that is the erroneous premise of "The Old Die Rich" (Gold) — available today as an "X-Minus One" radio tape — that nothing can exist either "before" or "after" it exists. Thus a time traveler cannot travel into the past further than his own birth, and if he eats anything while in the past, he will be hungry when he returns to the present because the food will have decomposed before being digested! (The author of this astonishing interpretation of time travel was so proud of its "brilliance" that he published his working journal notes on his reasoning as an afterword to the story.)

As for the second basic paradox, we find it in Smart's paper, directly following his claim about rockets being unable to visit the past. He begins with "Suppose it is agreed that I did not exist a hundred years ago. It is a contradiction to suppose that I can make a machine that will take me to a hundred years ago. Quite clearly no time machine can make it be that I both did and did not exist a hundred years ago." I wonder if Smart wrote that outrageous argument with tongue in cheek: The first sentence is simply an initial hypothesis equivalent to denying backward time travel in the first place. How does he *know* he didn't exist a hundred years ago? Why does he so quickly agree with that assertion? If time travel to the past of a hundred years ago is possible, and if his time trip to the past won't start until next year, then it is quite clear that Smart makes a mistake in so readily agreeing to his earlier non-existence. If his argument proves anything, it is just that if a time machine can be made, then he well *might have* existed a hundred years ago! Even the British philosopher Jonathan Harrison (1971) a non-enthusiastic analyst of time travel, has admitted the force of that position.

A variant of Smart's flawed example is found in the well-known college philosophy text by Hospers.[3] Hospers argues against the logical possibility of time travel, writing "We can imagine ourselves as having been born in a different era and being with the Egyptians building the pyramids. But can we imagine ourselves, *now,* in the 20th century A.D., *being* (not merely in our imagination) in 3000 B.C.? How can we be in the 20th century A.D. and the 30th century B.C. *at the same time?*" That so misstates what is meant by time travel to the past that I doubt Hospers would find many who would agree that his statement is even superficially plausible. A Socratic dialogue is a fine teaching tool, but one who employs it must create at least a façade of reasonableness in setting up the red herrings!

Hospers next tries to play devil's advocate and presents the obvious rational rebuttal against his ridiculous straw-man position: " 'But,' one may object, 'this is not the situation we are imagining. What we are imagining is being one day in the 20th century and then moving backward in time so that the next day we are in the year 3000 B.C.—and on that day we are no longer in the 20th century A.D.' " This is certainly true, but Hospers then refutes his reasonable restatement in a most astonishing manner. In essence, he claims that if the day before your trip is January 1, then the next day in your life *has* to be January 2 and certainly not some day in 3000 B.C. Like Smart, Hospers is begging the question of time travel to the past because he is saying nothing more than that time travel is impossible because it is not possible! Anything else, he claims, is a contradiction in terms and hence logically impossible. That, of course, is nothing more than a grammarian's disproof of time travel when what is really needed is mathematical physics.

Can the Present Change the Past? Can the Past Be Un-Done?

Hospers unleashes what he thinks is the ultimate argument against time travel to the past, with an opening line equivalent to asserting the impossibility of what he claims to be proving:

> Many centuries B.C., the pyramids were built, and when all this happened you were not there—you weren't even born. It all happened long before you were born, and it all happened without your assistance or even your observation. This is an unchangeable fact: *you can't change the past*. That is the crucial point: the past is what has happened, and you can't make what has happened not have happened. Not all the king's horses or all the king's men could make what *has* happened *not* have happened, for this is a logical impossibility. When you say that it is logically possible for you (literally) to go back to 3000 B.C. and help build the pyramids, you are faced with the question: did you help them build the pyramids or did you not? The first time it happened, you did *not*: you weren't there, you weren't yet born, it was all over before you came on the scene. All you could say, then, would be that the *second* time it happened, you *were* there—and there was at least a difference between the first time and the second time: the first time you weren't there, and the second time you were.

That, as you may have guessed, is a textbook example of "say it enough times and everybody will agree I am right just to get me to stop saying it!"

One science fiction fan long ago summed up Hospers' position in a letter to the editor of *Astounding Stories* (August 1931): "It is said that the past cannot be changed, and that any effort to do so would be useless. In my belief, no matter where or when a man goes in the past, if he appears in a year or day that has already gone by, *he is changing the past*. Then there should be no room for doubt: time traveling is impossible. It will never be done." Both Hospers and that fan claim that the past is unchangeable, a claim that is not in dispute. As Kotarbinski (1968) puts it, "If impossibilities could have degrees then it would be more improbable to undo yesterday's flight of a gnat than to send the moon adrift from its usual orbit tomorrow." But it simply does not follow, as Hospers and the fan assert, that backward time travel is therefore impossible. Each has made the same logical error in reasoning.

Hospers' puzzle is, of course, simply the grandfather paradox in disguise. Hospers' error in his argument against time travel occurs, in particular, precisely at the point where he states his belief that 3000 B.C. occurs twice. In fact, there is no reason for believing that—3000 B.C. (or any other year) happens just once. If you *will* go back to 3000 B.C., then you *were* there; and if you *weren't* there, then you *won't* go back. You don't remember 3000 B.C. even if you were there (and even though that year is in the global past), because your time trip is not in your local past but rather in your personal future. This may all seem odd, of course, but it is not illogical. Indeed, even as Hospers wrote his book, it was understood that his form of reasoning is faulty. Certainly, today, the philosophical consensus is that the idea of backward time travel is perfectly consistent with the four-dimensional block universe discussed in Chapter Two.

Hospers' error is repeated by Herbert (1988), a physicist who claims that with the existence of a time machine, "no longer would there have to be a 'road not taken'—we could simply travel back in time and make some other choice. And if that choice didn't work out, we could go back into the past and try again. In a time-traveling society, our actions would no longer be irreversible." In fact, none of these statements is true; they are false precisely because, as Hospers himself says, the past is unchangeable.

Early science fiction writers were just as puzzled by the grandfather paradox as Hospers and many of his fellow philosophers, but sometimes the writers were more open in admitting so. For example, in "Dark Interlude" (Reynolds and Brown) we hear the following from the inventor of the first time machine:

Figure 4.1 Illustrator Jack Binder was the author of a continuing series called "IF—" in *Thrilling Wonder Stories*. In each issue, the ellipses would be replaced with some phrase such as "the Sun exploded!" "there was another ice age!" or "there was no friction!" The installment shown here appeared in December 1938 and asserted that the past could be changed by a time traveler. Binder was the brother of writers Earl and Otto, who, under the fused pen name of "Eando," wrote some of the more literate time travel stories of the 1930s and 1940s, such as the 1940 "The Time Cheaters."

Illustration for "IF—You Were Stranded in Time!" by Jack Binder © 1938 by Better Publications, Inc.; reprinted by arrangement with Forrest J. Ackerman, Holding Agent, 2495 Glendower Ave., Hollywood, CA 90027.

Figure 4.1 Continued

"I have devised a method [for travel] into the distant past. The paradox is immediately pointed out—suppose [the time traveler] should kill an ancestor or otherwise change history? I do not claim to be able to explain how this apparent paradox is overcome in time travel; all I know is that time travel *is* possible. Undoubtedly, better minds than mine will one day resolve that paradox, but until then we shall continue to utilize time travel, paradox or not." Admirable courage, but certainly not so risky as it sounds, now that we realize there is no paradox involved at all. Less admirable is the way out invoked in the 1942 tale "Time Dredge" (Arthur): The heroes simply decide not to think about the "changing history" paradox because it makes them dizzy!

More modern writers continued to be puzzled by that "paradox" long after they shouldn't have been. Consider a fanciful work with no patience for physics, the novel *Morlock Night* (Jeter). The hero (who turns out to be a resurrected King Arthur!) learns he must defeat the Morlocks who have killed Wells' Time Traveller in the far future and have used his time machine to travel back in time to destroy the world. This rightfully causes some bewilderment. As he says, "Wait a moment. There's something wrong here. . . . If the Morlocks come back in time to their own past and wreak such havoc, aren't they endangering the chain of events that leads to their own existence? Why, they might be conquering and *then eating* their own ancestors! And thus obliterating their own nasty lives scores of generations before their own births!" That reasonable objection is brushed aside by his companion—none other than Merlin, of course—who replies, "I admire your astuteness. . . . Not many . . . could follow that, let alone come up with it themselves. Indeed, it *is* a violation of the Universe's natural order. This whole business of Time Travel is shot through with cosmic blasphemy, I'm afraid." Not much of an answer, either, I'm afraid, especially when you consider that four centuries before Christ the answer was clear to Aristotle. In his *Nicomachean Ethics,* in fact, we find him declaring that the Greek poet Agathon had known the answer a century before that, and he quotes the poet as saying, "For even God lacks this one thing alone, To make a deed that has been done undone."

Agathon and Aristotle not withstanding, some medieval theologians argued passionately that the past *could* be changed (but only by God). The eleventh-century Italian cleric Peter Damian (who became a Christian saint) is a famous exponent of that radical view—see McArthur and Slattery (1974), Remnant (1978), and Gaskin (1997). Writing in his *De Omnipotentia Dei* ("On the Divine Omnipotence in Remaking What Has Been Destroyed and in Undoing What Has Been Done"),[4] Damian made it clear that he believed nothing could withstand the power of God, not even the past. Ralph Waldo Emerson's beautiful poem "The Past" ("All is now secure and fast, Not the gods can shake the Past") would have been blasphemy for Damian. The following words from Damian testify to the strength of his commitment to a belief in the possibility of changing the past:[5] "Just as we can duly say 'God was able to make it so [that] Rome, before it had been founded, should not have been founded,' in the same way we can equally and suitably say, 'God can make it so that Rome, even after it was founded, should not have been founded.' "

Two centuries after Damian, Aquinas argued the contrary view that changing the past is *not* within God's power. Whereas Damian felt it impossible to deny any act to God, Aquinas took the far more moderate position that part of God's law is that there be no contradictions in the world and that certainly God would be bound by his own law. As he wrote, "It is best to say that what involves contradiction cannot be done rather than that God cannot do it." In his *Paradise Lost,* John Milton's God is constrained even more; he is free to act or not, but if he does freely decide to act, it can only be to "do right." That might seem to preclude causing contradictions, as in changing the past, but perhaps not. Milton's contemporary, Thomas Hobbes, declared that there is no *a priori* standard of goodness, and thus (for Hobbes) there are no constraints on God's powers. For Hobbes, therefore, it would seem that God could change the past.

Theological changing of the past leads, as might be expected, to all sorts of mind-boggling, logical puzzles. Because of such puzzles, theology would certainly be influenced by time travel, but just as certainly theological reasoning will not answer the question of the possibility of time travel. Philosophers such as Hospers, who incorrectly argue that similar change-the-past puzzles occur with backward time travel, have simply failed to grasp that time travel is a question for mathematical physics, not theology or grammar. Modern philosophers, who understand relativistic physics, agree that the past can not be changed and that backward time travel in no way implies that it could be changed. Yet perhaps we should not be too critical of Hospers and those of like persuasion; recall from Chapter One that even the man who started serious time travel analyses—Kurt Gödel—erroneously believed that backward time travel would allow changing the past.[6]

One philosopher, however—Anscombe (1971)—has advanced a non-relativistic argument for why it is not possible to change the past. The stage is set with the seemingly benign words "Things have taken a certain course, which perhaps can and perhaps cannot be reversed; some actions can be undone. But it makes sense to wish they had never been done, and when one says 'The past cannot change' one is stating that this is [impossible]." Anscombe then toughens her stance with "But 'a change in the past' is *nonsense,* as can be seen from the fact that if a change occurs we can ask for its date. If the idea of a change in the past made sense, we could ask the question 'When was the battle of Hastings in 1066?' and that not in the sense 'When in 1066 was the battle of Hastings.' *The idea of change in the past involves the idea of a date being dated* [my emphasis]." Finally, in case the point of her argument has been missed, she hammers her thesis home with "This consideration helps to remove the impression that when one says 'the past cannot change' one is saying something *intelligible* that is an impossibility." That is, the phrase "changing the past" is not an issue that may or may not be possible, but rather it is simply a silly sequence of words devoid of meaning. For Anscombe, philosophers who speak of changing the past are equivalent to veterinarians debating how best to perform a medical procedure on a unicorn.

I personally happen to agree with Anscombe's conclusion, but, as might be expected for an argument supported by the nuances of the meaning of words, there is always someone who interprets the words differently. For example, consider the rebuttal by Jack Meiland (the advocate of multidimensional time from Chapter Three who used that concept to argue *for* the logical possibility of changing the past). He claims there is a perfectly sensible and possible response to Anscombe's example: "Suppose a change occurs in the battle of Hastings after the battle is over. For example, suppose that Harold took part in the battle when it occurred, but after the change in the past battle he was no longer a participant in that battle. At some point in time Harold ceased to take (or to have taken) part in the battle of Hastings. This means that up to a certain point in time, say up to July 20, 1955, the proposition 'Harold participated in the battle of Hastings' is true and after that date that proposition is false. In this way we can date the change in the past battle of Hastings: the change occurred on July 20, 1955."

Curiously enough, Meiland goes on to present what he thinks Anscombe's reply to his claim might be but then *fails to refute it!* That is, Meiland destroys his own position, and, I might add, he does a rather thorough job of it. The problem with dating the change of a date is that, as he himself points out, "this method does not show that changes in past events can, even in principle, be dated, because this method is not a method that can in principle be used; we could never find out that the truth value of a proposition about a past event had changed; so we would never be able to assign dates to alleged changes in past events." That is, if an event in the past has changed, there could be no surviving evidence of its prior value. If there were such evidence, you see, then what could it mean to say that the event had *changed?*

Meiland tried to offer an example of how such evidential issues might occur by using a philosopher's story attached to a version of Damian's thesis that God can change the past. He asks us to consider the situation where a Roman ruin suddenly vanishes one day. With no other explanation available, and assuming that one believes in God, Meiland asserts that it is reasonable to conclude that God has changed the past, and thus the ruin no longer plays the role it once played in past events. This story begs the question about evidence, however, because *if* we accept Meiland's little fable, *then* we remember the now non-existent ruin, and that memory itself is all the evidence needed to assign a date to the change in the past. But why would our memories be unaltered? Of course, we might also wonder why they shouldn't remain unchanged.

The problem is that in Meiland's example, we are dealing with a philosopher's fairy tale and not with physics. There simply isn't anything logical *at all* that can be deduced from his story about the possibility or impossibility of changing the past. Anscombe's objection that changes in the past should be able to be dated stands, and Meiland's example is an empty (if entertaining) one. Changing the past, whether by God's agency or that of a mortal time traveler, is *logically* impossible in a universe with a single timeline. As a time traveler in the 1954 story "The Poundstone Paradox" (Dee) is told, "You can't possibly create a paradox in time, you see, because anything you do in the past must have been done already or you couldn't have been there to do it in the first place."

A less grandiose example of Meiland's idea occurs in the opening scenes of the 1989 movie *Time Trackers*. There, a laboratory coffee cup is broken just before the first time machine trip into the past. The laboratory staff later realizes that the experiment is a success, even before the traveler returns, when the cup is no longer broken—the time traveler caught it before it hit the floor! The past has been changed, then, but the *broken* cup is still remembered. Why? Making the same mistake the following year is the film *Future Zone,* in which a time traveler from thirty years in the future journeys back in time to save his father from being killed (as happened the "first" time). After he is illogically successful in this mission, the time traveler's original memories (of what?) are still intact.

And finally, the same error appears in "Not of an Age" (Benford), in which an historian uses a time machine to visit great writers of the past mere hours, or even minutes, before history records their deaths. She does this to tell them that the future holds their works in great esteem. Her appearances so unnerve her targets, however, that either they continue to write, producing works that "future history" never knew, or they save manuscripts that "future history" had declared to be lost. In short, she has changed the past. The special irony in the story comes at the end, when she herself is visited by an historian from *her* far future, who thanks her for the spacetime disturbances she has caused because they have benefitted literary history! This tale, charming as it is, is absolutely shot through with logical puzzles, not the least of which is how the future would know the past has been changed (compared to *what?*)

One philosophical writer, Fraser (1978), has referred to "the vicious circle of time travel" and has explained why he has that negative reaction by using a sketchy version of the faulty argument (à la a greatly abbreviated Hospers) that time travel implies changing the past. "Time travel into the past. (a) Misdirect your grandfather so that he will never meet your grandmother. You do not exist. (b) Find that it is impossible to interfere with the past. You are then a historian, not a time traveler." Fraser's points are not nearly so telling as he seems to think. His point (a) is simply not accepted as logical by modern students of time travel, because you *do* exist and so your grandfather *did* meet your grandmother. To claim otherwise is like claiming black is white, and if you are going to argue that, then there really isn't much left to say. His point (b) is somewhat more interesting, but it represents a common misunderstanding of what time travel would mean, a misunderstanding that should be cleared up by the end of this chapter.

Even though the consensus today is that the past cannot be changed, science fiction writers have used the idea of changing the past for good story effect. Consider this passage from *The Fall of Chronopolis* (Bayley), a novel about a "time-war." Just after the detection of temporal invaders, we read of them that "They had come in from the future at high speed, too fast for defensive time-blocks to be set up, and had only been detected by ground-based stations deep in historical territory. If the target was to alter past events—the usual strategy in a time-war—then the empire's chronocontinuity would be significantly interfered with." And in *Time of the Fox* (Costello), American physicists battle KGB physicists in a war of time travelers in the past, each side attempting to change history to its advantage. In this novel the history changers isolate themselves from all the alterations taking place outside of their Time Lab, and they compare their stored historical records with those of external libraries. That allows the staff historian to adjust for each new round of changes. As the historian explains, outside of the Time Lab "History might change, but here [in the Time Lab] the past lives on."

In a novel of a galaxy-wide confrontation between humans and androids—*Time and Again* (Simak)—the use of time travel to alter history is central: "A war in time . . . would reach back to win its battles. It would strike at points in time and space which would not even know that there was a war. It could, logically, go back to the silver mines of Athens, to the horse and chariot of Thutmosis III, to the sailing of Columbus. . . . It would twist the fabric of the past." Contrary to Simak's claim (and to the underlying claims in Bayley's and Costello's novels, as well), the modern view is that such actions are *not* logical. The rest of this chapter will present arguments to support that claim.

Despite the lack of support from either physics or logic for claims that changing the past makes sense, science fiction writers have been quite inventive in devising possible altruistic reasons for a desire to change the past. For example, in "Host Age" (Brunner) operatives of the Corps of Temporal Adjustment from the year 2620 intentionally change the past by infecting our present with a terrible disease against which there is no defense. The reason for this seemingly irrational act is, however, clever in its simplicity: Over the centuries medicine has triumphed over all known ills, so when alien invaders from the stars infect humankind with a new and vicious plague, there is no natural immunity. Accordingly, back in time travels the Corps, to alter history so that humans *will* have developed natural antibodies *before* the invasion occurs in the far future.

Consider also "Jon's World" (Dick), in which time travelers from Project Clock journey back to before an atomic war will devastate earth. They plan to steal the research papers of the scientist who discovered the secret of the artificial brain (!) and then to use the brain to build robots to help rebuild the blasted surface of the planet in the future from which they came. By accident, however, they kill the scientist and thus so completely change all of subsequent history that when they return to the future, it is nothing like what they remember. It is, in fact, idyllic, a world in which all the inhabitants spend their time discussing philosophy. The last lines of this story are "Let's go find some of the people [of the new, better future]. So we can begin discussing things. Metaphysical things. I always did like metaphysical things." Well, any real discussion of the metaphysics of time travel would, I think, convince one that the past could *not* be so changed. The lesson of the unalterability of the past is painfully learned by the history student in "Fire Watch" (Willis). Sent back a hundred years to the London Blitz of 1940, he becomes part of the fire watch team that saves St. Paul's Cathedral from destruction. But just as the salvation of the church in that war could not be changed, he knows that the later vaporization of the same church in a nuclear war in the early twenty-first century also cannot be avoided. This is an example of how a good story can be written without trampling over logic. Other works of science *fantasy* that assume the changeability of the past are (in rapidly increasing order of implausibility) *Time Tunnel* (Leinster), *Black in Time* (Jakes), and *Blake's Progress* (Nelson).

The movies have often illustrated the supposed use of time travel to change the past, with varying degrees of success. In the 1960 film *Beyond the Time Barrier,* a test pilot cracks both the speed of sound and the "time barrier" and flies into the year 2024 to a post–nuclear war world. He later returns to the present in an effort to prevent the war from ever occurring. The 1987

Timestalkers gets changing the past terribly wrong; the hero first sees his family killed in an auto accident and then uses a time machine to make the crash not have happened. The 1994 *Timecop* so completely mangles logic with its changing and unchanging of the past that it seems no subsequent film could do a worse job of treating time travel.

Other films, also illogical, at least have managed to be charming in their ignorance. For example, there is a subtle change-the-past sequence in the original *Back to the Future* film that is easy to miss. When the hero, Marty McFly, returns to 1955, he leaves from the parking lot of the Twin Pines Mall, so named because of the two pine trees that stand nearby. Marty arrives in the past literally with a bang, inadvertently destroying one of the (then) young pines. Near the end of the movie, when he returns to the future of 1985, he finds that the mall is now called the Lone Pine Mall. This is charming and fun—indeed clever, but modern scholars of time travel reject it, and other claims of changing the past, as not logically possible. What Marty's trip *would* explain is why the mall would *always* have had the name the Lone Pine Mall. As Lady Macbeth coolly declares, concerning the murder of Banquo, "What's done cannot be undone: to bed, to bed, to bed."

Changing the Past vs. Affecting It

The fear of time travelers from the future attempting to alter the past has led some philosophers (and no few physicists) to assert that time travel is impossible because it would mean the impossible could happen, i.e., changing the past. Hospers and Fraser are two such philosophers, but one of their professional colleagues, Geach (1968), identified the error in that reasoning in the conclusion of his paper: "[S]quandering vast sums on foolish enterprises is an everyday occurrence. [For example], will the U.S. time explorer get back and eliminate Lenin before his Russian rival gets back even earlier and eliminates George Washington? . . . If such spectacular folly once gets under way because governments have been convinced of some nonsensical theory, a logician will not . . . lose any sleep about who is going to succeed. . . ." The unchangeability of the past was the lesson learned in the 1923 tale "Where Their Fire Is Not Quenched" by the English novelist May Sinclair, in which the reader follows the heroine right into hell after her death; she ends up there because of an immoral life. She then wanders through time, into her past, but finds that she can change nothing. As she is told, "You think the past affects the future. Has it never struck you that the future may affect the past? . . . You *were* what you *were to be*."

Geach would have approved of that, and in fact he is correct (though for the wrong reason, his paper being a misguided attack on the Minkowskian view of spacetime); you cannot travel anywhere into the past unless you've already been there, and when you do make the trip you will do what you've already done there. You could not, like the time traveler in the 1990 story "Ben Franklin's Laser" (Beason), change the course of history by revealing twentieth-century physics in the eighteenth century. That does not mean you would necessarily be ineffectual during your stay in the past, however. Not being able to change the past is not equivalent to being unable to *influence* or *affect* what happened in the past. Science fiction writers have used this distinction to good effect.

Consider the case of the time-traveling submarine in "Uncommon Castaway" (Bond). When it appears in Biblical times, the crew soon rescues a strange man who has been set adrift. It is only after they return to the present that they realize the man they took aboard was Jonah, which of course explains the origin of the famous "swallowed by a whale" story! An ability to play a role in history is not without its constraints, however. You cannot prevent either the Black Death in the London of 1665 or the Great Fire the following year, but it *is* logically possible that you—a careless time traveler—could be the cause of either event or perhaps of both. That was the fate of the time-traveling historian from A.D. 2461 in "The Misfit" (Edmondson), who was the cause of the plague in A.D. 562 Rome, as well as of that in England nearly 800 years later.

The entire point of the beautiful 1958 short story "The Day of the Green Velvet Cloak" (Clingerman) is how something the modern-day heroine does in the present affects events in 1877. In my own 1979 story "Newton's Gift" it is the visit of a time traveler from the future that causes Newton's descent from first-rate physics to third-rate theology, a tragic misapplication of talent about which the time traveler knew but did *not* know the cause. See also "The Past Master" (Bloch), in which the mysterious time machine of a visitor to our near future from the thirtieth century is mistaken for a secret Soviet weapon and thereby triggers nuclear war. The time traveler had journeyed back, in fact, to save masterpieces of fine art from being destroyed in that very war (which he knew from history had been caused by some "trivial incident, unnamed").

Not all writers of early science fiction got the distinction between changing and affecting the past right. For example, a failure to understand this distinction is the rock on which "Sidetrack in Time" (McGiven) crashes. It is the story of the inventor of a time machine whose assistant plots to do away with him and steal the gadget. Emotionally unable just to shoot the inventor and be done with it, the assistant waits until their first test trip, a 5000-year journey into the future. There he knocks the inventor to the ground, leaps into the time machine, and returns alone. Stranded in the far future, thinks the assistant, his victim is as good as dead. The assistant's plan develops a fatal twist, however. Upon his return to the present he sees and hears the inventor—he has accidently returned to the day *before* the start of the test! His nerve fails under the shock, and pulling a gun, he shoots himself dead; "the blasting report reverberated through the lab." But this did not happen in the initial description of that day (after all, it would have been hard for the assistant to have overlooked his own dead body on the floor of the lab!) and so it cannot happen.

The distinction between changing the past and affecting the past has been understood only in relatively recent times. Recall that even Gödel slipped on this point, when he wrote of a time traveler being able to visit himself in the past (which is logically possible if such a visit *did* occur) and then doing something that he recalls did not occur (this is what is illogical). To illustrate Gödel's view from a different angle, consider the 1951 story "Journey" (Hunter), in which we find a thirteen-year old boy going forward in time from 1935 to 1950, to meet himself. It happens on a day he plays hookey by taking a streetcar into Los Angeles—suddenly he finds himself in the future. After seeing a movie, he finds his adult self and discovers he will be (is) unhappily married with all of his ambitions unfulfilled. It is then that the Gödelian objection occurs to him, so he asks his older self, "Wouldn't *you* remember [all this] happening to you when *you* were thirteen?" In reply, with an answer that Gödel himself apparently overlooked, we read, "One time when I was your age I can remember ditching school and hopping a streetcar to L.A. I know I went to a movie." He adds that on the way home that night, he couldn't remember what had happened after the movie. Then, in an attempt to change his life, the adult self puts his younger self on a streetcar back home (and back to the past) with the admonition "Damn you, *don't forget!*" And the happy—though illogical—ending is that he doesn't forget, and so presumably the future is changed. Isaac Asimov tripped over the mirror image of this same error when he wrote (1984b) that "to go into the past and do *anything* would change a great deal of what followed, perhaps everything that followed." Not true.

A fascinating treatment of the "meet yourself in the past as a child" idea, with a clever twist that avoids the mistake of changing the past, unfolds in "A Ticket to Childhood" (Kolupayev). A forty-year-old time-traveling medical researcher, who has lost all of his personal memories except those for the most recent fifteen years, travels thirty years back in time to see the ten-year-old version of himself. It is a strange meeting. The old self is a strong, healthy man, whereas the young self is a sickly boy who seems to think he will die young. And so he actually does: At the end of the story we learn he died at age twenty-five when on the verge of a promising medical research career, on the very day that the old self lost all of his prior memories. Indeed, the old-self is *not* the old-self at all, but rather the artificial creation of the young man, brought into being to continue the research the young man had no time left to do himself. The time traveler therefore exists only because the youngster he thought was his young self existed, and yet the time traveler had no childhood.

Robert Heinlein was one science fiction writer who clearly understood time travel paradoxes, both what they mean and what they do not mean. In his cold-war novel *Farnham's Freehold,* for example, the story of a family that is literally blasted twenty-one centuries into the future when their bomb shelter receives a direct hit from a Soviet nuclear warhead, we find the following exchange as two of the characters are about to return to their original time via a time machine.

> "The way I see it, there are no paradoxes in time travel, there can't be. If we are going to make this time jump, then we already did; that's what happened. And if it doesn't work, then it's because it didn't happen."
> "But it hasn't happened yet. Therefore, you are saying it didn't happen, so it can't happen. That's what I said."
> "No, no! We don't know whether it has already happened or not. If it did, it will. If it didn't, it won't."

Modern philosophers, and many physicists as well, who have examined the concept of time travel in depth agree with that explanation from Heinlein's character, and indeed they now invoke the so-called *principle of self-consistency* (it is generally attributed to the Russian physicist Igor Novikov); as it is stated in Friedman *et al.* (1990), "The only solutions to the laws of physics that can occur locally in the real Universe are those which are globally self-consistent." That is, strict causality is not demanded to determine what is and isn't allowed concerning time travel. All that is required is that a logical consistency exist between events at different times.

Interestingly, statements equivalent to the principle of self-consistency have been around in the physics literature for decades; Driver (1979) traces it back to 1903! And at least an intuitive understanding of the principle can be found in the mainstream literature from nearly as long ago. For example, in Lord Dunsany's short 1928 play *The Jest of Hahalaba* (the inspiration for the 1944 movie *It Happened Tomorrow*), a man obtains (via supernatural means) a copy of tomorrow's newspaper. In it he reads his own obituary, which so shocks him that he promptly expires—thus explaining the obituary notice.

The principle of self-consistency also appeared long ago in science fiction. An example is "Time Wants a Skeleton" (Rocklynne), originally published in 1941 and very possibly inspired by the last lines of T.J.D.'s 1927 letter to *Amazing Stories,* quoted at the beginning of this chapter. In it one character, after puzzling over a time travel paradox, realizes that "Future and present demanded co-operation, if there was to be a logical future!" A nice lecture on the principle (that pre-dates by three years the dialogue quoted above from Heinlein's *Farnham's Freehold*) is given by a character in "The Dandelion Girl" (Young). That story is particularly interesting because it was originally published not in a specialty science fiction magazine catering to an audience with "genre knowledge" of time travel but rather in that icon of general American culture, *The Saturday Evening Post.*

A more modern hint at the principle is dropped in "Hunters in the Forest" (Silverberg), a hint given without elaborate explanation, perhaps because the principle is now part of the accepted background knowledge that writers of time travel science fiction simply assume in their readers. A man from August 2281 meets, in the past, a woman from September 2281 while each is on a different one-day dinosaur-watching vacation courtesy of "Cretaceous Tours." He thinks, "When we get back [to 2281] maybe I'll look her up. The September tour, she said. So [I'll] have to wait a while after [my] own return." He concludes that because she does not know him in the past, and thus had not met (won't meet) him before her departure into the past, he *must* wait until her return to 2281 if the future and the past are to be consistent.

Not all science fiction writers, of course, have understood the requirement for consistency around a loop in time. In the 1942 story "The Message" (Wilson), for example, a man meets the inventor of a time machine and agrees to use it to travel into the future. Once he is in the future—which turns out to be quite dismal—the machine breaks. The man then finds another machine that, though it is too small for him to fit in it, is able to hold a recording he sends back into the past to himself, to *before* he began his journey. The message on the recording (which he did not receive the "first" time) is, of course, *not* to meet the inventor. This advice he follows, so the principle of self-consistency is violated twice in this story.

One writer has turned the self-consistency principle on its head and has used it to deny the possibility of time travel. As Waelbroeck (1991) writes, "A closed timelike curve is a time machine. A time traveler may follow this path in space-time and, at the end of his trip, find himself ready to depart. The argument goes that he may then pull out a gun and shoot his younger image, forbidding himself to make the journey. The solution to this paradox is that the end of the time traveler's journey must be consistent with its start. Thus, a closed timelike curve introduces a periodicity condition on the time evolution. Considering the experimental evidence against such periodicity, we will take the conservative view that one should reject any initial data that leads to closed timelike curves." Waelbroeck, however, tells us neither what "experimental evidence" he is referring to nor how we would ever even recognize any such evidence.

It is important to realize that there are ways to affect the past that avoid any sense of time travel but are in fact mere word games; Swinburne (1966) describes several such cases, and more are given in Ni (1992). For example, the truth of a statement about the past such as "the atom bomb dropped on Hiroshima was the first of the only two such bombs to destroy populated cities" can be altered by the later actions of statesmen and soldiers. The truth of whether Father Jones baptized, in 1997, the greatest pianist of the twenty-first century depends on the subsequent career of Baby John. These are trivial examples that do not capture what is meant by a claim that it is (or is not) possible to affect the past.

Swinburne says that what we really mean by talk of affecting the past is something like "Nothing anyone can do now can make it not have rained yesterday if, in fact, it *did* rain yesterday." That example is correct, but it is not what is meant by *affecting* the past; rather, it is an example of the impossibility of *changing* the past. What Swinburne should have written—given the title of his paper and the position he takes in it—is "Nothing anyone can do now can be the *cause* of it not having rained yesterday." If it rained, it rained, and that's final. The philosophical issue still at hand is merely *when* the cause of the rain occurred, before or after the rain?

Exploration of that question can be found in the science fiction literature from more than sixty years ago. In the 1937 story "The Time Bender" (Saari), for example, a time traveler leaves the Chicago of 1942 for the year 3000. Much later, in the year 2564, another time traveler interested in history journeys back to 2253 in an attempt to learn the cause of the great Chicago explosion of that year. The explosion was centered on the site of an ancient laboratory once used by a scientist who mysteriously vanished in 1942. The second time traveler begins his journey on the same spot, with plans to go back to the day before the explosion. At the end we learn that disaster was the result of the two time travelers colliding as both "passed through" 2253. The backward-traveling historian, therefore, by pushing a button in 2564, is the cause of an event that happened 311 years earlier.

A special, historically interesting case of a belief in the possibility of affecting the past is the retroactive petitionary prayer. (An "ordinary" petitionary prayer, like the Lord's Prayer in *Matthew* 6 and *Luke* 11, asks for something in the present or the future.) Examples of retroactive prayers include that of the surgical patient who prays, just before an exploratory operation, for his tumor to be non-malignant and that of the soldier's wife who prays that her husband was not among those killed in yesterday's battle. These prayers are for a happy outcome to an event that is over and done with at the time of the prayer. One might accept the rationality of praying about the future ("Please, God, let me survive tomorrow's battle and I'll be good for the rest of my life"), but are prayers about the past even sensible?

In an appendix titled "On 'Special Providences' " in his book on miracles (1978), C. S. Lewis answers that question as follows:

> When we are praying about the result, say, of a battle or a medical consultation, the thought will often cross our minds that (if only we knew it) the event is already decided one way or the other. I believe this to be no good reason for ceasing our prayers. The event certainly has been decided—in a sense it was decided "before all worlds." But one of the things taken into account in deciding it, and therefore one of the things that really causes it to happen, may be this very prayer that we are now offering. Thus, shocking as it may sound, I conclude that we can at noon become part causes of an event occurring at ten A.M. (Some scientists would find this easier than popular thought does.)[7]

Here we see Lewis, a prominent lay theologian, arguing for the present influencing (but not changing) the past. What can we make of that? Was Lewis arguing for backward causation? I think perhaps so—the last sentence in that excerpt makes it seem that he may have had this in mind. It is a view that does find much support in the block universe interpretation of Minkowskian spacetime. Lewis never mentions the block universe concept by name, but it is clear that he believed in the idea of God being able to see all of reality at once (recall the words of Aquinas and Boethius from Chapter Two). Lewis believed, therefore, that God knew of the petitionary prayer before it was made; or, even stronger, if God is not a temporal being but rather is eternal and knows all of time "at once," then God knows of the prayer and the event being prayed about "at the same time."

There have been all sorts of opinions expressed through the ages in reaction to the idea of affecting the past via the retroactive petitionary prayer, and many of them have centered on the old concern of free will versus determinism discussed in Chapter Two. Brown (1985) gives a good summary of those opinions. Most theologians want to retain free will, as did Lewis—the philosopher

Michael Dummett discusses (1964) Lewis' concept of retrospective prayer with great sympathy—and backward causation lets them do so, in addition to keeping divine omniscience. That is, it is not God's foreknowledge that causes our later actions, that forces our behavior and turns us into automatons. Rather it is our later actions that cause God's foreknowledge!

We find the origin of the modern philosophical debate on the issue of affecting the past in the companion papers by Dummett and Flew (1954). Dummett, who believes in the logical possibility of backward causation, unfortunately presented his case in the questionable form of philosophers' stories (in Chapter Two I gave my reasons for dismissing this approach as having no merit other than as entertainment). Specifically, Dummett asks us to imagine a man who always wakes up in the morning three minutes before his alarm clock goes off (unless he forgot to wind it, and then he sleeps late); a magician who finds the weather is fine in Liverpool yesterday after he recites a spell today (assuming he does not know ahead of the recitation what the weather actually was); and a man who finds that if he says "click" before opening an envelope, it never contains a bill.

Dummett argues that by broadening our idea of cause, we can retain such occurrences in a logical world. But because there is no evidence for those curious events—they are simply the offspring of Dummett's imaginative mind— *why should* we be willing to broaden what appears to be an already complete concept of causation? Dummett never offers us the slightest physical explanation, or even hints of a physical mechanism, for his tales. Whatever may be the truth of his conclusions, his method of reaching them has no compelling power to persuade.[8]

In his reply rejecting Dummett, however, Antony Flew was no more successful in convincing the reader of the logic of *his* argument. He was not even willing to entertain the logical possibility of backward causation, which is all that Dummett was claiming. Flew declared, "that the cause must be prior to . . . the effect is not a matter of fact but a truth of logic." That sounds very much like Aristotle, who could have simply dropped two unlike weights, as Galileo did, to *observe* what actually happens but who preferred to argue "logically"—and falsely—and with such a declaration there is nothing left to say (at least not to Flew).

Flew, of course, convinced few with his begging of the question. Indeed, he had completely missed the point, as was noted some years later by Brier (1973), who correctly charged Flew with having blundered by confusing changing the past with affecting the past. In a reply immediately following Brier's paper, Flew showed that he continued to make the same error when he wrote that he found the distinction between the two concepts "quite breathtakingly perplexing." Compounding that admission of ignorance, Flew went on to charge incorrectly that the distinction between changing and affecting the past was "one with which any faithful reader of the philosophical journals is [not] acquainted."

That prompted a reply from Dwyer (1977), who took as his goal the education of Flew on the difference between changing and affecting. Referring to an earlier critique of Hospers—Dwyer (1975)—Dwyer wrote of a Gödelian time traveler visiting the past in a rocket:

> Time travel, entailing as it does backward causation, does not involve changing the past. The time traveler does not undo what has been done or do what has not been done, since his visit to an earlier time does not change the truth values of any propositions concerning the events of that period. Thus even before the time traveler enters his rocket in 1978 to begin his successful mission to the year 3000 B.C., an accurate catalogue of all the events occurring in Ancient Egypt that year would include an account of his arrival from the sky, as well as an account of his various actions and reactions in that new environment. The contents of such a catalogue may never be revealed but that is beside the point. And yet, while the time traveler thus does not *change* the past when he goes back to it (for he cannot *do* anything in 3000 B.C. that was not *done* in that year) he does *affect* the past in that the arrival of the rocket, as well as his pyramid-building activities, etc., are members of the class of events that characterize the (unique) year 3000 B.C. It seems to me that there is a clear distinction to be made here, between the case where a person is presumed to change the past, which indeed involves a contradiction, and the latter case where a person is presumed to affect the past by dint of his very presence in that period.

Dwyer's paper concludes with the statement that "certain criticisms of backward causation theories, such as Flew's response to Brier, are . . . easily seen, in the context of time travel, to be based on misunderstandings." Another philosopher Dwyer would surely accuse of similar misunderstandings is J. R. Lucas, who in writing about time machine stories says,[9] "Often we are invited to imagine ourselves witnessing past events. Provided we are inactive and invisible (recall Gernsback's similar view), no paradox need ensue. If we are merely passive spectators, we are not altering the unalterable past; and provided we cannot be seen, heard or felt by any of the actors of the events, our presence makes no difference, and the purity of the past is preserved untouched." Like Hospers, Lucas appears to think the past happens twice, once without the time traveler and then again with the time traveler. This is a view that few modern students of time travel accept.

For Dwyer's presentation to be persuasive, of course, we must be willing to accept at least the *possibility* of time travel; else we are reduced to the telling of a new philosopher's story that tries to explain one extraordinary idea (backward causation) in terms of another, even more extraordinary one (time travel). For the thesis of this book, the importance of Dwyer's 1977 paper is that in it he explicitly gives us a rational mechanism: "One can provide an *explanation* for time travel in terms of the field equations of General Relativity, together with initial conditions on the distribution of mass-energy in a certain region of space-time."

I agree with Dwyer, but of course not everybody does. For example, in Spellman (1982) we find "I do not intend to enter the debate over whether time travel is conceivable. What I do want to hold, however, is that there is no reason to call Dwyer's example *backward* causation. Dwyer describes pyramid building as in the [time traveler's] 'causal future' but part of his 'chronological past.' But if time travel is possible, what reason is there for saying that pyramid building *precedes* the rocket's firing? Rather, from the [time traveler's] point of view, what happens is perfectly ordinary causation—an earlier event (rocket firing) causes a *later* one (arrival at one's destination)." Spellman is certainly correct in his logic, as far as it goes, but what of the point of view of a non–time-traveling observer who watches the entire process linearly from 3000 B.C. to A.D. 1978? For him (and if such a lengthy life span is bothersome, then just replace 3000 B.C. with A.D. 1950) the process certainly *does* involve backward causation.

As we saw in Chapter One, many science fiction stories have gone to great lengths to show how time travelers would have to be excruciatingly careful while in the past; an example is Ray Bradbury's "A Sound of Thunder." At the other extreme is Clifford Simak's "Over the River & Through the Woods," which features an intentional effort to change the past. Although modern philosophers may find such stories charming, they rightfully reject them as illogical. These stories are illogical because they erroneously liken time travel into the past to ordinary travel through space. With the latter, you can indeed visit places you have never been before, and once there you can do things you have never done before. Not so, however, with time travel to the past. When the narrator in the novel *A Time to Remember* (Shapiro)—the basis for the 1990 made-for-TV movie *Running Against Time*—wondering whether he can save Kennedy from Oswald, asks, "Can I really go back to 1963 and stop the unspeakable crime, to unstitch historical fabric and resew the past?" the answer is no. You can't save Joan of Arc with a fire extinguisher either, or Jesus with a rifle, as in "The Rescuer" (Porges), because either you weren't there then and so you can't *be* there then, or you were (are going to be) there then, in which case we already know you failed for some reason. Trying to intercept John Wilkes Booth outside of Ford's Theatre and warning the Nazis of D-Day are equally doomed. And as a special, personal case, if some day you should come into possession of a time machine, it will be similarly futile to attempt to visit yourself at any time you know you were alone. Go back to a time when there were lots of people around, but don't try to talk to yourself *unless* you recall a time when a mysterious yet somehow familiar stranger approached you with talk about traveling through time!

In Poul Anderson's "time war" novel *The Corridors of Time,* a twentieth-century man recruited by a visitor from the future exclaims, "Huh? Wait! You mean you people *change* the past?" The logical reply, from his visitor, is "Oh, no. Never. That is inherently impossible. If one tried, he would find events always frustrated him. What has been, is. We time travelers are ourselves part of the fabric." If you *weren't* then, you can't *be* then — and there is simply nothing to be done about it. As Shakespeare's Pericles observes, "Time's the king of men; He's both their parent, and he is their grave. And gives them what he will, not what they crave." In the 1941 story "Beyond the Time Door" (O'Brien) we have a tale that, at first, appears to misunderstand this point but that, in fact, presents a logical treatment. A convicted killer suddenly disappears from the electric chair in 1940 and finds himself in the year 3000. As the inventor of the time machine that has made good this astounding escape explains to the stunned criminal, if the killer will carry out a "contract hit" for the inventor, then "I'll send you back one year before the time you were caught and sentenced to death! . . . You can see to it then you don't take the step that led you to the chair!"

That illogical claim seems to indicate a primitive change-the-past story, but that impression is weakened a few sentences later when the inventor further claims that history records the time and date of the killer's execution. As the story unwinds, we find that although the killer is a brutal man, he is not stupid. Before the time machine controls can be changed from the instant at which the killer was plucked from the electric chair, the killer changes places. Thus it is the evil inventor who is sent back through time — and it is *his* execution, not the killer's, that is graven in history's unchangeable record.

The distinction between changing the past and affecting it also provides an answer to the "cumulative audience paradox" in Robert Silverberg's novel *Up the Line.* That paradox claims that as time travelers to the past continue to visit certain historically interesting dates and places, there will be an ever-increasing number of people present. As stated in the novel, "Taken to its ultimate, the cumulative audience paradox yields us the picture of an audience of billions of time-travelers piled up in the past to witness the Crucifixion, filling all the Holy Land and spreading out into Turkey, into Arabia, even to India and Iran. . . . Yet at the original occurrence of [that event] *no such hordes were present!*" (Is this where Hawking got the inspiration for his 'no hordes of tourists from the future' concept that he claims is "experimental evidence" for his Chronology Protection Conjecture?) And later in the same work, more poetically, "A time is coming [when we] will throng the past to the choking point. We will fill all our yesterdays with ourselves and crowd out our own ancestors."

The modern view of Silverberg's "paradox," of course, is not that the above absurd situation means there could be no time travelers present at the Crucifixion because "obviously" time travel must be impossible, but rather that *all* the time travelers who *were* (will be?) there are in the "original" hordes—Silverberg's use of the word *original* is not correct because it repeats Hospers' error in claiming the past happens more than once.

"The Ring" (Hudec) is a clever story that also has its subtle but quite fatal flaw in a failure to appreciate the difference between changing and affecting the past. It also illustrates object duplication, a time travel effect that Silverberg calls a special case of the cumulative audience paradox. In this story we are told that a time traveler makes a huge amount of money by selling the same diamond ring 94 times; that is, he sells 94 different rings yet the rings are "identical, definitely and absolutely, down to the last atom." There is no crime involved here, no fraud, because the rings are each worth every penny charged. But a police inspector is intrigued and so investigates. What he discovers is that the time traveler simply went back to seventeenth-century Amsterdam and bought the same ring over and over, each time one hour earlier than before, until the time when the jeweler had not yet himself acquired the ring! From the Minkowskian spacetime view, however, when he presented himself the first time (the first for him but the last for the jeweler), he would have been thought crazy. "But sir," the jeweler would cry, "you have already bought the ring of which you speak!" The ring would simply not be in the jeweler's shop after the purchase furthest back in time—indeed, its absence would be evidence to the time traveler on his first visit to the jeweler that he would later go back even further and buy it. The same idea (and flaw) can be found in "The Fourth-Dimensional Demonstrator" (Leinster) and "Gravesite Revisited" (Moon).

Perhaps the classic of object duplication stories is the enormously clever, diabolical "My Name Is Legion" (Del Rey). In this story the object duplication is accomplished by "reaching into the *future*," not into the past. The inventor of the gadget that does this, whose wife and children had been murdered by the Hitler Youth, explains, "I pull an object back from its future to stand beside its present. I multiply it in the present. As you might take a straight string and bend it into a series of waves or loops, so that it met itself repeatedly." The "string" is certainly Del Rey's metaphor for the world line of the object. What gives this tale particular power is that the duplicated object is Adolf Hitler, duplicated 7000 times (one from each of the subsequent 24 hours for the next 20 years), each increasingly older copy being slightly more decrepit than the last. (The story originally appeared in 1942, before Hitler's real fate in the Bunker was known.) The mind of Hitler is thus forced to live through the same 24 hours 7000 times. (Del Rey avoids Silverberg's error of repeating history 7000 times by having 7000 copies of Hitler present at once, rather than one additional Hitler for each repetition.) We have discussed other stories that have time travelers duplicating themselves via time travel. They include "The Trouble with the Past" (Eisenstein), "Obituary" (Asimov), and "We're Coming Through the Window" (O'Donnell).

Some writers of fiction find object duplication so objectionable that they simply deny its possibility. In *Hot Wireless Sets, Aspirin Tablets, the Sandpaper Sides of Used Matchboxes, and Something That Might Have Been Castor Oil* (Compton), for example, when Roses Varco (the mentally retarded time traveler we met in Chapter One) returns to the year he was (is) eighteen, he is "found [by the equivalent of Varley's cosmic disgust theory] to be philosophically impossible." Thus he vanishes with "a roar like that of an express train," leaving behind no trace other than the strange odor described by the novel's curious title. (In 1970 Ace Books in New York published the novel under the somewhat more "traditional" science fiction title of *Chronocules*—a chronocule is described as a "particle of time.")

An early (1943) but scientifically sophisticated science fiction, spacetime presentation of the basic ideas behind all "change-the-past" stories is found in "Forgotten Past" (Morrison). There one character explains it all to another: "Most of this talk of time travel is rot. Notice that I don't say all of it, but most of it. You can't travel into the past. The past is an infinite region in the four-dimensional space-time continuum whose nature has been completely determined. All the world-lines, as Minkowski put it, the world-surfaces, the world-volumes, are completely known. Traveling in the past would change them. It can't be done." This argument loses its force, of course, given the position taken in this book (that is, if one accepts the idea that the world lines of the time traveler are bent back into the past), so time travel does not change the past but rather is built into the past.

John Wyndham's "The Chronoclasm" fails to grasp the distinction between changing and affecting the past, and all the characters exhibit erroneous concerns—concerns that time-traveling historians are changing the past when, in fact, all the quirky historical happenings cited (such as Hero demonstrating a steam turbine in Alexandria years before Christ, Archimedes using napalm—"Greek fire"—at the siege of Syracuse, and Da Vinci's drawings of parachutes) *are* history. The leakage of information by loose-tongued time travelers is not changing the past but rather is affecting it just as history records it. It is, indeed, *necessary* for such leakage to occur to avoid altering the past (in this story) as we know it.

Other stories have done a good job of correctly distinguishing affecting the past and changing the past. For example, in the 1940 story "The Time Cheaters" (Binder) a friend tries to talk two time travelers into having a little fun after their successful initial test trip. Just before the time travelers leave 1941 for the future, he says, "Why not take in the World's Fair before you leave again? You won't have another chance. Or—uh—will you?" (This shows the risk of trying to predict the future. The story appeared March 1940, while the Fair was still open, but it closed just a few months later, in October, after declaring bankruptcy.) One of the chrononauts answers with a laugh, "We will. We'll have the distinction of going into the future, reading about the Fair being over, and then come back to see it!" Replies the friend, "I don't believe it! You can't do that. . . . If the Fair is over without you two being in it, you never *were* in it." Absolutely correct.

Another such tale that gets things right—indeed, it is a classic of the genre—is "The Biography Project" (Gold), about the Biotime Camera that can film the past (alas, no sound). Using this wonderful gadget, the Biofilm Institute funds teams of biographers to study the lives of past notable personages. In particular, the lives of those who developed neurotic psychoses, such as Robert Schumann, Marcel Proust, and Isaac Newton, are of great interest. And, indeed, the Biotime Camera does capture these individuals' images as they begin to display increasingly disturbed behavior. We see Newton, for example, begin to peer into dark corners, looking for those he has come to believe are spying on him. On his death bed, the biography team assigned to him reads his lips and discovers that his final words are "My guardian angel. You've watched over me all my life. I am content to meet you now." It is then that the Biofilm Institute realizes what it has done. Newton *was* in fact being spied upon—by the Biotime Camera, which has not changed the past but *has* affected it. (This is, of course, really just a stimulating exercise in speculative fiction; present medical opinion is that Newton's odd behavior was actually due to mercury poisoning from alchemy experiments, not from being time-viewed!) The same idea is used in "One Time in Alexandria" (Franson), in which an archeologist uses a time viewer to read the lost manuscripts in the ancient library at Alexandria before it burned. The viewer uses an infrared beam—and it is the heat from that beam from the future that proves to be the origin of the fire in the past.

The idea of time viewing has even appeared in the popular, non-fiction literature. For example, John L. Cotter, curator emeritus of The University Museum of the University of Pennsylvania, was so inspired by reading about the wormhole time machine in the 1988 paper by Morris, Thorne and Yurtsever that he responded as follows:[10] "Thorne and his colleagues . . . say that if travel into the past is theoretically possible, this possibility would have 'profound philosophical consequences.' It sure would. It could put archeologists out of business. It would also offer a dilemma for historians and theologians. Imagine reviewing the actual lifetimes of Moses, Jesus, Mohammed, or Gautama Buddha, to say nothing of Lucy, Neanderthal Man, the artists who painted Lascaux Cave, Sargon, and Julius Caesar." Cotter then offered up an unsettling observation, which was probably new for most of his readers but which we recognize as the idea of the Biotime Camera: "It must occur to us that *we* may be viewed by those living 2000 or more years in the future." As the narrator warns us at the end of "I See You" (Knight)—a cautionary tale about the full implications of being able to view the past—"You realize that there are no secret places. And beyond you in the ghostly future you know that someone is watching you as you watch, and beyond that watcher another, and beyond that another. Forever." Ponder that the next time you *think* you're alone before you do anything you'd hate to see show up in a graduate student's doctoral thesis in the fortieth century—although for some of us it may already be too late to worry about that!

The two *Terminator* movies cleverly incorporate both affecting and changing the past. The central idea common to both movies is that after a self-aware military computer intentionally starts the nuclear holocaust of 1997, killer-machines enter into a continuing conflict with the human survivors who are inspired by a charismatic leader to resist defeat. In the original *Terminator,* the machines send a killer-robot back to 1984 to change the past by terminating the mother-to-be of the leader-to-be *before* he is born. To counter this, the leader sends a friend back to warn his mother—and this friend becomes his father. That is, the leader exists in 2029 because in that year he does something that *affects* 1984; he arranges for his father to meet his mother!

In the 1991 sequel, *Terminator 2,* we learn how the self-aware military computer came into existence. At the end of the original film, the killer-robot is destroyed before it can complete its (logically impossible) task, but a single computer chip from its 'brain' is salvaged; the high-tech innovations spawned by the secrets of that chip lead to the invention of the self-aware computer and, thus, to the war of 1997. In the second film a second killer-robot is again sent back in time, now to kill the leader himself when still a child (another logically impossible task, yes, but after all, this is Hollywood!), and again the killer is (logically) thwarted. Indeed, that machine fails because again the leader sends back a counter-warrior (a "good" robot, this time) to protect the boy-version of himself. In addition, the good robot finds and destroys the advanced computer chip left in the past of 1984 before it can be deciphered; this is supposed to prevent the construction of the self-aware computer in the first place, which is of course logical nonsense.

Does this mean that *Terminator 2* is logical nonsense, then, after having done a pretty good job with time travel until the final scene? I don't think so. In the final battle scene of the sequel, the good robot has an arm wrenched off (just as happened to the "bad" robot at the end of the first movie). Thus, despite the self-sacrifice of the good robot in a vat of molten steel to destroy the chip in *its* brain, there is still future technology existing in the present. Who can say what may be in that surviving arm? My guess is that it will be the basis for *Terminator 3*.

Both *Terminator* movies send the message to their audiences that the future can be changed by changing the past, an assertion that most physicists and philosophers today would say has no meaning. The general conclusion, then, is that you cannot change the past, but you might well affect it. That view rejects, for example, the position taken in "Pebble in Time" (Goldstone and Davidson), which relates how an elder of the Church of Latter-Day Saints invented a time machine so he could travel back to 1847 to watch Brigham Young declare, *"This is the place!"* at what would become Salt Lake City. Inadvertently interfering with the past, however, the time traveler is shocked instead to hear *"This is not the place! Onward!"* and to see Young continue on to San Francisco. As home of the Mormon Church in this altered history, San Francisco becomes associated with the initials L.D.S. (the story appeared in 1970; this constitutes a play on the initials associated with the L.S.D. drug culture of that time and place). In contrast, Michael Moorcock's brilliant 1969 novel *Behold the Man* gets the impossibility of changing the past, and the possibility of affecting it, right. When a disturbed man journeys backward in time to ancient Galilee to meet Christ, only to discover that there is no such person, *he* assumes the role and lives out the Biblical accounts up to and including dying on the Cross. He has not changed the past, but he certainly plays an important role in it! This same theme was used more recently in the 1994 story

"The Tetrahedron" (Harness), in which the Mona Lisa is the historically "explained" character—an idea that runs counter to Chapman (1982), a philosopher who incorrectly argues that a time traveler could not "causally interact" with anything in the past and who, echoing Gernsback, says a time traveler must be "disembodied."

Physicists who worry about cosmic disaster if a time machine were actually to be built, and those who reject physical theories simply because they allow time travel to the past, are worrying about a non-problem.[11] Even if time travel to the past is possible, you cannot go back and kill your grandfather before your father is conceived—or, even "better," you cannot kill yourself in the past. But it *is* logically possible that you could be the one who introduces your grandfather to your grandmother.[12] Even the funny conversation two time travelers have with themselves in a time loop (we see it twice, once from each side of the exchange) in the 1989 movie *Bill & Ted's Excellent Adventure* is logical.

Why Can't a Time Traveler Kill His Grandfather?

Even when all has been said about the impossibility of changing the past, and even when they are finally willing to concede that point, most people still cannot help wondering *why* the time traveler can't kill his grandfather. There the time traveler is, after all, just two feet away from the nasty young codger (I assume he is nasty to make the whole unpleasant business as palatable as possible). A perfectly functioning and well-oiled revolver is in his hand, cocked and loaded with powerful, factory-fresh ammunition that even Dirty Harry would find excessive. What can possibly prevent the time traveler from simply raising his arm and doing the deed? Indeed, the opening illustration to the 1944 story "Thompson's Time Traveling Theory" (Weisinger) shows this act in detail, including the smoking gun in the hand of the time traveler who has just taken a shot at grandpop. And if that still leaves open the remote possibility of an aiming error through nervousness, then why can't a suicidal time traveler just wrap his entire body in factory-fresh dynamite and blow-up grandad—along with himself and everything else within a hundred feet?

As the rest of this section will argue, killing your grandfather is *logically* impossible. No one will ever find a note in the empty laboratory of a missing traveler who, skeptical of the grandfather paradox, has written, "To prove the falsity of the grandfather paradox, I will take my time machine back fifty years and kill my grandf. . . ." Nor will the inventor of a time machine have to be concerned about the pretty twist in Fredric Brown's "First Time Machine." There, the inventor of a time machine shows the gadget to three friends; one of them steals the machine to go back sixty years to kill his grandfather; the story closes with a *near* repeat of the opening, with the inventor showing the gadget to *two* friends.

It is a shame that the classic time travel paradox takes such a murderous form, but that is the historical origin of the idea. Stewart (1994) leaves the impression that this paradox is only about fifty years old when he writes of "the 'grandfather paradox,' which goes back to René Barjavel's story *Le Voyageur Imprudent*." That tale dates back to 1944, but in fact we can find the grandfather paradox being discussed more than a decade earlier, in a letter to the editor at *Astounding Stories* (January 1933). The author of that letter wrote, "Why pick on grandfather? It seems that the only way to prove that time travel is impossible is to cite a case of killing one's own grandfather. This incessant murdering of harmless ancestors must stop. Let's see some wide-awake fan make up some other method of disproving the theory." As we proceed, you'll see just how clever some of those who responded to that writer's plea have been, but even today, as it stands revealed as a red herring, the grandfather paradox is preeminent in most people's imaginations.

If a solution to the grandfather puzzle escaped an early science fiction writer, then he would generally just mysteriously refer to it and then hasten on to other matters. For example, in the 1942 story "The Time Mirror" (South) we find the following exchange between the stock pulp-fiction characters of a young hero and a brilliant old scientist.

> "You mean that time travel really is possible? That men can be transported into the future or the past—."
>
> The other held up a restraining hand. "Yes. Time travel *is* possible . . ."
>
> "But professor! Think of what you're saying! You're telling me that I could go back and murder my own grandfather. That I could prevent myself from being born—."
>
> Again the elder man sighed. "I was afraid of this," he said. "I knew you could not understand." He hesitated. Then: "At any rate, take my word for it that time travel is possible. Also, I assure you that there are any number of perfectly sound theoretical and practical reasons why you never could hope to murder your grandparents."

We are, however, not told just what these reasons might be.

The earliest story that I've found in which a time traveler specifically kills his grandfather in the past is the 1934 tale "The Time Tragedy" (Palmer), but the paradoxical aspects of the act are not developed. The grandfather paradox is also mentioned, in passing, in the earlier 1929 "Paradox" (Cloukey). In 1933, however, the far more sophisticated story "Ancestral Voices" (Schachner),

Figure 4.2 In this illustration from Raymond A. Palmer's "The Time Tragedy" (*Wonder Stories,* December 1934), the inventor of a time machine demonstrates it by sending the family cat on a trip. The inventor travels back to 1901, where he accidently kills his grandfather in an early, non-paradoxical version of the famous riddle.

Illustration for "The Time Tragedy" by Frank R. Paul © 1934 by Continent Publications Inc.; reprinted by permission of the Ackerman Science Fiction Agency, 2495 Glendower Ave., Hollywood, CA 90027 for the Estate.

based on the concept of killing more ancient ancestors, caused a brief stir. In this tale people vanish by the tens of thousands because *one* man is killed fifteen centuries earlier. Schachner himself seems to have believed that this makes sense: In a letter to the editor of *Astounding Stories* (December 1933) he wrote that " 'Ancestral Voices' attempts the logical unfolding" of the grandfather paradox. A subsequent flurry of fan letters to the magazine, however, showed that many readers did not find the story at all logical. Other writers, however, found the grandfather paradox irresistible; twenty years later, the time traveler in "Time Goes to Now" (Dye) topped Schachner's story by accidently killing the original "intelligent baboon" in the ancient past, thereby wiping out the entire human species!

The grandfather paradox nags at all students of time travel. Some believe it is unresolved even today. For instance, Schmidt (1998) claims that "there are as yet no generally accepted solutions" to the grandfather paradox. As a character in the 1950 story "Typewriter from the Future" (Worth) says, "The resolution of [the grandfather paradox] is the key to time." And it *is* troublesome, if not quite so central to the core of time as Worth declares. As Gorovitz (1964) somewhat less melodramatically puts it, time travel and its concomitant backward causation—the apparent possibility of a time traveler being able to do away with both his grandfather and himself—gives "rise to such puzzles that we are forced to question its [time travel's] intelligibility."

Gorovitz correctly believes that the past cannot be changed, and so the potential violent act of a time traveler murdering his grandfather or committing suicide simply can't happen. But then he goes on to say that the inability of the time traveler to kill either himself or his grandfather results in our being "faced with the problem of explaining why it is [that he] cannot fire the gun or, if [he] can, why it is [that he] can fire only in certain directions [i.e., the ones that miss]." Gorovitz asserts that there can be only two possible answers: "Either the gun is not behaving as the normal physics object we take it to be, or the notion of voluntary action does not apply in the usual way." And so here, at last, we have an explicit concern over the issue of free will. Gorovitz isn't alone in his unhappiness over the grandfather paradox; Herbert (1988) calls the paradox the strongest argument he knows against time travel. Both would agree with one of the characters in the 1944 story "Thompson's Time Traveling Theory" (Weisinger), who uses the grandfather paradox to conclude that "all time traveling stories are one hundred percent sheer oil of over-ripe bananas!"

Gorovitz has gotten himself into this logical quagmire precisely because, like Hospers, he thinks of the past as happening twice—once without the time traveler and his gun, and then again with him and it. With that second chance,

Gorovitz argues, the time traveler should have the ability to do something he didn't do on the first try. So why can't he? This puzzle is all of Gorovitz's own making, because he is violating his own fundamental belief in the unchangeable nature of the past. Assuming that the time traveler did once confront his grandfather (or himself), then he *must* fail because he *did* fail. To demand an accounting for the specific "why of failure" before accepting the failure is as misguided as a stranded motorist refusing to believe his car won't start until he knows *why* it won't start.

Some science fiction stories have missed that point and have invoked forces (mysterious and otherwise) to protect the fragile past from future tampering; an example is the so-called time police. These time commandos roam the corridors of time, disrupting the plans of those who would change history to suit their personal desires. Stories of temporal cops are simply westerns, mysteries, police procedurals, or some similar type of specialty story form wearing thin camouflage, and this particular story device can be, as Lewis (1976) calls it, "a boring invasion."[13]

So why do we so often find, in fiction, various incarnations of the time police, such the Time Security Commission in "A Star Above It" (Oliver), which prevents a rogue historian from changing the past to allow the Aztecs to defeat Cortez, and the operatives in Poul Anderson's many stories about the Time Patrol? These "chaperones" have appeared in the movies, too; see the time police films *Trancers* (1985) and *Timecop* (1994). The short answer is that the writers of such stories—and many philosophers, as well—believe this is the only way to have both time travel and free will. Let's look at two examples of this supposed conflict.

Returning one last time to Hospers, we find the following tale:

> Our hero in 1900 pulls the lever [of his time machine] and finds himself in . . . the future. There he meets a girl, marries her, and takes her back with him in the time machine to the year 1900. The girl wasn't born until A.D. 40,000, yet she gave birth to his child in 1900, long before she was born. One is tempted to speculate: What if he had decided, in the year 40,000, *not* to marry her and bring her back after all? Then her child (born in 1900 though the mother wasn't born until 40,000) wouldn't have been born either; and yet after 1900 he had already been born. Indeed, that child might have become the prime minister of Britain, and affected the course of the world in such a way that no human beings would have existed on the Earth in the year 40,000. What if there had been a nuclear explosion in 1990 that obliterated life forever from the Earth?

All those questions tell us more about Hospers' badly mangled concept of what constitutes a paradox than they tell us about time travel. To complain about the girl giving birth before her own birth is to beg the question of time travel itself; this is simply backward causation, which is inherent with time travel. The issue of free will arises when Hospers wonders "what if" the hero does not bring the girl back to 1900. The answer is simply that then the time traveler's son would not have been born in 1900 and he would not have gone on to become the prime minister responsible for wiping all life out in A.D. 40,000. Indeed, if all life is so wiped out, it is *Hospers'* obligation to explain to *us* why the girl is alive then! On the other hand, if his son was born in 1900, then the hero *must* bring the girl back—it is this *must,* which is required by consistency, that bothers Hospers.

Thus Hospers has created his puzzles for himself by demanding to have matters all possible ways all at the same time (so to speak). If there was a nuclear explosion in 1990 that obliterated all life, then the hero simply would not find a girl in 40,000. Again, Hospers has manufactured his own puzzle by claiming both that the girl is there in A.D. 40,000 *and* that all life vanishes forever in 1990. To declare that a paradox is tantamount to first declaring a gun is empty, without having any evidence for that claim, and then being surprised when you shoot yourself in the foot.

As our second example, Martin Gardner (1982) makes the same kind of argument, and the same error, in his tale of Professor Brown, who goes forward in time by 30 years, carves his name on an oak tree, and then returns to the present—where he promptly cuts the tree down! How, asks Gardner, if Brown cuts the tree down now, can it be there 30 years in the future for him to carve his name into? The answer is simply that if the tree exists in the future, then the Professor simply didn't (won't) cut it down—or if he does cut it down, then it won't be there 30 years later. Gardner forces the "paradox" himself by demanding both that the tree be there *and* that it not be there. Here is what I believe is really bothering Gardner: If the Professor does find the tree in the future, then why can't he cut it down upon his return? This, of course, is the mirror image of the grandfather paradox. We don't know why; we just know that the Professor won't (perhaps his axe will break or he will be arrested for attempted vandalism, etc., etc.) As a character in Robert Forward's novel *Timemaster* puts it, "Once time machines exist, no event is low probability if it is needed to make the past consistent." And there is no second or third or fourth chance—the Professor's world line in spacetime brings him to the tree *once* and he fails *once* (for some reason) to cut it down.

Figure 4.3 A time machine inventor makes an experimental test of the grandfather paradox!

Illustration for "Thompson's Time Traveling Theory" by Malcolm Smith, © 1944 by Ziff-Davis Publishing Co.; reprinted by arrangement with Forrest J. Ackerman, Holding Agent, 2495 Glendower Ave., Hollywood, CA 90027.

In the novel *The Unknown Soldier* (Reichert) a madman in the year 2058 plans to destroy the existing world by setting off widespread nuclear explosions. Meanwhile, his followers will travel into the past, retrieve the madman when he is ten years old, and take him into the very far future to create a new world. Replace Gardner's tree with Reichert's madman and you have the same paradox—and the same logical error. As the hero of Reichert's potboiler thought about this scheme, "The idea seemed utter madness. . . ." Indeed.

One science fiction story that does get the grandfather paradox right is the clever 1944 "Thompson's Time Traveling Theory" (Weisinger). A time traveler journeys back from 1943 to 1870 and shoots his then fourteen-year old grandfather in the head. Leaving his victim lying on the ground with "blood oozing all over the youth's forehead," the would-be killer returns to 1943. Once back, however, he finds himself in a strange place where he learns from two men that the Germans destroyed New York in 1920 with poison gas! Suddenly realizing that the death of his grandfather has apparently changed history (a curious oversight for anyone smart enough to invent a time machine and then actually to force the "grandfather paradox"), he decides he'd rather be dead than be cut off for all time from *his* world. Accordingly, he shoots himself. As he lies dead, we learn that the two men are actually inmates in an asylum who like to make up stories for unsuspecting strangers. We also learn that the time traveler's grandfather's photographs always did show him with a "white, furrowed scar on his forehead that might have been caused by a glancing bullet."

In an afterword to his 1946 story "Dead City," Murray Leinster observed, "You've heard the old argument that a man can't travel backward in time because he might kill his grandfather. I've wondered why nobody has argued that a man can't travel forward in time because he might be killed by his grandson." Perhaps nobody argued that because it really isn't much of a puzzle! The answer to Leinster is simply that if, at the moment our forward-bound time traveler departs, he has not yet sired a child, then there simply won't be a murderous grandson waiting for him in the future. Or, if he has sired a child before starting his trip, then there *could* be an ungrateful descendent who, without paradox, *could* indeed kill him.

The grandfather paradox was most recently illustrated in the 1993 "change-the-past" movie *The Philadelphia Experiment 2,* in which the winner of World War II is changed from the Allies to the Nazis and then back to the Allies. In the course of this illogical film, the hero, while in the past, kills the father of the villain before the father has conceived his son. And so, amid much spectacular electrical sparking, the bad guy noisily winks out of existence. And although the film never actually shows it happen, the villain in the 1994 *Timecop* routinely threatens to kill the ancestors of his unwilling helpers unless they do his bidding. Half a century after Weisinger got it right, Hollywood still has a lot to learn about the grandfather paradox.

In an attempt to analyze the free-will issue in time travel, in the context of the grandfather paradox, Thom (1975) presents an interesting grammatical argument. Thom asks us to imagine the usual situation: A time traveler as a mature man travels into the past and confronts himself as a boy. Can the time traveler kill the boy (himself)? As I have argued up to this point, the answer is yes but also that he won't because he didn't. The fact that he won't (didn't) doesn't mean he can't (couldn't). But Thom pursues this and says that if the time traveler *can,* then one would appear to be on safe ground in assuming that there is no logical inconsistency in imagining this "could happen" event actually occurring. But, of course, the time traveler killing his younger self *would* lead to a logical paradox, thus seeming to refute the possibility of a possible event's actually occurring! If this is so, then what does—what *could*—*possible* possibly mean?

Meiland (1974) has made a very pointed reply to that "paradoxical" question. In his paper he writes of time travel suicide, an unusual theme treated fictionally in the 1953 story "Mission" (Neville): "If we assume that it is impossible for [a time traveler to kill his younger self], some people are inclined to ask such questions as this: 'But how can the laws of logic prevent him from killing his younger self? Do they cause his finger to slip on the trigger or the bullet to fly apart in mid-air?' The implication of such questions is that the laws of logic cannot prevent such actions. But such questions are like asking: 'How do the laws of logic prevent the geometer from trisecting the angle or squaring the circle? Do they, for example, cause his ruler to slip at a crucial moment every time he tries it?' " A similar point was made later, with a different example, by Arntzenius (1990): "Surely it is not an impairment of 'freedom of action' . . . that, e.g., you cannot push another person harder than he/she pushes you. Just as one would explain this is the case by reference to Newton's third law, one could explain the impossibility of [causing a paradox] by reference to the laws which imply such an impossibility. If this explanation is taken to be unsatisfactory, it would seem that one is saddled with a general problem concerning the reconciliation of physics and 'freedom,' and not with a specific argument against [paradoxes]."

The short (29-minute) 1963 French film *La Jetée* accurately captures the concept of a time traveler moving along a closed, timelike curve that brings him together, fatally, with a younger self. After World War III has rendered Earth's surface so radioactive that the survivors are driven underground, experiments in time travel are started in an attempt to escape the horrors of the present. The first trial uses a man obsessed with a childhood memory of seeing, while he is watching the planes at Orly airport, a running man shot dead on the runway. The initial test sends him into the past, where he falls in love with a beautiful woman. Then he is pulled back to the present to prepare for a second test, this one a trip to the future. That test succeeds, too, but while he is there he asks the denizens of the future to use their advanced technology to send him directly back to the past, to his love. This they do, and he arrives at Orly airport. Seeing the woman, he runs toward her but is shot by one of the time travel experimenters from the "present" (who has followed him into the past to execute him for his selfish attempt at a personal escape from the awfulness of the "present.") As he falls dead on the runway, his final thought is that "one cannot escape time," because he realizes that *he* was the man he saw shot when he was a child. Indeed, he realizes that his child-self is even then watching him as he dies. This film was the inspiration for the 1995 Hollywood film *12 Monkeys* (complete with its own airport scene). Both films are subtle arguments against free will.

A blatant, explicit rejection of free will, combined with a passionate embrace of the block universe view, is offered in "Beep" (Blish). There one character tells another, after discovering how to receive radio signals from the future: "I *was* going to do all those things. There were no alternatives, no fanciful 'branches in time,' no decision-points that might be altered to make the future change. My future, like yours . . . and everybody else's, was fixed. It didn't matter a snap whether or not I had a decent motive for what I was going to do; I was going to do it anyhow. Cause and effect . . . just don't exist. One event follows another because events are just as indestructible in space-time as matter and energy are." Blish's tale is actually a sophisticated elaboration of the much earlier, 1930 story "The Time Annihilator" (Manley and Thode). In that story, time travelers from 1945 see the destruction of humanity in 2250—and realize there is nothing they can do to alter it. In his editorial introduction to the story, Hugo Gernsback wrote that "we cannot change the future by one iota. No matter how we strain against and battle against the events of some future era, we cannot alter them the least bit. They are written indelibly in the book of fate."

To many, the philosophical view that rejects free will, as expressed by Gernsback, in the stories by Manley and Thode, and by Blish, may seem awful, whereas to others it may eliminate a lot of guilt for past poor decisions. It certainly contrasts greatly with the statements made at the conclusion of the 1990 movie *Back to the Future III* (statements obviously and specifically aimed at the film's youthful audience) that the future is theirs to make. This film assumes the view expressed in Melville's *Mardi:* "The future is all hieroglyphics." The same view is expressed in the 1991 film *Terminator 2,* in which the past is changed to change the future—at one point a character carves the words "No Fate" into a table top.

One writer who would no doubt disagree with nearly everything I have written in this section is Martin Gardner, who wrote (1979), "In all time-travel stories where someone enters the past the past is necessarily altered. The only way the logical contradictions created by such a premise can be resolved is by positing a Universe that splits into separate branches the instant the past is entered."[14] As I have argued, I think Gardner is mistaken, and I think his confusion originates in a failure to distinguish properly between changing and affecting the past. But his last line *is* intriguing. Indeed, it leads us into the next section, where we will discover how some physicists have invoked quantum mechanics, as well as general relativity, in their analyses of time travel. In fact, the present-day view is that in any complete analysis of time machines and time travel, quantum theory is essential.

Quantum Mechanics and Time Travel

One early science fiction technique for allowing backward time travel and a changeable past, while still avoiding paradoxes, is that of alternate universes.

According to this idea, if a time traveler journeys into the past and introduces a change (indeed, his very journey may be the change), then, as Gardner stated, reality splits into two versions, with one fork representing the result of the change and the other fork being the original reality before the change. (To a fifth-dimensional observer, of course, all conceivable forks, all possible four-dimensional spacetimes, have always existed.) Indeed, according to this view the entire universe is splitting, at every microinstant, along every alternative decision path for every particle in the cosmos! This is often called the theory of *alternate realities with parallel time tracks.*

This apparently fantastic view seems actually to have some scientific plausibility because of the so-called many-worlds interpretation of quantum mechanics, pioneered by Hugh Everett III in his 1957 Princeton doctoral dissertation. For this book, however, the underlying scientific theory of time travel is classical (that is, non-quantum) general relativity, and that theory has nothing to say about alternative time tracks. For most time travel theoreticians there is *one* time track, and the past is unique and inviolate. I agree with the great quantum physicist J. S. Bell, who, in the essay "Quantum Mechanics for Cosmologists" included in Bell (1987), wrote of the Everett theory that "if such a theory were taken seriously it would hardly be possible to take anything else seriously."[15] As Bell points out in that paper, in the many-worlds interpretation "there is no association of the particular present with any particular past," a bizarre idea that appeared in science fiction as long ago as 1954, in "Mr. Kinkaid's Pasts" (Pierce). Visser (1993) has repeated Bell's observation in the more recent physics literature.

Still, to be complete, I must admit that although most early time travel analysts did base their work just on classical general relativity, there *are* now many more who think quantum mechanics itself, independent of its interpretation, has much to contribute as well. Perhaps it may make an absolutely crucial contribution. One analyst who believes this (along with the even stronger belief in the many-worlds interpretation of quantum mechanics) is David Deutsch at the Oxford University Mathematical Institute. Deutsch holds (1991) that general relativity is not the proper theory with which to study the physical effects of closed timelike lines. He believes that the traditional mathematical machinery of general relativity actually obscures, rather than clarifies, the difficult task of separating the merely counter-intuitive from the unphysical. Indeed, Deutsch calls the conventional spacetime methods, based on general relativity and differential geometry, perverse. He also does not like the technical and conceptual problems of general relativity's wormholes and singularities. Any non-quantum mechanical discussion, he says, of the "pathologies" of backward time travel is simply not adequate. In particular, Deutsch separates these pathologies into two classes: (1) paradoxical constraints, e.g., the free-will problem in the grandfather paradox, and (2) causal loops that create information, e.g., a mathematician who is visited in his youth by a time traveler from the future who gives him the proof of a theorem for which the mathematician is (will be) famous in the future. (*Where,* then did the proof come from?)

Deutsch claims his quantum analyses show that the first class of pathologies simply does not occur. This is because the past that the time traveler enters is (according to Deutsch) the past of a parallel world (see Note 15 again), not that of the world the time traveler has left. Deutsch has also arrived at certain preliminary results that lead him to believe that pathologies of the second class may be "avoidable." This is not the orthodox view among most time travel students, however (though of course this does *not* mean that Deutsch is wrong), and general relativity *is* the standard tool used by the majority of time travel theoreticians. When quantum mechanics does enter the calculations of most analysts, it is generally on an *ad hoc* basis.

Deutsch's position does raise the obvious question of what motivates a quantum theoretician to study closed timelike lines at all, given that they originate in general relativity and *not* in quantum mechanics! Deutsch's response is that although closed timelike lines did indeed originate in classical Einsteinian general relativity, the still incomplete theory of quantum gravity—see Sorkin (1997) for a tutorial—*does* predict closed timelike lines, too. And that is an exciting observation, because as Deutsch writes, the results of his quantum study of closed timelike lines show that "contrary to what has usually been assumed, there is no reason in what we know of fundamental physics why closed timelike lines should not exist." That view was later endorsed by Goldwirth *et al.* (1994), who wrote, after a quantum mechanical study of how a particle could transit a time machine spacetime in a physically consistent manner, "there is no contradiction between the postulates of quantum mechanics and the possible existence of causality violation in general relativity."

To address his concern about how to distinguish between the merely counter-intuitive effects of time travel and the "downright unphysical ones," Deutsch's interesting approach is that of studying various elementary computational circuits with negative time delays. (All ordinary physical systems, of course, the ones we call *causal,* have non-negative time delays.) Deutsch's circuits not only have information-processing loops in space, as ordinary networks do, but also loops that extend backward through time. He calls such a temporal loop a *chronology-violating link.* Deutsch's reason for adopting this approach is that it makes possible the study of a well-known physical system—such as basic computer gates or implementing the exclusive-OR logic function—under the unfamiliar situation of chronology violation, i.e., time travel.

With such simple computer gates, Deutsch is able to construct computational analogues to the classic time travel paradoxes, such as a time traveler going into the past to interact with an earlier version of himself. Deutsch's general conclusion is that if classical physics is assumed for the operation of the gates, then certain logical, yet what he calls paradoxical, constraints are required, such as the principle of self-consistency; but if the gates are quantum devices, then there are no required constraints. And, of course, the world is, at a fundamental level, quantum, not classical. Deutsch does leave unanswered one interesting question: How does one actually make a negative time delay, which is simply another name for a time machine? Hans Moravec, a computer scientist at Carnegie Mellon, has similarly used negative time delays and quantum mechanics to build conceptual systems with time machine logic—that is, *chronocomputers*. With such computers, Moravec shows (1991) how some famous computational problems, such as the traveling salesman puzzle, and chess, could be solved in zero time.

A quantum mechanical approach to time travel very different from Deutsch's and Moravec's has been described in Vaidman (1991). There we read of a quantum time machine that "is not for time travel," but rather for modifying the *rate of time flow* for an isolated system. In this way the machine resembles the entropy time-cabinet of Farley (1950), described in Chapter Three, more than Wells' time machine. Vaidman's machine can alter the time flow by either a positive or a negative factor, and it is the second possibility that allows his machine to return to a previous quantum mechanical state (the "past"), which is what Vaidman means by time travel.

There are at least three problems with his device. I have already mentioned the need to isolate Vaidman's machine from the rest of the world, a difficult, perhaps even impossible, requirement. If it is not isolated, then the machine returns to a *counterfactual* past—not to its actual past, but to the past it would have experienced if it had been isolated. Second, it works by surrounding a quantum mechanical system (such as an atom in an elevated state) with a massive spherical shell of variable radius; building such a machine would be an engineering nightmare. But worst of all is the random nature of the machine's operation; it works only *some* of the time! In fact, the bigger it is, the less likely it is that it will work. This is not the feature of any practical time machine, and in this case quantum theory is disappointing to those who hope for theoretical support for time travel.

Indeed, some recent quantum mechanical analyses of time travel, based on a mathematical theory that was developed in the early 1940s by Heisenberg but that has its origins in a 1937 paper by the pioneering master of black holes, John Wheeler—see Cushing (1990)—are cautionary. Called S-matrix theory, it allows the study of how interacting quantum fields change states because of their scattering interaction (the German name for this theory, *streung,* means "scattering"). From the elements of the S-matrix for a quantum system, it is possible to calculate the conditional probabilities that a given *before* (the interaction) state of the system will transition to each of its possible *after* states. For there to be self-consistency to quantum mechanics, it is clear that the sum of these conditional probabilities should be unity. There is such a sum, of course, for each of the possible *before* states. This summing property, which physicists call the *conservation of probability,* implies that the S-matrix has the mathematical property of unitarity. Unitarity, however, as shown in Friedman *et al.* (1992) and Politzer (1992), fails in the curved spacetimes that contain the closed timelike lines that come with time travel!

Those two papers, however, hold out at least a little hope for time travel by concluding that the failure of the conservation of probability can be salvaged via Feynman's famous "sum-over-histories" path-integral technique—see Feynman and Hibbs (1965). Almost simultaneously, Goldwirth, Perry, and Piran (1993) reached similar conclusions using the path-integral. Ironically, however, Goldwirth *et al.* concluded that *if* the idea that quantum gravity allows the metric of spacetime to fluctuate is accepted, "then the obstacle [failure of conservation of probability] for the construction of time machines that we raise here is removed." This is ironic because the authors cite a paper by Hawking (1982) to support their conclusion, and Hawking is no friend of time machine enthusiasts (recall his chronology protection conjecture).

The failure of unitarity in a time machine spacetime does not necessarily imply any violation of the laws of physics, but it does have what Politzer (1994) calls "decidedly eerie consequences." As Politzer points out, quantum mechanics requires some sort of generalization in the presence of closed timelike lines, and several different generalizations (each logically consistent—one example is Deutsch's) have been proposed. All of them reduce to standard quantum mechanics when there are no closed timelike lines present, but each makes different, often quite bizarre predictions when such entities are present. In the path-integral formulation, for example, the modified quantum mechanics is non-linear (see Note 42) and this non-linearity should have detectable consequences *before* the closed timelike lines are created! That is, as Politzer writes, "with this sort of mechanics, suitable experiments before the

CTC [closed timelike curves] epoch could determine that CTC's will be in their future." As I write, quantum mechanics appears to be absolutely linear, but of that Politzer writes, "the observed linearity of quantum mechanics, although verified in experiments of enormous precision, does not in practice tell us much about future CTC's, except that they are not in our immediate future." But see Stapp (1994) for a discussion of an experiment involving backward causation that Stapp suggests can be explained by non-linear quantum mechanics. And Hawking (1995) argues that it is not really a problem because in a time machine spacetime, the S-matrix loses its probability interpretation. See also Mensky and Novikov (1996b).

Most recently, Rosenberg (1998) has extended Politzer's idea of testing *now* for the possible effects due to the *future* operation of a time machine. His idea was to calculate the cross section of a two-particle scattering experiment, both with and without closed timelike curves in the future of the experiment. His results were that, yes, the total scattering cross section *now* does indeed depend on the future operation of "manmade time machines, which could be of a size traversable by humans"—which means on the order of one-meter spatial extent and involving one-second time loops (the temporal separation between the time machine and the experiment is related to the speed of the interacting particles). But, for electron–electron scattering and the time machine just one meter distant from the experiment, the alteration of the cross section compared to the cross section with no time machine was four orders of magnitude below what could be detected with the best technology now available. Another remarkable result is Rosenberg's calculation that, in a spacetime in which closed timelike curves pass through every point, there would be (because of a cancellation effect) *no* alteration at all in the scattering cross section!

Fiction has enthusiastically embraced quantum mechanics and its possible connection with time travel. The first such tale in science fiction was probably the 1935 story "The Branches of Time" (Daniels), which also contained the observation that although alternate time tracks may allow changing the past for the better (something that can't be done, for better *or* for worse, with just one time track), in the end any such change may still be futile.[16] As Daniels' time traveler puts it, sadly, "I did have an idea to . . . go back to make past ages more liveable. Terrible things have happened in history, you know. But it isn't any use. Think, for instance, of the martyrs and the things they suffered. I could go back and save them those wrongs. And yet all the time . . . they would still have known their unhappiness and their agony, because in this world-line those things have happened. At the end, it's all unchangeable; it merely unrolls before us." Many years later, in Shimony (QCST), which is a critique of the many-worlds idea, a philosopher/physicist echoed Daniels: "[In the world] that I (subjectively) experience I may blunder, but [in another world], with equal actuality, I triumph gloriously. The Everett interpretation can be used this way to mitigate sorrows, but this use is two-edged, for it equally well implies the speciousness of happiness." The recent novel *Branch Point* (Clee) makes a similar observation.

The theme of Daniels' story is what Lem (1974) calls the *ergodic theory of history:* Nearly all possible histories have the same general characteristics, differing only in the details. Another example of Lem's ergodic theory is the 1950 story "Ounce of Prevention" (Carter). As the last man alive (on Mars) watches Earth burn in the Final Atomic War, he is offered a deal by scientifically advanced Martians. They will send him back to Earth, back in time a billion years, along with a gadget to eliminate all fissionable elements. Thus no Final Atomic War. Once back, however, he is confronted by another version of himself, one sent back by the Martians from the new reality *he* will create, to prevent the Final Biological War. That war will be caused by poverty and economic unrest occasioned because, for some reason, all fissionable materials had disappeared from Earth in the ancient past! In this (illogical) story the details of history change, but not the big picture.

Perhaps one of the most imaginative alternate time track stories is "Sail On! Sail On!" (Farmer), which originally appeared in a 1952 issue of *Startling Stories*. In this story we read of an alternate world where radio was discovered rather earlier than in our world; one of Columbus' crew is a Friar Sparks, who sends Morse code–like signals in Latin! And "The Cliometricon" (Zebrowski), a story that describes the title gadget (named after Clio, the muse of history) that displays the alternate worlds of quantum mechanics, actually opens with a quote from a 1970 article in *Physics Today* on the quantum splitting of universes.

Splitting universes have been used in literary works outside the genre of science fiction, as well. Examples include "The Garden of Forking Paths" (Borges), the first play J. B. Priestley wrote, the 1932 *Dangerous Corner,* John Updike's 1997 novel *Toward the End of Time,* and Gore Vidal's 1998 novel *The Smithsonian Institution.* Typical of these fictional fantasies about splitting universes is "Lost" (Dunsany), the tale of a man who goes back in time to correct "two or three mistakes he had made in his life." This he successfully does, but the result is a new, subtly different subsequent history. But the differences are not infinitely subtle; after the changes, he finds that his home, his wife, and all the delicate details of his life have vanished. As he relates to a visitor at the lunatic asylum he is now confined to, as the result of his despair, "I tell you I'm lost. Can't you realize that I'm lost in time? I tell you that you can find your way traveling the length of Orion, sooner than you shall find it among the years. . . . Don't go back down the years trying to alter anything. . . . Don't even wish to. . . . [T]he whole length of the Milky Way is more easily traveled than time, amongst whose terrible ages I am lost." Decades later Isaac Asimov used Dunsany's idea in the tragic "Fair Exchange?" After a time traveler journeys back to 1871 London to retrieve a lost Gilbert and Sullivan operetta, he returns to the present to find that his wife (who was alive when he left) has been dead for a year on the new time track his actions in the past have created. (As an early hint of this tragedy, Asimov tells us at the very beginning of the story that the "lost" operetta is called *Thespis*.) As the story ends, the devastated time traveler thinks, "I had changed history. I could never go back. I had gained *Thespis*. I had lost Mary."

Getting lost in time is also the fate of the historian in "Remember the Alamo!" (Fehrenbach). Traveling back to 1836 to observe the attack by General Santa Anna on the famous Texas fort, the historian is puzzled when events seem oddly at variance with what he remembers reading in history books. Jim Bowie and Davey Crockett talk more like politicians than rustic backwoodsmen, and in the end they ride away from the battle rather than gloriously dying. It is then that the historian realizes why he had such a rough ride into the past on his time machine: "Oh, those ponces who peddled me that X-4-A—the *track jumper!* I'm not back in my own past. I've jumped the time track—*I'm back in a screaming alternate!*" More eloquent in expressing the same thought is the beautifully written 1987 story "The Forest of Time" (Flynn), which tells of a traveler lost in an infinitude of time tracks with no hope of ever finding his way home: "In all of time, how many, many worlds there must be. How to find a single twig in such a forest?"

Other interesting stories on the alternate-world theme have also been written with the twist that, given an infinity of such worlds, it may be awfully hard to find your way back to your original world. Examples include "One Way Street" (Bixby), "Rumfuddle" (Vance), "Trips" (Silverberg), and "Worlds Enough" (Thompson). Robert Heinlein's novel *The Number of the Beast* has only a finite number of alternatives, but the difficulty twist still applies because the number is 6 raised to the sixth power, all raised to the sixth power. More precisely, the number of parallel worlds is 10,314,424,798,490,535,546,171,949,056 (ten million sextillion!)—pretty big, but if Heinlein had written the "number of the beast" in the alternative way of 6 raised to the power of six raised to the sixth, he would have arrived at the far more stupendous number of 2.659×10^{36},[305]. But that still wouldn't have been the science fiction record; in the paratime stories of H. Beam Piper, the number of alternate universes is even larger: $10^{100,000}$.

Jack Finney's "The Coin Collector," however, takes just the opposite position. In that story a man enjoys the company of two wives, each in one of just two parallel universes, and he flips back and forth between them with ease. In "What Time Do You Call This?" (Shaw) there are also just two parallel worlds, but the consequence is far less happy than in Finney's tale. When a bank robber in one world tries to make his escape into the other one, he literally runs into "himself" trying to escape after robbing the "same" bank in the parallel world!

In writings for the masses, rather than for the more limited science fiction and fantasy audiences, perhaps the best-known literary work of alternate history is the classic 1953 novel by Ward Moore, *Bring the Jubilee,* in which Lee wins the Battle of Gettysburg and the South wins the Civil War. Moore's book beautifully explores the implications of that, with the *ending* (read on, and this will soon be clear) describing the invention of a time machine, the HX-1. By using this machine, a historian travels from 1952 (of the world in which the South wins) into the past of 1863 to study the battle, where he inadvertently disrupts events to the point that the North wins; that is, reality splits, the new fork representing the time track of *our* world. The historian is trapped on this new fork, cut off forever from his original time track and its future. Indeed, the entire novel is in the form of a discovered manuscript, written in 1877 and found in 1953, and the pathos of what must be the ultimate isolation endows it with great emotional impact. Lem (1974) incorrectly associates this story with a grandfather-type of paradox using a single time track, even though it is quite clear (I think) that Moore had multiple tracks in mind.

A more recent novel on the same theme, *The Guns of the South* (Turtledove), begins with a fascinating premise but then completely misses the crucial issue of single versus multiple time tracks. In this work racists from the year 2014 arrive by time machine at Lee's 1864 winter camp. They bring with them AK-47 automatic rifles and offer to supply Lee's army with all it can use. Lee accepts and the South wins the Civil War. The future, of course, changes—or does it? The time travelers have brought back books from the future showing that the South lost the war, so the implication is that history must have forked. But all through the novel, the time travelers move back and forth between the nineteenth and twenty-first centuries, apparently finding their own time unchanged. And if that is so, then the whole point of the story vanishes. Why all the effort to change history when it is clear that nothing has changed? The novel *is* entertaining reading, but (in my opinion) Moore's

Bring the Jubilee remains the superior work of science fiction. See, too, "An M-1 at Fort Donelson" (Fortenay) for a "try to change the Civil War by time travel" story that is logically consistent without the use of splitting time tracks.

The idea of an alternate time track being created by the disturbance of a time traveler is treated in an interesting way in *The Trinity Paradox* (Anderson and Beason). In this novel a nuclear weapons protestor causes an explosion while destroying a test site, and the explosion blasts her back in time, to the Los Alamos of 1943. There she inserts herself into the Manhattan Project, where scientists are trying to build the first atomic bomb, and thus finds herself in the perfect position to disrupt the history of her original time track. Before she is through, her efforts result in the radioactive destruction of New York City; the deaths of Edward Teller, General Groves, and Robert Oppenheimer; and the atomic bombing of the Nazi rocket base at Peenemunde. Somewhat more "fortunate" is Richard Feynman, who merely ends up in a wheelchair. Whether all this tragedy has resulted in a better future is, of course, open to debate, in agreement with Lem's ergodic theory of history.

Filmmakers seem to have mostly overlooked the dramatic plot device of alternate time tracks; the only attempts I know of are the 1971 British movie *Quest for Love,* which is based on the short story "Random Quest" (Wyndham), the 1987 film *Julia and Julia* and the 1998 *Sliding Doors.* The 1993 television movie *Time Trax* does hint at the related idea of parallel worlds, with its interesting treatment of criminals from 2193 fleeing two hundred years into the past. Ultimately, however, the film devolves into a routine cops-and-robbers chase movie. It presents the idea of a *single* additional world parallel to ours, presumably to avoid having to deal with the time travel paradoxes that arise with a single time track, but even that minimal view is bungled. For example, when the police officer who chases the criminals wants to communicate from 1993 to his superiors in 2193, he simply places a (cryptic) notice in a newspaper. But why would a notice printed in one parallel world necessarily appear in the newspaper of the other world (assuming, of course, that there *is* such a newspaper)?

I'll end this discussion on splitting universes with a startling theological issue raised by Smith (1989). Arguing that God cannot branch into multiple time tracks because God is unique, Smith states that God therefore exists in exactly one out of how ever many different time tracks there may be. What if the chosen time track isn't ours? Then, concludes Smith, Nietzsche's nineteenth-century metaphorical claim "God is dead" (for us) might be literally true! Smith himself admits that this is "fanciful," but still. . . .

Causal Loops

Dwyer (1978), in offering a possible reason why his time traveler journeys back to 3000 B.C., discusses a paradox with far more mystery than the grandfather paradox has: "In our time travel story it just may be that the traveler's interest in going back to ancient Egypt is stimulated by recently discovered documents, found near Cairo, containing the diary of a person claiming to be a time traveler, whereupon our hero, realizing it is himself, immediately begins . . . construction of a rocket in order to 'fulfill his destiny.' " In other words, (1) he builds a time machine and goes back to the past because of the discovered diary, and (2) the diary is discovered because he goes back to the past. Each of these points by itself has logical clarity, but together they form a *closed time loop* of enormous mystery.

The discovery of an anachronistic artifact (such as the diary), even without explicit mention of the closed-loop aspect, is certainly an intriguing idea. In "The Marathon Photograph" (Simak), for example, the discovery of a holographic photograph of the Battle of Marathon, 490 B.C., opens the discussion of time travel. And it is the later finding of another such three-dimensional image, a shocking view of the Crucifixion, that sends a time traveler back into the past. But it is the closing of the time loop that adds the real mystery. Simak clearly liked the idea of a closed time loop, and he used it more than once. In his "The Birch Clump Cylinder," a private college, endowed decades before by a generous but mysterious benefactor, experiments with a time machine. Suddenly, one of the college's graduates is accidently sent a hundred years into the past—where she becomes the benefactor. The college comes into existence, therefore, because it will exist. For this reason closed time loops are called *causal loops.*

Similarly, in the 1931 story "The Time Hoaxers" (Bolton) a time traveler journeys a century into the past *because* she finds an old, yellowed newspaper story describing her arrival. In my own 1979 story "Old Friends Across Time," I used an old photograph to convince the time-traveler-to-be that a trip to the past is in his future—it shows him in a group picture with men who died long before he was born. (Like Simak, I liked this idea enough to repeat it in my 1985 tale "The Invitation.") The hero-physicist in the 1953 story "The Gadget Had a Ghost" (Leinster) knows *something* odd lies in his future when he is confronted with a 700-year-old museum copy of a book. The puzzle is how to explain a message penned in ancient, faded ink, in modern English *and in his handwriting,* on the back side of one of the recently unglued endpapers! How also to explain his own fingerprints all over the same endpaper? How, indeed to answer these questions is his problem when he is presented with all of this and is asked, "Have you, by any chance, been visiting the thirteenth century?" At the end of the story the time loop is closed when the hero finds himself writing that same message on a *brand new* copy of the book that has been sent from the past (and that he returns to the thirteenth century via a "time portal").

The general plot device of a closed-loop artifact has a long history; as long ago as in the 1931 story "Via the Time Accelerator" (Bridge), for example, we find one of the first sophisticated treatments of causal loops in science fiction. A time traveler in 1930, about to start his journey into the future in an airplane/time machine, wonders at the last moment if he should really go—then he sees himself returning and thus *knows* he will successfully make the trip. As he later tells a friends, "That decided me. . . . Paradoxical? I should say so! I had seen myself return from my time-trip *before* I had started it [just like Marty McFly in the original *Back to the Future* film]; had I *not* seen that return, I would *not* have commenced that strange journey, and so could *not* have returned in order to induce me to decide that I *would* make the journey!" And later, when he finds himself in a dangerous situation in the future, he draws hope from that initial experience: "I *would* escape. . . . It was so decreed. Had I not, with my own eyes, seen myself appear out of the fourth dimension back there in the Twentieth Century, and glide down to my landing-field? Surely, then, I *was* destined to return to my own age safe and sound."

Even more dramatic is the second, internal time loop that ends the story. When the time traveler arrives in a ruined city in the year A.D. 1,001,930 he is greeted, *by name,* by an old man who says he is the Last Man alive. He knew the time traveler was coming because an ancient history book had said the Last Man had, in fact, appeared in the year A.D. 502,101 in the very time machine out of which the time traveler has just stepped. The time traveler is so startled by all this that he decides to mull it over until the next day. As he wakes up in the morning, he is just in time to watch the Last Man depart for 502,101. Stranded in the future, the time traveler wanders the empty city in despair until he chances upon a museum. And in the museum, sealed in a glass case, is his time machine (!)—it has been there for half a million years, since the end of the Last Man's journey. And so the time traveler is saved; he merely adds some oil to the still-functional engine (if you can accept time travel, I suppose this is no more difficult to believe) and returns to 1930—just as he saw himself do at the beginning of the story.

Since Bridge, the idea of a causal loop in time has been used many times in science fiction stories. In the 1941 tale "Weapon Out of Time" (Blish), for example, we find armed time travelers returning to the Triassic age to uncover the secret of a mysterious artifact; at the end we learn that it is the remains of their own automatic rifle. In the 1948 story "The Tides of Time" (Chandler) a time traveler journeys backward 500 years, suffers an accident that results in his being "agelessly stuck" in his time-traveling gadget until he is freed—by himself, 500 years later. He then gets into the gadget to journey backward 500 years. In "A Two-Timer" (Massor) a time traveler from 1964 is secretly observed by one of the "locals" when he arrives in 1683. The oddness of the sudden appearance of the time traveler and his time machine ("It were a kind of Dazzle") makes the local think it might be that the stranger is the man who stole some items from his home the previous night, the same night he had an "ill Dream." Stealing the time machine after the time traveler has gone exploring, the local travels to 1964, where he learns how valuable antiques are. So back he goes to 1683, to the night before the time machine first arrived, to get some "antiques" from his house. And thus he realizes who the thief *really* was. Before leaving, he enters his own bedroom first to see himself asleep and then to awaken. And so he also learns the cause of what he called his "ill Dream."

The 1950 story "A Bit of Forever" (Sheldon) opens with the world suddenly losing five minutes, an astonishing event that comes to be called "the time drop." After two weeks of investigation, a reporter traces this event to a reclusive (but brilliant, of course) scientist who tells the reporter that he has invented a time machine. The reporter decides to test this claim by using the machine to return to just before the time drop, to observe precisely what caused it—and it is, in fact, a malfunction of the machine that is at fault. And so the reporter finds himself caught in a two-week-long causal time loop. In "The Man Who Went Back" (Knight) a man gets the money to support his experiments in time travel by selling a large collection of old, rare comic books he has discovered in his late mother's attic. Later we learn how the comic books came to be there; after his experiments are successful, the inventor himself travels back into the distant past, buys the *newly* published comic books right off newsstands, and stores them in his mother's attic where, decades later, he knows his younger self will grow up and find them (and thus get the money to make it all happen).

There are a nearly infinite variety of such causal loops, and they have continued to form the basis for some of the most puzzling of science fiction stories. For example, in Harry Harrison's classic novel *The Technicolor Time Machine,* we read of a movie production crew that goes into the past to make a film. At the end of the story, it becomes clear that their presence in the past was not an insignificant event, as one character realizes after seeing the evidence of how they affected (*not* changed!) the past: "If this is true, then the only reason that the

Vikings settled in Vinland is because we decided to make a motion picture showing how the Vikings settled in Vinland." The final scene of this clever novel has the head of the movie studio, who is thrilled at how well things have worked out, planning a new project. It will involve traveling into the past once more to make "the absolutely [greatest] religious picture of all time," an idea that doesn't excite the production chief. He is almost certainly thinking of how a causal time loop might again be created (recall the very nasty loop in Michael Moorcock's explanation of the origin of Christ, in *Behold the Man*).

In *There Will Be Time* (Anderson) time travel is not by machine but by will power in those who have suffered a special genetic mutation. The mutation is caused by a virus. And where did the virus come from? From time travelers in the far future who traveled back to release the virus into humankind—thus time travelers exist because they exist. And in "Willie's Blues" (Tilley) we have the story of a time traveler from 2078 who journeys back a century and a half to learn the details of the short career of a famous jazz musician, and of his early death. On his trip the time traveler brings with him a recording of the musician's most brilliant performance, made the very night of his death. Arriving eighteen months before that night, the time traveler finds the musician a dispirited man. But when the musician hears the recording of his own yet-to-be-written music, his talent is awakened, to flourish until it results in making the recording that is its own cause.

In Fulmer (1980) a philosopher tells a causal-loop tale that puts the grandfather paradox to shame: "If James cannot decide whether to marry Alice or Jane, he simply travels to the future and learns that he is to choose Alice; he then chooses her for this reason. One wants to object that the decision to marry Alice was never really made at all! But this is not true; the decision was made—as a result of the knowledge that this was the decision. . . . It is not the case that the prospective bridegroom could visit the future and compare the results of marrying Alice with those of marrying Jane in order to decide between the alternatives. For if he visits the future, he will learn only that in fact he chose Alice, for better or for worse!" Similarly, when the time travelers in the 1940 story "The Time Cheaters" (Binder) arrive in the forty-sixth century, they find they are expected. Their host tells them why: "I have been awaiting your arrival, from the past. I have a written record of your coming. You see, I have a time machine myself. . . . With my time machine, I recently went a year into the future, and read the written account I had made, or will make after you leave. Then I came back, awaiting your arrival."

Fulmer elaborated on these astonishing "uses" of causal time loops in a subsequent paper (1981). He asked, "What if time travel becomes commonplace, so that we must deal with a constant stress of time travelers returning from the future to reveal what they have seen?" His answer is "I think it is clear that the . . . causal loop we have been discussing would become very common, and would play a prominent role in human affairs." Fulmer, however, denies that such causal loops would mean the loss of free will. As he explains his position, knowledge of a rigged roulette wheel will not prevent you from putting your money on the table *if you want to,* but perhaps that knowledge will influence your *freely* made decision making. Whether you learn that the roulette wheel is rigged by traditional means (perhaps you see magnets being installed under the table) or by means of time travel is irrelevant—even with this knowledge, you act freely.

Other philosophers, however, are so puzzled by causal loops (recall from Chapter Three that Murray MacBeath thinks they have a "queer smell") that they believe the consequent possibility of such loops directly reveals the impossibility of time travel (Fulmer and MacBeath, I should add, are *not* among that group). Mellor (1981), for example, is of the school that takes causal loops as simply impossible, but see Riggs (1991) for a strong rebuttal. Some philosophers who dislike causal loops as much as Mellor does have been somewhat more sophisticated with their reasons for rejecting them. Capek (VOT), for example, rejects time travel because causal loops necessarily have spacetime diagrams that bend back on themselves. Declaring any idea of "time travelers visiting their own past or the past of their ancestors" as nothing but "Wellsian fantasies," Capek continues with "there would be some events which, besides being simultaneous with themselves would also be simultaneous with other instants of time! In other words, a certain event corresponding to a single point in which the corresponding world line recrosses itself would be simultaneous with a remote future instant. In such a case we would be clearly on the brink of magic." Here Capek is failing to distinguish the proper time of the time traveler from the time of the rest of the world.

Hospers makes the same error when he asserts that 3000 B.C. happens twice—once without a time traveler and then once again when the time traveler arrives from the future. In fact, 3000 B.C. (or *any* point in spacetime where a world line comes close to itself—Capek makes a very serious, indeed *explosive,* error when he imagines a world line "recrossing itself") happens *once*. In the case of a time traveler who goes back to talk with his younger self, there are of course two sets of memories in a single brain: one set as the younger self and one set as the older, time traveler self.[17] This idea has often been used in science fiction, because it offers obvious avenues into powerful stories, including the 1935 story "A Thief in Time" (Young) as well as "Who Else Could I Count

On?" (Wellman), ". . . And It Comes Out Here" (Del Rey), "By His Bootstraps" (Heinlein), *The Man Who Folded Himself* (Gerrold), and *Who Goes Here?* (Shaw). With the four-dimensional block universe concept, all world lines lie tenseless in spacetime, so any self-encounter happens just once in spacetime. The older version speaks the same words he heard (even if he has forgotten them) when his older set of memories formed. He has to; otherwise the past would be changed.

A story that makes the mistake of representing a time loop as a continual, cyclical process, with events repeating with every passage around the loop (and a consequent duplication of the time traveler each time around), is "The Trouble with the Past" (Eisentein). That error is coupled with yet another: Each time around the loop, the past changes. This continues in the story until the duplicated time traveler(s) figure it out and put a stop to it after the ninth passage. It's a nice story, but illogical. Similar presentations, equally flawed in their logic, can be found in "Me, Myself, and I" (Tenn), "Of Time and Cats" (Fast), "A Little Something for Us Tempunauts" (Dick), "One Giant Step" (Stith), and "Hunter Patrol" (Piper).

Another quite interesting science fiction tale that makes the error of a closed time loop going around and around, endlessly repeating, is the 1946 "Find the Sculptor" (Mines). Here the inventor of the first time machine travels five hundred years into the future where he finds a bronze statue of himself that honors his discovery of time travel. Suddenly injured, fatally, he returns to the present with the statue and then dies. As a memorial, the statue is placed in the very spot where the inventor found (will find) it. Just as we wonder who made the watch in the film *Somewhere in Time* (see Note 11 for Chapter Three), we are left with the mysterious puzzle of who made the statue. This is fine, but then the author goes just a bit too far. As the tale ends, the late inventor's lab assistant wonders to himself what will happen five hundred years later: "Suddenly a strange machine will come out of the past and [the inventor] will be here again—although he is dead and has been dead five hundred years. [He will take the statue] and go back to the past . . . to die. And once again that maddening cycle will begin, to go on and on forever as long as time spins its threads." Robert Silverberg's "Mugwump 4" makes this same error.

Even philosophers who are sympathetic to time travel and causal loops can make the error of thinking such loops are cyclical. For example, Reichenbach (1956) tells us that

> there is nothing contradictory in imagining causal chains that are closed, though the existence of such chains would lead to rather unfamiliar experiences. For instance, it might then

> happen that a person would meet his own former self and
> have a conversation with him, thus closing a causal line by
> the use of sound waves. When this occurs the first time he
> would be the younger ego, and when the same occurrence
> takes place a second time he would be the older ego. Perhaps
> the older ego would find it difficult to convince the younger
> one of their identity; but the older ego would recall an iden-
> tical experience long ago. And when the younger ego has be-
> come old and experiences such an encounter a second time,
> he is on the other side and tries to convince some "third" ego
> of their physical identity. Such a situation appears paradoxical
> to us; but there is nothing illogical in it.

What Reichenbach has erroneously described is the beginning of an endless
succession of encounters around a closed causal loop.

There is just *one* encounter on such a loop in spacetime (but the mind of the
time traveler experiences it twice, of course), subject to the constraint of self-
consistency. As discussed earlier, the so-called principle of self-consistency has
been around in physics for decades, certainly long before the modern, serious
physics treatments of time travel began in the late 1980s. In the context of
time machines, however, the Russian astrophysicist Igor Novikov has become
particularly identified with it, probably because he discussed it before other
physicists had the nerve to do so. In his book *Evolution of the Universe* (originally published in Russian in 1979) he wrote, "The closure of time curves
does not necessarily imply a violation of causality, since the events along such
a closed line may be all 'self-adjusted'—they all affect one another through the
closed cycle and follow on another in a self-consistent way." Novikov (1989)
reiterated that view in one of the first time machine papers in the physics literature.

Some physicists are so concerned about multiple trips around closed timelike curves (CTCs), because they think such trips would allow the past to be
changed, that they have felt it necessary specifically to forbid such a possibility.
As Friedman *et al.* (1990) explains, "That the principle of self-consistency is
not totally tautological becomes clear when one considers the following alternative: The laws of physics might permit CTCs; and when CTCs occur, they
might trigger new kinds of local physics which we have not previously met.
For example, a quantum-mechanical system, propagating around CTCs,
might return to where it started with values for its wave function that are inconsistent with the initial values; and it might then continue propagating and
return once again with a third set of values, then a fourth, then a fifth. . . . The
principle of self-consistency by fiat forbids changing the past." This last statement is, of course, in agreement with the position of this book, a position that

has generally been accepted by most philosophers for over thirty years, but the proponents of the principle of self-consistency seem to have been driven to it by a fear of the past "happening again" over and over. David Deutsch at Oxford, on the other hand, flatly rejects any need for the principle, calling (1991) it simply redundant.

Princeton University philosopher David Lewis has written with particular insight on causal lops, especially ones that involve just information transfer, such as a time traveler going back in time to tell his younger self how to build a time machine so that once it is constructed he can go back and tell himself how to do it. This idea is found in the 1931 Nicholson letter quoted at the beginning of this chapter and as the second of Deutsch's backward time travel "pathology classes."[18] As Lewis (1976) explains the puzzle, "But where did the information come from in the first place? Why did the whole affair happen? *There is simply no answer* [my emphasis]. The parts of the loop are explicable, the whole of it is not. Strange! But not impossible, and not too different from inexplicabilities we are already inured to. Almost everyone agrees that God, or the Big Bang, or the entire infinite past of the Universe, or the decay of a tritium atom, is uncaused and inexplicable. Then if these are possible, why not also the inexplicable causal loops that arise in time travel?"

A few years later, Levin (1980) gave a similar response to a paradox involving a causal loop similar to Lewis'. Levin's tale is of a book containing instructions on how to make a time machine, a book that travels into the past on the machine so it can be read—in order to make the machine! In answer to the question "Who wrote the book about building a time machine?" Levin says that this question is "no different from questions about where *anything* originally came from. We can ask about the origin of the atoms . . . their time line is not neatly presented to us. The atoms either go back endlessly, or if the Universe is finite, they just start. In either case the question of ultimate origin is as unanswerable as the question of the book's origin. What makes us think that when such questions are asked about the loop they are different and *ought* to be answerable is that the entire loop is open to inspection. *Sub specie aeternitatis* this difference disappears."

An analyst who takes strong exception to Lewis' and Levin's views on causal loops is David Deutsch, who, you may recall from the previous section, illustrated his second class of time travel pathologies with a causal loop. Deutsch (1991) writes that "the real problem with closed timelike lines under classical physics is that they could be used to generate knowledge in a way that conflicts with the principles of the philosophy of science, specifically with the evolutionary principle." What Deutsch is referring to is the metaphysical claim, attributed to the philosopher Karl Popper, that *knowledge comes into existence only by evolutionary, rational processes* and that solutions to problems do not spring fully formed into the universe. One might call this the physics version of the work ethic—the creation of knowledge demands hard work!

One recent philosophical paper—Smith (1997)—has offered a quite interesting response to Deutsch's concern. Smith writes, of information "appearing out of nowhere," that "These cases are puzzling, but they by no means show that the time travel scenarios in question are impossible or incoherent, *or even improbable*. We think it very improbable that . . . information should come from nowhere—but only because this does not happen very often. It does happen *sometimes*—for instance, when you say something and I mishear you. I think that you said something very profound—something which neither of us would, in fact, ever have thought of. Where does the idea come from? If this sort of thing were to start occurring regularly [as via causal loops], then we would simply accept it without raising an eyebrow." Two Russian physicists, Lossev and Novikov, have actually given a dramatic physical example of how an information-creating time loop might be constructed using a wormhole time machine; that example is described in detail in Tech Note 9. Deutsch, however, uses the evolutionary principle to (for example) reject creationism, the anti-evolution theory that "explains away" fossils millions of years old with the assertion that they were created, fully formed by God just a few thousand years ago. Deutsch's position is, of course, considered by nearly all scientists to be correct for the specific case of creationism, but the evolutionary principle does not tell us much about causal loops—other than to declare them impossible because they are not possible.

The idea of using time travel to acquire knowledge from the future received an interesting fictional twist in "Trespass" (Anderson and Dickson). In this story time travel, discovered in the year 2007, is found to have a limited future temporal reach of fifty years. In addition, all technical progress comes to a halt. The reason for the time restriction on time journeys is that fifty years in the future there is a puzzling law banning time travelers from the past; all time travelers to 2057 and beyond are promptly arrested and "deported" to their own time. The story eventually explains that the law was passed precisely because of Deutsch's concern. As one character in the tale explains, "Suppose [that one could travel more than fifty years ahead], then a time traveler from the past could get [new inventions], carry them back to his own time, and give them to scientists—which [would] cancel all the long period of invention which [produced the inventions]. Which [would] violate causal laws." It takes fifty years to figure this out, but once the time travel law is passed, technical progress resumes. Deutsch (and Popper) would, I presume, approve. And the legal profession can take pride in how physics is saved by a smart lawyer!

I believe Lewis and Levin have the causal loop issue correct, but one can actually find similar ideas in science fiction considerably before the appearance of their papers. For example, "Absolutely Inflexible" (Silverberg) describes the world of A.D. 2784 as one that has eradicated all disease; as a consequence, humanity has lost its immunity to even the most ordinary illness. Thus when time travelers from the past show up, they are immediately sealed into bulky space suits, screened by an administrator called "absolutely inflexible" Mahler, and then quarantined on the Moon. Mahler has earned his nickname by being so emotionally hardened that he has never let a time traveler go free. Time travelers cannot be sent back home to the past because (so we are told) two-way time travel is impossible. At least it *was,* until one day a time traveler arrives with what he claims is a two-way time machine. Mahler doesn't believe him (oddly enough, the time traveler understands immediately that it is useless to argue), and Mahler promptly sends him off to the Moon. Then, curious, Mahler tries the gadget, and indeed he goes into the past. Upon his return, however, he is seized, sealed in a space suit, and processed—by himself because he has not come quite all the way back to the moment when he left for the past. As the story ends, "Suddenly Mahler saw the insane circle complete. . . . But how did the cycle start? Where did the two-way rig come from in the first place? He had gone to the past to bring it to the present to take it to the past to. . . ." Knowing it is useless to argue with himself, Mahler goes off to the Moon without resistance.

Mahler's fictional puzzle, of course, is precisely the same question about the origin of knowledge in a time loop that Lewis asks. At least two science fiction writers have hazarded answers to the "origin problem." In *The Technicolor Time Machine* (Harrison) one character is perplexed over a piece of paper with a diagram on it: "Then no one ever *drew* this diagram. It just travels around in this wallet and I hand it to myself. Explain that." His friend replies,

> "There is no need to, it explains itself. The piece of paper consists of a self-sufficient loop in time. No one ever drew it. It exists because it is, which is adequate explanation. If you wish to understand it, I will give you an example. You know that all pieces of paper have two sides—but if you give one end of a strip of paper a 180-degree twist, then join the ends together, the paper becomes a Möbius strip that has only one side. It exists. Saying it doesn't cannot alter the fact. The same thing is true of your diagram; it exists."
> "But—where did it come from?"

> "If you must have a source, you may say that it came from
> the same place that the missing side of the Möbius strip has
> gone to."

In the "The Man Who Met Himself" (Farley) we find much the same explanation. As a tribute to the mystery of causal loops, Farley has a Cambodian Buddhist monk explain matters to the hero (and, of course, to us):

> "But you haven't yet told me where the time-machine
> came from!"
> "You yourself brought it here out of 1935."
> "But where did it come from *originally?*"
> "There never was any 'originally.' You flew the time ma-
> chine backward through time from 1935 to 1925. And this is
> why I now have the machine."
> "To fly back again to 1925?"
> "No. Nothing of the sort. There is no round-and-round
> circle of events; no repetition. Merely *one* closed cycle. *One*
> overlapping of events for only ten years. . . . One time-
> machine, found in 1935 and brought back to 1925—found
> in 1935 *because* brought back to 1925. That is all."
> "But who made it in the first place?—Oh, skip the 'in the
> first place.' Just plain: who made it?"
> "No one. It was never made. . . . It is here because it is
> here."

This is, of course, the issue young Jim Nicholson raised as item 1 in his 1931 letter to Hugo Gernsback, quoted in the second section of this chapter.

The causal loops discussed by philosophers are usually of the type that loop from the present into the past and back to the present, as do Farley's and Harrison's. This need not be so. One science fiction treatment that describes how a loop might extend into the future is "Dominoes" (Kornbluth). A wealthy stock market wheeler-dealer travels two years into the future to get an edge. While there, he quickly learns from newspaper archives that a stupendous crash occurred just hours after the start of his time trip. He then returns to the present and unloads all of his holdings—and thus precipitates the very crash he read (will read) about. (The inventor of the time machine, who lost all his wealth in the market plunge, strangles his greedy patron, thereby showing that poetic justice can prevail even in the strange world of time travel and causal loops!)

Another example of a future-extended loop, one that explicitly acknowledges the chaos that would be caused by one event on a closed time loop changing another event on the loop, is the 1947 story "Collector's Item" (Long). When a time-traveling phonograph recorder from the mid-twentieth century silently appears in the middle of the twenty-sixth century, it captures the secret of a fearsome nuclear explosive (a madman who has just used the explosive is conveniently mumbling the secret formulas to himself at the moment when the recording gadget materializes, unseen). The weapon is so powerful that it can destroy the force-field bubble shields that have protected all of Earth's cities for six hundred years. After the time-traveling phonograph returns to the mid-twentieth century, its inventor goes crazy and the recording inadvertently ends up in a used-records shop. There it is found by the discoverer of the force field, who, after listening to the record, realizes his newly invented defensive bubbles are not impregnable—the shields are therefore *not* deployed! As the author writes near the end of this illogical tale, "It was like a telescoping of a tomorrow back into yesterday that had made that tomorrow impossible."

A loop that involves just information is strange enough. But when there is a physical artifact in the loop, there is an additional puzzle that nearly all writers seem to have missed; I have found this puzzle mentioned in the philosophical literature only by Nerlich (1981) and MacBeath (1982). Consider once again, then, the watch in the central causal loop of the film *Somewhere in Time* (see Note 11 for Chapter Three). Assume that the watch received by the man in the present is bright and shiny. He then takes it back into the past and gives it to his love. It remains with her after his return to the present until, decades later, she gives it to him—bright and shiny. Why didn't it tarnish? Is there some peculiar anti-tarnish property to the watch? Well, probably not. Perhaps she simply polished it just before giving it to him.

For a watch this seems an acceptable explanation (even though it smacks of *deus ex machina*), but now replace the watch with a crisp, new book. He receives it, takes it back, leaves it with her, and gets it back years later—crisp and new! Why haven't the pages turned yellow and brittle and dark with decades of fingerprints? Or, if they are yellow and brittle and dark, then the unalterability of the loop is lost because the book was crisp and new the "first" time around. This is all much harder to explain than is the shiny watch.

Somewhere in Time contains two additional, intertwined and far more subtle loops than the one with the watch, loops that are easy to miss on a first viewing of the film. In one, the movie begins with the hero falling in love with the heroine when he views an old, enigmatic photograph of her—in it she appears to be looking specifically at *him*. It is this photograph that largely prompts him to make the journey into the past. Later, when he is in the past, there is a scene in which she is having a prim and proper photograph taken when he appears; she then smiles her beautiful smile, and, in fact, she *is* looking specifically at him.

For the second loop, at the same time he first sees her photo in the present, he also happens upon a music box she once owned. He opens it and finds it plays his favorite piece by Rachmaninov. This link between him and her also spurs him on in his quest to travel back in time. Later, in the past, when he hums his favorite piece for her, he discovers she knows *nothing* of Rachmaninov's music. After his disappearance, when he returns to the present, she acquires the music box as a link to her memory of her lost love. In both of these loops, then, the artifacts (photograph and music box) that motivate his trip into the past are created *because* of his trip into the past. Unlike the watch loop, however, there is no puzzle about the origin of the artifacts.

The mystery of such causal loops is part of MacBeath's explanation of why, although he believes in the logical possibility of time travel, he nevertheless views temporal loops with much suspicion. An extremely clever story that properly handles a causal loop is "Paycheck" (Dick), which has an ending (that closes the loop) so spectacular that MacBeath would surely have rolled his eyes at it if he had read it before his recent death. Nerlich (1981) is of the same persuasion as MacBeath concerning causal loops. He notes that "despite [strong] arguments for the consistency of time travel stories, the impression is apt to remain that something is wrong with them. I think this impression is correct." He then tells us a tale (he attributes it to one of his students) that he claims proves his point.

Nerlich's tale is remarkably similar to the central puzzle in the well-known "As Never Was" (Miller), of which Nerlich and his student were apparently unaware (no citation is given). In Miller's story (originally published in 1943), a knife brought from a museum of the future arrives in the present with a flawless blade, but soon thereafter it gets a nick in the blade. How, wonders the narrator, can the time loop be completed "again"? I do not find this quite the puzzle Miller and Nerlich do. It is simply a variation of the grandfather paradox. *If* the knife is found flawless in the future, *then* it was not (will not) be nicked in the past. To demand otherwise is not to show a flaw in time travel but simply to be inconsistent!

In his imaginative 1984 novel *Them Bones*, Howard Waldrop makes the "nicked knife" mistake in a causal loop containing a living creature. In this poignant tale of disaster, time travelers from a world devastated by a nuclear war leave the year 2002, and aim for the 1930s, in an attempt to change the future course of history and thus to undo the war. They end up too far in the past, hundreds of years "off target," however, and the subsequent story is alternately told from three different viewpoints: that of the main body of time travelers (which is eventually massacred by Indians), that of one time traveler who winds up on a parallel time track, and that of the archeologists who in 1929 dig up the long-dead bones of the main body. The novel is a terrific story, but it makes a major blunder in its science.

The blunder centers on a causal loop that is supposedly entered by the main body as soon as it arrives in the past. A stray, hungry dog is captured and treated for a damaged paw; in particular, one of its dewclaws has to be amputated. A few days later, one of the scientists in the group sends the dog backward a short time interval in order to calibrate his equipment. In the struggle to shove the dog into the time portal, the dog hurts a dewclaw. The dog then goes back to sometime before he will be captured, and so we have the explanation for the "original" appearance of the dog, hungry with a damaged dewclaw. Or do we? Waldrop apparently forgot that the dog "originally" appeared with both dewclaws still attached! This self-healing time loop is inconsistent, and beyond being counter-intuitive (as all such loops are), it is also *biologically* flawed.

To say that causal loops are counter-intuitive is barely to hint at their mystery, but that is *not* an argument against time travel. Rather, *if* time travel is possible, *then* it would seem that we will have to accept causal loops, too; and *if* it is somehow shown that time travel is not possible, then for that same reason (whatever it might be), so too would causal loops thus be shown to be impossible. One time traveler in *Great Work of Time* (Crowley), a novel that is one gigantic causal loop from start to finish, certainly has matters right when he says, "It *is* mighty odd. The paradox is acute: it is. Completely contrary to the usual cause-and-effect thinking we all do, can't stop doing really, no matter how hard we try. . . . Strictly speaking it is unthinkable. And yet there it is."

There is, I should add, one possible way to explain causal loops: through the use of parallel universes. The best way to illustrate that is, I think, with the example of the extremely clever 1938 story "Other Tracks" (Sell). To improve the performance of his time machine, the inventor in this tale needs batteries with tremendous energy density, a density far in advance of the batteries in the present. Unable to travel far into the future—if he *could* obtain them there, then of course a causal loop (the very entity we wish to avoid) would be created upon his bringing them back to the present—his assistant first travels back to 1851. There he leaves a note on the desk of a well-known experimenter, with a plea for him to devote his life to battery research; a copy of the 1937 *Electrical Handbook* is left with the note as proof that there really has been a visit from the future! Before returning to the present, the assistant takes a sheet of (new) 1847 five-cent stamps from the experimenter's desk.

Returning to the present, which is now different (a new time track, in accordance with the splitting-universe idea), powerful batteries *are* readily available *because* the experimenter believed the note. Buying several of them, using money obtained by selling the pristine 1847 stamps to a collector, the assistant returns to a slightly earlier 1851 than before (to *before* the fork in time!), watches himself appear and leave the note and the handbook, and *then, unobserved, the assistant removes both:* Thus, upon returning once more to the present, he finds all is as before—except now they have the powerful batteries. As before, one might ask where the batteries came from, but unlike the previous mystery of information-creating causal loops, the answer is clear and non-mysterious. They came from the experimenter on a *different time track*. This science fiction answer is, in fact, the answer that David Deutsch rediscovered decades later—see Deutsch and Lockwood (1994).

Shuttling back and forth between time tracks is the signature of what is called a *cross-time* story, a device to avoid paradoxes while still allowing for changing the past. The first example of this time travel sub-genre is the 1934 "Sidewise in Time" by Murray Leinster, who liked the idea so much that he returned to it over the decades of his long writing career. See, for another example, his "The *Corianus* Disaster," in which a spaceship traveling over eleven hundred times the speed of light bounces into a parallel time track after running into drifting space debris. Such stories reached a peak with "The Other Inauguration" (Boucher), in which two professors "rotate" themselves out of our world into an alternative one after the wrong candidate (i.e., not their candidate) wins the U.S. Presidential election. Written as a response to the hysteria of McCarthyism in the early 1950s, it is an eerie tale of how alternatives aren't always improvements. A similar thesis is advanced in "Minor Alteration" (Richards), in which an alteration in the activities of Booth on the night of April 14, 1865, leads to a world far more awful than the one that follows Lincoln's assassination (although in this tale, unlike in Boucher's, the damage is eventually undone). *Vestiges of Time* (Meredith) has the interesting idea of the hero traveling cross-time from parallel world to endless parallel worlds in search of one in which a *true* time machine has been invented, a machine that travels forward and backward along *one* time track.

And finally, in "The Probable Man" (Bester), originally published long ago in a 1941 issue of *Astounding Science Fiction,* we find a logical mixing of a self-consistent closed time loop and the splitting of time tracks. In this tale each journey into the past causes the future to fan out into an infinity of new tracks, and so (it is argued) a time traveler cannot hope to return to *the* present he knew. Each such future has the same root past, however, as in Sell's story. That allows the story to end with the closing of a loop by having the time traveler return to the past he left at the beginning of the tale—but he goes back just a little bit too far and so sees himself leave. (At the start of the story there is a subtle hint that he did, indeed, feel as though somebody was watching him!)

This is precisely the same ending used decades later at the end of the 1985 film *Back to the Future.* Spielberg's movie is fun, but Bester did it first!

Sexual Paradoxes

There are causal loops even stranger than the ones we have already discussed, hard as that may be to believe. These are the sexual paradoxes, first mentioned in 1931 in the Nicholson letter to Hugo Gernsback. As a character in *Up the Line* (Silverberg) declares with interesting enthusiasm, "You haven't lived until you've laid one of your own ancestors." Philosophers have found these particular paradoxes full of dramatic appeal, too. For example, as a challenge problem to the readers of the scholarly journal *Analysis,* in 1979 British philosopher Jonathan Harrison posed the bizarre astonishing situation. A young lady, Jocasta Jones, one day finds an ancient deep freezer containing a solidly frozen young man. She thaws him out and learns that his name is Dum and that he possesses a book that describes how to make both a deep freezer and a time machine. They marry. Soon they have a baby boy and name him Dee.

Years later, after reading his father's book, Dee makes a time machine. Dee and Dum, taking the book with them, get into the machine and begin a trip into the past. Running out of food during the lengthy journey, Dee kills his father and eats him. (In this story, time travel machines are like the reverse entropy machine in "The Very Slow Time Machine" described in Chapter Three.) Arriving in the past, Dee destroys the time machine, builds a deep freezer (again, using the book), gets into it, and . . . wakes up to find that a young lady, one Jocasta Jones, has thawed him out. When asked his name he replies Dum and shows Jocasta his book; they marry, and. . . .

Harrison concluded this astonishing tale with a challenge question for his readers: "Did Jocasta commit a logically possible crime?" That issue is just the surface of an ocean of puzzles found in this story! Jocasta's crime, of course, is that she has (if unwittingly) committed incest; readers who remember their Greek myth of Oedipus, and who his mother–wife was, will understand why Harrison named his heroine as he did. But what of Dee's crime? He has, after all, eaten his father! But perhaps it isn't a crime after all, because Dee and Dum are one in the same, and is it really a crime to eat yourself? According to MacBeath (1982), Harrison's story is "a story so extravagant in its implications that it will be regarded as an effective *reductio ad absurdum* of the one dubious assumption on which the story rests: the possibility of time travel."[19]

Harrison's challenge provoked nearly a dozen replies. Most thoughtful was Levin (1980)—quoted earlier on causal loops—which makes the thought-provoking observation that not only has Jocasta committed incest but she has also done so with a single act of intercourse. As explained previously, the events on a causal loop do not happen endlessly but rather only once; thus Jocasta thaws Dum (Dee) out just once, she marries him just once, and the two consummate their marriage just once. Ordinarily we think it takes two sexual acts to commit incest, the first resulting in the birth of a child, and the second being a parent's union with the child, but this is not so in a causal loop. Time travel *is* an odd business.

And finally, in the only other printed solution to Harrison's challenge, Godfrey-Smith (1980) makes the telling point that, irrespective of physics, the story is biologically flawed and fatally so. As he writes, "The biological problem is the following. Dee is the son of Dum and Jocasta. So Dee obtained half his genes from Dum and half from Jocasta. But Dum is diachronically identical with Dee and is therefore genotypically identical with him (i.e., himself). That is, Dee is both genotypically identical with and distinct from Dum, which is absurd." Harrison seems amazingly unperturbed by that valid objection, dismissing it as a mere "law of nature, not logic." If one is going to ignore airily the laws of nature, then why bother debating time travel (or anything) in physics? Conservation of momentum is not a law of logic, either; should we overlook it, too, when it suits us? It is interesting to note that Godfrey-Smith's biological objection had actually been raised some years earlier by a physicist—see Schulman (1971a)—and by a science fiction writer in the 1955 story "Time Patrol" (Anderson).

While certainly instructive, the sexual paradox stories by Harrison and MacBeath (see Note 19 again) are remiss in not indicating that the concepts they are dealing with have long been a staple of science fiction and that the sexual paradoxes received much critical analysis in that genre long before philosophers (or physicists, for that matter) discovered them. The philosophers' stories are clever but derivative. From science fiction, for example, the 1954 story "The Poundstone Paradox" (Dee) is the tale of a young man who travels backward in time 1250 years, from A.D. 3207 to 1957, to become his own grandfather fifty generations removed. And even that is tame compared to the sexual paradoxes other science fiction writers conjured up before philosophers discussed them.

In "Child by Chronos" (Harness), for example, a tale written decades before Harrison's and MacBeath's, we meet a young lady caught up in a mind-twisting affair in which the mystery of a causal loop is the least of her troubles. In 1957 a girl is born, and after twenty years of intense competition with her mother (who has an uncanny ability to predict the future), she travels back from 1977 to a few months before her own birth. She becomes pregnant (by a man who she later discovers is *her* father) and gives birth to a girl. The new mother has, of course, knowledge of all that will happen during the next twenty years, including the fact that she will have an intense competition with her rebellious daughter. . . .

In 1959 Robert Heinlein extended the sexual paradox with his acclaimed classic "All You Zombies—," generally acknowledged as the best sexual paradox tale ever written. We are given only a hint of what is to come when a character listens to a song called "I'm My Own Grandpaw!" In 1945 a newborn girl, Jane, is found on the steps of an orphanage. At age 18, in 1963, she has a one-night affair with a mysterious stranger that leaves her pregnant. Some months later, during the birth of a daughter, it is discovered that Jane actually has a double set of sexual organs, and because the female set has been ruined by the pregnancy, the doctors restore her as a man. Soon after, the baby girl is kidnaped from the hospital ward. Years later, in 1970, Jane (now a man, of course) meets another stranger who uses a time machine to transport both of them back to April 3, 1963. By April 24 male-Jane meets female-Jane and impregnates her; now we know who *that* mysterious stranger was! Meanwhile, the stranger with the time machine travels forward to March 10, 1964, a little after female-Jane has given birth, kidnaps the baby from the hospital, takes her back to September 20, 1945, and leaves her on the steps of an orphanage. And so we see that Jane is her own mother *and father,* thus out-doing Harness in the self-parenting game.

This is pretty impressive, but Heinlein still has one more twist for us. After leaving baby-Jane in 1945, the time machine stranger returns to April 24, 1963, retrieves male-Jane (who has just kissed female-Jane goodnight after fathering her-himself in herself), and takes him to 1985 where he recruits him into the Temporal Service—and finally, the stranger jumps forward to 1999, his "real time." At the end we at last learn that the stranger is, in fact, an even older version of male-Jane—*all* the characters in the entire story are the *same* individual at various points along a single, highly twisted world line. The lone character in Heinlein's tale is truly a self-made man/woman in every sense of the phrase! This ultimate act of *creatio ex nihilo* has been called by Lem (1974) "smaller than the minimal loop." In a December 1958 letter to his literary agent, Heinlein wrote of this amazing tale, "I *hope* that I have written in that story the Farthest South in time paradoxes."[20]

The sexual paradox has continued to fascinate science fiction writers up to the present day. Indeed, time travel sex even without paradox fascinates. In Robert Forward's 1992 novel *Timemaster,* for example, the hero at one point spends a night with his wife—and with two versions of himself from the future. He will, of course, live through that night two more times! Later, he becomes upset when his wife runs off with one of the older versions, but he quickly calms down when he considers that eventually *he* will be the older version. Now, add paradox to the sex and it all becomes nearly too much to handle.

Consider Gregory Benford's beautifully written story "Down the River Road," which tells of a young man hunting the father who, years before, had abandoned him in a burning house. The death of the young man's mother in the flames has sent him on a ten-year quest for revenge up and down what is literally a river of time, where to travel in one direction ("up time") is to move into the past, whereas moving "down time" leads to the future (and to the legendary waterfall at the end of eternity). Eventually he corners the father and, despite the man's pleading, kills him. It is only later, after examining papers he finds in his father's pocket, that the boy realizes he has killed his *future* self (Benford, a physicist, knows the pitfalls of time travel, and there is no grandfather paradox here, you'll notice!) Benford's tale is reminiscent of Huck Finn on the Mississippi, as well as of the 1954 Czechoslovakian film *Journey to the Beginning of Time,* in which four boys raft down the river of time into the prehistoric past. In Benford's story, time is monstrously convoluted, and the ending hints at an explanation for an earlier, mysterious encounter the boy has with an older self in a futile attempt to avert his (then) future murder (suicide?). By having the young man kill his future self, this tale has introduced the clever twist of *inverting* the grandfather paradox.

The biological objection raised by Godfrey-Smith to Harrison's story is, of course, equally valid criticism of those of Harness, Heinlein, and Benford. David Gerrold avoided Godfrey-Smith's objection in his classic novel *The Man Who Folded Himself* by having the protagonist become involved in a homosexual affair with himself—indeed, with multiple copies of himself ("The night when six of us, naked and giggling, discovered what an orgy *really* was" has to be one of the most unusual lines in all of time travel fiction). Other than that curious story device, however, Gerrold's novel holds little additional interest for us in this book, because its underlying logic is that the past can be changed at will. Stepping around that error, as well as avoiding the genetic objection, is *The Janus Equation* (Spruill). There, a brilliant mathematician who is attempting to unravel the theory of time travel undergoes a sexual identity crisis—and then he meets a beautiful woman who breaks him out of his slump and straightens him out. Or, at least she does until he learns that she is himself, time traveling to the past after a sex change operation and other sorts of plastic surgery. The story title refers to the two-faced Roman god Janus, who faced both the past and the future, but undoubtedly the author had this curious time travel coupling in mind, as well.

All of the stories discussed in this section are written in a "serious" tone, but there is of course a sense of the absurd to the sexual paradoxes. This has not been overlooked by science fiction writers. "Genetic Coda" (Disch), for example, is a very funny spoof of the idea of being one's own father; although Disch makes it quite clear that he understands the genetic objection to self-parenting, he doesn't let that stop him from having a good time with the idea. The only film I know of that explicitly uses the idea of a time traveler being her own (remote) ancestor is the 1989 *Time Trackers*. However, this movie also argues illogically throughout for the possibility of changing the past (and in one scene, it actually does so).

Maxwell's Equations and Advanced Effects

Every physicist and electrical engineer knows that the mathematical description of the electromagnetic field is given by Maxwell's equations. In particular, radio engineers know that the waves of energy their antennas launch into space follow the predictions of these equations with astonishing accuracy. Indeed, when Einstein's relativity theory was completed, it was found that Maxwell's equations automatically satisfy relativity because magnetic effects *are* relativistic effects; in other words, relativity is built into Maxwell's equations. Whereas Newton's laws of dynamics had to be patched up, Maxwell's equations were untouched by the discovery of special relativity—but see the end of Note 39.

Thus it was a puzzle when physicists discovered that careful study of the seemingly perfect Maxwell equations, when applied to antennas, apparently results in the prediction of causality violation. It is found, in fact, that the equations admit two solutions. One, as expected, contains the feature of time delay; that is, creating an electromagnetic disturbance at the antenna *now* causes a detectable effect at a distant point in space *later*. This is the so-called *retarded solution,* and its common-sense physical interpretation is that of energy waves traveling away from the antenna as they also travel into the future. The shock was that Maxwell's equations also accept an *advanced solution:* energy waves *arriving* at the antenna from infinite space.

According to Beichler (1988), Paul Renno Heyl wrote perhaps the earliest scientific work discussing advanced effects, in his 1897 University of Pennsylvania doctoral dissertation with the provocative title "The Theory of Light on the Hypothesis of a Fourth Dimension." Heyl cited the guru of the fourth dimension, C. H. Hinton (see Chapter Two again), as his inspiration. The situation described by Heyl is something like that of a child standing at the edge of a pond. She throws a rock into the middle of the pond and watches ripples spread out and away from the splash. Suddenly she sees ripples appear all around the edge of the pond and then travel inward toward the center—where they all converge at once. A spout of water erupts from the surface of the pond at the simultaneous meeting of the inward-traveling ripples, and she watches as a rock is ejected from the spout to land back in her hand. She is, of course, open-mouthed with amazement! How absurd, you think as you read this, and who could blame you? Not I! This astonishing imagery of advanced effects we owe to the philosopher Karl Popper (1956), and it has come to be called "the fable of the Popperian pond."

Pursuing the mathematics of wave motion in the fourth dimension, Heyl wrote, "We are led to the curious conclusion that in Hinton's aether the nature of the central disturbance *after* a given instant can influence the form of the aether *before* that instant. In other words, the aether seems to be endowed with an uncanny faculty of foreknowledge." We can avoid such a counter-intuitive implication of advanced effects, but only at the price of something many physicists and philosophers consider equally unacceptable—information traveling from the future into the past. We can still think of the advanced solution as representing electromagnetic waves traveling *away* from the transmitting antenna—that is, as being broadcast, just like the retarded solution, rather than being received—*if* we also think of the waves as traveling backward in time.

Thus the advanced solution to Maxwell's equations holds out the possibility of sending messages to the past, a sort of poor man's time travel. It may seem that we have simply traded one problem for another, however, because just sending *information* into the past can cause many of the same paradoxical, causality-busting situations that physical time travel is claimed to cause. Even before the turn of the century, the self-described *enfant terrible* of Victorian literature and science, Samuel Butler (1835–1902), had a sense of the turmoil such transmissions backward through time might cause. In the "Imaginary Worlds" entry in *The Notebooks of Samuel Butler* he wrote, "Communication with a world exactly, to minutest detail, a duplicate of our own [but] twenty thousand years ahead of us might ruin the human race as effectively as if we had fallen into the Sun."

The well-known cosmologists Fred Hoyle and J. V. Narlikar share Butler's opinion that communication backward in time could cause trouble. Concerning Maxwell's equations, they wrote in their 1974 book *Action at a Distance in*

Physics and Cosmology that "the equations supply us with both advanced and re-tarded solutions (and, because of their linearity, with any linear combination of them). . . . With so many solutions theoretically possible, why does nature always select the retarded one? That this question cannot be answered within the framework of Maxwell's theory must be regarded as one of its intrinsic weaknesses."

Hoyle and Narlikar's branding of Maxwell's theory as having an "intrinsic weakness" because of its prediction of an advanced solution was, I think, un-warranted. Indeed, Schwebel (1970) and Stephenson (1978) have shown how the advanced solution can have a perfectly reasonable physical interpreta-tion in the context of Maxwell's theory. Stephenson, in particular, imagines a transmitting antenna sending electromagnetic waves to an identical receiving antenna. At any point in space between the two antennas there are electric and magnetic fields. Maxwell's equations allow us to calculate the fields produced by alternating currents in the two antennas. When we do an analysis of the re-lationship between the transmitting antenna's current and the fields, we use the retarded solution because the current is the cause and the fields are the ef-fect. But in the analysis of the relationship between the fields and the receiving antenna's current, the situation is reversed, and the fields are the cause and the current is the effect. That is, the *advanced* solution is simply the mathematics relating the current in the receiving antenna *now* to the fields *in the past.*

An acceptance of both solutions has, in fact, been in the physics literature for three-quarters of a century. As Yale physicist Leigh Page wrote in 1924, "While the advanced potentials, as well as the retarded potentials, satisfy the electromagnetic equations, the former has generally been discarded for the reason that it has been more in accord with the trend of scientific intuition to consider that the present is determined by the past course of events than by the future. However, if it is once admitted that the present state is uniquely deter-mined by any past state, it follows that the future is also so determined, and hence the employment of a future state as well as a past state in specifying the present marks no inherent departure from our accustomed methods of de-scription. . . ." And, indeed, Hoyle himself has indicated (1983) that he has had a change in his opinion of advanced field effects.

It may still be tempting, however, just to dismiss the advanced solution as a mere anomaly of the mathematics—to discard it on physical grounds. This is the traditional approach taken by physicists when confronted with non-causal solutions in any physical theory, and indeed that was what Swiss physicist Wal-ter Ritz (1878–1909) did with the advanced solution to Maxwell's equations. This approach, according to Zeh (1989), involved Ritz during his last year of life in a dispute with Einstein. For Ritz, the reversal of cause and effect was simply too violent to his intuition to be taken seriously. And Ritz's attitude *is* understandable; using his approach we might just *impose* causality on Maxwell's equations, a condition they clearly do not inherently contain, by *a priori* rejecting all advanced solutions—see De Beauregard (1971) on this.

Electrical engineers make similar kinds of judgments all the time, of course, as when the solution of a quadratic equation for a passive (energy-dissipating) resistor gives both a positive value (which "makes sense"), and a negative value that doesn't "make sense" and so is simply ignored.[21] A half-century ago, however, the eminent M.I.T. physicist Julius Adams Stratton echoed Leigh Page's warning against the temptation to make this sort of argument without careful consideration. He wrote,[22] of the disturbing advanced solution, "The familiar chain of cause-and-effect is thus reversed and this alternative solution might be discarded as logically inconceivable. However the application of 'logical' causality principles offers very insecure footing in matters such as these and we shall do better to restrict the [Maxwell] theory to retarded action solely on the grounds that this solution alone conforms to the *present* [my emphasis] data." And in a famous paper we will discuss later in this chapter, Wheeler and Feynman (1949) declared that "We conclude advanced and retarded interactions give a description of nature logically as acceptable and physically as completely deterministic as the Newtonian scheme of mechanics. In both forms of dynamics the distinction between cause and effect is pointless. With deterministic equations to describe the event, one can say: the stone hits the ground because it was dropped from a height; equally well: the stone fell from a height because it was going to hit the ground." For Wheeler and Feynman, the reversal of cause and effect inherent to backward causation and time travel to the past offered no conceptual difficulties.

The elimination of an appeal to causality, or to the "weirdness" of advanced waves, in arguing for the naturalness of the retarded solution to Maxwell's equations was first done in Aichelburg and Beig (1976). The only auxiliary condition applied to the equations was simply that of requiring the initial field energy to be finite—see also Anderson (1992). And, indeed, there is today still no experimental evidence for the physical reality of the advanced solution. Now and then one does run across speculations that the advanced waves of *something* traveling backward to us from the future might explain the so-called ESP talent of precognition, but that is all it is, speculation; see Hesse (1961).

Even before Wheeler and Feynman, advanced waves were discussed in science fiction. For example, one story that specifically invokes the advanced solution to Maxwell's equations is the 1938 tale "Pre-Vision" (Pierce), in which a gadget that makes use of what the author called the "anticipated potentials" can display the near future on a screen much like a television. (The author's by-line proudly gave his academic credentials as including a master's degree, and in fact John Pierce was a graduate student in electrical engineering at Caltech. He received his doctorate just months after this story was published and then went on to a highly distinguished career at Bell Telephone Laboratories and later at the Jet Propulsion Laboratory operated by Caltech for NASA. Peirce knew all about Maxwell's equations, of course, and actually opened his story with a quote from Page's 1924 article on the advanced solution. How many fictional pieces include quotes from *The Physical Review?*)

And it would probably take something like advanced waves to explain the funny doings in the 1938 "Anachronistic Optics" (Schere), the story of a man caught in a time machine accident. Nearly all of his body ends up four years in the future—but only *nearly* all, because his eyes remain in the present! As one of the puzzled observers of this odd business wonders, "Strange, that his eyes, now, can convey a message to his brain, four years hence, and his brain tells the eye muscles to move the eyeballs which are four years behind them—." This idea of a body and mind existing at two different times was later used by Philip K. Dick in his powerful 1956 political novel disguised as science-fiction, *The World Jones Made*. That work tells us of a prophet who can see a year into the future. As he says, "To me *this is the past.*" Later we are told, "He was a man with his eyes in the present [the world's future] and his body in the past [the world's present]." Was Dick aware of the advanced waves in Maxwell's theory?

Communication with the Past

Some of the most intriguing paradoxes of time travel involve no traveler, or at least no living one—only information. Of course, any information flow at all, independent of time travel, involves the flow of energy, and as Einstein showed, energy and mass are different aspects of the same thing. Accordingly, information time travel involves the transfer of mass/energy. Thus a man in the twenty-fifth century who sends a backward-in-time "temporal radio" message to a twentieth-century woman stating that he loves (will love?) her is sending much more than mere emotion.[23] One point should be clearly understood, however: The paradoxes of information time travel, as with all time travel, arise only when *backward* time travel is involved.[24]

Indeed, all forms of present-day communication are transmissions only to the future. If you speak to someone and if you send a radio message, there are delays depending on the distance of separation and the speed of transmission of sound and light, respectively. If you want to send a message to the 125th century, you can; just write a letter and seal it in a pressurized bottle of helium, a variation on a theme exploited in the novel *Time Bomb* (Dixon). The same idea is more dramatically presented in *The Dechronization of Sam Magruder* (Simpson). Magruder is a scientist who is accidently transported from 2162 back to the late-Cretaceous, eighty million years into the past. He leaves a written record on seven sandstone slabs of his subsequently brutal, lonely life among the dinosaurs—letters across time, if you will—found by twenty-second-century geologists some years after his disappearance. This is all quite ordinary (at least as far as the time direction of Magruder's letters goes!), but not so in the reverse temporal direction. What, for example, could be a more exciting message than the one received from the future by the young genius Cullen Foster, inventor of the first time machine in the 1942 story "How Much to Thursday?" (Stapleton)? After his initial experiment of sending the pilotless machine into the future, it returns with an envelope inside. Eagerly tearing it open, he finds that the note is from the National Academy of Sciences: "We know from old records and museum models that this is the Cullen Foster experimental machine. Fifty years looks down on you and says 'Good work.'"

Heady stuff, that, but lots of other possible messages are capable of competing with young Foster's in generating excitement. For example, suppose you had a gadget that is superficially similar to a telephone but that calls telephones in the *distant* future. You can hear the person (in the future) on the other end, but they can't hear you (in their past). That is, information can flow only from future to past.[25] It is then easy to imagine situations that at least seem paradoxical in the use of this device. For example, suppose you call your own private number one month ahead. You hear your future self first answer the phone and then recite the winning lottery for the "previous" day, which is a month in your present self's future. (Your future self does this somewhat odd recital because a month from now you will remember, when your private phone rings, *who* is calling!) So now in the present you can make a fortune winning the lottery a month later.

So far, of course, there is nothing paradoxical, or even illegal, in all of this. But what if, when the phone rings in the future the day after you won the lottery, you then decide *not* to read the winning number? This particular paradox has, in fact, already been treated with the aid of the block universe view of spacetime—that is, if the future you spoke the lottery number, then the present you *must,* inevitably, read it. To see how this particular idea has been treated in fiction, read "The Other End of the Line" (Tevis).

Now suppose that instead of calling your future self, you call the weather service and listen to the recorded message telling you the daily weather (thirty days hence, of course). You do this day after day, and after a while you get a reputation for being able to predict, *perfectly,* the weather for every day to come, up to a month into the future. Your reputation spreads far and wide, and after a while the weather service hears about you. Meteorologists check and find you are *never* wrong. Their computer models are only 80% accurate out to three days, and for a week's prediction and beyond, the general public might as well flip a coin on whether it will rain any particular day. But *you* are 100% correct out to ten times their range, and so they hire you—and as a secondary job, you also make the daily weather recordings. (The voice on the other end of the gadget *has* sounded sort of familiar!) Here, then, is the puzzle, the one we encountered earlier in causal loops carrying information: Just *where* is the information in the flawless weather predictions coming from?

One easy answer is that the question is meaningless because such a future-to-past information flow must be impossible. Indeed, if I am to avoid telling a "philosopher's fairy tale," like those I criticized in Chapters Two and Three, I must admit that one consistent, non-paradoxical answer is found in recognizing that I have *assumed* that those 30-day weather reports are correct. Maybe, however, they are no better than anybody else's predictions. And so you *don't* become famous, and so you *don't* get hired—and so there *isn't* any paradox. Is that the way to avoid paradoxes involving information flowing backward in time?

Perhaps not. As long ago as 1917 it was realized that special relativity does not preclude such an apparent backward flow. That is, if information could be transmitted faster than light, then messages could travel backward in time. This was the year that Richard Tolman, a professor of physical chemistry at the University of Illinois and later at Caltech, wrote, "The question naturally arises whether velocities which are greater than that of light could ever possibly be obtained." He then answered that question, his general conclusion being that if such velocities are possible, then a faster-than-light (FTL) observer could see the time order of two causally related events reverse. And thus the observer would see the result *before* the cause. Alternatively, a *sub*luminal (slower-than-light) observer could see the two events, which are connected via an FTL causal interaction, reversed in time order from what a stationary observer would see. Either situation has come to be called Tolman's paradox (see Tech Note 7), but Tolman himself was careful with his words: "Such a condition of affairs might not be a logical impossibility; nevertheless its extraordinary nature might incline us to believe that no causal impulse can travel with a velocity greater than that of light."

That was an astonishing statement, given that Einstein himself had specifically stated in his original 1905 paper on special relativity that such a thing simply could not occur. There is nothing, it would seem, to be "inclined" about! And even though Tolman's paradox is itself an astonishing result, it has in fact been extended. Tech Note 4 shows that the time order of two events can appear reversed even for a subluminal observer—but only if the two events are not causally related. Faster-than-light motion extends this reversal to causally connected events; that is, FTL motion, reversed causation, and time travel to the past go hand in hand in hand.

This rather technical connection between FTL speeds and backward time travel made the transition from theoretical physics to popular culture very quickly. It was in the British humor weekly *Punch,* for example, that the famous (but nearly always misquoted) limerick by A. H. R. Buller (1874–1944) first appeared:

> There was a young lady named Bright
> Whose speed was far faster than light;
>> She set out one day
>> In a relative way
> And returned on the previous night.[26]

Where *Punch* dared to go, Hollywood could not be far behind. Indeed, in this case it was actually there first, with the 1922 one-reel silent comedy movie *The Sky Splitter.* This was just a short film (feature pictures generally had at least four reels), so it is not clear how widely distributed and viewed it may have been. The story is that of a scientist testing a new spaceship; when it exceeds the speed of light, he begins to relive his life.

The linkage between time travel to the past and FTL motion continues to fascinate science fiction writers. In the 1953 story "Throwback in Time" (Long), a time machine experimenter in the twenty-seventh century wonders, "Was the speed of light the core of the mystery? At the speed of light did the past and the future become a shining, merging road down which men could walk—in their ears the thunder of time passing. . . ?" Not everybody is pleased with the idea of such a thing, however; Eddington (1929), for instance, declared that "the limit to the velocity of signals is our bulwark against the topsy-turvydom of past and future."

Other physicists have seen what caused Eddington's concern as an opportunity. The property of a wormhole that allows nearly instantaneous FTL travel between its mouths can be used to achieve time travel via a coupling of the mouths of a *pair* of wormholes together (see Tech Note 9). One can reverse (conceptually) the sequence, too, and achieve FTL speed via time travel.

To go from A to B at FTL speed, just follow this procedure: put yourself in hibernation at A; cruise as slowly as you'd like to B; and finally, wake up, get into your time machine, and travel backward in time by as much as the proper temporal duration of your cruise (as measured by your wristwatch). This lets you actually achieve *infinite* speed. And going back just a little further means, of course, that you arrive at B before you leave A![27] This idea was used in "Minimum Sentence" (Cogswell) and was hinted at in *There Will Be Time* (Anderson).

The obvious question at this point, of course, is whether it is even conceptually possible to build a gadget to send FTL messages backward in time? Einstein himself thought not, saying (1922), "We cannot send wire messages into the past." But was he right? One hint at the possibility of achieving FTL speeds is in Dirac's 1938 paper, discussed in Chapter Three. There, in his remarks about pre-acceleration, Dirac wrote, "Suppose we have a pulse sent out from place A and a receiving apparatus for electromagnetic waves at a place B, and suppose there is an electron on the straight line joining A to B. Then the electron will be radiating appreciably [because accelerated charges radiate] before the pulse has reached its centre and this emitted radiation will be detectable at B at a time . . . earlier than when the pulse, which travels from A to B with the velocity of light, arrives. *In this way a signal can be sent from A to B faster than light* [my emphasis]."

This exciting conclusion goes a step beyond the usual examples of things that go faster than light.[28] Dirac had an equally exciting reaction (and here the emphasis is his): "This is a fundamental departure from the ordinary ideas of relativity and is to be interpreted by saying that *it is possible for a signal to be transmitted faster than light through the interior of an electron. The finite size of the electron now reappears in a new sense, the interior of the electron being a region of failure, not of the field equations of electromagnetic theory, but of some of the elementary properties of space-time.*" This last line sounds very much like the things people say today about the singularity inside a black hole event horizon. And yet, Dirac was careful to point out that as weird as FTL speed may appear, special relativity is not violated because "in spite of this departure from ordinary relativistic ideas, our whole theory is Lorentz invariant."[29] Of course, like any scientific theory, Dirac's theory is not necessarily the last word, and we have to admit the possibility that at least some of its implications (in particular, the possibility of FTL speeds) just aren't so. We must also admit that it is one thing to talk of science fiction "Dirac radios," as mentioned in Chapter Three, and quite another to see how physics may actually enable one to talk to the past.[30]

Wheeler and Feynman and Their Bilking Paradox

In 1941, at a meeting of the American Physical Society, Princeton University physicist John Wheeler and his student Richard Feynman discussed a seemingly outrageous idea that provided a possible clue to how a Dirac radio might function. The idea was that the advanced wave solutions to Maxwell's equations are not mere mathematical curiosities but rather have profound physical significance. At the time, their talk received only a small abstract notice in the *Physical Review,* but after World War II they wrote it all up in a beautiful paper—see Wheeler and Feynman (1945).

Their primary motivation was to explain the origin of the force of radiative reaction discussed by Lorentz earlier in the century. This reaction force is the cause of the energy loss suffered by an accelerated, charged particle. Lorentz, who thought of charged particles as having a finite size, attributed this reaction force to the retarded (by the time required for light to cross the width of the charged particle) coulomb repulsion force of one side of the particle's charge on the opposite side. This view, however, leads to various conceptual and mathematical problems, including an arbitrary assumption on how the charge is distributed over/through the finite volume, as well as the problems of infinite self-interactions and the issue of what keeps the charge from blowing itself apart by internal coulomb repulsion.

Wheeler and Feynman's theory, on the other hand, avoided those problems by postulating point charges, because a point charge cannot repel itself. But whence the reaction force if there is no repulsion? Their revolutionary explanation was first to imagine the accelerated point charge as radiating retarded radiation outward, eventually to be absorbed by distant matter. This distant matter, which itself consists of point charges that are accelerated by the received radiation, then radiates *backward* in time, back toward the original charge that started the chain of events. This backward-in-time, or *advanced,* radiation arrives in the past of the original charge, and *it* is the cause of the observed reaction force. Indeed, Wheeler and Feynman proposed that an accelerated charge will not radiate unless there is to be absorption at some other distant place and future time. The *future* behavior of a distant absorber determines the *past* event of radiation; there is simply no such thing as just radiating into empty space. The entire universe, spatially *and* temporally, is a very "connected" place!

Astonishingly, that non-causal view of spacetime had been around in physics for at least twenty years before Wheeler and Feynman's talk. They had independently developed their views, but after their 1941 talk, Einstein (who perhaps recalled his 1909 debate about advanced effects with Ritz) brought a 1922 paper by Hugo Tetrode to their attention.[31] In that paper Tetrode had written that "the Sun would not radiate if it were alone in space and no other bodies could absorb its radiation. . . . If for example I observed through my telescope yesterday evening that star which let us say is 100 light years away, then not only did I know that the light which it allowed to reach my eye was emitted 100 years ago, but also the star or individual atoms of it knew already 100 years ago that I, who then did not even exist, would view it yesterday evening at such and such a time." None of this, of course, is obvious! As Berenda (1947) put it in a tutorial appearing just two years after Wheeler and Feynman's 1945 paper, "Any physical theory which seriously proposes that events in the future may be the efficient cause of events in the past may be regarded—at least at first glance—as rather revolutionary doctrine." Indeed!

To be sure the "doctrine" is clear, let me restate what Tetrode, and later Wheeler and Feynman, had in mind. Imagine we have an electric charge that we mechanically shake—that is, accelerate. This allows us to assign a definite *cause* to the charge's acceleration, which of course creates radiation. This radiation travels outward into space as observed retarded fields until they are eventually absorbed by distant matter. The charges in that distant matter are thus accelerated, and they in turn therefore radiate. This induced radiation again consists, according to Wheeler and Feynman, of *both* retarded and advanced fields. The advanced fields radiate outward but *backward* in time toward the original charge, collapsing upon it at the precise instant we first shook it, thereby producing the radiative reaction force. At any instant of time, at any point in space, the observed field is the sum of the retarded field traveling away from the source into the future and the advanced field traveling to the source in the past.

But, argued Wheeler and Feynman, there is one last point that has been left out of the picture—there is also an advanced field (traveling away from the source and backward in time) because of the original, mechanical shaking of the source charge. Equivalently, a field traveling *forward* in time will *converge onto* the source because we *will* shake it. Wheeler and Feynman showed that before the mechanical shaking that starts this whole process, the advanced radiation field of the source and the advanced radiation fields of the absorbers exactly cancel each other at every point in space and every instant of time (if there is total absorption in the future), which accounts for the experimental fact that we observe a zero total field before the shaking occurs.

Wheeler and Feynman showed that if we accept these ideas, then everything we actually observe is predictable: radiative reaction, the direction of the electromagnetic arrow from past to future (retarded-only effects), and the absence of infinite self-interactions.[32] That is, we gain these rewards *if* we accept backward time travel, a step too big for many in 1945 and for nearly as many today. This formulation of particle interactions is difficult for many physicists to swallow because it leads to apparent paradoxes. For the same reason, Tetrode's earlier work also went virtually unnoticed during the two decades before Wheeler and Feynman.

Tetrode published in a German journal, but Wheeler and Feynman's ideas were anticipated even in America. Back in 1926, G. N. Lewis had stated that "I'm going to make the . . . assumption that an atom never emits light except to another atom, and to claim that it is as absurd to think of light emitted by one atom regardless of the existence of a receiving atom as it would be to think of an atom absorbing light without the existence of light to be absorbed." Wheeler and Feynman were aware of Lewis by 1945 (but there is no evidence that Lewis was ever aware of Tetrode). Certainly Wheeler and Feynman must have been intrigued by Lewis' paradox: "I shall not attempt to conceal the conflict between these views and common sense. The light coming from a distant star is absorbed, let us say, by a molecule of chlorophyl which has recently been produced in a living plant. We say that the light from the star was on its way toward us a thousand years ago. What rapport can there be between the emitting source and this newly made molecule of chlorophyl?"

The paradox in that, of course, arises from the issue of what happens if, at some intermediate time and place, the star's light is blocked, thus preventing its absorption by the chlorophyl. Could refusing to look at a star *now* affect the emission of the star's light in the *past*? Lewis was obviously making a clear statement of backward causation when posing his bilking paradox. His very next words show that he understood the probable reaction of his readers: "Such an idea is repugnant to all our notions of causality and temporal sequence." Like Tetrode's work, Lewis' ideas were ahead of the times.

The astonishing puzzles of Tetrode and Lewis were actually not repugnant to everyone. In fact, similar puzzles were an inspiration to Wheeler and Feynman and almost certainly motivated them to create their own famous bilking paradox, which they presented in a 1949 paper. They opened the presentation of their paradox as follows: "If the present motion of *a* is affected by the future motion of *b*, then the observation of *a* attributes a certain inevitability to the motion of *b*. Is not this conclusion in direct conflict with our recognized ability to influence the future motion of *b*?" Here once again, the conflict between free will and determinism arises—see Newton (CPT)—and to sidestep this human concern, Wheeler and Feynman constructed a "paradox machine," one that operates totally automatically and that, in the subsequent literature represented by Fitzgerald (1970), has come amusingly to be called the "logically pernicious self-inhibitor"!

In their description of the paradox machine, Wheeler and Feynman ask us to imagine two charged particles, *a* and *b*, positioned five light-hours apart. As shown in Figure 4.4, *a* is attached to the arm of a pivoted shutter, toward which a pellet is moving from initially a great distance away. Now, normally we would think of what happens next in terms of just retarded fields. That is, the pellet hits the arm, knocking it downward and thereby accelerates charge *a*; this acceleration of charge *a* creates a retarded radiation field that arrives at charge *b* five hours later and thereby accelerates charge *b*; this acceleration of charge *b* creates a retarded radiation field that arrives back at charge *a* five hours later (ten hours after the pellet hit the arm).

The Wheeler and Feynman view, however, as I stated earlier, claims that this description leaves out half the story—the *advanced* fields. Specifically, suppose the pellet will hit the arm and accelerate *a* at 6 P.M. Then, *b* will be affected not only five hours later at 11 P.M. but also *earlier* at 1 P.M. This advanced acceleration of *b*, in turn, sends out an advanced field that arrives at *a* at 8 A.M. The paradox is now easy to see. As Wheeler and Feynman described events, we see *a* exhibit a premonitory movement at 8 A.M. Seeing this motion in the morning, we conclude that the pellet *will* hit the arm in the evening. We could then return to the scene a few seconds before 6 P.M. and block the pellet from

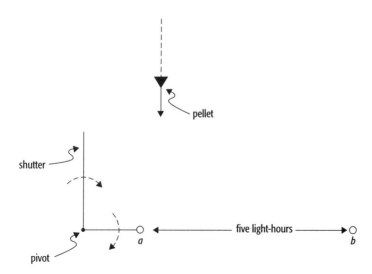

Figure 4.4 Wheeler and Feynman's paradox machine.

acting on a, a task automatically accomplished by the shutter in Wheeler and Feynman's machine. But then we are faced with the puzzle of explaining *why a* moved in the morning.

Wheeler and Feynman claimed they had resolved their bilking paradox by observing that discontinuous forces (more generally, *signals*) are never seen in nature. They concluded that the shutter does not completely block the pellet but rather that the shutter suffers a "glancing blow." That is, a very weak advanced signal is received by charge a, which moves the shutter just enough to induce the "glancing blow," and it is this partial interaction that results in the weakened signal in the "first place." This is the very same explanation that was rediscovered decades later to resolve similar bilking paradoxes that involve self-interacting billiard balls transiting wormhole time machines—see Tech Note 9.

Absorber Theory and Signaling to the Past

Wheeler and Feynman's argument *is* logically and physically sensible; it is simply an early statement of the principle of self-consistency. Still, I think it is amazing that few physicists challenged Wheeler and Feynman on it.[33] A philosopher did, however, and I think her analysis—see Hesse (1961)—is correct, as is her conclusion that "there is no solution in terms of classical physics," although Hesse does talk about a quantum mechanical explanation. Many years later Feynman wrote (in his autobiographical *Surely You're Joking, Mr. Feynman*) that at one time he and Wheeler thought it would not be too difficult to work out the quantum version of their theory. But first Wheeler failed in the task, and then Feynman tried his hand at it and, as he stated, "I never solved it, either—a quantum theory of half-advanced, half-retarded potentials—and I worked on it for years." Their paradox (if indeed it *is* a paradox; if advanced fields don't actually exist, then there is no problem) remains unsolved.

Wheeler (1979) himself summed matters up nicely years later when he wrote, "Interconnections run forward and backward in time in such numbers as to make an unbelievable maze. That weaving together of past and future seems to contradict every normal idea of causality. However, when the number of particles is great enough to absorb completely the signal starting out from any source, then this myriad of couplings adds up to a simple result: the familiar retarded actions of everyday experience, plus the familiar force of radiative reaction with its familiar sign."

Is Wheeler and Feynman's view of nature correct? Some physicists have simply assumed that the advanced field does exist and have explored where that assumption might take them. In Schild (1963), for example, the relativistic motion of two point charges (in concentric circular orbits in the same plane) interacting through time-symmetric fields is shown to be consistent with stationary (i.e., periodic) solutions. And in Driver (1979) the one-dimensional motion along the infinite x-axis of two point charges interacting through time-symmetric fields is shown to be unique, *provided* that the charges are initially "far enough" apart—and "far enough" doesn't have to really be very far: as 4600 classical electron radii will do. As Driver correctly observes, such few-body, time-symmetric interactions have received little attention.

Could we use advanced waves to send signals to the past? Wright (1979) argues rather strongly (if not always coherently) that we could. Or, if that requires some yet-to-be-developed technological breakthroughs in transmitters and if receivers are easier to construct, could we at least listen to the future (inasmuch as we are the future's past)? And if we could do that, could the future send us the details of the transmitter breakthrough (thus creating a causal information loop in time)?

John Cramer at the University of Washington has offered a clever suggestion on how to use such signals in principle, although he maintains a conservative skepticism about advanced fields, given the absence of any experimental evidence.[34] Observing that for retarded waves, there is about a 2.5-second round-trip delay in transmitting a radar signal to the moon and then receiving the echo, Cramer extends this observation to advanced waves; the echo for the advanced-wave case would be received 2.5 seconds *before* transmission. By using multiple reflections of the advanced wave between Earth and Moon, one could in principle extend the echo as far back into the past as desired. Whether this would be "good" is problematical for Cramer, who wrote, contrary to the arguments presented earlier in this chapter, that "if advanced waves [could be so used] then our grip on reality would become more tenuous. The past could never be considered over and done with, because anyone with the proper hardware could send messages back in time and alter what had already happened." That view, of course, is not the view I presented earlier in this chapter, nor is it the view of most physicists and philosophers, who believe that the past can be affected but *not* changed or altered.

A suggestion for conducting an experimental test of the absorber theory was actually made *before* Wheeler and Feynman's work, by Lewis (1926), a suggestion that inspired a quick (but sympathetic) rebuttal by Tolman and Smith (1926). The first experimental search for advanced waves seems to be the effort described by Partridge (1973). Flaws in that search process prompted Heron and Pegg (1974) to discuss an experiment designed to detect advanced waves (if they exist). As those two authors wrote, in a grand understatement, the exciting possibility of a positive result "would have such far-reaching consequences on our ideas of the unidirectionality of time and causality that . . . the experiment justifies a large amount of effort, even if no conclusive result is obtained for years." A more recent attempt to detect advanced waves is described in Schmidt and Newman (1980).

All of the searches for advanced waves have so far given negative results, and the world still awaits the first Dirac radio. However, attempts have been made to support Wheeler and Feynman's bilking explanation theoretically with analyses of the various paradoxes occasioned by sending messages to the past via hypothetical FTL particles called tachyons—see Schulman (1971a) and Peres and Schulman (1972)—or through the study of self-colliding billiard balls time traveling through the wormhole time machines described in Tech Note 9—see Friedman *et al.* (1990) and Echeverria, Klinkhammer, and Thorne (1991). Novikov (1992) replaced the billiard balls with exploding bombs, a twist that gives the ultimate in "time bombs"!

Over the years the Wheeler and Feynman view of nature has also been the object of many theoretical objections. Hogarth (1962), for example, rightfully complained that Wheeler and Feynman had assumed a static, time-symmetric spacetime for the universe, in which the properties of all past and future absorbers are identical. That is obviously not so in an expanding (or contracting) universe, and as Hogarth wrote, "No serious modern cosmological theory is framed in a static Universe." Another puzzle for Hogarth is that Wheeler and Feynman took a time-symmetric theory of half-retarded/half-advanced waves in a time-symmetric universe and arrived at a *non*–time-symmetric solution! As Price (1991b) aptly puts it, Wheeler and Feynman "plucked an asymmetric rabbit from a symmetric hat." They performed that trick by supposing that although the universe is static, it was created with asymmetric initial conditions of low entropy. Thus, for Wheeler and Feynman, the one-way thermodynamic arrow of time is the primary arrow, the electromagnetic arrow its consequence. (The *how* of the low-entropy initial cosmological boundary condition was left unexplained.) This ordering of the primacy of the temporal arrows was the position adopted by Einstein in his 1909 debate with Ritz and was also the position more recently (1985) taken by Stephen Hawking.

Many years later Wheeler (1979) explained, again, his and Feynman's 1945 viewpoint: "The particles of the absorber are either at rest or in random motion before the acceleration of the source. They are correlated with it in velocity after that acceleration. Thus radiation and radiative reaction are understood in terms, not of pure electrodynamics, but of statistical mechanics." Hogarth objected to this explanation, which he felt was unnecessary, believing instead that the expansion of our actually non-static universe provided the required asymmetry. In Hogarth's view it is the cosmological arrow that is the primary one, and the electromagnetic arrow is *its* consequence.

Wheeler and Feynman had shown that both the advanced and the retarded solutions are self-consistent in a static universe; Hogarth's interest was in whether the observed retarded solution, alone, would be self-consistent in an expanding universe. His conclusion? It depends (on the details of the expansion)! In 1964 Hoyle and Narlikar expanded on Hogarth's study and claimed to have shown that the retarded solution alone is self-consistent *if* the expansion is steady-state via the continuous creation of matter. That would be the case because if only retarded effects are to occur, then each emitter of radiation needs a large number of absorbers (such as ionized intergalactic gas) in its future light cone to provide for complete absorption. This, in turn, requires that the density of matter not decline "too fast" with the expansion. That is, the future universe must not be "too transparent."

That conclusion was embraced with enthusiasm by Hoyle, a British cosmologist whose name has long been identified with the idea of continuous creation of matter. Since then, however, continuous-creation cosmologies have fallen into disfavor because it was in 1965, just a year after Hoyle and Narlikar wrote, that the cosmic microwave background radiation was detected. That is now taken as very strong evidence for the existence of the primordial explosion, or Big Bang, that started the expansion of the universe—and as equally strong support for therefore rejecting a steady-state universe. Not by Hoyle, though, who has an almost fanatical devotion to non–Big Bang cosmologies; see Hoyle (1975) for his explanation of the microwave background radiation. Most recently, Gott and Li (1998) have shown that a universe with closed timelike curves is stable and self-consistent if only the retarded solution is allowed.

A real puzzle for the Big Bang universe, however, is that it expands from a dense, opaque past into a less dense, ever-more-transparent future, each emitter having a large number of absorbers in its past light cone. That should result, note Hoyle and Narlikar, in an observed advanced solution and thus in a reversed electromagnetic arrow that would allow communication with the past. The fact that we have not yet discovered how to perform such communication might be taken to mean that the advanced waves of Wheeler and Feynman are, in fact, just a bit of pretty mathematics devoid of any physical reality. One possible rebuttal to this, of course, is that the analyses of Hogarth and of Hoyle and Narlikar are simply wrong, as discussed by Davies (1972a).

But Davies' analysis has its puzzles, too. He showed that all of the standard expanding, relativistic cosmological models of the universe are too transparent in the future to explain the observed retarded-only waves in terms of the Wheeler and Feynman absorber theory. This conclusion has recently been given an interesting interpretation. One surprising result from the Hubble Space Telescope was that the oldest stars appear to be about twice as old as the universe itself. This seems paradoxical (to say the least!) but as Hoekzema (1996) observed, as the universe expands, the matter in it thins out, and so it will become increasingly more difficult for stars to "find" absorbers for their emitted radiation. That is, in the *past,* stars found it *easier* to radiate. This means that stars aged more rapidly in the past than they do now; because we implicitly assume an essentially constant aging process (at the rate we observe today), it is no surprise that we overestimate the age of the older stars. Making some reasonable assumptions about the physics of stars and relativistic spacetime, Hoekzema showed that it is not difficult to account for a factor of 2 in the apparent ages of the oldest stars (measurement errors are now thought to be the real answer).

There is also another problem with the Wheeler and Feynman absorber theory: the puzzle of neutrino absorption. Neutrinos are particles that interact so weakly with matter that a beam of them would have to travel through many hundreds of light years of lead for there to be a significant attenuation of the beam intensity. How can such particles find enough future absorbers to make possible their observed journeys into the future of the expanding universe? Narlikar (1962a) admitted that a continuous-creation universe "only 'just' manages" to explain retarded neutrinos, but at least it manages (again, the 1965 detection of the cosmic microwave background, believed to be the signature of a Big Bang origin of the universe, fatally wounded that analysis).

Is there any evidence for advanced waves outside of theoretical physics? What of precognition? Bob Brier, a philosopher who specializes in parapsychology, lists (1974) the various categories of the subject as follows: "Extrasensory perception" or ESP, which is knowledge of the environment gained by means other than the five usual senses, is divided into clairvoyance, telepathy, and precognition. The first is ESP of mental states in the present, and the third is ESP of physical objects or events in the future. As Brier points out, it is precognition that presents interesting philosophical questions, whereas the other two "talents" might have physical explanations in terms of ordinary, radiating brain waves (he does acknowledge that such waves are extraordinarily weak, but at least they exist without violating any known laws of physics).

Brier is also aware of the idea that precognition, too, could be included in the radiating-brain-wave hypothesis *if* brain waves can travel not only forward in time but also backward into the past. That view, endorsing backward causation, is debated just as hotly by philosophers as by physicists; see Mundle (1964), who rejects precognition precisely because he sees it as requiring acceptance of backward causation. Mundle's reason for rejecting backward causation is banal, however, because he *defines* a cause as having to occur before any of its effects, which simply begs the question. Brier, too, has difficulty with time-traveling radiation as the particular agent of precognition, although he does accept the idea of backward causation, itself, as logical. As he writes, "We find it difficult if not impossible to imagine waves that go into the future *and return to the present* [my emphasis] bearing information about where (and when) they have been."

For such an exciting idea as communicating with the past, I have been able to find only a few uses of advanced-wave radio in science fiction. In Poul Anderson's story "Earthman, Beware," first published in a 1951 issue of the magazine *Super Science Stories,* there is the hint of such waves with something called the "ultrawave effect": "While gravitational effects were produced by the presence of matter, ultrawave effects . . . did not appear unless there was a properly tuned receiver somewhere. They seemed somehow 'aware' of a listener even before they came into existence [i.e., aware of a future absorber, in Wheeler and Feynman's terminology]." And the evil scientific genius Lex Luthor, Superman's archenemy, invented such a radio while in prison in 1961. He constructed it from ordinary AM radio parts and used it to call for rescue by, and to listen to messages from, an organization with the interesting name of "Legion of Super-Villains" from the thirtieth century!

The more recent *Thrice Upon a Time* (Hogan) is a novel-length discussion of potential bilking paradoxes produced by sending messages backward in time. The puzzles presented are undeniably fascinating, but Hogan's answer to them is to allow the changing of the past. Indeed, the title comes from the plot device of twice changing the past by sending messages to the past to save the world from terrible disasters. (One of them is a swarm of micro-sized black holes buzzing around in the planet's interior!) Thus we read through entire time periods *three* times before finishing the novel. As one character blurts out, "We can monitor the actual consequences of our decisions and actions, and change them until they produce the desired result! My God . . . it's staggering!" Quite so.

Far less thoughtful is *The Forty-Minute War* (Morris and Morris), in which terrorists crash a plane carrying a nuclear bomb into Washington, DC. This sets off a thermonuclear exchange between the Soviet Union and the United States. In the end, however, the entire book renders itself moot because the hero sends a history-changing message backward in time to the CIA, which "then" eliminates the terrorists before they can crash their plane. No physical theory is advanced to explain how the message is transmitted, and there is nothing in today's time machine research to support such a concept (or Hogan's either).

The most interesting science fictional use of backward-in-time signaling is perhaps that in "Beep" (Blish). There the "Dirac radio" for instantaneous transmissions is described, and we learn that at the bèginning of each received message there is always an irritating audio beep (hence the title) that is seemingly a useless artifact of the mysterious workings of the gadget. Its only obvious characteristic is a continuous frequency spectrum from 30 hertz to well above 18,000 hertz. It is only at the end of the story that the main character learns that this spectrum is the "simultaneous reception of every one of the Dirac messages which [has] ever been sent, or will be sent."[35] There is no mention of advanced waves, but clearly Blish knew that instantaneous (infinite-velocity) signals would travel into the past. The story does a masterful job at presenting the mystery of listening to the future. At one point characters in the twenty-first century hear the commander of a time-traveling "worldline cruiser" transmit a poignant call for help from 11,000,000 light years away and from 65 centuries in the future. Most interesting of all, however, is Blish's statement of a technical issue that I have not seen raised before, although "Cambridge, 1:58 A.M." (Benford), the precursor story to Benford's classic novel *Timescape*, hints at it in passing: If signals arrive at a receiver simultaneously from all future times, how can they be separated? Blish resorts to some scientifiction babble-talk to answer that question, but I believe it remains a puzzle.

Tachyonic Signals, Spooky Actions, and the Bell Antitelephone

Science fiction writers have often used FTL motion to reverse time. "The Worlds of If" (Weinbaum) includes this throwaway line in a lecture on time travel from a curmudgeonly but lovable old physics professor named Haskel van Manderpootz: "And as for the past—in the first place, you'd have to exceed light speed, which immediately entails the use of more than an infinite number of horsepowers." This is the second reason, beyond the reversal of cause and effect, why many physicists reject backward time travel by FTL, irrespective of the logical paradoxes. That is, FTL motion implies time travel to the past, yes, but such motion is impossible because of the infinite energy required by special relativity just to get to the speed of light, much less penetrate it. Therefore, because you can't get through the "light barrier," time travel to the past must be impossible, too, or so goes this line of argument.

The professor's point is a good one—it comes straight from Professor Einstein; see the opening quote in Tech Note 7—but maybe there is a way around it. The key to a possible rebuttal depends on the existence of FTL particles, the now well-known *tachyons,* a name coined by the American physicist Gerald Feinberg (1967) from the Greek word *tachys* for "swift."[36] (Feinberg once admitted—see Benford (1993)—that his interest in such matters was originally sparked by Blish's FTL story "Beep.") The idea of the tachyon is actually a very old one that is even hinted at in the work of the Greek poet and philosopher Lucretius (who died twenty years before the birth of Christ). In his discussion of visual images in Book 4 of his giant (well over 7400 lines) science poem *De Rerum Natura,* we find the following words about particles of matter originating from deep inside the Sun:[37] "Do you see how much faster and farther they must travel, how they must run through an extent of space many times vaster in the time it takes the light of the Sun to spread throughout the sky?"

The first attempt at a relativistic treatment of FTL particles in the physics literature was not recorded until two thousand years later, in Tanaka (1960) and Bilaniuk *et al.* (1962), which observed that special relativity is *not* violated by FTL motion.[38] Relativity theory indeed precludes the acceleration of a massive particle up to the speed of light, but does allow a massless particle like the photon to exist just at the speed of light. Photons are emitted during various physical processes, and they move from the instant of their creation at the speed of light;[39] the only way to slow down a photon is to destroy it by absorbing it. Advocates of the possibility of the existence of tachyons make a similar argument when asking if there might not be particles, emitted during various (yet-unknown) physical processes, that move from the instant of their creation at speeds greater than that of light?

An affirmative answer would neatly avoid the "acceleration through the light barrier" problem, but then there are other concerns. For example, such FTL particles would have to have an imaginary rest mass—see Tech Note 7—if they were to carry real energy and momentum, and what could *imaginary mass* mean? That question is convincingly answered by the proponents of tachyons, who say that the rest mass of a superluminal particle would be unobservable because there is no subluminal frame of reference in which the particle could be at rest. That is, there is no frame of reference in which the mysterious imaginary rest mass could be measured, and it is only observable changes in the real energy and momentum that characterize particle interactions.

A more serious problem for tachyons, according to those who dislike the ideas of backward causation and time travel, is the observation that in some frames of reference, an FTL particle would appear to have *negative* energy. Feinberg (1967) explains why this is a problem: "By the principle of relativity, any state which is possible for one observer must be possible for all observers, and hence FTL particles can exist in negative-energy states for all observers. . . . The occurrence of negative-energy states for particles has always been objected to on the grounds that no other system could be stable against the emission of these negative-energy particles, an entirely unphysical behavior."

This kind of objection to FTL particles was raised early in the history of tachyons, even before they received that name, and it was addressed by Bilaniuk *et al.* (1962), who proposed the so-called *reinterpretation principle* or what I'll call here the RP. To see how the RP works, consider Figure 4.5, in which a source S_1 at x_1 emits an FTL particle at time t_1. This particle then travels to an absorber S_2 at x_2, arriving there at the later time t_2. S_1 and S_2 are in the same reference frame, and for an observer in that frame, the particle energy E is positive. However, it is always possible to find another observer in a relatively moving frame for whom this process would look as though t_2 is less than (that is, earlier than) t_1 with $E < 0$. In other words, for the moving observer, the particle would appear as a negative energy particle moving backward in time. (See Tech Note 7, where it is shown that the particle speed must be not just superluminal but the even faster *ultraluminal*.)

Note that for the moving observer, the emission by S_1 of negative energy *increases* the energy of S_1, and the absorption of negative energy by S_2 decreases the energy of S_2. S_2's energy decrease (for the moving observer) occurs before the increase in S_1's energy because, as we have noted, for the moving observer, $t_2 < t_1$. The moving observer naturally interprets this process as the emission of positive energy by S_2, followed by absorption by S_1. This reinterpretation would thus seem to preserve our common-sense idea of causality, as well as avoiding any mention of backward time travel.[40] The RP appears to have slipped around those problems merely by redefining which source is

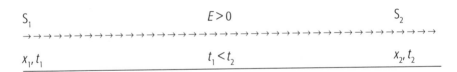

Figure 4.5 The emission of a positive energy particle, followed by absorption.

transmitting, and which receiving, the tachyon. Indeed, Feinberg (1967) claimed that the RP avoids the creation of causal loops and their associated paradoxes—a claim repeated by Italiano (1986), who used the RP to eliminate paradoxes from Gödel's time travel rotating universe.

Despite Feinberg's analysis, Newton (1970) argued that the effectiveness of the RP in avoiding causal loops is "illusory" and "irrelevant," while Thouless (1969) and Rolnick (1969) concluded that the causal paradoxes would actually preclude the possibility of tachyons interacting with ordinary matter, which is just a polite way of saying that tachyons have no more reality than unicorns! Tech Note 7 explains what Thouless, Rolnick, and Newton meant by their criticisms of the RP by giving an example of how a paradoxical time loop could be created via the exchange of FTL signals, a loop from which the RP cannot escape; indeed, the RP plays a crucial role in *creating* the paradox! That does not mean the loop couldn't logically exist—only that backward causation and time travel to the past *are* an odd business. And finally, as shown in Kamoi and Kamefuchi (1971), any so-called S-matrix, or *scattering* matrix, that describes how tachyons interact with ordinary matter necessarily violates unitarity if it is also required to be consistent with the RP. That is, an RP-consistent S-matrix violates the conservation of probability, a violation generic to spacetimes containing closed timelike curves and a law that most physicists are extremely reluctant to give up.

The RP's effect of flipping the roles of transmitter and receiver has attracted criticism, as well. Some analysts have pointed out that if one can modulate a superluminal signal to send a message into the past, then certainly one can *sign* the message. Because the RP cannot alter a signature, the origin of the message is consequently completely unambiguous. To quote a delightful example from Benford, Book, and Newcomb (1970), "If Shakespeare types out *Hamlet* on his tachyon transmitter, Bacon receives the transmission at some earlier time. But no amount of reinterpretation will make Bacon the author of *Hamlet*. It is Shakespeare, not Bacon, who exercises control over the content of the message." The last line of the quote is of central importance. The authors emphasize it by immediately observing that a signature is a relativistic *invariant* and that, indeed, it establishes a causal ordering quite independent of any temporal ordering. As Fitzgerald (1974) and Craig (1988) put it, respectively, in such cases the RP "is laughed to scorn" and the RP "sounds merely like the endorsement of what can only be characterized as a fantastic delusion."

Even more sophisticated scenarios than the one given by Benford *et al.* have been devised to show how backward causation can result from FTL signals, irrespective of one's view of the RP. One such example involves a closed chain of signal transmissions in one spatial dimension among four observers; the last observer to receive the signal is the one who first sent it. Bryce DeWitt, who created this example—it is called the "DeWitt Gambit" in Fitzgerald (1970)— showed how to arrange the spacetime geometry of the participants in the chain so that at each stage, there is no dispute over who is sending and who is receiving and thus the RP is *a priori* avoided. Yet when the signal reaches the last (first) observer, it is *before* he started (starts) the process! This sets up a potential bilking paradox, of course: What if "now" the first observer decides *not* to send the signal? It is the time travel version of Escher's famous visual paradoxes *Ascending and Descending* and *Waterfall.*

DeWitt's example was extended by Pirani (1970) from one spatial dimension to two, and the same conclusion was reached; causal paradoxes could not be avoided simply by invoking the RP. Parmentola and Yee (1971), however, argued that tachyon causal anomalies *could* be avoided. A summary of this issue, and of many related tachyon issues as well, is given in Caldirola and Recami (1981). For a theory of tachyons in which causal loops cannot occur, see Everett and Antippa (CPT).

Many physicists today reject the possibility of backward-in-time messages, not because of flaws in the RP but because such messages could create potential bilking paradoxes, as in the Dewitt Gambit and in the example described in Tech Note 7 that we owe to Yoshikawa. Sudarshan, one of the original advocates of tachyons, had actually admitted the possibility of such paradoxes when he wrote (1969) that "if FTL exists, we are provided with an *almost* [my emphasis] instantaneous communication channel. While distant observers can communicate by tachyons there is a physical limitation to the speeds that can be employed: two observers in relative motion with velocity $v < c$, can employ only [non-ultraluminal] tachyons." This speed limitation is, of course, precisely what is required to forbid the sending of messages to the past and to prevent the attendant possibility of creating a bilking paradox.

Arntzenius (1990) has observed that relativistically invariant, instantaneous action-at-a-distance theories do exist and that all such theories impose equations of motion on tachyons that are exactly what is needed so that any procedure that could imaginably lead to a paradox (such as a bilking paradox) is not allowed *by the physics.* That is, in such theories we would never be in the position of having to make a metaphysical argument about retaining free will so as to avoid the backward-causation issues of time travel. In an interesting follow-up thought experiment, Arntzenius (1994) expanded on this notion and, in fact, reversed it. That is, all of the tachyon puzzles are apparently caused because tachyons move faster than light, so if light were infinitely fast, would there be no paradoxes?

Arntzenius says no. He asks us to imagine that we have a flashlight pointed at a mirror. The on–off switch on the flashlight is wired through a sensor in such a way that we can turn the flashlight on only if there is no light incident on the sensor (a simple circuit to build). Further, imagine that the speed of light is infinity. Then the flashlight can be turned on only if it can't be turned on—a paradox! Arntzenius concludes that this does *not* mean the speed of light must necessarily be finite (that is an experimental fact) or that such sensor flashlights are impossible. Rather, he concludes only that certain physical conditions each of which is individually possible are not *logically* possible when taken in combination.

The old Wheeler and Feynman idea for explaining bilking paradoxes—that no signal is really discontinuous—was reexamined by Schulman (1971a) in the context of tachyons. Schulman's paper asks us to consider the following situation: A human (call him A) has a lamp on a table before him. The lamp is controlled by a tachyon receiver; in other words, the lamp is illuminated only when a tachyon signal (a pulse, let us say) is detected. At three o'clock A will send a tachyon signal to B (an echo-transmitter that immediately rebroadcasts everything it receives) if the lamp does not glow at one o'clock. Now, the spacetime geometry of A and B is arranged to be such that a signal sent by either A or B to the other travels one hour backward in time. Thus, if A sends at three o'clock, then B will receive at two o'clock (and immediately echo), and the echo will arrive at A at one o'clock. The paradox, of course, is that A sends a signal only if the lamp does *not* glow—that is, only if A does *not* send the signal!

Schulman next reminds us of the Wheeler and Feynman claim that every pulsed signal is actually continuous; this argument would include the illumination of the lamp. Therefore, the lamp is not just on or off but potentially at any level of illumination between those two extremes. So there sits A, and at one o'clock the lamp seems to glow dimly. Says Schulman, "A thinks it over, vacillating, finally sending a slightly late signal which isn't full strength." Of course, the echo isn't full strength either, which accounts for the original dim glow. This conclusion *is* consistent (but what if A's sending device is a toggle switch that *snaps* one way or the other—why then only a partial strength signal?), but it does seem to ask for a lot of supposing. Schulman himself is not so sure about the validity of universal continuity, writing at the end that "it is not clear that the Wheeler–Feynman assumption . . . ought to be made." Indeed, Kowalczynski (1984) disagreed with the assumption of universal continuity and specifically declared Schulman's analysis to be in error!

A generalization of Schulman's argument, to cover all signals from simple pulses to arbitrarily complicated signals, was presented by Donald (1978), the principle of universal continuity remaining central. Donald believes that his generalization resolves *all* potential paradoxes that one could imagine resulting from the existence of a closed timelike curve in spacetime. More recently, Kip Thorne at Caltech and his colleagues have tried to avoid bilking paradoxes and backward causation in wormhole time machines (see Tech Note 9) by using Schulman's approach—see Redmount (1990) and Friedman *et al.* (1990). Schulman's argument has even appeared in science fiction; see, for example, the 1980 novel *Timescape* (Benford), in which the future tries to change the past by sending tachyon messages warning of impending disaster.

Not surprisingly, the supposed ability of a modulated beam of tachyons to send a message into the past raised concerns among many about free will and fatalism. Suppose, say those who are concerned, that you receive a tachyon message from yourself from tomorrow, informing you that a man you plan to kill tonight is still alive (tomorrow). Does that mean it is beyond your power to kill him tonight? According to Fitzgerald (1970), the answer is no; you *could* kill him—but if you do, then the message from tomorrow would not have arrived. And because ignorance is not a precondition to free will, your newly acquired knowledge does not, by itself, suddenly limit your ability to kill the man. But, Fitzgerald went on, if you do not attempt to kill the man because you believe the message from your future self, then in fact the time message *has* limited you!

Fitzgerald's position was rebutted by Craig (1988), who argues that it is not your ability to kill that is altered but rather your motivation. Craig points out that such motivational changes can occur without invoking anything as radical as a message from the future. Suppose, he says, that just before you fire the fatal shot into your victim, you learn from him that he is, in fact, your beloved, long-lost uncle. Clearly your motivation for killing him is likely to be changed, but equally clearly, your ability to kill him is unchanged. The mechanism for obtaining genealogical information, whether via time travel or as a last-minute appeal from your intended victim, is (says Craig) simply irrelevant. One analyst who is not entirely convinced by Craig's argument is Schulman (1971a). He writes that "history is a set of world lines essentially frozen into spacetime. While subjectively we may feel strongly that our actions are determined only by our backward light cone, this may not always be the case. . . ." That is, Schulman appears open to the possibility that influences originating in the future might indeed have an impact on the present.

More recently, the romance of communicating with the past via superluminal speeds has passed from tachyons to quantum mechanics through a mathematical result called Bell's theorem. The late John Bell, a physicist at CERN, published his work "On the Einstein–Podolsky–Rosen Paradox" as an unimposing little article in an obscure, now-defunct journal. Since then it has become one of the most oft-cited physics papers of the 1980s. The paradox cited in Bell's title refers to a famous paper in which Einstein (1935b) challenged the conventional view of quantum mechanics, that there is no objective reality to anything that is observed. De Beauregard (1980) says Einstein framed the same challenge in different terms as early as 1927.

The conventional view of quantum mechanics formulates physics in terms of probability wave functions that collapse into specific realities only when measurements are made of states of systems (the system may be as elementary as a single particle—see Note 15 again). Until such measurements are made, says this view of quantum mechanics, a system has *no* specific state; instead it has a probability distribution over a set of possible states (but see Figure 4.6 for a time travel puzzle concerning this claim). Einstein and co-authors Boris

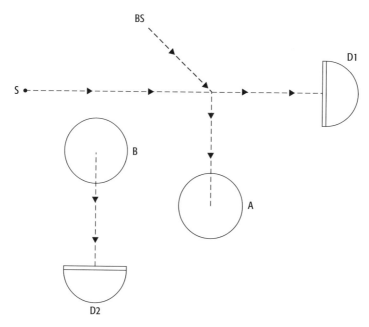

Figure 4.6 In this situation, adapted from Gonella (1994), S is a low-intensity source of photons, so weak that at any instant there is never more than one photon in the system. Each photon begins by traveling toward BS, a half-silvered beam-splitting mirror that, with equal probabilities, either passes a photon to the right, where it is detected by D1, or downward into mouth A of a wormhole time machine. Conventional quantum mechanics says that what happens at BS is determined *when the photon reaches BS*—that is, the wave function of the photon collapses at BS. But for a photon that is reflected into A, it emerges from mouth B *into the past*, where it is detected by D2 *before* the time of its arrival at BS. That means it is known what *will* happen at BS *before* the photon arrives at BS. So—when *did* the wave function collapse?

Podolsky and Nathan Rosen (the Rosen of the Einstein–Rosen bridge wormhole) rejected this probabilistic interpretation of nature. Recall Einstein's famous dictum "God does not play dice with the cosmos." Einstein and his colleagues asserted that quantum mechanics may be valid as far as it goes but that it leaves out something (as yet unknown) in describing reality. That is, they suggested that quantum mechanics is "incomplete" and that it incorporates "hidden variables." They expressed this view in the form of a paradox. The "EPR paradox" was posed as a thought experiment in which quantum mechanics declares that the properties of a spatially distributed system, when measured at point A, are forced to assume specific values at point B (*without* there being a measurement at B).

Thus, said Einstein, there are just two possibilities. Either the system properties at B must have been what they are from the very start (even if the measurement at A had not been done), which is the view he held, *or* there must have been a linkage between the system at A and the system at B such that the wave function collapse at A is instantly transmitted to B to allow the wave function to collapse there as well. Because A and B may be arbitrarily far apart, this second view obviously requires an FTL transmission mechanism, something Einstein called a "spooky action-at-a-distance"; the term eloquently expresses his opinion of the idea! Some translations replace *spooky* with *ghostly*, but the sentiment is clearly the same. For nearly thirty years the debate between proponents of these two alternatives remained at a metaphysical, nonquantitative level. Then came Bell's paper in 1964.

Bell's theorem mathematically poses the choice between Einstein's hidden-variables view and the conventional view of quantum mechanics through the use of an inequality involving certain measurable properties of a system. (The details are not important here, but Bell himself wrote (1987) a lovely exposition of it all for the lay person in an essay with the wonderful title "Bertlmann's Socks and the Nature of Reality.") If these measurements are such that the inequality is violated, then the conventional interpretation of quantum mechanics is vindicated, and Einstein's FTL spooky action-at-a-distance effect simply doesn't exist. Bell's great contribution, then, was to remove the debate about quantum mechanics from metaphysics and to place it squarely in the realm of experimental physics.

"All" that needed to be done was to make the required measurements. These technically difficult experimental measurements were eventually performed by the French physicist Alain Aspect and his colleagues (1981, 1982a,b), at the Institute of Applied Optics of the University of Paris, a decade and a half after Bell showed what had to be done. The results unequivocally supported the conventional view of quantum mechanics. (Earlier experiments had been done, but various technical details left some uncertainty about how to interpret the results. The Aspect experiments, however, are considered to be definitive. See also Sudberg (1997) and Bouwmeester *et al.* (1997) for the latest in such experiments.) Einstein was simply wrong, and his spooky action seems to exist. Does that mean we have, at last, experimental evidence of the possibility of information transfer at FTL speeds?

Well, it seems not. Ghirardi, Rimini, and Weber (1980), a paper that rejects the possibility of an FTL mechanism in quantum mechanics, identifies subtle loopholes in the arguments of those who thought they had found a way around special relativity. Indeed, those authors concluded their paper with the hope that their analysis would "stop useless debates on this subject." But alas! The majority of physicists today are actually more perplexed over what Bell's theorem is saying than they were over Einstein's original EPR paradox. In those early days one could agree with Einstein, who argued that quantum mechanics was valid *as far as it went* but that, even though all our measurements are stochastic, a deeper, more comprehensive theory would show the existence of hidden variables that would explain the reality of a system's specific properties.

Because of the work by Bell and Aspect, however, it is now known that quantum mechanics as it stands leads to correct predictions. In other words, there are no hidden variables—see Wigner (1970) for more on this point—and in addition, one seems to have to accept FTL transmission to explain the experimental results. Just what to make of all this is still the center of great controversy.[41] The consensus (among those who dare to venture an opinion!) seems to be that there is no FTL-information-transfer mechanism involved in quantum mechanics. In a play on the hard-to-miss coincidence in names, the tachyonic antitelephone cannot, according to present thought, be made in the form of a "Bell antitelephone."[42]

However, it is not obvious to everyone that this is a correct conclusion. For example, Cornell physics professor N. David Mermin has reported that some years ago he received the text of a letter written by a senior person at an unspecified California think tank (an organization such as, for example, the RAND Corporation) to the Under Secretary of Defense for Research and Engineering in the Pentagon.[43] In it he read, "If in fact we can control the FTL nonlocal effect, it would be possible . . . to make an untappable and unjammable command-control-communication system at very high bit rates for use in the submarine fleet. The important point is that since there is no ordinary electromagnetic signal linking the encoder with the decoder in such a hypothetical system, there is nothing for the enemy to tap or jam. The enemy would have to have actual possession of the "black box" decoder to intercept the message, whose reliability would not depend on separation from the encoder nor on ocean or weather conditions. . . ."

One can't help but wonder what might have been the Under Secretary's response to that letter and what sorts of top-secret experiments may have been conducted. It is almost certain that whatever they may have been, they failed. As Cramer (1988b) puts it, "Up to now nature has covered her tracks pretty well, blocking all possibilities for using the EPR effect for FTL communication."[44] Of course, the think tank letter actually represents a *failure* of imagination, because the backward causation effect of EPR's FTL effect is certainly a "quantum jump" beyond mere unjammable C^3 submarine communications.[45]

More recently, Price (1991a,b, 1994) has reintroduced an idea first put forth by Sciama (1958), with the suggestion that it may again be time to think about a modified interpretation of Einstein's hidden-variables view of quantum mechanics. This altered interpretation, which avoids the constraints imposed by Bell's inequality, says hidden-variable values depend not only on previous states of the system in question but also on its future states. Because the future states are unknown, any quantum theory developed will have an apparently indeterminate nature, and that is why, according to this idea, quantum mechanics "looks" statistical in nature. Price then observes that "the fact that the instantaneous influence approach is nevertheless regarded as less implausible is an indication of how strongly counterintuitive we find the idea of influencing the past. But what is the basis of this intuition?" Price believes it is simply "prejudice, grounded . . . in our own asymmetric nature and condition."

As a final example of the problems that quantum mechanics presents for time travel, I'll mention the issue of "quantum coherence." Imagine a lump of matter in a single "pure," or coherent, quantum state that then gravitationally collapses to form a black hole. The information inherent in describing that pure state is thus hidden from view by the hole's event horizon; *all* black holes have only three parameters (mass, angular momentum, and charge). Then, via Hawking thermal radiation, the black hole slowly evaporates by emitting energy representing a *mixture* of quantum states that displays none of the original matter's quantum coherence. There's the puzzle. Where did the original pure quantum state information go?

How this puzzle is related to time travel—see Page (1993) and Bekenstein (1993) for more on the original question—was summarized by Hawking (1995) in his final paragraph, which ends with one of his signature, somewhat mysterious one-liners:

> Personally, I do not believe that closed timelike curves will occur, at least on a macroscopic scale. I think that the chronology protection conjecture will hold and that divergences in the energy-momentum tensor will create singularities before closed timelike curves appear. However, if quantum gravitational effects somehow cut off these divergences, I am quite sure that quantum field theory on such a background will show loss of quantum coherence. So even if people come back from the future, we will not be able to predict what they will do.[46]

All of this chapter (indeed, this entire book) is concerned about controversial issues. What should readers think about the debates over FTL speeds, quantum mechanics, backward causation, black holes leaking information, and time travel? By 1974 the outpouring of papers on tachyons, for example, and the arguments about the causal loop paradoxes that tachyons seem likely to create, had reached the flood stage. Some of the exchanges become a bit heated, too, with opposing authors hinting—not always subtly—at the incredible obtuseness of the other side. For example, Fox, Kuper, and Lipson (1970) are relatively gentle: "The quantum theory of free tachyons . . . is an elegant but empty formalism," whereas Kowalczynski (1984) put matters more bluntly: "The [tachyonic] literature is infested with a large number of childish works." Indeed, with the publication in 1974 of a bibliography on FTL particle papers in the *American Journal of Physics*,[47] the journal's editor felt compelled to write, "Until tachyons are verified to exist by experiment, the prudent scholar will treat statements made in these references as hypotheses rather than statements of fact. Moreover, since these authors disagree among themselves, it is a logical certainty that at least some of them are wrong. Readers beware!"

Good advice, indeed, concerning tachyons, concerning the spooky actions of quantum physics, and concerning time machines, too!

We all from time to time indulge in dreams about traveling
back into time. Every time we recall something that happened
in the past we make a mental journey back in time. But
wouldn't it be grand if we could actually return physically to
the most delightful moments of our memories? Wouldn't we
all like to reexperience the bliss of childhood, the heydays of
youth, or return to make that big decision which we failed to
make then, for the lack of courage?

—A philosopher daydreams in Faye (1987)

I have been here before,
But when or how I cannot tell:
I know the grass beyond the door,
The sweet keen smell,
The sighing sound, the lights around the shore.
:
:
Has this been thus before?

—Dante Rossetti's "Sudden Light"

[Science fiction] cannot be good without respect for good
science. . . . This does not include time machines, space warps
and the fifth dimension; they will continue to exist in the
hazy borderland between [science fiction] and fantasy.

—Harry Harrison, in his essay
"With a Piece of Twisted Wire . . ." from *SF Horizons,* 1965

What can be said in closing a book like this one? As the opening quote indicates, we seem to have come full circle. Just as at the start of Chapter One, we are still faced with the obvious question: Is time travel to the past possible? I have tried hard not to proselytize blindly for such a possibility and have tried equally hard to present an even and balanced account on both sides of that question. The central philosophical arguments against backward time travel—causality violation and the logical paradoxes—are exactly what makes the question so interesting in the first place.

Many, naturally, find those negative arguments highly persuasive. In particular, many like to cite versions of the grandfather paradox, along with the claim that the only way out of that paradox is to give up free will, a position that Chapter Four suggests has no logical or physical basis. It is astonishing, indeed, that the power of that philosophical red herring has hardly abated since it first appeared in the science fiction pulps nearly three-quarters of a century ago. Just a few years ago, in an essay in *Science* on Gott's cosmic string time machine, for example, Travis (1992) called the grandfather paradox an "obvious problem" for time travel. And in the very recent philosophical literature we still find lengthy, quite detailed analyses of that phoney "paradox," as in Adams (1997), Riggs (1997), and Smith (1997). Science fiction writers, too, whatever they may write in their stories, often still labor under the skeptical position stated by writer and critic Kingsley Amis, who wrote in *New Maps of Hell* that "time travel is inconceivable."

Isaac Asimov echoed Amis when he flatly stated, in his introduction to Lafferty (1991), "I think scientists who think up methods of time travel are probably all wrong." Are Amis and Asimov right? Whether the answer is yes or no, the analyses published since Gödel's pioneering time travel paper of 1949 have taken a long time to reach even those who would surely find them high drama. Indeed, Gödel's time travel work seems, even several years later, to have been unknown outside of a limited mathematics and physics community. In 1952, for example, the anthologist Groff Conklin wrote, in his introduction to Murray Leinster's story "The Middle of the Week After Next," "In this tale we meet our first Mad Scientist. Just as in reality the thoroughly cracked pots used to be found inventing perpetual-motion machines, so in science fiction we find the lunatic fringe more often than not trying to perfect time-travel mechanisms." That same year H. L. Gold, the founding editor of *Galaxy Science Fiction Magazine,* declared (GRSF1), "Time travel requires a suspension of disbelief that is almost unbelievable. . . . [S]cientifically, time travel can't stand inspection."

Taking a somewhat less rigid position in those early days of 1952 than that of Conklin and Gold was Chad Oliver's non-fiction essay "The Science of Man," included in his novel about a teenager who travels in a "space-time machine" back to 50,000 B.C., to the start of Cro-Magnon time (*Mists of Dawn*). There we read, "It would be untrue . . . to present the idea of a time machine as anything but what it is, an intriguing literary device, part of the bag of tricks of the science fiction writer. . . . [T]here is no such thing as a 'science' of time travel." Conklin, Gold, and Oliver all wrote just three years after Gödel; perhaps it was simply too soon for his work to be widely known outside the physics community.

Paul A. Carter, in *The Creation of Tomorrow,* his excellent 1977 analysis of the first half-century of the science fiction magazines, was aware that time travel has a physical rationality, but even at that late date he apparently had not appreciated either Gödel's time travel analyses or the ones that followed. After remarking on how the conventional view has been that backward time travel is simply impossible and then citing the work of Tipler, Carter writes, "Only as recently as 1974, in the sober pages of the *Physical Review,* has a physicist been more bold. . . . For seventy years in the meantime, however, without waiting for Professor Tipler to solve his equations . . . writers had happily helped themselves to Mr. Wells' invention and sent their characters through time in every direction, forward, backward, and sideways." Given that *The Time Machine* was published in 1895, it is not clear how Carter arrived at the value of seventy, but in fact it was only fifty-four years between Wells' time travel fiction and Gödel's time travel mathematical physics.

In the 1980s, however, writers were apparently just as unaware of Gödel's time travel analyses (and of the much later ones of Frank Tipler) as were the early commentators. For example, in his marvelous book *The Past Is a Foreign Country,* David Lowenthal repeatedly refers to time travel as "fantasy" and to science fiction stories about time travel as "unbridled by common sense." The well-known science fiction writer and critic Alexei Panshin, for yet another example, agrees with Lowenthal; in his Foreword to Robert Heinlein's classic time travel story "All You Zombies—" Panshin wrote, "Time travel is a philosophical concept, not a scientific one. It is, in fact, as has often been pointed out, scientific nonsense. . . . That s-f writers . . . kept returning . . . to the scientifically unsensible theme of time travel indicates that their viscera knew something that their minds did not." Mr. Panshin, obviously, had not appreciated either Mr. Gödel or Mr. Tipler.

Even more recently, one of the best of the new writers of science fiction, Orson Scott Card, has referred (MIAM) to time travel as an impossibility and to time machines as simply a "magic trick" for writers of fiction. Somewhat more muted in his unhappy feelings about time travel was the poetic British science fiction writer Brian Aldiss. In the Foreword to his short-story collection

Man in His Time, Aldiss wrote, "Of course, things may seem plausible which are not really credible. Time travel has this quality." Hedging his bets even more was Robert Silverberg who, in the Afterword to the original 1974 publication of his story "Trips" (in *Final Stage,* Charterhouse Books), wrote of time travel that it is a "philosophically stimulating but physically unlikely concept."

And finally, consider the case of James Gunn, professor of English at the University of Kansas, past president of both the Science Fiction Writers of America and the Science Fiction Research Association, author of *The Immortal* (inspiration for the 1970–1971 TV series of the same name), and eminent scholar (see his 1975 book *Alternate Worlds*). His literary credentials are impeccable and his critical influence profound. And yet, thirty years after Gödel and five after Tipler, Professor Gunn wrote in *The Road to Science Fiction,* "Time travel has been an anomaly in science fiction. Clearly fantastic—there is no evidence that anyone has ever traveled in time and *no theoretical basis for believing that anyone ever will* [my emphasis]."[1] If you've read this far, in particular of the modern analyses by Thorne, Novikov and Gott, you know that what Gunn claims in those last words is simply not so.

The eminent British-born American physicist Freeman Dyson of the Institute for Advanced Study has commented (1979) on that sort of narrow mindset, with words quoted from the 1979 physics Nobel prize winner Steven Weinberg, words reminding us that rigidity is not limited to science fiction writers: "This is often the way it is in physics—our mistake is not that we take our theories too seriously, but that we do not take them seriously enough. It is always hard to realize that these numbers and equations we play with at our desks have something to do with the real world. Even worse, there often seems to be a general agreement that certain phenomena are just not fit subjects for respectable theoretical and experimental effort." As a case in point, consider that in 1993, Kip Thorne began adding a line in the customary acknowledgments paragraph at the end of his time machine papers stating that the National Science Foundation had refused to continue to support that particular aspect of his research program. Fortunately, internal money at Caltech was available, at least until Thorne's own long-time interests in gravity wave detection lured him away (at least for a while) from time machine studies.

The NSF's reaction is, of course, not difficult to understand. Time travel, with its double burden of very difficult mathematics and non-intuitive philosophical implications, is not easy to come to grips with, even for those who are at least open to the possibility of time travel. Just try to imagine some poor NSF official explaining the agency's budget to Congress—to people who often don't appreciate the distinction between pure and applied research—

when the line-item for "time travel" comes up! Thorne himself has some concerns. As he has written (1992), "[Time machine] studies are giving us glimpses of how [closed timelike curves] influence physics; but whether these glimpses are teaching us something deep and important, or we are just playing fun mental games, is far from clear." Thorne has not come suddenly to that cautious assessment. In an April 1991 letter to me, for example, he mentioned a recent visit with Stephen Hawking in Cambridge. They had met to review the current status of their work on closed timelike curves, and their joint conclusion concerning time travel was a sobering one: "We now think that we understand these issues less well than either he or I believed a few weeks ago."

Three reasons for why both Thorne and Hawking felt less sure than before about the physical possibility of time travel, and of wormhole time machines in particular, were given by Thorne in a paper published that same year, and as I write (early 1998), I think all three reasons still stand undefeated: (1) Spacetime may, in fact, be simply connected and so wormholes just don't exist. (2) Even if wormholes can exist, briefly, quantum mechanics may enforce the average energy condition and thereby forbid wormhole throats to remain open. (3) The Cauchy horizon may be unstable and may collapse into a singularity. Despite those concerns, however, Thorne thinks the definitive case against time travel is yet to be made. Hawking disagrees (although in the last two years or so—see his Foreword to Lawrence Krauss' *The Physics of Star Trek*—he does appear to be wavering just a bit on this!)

The split between Hawking's and Thorne's views about the plausibility of time travel is centered on the proper way to handle the quantum-gravitational details of wormholes and the stability of the Cauchy horizon. Hawking (1992), in fact, says his analysis shows that any mechanism for time travel, wormhole or otherwise, is forbidden (by his *Chronology Protection Conjecture.*) As he writes, "if you try to create a wormhole to use a time machine, you have to warp the light cone structure of spacetime so much that closed timelike curves appear anyway," and so the specific details of a particular wormhole are irrelevant. It is the creation of closed timelike curves by *any* means that Hawking denies. This therefore led Hawking to dismiss the cosmic string time machine described by Gott (1991).

On the other hand, Kim and Thorne (1991) disagree with some of the details of how Hawking does *his* quantum gravity! More recently, Kim (1992) has argued that the Cauchy horizon *is* stable at the quantum level, but on the other hand (again), Visser (1993) has generalized and endorsed, as has the Russian physicist Valery Frolov (1991), Hawking's conclusion forbidding time machines. Visser doesn't entirely agree with the details of Hawking's analysis, either, but Visser does agree with Hawking's general claim of the Chronology Protection Conjecture and with Hawking's humorous statement that the Conjecture "makes the Universe safe for historians." By the way, a fictional treatment of what Hawking is alluding to in that provocative statement is the 1950 story "Sunday Is Three Thousand Years Away" (Jones), in which a Master Historian and his students in a graduate course on 'Experimental History,' in the forty-sixth century, try to correct a problem created by previous tampering with the past. Hawking seems to be afraid that the assistant television producer's bold comment, in J. G. Ballard's "The Greatest Television Show on Earth," is exactly how the past would be viewed by a world with time travel: "History is just a first-draft screenplay."

Igor Novikov (a former colleague of Frolov's at the P. N. Lebedev Physical Institute in Moscow), however, takes the position that via the principle of self-consistency, the universe will remain "safe" even with time travel. As his fellow Russian A. Yu. Neronov (1998) has recently put it, "The global solutions of the dynamical [Einstein] equations must satisfy the 'self-consistency' principle, which eliminates paradoxes associated with the breakdown of causality in a space-time with closed time-like curves." Novikov also is not quite so ready to put his faith in quantum theory's ability, as we presently know it, to forbid time travel. As he and a colleague have written—see Lossev and Novikov (1992)—"our understanding of the fundamental structure of the vacuum and the effects of quantum fluctuations is so inadequate that from existing quantum theory only we should definitely exclude the possibility of the very existence of the Universe . . . but experimentally it exists."

But maybe some other mechanism(s) will enforce the conjecture; Visser has argued that physics always seems to come up with a new way to defeat the possibility of closed timelike curves every time some clever analyst thinks he has discovered a new way to make such entities. As Visser puts it, Nature appears to have a "defense in depth" against time travel. Such a position is not new, either for time travel in particular or for physics in general. All through this book, for example, we have seen how people have argued against Tipler's cylinder time machine because it is unphysically long, against Gödel's time travel universe because it requires an unphysical rotation, against wormhole time machines because they assume what some believe to be unphysical energy conditions, against cosmic string time machines because . . . and so it goes. Thorne (1992) replies, "I do not find such assertions at all satisfying," and then he goes on to give an historical list of similar arguments—arguments once made to prove how some process just could not be because it was "unphysical."[2] *All* the arguments on Thorne's list were subsequently shown to be wrong.

Another amusing example of such arguments, which Thorne did not mention but that I think particularly appropriate in this book, was the debate in the 1930s between the illustrious British astrophysicist Sir Arthur Eddington and the young Indian astrophysicist Subrahmanyar Chandrasekhar (see Note 16 for Chapter One). In his analyses of the life history of stars, Chandrasekhar had arrived at an astonishing conclusion. As Eddington sarcastically explained[3] in an address at Harvard University in the summer of 1936, "above a certain critical mass (two of three times that of the sun) the star could never cool down, but must go on radiating and contracting until heaven knows what becomes of it. That did not worry Chandrasekhar; he seemed to like the stars to behave that way, and believes that is what really happens." Eddington then went on to declare such behavior nothing less than "stellar buffoonery."

As far as Eddington was concerned, Chandrasekhar had simply made an error in combining relativity theory with non-relativistic quantum theory. Indeed, so appalled was Eddington at the thought of a star contracting "until heaven knows what becomes of it" (that is, until it gravitationally collapses into a black hole) that he had earlier, in 1935, stated, "there should be a law of nature to prevent a star from behaving in this absurd way!" Today, of course, no astrophysicist (least of all Hawking) feels the need for such a "star protection conjecture."

But Hawking remains convinced that we do need chronology protection. As he wrote of time travel his 1993 collection *Black Holes and Baby Universes and Other Essays,* "The best evidence that time-travel will never be possible is that we have not been invaded by hordes of tourists from the future." I must admit I am puzzled by Hawking's fascination with that particular argument, one that I don't think actually proves what he is asking of it.[4]

What can one conclude from all this controversy? Not much, I think, except that time travel is an open question and will remain the subject of ongoing study. As I write, nobody knows the answers to Thorne's list of concerns, or whether Hawking's conjecture is accurate, or whether Visser or Novikov is right or wrong, so the time travel debates among physicists continue. In a personal communication with to me (August 12, 1992) Tipler wrote that he tends to side with Novikov but also that the present quantum arguments are all "order of magnitude handwaving."

As the writing of the first edition of this book drew near its end, I wrote to the great American physicist John A. Wheeler at Princeton University, whose work has been discussed all through this book, and put the question to him. I received a delightful letter in response, and it is as appropriate for this edition as it was originally. Professor Wheeler's letter opened as follows:

The idea of going backward in time? Of it let me say what Thomas Brown said of the dean of his Cambridge college in 1680 when he threatened to expel Brown:

"I do not like thee, Dr. Fell.
The reason why, I cannot tell;
But this I know, and know full well,
I do not like thee, Dr. Fell."

The spectacular—and that seems the only proper word to use—nature of the original time travel analyses of Kip Thorne's theoretical astrophysics group at Caltech creates at least a *little* doubt about that assessment. Ironically, Thorne was Wheeler's doctoral student at Princeton in the early 1960s.[5] Hawking's and Visser's theoretical doubts that we discussed earlier shouldn't be forgotten, of course; and I think Thorne would certainly be the first to admit that even if no flaws appear in the theory of time travel, the engineering limitations on the actual construction of a time machine are light-years beyond the merely staggering.

Thorne admits—see Travis (1992)—that Hawking's analysis against time travel is "a very powerful result," and in a personal communication to me (August 4, 1992) he wrote, "the central issue in time machine research, now, it seems to me, is the issue of whether vacuum fluctuations of quantum fields AL-WAYS destroy a time machine at the moment one tries to activate it. I suspect the answer is yes, but I also suspect we will not know for sure until we understand the laws of quantum gravity." Thorne has always been careful to state that the driving force behind his work on the physics of closed timelike curves is to discover what the laws of physics allow, not actually to build a time machine.[6]

To get right down to it, I don't know whether it is possible to visit the past, and I fear the odds are that I'll learn the truth about Heaven (or Hell) in the next world before I learn the truth about time travel in this one. I am fairly certain that if time travel is ever achieved, it will be by means that we cannot today even begin to guess. Indeed, it required a mutant child genius with an IQ of 270 to fix the slightly broken time machine found abandoned in a cellar in "A Guest in the House" (Long)! But that isn't a view uniformly shared in much of science fiction. I very much doubt, for example, that things will be quite so elementary as in "Heritage" (Abernathy), wherein Nickolus Doody's time machine was so simple that "If it were taken apart or put together before you, your wife, or the man across the street, you would wonder why you didn't think of it yourselves." Not only that, but its power source was just two dry cells! The time machine in the 1933 story "In the Scarlet Star" (Williamson) is almost as simple, requiring (besides a strange piece of crystal) only a "little stack of dry cells, a Ford coil, a small brass switch, a radio 'B' battery, an electron tube, and a rheostat." Even Wells' marvelous *Time Machine* couldn't resist making it all look simple; as Pitkin (1914) observed, "The time machine, like all products of supreme inventive genius, was a remarkably simple affair. A few rods, wires, some odd glass knobs—nothing more!"

That sort of simplistic fiction reminds me of a passage from "The Twentieth Voyage of Ijon Tichy" (Lem): "There have been mountains of nonsense written about traveling in time, just as previously there were about astronautics—you know, how some scientist, with the backing of a wealthy businessman, goes off in a corner and slaps together a rocket, which the two of them—and in the company of their lady friends, yet—then take to the far end of the Galaxy. Chronomotion, no less than Astronautics, is a colossal enterprise, requiring tremendous investments, expenditures, planning. . . ." Lem would have snorted in derision at the statement made to a prospective graduate student by the head of a college physics department in the novel *River of Time* (West). The student is casually told that the college "has been awarded a million dollars to build [a time machine]. It means . . . a raise for me and maybe a doctorate for you, so we'll build one and have some fun doing it."

It is no wonder that Lem so readily dismisses stories that reduce space (and time) travel to weekend adventures in a home laboratory. As he wrote in another essay (1977), time travel and its close relation, faster-than-light space travel, have reduced much of science fiction to "a bastard of myths gone to the dogs." Because of precisely that, Harry Harrison (in the same essay from which I took the last of the opening quotes) wrote of the early science fiction magazines that published so much nonsense, "I used to moan over the fact that pulp magazines were printed on pulp paper and steadily decompose back towards the primordial from which they sprang. I am beginning to feel that this is a bit of a good thing."

Well, I do not know whether time travel to the past can actually be accomplished, but I do know that speculations once thought as outlandish as finding the philosophers' stone for turning base elements into gold, *have* eventually been realized (and come to think of it, with modern nuclear physics we *have* learned how to turn lead into gold, if only a few atoms at a time). Television, nuclear power, VCRs, home computers that run at 400 MHz in the bedrooms of high school students, and supercomputers that animate our movies and simulate the formation of black holes and galaxies—all these amazing developments would be pure magic to nineteenth-century science. The ghosts of not a few Victorian scientists have watched their reputations eat a lot of posthumous crow during the last hundred years. My personal position is described by the rejoinder to the skeptic in Heinlein's "By His Bootstraps," who, even after having done some time traveling, *still* argues against it by invoking paradoxes. He is sharply rebuked with "Oh, for heaven's sake, shut up, will you? You remind me of the mathematician who proved that airplanes couldn't fly."[7] I subscribe to the optimistic philosophy of British writer Eden Phillpotts, who wrote, in his 1934 novel *A Shadow Passes,* "The Universe is full of magical things, patiently waiting for our wits to grow sharper." Perhaps he had J.B.S. Haldane's famous lines from his 1928 *Possible Worlds* in mind: "Now my suspicion is that the universe is not only queerer than we suppose, but queerer than we can suppose."

Still, even if time travel is possible, the engineering phase will surely be tough going. I am certain that before we see a working time machine, there will be many, *many* episodes like the one described in the funny, novel-length spoof of academic research *Dr. Dimension* (De Chancie and Bischoff). All physicists and engineers who have tried to get some stubborn piece of apparatus to work, apparatus that *should* work and yet simply won't, will appreciate Professor Demetrios Demopoulos' frustration and will, I am sure, forgive him for his intemperate language:

> . . . the distinguished physicist took a step back and, arms akimbo, surveyed the complex and sophisticated machine that was the culmination of years of dedicated scientific research and pains-taking technological development.
>
> "What a pile of ****," he said.
>
> "Oh, no, Dr. Demopoulos, don't say that!"
>
> "Well, it is." A sneer formed on the professor's thin lips. "Time machine, my ***. This thing couldn't give you the time much less travel in it."
>
> "But we haven't incorporated all our latest test data yet," the pretty research assistant reminded him. "These last few adjustments might do it, Professor."
>
> "Hell, we've been tinkering with it for two years," Demetrios complained. "We've tried everything and it's all come to dog ****?"

Even before the practical nuts-and-bolts bugs in the Professor's machine are worked out, I think some adjustments are called for in our thinking about time travel.

I believe that present-day philosophers and science fiction writers are going to have to become knowledgeable about the new work by physicists on time travel. It simply won't do any longer for Philosophy Professor X to invoke the grandfather paradox during a discussion of causality and free will and airily declare them to be "obviously" incompatible with time travel to the past. And it simply won't do any longer for Famous S-F Writer Y to send his hero into the past to kill Hitler as a baby and thereby change recorded history. One might as well keep watching a videotape of the Challenger disaster in the vain hope that maybe, on the next viewing, the Shuttle won't blow up.

The principle of self-consistency around closed timelike curves is going to have to become as much a part of the science fiction writer's craft (or else she will be a writer of fantasy) as it will have to become part of the fundamental philosophical axioms.[8] The "time police," such as science fiction writer Andre Norton's "operatives of the Bureau of Time Exploration and Manipulation," will have to be put out to pasture with the unicorns and telepathic dragons of fantasy fiction. Just as the recent physics literature on time travel has displayed a growing awareness of what science fiction writers and philosophers have had to say on the subject, so too are writers and philosophers going to have to learn some more physics. Most people can enjoy a good fantasy tale now and then, but the use of "magic mirrors" to see through time is *not* physics. Such devices were popular and acceptable in medieval times (see "The Squire's Tale" in Chaucer's *The Canterbury Tales*) and later (see Act IV of *Macbeth*), but good science fiction needs much more than that today.

Time travel to the past is a beautiful, romantic idea, and some words written by two physicists in a technical paper[9]—words embedded in the midst of swirls of tensor equations—show that even hard-nosed physicists can share this dream: "In truth, it is difficult to resist the appealing idea of traveling into one's own past. . . ." The appeal of this dream is explained in Ray Bradbury's Foreword to a beautiful little 1989 book by Charles Champlin (*Back There Where the Past Was*). Bradbury, I think, has clearly illuminated *why* we want to go back to the past. It is for the same reason that we go, time and again, to see *Hamlet, Othello,* and *Richard III:* "We don't give a hoot in hell who poisoned the King of Denmark's semicircular canal. We already know where Désdemona lies smothered in bedclothes and that Richard goes headless at his finale. We attend them to toss pebbles in ponds, not to see the stones strike, but the ripples spread."

That's why a visit to the past is so mysteriously and marvelously fascinating. It would let us watch ripples spread through time. Our own visit, in fact, might even *be* the pebble in the pond that starts an interesting ripple or two that will one day sweep over—us! Who would want to miss that? Indeed, if modern philosophers are right, if the analyses discussed in Chapter Four are correct, you *can't* (didn't/won't) miss it. I think time travel appeals, irresistibly, to the romantic in the soul of anyone who is human. A time traveler does not exist either *here* or *then,* but rather *everywhen.* For a time traveler passing back and forth through the ages, history would be the ultimate puzzle, a chronicle described in John Crowley's brilliant novel *Great Work of Time* as beginning "not in one place, but everywhere at once. . . . It might be begun at any point along the infinite, infinitely broken coastline of time."

Romanticism doesn't preclude there being a dark side to the idea of visiting the past, of course, as the time traveler from 1989 in *A Bridge of Years* (Wilson) realizes when he takes up residence in 1962. Falling asleep on a hot summer night in that long-ago year, he thinks "JFK slept. Oswald slept. Martin Luther King slept. [I sleep and dream] of Chernobyl. . . . *I am a cold wind from the land of your children.*" But, I must admit, I personally am more attracted by happier descriptions of time travel. In his marvelous book *1939: The Lost World of the Fair*—which is proof that there are not enough Pulitzers to go to all of the books that deserve one—David Gelernter caught just the right spirit in his Prologue: "The best of all reasons to return to the fair is that travel is broadening, and time travel most of all. . . . The 1939 New York World's Fair is one amazing show. It still stands, undisturbed on Flushing Meadow, just over the edge of time; it would be an unforgivable shame to miss it." Trust me—if you read Gelernter's book, you'll come as close as you can in today's world to taking a ride in a "time machine"!

The eminent philosopher Sir Karl Popper opens his biography with a wonderful story about his apprenticeship as a young man in 1920s Vienna to a master cabinetmaker.[10] After winning the old man's confidence, the student learned his mentor's great secret: For years the master had been looking for the solution to perpetual motion. He knew the physicist's judgment on such machines (recall Groff Conklin's accurate assessment), but nonetheless he had never given up his dream: "They say you can't make it; but once it's been made they'll talk differently." He sounds just a bit like Gertrude Stein in her 1938 essay "Picasso," where she writes, "It is strange about everything, it is strange about pictures, a picture may seem extraordinarily strange to you and after some time not only it does not seem strange but it is impossible to find what there was in it that was strange." Might we one day say the same thing about time travel?

The theoretical basis for time travel is very different from that of perpetual motion, of course, so maybe some day, *just maybe,* the first time traveler will propose a toast such as the one in the story "Time's Arrow" (McDevitt), offered when the inventor of the first time machine and his no longer skeptical friend successfully arrive in the Civil War past.

> "To you, Mac," I said.
> McHugh loosened his tie. "To the Creator," he said, "who has given us a Universe with such marvelous possibilities."[11]

Prologue

1. An example of the sort of tale that gave an aroma of the sophomoric to early time travel science fiction is the 1941 "The Brontosaurus" (Thompson). It is the story of a young man of the far future, with access to a time machine, who wants to see a dinosaur before he dies. So back he travels, back, back, until he at last finds himself in a "subterranean cave, dark and foul-smelling." At first he is puzzled (did dinosaurs live underground?), but then suddenly he hears a thundering roar and sees a huge black shape in the gloom. There can be no doubt now; it *is* a dinosaur, and he sees its red, gleaming eyes just as it crushes him into a pancake. But that's okay; he saw a dinosaur before he died. Then comes the dénouement. He hadn't really gone back quite as far as the Jurassic period but only to the twentieth century, where he has been run down by the local express in a subway tunnel.

Chapter One

1. Many readers, I know, will fondly recall Mr. Peabody, that nice but slightly stuffy, professorial white beagle (don't all dogs wear glasses and a bow tie?) who, with his brainy adopted son Sherman, routinely traveled into the past in his "Way-Bac" machine to see what really happened. A derivative of this idea was the basis for the 1982–1983 television series "Voyagers"; in each episode, two time travelers would "help" keep history from going astray. At the end of each show, young viewers would be encouraged to learn about that week's "history lesson."

2. It is not just radio tapes of long-ago broadcasts that sell well. One can now buy authentic reproductions, with modern electronics, of the old-time radios, too. As one manufacturer declares on its shipping boxes, "We Sell Yesterdays." Nostalgia, itself, is a subject that has received serious study; see F. Davis, *Yearning for Yesterday: A sociology of nostalgia*

(New York: The Free Press, 1979), which never mentions time travel, even in passing.

3. Lucifer doesn't always win, of course, and the stories "The Brazen Locked Room" (Asimov) and "Time Trammel" (DeFord) are examples of how Mr. Scratch sometimes loses. Both involve time travel but, I am sorry to say, in not very innovative ways. Much more interesting are "Enoch Soames" (Beerbohm) and "The Hell-Bound Train" (Bloch). Also excellent is T.R. Cogswell, who wrote two devilish stories about time travel (found in the anthology DD). In particular, his "Three-sie" shows just how careful one has to be in reading the fine print in a contract with Beelzebub. Joseph Cruthers advertises his soul for sale in the local newspaper, Satan replies, and the standard deal is struck: Joe's soul for three wishes. The first two are for the usual wealth and immortality, but the third is for three *more* wishes. And Joe gets them, too. He is sent back in time to the moment he places his advertisement, and so he gets wishes forever, the same three, in an endless, looping temporal treadmill that becomes Joe's own personal hell. For more on that idea, see "The Time-Wise Guy" (Farley) and also *The Third Policeman* (O'Brien), which tells us that "Hell goes round and round. In shape [in time] it is circular and by nature it is interminable, repetitive and very nearly unbearable." Jean Paul Sartre's 1944 play *No Exit* presents a similar image of Hell. The 1993 film *Groundhog Day* is based on the idea of circular time, in which a TV weatherman is, for an unexplained reason, forced to loop over and over through February 2. This is not a true time loop, however, because the weatherman is aware of his situation and, indeed, can change events within the loop at will. A 1990s made-for-television film short, *12:01 PM,* is essentially a nightmare version of *Groundhog Day.* The time loop, or *time bounce* as it is called in the film, is one hour in duration, and the only man in the world who realizes what is happening slowly goes crazy. There is a weak attempt to explain the time loop—the universe is colliding with an anti-matter universe, but see the end of Note 16 for Chapter Two for what would be the far more likely result. Another 1993 made-for-television movie, with a similar title (*12:01*), utilizes a malfunctioning "time-field generator" rather than colliding universes to explain its time loop. It makes the same errors as *Groundhog Day* and *12:01 PM,* however, with a lone character who realizes he is in a loop and who repeatedly alters the past, whatever that might mean. The 1990s obviously found time loops quite interesting, the 1997 *Timequake* (Vonnegut) being yet another (not very good) try at it. This is an old fantasy concept, of course, and it can be found in pulp fiction from at least half a century ago; see, for example the 1943 "Yesterday's Clock" (O'Brien). In that story there is also a hint of *why* the central character lives in a day-long time loop—a mysterious clock loops him back twenty-four hours.

4. Perhaps the definitive analysis of Jack's identity can be found in A.G. Kelly's *Jack the Ripper: A bibliography and review of the literature* (London: Association of Assistant librarians, S.E.D., 1972). Kelly specifically recommends Bloch's story, which has the Ripper brought into the future to serve as a sexual playmate for a spoiled woman who has the tables turned on her in a most gruesome manner. Ellison's effort, which tells about Jack's ultimate fate in the year 3077, is a sequel to Bloch's. Neither story is for the faint of heart or the weak of stomach.

5. N.P. Davis, *Lawrence and Oppenheimer* (New York: Simon and Schuster, 1968), 95.

6. For more on the possibility of sound recordings from the ancient past, see D.E.H. Jones, *The Inventions of Daedalus* (San Francisco: CA: W.H. Freeman, 1982), 26–27. A letter from Woodbridge, in which he reveals that his paper was initially "perfunctorily rejected is being 'too specialized' " by *Nature*, is reprinted in Jones' book. The 1947 "Collector's Item" (Long) goes in the other direction; there we find a phonograph recorder that travels six centuries into the *future* and then re-

turns with a momentous audio message that changes the future, an illogical result that will be discussed later in this book.

7. N. Berdyaev, *The Bourgeois Mind and Other Essays* (Freeport, NY: Books for Libraries Press, 1966 [originally published in 1934], 50. Even philosophers are human, of course, and the story is told of Berdyaev once pleading with intensity and passion for the insignificance of time and then, suddenly, stopping in mid-sentence, first to look at his watch and immediately after that to rush from the room crying, "I'm late, I'm late, I should have taken my medicine two minutes ago!" When Pierce wrote his story on the vagueness of the past, he almost certainly had in mind a paper written by Einstein (1931) and two colleagues, some years earlier, on just this point.

8. Reversing the idea in Anderson's tale is *The Way Back* (Chandler), which has its characters return from the past to their proper time by traveling even further backward, right through the Big Bang and into the previous and identical cycle of time. Larry Niven (1971) argues that this is a conceptually valid way to travel into the past, but he does raise one cautionary note: "Removing your time machine from the reaction of the Big Bang could change the final configuration of matter, giving an entirely different . . . history." In *Nature* 66 (July, 3 1902):223, there is a curious letter that bears on Niven's point written by Hiram S. Maxim, the American-born British inventor of the first fully automatic machine gun. In one place it reads, "The grandest words ever uttered by any man on this planet were spoken by Lord Kelvin when he said that if all the matter in the Universe were reduced to its ultimate atoms and equally divided through all space, the disturbance caused by the beating of the wing of one mosquito would bring about everything that we find in the material Universe today." Presumably Maxim and Niven believed that any deviations in such a disturbance would result in different Universes. This idea has become popularly known today in mathematical chaos theory as the Butterfly effect, after Bradbury's use of it in his time travel story "A Sound of Thunder"—the story Isaac Asimov declared to be his favorite of all of Bradbury's tales—but as the *Nature* letter shows, it should more accurately be called the "mosquito effect."

9. Originally published in 1748, *Enquiry* has been reprinted many times over the years. I used the work as published by The Open Court Publishing Company (La Salle, IL: 1963).

10. P. Health, "The Incredulous Hume," *American Philosophical Quarterly* 13 (April 1976):159–163.

11. *New Scientist* 77 (January 26, 1978):239.

12. Gödel's universe is a rotating, infinite, static one, with a non-zero cosmological constant; see Tech Note 8 for the implications of this. The observable universe is non-rotating and expanding—that is, it has a red-shift—so although Gödel's analysis satisfies the general relativity equations, its time travel property does not hold in our universe. Gödel may have been inspired to study his model by a letter written three years before by George Gamow (1946). See also Narlikar (1962b). In Birch (1982) astronomical data are presented indicating that the universe is rotating no faster than once each 60,000 billion years, which is much slower than the minimum required by Gödel's solution (see Tech Note 8). The question asked by the title of Birch's paper has recently been answered, in the negative, by Panov and Sbytov (1992). See also Korotkii and Obukov (1991) for more on the relationship between a shear-free rotation of the universe, as in Gödel's analysis, and time travel. For technical analysis of Gödel's work, see Chandrasekhar and Wright (1961), a faulty study cited as late as 1962 by MacKinnon in support of the mistaken view that Gödel had made an error, and Stein (1970), which shows how

Chandrasekhar and Wright erred. The critical technical point, elaborated on by Stein, was actually mentioned some years earlier in Earman (1967a). So far as I know, no one has written a time-travel story using the rotating-universe idea but, if he had lived, perhaps the science fiction writer James Blish would have. In David Ketterer's biography of Blish, *Imprisoned in a Tesseract* (Kent State University Press, 1987), there is this comment by Blish from a 1970 letter: "I am especially intrigued by the spinning-universe form of time travel, especially since . . . nobody has touched it. . . . But I should really stop mentioning the spinning-Universe in public, or somebody will nobble onto it before I can get into it!" Gödel's time-travel idea still waits for its first fictional use. Dawson (1989) reports that a much longer version of Gödel's paper (1949a) on time travel in rotating universes remains unpublished in the archives of the Institute for Advanced Study. For more on Gödel, based in part on unpublished material of his at the IAS, see Yourgrau (1991) and in particular his Chapter Three, "Time Travel and the Gödel Universe."

13. In a footnote Gödel says that the time traveler would have to move at least as fast as nearly 71% of the speed of light and that if his rocket ship could "transform matter completely into energy," then the weight of the fuel would be greater than that of the rocket by a factor of 10^{22} divided by the square of the duration of the trip (in years, in rocket time). A trip to the past in Gödel's universe would require a time machine that looked something like Dr. Who's telephone booth attached to a fuel tank the size of several hundred *trillion* ocean liners. Capek (BST) calls the whole business "fantastic." These *are* formidable numbers—see Purcell (1963) for an analysis of a fantastic rocket powered by matter/anti-matter, a process that satisfies Gödel's vast energy requirement—but they require no violation of physical laws, and that is what really counts if time travel is to be disproved. As pointed out in Earman (1972), Gödel's use of engineering limitations for explaining away backward time travel is worse than simply being wrong, because the puzzle is not in practicality but rather in showing, assuming that general relativity is correct, how correct mathematics and physics can lead to what seems to be a paradoxical conclusion. More recent analyses by Chakrabarti *et al.* (1983) and Malament (1985, 1987) have established that there must be a lower bound on the total integrated acceleration of a time traveler in order for her to move backward in time—see also Pfarr (1981). In particular, if TA is the total integrated acceleration of a rocket ship with a total initial mass equal to $m_p + m_f$—that is, with a payload mass of m_p and a fuel mass of m_f, then Malament showed that $m_p/(m_p + m_f) \le \exp(-TA)$, which says that as the total integrated acceleration increases, the fraction of the rocket mass that can be transported into the past in Gödelian spacetime *decreases very quickly*. As Malament put it, "Any 'time traveler' who would return to an 'earlier' point on her own world line must undergo some acceleration, sometime during the trip." That is, such a time traveler could not just fall freely through spacetime at a high speed (such a trajectory is called a *geodesic*), but rather would have to experience some force. It was on this point, in fact, that Chandrasekhar and Wright (see Note 12 again) went wrong; see also Chandrasekhar (NT). There are other universes, however, in which free-fall geodesic time travel would be possible—see Soares (1980) and Paiva, Reboucas, and Tiexeira (1987).

14. This isn't to say that there aren't some problems with quantum mechanics, too. In particular, there is much confusion today over what seem to be FTL interactions in certain quantum systems, in direct violation of special relativity. Such interactions, if they really exist, hold out the tantalizing possibility of being able to transfer information into the past (see Chapter Four and Tech Note 7). This is, of course, of special concern to physicists, who are either horrified or fascinated, or both, by such a prospect.

15. In actuality, the collapse may stop short of the singularity. See, for example, Israel (1967) and Donald (1978), which suggest that a collapsing body will rebound after reaching a minimum non-zero volume, but even then it is also suggested that causality may still be violated. The infinity of the singularity may just be the mathematics telling us that general relativity has finally failed. And yet, as Chapter Three explains, one modern view is that it is the occurrence of the singularities at the start, and perhaps at the end, of the universe that forces time to be linear rather than circular. What is meant by a singularity in relativistic gravitational field theory is given an excellent treatment in Geroch (1968), the general conclusion being that it is not obvious what the term means. This essential difficulty with relativistic gravitational field theories does not occur in other field theories, such as electrodynamics, because such theories use spacetime as a background; that is, a singularity in electrodynamics is simply a place where the electromagnetic field is undefined, with the "place" referenced to the background spacetime in which the field is embedded. In gravitational field theory, however, it is the very background spacetime itself that may be singular, and there seems to be nothing more fundamental to use as the reference; that is, what could *spacetime* be embedded in? For an explanation of the term *naked singularity,* see Note 22 for this chapter.

16. Curiously, Schwarzschild's result had been discovered earlier using Newton's theory of gravity, and the result was reported in a letter written in 1783(!) by the Reverend John Michell to the English scientist/recluse Henry Cavendish. Somewhat later, and independently, the great French scientist Pierre-Simon Laplace arrived at the same result, writing that "it is therefore possible that the greatest luminous bodies in the universe are . . . invisible" [see Schaffer (1979)]. Laplace called such objects "dark bodies." The idea of black holes appeared again in twentieth-century technical literature only a few years after Einstein published the general theory. For example, in the general relativity analysis by Anderson (1920), we read "We may remark, though perhaps the assumption is very violent, that if the mass of the sun were concentrated in a sphere of diameter 1.47 kilometeres [this is wrong—see below], the index of refraction near it would become infinitely great, and we should have a very powerful condensing lens, too powerful indeed, for the light emitted by the sun itself would have no velocity at its surface. Thus if, in accordance with the suggestion of Helmholtz, the body of the sun should go on contracting [before nuclear reactions were understood, gravitational contraction was thought to be the source of a star's energy], there will come a time when it will be shrouded in darkness, not because it has no light to emit, but because its gravitational field will become impermeable to light." The very next year the quintessential Victorian physicist, the seventy-year-old Oliver Lodge, showed that he had been thinking along similar lines when he wrote (1921), "If light is subject to gravity, if in any real sense light has weight, it is natural to trace the consequences of such a fact. One of these consequences would be that a sufficiently massive and concentrated body would be able to retain light and prevent its escaping. *And the body need not be a single mass or sun, it might be a stellar system of exceedingly porous character* . . . [my emphasis]. If a mass like that of the sun could be concentrated into a globe about 3 kilometers in radius, such a globe would have [the light-trapping property]; but concentration to that extent is beyond the range of rational attention. The earth would have to be still more squeezed, into a globe 1 centimeter in diameter. But a stellar system—say a super spiral nebula—of aggregate mass equal to 10^{16} suns, say 10^{49} grammes, might have a group radius of 300 parsecs, or 10^{21} centims., with a corresponding average density of 10^{-15} c.g.s., without much light being able to escape from it. This really does not seem an utterly impossible concentration of matter." Thus Oliver Lodge, born in the middle of the nineteenth century nearly thirty years before Einstein's birth, was apparently the first to imagine super-massive black holes with very small average density. A very

nice technical discussion of the historical evolution of Newtonian dark bodies into Einsteinian blackholes, and of the curious reluctance of theoreticians, including Einstein, to take seriously the prediction by general relativity of gravitational collapse, is given in Eisenstaedt (1993). See also Thorne (1994). Pioneering analysis on the gravitational collapse of stars was done by the young Indian astrophysicist Subrahmanyan Chandrasekhar (mentioned in Note 12), who in 1931 combined special relativity and quantum mechanics to show that non-rotating stars above a certain mass (the *Chandrasekhar limit*) will experience a different fate from less massive ones; we shall have more on this story in the Epilogue. Beyond this limit of about 1.4 solar masses, a star cannot simply evolve into a white dwarf, something that before Chandrasekhar had been believed to be the fate of all stars. For this work Chandrasekhar shared the 1983 Nobel prize in physics. Interestingly, about the same time as Chandrasekhar's work, the classic science fiction tale "Sidewise in Time" (Leinster) in the June 1934 *Astounding Stories* expressed similar ideas. This is primarily a parallel-universe story, about which more is said in Chapter Four, but near its conclusion in the obligatory genius-explains-it-all dénouement, we read, "We know that gravity warps space. . . . We can calculate the mass necessary to warp space so that it will close in completely, making a closed universe. . . . We know, for example, that if two gigantic star masses of certain mass were to combine . . . they would simply vanish. But they would not cease to exist. They would merely cease to exist in our space and time." As another character sums it up, "Like crawling into a hole and pulling the hole in after you." The explicit use of the term *black hole* for a region of spacetime operating as a temporal and/or spatial portal can be found in science fiction before Wheeler's use of the term in the technical literature. See, for example, the 1950 story "Typewriter from the Future" (Worth).

17. A non-rotating black hole, which has a point singularity, cannot be used for time travel, but because angular momentum is conserved and because it is most unlikely that any precollapsed star would have exactly zero spin, it is very unlikely that a black hole would not be rotating. Indeed, it is thought most black holes would spin at nearly the maximum rate—the speed of light at the "surface"! A rotating Kerr–Newman hole has a *ring* singularity through which a time traveler can theoretically pass and still avoid the deadly infinity to enter other universes and/or to time-travel in this one. See Weingard (1976b) on *how* to time-travel through this singularity. An analysis similar to that of Chandrasekhar and Wright's (see Note 12), for the special case of free-fall time travel through a Kerr–Newman black hole spinning so fast that its singularity is "naked" (see Note 22), appears in Calvani *et al.* (1978). The popular use of rotating black holes for interstellar space and time travel, as discussed in science fiction—see, for example, "Singularities Make Me Nervous" (Niven)—was at first flatly rejected by Caltech physicists Michael Morris and Kip Thorne (1988), with just one of their objections being a causality-violation paradox, i.e., the grandfather paradox. However, in a later analysis (1988) they, along with their colleague Ulvi Yurtsever, seemed to be more willing to accept the possibility of backward time travel when they reported on their discovery that an advanced civilization might, indeed, be able to construct a machine for backward time travel. As they observed, "Some readers will wish to await a definitive answer from future research as to what the laws of physics prevent and what they permit." The idea that the interior of a black hole might "lead to other places" was originally seen as a way to avoid the singularity problem. That is, the collapse of a star into a black hole would not proceed all the way down to a point. Rather, the collapse would stop before reaching the singularity by gravitationally rebounding (see Note 15 again). This "bounce" would occur after the star was inside its event horizon, so an external observer, watching the initial collapse, would not see the later expansion. This expansion would be outward through the event

horizon radius into a *different region of the universe!* The first paper I know of that developed this astonishing concept is Novikov (1966). When Novikov's work was generalized by De La Cruz and Israel (1967), the authors clearly had a hard time believing this imagery despite their own mathematics, concluding with, "It then appears necessary to believe in the existence of other [regions of the universe] similar to but distinct from ours, which will accommodate the re-expansion. This seems at least as fantastic as the alternative of irreversible collapse to virtually point-like dimensions." See also Frolov *et al.* (1989). Even more majestic would be to have the *entire universe* collapse, as is allowed in some cosmological models, and this ultimate disaster can be found in early pulp science fiction, such as "The Tides of Time" (Williams). In that 1940 story, humanity escapes annihilation by crossing a "dimensional bridge" into another universe, just as discussed by some modern-day physicists. See, for example, John Cramer's novel *Einstein's Bridge.*

18. In 1974 Stephen Hawking announced (1974, 1975) an astonishing partial connection of quantum mechanics with general relativity's black holes. Hawking showed that, contrary to the usual image of black holes as being one-way trap doors to . . .?, black holes actually *must* radiate energy. His analysis, which stunned physicists by its beautifully simple arguments, invokes Werner Heisenberg's uncertainty principle, one of the cornerstones of quantum mechanics. Hawking himself found the result "greatly surprising." He also cautioned (in his 1975 paper) that the following picturesque imagery is "heuristic only and should not be taken too literally," but it has now been in physics for over twenty years and appears to be here to stay. The uncertainty principle states that there are certain pairs of variables associated with particles, variables that cannot simultaneously be measured exactly. Time and energy are such a pair; that is, a non-zero time interval is required to measure a particle's energy, and the product of the uncertainty in both the interval and the energy must be equal to or greater than a certain non-zero constant. That is, if h is Planck's constant, then $\Delta \bar{E} \Delta t \sim h$. This allows the process of *virtual particle creation,* the appearance of particle/anti-particle pairs just outside the surface of a black hole. The uncertainty in the energy is what gives the combined mass of the particles; this uncertainty in the energy is the quantum fluctuation energy of the intense gravity field around the hole. The only constraint is that the energy be returned to the field, via mutual annihilation of the matter/antimatter pair, within the time uncertainty dictated by Heisenberg. But, as Hawking showed, this time interval, although incredibly short, is still long enough for the two virtual particles to separate before annihilation, one falling into the hole and the other escaping. Indeed, in Tryon (1973) it is argued that the *entire* universe might have been created by this process, out of nothing, a so-called *vacuum-fluctuation.* The explanation for why the universe does not then disappear—and *very* quickly indeed, because the energy for all the mass in the universe is quite large—is that the *negative* energy in the gravitational fields of the newly created matter cancels the positive mass-energy and so the total energy is zero, just as it was *before* the fluctuation. $\Delta E = 0$, so $\Delta t \approx \infty$ is allowed by $\Delta \bar{E} \Delta t \sim h$. As Tryon whimsically put it, "I offer the modest proposal that our Universe is simply one of those things which happen from time to time." [Gott and Li (1998) have taken this one step further by suggesting that the universe, via time travel, may have caused itself. As they put it, "the laws of physics may allow the Universe to be its own mother."] Hawking radiation has the interesting side effect of tending to neutralize the electric charge of an evaporating black hole. That is, if the particle/antiparticle pair is an electron/positron, then a negatively (positively) charged black hole will tend to attract the positron (electron), in either case driving the net electric charge of the hole toward zero. By this incredible quantum process, then, the black holes of general relativity slowly evaporate as they glow with what is now called *Hawking radiation.* That is, black holes appear to be hot bodies. But *hot* is

relative; as Hawking shows, a black hole with the mass of the Sun acts as a body with a temperature just sixty *nano*degrees Kelvin above absolute zero! At the other end of the spectrum, so to speak, are the micro-black holes with a mass of "only" a billion tons or so and a Schwarzchild radius of about 10^{-16} meter. The temperature of such a micro-hole is well over one hundred *billion* degrees, and it radiates energy at the rate of six thousand megawatts, the output of a good-sized nuclear power plant with a wide-open throttle! By the time this "tiny" black hole is down to its last thousand tons, its temperature has risen to 10^{17} degrees and it has just 0.1 second to go before it is completely gone. Star-sized black holes take a long time to reach this final state, however; a black hole with the mass of the sun will last 10^{66} years. And even more interesting for our purposes in this book, Hawking (1977) showed how the particle *entering* the hole could alternatively be thought of as an *emitted* particle traveling backward in time. The uncertainty principle has long been used in time travel science fiction. In the 1951 story "Ambition" (Bade), for example, a character is transported from 1950 to 2634 by a scientist of the future. Once there, this character decides he'd like to remain permanently in the twenty-seventh century. He is told he can't because he is like an atom excited into an elevated energy state. And just as quantum mechanics predicts that eventually an electron in such an atom will drop back down into a lower state, so do the "laws of time travel" require that he drop back to his normal time. How long he can remain in the future, he is told, "depends on the mass [energy] of his body and the number of years the mass [energy] is displaced." This is simply the uncertainty principle. Later, this quantum process was found to be useful in stabilizing hyperspace wormholes (see Chapter Two), which can be used, in principle, to build time machines (see Tech Note 9). For a quite readable account of the search for a theory of quantum gravity, see Isham (NP).

19. See also Starobinsky (1980). One paper, Birrell and Davies (1978), however, asserts that even without detailed knowledge of quantum gravity, quantum effects "would smash the idealized interior geometry" of a rotating, electrically charged black hole, thereby eliminating any possibility of using such a hole for time traveling, and Morris and Thorne (1988) support this conclusion. But, more recently, Mellor and Moss (1990) disagreed. Details of the supposed time-traveling properties of spinning black holes can be found in many of the books and papers on them, such as Weingard (1979a), the excellent presentations in Kaufmann (1977) and Shapiro and Teukolsky (1983), and Parker (1991). However, because black holes are bizarre objects—nearly as bizarre as time travel—it seems risky to try to understand one in terms of the other. Whether or not black holes actually exist is now taken by theoreticians as a crucial test of general relativity under conditions of strong gravity. (All of the famous early tests of general relatively, such as the advance of the perihelion of Mercury's orbit, are for weak gravity, on the order of a million or more times weaker than one would find at the spacetime horizon of a small black hole.) If general relativity fails the black hole test, then, of course, all of its predictions under extreme conditions, including time travel, become suspect. To date, no black holes have been directly observed, but there are three excellent indirect candidates, each as one half of an X-ray binary system. The first, nearest, and best known is Cygnus X-1, discovered in 1971 at a distance of ten thousand light-years from Earth, with a suspected black-hole component of sixteen solar masses. There is also speculation that any large number of closely spaced stars, such as occurs at all galactic cores, would almost certainly form a black hole with an average density of something like that of water. (The required average mass density of a black hole decreases with increasing hole size. Thus, if the radius were big enough, a black hole could have the density of air!) Galactic-core holes would be very massive black holes. The one at the center of the Milky Way, for example, may have the mass of five million Suns, whereas the core of Andromeda (two mil-

lion light-years distant) may be home to a black hole with a mass of fifty million Suns. Dwarfing this is the suspected black hole at the center of the giant radio galaxy M87, with a mass of five billion Suns! John Wheeler calls such monsters *Nimmersatt* (German for "glutton"). The more matter they take in, the bigger they get; the bigger they get, the more matter they absorb. Such behemoths would literally "eat stars." Jonathan Swift caught the spirit of this image in his poem "On Time" in *Riddles* (circa 1724). Written as his view on the destructive nature of time, his words apply equally well to Wheeler's Nimmersatt: "Ever eating, never cloying/ All-devouring, all-destroying/ Never finding full repast/ Till I eat the world at last."

20. A few years before Tipler, Brandon Carter (1968) used the term *time machine* in his analysis of the Kerr solution. As noted in the text, Carter was not nearly so willing to entertain time machines as was Tipler, but he did write that under certain conditions the solution "has the properties of a time machine" and that "it is possible, starting from any point in the outer regions of space far away from a rotating mass, to travel toward the mass and move backward in time as far as desired." However, neither Carter nor Tipler was the first to describe the details of how to make a time machine if we count non-technical publications. This was apparently done in 1899 by the notorious French novelist, poet, and artist Alfred Jarry (1873–1907) in the literary journal *Mercure de France*. See Jarry (1965) for a reprint and history of this intriguing essay and Henderson (1983) for a discussion of Jarry's interest in time travel. Jarry described his time machine via an unusual analogy with a gyroscope—an analogy repeated decades later in the 1976 story "Kelly Country" (Chandler). As Doctor Siebert explains his time machine in that tale, "Imagine a spinning gyroscope. You press down on one end of the axis and it resists the downward pressure. But it does move. It precesses, swings to one side at right angles to the applied force, in the direction of rotation. . . . The rotors of my machine precess, but not through any of the dimensions of normal space. But they precess, nevertheless, within the Space-Time continuum. . . . Temporal precession." Jarry's essay was brought to the attention of the well-known British physicist William Crookes, who then wrote to Oliver Lodge about it (I thank the Society for Psychical Research in London, to which both Crookes and Lodge belonged, for providing me with a photocopy of this interesting letter, dated July 7, 1899). Crookes seems to have thought Jarry was serious about how to build a time machine, even though Jarry signed his essay as "Doctor Faustroll," which is a dead give-away to the true poetic intent of the essay! Obviously tongue-in-cheek on the same topic is Brunner (1965). A more recent time machine proposal, which has a certain rugged charm to it, graces Platt (1974), printed the same year as Tipler's paper. Two years later, in 1976, Tipler's Ph.D. dissertation (1976b) was accepted, and in it he was even more direct about his interest in time machines than he was in his 1974 paper. This isn't to say that other physicists hadn't seriously talked about backward time travel before Tipler. Richard Feynman, for one, did in his Nobel prize lecture, "The Development of the Space-Time View of Quantum Electrodynamics," which is reproduced in *Science* 153 (August 12, 1966):699–708. Feynman, who was John Wheeler's student, recalled how Wheeler called him up one day with his "proof" for why every electron in the universe has exactly the same charge (using the argument that there is only *one* electron, weaving back and forth in time), with positrons being the backward-traveling electrons. Feynman then said, "I did not take the idea that all electrons were the same one . . . as seriously as I took the observation that positrons could simply be represented as electrons going from the future to the past. . . . That, I stole!" Indeed, he wrote (1948) that "this idea that positrons might be electrons with the proper time reversed was suggested to me by Professor J.A. Wheeler in 1941." In an astonishing coincidence, even as Feynman and Wheeler were talking, two science fiction writers also came up with the same idea.

Writing as "Will Stewart," Jack Williamson and John W. Campbell (the editor of *Astounding Science Fiction*) identified anti-matter with backward time travel in "Minus Sign." Later, Stanislaw Lem took this idea, combined it with the concept of energy fluctuation, and in "The Eighteenth Voyage" came up with one of his typically outrageous ideas—shooting a single positron out of an accelerator back to the very beginning of time! He called this fantastic machine the "Chronocannon" and claimed that's what started the universe. Soon after Lem, the philosopher Fulmer (1983) used a variant of this idea (in which the Big Bang creation of the universe was caused by a time traveler from the future who saw a need—his own existence—to generate the Big Bang) to speculate on the cosmological implications of God as a time traveler. Isaac Asimov used a variation of this idea in his 1989 story "The Instability."

21. The restriction does prevent at least one particularly odd paradox from happening: a traveler going backward in time to tell the inventor of the time machine (perhaps an earlier version of the time traveler himself) how to build the time machine. A story using this idea is "Package Deal" (Franson), and a minor variant of it is amusingly illustrated in the 1985 film *Star Trek IV: The Voyage Home* (when you next watch the movie, ask yourself who actually invented "transparent aluminum"?) This paradox, which appeared as long ago as 1929 in "Paradox" (Cloukey), belongs to a class of puzzles called *causal loops;* such loops are very mysterious and still befuddle philosophers. They are discussed in greater detail in Chapters Three and Four. One could use a Tipler cylinder to travel forward in time, too; one would just orbit around it in the opposite sense. Again, one could not travel into the future beyond the time the cylinder ceases to exist. Science fiction writer Oliver Saari anticipated this point as long ago as 1937 in his story "The Time Bender." In an astonishing coincidence with Tipler's incredible cylinders, Saari uses a time machine that works by warping spacetime via a plate of superdense material. As the story explains, the time traveler "could not travel into the past, for the plate had to exist in all ages traveled, and it had not existed before he had made it." An even earlier story, "The Time Deflector" (Rementer), also makes this argument, but in that tale the means of time travel is via the hocus-pocus of "rays" emitted by a newly discovered element called *tempium*. Kip Thorne (1991) has rebutted Hawking's experimental evidence for his *Chronology Protection Conjecture* by using Saari's kind of argument. Thorne indicates that (1) time machines, if possible, must have the property of not being able to travel back before their creation (as in the case of a Tipler cylinder), and (2) no time machine has yet been created. "Azimuth 1,23,3 . . ." (Knight) repeats Saari's argument in a science fiction setting. Published in 1982, this story therefore also anticipated Thorne's rebuttal. More on the Chronology Protection Conjecture can be found in Tech Note 9. In the popular press, Robert Forward (1980) wrote with enthusiasm about Tipler cylinders, for which he was politely but firmly rebuked in Rothman (1985) on the technical issues of self-gravitational collapse. When Forward (1988) later wrote again on the subject, he ignored the criticism, leaving the technical objection both unmentioned and unanswered.

22. A strange theoretical possibility is that at such incredible matter densities, the propagation speed of low-frequency sound waves (*density fluctuations*) might exceed the speed of light—see Kirzhnits and Polyachenko (1964), Cutkosky (1970), and Bludman and Ruderman (1968, 1970). An excellent book on the details of how matter crushes (that is, on the so-called *equations of state* that relate the internal pressure response of matter to the compression of gravity), along with some fascinating history of the subject, is Harrison, *et al.* (1965). More on this can also be found in Thorne (1965a) and in Thorne (MWM), which presents the curious result that there cannot be a spacetime horizon around a gravitationally collapsed

non-rotating infinite cylinder; in this physically implausible situation, it seems that the threadlike singularity would be visible to a remote observer. As mentioned in the text, such a singularity is said to be *naked*. This particularly unphysical geometry has continued to fascinate Thorne for decades. Most recently, one of his graduate students at Caltech studied the formation of a naked singularity by the collapse of an infinite, non-rotating cylinder in great detail, both analytically and by computer simulation, i.e., Echeverria (1993). Many physicists believe a naked singularity cannot actually happen in any real, physical case and declare its impossibility via a metaphysical law called *cosmic censorship*, put forth in 1969 by Roger Penrose—see Wald (1974) and Kosso (1988). It is interesting to note, however, that several years after stating his censorship law, Penrose still wrote (1972), "The possibility that naked singularities may sometimes arise must be considered seriously." The very next year the theoretical discovery of a spinning naked singularity was reported in Gibbons and Russell-Clark (1973) (which the authors then dismissed as having limited "astrophysical application" because of associated closed timelike curves with "pathological" consequences, i.e., time travel). Thorne, however, conjectures that a naked singularity might form in a sufficiently aspherical collapse of a finite mass even if rotating, with the infinite non-rotating cylinder simply being one special limiting case; see Apostolatos and Thorne (1992). Shapiro and Teukolsky (1991a,b), who describe computer simulations of the general relativistic collapse of prolate spheroids of gas, have seemingly confirmed Thorne's idea. See also Clarke and de Felice (1982), Christodoulou (1984), and Gott (1991). An interesting "explanation" for the naked singularities that have been discovered (on paper or in computer simulations) in spherically symmetric collapses is that they are *massless*—see Lake (1992). This means that they do not gravitationally interact with matter or radiation, so in fact they do not violate cosmic censorship (because they have no causal influence on anything). The situation for non-spherical gravitational collapse is still far from clear. Wald (1992), however, is simply skeptical of all these simulations, arguing that their results may be nothing but artifacts of the computer codes. Finally, a form of singularity that is perhaps even stranger than the naked is the so-called *thunderbolt*. This relatively recent discovery arises under certain conditions in the mathematics of evaporating black holes. In the words of Hawking and Stewart (1993), the thunderbolt "is a singularity that spreads out to infinity on a space-like or null path [i.e., at least at the speed of light—see Tech Note 4]. It is not a naked singularity because you do not see it coming until it hits you and wipes you out." As bizarre as this is, the thunderbolt was anticipated in science fiction by more than half a century. In the 1940 story "The Tides of Time" (Williams), the universe is collapsing at faster than the speed of light. Human scientists learn this when fleeing aliens stop their faster-than-light space ships to warn them. One of the human characters then looks out into the night sky and, in words that sound like those of Hawking and Stewart, realizes that "There would be no warning, for the rolling tide was traveling faster than light. . . . It would come faster than the flicker of an eye. No one would see it come. One instant the world you knew would be around you. The next instant, there would be nothing. You would not even have time to know what had happened. Death, faster than the lightning flash!"

23. It should be mentioned that Tipler was not the first to study rotating cylinders in the context of general relativity. Such cylinders had been around for decades, as Tipler himself notes, and a good reference is M. A. Mashkour, "An Exterior Solution of the Einstein Field Equations for a Rotating Infinite Cylinder," *International Journal of Theoretical Physics* 15 (October 1976):717–721. This paper contains citations to the earlier literature on rotating infinite cylinders dating back to 1932. The first-analyzed configuration of matter that generates closed timelike lines, solved in all its general relativistic detail, was in fact the infinite rotating cylinder

studied by Van Stockum (1937). This is particularly interesting because Van Stockum's cylinder is made of ordinary, not exotic, matter. Van Stockum himself, however, did not spot the presence of closed timelike lines in his solution. The details of Van Stockum's solution are discussed by Bonner (1980).

24. More recently Niven mentioned a Tipler cylinder in a very funny story entitled "The Return of William Proxmire," which tells about how former Senator William Proxmire (whose claim to fame was his supposed outrage at "silly government spending"—except when the spending was done in his home state) might use a time machine to undo the expensive American space program.

25. To be absolutely sure I don't put my words in Tipler's mouth, I should mention a second paper of his (1976a) in which he somewhat more cautiously wrote, "There are many solutions to the Einstein equations which possess causal anomalies in the form of closed timelike lines (CTL). It is of interest to discover if our Universe could have such lines. In particular, if the Universe does not at present contain such lines, is it possible for human beings to manipulate matter so as to create them? I shall show in this paper that it is *not* [Tipler's emphasis] possible to manufacture a CTL-containing region without the formation of naked singularities, provided normal matter is used in the construction attempt."

Chapter Two

1. Dummett (1969) has a very nice way of systematically generating this infinite regress of complex predicates: "Let us call 'past,' 'present,' and 'future' 'predicates of first level.' If, as McTaggart suggests, we render 'was future' as 'future in the past,' and so forth, then we have the nine predicates of second level, where we join any of the three on the left with any of the three on the right:

past		past
present	in the	present
future		future

Similarly there are twenty-seven predicates of third level. . . ." Indeed, this construction clearly shows that at the Nth level there are 3^N predicates, most of which are incompatible.

2. Quoted from B. Hoffmann, *Albert Einstein: Creator & rebel* (New York: New American Library, 1972), 257–258.

3. An abridgement of Wells' thesis can be found in *Nature* 153 (April 1, 1944):395–397. The miserably brief duration of human history is nicely demonstrated in the well-known metaphor of the "cosmic year." If we imagine that the entire history of the universe—about fifteen billion years—is compressed into just one year, and that the Cosmic Clock has just struck midnight on December 31 to ring in the second Cosmic Year, then dinosaurs were walking the Earth until the middle of yesterday and Jesus died on the Cross four seconds ago.

4. See Dyson (1979) for fascinating analyses of how physics predicts the way the evolution of an open universe would go. Such a universe does not eventually collapse into a Big Crunch, and thus it would have an infinite future. As Dyson writes, "The main conclusion I wish to draw from my analysis is the following: So far as we can imagine into the future, things continue to happen. In an open cosmology, history has no end." Dyson, who assumes absolute stability for the proton, wrote his paper partly in response to a paper co-authored by time machine pioneer Frank Tipler—see Barrow and Tipler (1978). In that paper Barrow and Tipler, who assume that electrons and protons eventually decay into radiation, are far less sanguine about the future than is Dyson. See also Page and McKee (1981).

5. J. Swift, *Gulliver's Travels* (New York: Heritage Press, 1940), 22–23.

6. An example of this attitude appears in the excellent essay by N.J.A. Sloane, "The Packing of Spheres," *Scientific American,* January 1984. This paper, which addresses such problems as how to pack spheres most tightly in spaces of dimensions up to 100,000 (the solution to this problem has enormous practical value in the design of electronic communications systems), says (correctly) that "there has been a great deal of nonsense written in science fiction and elsewhere about the mysteries of the fourth dimension." The author is a mathematician who takes a particularly skeptical view of assuming "as the physicist does, that the fourth dimension represents time."

7. The boring repetition of masculine heroes saving weeping women from ungodly horrors was not always appreciated by the male readers of the science fiction magazines. As one wrote in a letter to the editor of *Astounding Stories* (June 1931), "Just why do you permit your Authors to inject messy love affairs into otherwise excellent imaginative fiction? Just stop and think. Our young hero-scientist builds himself a space flyer, steps out into the great void, conquers a thousand and one perils on his voyage and amidst our silent cheers lands on some far distant planet. Then what does he do? He falls in love with a maiden—or it's usually a princess—of the planet to which the Reader has followed him, eagerly awaiting and hoping to share each new thrill attached to his gigantic flight. But after that it becomes merely a hopeless, doddering love affair ending by his returning to Earth with his fair one by his side. Can you grasp that—a one-armed driver of a space-flyer! . . . We buy A.S. for the thrill of being changed in size, in time, in dimension . . . not to read of love . . . I wish . . . for plain, cold scientific stories sans the fair sex."

8. See "Like a Bird, Like a Fish" (Hickey), for example, for what happens to time-traveling monsters from another dimension when they run afoul of a rural mechanical genius who fixes their machine after it breaks down. In my own attempt at this subgenre, "Twisters," I had a small-town doctor defeat (sort of) the insidious attack of other-world invaders who were using four-dimensional doughnuts as combination "bait and trap" devices to capture humans for their gourmet food shops back home. I am willing to admit that the only possible redeeming value in my story is that its description of a Klein bottle (a four-dimensional object of central importance to topologists) might stimulate young readers to look elsewhere for more information (see also Note 20 for this chapter). Long after I wrote my story, I discovered that British writer John Collier (1931) had used the idea more than fifty years before, in his *No Traveller Returns.* In that tale a Professor of Quantum Theory stumbles through a temporal gate, and it is as though he has trod upon a "mathematical banana skin." "After an eternal instant of slip and twist," the Professor finds himself in the far future of Earth, where the inhabitants eat their own ancestors that have been caught in the time trap. This story also has what reads like an early hint at a wormhole in fiction. As the Professor is told, "Time doubles and redoubles upon itself. . . . It happens that that point in time at which we are now arrived is, in another dimension, closely adjacent to the point you recently left. It was thus a very simple matter for us to *burrow through* [my emphasis] and set up a little trap, which has enabled us to supply the atavistic members of our community . . . with a little fresh butcher's meat." Even earlier was the time tunnel that appeared the year before in "In 20,000 A.D." by Nat Schachner and Arthur Zagat. In that 1930 tale from *Wonder Stories* there is a pipeline to the future that exists in a stand of trees; "the trees were acting funny. Instead of standing up tall and straight [the trees leaned] way over in a sort of double curve. There was a path in between, and on each side the trees leaned away from it, like . . . it was a funnel."

9. G.F. Rodwell, "On Space of Four Dimensions" *Nature* 8 (May 1, 1873):8–9.

10. P.G. Tait, "Zollner's Scientific Papers," *Nature* 17 (March 28, 1878):420–422. Thus Clifford Ashley's 1944 classic book on knots (advertised as "7,000 drawings

for 3,900 knots") would be as meaningful to a four-dimensional being as a book on unicorns. Three years before, however, Tait had suggested (along with fellow Scottish physicist Balfour Stewart), in their book *The Unseen Universe,* that a human soul *is* a four-dimensional knot in the ether. James Clerk Maxwell, Tait's friend since childhood (as well as one of the greatest theoretical physicists of all time), took a decidedly skeptical view of such speculations in his *Encyclopaedia Brittanica* essay "Ether." As for Zollner (1834–1882), he was a well-known and somewhat unconventional professor of physical astronomy at the University of Leipzig. After meeting William Crookes in 1875, he became intensely interested in Spiritualism, as well as being convinced that a fourth dimension to space would provide a scientific explanation for certain psychic phenomena. As examples of such phenomena, Zollner (1878) used the claims of a then-famous medium who was later convicted of fraud. Because of the medium's fall from grace, Zollner's own reputation fell, and for a while both his name and the fourth dimension became identified in the popular mind with deceit. For more on this, see the amusing tale "The Church of the Fourth Dimension" by Martin Gardner. In 1893 Ambrose Bierce explicitly used the scientific idea of non-Euclidean space in his "Mysterious Disappearances," in a context that earlier times would have called supernatural.

11. I.M. Freeman, "Why Is Space Three-Dimensional?" *American Journal of Physics,* 37 (December 1969):1222–1224, and L. Gurevich and V. Mostepanenko, "On the Existence of Atoms in *n*-Dimensional Space," *Physics Letters A* 35 (May 31, 1971):201–202. See also J.D. Barrow, "Dimensionality," *Philosophical Transactions of the Royal Society of London A* 310 (December 20, 1983):337–346. A more philosophical presentation is G. J. Whithrow, "Why Physical Space Has Three Dimensions," *The British Journal for the Philosophy of Science* 6 (May 1955):13–31. See also Abramenko (1958) and Good (1965) for what Good himself describes as a "partly-baked idea" on how spacetime might be eight-dimensional. Less speculative arguments for a four-dimensional spacetime, based on field equations of equal mathematical strength, can be found in R. Penny, "On the Dimensionality of the Real World," *Journal of Mathematical Physics* 6 (November 1965):1607–1611. Penny admits that some may find the strength criterion (which we owe to Einstein himself) to be "slightly metaphysical." A radically different argument is presented in Rosen (1968), where it is shown that only in a spacetime with an even number of dimensions does the TCP theorem hold. This theorem says that the "mirror-image" of a physical process is a legitimate process if the "mirror" reverses time (T), electric charge (C) (so that particle and anti-particle are interchanged), and parity (P) (the reversal of left and right that one observes when looking in a mirror); see E.P. Wigner, "Violations of Symmetry in Physics," *Scientific American* (December 1965). If we let the symbol * denote *reversal,* then under successive P*, C*, and T* operations, the TCP theorem says that if the particle interaction on the left of Figure R2.1 is legitimate, then so is the interaction on the right. This example shows that the 1948 experimental observation of the beta decay of a free neutron into a proton and an anti-neutrino (decay by the ejection of an electron) immediately implied that the creation of an anti-neutron by a neutrino, a positron, and an anti-proton must also be possible. There is strong reason to believe the TCP theorem, because quantum field theory is compatible with special relativity only if the TCP theorem holds. A mathematical proof of the theorem is given in the essay by W. Pauli, "Exclusion Principle, Lorentz Group and Reflection of Space-Time and Charge," *Niels Bohr and the Development of Physics* (New York: McGraw-Hill, 1955), 30–51. For a more recent philosophical discussion, including references to the newer physical literature (including the possibility of a fractal dimensionality to spacetime), see F. Caruso and R.M. Xavier, "On the Physical Problem of Spatial Dimensions: An Alternative Procedure to Stability Arguments,"

Fundamenta Scientiae 8, 1 (1987):73–91. Most recently, Bechhoefer and Chabrier (1993) have argued that only in three (extended) spatial dimensions can there be a non-zero Chandrasekhar limit for stars (see Note 16 for Chapter One). That is, in spaces with four or more spatial dimensions, *all* stars either collapse into black holes or disperse. Only in a universe like our four-dimensional one with three spatial dimensions can there be the diversity of white dwarfs, neutron stars, and massive planets such as Jupiter. Taking a different approach is the paper by Lämmerzahl and Macias (1993). Rather than extending particular three-dimensional theories into higher dimensions to see whether they then give rise to non-observed effects, the authors assume only a small number of fundamental *principles* that are independent of both the dimensionality and the geometry of spacetime. Their conclusion is, as before, that our world is necessarily four-dimensional.

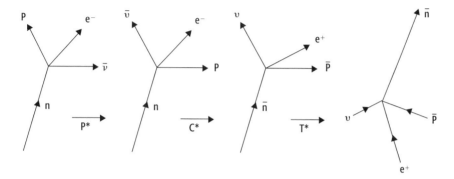

Figure R2.1   The TCP theorem predicts the anti-neutron.

12. K. Pearson, "Ether Squirts," *American Journal of Mathematics* 13 (1891):309–362. The luminiferous ether was a mysterious substance thought to fill all space; its only role was to give light something through which to move. The response came from Arthur Schuster in *Nature* 58 (August 18, 1898):367 who claimed in a subsequent letter to *Nature* not to have been aware of Pearson's paper when he first wrote. Pearson certainly came to his views on the fourth dimension at least partly through the influence of the work by William Kingdon Clifford (1845–1879), who anticipated Einstein's curved space by decades (see Tech Note 4). Pearson edited Clifford's lone book (*The Common Sense of the Exact Sciences*), published posthumously in 1885, in which Clifford wrote "We may conceive our space to have everywhere a nearly uniform curvature, but that slight variations of the curvature may occur from point to point and themselves vary with time. These variations of the curvature with time may produce effects which we not unnaturally attribute to physical causes independent of the geometry of our space. We may even go as far as to assign to this variation of the curvature what really happens in that phenomenon which we term the motion of matter." And as an example of how (perhaps) there is something (more or less) in nearly all speculations, see the paper on spacetime wormholes by Ellis (1973), which has a title that Pearson would have loved. Ellis' paper, in fact, anticipated part of the wormhole analyses in the 1988 papers by Morris, Thorne, and Yurtsever (see Tech Note 9).

13. A very nice treatment of Hinton's life and work appears in *Speculations on the Fourth Dimension: Selected writings of Charles H. Hinton,* ed. R. Rucker (New York: Dover, 1980). All of Hinton's important 4-space essays are reprinted in that book.

14. *Nature* 31 (March 12, 1885):431.

15. This story appeared in 1931, just a few years after the German physicist Theodor Kaluza had expanded four-dimensional spacetime to five dimensions. He did this in a successful attempt to unify gravity and electromagnetism. A brief, readable exposition of this theory can be found in H. T. Flint, *The Quantum Equation and the Theory of Fields* (New York: John Wiley, 1966). Perhaps Kaluza's work was Leinster's inspiration. The story was liked well enough by science fiction fans to encourage the writing of a sequel, "The Fifth-Dimension Tube." Soon thereafter, in 1936, *Astounding Stories* published "Infinity Zero" (Wandrei), in which an explosion at a chemical plant (it "contained all the elements of the universe") ends up destroying the world. As a mathematician explains to a reporter, "At the moment of explosion, the whole universe was concentrated [in the plant] in miniature, thus creating the new, fourth dimension of ultra-space. And since ultra-space cannot exist in our universe, the ultra-space is . . . blotting out . . . the three space dimensions and the time factor." With the "new, fourth dimension of ultra-space," then, Wandrei seems to have a total of five dimensions in mind. Kaluza's five-dimensional spacetime, in which gravity also appears as electromagnetism in our apparent world of four dimensions, leaves open the question of where the extra dimension is. For that reason, as Dobbs (1956) observes, many physicists came to regard five-dimensional spacetime as "an artful dodge having no physical significance." In 1926, five years after Kaluza, the Swedish physicist Oscar Klein put forth an astonishing speculation of where Kaluza's fifth dimension might be. The eminent theoretical physicist Bryce S. DeWitt has commented on this ("Quantum Gravity," *Scientific American,* December 1983): "The reason space *appears* to be three-dimensional is that one of its dimensions is cylindrical . . . the circumference of the universe in the cylindrical direction is only a few (perhaps 10 or 100) Planck lengths. As a result an observer who attempts to penetrate the fourth spatial dimension is almost instantly back where he started. . . . The fourth spatial dimension is simply unobservable as such." That is, the fifth dimension is periodic, and a particle moving along just this dimension would follow an incredibly tiny circular path. More on this can be found in Chodos and Detweiler (1980), Chyba (1985), Davidson and Owen (1986), and Davidson (1987). Dobbs, however, suggested that the fifth dimension is a second time dimension, and this suggestion is repeated in the physics literature by Sakharov (1984). The idea of a unified field theory quickly entered science fiction. In "The Meteor Girl" (Williamson), published in 1931, the hero-genius is asked, "What's the good of Einstein, anyhow?" and before he can reply, his interrogator goes on. "I know that space is curved, that there is really no space or time, but only space-time, that electricity and gravitation and magnetism are all the same. But how is that going to pay my grocery bill—or yours?" (We must remember that these were the early days of the Great Depression.)

16. The analogy involves a comparison with a prison in 2-space *flatland,* which would merely be a circle around the captive. Knowledge of the third dimension would make it possible to escape by moving along that new direction, over the circle, and then back into the plane. To a flatlander guard it would seem that the prisoner suddenly vanished from view inside the circle and just as suddenly materialized again outside it. Similarly, to escape from a 3-space prison, one would merely move along the fourth dimension. In the same way, one could remove the yolk from an Easter egg without damaging either the shell or the white; indeed, one could remove the yolk directly from the chicken without damaging the chicken. In fiction, this way of using the fourth dimension appears in "The Appendix and the Spectales" (Breuer), where it is used to perform a wound-free appendectomy. The concept was mentioned even earlier, in passing, in the previous year's 1927 story "The Four-Dimensional Roller-Press" (Olsen); see also his "Hyper-Forceps," which were invented in "Four Dimensional Surgery" and later used in "The Great Four Dimensional Robberies" to rob locked safe deposit boxes. Still later, in "The Four

Dimensional Escape," a man sentenced to die by hanging at San Quentin Prison is rescued while standing on the gallows' trap by an inventor who pulls him through the fourth dimension! Such ideas are far from new. The German mathematician and astronomer August Möbius (1790–1868) discovered in 1827 that any three-dimensional object can be converted into its mirror image by flipping it over through the fourth dimension; for instance, a left-handed glove can thus be made by pure geometry (no scissors, thread, or needle required) into a precise copy of its right-handed mate. Later (but still in the nineteenth century), in his Presidential Address to the American Mathematical Society, Simon Newcomb made this very point. This idea has been used in fiction; see "Left or Right?" (Gardner) as well as "The Plattner Story" (Wells). Both of these stories, which involve flipping living organisms (humans) over through the fourth dimension, have a literally fatal flaw. Everything in the body would be reversed, including the optically active organic molecules discovered by Pasteur in 1848, which are involved in vital biological processes. These molecules, called *stereoisomers,* exist in two versions in nature (the left-handed and right-handed versions, if you will), but our bodies have developed the biological ability to use only one version. To be flipped through the fourth dimension would, for example, make some reversed stereoisomers unable to participate in the digesting of food and we would starve to death. This problem was first fictionally addressed in "Technical Error" (Clarke), published in 1946, where the solution is to flip the victim through the fourth dimension a second time. (When *Thrilling Wonder Stories* reprinted this story in June 1950, the title was changed to the more appropriate "The Reversed Man," but for some reason the editorial lead-in spoke of the fifth dimension!) Later, "The Heart on the Other Side" (Gamow) repeated the idea. There is more to worry about with dimensional flipping than eating, however. Indeed, because everything flips, matter should become anti-matter, and Lewis Carroll was sharper than he could have imagined when he had Alice wonder in *Through the Looking-Glass,* "Perhaps Looking-glass milk isn't good to drink." Unless Alice is an anti-Alice, she and the anti-milk would mutually become pure energy via a spectacular 100% efficient explosion.

17. This surface, discovered in 1858 by Möbius, is easily made by taking a long strip of paper, giving it a half-twist, and taping the ends together. You can convince yourself that it is one-sided by coloring it with a crayon. During the coloring, *do not lift the crayon from the paper.* When you can color no more, you will find that every last bit of the paper is the same color. You cannot do this with a flat piece of paper without lifting the crayon and turning the paper over because that paper is two-sided. A hilarious story based on this property is "A. Botts and the Möbius Strip" (Upson), published in 1945. A similar idea appeared earlier in "The Geometrics of Johnny Day" (Bond). As these stories show, most people refuse to believe such an object actually exists until they see a demonstration: Even our ordinary three-dimensional space has some surprises left in store. Indeed, Sloane (see Note 6 again) makes this very point when he states that "all physicists know" what the maximally dense sphere packing in 3-space is, but nobody can prove it! This indicates that "the mathematical understanding of ordinary, three-dimensional Euclidean space is far from complete." I have heard of at least one practical use for the Möbius strip—as a conveyor belt in mines. The belt is one-sided, so it lasts twice as long as a normal two-sided belt because every bit of its total surface gets worn by the rocks, instead of just half. This idea has been used in fiction, too, as in "Paul Bunyan versus the Conveyor Belt" (Upson), but with, if I may use the word, a twist. This charming tale introduces the reader to one of many astonishing properties of the Möbius band. Cut the band lengthwise with a scissors. Most people believe you will get two strips, each the length of the original strip but each only half as wide. Actually, you get one strip with a full twist, which means that the strip

is a two-sided surface. And if you cut the new strip lengthwise once more, you get two separate loops, linked together. Try it and see. But be careful. Science fiction writer Cyril Kornbluth warned of the possible danger of unschooled experimentation in such matters when he wrote

> A burleycue dancer, a pip
> Named Virginia, could peel in a zip;
> But she read science fiction
> And died of constriction
> Attempting a Möbius strip.

18. To get an idea of how Heinlein's four-dimensional house would look, see Farley (1939), which provides a fold-and-paste cut-out pattern to let you make your very own three-dimensional model of a tesseract. Heinlein's story appeared in 1940, so Farley's article in *Scientific American* just the year before may well have been at least part of its inspiration. As the Bibliography indicates, Farley (the pen name of Roger Sherman Hoar) was also a prolific writer of time travel stories.

19. S., "Four-Dimensional Space," *Nature* 31 (March 26, 1885):481. In his 1934 *Experiment in Autobiography,* Wells wrote, "In the universe in which my brain was living in 1879 there was no nonsense about time being space or anything of that sort. There were three dimensions, up and down, fore and aft and right and left, and I never heard of a fourth dimension until 1884 [when Wells was eighteen] or thereabout. Then I thought it was a witticism." He had, in fact, said this before. In a 1931 edition of *The Time Machine* (New York: Random House), for example, the Preface is by Wells himself. Of the idea for the novel, he says, "It was begotten in the writer's mind by students' discussions in the laboratories and debating society of the Royal College of Science in the eighties and already it had been tried over in various forms by him before he made this particular application of it." In the biography by Geoffrey West, *H.G. Wells* (New York: W.W. Norton, 1930), we learn that "the idea for a time machine story had come to him one day . . . during the discussion following the reading of a paper—by another student in a debating society—on the fourth dimension." The best candidate for this other student is given by Bergonzi (1961) as E. A. Hamilton-Gordon; on January 14, 1887, Hamilton-Gordon read a paper on the fourth dimension to the debating society. As to the nature of the fourth dimension, however, time was not the only possibility mentioned. So, too, were life, Heaven, and velocity! The very next year, however, there was the very positive declaration of "a fourth dimension of space, which we call homogeneous time" in the French philosopher Henri Bergson's 1888 doctoral dissertation. But space held on to the fourth dimension for a long time after *The Time Machine*. For example, when Alfred Taylor Schofield's 1888 *Another World* (which suggested that God exists in a hyperspace of four spatial dimensions) was reprinted in 1905, he added a note indicating that he was aware of Wells' use of time as the fourth dimension—but Schofield himself remained unconvinced.

20. The Klein bottle was discovered in 1882 by the German mathematician Felix Klein. It is closely related to the Möbius strip, but unlike the strip, it cannot be made in 3-space. A 3-space bottle has an inside and an outside. This simply means that it has an *edge*—that is, the mouth of the bottle—which separates the two surfaces just as the edge of a piece of paper separates its two sides. A Klein bottle has no such edge, and its inside flows smoothly into its outside. To make such a thing in 3-space requires an intersection in the side of the bottle, but in 4-space there would be no intersection. Yet even though inside and outside are meaningless distinctions for the bottle, it could indeed hold water. And, paradoxical as it may

seem, the Klein bottle can be made from two Möbius strips that *do* exist in 3-space. An anonymous wit put it thus:

> A mathematician named Klein
> Thought the Möbius band was divine.
> Said he: "If you glue
> The edges of two,
> You'll get a weird bottle like mine."

The trick, unfortunately, is that before you apply the glue, you have to twist some edges through the fourth dimension. Never one to miss a chance to slip some mathematics into his children's stories, Lewis Carroll in *Sylvie and Bruno* describes a Klein bottle in everything but name in the guise of the "purse of Fortunatus." Stories involving a Klein bottle are not common, but two are "The Last Magician" (Elliott) and "The Island of Five Colors" (Gardner). As a final comment on Klein, in 1896 (less than a year after publication of *The Time Machine*), he gave a series of lectures at Princeton University. Published the following year as *The Mathematical Theory of the Top,* this book presents a four-dimensional non-Euclidean space with time as the fourth dimension.

21. The Time Traveller is never named. An earlier (1888) attempt at a time machine story with the awful title *The Chronic Argonauts* [reprinted in both Bergonzi (1961) and Wells (1987)] so embarrassed Wells that he later called it "imitative puerile stuff," "clumsily invented, and loaded with irrelevant sham significance," and "inept" and destroyed every copy of it that he could find. The hero is named Dr. Moses Nebogipfel. There is one passage in *The Time Machine* that does tantalize; as the Time Traveller explores a museum of "ancient" artifacts in the Palace of Green Porcelain (they are, of course, artifacts of our *future*), he reveals that "yielding to an irresistible impulse, I wrote my name upon the nose of a steatite monster from South America that particularly took my fancy." Thus the Traveller *has* given his name, but his signature exists in the future in a museum of the past that is yet to be built.

22. And so Newcomb actually was. Wells was a trained scientist (B.Sc. with first-class honors in zoology and second-class in geology in 1890 from the Royal College of Science), and he quite clearly kept up with technical developments. Certainly he read *Nature* (one of his college friends at the Normal School in South Kensington, Richard Gregory, eventually became editor of *Nature,* and one of their teachers, Norman Lockyer, was *Nature's* first editor) and found Newcomb's address on December 28, 1893, to the New York Mathematical Society reprinted there in full [*Nature* 49 (February 1, 1894):325–329]. This address must have struck a responsive chord with Wells, because only weeks after its appearance in *Nature,* he published what would be the opening to *The Time Machine,* containing the reference to Newcomb, as an unsigned essay in the March 17, 1894, issue of the *National Observer,* under the title of "Time Travelling. Possibility or Paradox?" The Time Traveller's dinner party must have taken place, therefore, in January or February of 1894. (It is amusing to note that *Morlock Night* (Jeter), one of the many sequels to *The Time Machine,* gets the date of the dinner party remarkably wrong, placing it in 1892 and so *before* Newcomb's address. *The Time Ships* (Baxter) goes that error one better and puts the dinner party in 1891). Several years later Newcomb expanded on his ideas in his Presidential Address to the American Mathematical Society (1898). He concluded his Address with the words "We must leave it to our posterity to determine whether . . . the hypothesis of hyperspace can be used as an explanation of observed phenomena." It was less than two decades later that Einstein did precisely that, explaining gravity in terms of curved four-dimensional hyperspace (spacetime). Newcomb was no Johnny-come-lately to the

fourth dimension; he had published on the topic, including the concept of curved hyperspace, sixteen years (1877) before his address to the New York Mathematical Society. In his 1893 address he called four-dimensional space "the fairyland of geometry" (he repeated the phrase in his 1898 Presidential remarks, as well) and hinted at the modern idea of parallel universes when he said, "Add a fourth dimension of space, and there is room for an indefinite number of Universes, all alongside of each other, as there is for an indefinite number of sheets of paper when we pile them upon each other." Newcomb's idea appealed to Wells' fancy so much that he built two novels, *The Wonderful Visit* and *Men Like Gods,* around it. In the first novel there is explicit mention of multiple worlds "lying somewhere close together, unsuspecting, as near as page to page in a book," and the second book speaks of one parallel universe being rotated into another. Other versions of the idea are in "The Ship That Turned Aside" (Peyton), "The Invisible Bomber" (Farley), "The Magic Staircase" (Bond), and "Carbon Copy" (Simak). More recently, John Cramer (a University of Washington physicist) repeated Newcomb's and Wells' parallel universes/pages-of-a-book/rotation imagery almost word for word in his novel *Twistor.*

23. "Zoological Retrogression," *The Gentleman's Magazine,* 271 (September 7, 1891):246–253.

24. "On Extinction," *Chamber's Journal* 10 (September 30, 1893):623–624.

25. *Nature* 65 (February 6, 1902):326–331.

26. *Nature* 52 (July 18, 1895):268.

27. An old but still very important historical analysis of Minkowski's spacetime appears in Holton (1965). For a more recent study that includes the original German text, careful English translations, and photographs of Minkowski's agonized corrections to his pre-address manuscript, see P. L. Galison, "Minkowski's Space-Time: From Visual Thinking to the Absolute World," *Historical Studies in the Physical Sciences* 10 (1979):85–121. English texts can also be found in Einstein (TPR) and Minkowski (BST).

28. W. G., "Euclid, Newton, and Einstein," *Nature* 104 (February 12, 1920):627–630. As is the case with the mysterious "S." in Note 19, the editorial staff at *Nature* has informed me that there is no record of the identity of W.G. in the journal's archives. In "The Dimension Segregator" (Click) of 1929, we have an early fantasy tale that shows the fascination Abbott's world of two dimensions had for readers in the formative years of magazine science fiction—a scientist discovers how to project three-dimensional objects (including himself) onto a two-dimensional surface. A series of fantastic adventures follow until disaster strikes. More recently, Abbot's satire inspired an almost unbelievably awful 1964 episode of television's "The Outer Limits." In "Behold, Eck!," a two-dimensional creature arrives on Earth through a time warp. Unfortunately, it has poor vision and loses its way. Eventually a helpful optometrist helps Eck find his way home, but only after the sharp-edge two-dimensional visitor accidently cuts a 37-story building in half! Abbott's story has had more important modern influences, too. For example, some interesting analyses of the details of what science would be like in a universe with two spatial dimensions (a three-dimensional "toy" spacetime, as physicsts call it) were published in *Scientific American* by Gardner (1980). This essay reported the work of Alexander K. Dewdney, a computer scientist at the University of Western Ontario, and included a challenge for scientists of all disciplines to continue such work. Gott and Alpert (1984) took up that challenge for gravity and studied Einstein's general theory in Flatland (or the *Planiverse,* as Dewdney called it). They found some curious results: Although Maxwell's equations support electromagnetic waves in Flatland (and so there are photons and Flatlanders can see!), there are no gravitational

waves, no gravitons, and no attraction between masses, and in the case of weak gravity, general relativity does *not* approach Newton's theory in the limit as it does in our four-dimensional spacetime. Perry and Mann (1992) continued the study of gravity in a three-dimensional toy spacetime with their analysis of time machine wormholes (for more on wormholes as time machines, see Tech Note 9), as did Headrick and Gott (1994) with respect to string time machines (for more on cosmic strings as time machines, see Tech Note 10).

29. *Seventh International Congress of Logic, Methodology and Philosophy of Science,* Vol. 4 (Salzburg, Austria, 1983), 176. Popper describes his early discussions with Einstein on the reality of time and the four-dimensional Parmenidean block universe in some detail in his autobiography; see volume 1 of *The Philosophy of Karl Popper,* ed. P.A. Schilpp, The Library of Living Philosophers (La Salle, IL: Open Court, 1974), 102–103. More recently, Ben-Menahem (1993) has echoed Popper in claiming Einstein was a determinist. Although that paper is concerned with Einstein's ideas on causality and general relativity, it curiously never mentions Gödel's backward-time-travel result from general relativity. I think Popper and Ben-Menahem mislabel Einstein with the term *determinist.* Determinism says, "If you do A, then B will happen, and if you do not do A, then (perhaps) something other than B will happen." A deterministic universe has plenty of room for free will, because you can *choose* to do A or not to do A, and what you decide makes a difference. Anybody who believes in classical (non-quantum) science is a determinist. A fatalistic universe, however, like the block universe, simply says, "You will do A, and B will happen." To accept the block universe, as Einstein did, is to be a fatalist, not a determinist. Einstein's final position on this matter might have been that of the time traveler in "Throwback in Time" (Long). After taking a little girl 25,000 years back into the past, where she sees an ancient ancestor of humanity, she asks if he is really alive. Replies the time traveler in this 1953 tale, "Every man who ever lived is still alive, child. In time there is no real death. When a man dies he's still alive ten minutes ago, ten years ago. He's always alive to those who travel back through time to meet him face to face." Did Einstein really believe this? Not everybody thinks so, and physicist-philosopher Milic Capek (VOT, BST) cites an early Einstein comment to support his case. At the 1922 meeting of the French Philosophical Society, the philosopher of science Emile Meyerson (1985) asked Einstein whether the spatialization of time [the idea of time as a dimension on the same footing as the spatial ones; see Christensen (1981)] is a legitimate interpretation of Minkowski's spacetime. Einstein's terse answer (1922) was that "it is certain that in the four-dimensional continuum all dimensions are *not* [my emphasis] equivalent." Of course, although this reveals Einstein's position in 1922, it may have nothing to do with his views thirty-three years later when he wrote to Besso's family. Certainly Einstein's later willingness to ponder Gödel's time travel solution to the general relativity field equations shows that his views had undergone some changes, and even Capek (1965, BST) admits this. Even before the Besso letter, Einstein had said much the same thing. For example, in an appendix on Minkowski's spacetime in Einstein (1961) he wrote, "From a 'happening' in three-dimensional space, physics becomes, as it were, an 'existence' in the four-dimensional 'world.'"

30. H. Weyl, *Philosophy of Mathematics and Natural Science* Princeton, NJ: Princeton University Press, 1949, 116. Sir James Jeans had said the same, somewhat less elegantly, in his 1935 Sir Halley Stewart Lecture: "The tapestry of spacetime is already woven throughout its full extent, both in space and time, so that the whole picture exists, although we only become conscious of it bit by bit—like separate flies crawling over a tapestry. . . . A human life is reduced to a mere thread in the tapestry." Jeans then immediately *rejected* this fatalistic view; see his *Scientific Progress* (New York: Macmillan, 1936), 20.

31. From a book review of Whitrow (1980) in *Scientific American* (April 1962):179–185.

32. L. Silberstein, *The Theory of Relativity* (London: Macmillan, 1914), 134.

33. E. Cunningham, *The Principle of Relativity* (Cambridge, England: Cambridge University Press, 1914) 191.

34. Most recently, it has been shown that time travel does not imply any fatal violation of conservation of energy—see Friedman *et al.,* (1990) and Deutsch (1991). Indeed, global energy–momentum conservation does not play the central role in general relativity that it does in classical physics. For example, nearly sixty years ago Richard Tolman showed how an oscillating universe (alternating Big Bangs and Big Crunches separated by "bounces") creates ever-increasing total energy with each cycle of oscillation; in such a universe, energy is not conserved. For more information, see Tolman's book *Relativity Thermodynamics and Cosmology* (Oxford, England: Oxford University Press, 1934) and Trautman (GRAV). Nearly thirty-five years ago Kip Thorne wrote (1965b), "One of the greatest unsolved puzzles of the [general] theory has been the nature of energy within the framework of relativity," and I think most physicists would say that most of the puzzle is still with us. More recently, the local and global conservation of energy and momentum in general relativity, in the context of FTL travel, has been discussed in Bishop (1984, 1988). Science fiction has shown increasing sophistication on this issue, too. For example, even in the mostly silly *Time Echo* (Lionel) the mass/energy issue is addressed with the idea of a time traveler changing places with an equal amount of mass/energy at the target point in spacetime. And in his "Ripples in the Dirac Sea," physicist Geoffrey Landis uses Dirac's theory of negative energy particles to explain the time machine in his story. The problem of the conservation of mass/energy is partially avoided in this tale by Landis' "rule of time travel" that requires each visitor to the past to return to the exact moment of his departure (and this, in turn, is the basis for the development of a highly original "trapped in time" idea, one with enormous emotional tension). This "rule" means, of course, that in a certain sense the mass/energy never left the present. And as for the arrival of the mass/energy in the past, Landis explains that problem away with reference to the temporary borrowing of energy from the infinite Dirac sea of negative energy particles via the Heisenberg uncertainty principle. (This is similar to what Hawking did for virtual particle creation at the spacetime horizon of a black hole.).

35. Hollis' story is, in fact, a very funny summary of all the previous storytelling, presented as a dialog between two men, Box, and Cox, who discover they have been living in the same space (a bed-sitter at 2 Temporal Crescent, rented from a landlady named Mrs. Chronos!) but in different times. One of the most curious of philosophical tales concerning time can be found in Schlesinger (1982). There we are asked to imagine that scientists have just discovered X, a solar system "many billions of light years away. . . . We cannot interact with X except that we are able to communicate with them via some strange medium which transmits signals instantaneously." The author of this paper, who was attempting to explain the sense we all have of the passage of time, says nothing about the extraordinary conclusion concerning time travel that follows from this assumption (see Tech Note 7). Equally odd was the rebuttal from a fellow philosopher—MacBeath (1986)—that began by observing, "To sidestep any relativistic explanation [we are asked] to accept that communication between us and the Xians is instantaneous." Rather than sidestepping relativity, this assumption of course throws the analyst *up to his neck* into relativity! MacBeath apparently did realize something was amiss: He declared the assumption that "signals might travel billions of light years in no time at all" is absurd, but he said nothing of why he felt it was absurd.

36. Considerably predating both Putnam and Rietdijk with the same argument was Philipp Frank, in a section of his book (1957) called "Is the World 'Really Four-Dimensional'?" Frank attributes the argument to even earlier writers—see Frank (BST), written in 1938, for specifics—and correctly concludes that it has no merit.

37. Putnam's argument for a non-empty future was restated a few years later in Sklar (1974). Sklar began by claiming that "the philosophers who maintain that past and future objects are not real existents, or that future events do not have determinate reality, are refuted out of hand by special relativity." Then he repeats Putnam's argument but says nothing of the prior objections to it. He does, however, waffle a little on how convinced he is by it all: "But when we put the argument this badly, the inability of the physical theory to resolve the philosophical dispute becomes obvious." He continued to hold this position in Sklar (RTR). The reason for the so-called failure is, of course, that special relativity is the wrong theory to use! To argue about the reality of the future from what special relativity says is to go hunting for tigers armed with a prayer book. Special relativity applies to spacetimes of a decidedly empty nature—that is, no mass-energy present in significant amounts (else spacetime will be curved rather than flat, and thus in the domain of the *general* theory). To wonder about future events in an empty spacetime is to wonder, literally, about nothing.

38. This is the issue of how we measure the speed of light. All the traditional techniques invariably measure the average speed over a two-way trip, with a reflection of the light beam at the far end of the path back to the source. This allows a single clock at the source to make all the time measurements. A one-way measurement assumes two distant clocks are synchronized, but how to achieve this presents unexpected problems. Einstein himself was well aware of this, and he commented on it in his 1905 paper. That was the starting point of the sporadic debate (mostly among philosophers) on the so-called "conventionality of simultaneity." For a treatment of the subtle points, see Salmon (1977) and Petkov (1989).

39. Even in the otherwise stunningly absurd and juvenile 1941 story "Bandits of Time" (Cummings), in which an Attila the Hun character roams up and down time kidnaping beautiful women for his harem, there is an "editorial" footnote telling the young readers of *Amazing Stories* that "scientists—especially the new order of meta-physical-scientists—are agreed on the principles of Space-Time. The future is not a thing which *will exist*. Rather it is a thing which *does exist*—all events from the Beginning to the End, spread in a record upon the scroll of Time."

Chapter Three

1. The same argument appeared earlier in Smart (1954), and as Webb (1977) observes, one can find this particular objection to a flowing time as far back as Kant. Not every philosopher finds this a meaningful objection. For example, Christensen (1976) observes that the answer to how fast time is passing is just "one second per second" and makes it plain that he thinks not to understand this is to be dense. I don't think that position is solid, and neither do others; Morris (1984) calls such a claim "as meaningful as defining the word *cat* by saying 'A cat is a cat.'" In an amusing spoof on controversies in physics over matters that are essentially hypothetical conjecture at best, Kate Wilhelm has two scientists confront each other at a conference in her story "O Homo; O Femina; O Tempora." One claims to have shown that time is slowing down; the other says it is speeding up. Things get pretty hot for a while, but then the wife of one asks, "Will we experience anything differently? . . . Nine months will still seem like nine months." With that settled, all concerned head for the conference center bar! A discussion of the

self-referential nature of McTaggart's infinite regression of times can be found in Löfgren (1984), including Gödel's paradox of a time traveler meeting himself.

2. An extended discussion and analysis of Lewis' ideas on time are offered in R.W. Stuewer, "G.N. Lewis on Detailed Balancing, the Symmetry of Time, and the Nature of Light," *Historical Studies in the Physical Sciences,* Vol. 6, Princeton University Press, 1975. Lewis was not a physicist but a chemist at the University of California at Berkeley (we owe the word *photon* to him), and his ideas on advanced waves predated those of the physicists Wheeler and Feynman by twenty years, which they acknowledged. (Wheeler and Feynman's advanced wave work is discussed in detail in Chapter Four.)

3. Quoted from *The Philosophy of Rudolp Carnap* ed. P.A. Schlipp, The Library of Living Philosophers (La Salle, IL: Open Court, 1963), 37–38.

4. F.H. Bradley, *Appearance and Reality,* 2nd ed. (London: Oxford University Press, 1897), 190.

5. J.N. Findlay, *Philosophy* 25 (1950):346–347.

6. One submission that also argued against Findlay was easy for him to rebut. The analyst argued that a time-reversed world would be analogous to a world in which everything doubled in size, which would be undetectable. This erroneous reasoning was rejected by Findlay: "Surely the analogy is with a world in which size-relations are inverted, in which the least is largest and the largest least? And that, surely, *would* make a difference." The terror aspect of living backward in time is nicely captured in the 1949 story "Reversion" (Pease). A scientist who is involved in an accident with radioactive materials has his sense of time flow reversed, and the story carefully and logically analyzes what his life would be like. For example, the scientist can talk (backward for others), so he is understandable only if his words are recorded and then played in reverse. He cannot eat, because for him that would involve the regurgitation of food. He cannot answer questions because "if he should answer any questions we put to him, it would mean he was giving the answer before he heard the question, on his time scale." And finally he can't pick anything up because the normally stable position and velocity error-correction mechanism between hand and eye, which is a negative feedback system in normal time, becomes an unstable positive feedback system in reversed time. The horror of his existence is contained in the only words the man utters (deciphered after reversed playback): "Where am I? What's happened? Why are things so different? Why? Why?"

7. This idea can be found quite long ago in science fiction, as in "Minus Sign" (Williamson) which connects anti-matter not only with physical time travel to the past but also with telephone calls backward in time. Not everybody likes the association of anti-matter with backward time travel. For example, Price (1991a) refers to "Feynman's rather loose talk of particles 'travelling' backward . . . in time," and in Earman (1967a) we read that "it is true that Feynman uses the slogan 'Positrons are electrons running backward in time,' but it is dangerous to draw conclusions from slogans." I am not sure what Earman means by slogans; a reading of Feynman indicates that he took the matter quite seriously. In Feynman (1949a), for example, he wrote that "the idea that positrons can be represented as electrons with proper time reversed relative to true time has been discussed by the author and others" and also that "Previous results suggest waves propagating . . . toward the past, and that such waves represent the propagation of a positron." Feynman declared the idea to be of value—but see Selleri (1997) for an interesting objection to Feynman's view—in understanding negative energy states in his famous book *Quantum Electrodynamics* (New York: W.A. Benjamin, 1961), 68. See also Note 20 for Chapter One for Feynman's Nobel Lecture comments on this topic. I see no slogans or "loose talk"

in all this. The "others" Feynman had in mind included, in particular, the physicist Ernest C.G. Stückelberg (1905–1984), who in a 1942 article in the Swiss journal *Helvetica Physica Acta* also wrote of waves scattering backward in time.

8. J. R. Lucas, *A Treatise on Time and Space,* London: Methuen 1973, 43–47.

9. MacBeath spends some time in his paper discussing the nature of this window, which he shows is essential. Indeed, recalling the anti-matter nature of Midge's world, it is absolutely *vital* to keep her and Jim apart. MacBeath observes that the window is actually double-paned, with a perfect vacuum in between. The exchange of light between the two worlds presents no problem (or does it?—see Note 13 for this chapter) because photons are their own anti-particles. In particular, there is no difference between the time sense of photons in either world, because the flow of proper time for a photon—traveling at the speed of light—is zero (see Tech Note 5).

10. In a direct, vigorous attack on the possibility of time travel, Smart, a philosopher, cities (1963) this very problem of communication. Professor Smart believes that real travel back in time means that one should be able to "have normal experiences," and of course this would not be so because one could not communicate. He mentions Norbert Weiner's position in *Cybernetics,* 2nd ed. (Cambridge, MA: M.I.T. Press, 1961), 34–35, in which Weiner calls the idea of a reversed-time being a "fantasy" and concludes that "within any world with which we can communicate, the direction of time is uniform." This is doubly irrelevant. First, as we have seen, communication is logically possible between reversed-time individuals, and second, once back in time (assuming one can actually get there), the time traveler's time direction *would* be that of the rest of the world.

11. Causal loops have been used in the movies with some effect. In the 1980 *Somewhere in Time,* for example, we see the following loop: The hero in the present is visited by a mysterious old woman who gives him an antique watch. Some time later he travels back to 1912, where he meets a girl to whom he gives the watch. He then returns to the present, and she lives out her life until she too reaches the present, where we discover she is the (now old) woman who gives the hero the watch. At every instant of its existence, the watch is in the possession of either the hero or the woman—so when was it made? A similar loop is used in the 1983 film *Timerider.* A related *information* time loop is illustrated in the 1980 movie *The Final Countdown.* The designer of a modern naval warship that temporarily travels back to the Pearl Harbor of December 6, 1941, turns out to be a crew member who was accidently left behind in the past. In the past he *will be* able to design the ship because he already knows how it *was* designed—by himself!

12. H. Putnam, *The Journal of Philosophy* 59 (12 April 1962):213–216.

13. This point was elaborated on in Swartz (1973). Here we read, "We have uncritically imagined someone looking in on . . . two worlds having opposite time directions. . . . Part of the story we tell, of the process of seeing, involves the emission of photons from objects [e.g., computer screens] and the subsequent impinging of these photons on our retinas. But this process is obviously directed in time. In a world where time ran oppositely to ours, we could not see objects at all: objects would be photon-sinks, not photon-emitters." In true Faustian fashion we would be unable to photograph such a universe, and a being made of Faustian matter would pass one of the traditional tests for being a vampire—no reflection in a mirror—in our world.

14. Dick's first attempt at this idea can be found in the rather muddled 1966 short story "Your Appointment Will Be Yesterday." Its only bright spot is the wonderful line "The tasks of tomorrow become the worse tasks of today." Many of the problems of a backward-running world, poorly handled in this first try, are given better

treatment in the novel. In his excellent introduction to the novel *Counter-Clock World,* David Hartwell writes that it is neither a tragedy nor a comedy. Rather, "It is a complex, ironic novel, continually mixing the colloquial and personal with the metaphysical." Hartwell is a far more astute science fiction critic than I could ever claim to be, and he might well take exception to my use of the term *horror.* I can only say I was horrified as I read the novel; the description of what it would be like for a dead body to come alive again in a buried coffin is the equal of anything Poe wrote. The *most* terrifying story about a reversal in the direction of time is, I think, Philip Jose Farmer's "Sketches Among the Ruins of My Mind." This is the tale of how the world goes to bed each night and wakes up each morning with three more days of memories gone. People physically age but mentally grow younger. What this does to families as parents watch their young children regress is pure and simple horror. A recent work that combines the reversal of cause and effect with horror is the 1991 novel *Time's Arrow* (Amis), which is not actually a time travel or a reversed-time tale; its narrator tells us, backward, the story of his life as a former Nazi doctor at Auschwitz. Since the result of any action is related before its cause is presented, this novel has all the logical puzzles of *Counter-Clock World* without asking the reader to accept reversed time. The 1983 British film *Betrayal* did the same thing, with its unfolding of an adulterous affair backward in time. Again, this is not truly reversed time but simply the visual equivalent of Amis' literary device reversing cause and effect. Still, when the picture was reviewed by the *Monthly Film Bulletin* (October 1983), the critic could not resist a little fun: "The fact that *Betrayal's* narration advances backward rather than forward . . . is neither there nor here."

15. Ignoring how such a thing might actually occur, one philosopher has argued it is at least logically possible Woodhouse (1976). I must admit that this is one of those philosophical papers that it has proved impossible for me to appreciate fully. More analytical is Schmidt (1966), which shows how, under certain initial conditions at the Big Bang, one possible solution to the gravitational field equations is an oscillating universe that temporally runs backward during the contraction phase; and Walstad (1980) further elaborates on this idea. Matthews argues (1979) that the direction of time is local, not global, and that the arrow of time can point in opposite directions at different locations (at the same time?). In Cocke (1967) we find a physicist speculating on the possibility that if an observer in the expanding phase could survive into the contracting one ("perhaps by shutting himself in a vault, so that his own time sense remained unchanged"), then maybe he might disrupt things sufficiently to create large regions of the universe with different directions of time. This imagery is repeated in Hawking (1985). John Wheeler is not at all happy with any of this, and he certainly would be appalled at Dick's reversed-time world. As Wheeler stated in the General Discussion at the end of [NT], "Most of us would probably agree that the universe has not contained and will not contain any backward-looking observers. We do not expect to see caskets with corpses in them coming to life, nor do we expect to find bank vaults in which a gram of radium will integrate rather than disintegrate." The idea of time changing its direction after each bounce in an oscillating universe (but pointing in the same direction during both the expansion and the contraction phase of a given cycle) is described in Davies (1972b). Davies shows how, with such a model, one can then derive the specific value of the observed Big Bang background radiation temperature. The following year Albrow (1973) showed how Davies' model also provides complete symmetry between matter and anti-matter (if one assumes the TCP theorem always holds, even during bounces), even though during any one cycle one type of matter is dominant. But what if the TCP theorem, somehow, does not always hold? A theoretical analysis of the implications of this assumption is done in Ne'eman (1970), which concludes that the ex-

panding and contracting phases of an oscillating universe obey *different laws of nature!* An experiment using kaons to test this interesting hypothesis is described in Aharony and Ne'eman (1970), an experiment that (so far as I know) has not been performed.

16. The same year that Dick published his work, Brian Aldiss published the non-machine time travel novel *An Age*, which finds its dramatic revelation in backward-running time. Aldiss is quite explicit in dealing with the human digestive tract operating in reverse, but I find his technical presentation of reversed time to be at least as flawed as Dick's. Aldiss has seen fit to fill his tale with mysterious, godlike beings who periodically arrive like cavalry to save the day, and scientific laws (such as the second law of thermodynamics) are brushed aside, when convenient, with no explanation.

17. For example, obtaining food from a grocery store was easily accomplished once Hull mastered the art of walking backward. As he put it, "A man walking backward from 12:00 to 11:55 looks like a man walking forward from 11:55 to 12:00." Thus "A man moving in this way who enters a store empty-handed at 12:00 and leaves loaded with food at 11:50 looks like a normal man who comes in with a full shopping bag at 11:50 and leaves without it at 12:00—a peculiar procedure, but not one to raise a cry of 'Stop thief!' " For this to make sense, Boucher did have to impose one extra condition of his own: As long as Hull grasped something, its time sense reversed, too.

18. One of the institutional villains in *Counter-Clock World*, for example, is The Library, which is controlled by a group called the Erads. This is a backward-running library that absorbs the written word, and its only goal is to eradicate knowledge. The Library, one of the groups desperate to censor the novel's resurrected prophet (indeed, the Erads finally murder him), is clearly a metaphor for the book burners of our world.

19. The *H*-theorem was a direct continuation of the pioneering work by the genius James Clerk Maxwell on the statistical properties of gas molecules (specifically, the calculation of the probability density function of the molecular speeds). In 1859 Maxwell found this function for a special case, and later, in 1866, he showed that his solution holds in the particular case of thermal equilibrium. In 1872 Boltzmann found the differential-integral equation the function satisfies in general, even out of equilibrium. From this Boltzmann was able to define a quantity *H* that he showed evolves in time such that the general solution always approaches Maxwell's equilibrium solution. That is, the *H*-theorem says that *H* always decreases in systems not in equilibrium and is at a minimum in systems in equilibrium. Boltzmann's brilliant work can be found in his 1895 book *Lectures on Gas Theory*, which was translated into English by S.G. Brush (Berkeley, CA: University of California Press, 1964).

20. For more on Boltzmann's view on this matter, see the end of his letter "On Certain Questions of the Theory of Gases," *Nature* 51 (February 28, 1895):413–415. He describes this view, in an enigmatic aside, as "an idea of my old assistant, Dr. Schuetz." The philosopher Karl Popper called Boltzmann's willingness to consider the possibility that different regions of the universe could have different directions of time "staggering in its boldness and beauty"; see Vol. 1 of *The Philosophy of Karl Popper*, ed. P.A. Schlipp, The Library of Living Philosophers (La Salle, IL: Open Court, 1974), 127–128. But Popper went on to assert that Boltzmann must be wrong because "it brands unidirectional change an illusion. This makes the catastrophe of Hiroshima an illusion." This is a metaphysical argument, of course, and although it has great emotional power for the human soul, I fail to see how it is related to physics. From time to time one can find, in the modern physics literature on relativity, serious speculations about reversed-time beings. An example is

Belinfante (1966), whose author tells us in a footnote that it took two years to get the paper published, because two referees could not understand it!

21. See Haldane (1928). The googol is a gigantic number, far greater than the number of raindrops that have fallen on the entire Earth during its entire history. And the googolplex is light years beyond that. Using a wonderful bit of imagery that I have not seen repeated since, R.B. Braithwaite wrote in his critical analysis of Eddington (1929) entitled "Professor Eddington's Gifford Lectures," *Mind* 38 (October 1929):409–435, "If a man shuffled just a single pack of cards as rapidly as an individual molecule hits other molecules in air, and if a snail started to crawl around the universe . . . at the rate of one centimeter *during the life of the sidereal system* [my emphasis], the snail would have got round the universe many millions of times before it would become at all likely that the man would have got the pack back to the original order." If this is what it takes to get a pack of cards back to its initial state, then try to conceive of the time required to restore the world to December 7, 1941. Stanley G. Weinbaum and Lyle D. Gunn were two authors who were not discouraged by such calculations, and they used the certainty of recurrence over infinite time in their fiction; see "The Circle of Zero" (Weinbaum) and "The Time Twin" (Gunn). The notion of eternal recurrence considerably predates Poincaré, as shown in Small (1991), which traces its scientific (as opposed to astrological) study back to the fourteenth century. The claim for eternal recurrence, based on scientific arguments (such as the conservation of energy), can be found in many places in the writings of the German philosopher Friedrich Nietzsche (1844–1900), again predating Poincaré. See, for example, his *The Gay Science* (1882) and *Thus Spake Zarathustra* (1883). All of Nietzsche's arguments are flawed—see Krueger (1978)—but they are rational, physical arguments, as opposed to arguments based on metaphysics or theology. In fiction, the glimmer of the idea of a repetition of human affairs preceded Poincaré by some years, too. For example, see "Human Repetends" (Clarke), a story originally published in 1872. (In mathematics, if a fraction has an endlessly repeating decimal expansion, the repeating part is called the *repetend*.) Of course, as argued by Theobald (1976), it would seem as though we could never *know* of the event of a recurrence, because the state of all conceivable records (geological evidence, memories, books, films, and so on) would, as part of the physical state of the universe, also recur. And those records could contain no knowledge of the recurrence because it had not happened "yet"! This is the flaw in the time loop in the film *Groundhog Day*—see Note 3 for Chapter One. A "time loop" pulp science fiction story that specifically avoids this error (and quotes Nietzsche, to boot) is the 1946 "Forever is Today" (Ksanda).

22. Indeed, this connection provides a plausible link between the identity of our psychological direction of time and that of increasing entropy. Psychological time is in the direction of increasing memories (increasing information), and the direction of increasing information is in the direction of increasing entropy. The universe is still young in terms of its capability to increase its entropy; it is estimated by Page and McKee (1981) that over the next 10^{116} years the entropy of the universe will increase by a factor of at least 10^{14}. See also Frautschi (1982).

23. The fuzziness of the relationship between entropy and time was captured by one physicist who asked, "If it were found that the entropy of the universe were decreasing, would one say that time was flowing backward, or would one say that it was a law of nature that entropy decreases with time?" See P.W. Bridgeman, *Reflections of a Physicist* (New York: Philosophical Library, 1955), 251. Even in a story in which we are told the physics of time travel "would have no meaning to you," the connection between time and entropy is made. When a character in *The Corridors of Time* (Anderson) asks how a time corridor works, he is told, "Think of it as

a tube of force, whose length has been rotated onto the time axis. *Entropy still increases inside; there is temporal flow* [my emphasis]."

24. Bunge (1958) asserts that two-dimensional, complex time is old hat in the theories of spinning particles. Indeed, in 1889 the Russian mathematician Sophie Kowalevski (1850–1891) used complex time to study the mechanics of a rotating mass. Her work is described in Michèle Audin, *Spinning Tops* (New York: Cambridge University Press, 1996). There has also been some discussion of multidimensional time in the physics literature concerning the superluminal Lorentz transformations associated with FTL particles (issues discussed in Chapter Four and Tech Note 7), but advocates of this have been strongly rebutted by other physicists. For a short bibliography of the physical literature and a concise summary of the objections to multidimensional time, see Kowalczynski (1984).

25. A.S. Eddington, *The Mathematical Theory of Relativity*, 2nd ed. (Cambridge, England: Cambridge University Press, 1924), 25.

26. Mirman (1973) asserts that this is the most fundamental property of time and argues that there can be only one-dimensional time.

27. Other papers on two-dimensional time include Thomson (1965), Nusenoff (1976), and Wilkerson (1973, 1979), but none treats the issue of changing the past. Coming at two-dimensional time from another angle are Oaklander (1983) and Schlesinger (1985), which discuss its applicability and use in the context of McTaggart's arguments about temporal becoming. Most recently, MacBeath (1993) has done a nice job of summarizing many of the earlier arguments against multidimensional time, but he concludes with "I would not want to rule out the possibility . . . that time is three-dimensional. Or worse."

28. See Dobbs (1951a,b; 1956), which suggests that the specious present may be the physical manifestation of two-dimensional time, and Broad (1937) and Broad and Price (1937), which discuss two-dimensional time as an explanation for precognition. (Broad, who was Dobbs' teacher, wrote (1959) more generally on two-dimensional time.) Dobbs' theory was rejected by Mundle (1954), but that didn't discourage Dobbs, who later advocated (1965) two-dimensional time as a rational mechanism for ESP. An indirect experimental test, claims Yndurain (1991), can (and has) been done for compact multiple-time dimensions. This paper discusses the implications of extra timelike dimensions that are "curled up," much like the spatial dimensions mentioned in Note 15 for Chapter Two. The author's conclusion is that such a situation would lead to the decay of quarks and protons at an experimentally detectable rate; because such decays are not observed, these compact additional time dimensions do not exist. Yndurain also includes this remark, however: "If there are extra time dimensions we get violations of causality, because one could sneak to yesterday through the extra dimensions, and . . . if you had sneaked to yesterday, you would have disappeared from today." As Meiland's analyses show, however, there is nothing at all illogical about such things with two-dimensional time. A discussion on the use of three-dimensional time in the analysis of superluminal problems appears in Gunawant and Rajput (1985).

Chapter Four

1. This experiment does inspire a curious question, however: Does the cube travel through time or is its journey "instantaneous," so to speak? Sometimes this distinction is made by calling the first *sliding* and the second *jumping*. That is, does the cube travel around time? If through time, then the cube is present at every instant after the start of its trip, so it should not vanish. The cube gets to each

instant before the observers do, but why this should produce the visual effect of disappearing is unclear. Brown's description implies that the cube traveled five minutes into the future without existing at any of the in-between instants. Wells tried to have it both ways in *The Time Machine* by invoking "diluted presentation" as the reason why we cannot see "the spoke of a wheel spinning, or a bullet flying through the air." If the cube "is traveling through time fifty times or a hundred times faster than we are . . . the impression it creates will of course be only one-fiftieth or one-hundredth." This explanation breaks down, of course, when one remembers that even if you cannot see the spoke or bullet, they are still there and you can get in their way—Wells, unfortunately, has one of his characters stick a hand into the space where the time machine was last seen. This objection to Wellsian time travel was raised soon after the 1895 publication of the novel—see the unsigned *Daily Chronicle* review of the book in Parrinder (1972)—and then again in 1914, by Pitkin. Wells, it is only fair to note, seemingly anticipated Pitkin when he had the Time Traveller say, "So long as I travelled at a high velocity *through time* [my emphasis] . . . I was, so to speak, attenuated—was slipping like a vapor through the interstices of intervening substances!" For the Time Traveller to stop "inside" anything (Pitkin's example was the pile of bricks the Time Traveller's laboratory is certain one day to become) would, however (as Wells wrote), cause "a profound chemical [actually nuclear, but Wells wrote in 1894] reaction—possibly a far-reaching explosion—[that would] blow myself and my apparatus out of all possible dimensions." Just why this awful fate doesn't befall it when the machine simply stops *in air,* never mind Pitkin's pile of bricks, is never addressed. In any case, it seems from all this that Wells' machine travels *through* time, just as the Time Traveller claims. But Wells, himself, raises doubt about this when he describes the observable effects of a departing time machine. At the beginning of the novel, when the Time Traveller sends his model machine into the future, we read, "There was a breath of wind, and the lamp flame jumped. One of the candles on the mantel was blown out. . . and it [the model time machine] was gone—vanished!" And at the end, when the Time Traveller makes his final exit, the narrator of the tale just misses seeing the departure but tells us, "A gust of air whirled around me as I opened the door, and from within came the sound of broken glass falling on the floor. The Time Traveller was not there. . . . Save for a subsiding stir of dust, the further end of the laboratory was empty. A pane of the skylight had, apparently, just been blown in." Both of these descriptions read as *implosions,* air rushing in to fill a spatial void, as though the time machine had *jumped* in time. Is there an inconsistency here? Well, perhaps not, if one accepts the curious idea of "slipping like a vapor" for an operational Wellsian time machine.

2. Brown thought of everything in this tale. At this point in the story one of the colleagues, puzzled by how the inventor will be able to place the cube into the time machine at three if it has already vanished from his hand and appeared in the machine, asks, "How can you place it there, then?" Replies the inventor, "It will, as my hand approaches, vanish from the [machine] and appear in my hand to be placed there." Although I am an admirer of Brown's work, I certainly am not an uncritical one. For example, his time-reversed story "The End" is a simplistic, nearly trivial presentation of a deep issue.

3. J. Hospers, *An Introduction to Philosophical Analysis,* 2nd ed. (Englewood Cliffs, NJ: Prentice-Hall, 1967), p. 175.

4. This work is in the form of a letter to his friend Desiderius (who later became Pope Victor III), in which Damian rebutted Desiderius' defense of St. Jerome's claim that "while God can do all things, he cannot cause a virgin to be restored after she has fallen." Desiderius thought the reason God could not restore virgins is that he does not want to, to which Damian replied that this meant God is unable to do

whatever he does not want to do, but this meant that God would then be less powerful than men, who are able to do things they don't want to do (such as go without food for a month). This is a good example of the dangers involved in getting into debates with theologians.

5. Borges was so inspired by Damian's view that the past could be changed that he wrote a short story based on it ("The Other Death") and put a character named after Damian in it.

6. Two eminent physicists after Gödel have also taken this position in their very influential book; see S.W. Hawking and G.F.R. Ellis, *The Large Scale Structure of Space-Time,* (London: Cambridge University Press, 1973), p. 189. See also Note 11 for this chapter. The same view was earlier endorsed by Whitrow as "logically irrefutable" in the 1961 edition of his impressive book (1980). The 1980 edition has correctly dropped this endorsement. In an equally curious footnote to this (found on page 306 in 1980 and on page 260 in 1961), Whitrow writes, "It is interesting to note that in Wells' novel the Time Traveller returns from his trip into the future but not from his trip into the past!" It is amusing that Whitrow, who rejects time travel, finds Wells of all people a source of comfort. In any case, a careful reading of the novel's Epilogue shows that the direction of the last trip is an open question. In an appendix to Wells (1987), however, Harry Geduld shows that Wells toyed with the idea of the Time Traveller visiting the past. Two draft chapters in Wells' hand (dated April 10, 1894), which do not appear in the published *Time Machine,* are reproduced in typescript by Geduld; they have the Time Traveller in 1645 and at even more remote times. Modern novel-length sequels to Wells' classic that hazard explanations of the fate of the Time Traveller include *The Time Ships* (Baxter), *The Return of the Time Machine* (Friedell), and *Time Machine II* (Pal and Morhain). *The Space Machine* (Priest) is a sequel of sorts. It tells us how Wells' other great scientific romance, *The War of the Worlds,* was intimately tied to *The Time Machine* and how the invisibility of an operating Wellsian time machine (perhaps a reminder of Wells' *The Invisible Man*) was used to help defeat invading Martian monsters in 1903.

7. Lewis makes it clear that he believes it is a sin to pray for something known not to have occurred—for example, to pray for the safety of someone known to have been killed yesterday. As he writes, "The known event states God's will. It is psychologically impossible to pray for what we know to be unobtainable; and if it were possible, the prayer would sin against the duty of submission to God's known will." Taking a less judgmental position (but essentially agreeing with Lewis) are Stump and Kretzmann (1981), who write of the battle of Waterloo, "for one who knows the outcome of the battle more than a hundred and fifty years ago, [a retroactive petitionary] prayer is pointless and in that sense absurd. But a prayer prayed in ignorance of the outcome of the past event is not pointless in that way." And, in support of backward causation, they also write that "to pray in 1980 that Napoleon lose at Waterloo" is logical because "why should your prayer not be efficacious in bringing about Napoleon's defeat?" Disagreeing is Geach (1969), who declares, "A prayer for something to have happened is simply an absurdity, regardless of the utterer's knowledge or ignorance of how things went." This is a position of faith, of course, about which mathematical physics has nothing to say. But see Tipler (1994), who, in perhaps the twentieth century's most bizarre "science" book, claims to have reduced theology to a branch of physics. Tipler agrees with Geach on the lack of utility in praying, but his book also claims to answer the question of whether there will be sex in Heaven, so let that be an indication to you of its nature.

8. Dummett is fond of making up fairy tales to illustrate his belief in backward causation. Ten years later, for example, he was at it again (1964), this time with one

that has become famous in philosophical circles—a tale about a tribal chief who finds that he can make young warriors *have been* brave on lion hunts if he dances during their entire absence from home. The details of this story are unimportant for us, but other philosophers have responded, including Gorovitz (1964) and the curious Zetterberg (1979), who attempts to use relativistic reasoning but seems to miss the distinction between affecting and changing the past. My view is that it is all beside the point. There is no evidence for Dummett's alarm clocks, magicians, and dancing tribal chiefs. They are all made up—and we can make up anything we'd like! What *is* important is the theoretical prediction of backward causation and time travel by the equations of physics. I think Dummett was correct in his conclusions, but I consider this method of analysis a distraction; indeed, it is irrelevant.

9. J.R. Lucas, *A Treatise on Time and Space* (London: Methuen, 1973), p. 50. Many of Dwyer's arguments can also be found in Dwyer (1975).

10. "Of Warps and Wormholes," *Archaeology,* March/April 1989, 80.

11. See, for example, pages 238–240 in Kaufmann (1977). Here we read that "the idea that time machines could exist is extremely troublesome to the scientist. Truly disturbing things could happen.... [T]ime machines violate *causality.*... The idea that effects could occur before their causes is denied by the rational human mind." Kaufmann is also worried, I presume, by the possibility of changing the past in the way described in the 1945 story "Delvers in Destiny" (Kummer). There, when a time traveler from the twenty-first century journeys back five centuries, he cuts down an oak sapling to make a club and uses it to kill a lamb for food. Because that sapling will (no longer?) grow to make future acorns and the lamb will "no longer" have descendants, certain alterations are induced. According to the story, for example, homes in the future vanish (no acorns in the past means no trees in the future and thus no lumber), and because wool isn't readily available anymore, trousers disappear off men in mid-stride! A similar story published a decade later in a 1954 issue of *Orbit Science Fiction* magazine, "The Penfield Misadventure" (Derleth), commits the same gaffe; a time traveler kills his uncle in the past and so causes his aunt to vanish from the present (though memories of her remain). See also the verbose novel *Beyond Time* (Johns), in which accidental time travelers to the future are arrested by the "time patrol" for unavoidably killing in their own defense. As the leader of the time patrol tells the shocked time travelers, "Had you killed my remote ancestor, you would also have killed me. I should have gone out of the existence...."

12. In opposition to this assertion, the final scene in *Time and Again* (Finney) has the hero return to the past and prevent the parents of another character from meeting. Fredric Brown directly tackled the grandfather paradox in "First Time Machine," also with the idea that the past can be changed. That, as argued in this chapter, is not logical. More logical is "Don't Live in the Past" (Knight), in which a time traveler voyages back four centuries in a misguided attempt to prevent a "temporal accident" in the present from changing the past. Once in the past, he discovers he is actually an historically important personage, so his presence does not change the past (but certainly does affect it). Taking this idea to its ultimate is *Chorale* (Malzberg), in which *all* of the personages in history important enough to appear in history books turn out to be time travelers from the twenty-third century. Even the inventor of time travel is a time traveler from the future! In this novel, therefore, the past is simply the present creating what it has found preserved in musty archives.

13. But not always. For example, in both *The End of Eternity* (Asimov) and "Time Enough" (Knight) these forces (be they of human origin or just "fate") are the central point of the story; they generate the tension and conflict that are vital to good fiction.

14. Playing "change the past" with the Civil War by using a splitting universe is one of the more powerful time-travel sub-themes (see the anthology FCW). Many writers in addition to authors of science fiction have amused themselves with this sort of speculation. Gore Vidal's 1955 play "Visit to a Small Planet" has a hint of it at the end. See also *IF: or History Rewritten*, ed. J.C. Squire (New York: Viking, 1931). One of the included essays, by Winston S. Churchill, is a double twist, written as an analysis of future events if Lee had not won the Battle of Gettysburg! Playing with so-called *counterfactual pasts* even has a respectable role in academic historical scholarship; see the quite interesting paper by Gould (1969).

15. See DeWitt (1973) for the details of Everett's work, and see Wolf (1990) for an interesting treatment of it. Healey (1984) calls Everett's theory "highly controversial" and declares that "few working physicists take it seriously." Stein (1984) calls the theory "a bizarre notion." DeWitt himself writes, "the idea of 10^{100} + slightly imperfect copies of (the universe) all constantly splitting into further copies . . . is not easy to reconcile with commonsense. Here is schizophrenia with a vengeance." Even the defense (sort of) of Everett in Geroch (1984) admits that the theory is at best "incomplete." An indirect argument that has been used to support the many-worlds idea is the anthropic one. Smith (1986b) outlines the puzzle of why many of the physical constants of nature (such as the weak fine structure and the strong coupling constants) have just the values required to produce a universe that supports life. Some have argued that all possible universes exist, with all possible variations in the constants, and so there must be universes (like ours) with life. Everett's theory is then invoked to explain all these universes, but Smith argues that in fact Everett's theory offers no such explanation. For the theological aspects of the many-worlds idea, see Craig (1990), which calls the concept "ontologically bloated" and "outlandish." John Wheeler, Everett's dissertation advisor, has changed his own mind about the concept (1979): "I once subscribed to it. In retrospect, however, it looks like the wrong track. . . . [I]ts infinitely many unobservable worlds make a heavy load of metaphysical baggage." At least one science fiction writer agreed even earlier (1971) in "What Time Do You Call This? (Shaw). In that tale the inventor of the "chronomotive impulse belt" (which allows moving between the *two* parallel worlds that are all that exist) calls the many-worlds idea the "Doctrine of Infinite Redundancy—which is, of course, utter nonsense." Tipler (QCST) disagrees, claiming that the enlarged ontology is a plus, not a minus! See also Tipler (1986). A rebuttal to Tipler is Shimony (QCST). Everett's theory is the antithesis of what is commonly called the collapse of the wave function, the idea that all potential possibilities have a non-zero reality until a consciousness actually decides or observes which one will be. This quantum mechanical concept gets its name from the probability wave equation formulated by the physicist Erwin Schrödinger. Before the observation, all possible futures have various values of probability; after the observation (which collapses the wave function), however, one of the futures (*the* future) has probability 1 and all the others have probability 0. The earliest stories I know of that treat the collapsing-wave-function concept are *The Legion of Time* (Williamson) and "Tryst in Time" (Moore), both published nearly twenty years before Everett's dissertation (in 1938 and 1939, respectively). In Everett's many-worlds interpretation, the wave function of the universe does not collapse. Indeed it *couldn't*, because there is no observer external to the entire universe to observe the universe; instead, the wave function "splits" at every decision point in spacetime. Although this leads to a multitude of realities beyond comprehension, cosmologists like it because it avoids the puzzle of an observer outside the universe. Writer L. Sprague de Camp was an early pioneer in the exploration of this idea in science fiction. In his 1941 novel

Lest Darkness Fall he uses the analogy of a tree (the "main time line") that is always sprouting new branches. The parallel-universe idea is different yet again in that *all* the possibilities *always* exist, independent and parallel in time. Most recently, this approach was used in *The Hemingway Hoax* (Halderman), which stated that "there is not just one universe, but actually uncountable zillions of them," and in *Time and Chance* (Brennert), in which a man and his counterpart in a parallel world briefly exchange places. Other interesting parallel-world stories can be found in Lawrence Watt-Evans' 1992 short-story collection *Crosstime Traffic*. Among them is the particularly clever "The Drifter," which describes the horrible fate of a volunteer for a crosstime experiment at the Princeton physics department. It is discovered, too late, that parallel time tracks are not simply grooves into which you drop, like a ball, after leaving the time track of our world. Each version of a person in each world is not like a ball rolling down a groove from past to future. Rather, each world's time track is just a line on a smooth surface; as the volunteer is told during a temporary stay in a world still close to his (our) original one, "We gave you a push sideways, and you moved off your original line—but instead of dropping into the next groove, you've just kept on rolling across the surface, from one line to the next, at an angle. There are no grooves, nothing to stop you from sliding on across the different lines forever. You have the same futureward vector as you started with, but you've added a small crosstime vector, as well." And so the volunteer drifts crosstime, and gradually the worlds he experiences grow ever more alien.

16. The editorial introduction to this pioneering tale is interesting; the opening line is "To say that this short story contains some revolutionary time-travel theories would be putting it exceedingly mild." The editor then goes on to tell us enthusiastically that "when the author . . . submitted this story to us, his accompanying letter stated that in it he had settled the time-travel question once and for all. We must admit that a broad, unbelieving grin spread over our countenances when the author dared make this assertion. BUT—the smile soon left our faces. . . . [T]o our chagrin, Mr. Daniels had really propounded so many brand new ideas about time and time-travel, and such logical ones—*that he has not left one loophole in his argument!*" John W. Campbell, the first (and only) editor of *Astounding Science Fiction* (today's *Analog*), called alternate-time-track stories "mutant" because they represented the first new innovation (or mutation) in the time travel concept since H.G. Wells. Campbell claimed that *The Legion of Time* (Williamson), originally published by *Astounding Science Fiction* in 1938, was the first such tale (see Campbell's editorial in the May 1938 issue) and that "Other Tracks" (Sell)—this story appeared in the magazine in October 1938—was the second (Groff Conklin, editor of SFAD, wrote that "Other Tracks" was the first and called it "precedent breaking"), but in fact it was Daniels who used the idea first. After Daniels the concept quickly became part of standard science fiction lore and could be used by writers with little explanation. For example, in the 1947 story "A Hitch in Time" (MacCreigh) the author did not have to say much about his "First Law of Chronistics," which determines the development of "the branches of Fan-Shaped time." It was sufficient for his readers to learn that should a time traveler to the past change anything, a parallel branch of time would be created on which the time traveler would be trapped. "The man who interfered with the space-time matrix, displacing even a comma in the great scroll of time, would be cut off from his origin forever." See also Note 42 for this chapter and comments there on "Parallel in Time" (Bond).

17. Self-encounters in mainstream literature have long pre-dated science fiction. One can find a brief description of almost such an occurrence, for example, in Goethe's autobiography from more than a century and a half ago. (*Almost,* because the experience of the young Goethe and the memory of the older Goethe eight years later are not quite the same). And in Osbert Sitwell's 1929 travelogue-disguised-

as-a-novel, *The Man Who Lost Himself,* the author used 280 (!) pages to get to its then-shocking conclusion—a man is apparently killed by his younger self. The 1989 *Back to the Future II* film takes great advantage of the weirdness of an encounter between a time traveler and a second version; in it, such self-meetings occurring for four different characters. Self-encounters also occur in the 1994 movie *Timecop.* Such self-encounters can even be misses and still have great impact. For example, in the 1942 story "Minus Sign" (Williamson) a spaceship fights a battle with itself as it self-interacts while traveling backward in time. In "Let the Ants Try" (Pohl) a time traveler journeys back forty million years. Upon stepping out of his time machine, he hears a "raucous animal cry" from somewhere in the nearby jungle. Later, after other adventures in time, he returns to near the same point in spacetime. After stepping out of his time machine, he sees himself in the distance—the version of himself during the first trip. Then suddenly, the time traveler meets a violent death: "As his panicky lungs filled with air for the last time, he knew what animal had screamed in the depth of the Coal Measure forest." And the time traveler in "The Uncertainty Principle" (Bilenkin) happens across his own grave while on a visit to the Middle Ages; he will die after another trip (in his personal future) even further back in the past. See also the 1950 story "Time Is a Coffin" (Mead). The original *Back to the Future* movie also has a near miss; at the end Marty McFly returns to the future (his present) ten minutes before he left for the past, making it possible for him to watch himself leave.

18. Lucas (see Note 9) had this particular puzzle in mind when he wrote of a loop that avoids direct human self-interaction: "It is very important, not only for reasons of modesty, that I should not be able to use a Time Machine to go into a public library and read my own biography." The protagonist in *Time and Again* (Simak) follows this advice when he discovers and refuses to read a book brought from the future by a time traveler, a book that he *will* write. Robert Heinlein did not agree with this view; in *The Door Into Summer* the protagonist, an inventor, travels thirty years into the future, where he reads some patent disclosures for inventions that he does not remember—but that are under his own name. He returns to his own time and promptly files the patents! And in the earlier "By His Bootstraps" the whole point of the story is the inexorable repetition of events as a time traveler doubles and triples (and more) back on his own world line. "Sam, This Is You" (Leinster) is a very funny story (available today on audio tape as an "X-Minus One" radio drama) about a telephone lineman who starts getting telephone calls from himself from ten days in the future. The first call tells Sam how to make the gadget to transmit such calls. This idea appeared earlier in the 1942 tale "Forever Is Not So Long" (Reeds). In the suspenseful "Party Line" (Klein) a man receives telephone calls from *two* versions of himself, one ten years in the future saying he absolutely must accept an invitation to fly to the Bahamas that very day and another version calling from tomorrow that insists the plane will crash. What should he do? The time traveler in "A Thief in Time" (Sheckley) is constantly running into people who tell him what he will do (because for them he has already done those things). So, too, does the time traveler in "The Best Is Yet to Be" (Murray). A time machine experiment gone wrong allows thirteenth-century Roger Bacon to meet twentieth-century scientists in the 1937 story "Lost in the Dimensions" (Schachner), an encounter that explains the amazing forecasts in his *Opus Maius.* In "Fool's Errand" (Del Rey) we discover how Nostradamus came to make *his* predictions: A time-traveling historian on a visit to A.D. 1528 from A.D. 2211 accidentally gives a copy of the predictions to the prophet while he is still a wild young student. In the film *Bill & Ted's Excellent Adventure* (where we learn that even the not very bright can be time travelers), a set of missing keys is necessary for the successful completion of a task. The two time travelers decide that after the task is done, they will go back in time, steal the keys (*that's* why they're missing!), and hide them so they can use

them *now*. Where should they hide them? Why, "over there," says one of the boys, pointing at a hiding place—and sure enough, when they go over and look, the keys are there. They agree that once they have finished with the keys, it will be *most* important that they really do put the keys in the hiding place! Isaac Asimov made a similar causal loop the centerpiece of his major time travel novel *The End of Eternity*. In it a time traveler journeys back to teach the "inventor" of time travel how to do it, because ancient documents show that this is what happened (will happen). Asimov takes the position in his work that the past can be changed, that new and better realities can be engineered by wise people who guard the centuries, and that it is possible to "change yesterday today for a better tomorrow," a position that this entire chapter argues is illogical. The idea of information "circulating" in a closed time loop is actually an old one. In the 1904 *The Panchronicon* (Mackaye), for example, we learn how Shakespeare came to write his plays: A time-traveling fan from 1898 whispers the magic lines she has memorized (for her literary club meetings) into his ear. Would this make Shakespeare a plagiarist? Of himself!?

19. This isn't to say MacBeath asserts that time travel is impossible. Indeed, he boldly declares that he believes in the logical possibility of time travel, and his paper is devoted to discovering what he thinks is incorrect in Harrison's story. MacBeath does this by retelling the story with what he believes are crucial modifications to make it sufficiently less outrageous to be taken seriously. Our new hero, thawed out from a deep freezer, is Arthur. Arthur is, unfortunately, suffering from total amnesia (this is MacBeath's way of avoiding the problem of Dee remembering he ate Dum and even worse to come), and so when asked his full name, he is himself sufficiently puzzled to reply, "Arthur who?" He is finally called (what else?) Arthur Who. And as you can now no doubt guess, his son becomes Dr. Who! We are told of the son's genius (a 1999 Ph.D. at age 14, with a dissertation on the principles of time travel, is followed by a trip back into the past with his father, by the eating of the father by the son, by the entering of the deep freezer by Dr. Who, etc., etc.) The whole business is quite entertaining and *at least* as complex as Harrison's story. Just how complex is summed up in MacBeath's last, wonderful line: "The Who who was Dr. Who's father was not Dr. Who—that is, not the Dr. Who whose father he was." See Gordon (1987) for a pop-psychology analysis of the "incest urge" as the supposed reason for the fascination of time travel to the past. Karen B. Mann has written on the supposed link between sex and time travel to the past in "Narrative Entanglements: *The Terminator*," *Film Quarterly* 43 (Winter 1989–90):17–27. Mann sees, I think, far more sexual imagery in time travel movies than is warranted, and in at least one case she completely misses the central point of the movie. Specifically, she calls the sexual paradox at the heart of *The Terminator* an "unnecessarily bizarre twist on time travel" and offers instead the banal theory that it is simply an excuse to send "as violent a creature as Arnold Schwarzenegger after a woman." In fact, the movie would be utterly pointless without the sexual paradox. For a very funny spoof of the coupling of time travel and incest, see "Klein's Machine" (Weiner).

20. I agree with the assessment that Heinlein's is the best short time-travel story (although it, too, is biologically flawed), but certainly the most complicated time travel short story has to be one by Pierre Boulle (who is best known as the author of the book that inspired the time travel film *Planet of the Apes*). This story, "Time Out of Mind," describes how two time travelers manage to murder each other, even though one was born 6000 years before Christ and the other 18,000 years later. Boulle takes the position, for which I have argued here, that the past is unalterable (and his tale includes some interesting discussion on free will and determinism). The world lines of the men are, as you might imagine, horribly twisted. As one of them accurately declares, "Our lives are so intertwined, my past with his future, my future with his past, that the gods themselves can't make head or tail of

it any longer!" Certainly I couldn't. As a final twist, the story is told by a narrator who at the end gets caught (along with the reader) in a causal loop.

21. A similar example is given in Horwich (1975). To the claim that we can "rule out Gödel's solutions in the way that we often reject unacceptable mathematical solutions to mathematical problems," Horwich replies in a footnote, "Suppose we have to solve a quadratic equation in order to solve a problem about how many men would be required to dig a ditch, etc. If one of the roots is the square root of minus two it is instantly rejected." This is a poor example, however, because a fundamental result from algebra tells us that when we have a quadratic with real coefficients, if one root is imaginary, then so is the other. The proper conclusion in Horwich's example would be either that the original data are inconsistent or that we have made a mistake in constructing the quadratic equation itself. It is significant, I think, that when Horwich rewrote this essay for inclusion in his excellent book (1987), he deleted that footnote.

22. J.A. Stratton, *Electromagnetic Theory* (New York: McGraw-Hill, 1941), p. 428. Like all the other laws of physics (except for the rare K mesons mentioned in Chapter Three), Maxwell's equations are indifferent to the sign of time. However, as the Stanford physicist Leonard Schiff has so nicely demonstrated (NT), the Maxwell equations lose this indifference if there is such a thing as a magnetic monopole (either a north or a south pole *alone,* without the other kind attached). As I write, monopoles have not been observed, even by those who have looked very hard for them. In fact, Tipler (1975) has shown that the absence of monopoles can be interpreted as experimental *support* for the direct-interparticle-action theory of Wheeler and Feynman (1949).

23. See "The Love Letter" (Finney) for a masterful demonstration of the emotional impact that such backward-in-time communications might have for lovers irrevocably separated in time. "When You Hear the Tone" (Scortia) does the same with telephone calls to the ever-more-distant past, with a clever twist at the end that lets the lovers finally meet. "Line to Tomorrow" (Padgett), on the other hand, uses a similar plot device for a grim conclusion leading to insanity and suicide.

24. Not everybody appreciates this. For example, in *Utopia 239* (Gordon) three time travelers from 1958 journey a century into the future to escape political oppression (and a coming atomic war). At no time is there any discussion of paradoxes resulting from this trip, and yet the writer of the dust jacket blurb for this novel declared, "*Utopia 239* is a novel about the future and about time travel. Time travel raises many problems and paradoxes, but the author solves them ingeniously." I didn't see it. The one occasion where Gordon even hints at a paradox is early in the story, where he has one of the characters mention the issue that Hawking has used as "experimental evidence" for his chronology protection conjecture against time travel to the *past:* "If you imagine I've invented a time-machine which could take us back a hundred years, you are [wrong]. Even a layman knows that no one has ever come back from the near or distant future." This isn't to say one can't imagine a paradoxical situation during a forward-directed time trip. Suppose a time traveler goes into the future and, before returning to his normal time, finds himself. Suppose further that the two quarrel and the time traveler kills his future self. This is not paradoxical, but rather a delayed form of a sort of suicide. But suppose it is the time traveler who is killed. This is then just a variant of the grandfather paradox, and all the previous arguments against it apply here, too. It is crucial to note that the assumption that the time traveler does in fact find an older self in the future implies that a backward journey will take place. If a time traveler journeys forward and stays in the future, then he will *not* find another version of himself there.

25. My example is somewhat flawed, of course. For the gadget to call ahead in time means that some sort of signal, as yet unspecified, has traveled into the future (because *something* made the future phone ring). For the present discussion I have ignored this issue for the sake of the dramatic impact of the example. Soon, however, I will describe in some detail how one might, in principle, actually build this gadget, which in the physics literature is called an antitelephone. (Such a device is an *anti*telephone because the person who is the receiver is in the sender's *past,* the opposite of the case for an ordinary telephone.) For an interesting science fiction use of this idea, told in first-person narrative as a hard-boiled detective murder mystery, see "Worthsyer" (Schmidt). The gadget in this story is called a time telephone, but no theory for its operation is given. It is merely a "straightford application of an impressive, but limited, technology." (A grand, impressive understatement, I think!) This sort of signal transmission was also used in "You Want It When?" (Dalkey). In that story a character who discovers how to send messages backward in time soon thereafter receives a congratulatory note from herself, from ten years in the future. In Forward (1980) we are told that we can send messages into the past by compressing a 15-billion-ton asteroid into the volume of an atomic nucleus (!), spinning it, and then aiming frequency-coded gamma ray bursts through the nearby region of "unhinged time." This is, of course, an artificial Kerr–Newman black hole telegraph. Forward, evidently an optimist of the first rank, thinks we will be able to do this before the end of the twenty-first century. It would seem, then, that what should be done now is to build gamma ray receivers (well within present-day technology) and listen for such messages from the future. The technical details of such receivers would not matter, of course, as long as they are put into the public domain. That way the future will know what those details are merely by reading about them from a musty library book and it will build its transmitters to be perfectly compatible!

26. *Punch* 165 (December 19, 1923): 591.

27. Oddly enough, Isaac Asimov dismisses FTL motion (1984a) without once mentioning the time travel connection, which he had totally rejected just a few months before (1984b). Some discussion on a non–time travel method for achieving FTL motion has appeared in the philosophical literature. Resurrecting the matter-transmitter idea of Weingard (1972a) is the suggestion in Elliot (1981) that "one's biochemical details are recorded and one's body is then disintegrated. The recorded biochemical blueprint is electro-magnetically transmitted to the arrival center at the desired destination." (Seeming to be unaware of this philosophical analysis, many later writers on this idea refer instead to the beam transporter in television's "Star Trek"). Elliot's idea is plagued by all sorts of obvious problems (not counting that "disintegration" business), such as how the "arrival center" got there in the first place, the failure of distant simultaneity, and personal identity (the arrival center will reconstruct you from its element bank). For more on the brief debate that Elliot's suggestion sparked, see Brennan (1982), Ray (1982), and Elliot (1982). Some years later, an analysis of personal identity crises via time travel appeared in Ehring (1987), which showed how our normal ideas of what constitutes distinct persons (different bodies, causal independence of those bodies, and the lack of a shared consciousness) fail in the case of a time traveler who journeys backward to talk to himself. To an observer this may look like two people talking, but in the spacetime view it is just one bent-back world line—that is, *one* person. A similar view is expressed by Dainton (1992), who argues that a time traveler meeting himself in the past is, despite appearances, "the same embodied person." Lawyers may disagree; the potential legal problems of a duplicated time traveler are explored in "Double Indemnity" (Sheckley).

28. See Rothman (1960) for the usual "things," such as the intersection point of two very long, closing scissor blades. The explanation for such things is that they are

massless, do not participate in a causal chain, and carry no information, so special relativity is not violated. Much more perplexing, at first glance, are the FTL phenomena that astronomers have observed in space; for instance, Supernova 1987A in the Large Magellanic Cloud displays apparent expansion speeds of up to twenty times the speed of light. Until recently all such superluminal appearances were in other galaxies, but in 1994 the first one in our own Milky Way was observed at a distance of about 42,000 light-years. Such events are thought to be light-echo optical illusions. For more on this and other proposed explanations, all within the bounds of relativity, see Rees (1966), which explained the mechanism of apparent superluminal speeds *before* the phenomenon was first observed! See also Blandford, McKee, and Rees (1977) and Sheldon (1990). A less well-known superluminal phenomenon is the "tunneling photon" wave packet in quantum optics. Experimentally, photons that pass (tunnel) through a nearly perfect reflective mirror are observed to have an average tunneling speed significantly greater than that of light. See Chiao *et al.* (1993) for a prose explanation of how this can be so without violating special relativity, and see Deutch and Low (1993) for the *mathematics* of it all.

29. That is, even though "faster than light" means "backward in time," which means "causality failure," special relativity still holds true and nothing awful happens to physics, only to our intuitions. The reason for this is simply that causality, contrary to common belief even among many physicists, is not a premise or starting point of special relativity. See Nerlich (1982) for an extended discussion on just this point. Kirzhnitz and Polyachenko (1964) had vigorously made this point even earlier, before studying the possibility of superluminal sound in superdense matter.

30. The possibility of FTL "cosmic communication" was discussed by Herbert Dingle (see Tech Note 5) in a letter to *The Observatory* 85 (December 1965): 262–264. Dingle rejected special relativity (he called it "no longer tenable") and claimed that "the Doppler effect affords a means of instantaneous communication over any distance at all." Dingle was rebutted by R.F. Griffin in a reply immediately following, and Dingle countered in the same journal [*The Observatory* 86 (August 1966): 165–167]. Dingle had earlier (in 1960) written in much more detail on this in *The British Journal for the Philosophy of Science* 11 (May 1960): 11–31 and 11 (August 1960): 113–129. I know of no one who thinks Dingle was onto anything, and in any case he said nothing about communication with the past, a concept that, given his conservative, classical approach to physics, would surely have rendered him unconscious on the spot. Poul Anderson, trained as a physicist and usually careful with the science in his science fiction, ignores the backward-time-travel effect in an instantaneous communication system in his story "Dialogue." So does the 1939 story "Lightship, Ho!" (Bond), wherein a man on Pluto invents a way to send messages to Earth at twice the speed of light. (A would-be dictator is on his way to Earth in a light-speed rocketship, and only an FTL message can warn Earth in time.) Explaining the flaw in this signaling scheme, which Bond describes in some detail, would make a good question on a Ph.D. qualifying examination in physics or electrical engineering. See Diener (1996) for the mathematics of why information cannot be transmitted at FTL speeds by a modulated signal. A short, elegant horror story that plays with the implications of being able to outrun light but that also overlooks the time travel angle is "Time Fuze" (Garrett). In this tale, FTL spaceships have a fatal flaw—their use induces nearby stars (including the sun) to go nova. Because the first such ship outflew the light of the sun's explosion upon the ship's departure, the crew don't see the destruction (of the entire solar system!) until their return.

31. H. Tetrode, "Uber den Wirkungszusammenhang der Welt. Eine Erweiterung der Klassischen Dynamik," *Zeitschrift Fuer Physik* 10 (1922): 317–328. Tetrode's imagery was, curiously, captured decades before in the words of the nineteenth-century English poet Francis Thompson in his "The Mistress of Vision."

> All things . . . near and far,
> Hiddenly to each other linked are,
> That thou canst not stir a flower
> Without troubling of a star.

It is interesting to note that Einstein apparently said nothing to Wheeler and Feynman about a paper that pre-dated Tetrode's by three years. In 1919 the Swedish physicist Gunnar Nordström had suggested that the advanced solution might offer an explanation for a perplexing problem in atomic theory. Maxwell's theory says that an accelerated electric charge radiates energy, which implies that the orbital electrons in the classical (Bohr) model of the atom should quickly spiral in toward the nucleus; i.e., all matter should collapse. This cataclysmic event, of course, has not happened, and Nordström's idea was that if one took into account not only the usual retarded solution but the advanced one as well, then perhaps things could be understood. Indeed, Nordström was able to show that such an analysis does indeed give zero for the *average* energy radiated by an orbiting electron. Later, however, Page (1924) showed that the instantaneous radiated energy is not zero, and that this would lead to observable effects that in fact are *not* observed.

32. Yet, as pointed out in Gold (1962) and Cramer (1980), it is not clear that Wheeler and Feynman's claim to have avoided the self-interaction problems by referring to advanced fields is a step forward. Indeed, the self-interaction of the electron is needed to explain the 1947 experiment by Lamb that measured the deviation (the "Lamb shift") of the spectrum of hydrogen from what Dirac's theory of the electron predicted. Ironically, it was this experiment that helped inspire the renormalization of quantum electrodynamics to get rid of the infinities then plaguing it, which led to Feynman's share of the Nobel Prize in 1965. In fact, just four years after Wheeler and Feynman's first joint paper, Feynman (1949b) expressed his revised view that self-interaction could not be avoided. See Teitelboim (1970) for an explanation of how to get the radiative reaction without invoking advanced fields.

33. In a footnote, Earman (1969) cites a French source that he says does contain negative reactions by physicists to Wheeler and Feynman's "solution" to their bilking paradox. Earman himself calls their solution "brazen."

34. This isn't to say that Cramer dismisses advanced waves as a logical impossibility. Indeed, he has written (1983) on the Wheeler and Feynman absorber theory and has proposed an ingenious idea explaining why we see only the retarded solutions to Maxwell's equations and not the advanced ones. When we visualize the singularity of the Big Bang as a reflector of advanced waves back to the future (after all, you cannot go back in time further than the Big Bang), advanced waves cancel themselves, leaving only the retarded ones. As Cramer has noted, this idea has an immediate cosmological implication: The universe must not be fated to suffer a collapse in the distant future into a Big Crunch, because then that singularity would be a reflector of the retarded waves (and thus cancel them out too). In such a case, Cramer's theory says that we would observe *no* solutions to Maxwell's equations. Of course, in that case, we would not be here not to observe anything. Cramer has also put forth an ingenious idea that explains how the thermodynamic arrow of time is a consequence of the electromagnetic arrow (1988a). Observing that Boltzmann assumed, in his proof of the H-theorem (see Note 19 for Chapter Three), that the motions of molecules of gas are correlated *after* they collide, Cramer says that Boltzmann's assumption is consistent with a retarded electromagnetic arrow—that is, with delayed electromagnetic interactions between molecules. If, on the other hand, we lived in a universe with advanced waves, then molecular motions would be correlated before collisions, which would result in entropy decreasing with time. Thus, for Cramer, the expansion of the universe (cosmological arrow) via its lack of a Big Crunch allows for the (retarded) electro-

magnetic arrow, which in turn results in the observed thermodynamic arrow of increasing entropy. And finally, it is this thermodynamic arrow that gives rise to the one-way chemical reactions in our brains that result in our sense of the subjective arrow of time (the "moving now"). As stated in Schulman (1977), "consciousness and the thermodynamic arrow ought to go in the same direction." And yet Schulman had to admit that although this is "reasonable," it is also a "difficult to prove hypothesis."

35. Blish is actually pretty close to the mark here. A composite signal with a continuous spectrum (with the energy distributed uniformly in frequency), such as one might expect the overlay of many independent signals to be, does indeed have a narrow time structure. If applied to a loudspeaker, such a signal would sound like a sharp pulse or click—or even a *beep*. In the limit of an infinitely wide spectrum, the time signal becomes one of infinite amplitude and zero duration, a singular impulse function called (by theoretical physicists and radio engineers alike) the "Dirac delta function."

36. In fiction, Edward Page Mitchell, the Victorian pioneer in the time-travel-paradox story genre, who was mentioned in Chapter One, described a gadget (called, in anticipation of Feinberg, the *tachypomp*—literally "quick sender") for reaching any speed, no matter how great. This story originally appeared in the March issue of *Scribner's Monthly*—for 1874! The usual massive particles of physics, moving at less than the speed of light, are often called by the obvious (if vulgar) name *tardyons*. Disliking this and observing the classical origin of the word *tachyon*. Fox, Kuper, and Lipson (1970) and Bers *et al.*, (RAG) suggest that normal particles be called *ittyons*, from the Hebrew for "slow." This suggestion has not to my knowledge had much success. The tachyon, also called the *psitron*—see Dobbs (1965)—by psychical researchers, is well known on paper, but it has yet to be observed experimentally, although there have been extensive searches for it. See, for example, Alvager and Kreisler (1968), which describes an unsuccessful attempt to create electrically charged tachyon-pairs by bombarding a lead target with gamma rays and then looking for Cerenkov radiation (electromagnetic radiation emitted by an electric charge exceeding the speed of light in the local medium). The theory behind this experiment was declared to be faulty in Jones (1972). The idea of looking for Cerenkov radiation as a signature for tachyons has already appeared in time travel science fiction, such as *Time of the Fox* (Costello). Lapedes and Jacobs (1972) suggest that the same effect should exist for gravitational radiation, but at a much reduced rate; a tachyonic electron would electromagnetically lose all its energy in a tenth of a nano-nanosecond, whereas a tachyonic neutron would take two million years to radiate its energy gravitationally. How to detect this faint gravitational radiation is an open question. There have been analysts who have questioned the perhaps too-facile assumption that tachyons would be small. For example, Schulman (1971b) concludes that macroscopically sized tachyons with nuclear densities should produce gravitational shock waves detectable "across the entire solar system." Several years later, Clay and Crouch (1974), following the lead of Murthy (1971), reported the possible detection of tachyons via the precursor indication of subsequently measured Earth-surface air showers induced by cosmic rays in the upper atmosphere. That is, FTL tachyons produced high above the surface reached the ground before the main burst of subluminal particles. Later researchers, however, have been unable to repeat this work. There have been various explanations put forth for these failures, in addition to the direct one that tachyons simply do not exist. To explain their non-existence, Bers *et al.* (RAG) argue that tachyons are *unstable* solutions to the Klein–Gordon wave equation that describes the motion of particles with the tachyonic properties of superluminality, imaginary mass, and zero spin. In contrast, Peres (1969) asserts that the quantum wave function of a tachyon is such that tachyons cannot be localized in space and so are

simply unobservable. See, for example, the infinite-velocity tachyons in Tech Note 7, which are "everywhere at once." Drawing the same conclusion from a different angle is Kirch (1975), who asserts that the mass of a tachyon is so low as to be virtually zero, whereas Wimmel (1972) and later Corben (1976) claim that the failure to observe Cerenkov radiation is of no significance because a charged tachyon would not be a Cerenkov emitter, either electromagnetically or (even if uncharged) gravitationally. Despite that, Kowalczynski (1978) proposed an experiment using photons to detect the "electromagnetic shock surface" (Cerenkov radiation) of a charged tachyon. Parker (1969) suggests that the scattering of photons by photons could be used to test indirectly for tachyons. Because of the very low cross section of photons, however (on the order of 10^{-65} cm^2), the probability of a photon scattering off another photon is very small, and so very intense photon beams would be necessary to produce observable results. See Csonka (1967) for several imaginative suggestions on how sufficiently brilliant beams might be generated. Davidson and Owen (1986) have suggested that all observed particles are actually the projections of tachyons in five-dimensional spacetime into our four-dimensional world, and Davidson (1987) has extended this idea to explain the extreme smallness of the periodicity of the fifth dimension (recall Note 15 for Chapter Two). Cramer (1993) discusses the possibility that the electron neutrino (so called because it is created when tritium decays via the emission of an electron–neutrino pair) may be a tachyon, a suggestion first made in Chodos *et al.* (1985). In any case, if tachyons are one day discovered, then, as an old joke puts it, the day before that momentous occasion a notice from the discoverers should appear in newspapers, announcing "Tachyons have been discovered tomorrow."

37. My source for Lucretius' words is the translation of Book 4 by John Goodwin (Wiltshire: Aris and Phillips, 1986), p. 23. A more poetic allusion to faster than light appears in Shakespeare's *Romeo and Juliet*. In Juliet's words (Act II, scene 5) " . . . love's heralds should be thoughts, Which ten times faster glide than the sun's beams, Driving back shadows. . . ."

38. The British electrical engineer Oliver Heaviside theoretically studied FTL charged particles as long ago as 1888, using nonrelativistic classical physics; see my book *Oliver Heaviside: Sage in Solitude* (New York: IEEE Press, 1988), pp. 124–126. Tanaka's work was later discovered not to be relativistically invariant. The question of invariance was raised first in Broido and Taylor (1968) and later in Jones (1972), which pointed out that the original quantum theory of tachyons (which we owe to Feinberg) is not relativistically invariant. Attempts by tachyon advocates to address this failure can be found in Arons and Sudarshan (1968) and in Dhar and Sudarshan (1968). Just before his death in April 1992, Feinberg coauthored (with two philosophers) a paper in which he seems to have abandoned his support for tachyons as possible carriers of information backward in time. The conclusion in Feinberg, Albert, and Levine (1992), which never mentions tachyons, is that knowledge of both past and future is *necessarily* limited.

39. Some recent theoretical work hints at the possibility of FTL photons! The effect is very small, however—well below any present or near-future experimental means of detection; the increase in speed is predicted to be at best one part in 10^{36}. See K. Scharnhorst, "On Propagation of Light in the Vacuum Between Plates," *Physics Letters B* 236 (February 22, 1990):354–359. More abstract was the earlier work by Drummond and Hathrell (1980), where they show how the *failure* of general relativity's principle of equivalence (all local inertial frames are on equal footing) for quantum fields in curved spacetime leads to FTL photons. The authors present convincing explanations why the classical FTL communication paradoxes would *not* occur. Their work was later extended to the generation of FTL photons in the vicinity of a charged black hole, in Daniels and Shore (1994). Even more provoca-

tive are the calculations in Band (1988a,b) showing how to achieve an electromagnetic (i.e., photonic) wave packet with an FTL group velocity. Electrical engineers and physicists almost intuitively associate the group velocity with the speed at which information is transferred, but Band reminds his readers that this speed is really the signal speed (the speed at which a "well-defined shape" propagates). The group velocity in an absorption band in a dispersive medium can indeed be faster than light, but in this case the signal speed is still subluminal. These points are discussed in Leon Brillouin's beautiful little book *Wave Propagation and Group Velocity* (New York: Academic Press, 1960). See also Fox, Kuper, and Lipson (1969, 1970), which explains why group velocity has no physical significance for tachyons. Band's analysis, however, is interesting because he arrives at his result using a non-dispersive medium, so the signal speed does appear to be truly FTL. But then he pulls the rabbit out of the hat—his entire analysis depends on a geometry that has cylindrical symmetry (such as a coaxial cable or a transmission line). Thus, if such an FTL cable connects A to B, then the addition of a return cable from B to A would ruin the symmetry; it is only through the presence of a *closed* communication path that paradoxes could occur. Band states that "there is a fascinating similarity" between his superoptic wave packet and a tachyon, but he is not ready to go so far as to claim that the wave packet is a realization of the tachyon. In 1986 Henning F. Harmuth, a professor of electrical engineering at the Catholic University of America, caused a minor stir by publishing three papers that he claimed *correct* Maxwell's equations! In particular, he devoted much space to a discussion of the signal speed. All three papers were published by the Institute of Electrical and Electronics Engineers (IEEE) over the objections of referees who unanimously recommended rejection. The reviewers' comments (and the author's replies) were printed at the end of each paper. So far as I know, these papers had no lasting impact on the electrical engineering or physics communities, but they do make fascinating reading. See the *IEEE Transactions on Electromagnetic Compatibility* 28 (November 1986).

40. There is a curious implication in this process. Even if we grant that the reinterpretation principle may avoid causal paradoxes (but see Tech Note 7 for reasons why most physicists in fact think it doesn't), the fact is that physics isn't fooled as easily as a human observer. That is, the receiver does actually lose energy upon the arrival of the tachyon, which is the opposite of what normally happens in a radio receiver when it receives a photon. Thus the receiver must be in an elevated energy state prior to that arrival; it must be prepared beforehand to receive a message. If the receiver were passively sitting in its lowest energy state, it could not accept (or *eject,* according to the RP) the tachyon.

41. Even before Bell's theorem and the EPR paradox, however, the potential link between quantum systems and faster-than-light signaling had attracted attention. The possibility that two atoms (A and B) might be able to "communicate" (by A emitting a photon and then B absorbing it) at superluminal speed was first considered by Enrico Fermi in 1932. Probably because it seemed to confirm the common-sense view that such a possibility was actually an impossibility, his work attracted little notice. Later, however, Fermi's mathematical analysis was found to be in error. Most recently, Hegerfeldt (1994) has proved (in the context of standard quantum mechanics) a very general theorem that establishes the non-zero probability of B responding to A's emission *as soon as* the emission occurs. This hint at a possible failure of causality so stunned the editor of *Nature* that he quickly wrote a calming reply entitled "Time Machines Still Over Horizon" (February 10, 1994, p. 509). An excellent summary of the controversy in quantum mechanics in general and of the work by Bell and Aspect in particular (including interviews with both men) is *The Ghost in the Atom,* edited by P.C.W. Davies and J.R. Brown (New

York: Cambridge University Press, 1986). The EPR paradox has not been the only source of quantum mechanical speculations about nonlocal effects. The Aharonov–Bohm effect, a quantum mechanical phenomenon that allows the motion of a charged particle to be influenced by electromagnetic fields existing where the particle never goes, has also sparked a decades-long controversy. The history of this effect, from its 1959 prediction to its experimental verification in 1986, can be found in M. Peshkin and A. Tonomura, *The Aharonov–Bohm Effect* (New York: Springer-Verlag, 1989).

42. Polchinski (1991) uses the somewhat inconsistent name of EPR phone for the Bell antitelephone. Unlike the tachyonic antitelephone, the EPR version is a quantum antitelephone, and, as both Gisin (1990) and Polchinski have shown, if quantum mechanics is actually slightly non-linear (as are many other phenomena in physics at sufficiently high energy levels) and if one adopts the collapsing-wave-function interpretation of quantum mechanics (recall Note 15), then the EPR antitelephone will work. Polchinski has even suggested a far more radical quantum antitelephone called the Everett–Wheeler [anti]phone. Again, if one supposes that quantum mechanics is actually non-linear and if one accepts Everett's many-worlds interpretation of quantum mechanics, then Polchinski has shown that one could communicate not just with the past, but also with the infinitude of the many-pasts in the ever-splitting branches of the many-worlds! A very funny illustration in Cramer (1991a) shows what that might be like. Polchinski is well aware of the spectacular nature of his results. As he writes, "Do the results imply that nonlinear quantum mechanics is inconsistent, and thus 'explain' the linearity of the theory? Communication between branches of the wave function . . . seems even more bizarre than FTL communication . . . but it is not clear that it represents an actual inconsistency." See "Trans Dimensional Imports" (Farber) and *The Quicksilver Screen* (DeBrandt) for science fiction uses of this idea. And just to show that even with this seemingly outrageous concept, there is little that wasn't anticipated in *early* science fiction, see the 1940 story "Parallel in Time" (Bond). In this tale an inventor, trying to build a radio with which to signal Mars, accidently stumbles on the "temporal-aberrant carrier wave" and thus establishes communication with a parallel universe, a universe that forked off ours when Lee *won* the Battle of Gettysburg in 1863. The possibility of communication between splitting branches is taken seriously by others besides Polchinski. For example, the Oxford theoretician David Deutsch (1991) writes that "closed timelike lines would provide 'gateways' between Everett Universes." Deutsch has believed this for some years, in fact; earlier he described how to design super-fast quantum computers that solve different parts of a problem in different Everett worlds and that then collect the partial results from the different worlds by effectively photographing the other worlds—see Albert (1988)! As Whitaker (1985) has written (without exaggeration, I think), such an achievement would be "perhaps the most amazing discovery in the history of science, indeed in the history of mankind, access being gained in principle and to at least a reasonable extent in practice, to other possible modes of development of the universe, and to other conceivable types of existence on the Earth. . . ." More recently Plaga (1997) wrote, "interworld communication would lead to truly mind boggling possibilities." Some of these possibilities have been described in *Paths to Otherwhere* (Hogan), a novel dedicated to Deutsch. Some physicists are using quantum mechanics in time machine analyses in somewhat less (but not much less!) exotic ways. Klinkhammer (1991) makes an attempt at applying quantum field theory to the averaged energy conditions that play an essential role in wormhole time machines; Echeverria, Klinkhammer, and Thorne (1991) redefines the Cauchy problem in terms of quantum mechanics (so that "well-defined" is given a stochastic interpretation) and thereby shows that a simplified form of the grandfather paradox is *always* self-consistent; and Kim and Thorne (1991) and

Frolov (1991) examine the possibility that fluctuations in quantum fields might destroy the Cauchy horizon that such time machines require. For more discussion on these points, see Tech Note 9.

43. N.D. Mermin, "Is the Moon There When Nobody Looks? Reality and Quantum Theory," *Physics Today*, April 1985, 38–47. For more on the enigmatic letter on FTL effects, see Jack Sarfatti's letter to *Physics Today*, September 1987, 118, 120.

44. Cramer was specifically discussing a suggestion in Datta, Home, and Raychaudhuri (1987) for an EPR/FTL communication system. That paper received several highly critical replies pointing out crucial errors, all of which responses appeared in the December 1987 issue of the same journal. Earlier, Herbert (1982) described FLASH, the "First Laser-Amplified Superluminal Hookup." Years later, in Herbert (1988), the author admitted, "The FLASH scheme . . . simply doesn't work" as an FTL communication system. But all is perhaps not lost concerning the military usefulness of quantum communication, as shown in Ekert (1991). This paper offers a theory on how the experimental setup used by Alain Aspect and his co-workers to test Bell's theorem might be used to send cryptographically unbreakable coded messages.

45. A discussion that specifically addresses the possibility of explaining the "spooky actions" of quantum mechanics by backward causation, and certain causality paradoxes, appears in Sutherland (1983).

46. The last line puzzled me enough that I wrote to Hawking about it. On March 7, 1996, he replied: "It was a joke. I wasn't referring to time travellers bringing back information from the future. I take it the future has to be a self-consistent solution of the field equations even in the presence of CTCs. Rather it is that our ability to predict the future is reduced if there is loss of quantum coherence." Because Hawking did write of people returning from the future (which sounds like time travelers to me), and if they tell us what they did in the future, then I still wonder what is wrong with their memories (which is what I asked Hawking in my letter).

47. L.M. Feldman, "Short Bibliography on Faster-Than-Light Particles (Tachyons)," *American Journal of Physics* 42 (March 1974):179–181.

Epilogue

1. There is *perhaps* one possible exception to this statement: the case of two English academics from St. Hugh's College at Oxford. The Misses C.A. Moberly and E.F. Jourdain, while on holiday at Versailles in 1901, inadvertently made (or so they came to believe) a trip back to the year 1789. Moberly wrote the book *An Adventure* (London: Macmillian, 1911; first published under the pseudonyms "Elizabeth Morison" and "Frances Lamont") describing their experience, and an enthusiastic summary and evaluation of the affair can be found in Phillips (FSFS).

2. It is amusing to note that this same sort of argument occurred in the literary world as well. The two great masters of the "scientific romance" in the nineteenth century were of course H.G. Wells and Jules Verne. Verne was the "practical engineer," Wells the visionary who looked far beyond just the next few decades. Verne, for example, got his characters *From the Earth to the Moon* in 1872 in a spaceship powered by "conventional means" (400,000 pounds of guncotton stuffed into a 900-foot cannon!), whereas Wells used "Cavorite" (a metallic alloy that blocks gravity) in his 1901 *The First Men in the Moon*. In a 1903 magazine interview, Verne revealed how he felt about the difference between his and Wells' work: "It occurs to me that his stories do not repose on very scientific bases. . . . He invents. I go to the moon in a cannon-ball, discharged from a cannon. . . . He goes to Mars [sic] in an airship, which he constructs of a metal which does away with the law of gravitation. *Ça c'est*

très joli, but show me this metal. Let him produce it." Today the cry from those who dislike wormholes is, of course, "show us the exotic matter!"

3. S. Chandrasekhar, *Eddington: The Most Distinguished Astrophysicist of His Time* (Cambridge, England: Cambridge University Press, 1983), p. 48.

4. I have made this point before in this book, but Hawking is such an enormous presence in physics today that I think this particular, erroneous argument of his needs to be firmly rebutted. To quote the concluding paragraph of Headrick and Gott (1994), "Finally, we would like to point out that, with any general-relativistic time machine, events prior to the construction of the time machine, i.e., prior to the Cauchy horizon, will be inaccessible to time travelers. *Since we have constructed no time machine yet* [my emphasis], 'the fact that we have not been invaded by hordes of tourists from the future' does not constitute evidence for or against the possibility of future time-machine technology."

5. Thorne was, at one time, the leading American researcher on time machines (he has at least temporarily dropped that work in favor of research in gravity wave detection), but Wheeler's letter shows that there is a certain irony in this. In the early 1970s Thorne collaborated with Wheeler (and with Charles Misner of the University of Maryland) on their gigantic *Gravitation* (W.H. Freeman, 1973). A desk-thumping 1279 pages overflowing with mathematics, this book has the heft of the New York City telephone directory and speaks to every aspect of the general theory—except one. As Wheeler wrote me, one day he took Thorne around to meet Gödel at the Institute for Advanced Study, and they found that Gödel wanted to talk only of the rotating-universe idea. Nearly a quarter of a century after the fact, his discovery of a time travel universe ("populated," wrote Wheeler, who had heard Gödel's original 1948 lecture at the Institute, "by entities of a strange new character, closed timelike lines") still fascinated the old master. To their chagrin, however, Gödel's visitors "had to confess that we had not discussed the subject in our book; had not even considered it part of our task to look into any astrophysical evidence on the point." When he met Gödel, Thorne would not have believed what part of his own research would be less than twenty years later.

6. See the cover story on Thorne's time travel research in *California Magazine* (October 1989). Thorne has been careful to state clearly the motivation behind his research in order to avoid the fate of the time traveler in "Forever Is Not So Long" (Reeds), who laments the fact that many people associate time machines with crackpots. As Reeds writes, "For time machines . . . were things to be left to H.G. Wells. . . . It was no doings for a man of action and, above all, for a man of science." Even more blunt is the character in "Advice from Tomorrow" (Reynolds) who describes another character, a physicist, as "a crackpot. He used to have quite a reputation, but the last couple of years he's been working on time. . . . You know, time travel, that sort of rot. An A-1 crackpot." A non-fictional example of such negative reactions can be found in William D. Gray's *Thinking Critically About New Age Ideas* (Belmont, CA: Wadsworth, 1991). Gray is a philosopher who seems surprisingly unfamiliar with the time travel literature in both physics and philosophy. For example, to support his claim that time travel is *a priori* illogical nonsense (and so anybody who studies time travel is a fool), he cites the grandfather paradox as though it actually proves something. ("If I went back to a point in time before I was born . . . I could arrange it so my parents would never meet!") To be fair to Gray, Hawking (1992) makes this same gaffe. But then, unlike Hawking, Gray declares that he finds time travel to be unintelligible (his earlier statements have convinced me that he is indeed confused about time travel), and he concludes with the chilling words "If a concept is unintelligible, scientific efforts to validate it are a waste of valuable resources."

7. Heinlein was referring to Simon Newcomb (1835–1909), who, you may recall from Chapter Two, was specifically named by H.G. Wells in *The Time Machine* as the expert who had lectured "to the New York Mathematical Society only a month or so ago" on the fourth dimension. At the turn of the century, Newcomb published "proofs" that it would be impossible with known science to build a "practicable machine by which men shall fly long distances through the air." See, for example, Newcomb's essays "Is the Airship Coming?" *McClure's Magazine* 17 (September 1901): 432–435 and "The Outlook for the Flying Machine," *The Independent* 55 (October 22, 1903): 2508–2512. Frank Tipler could not resist concluding his pioneering doctoral dissertation (1976b) on time travel around rotating cylinders with an amusing, cautionary reference to Newcomb's proofs. The physicist Gregory Benford has described *his* fear of repeating Newcomb's mistake when his 1970 *Physical Review* paper "The Tachyonic Antitelephone" was in the germinal stage. As he wrote (1993), "My two coauthors were David Book and William Newcomb. Newcomb was the grandson of the famous Simon Newcomb, an astronomer who wrote the infamous paper showing why airplanes could not fly. When he happened to mention this over a beer, my alarm bells went off. Was I signing onto a similar blinkered perspective, to be cited with ridicule generations later?" Perhaps one shouldn't be too hard on Newcomb, however. Even H.G. Wells had some reservations about airplanes. In 1901 he believed they would be developed, eventually, but probably not for a long time. Writing that year in his *Anticipations,* he said, "long before the year A.D. 2000, and very probably before 1950, a successful aëroplane will have soared and come home safe and sound." The concept of world-girdling commercial airlines decades *before* 1950, however, would surely have struck even the "inventor of time machines" as absurd. But we all know what happened at Kitty Hawk just two years after Wells wrote his words.

8. In a recent work, Bud Foote (1991), a professor of English at Georgia Tech, has stated that consistency is simply a well-used plot device: "the attempt of the time traveler to prevent something or take advantage of it [and so causing] the event in question, is so popular and so ubiquitous that it seems to be about worn out." Worn out or not, I believe that plot device to be correct science. A few years ago the novella *Hawk Among the Sparrows* (McLaughlin) was considered science fiction. Today this work would, as a violator of consistency, be labeled fantasy. The story is of a nuclear bomb test that tosses a Mach 4 jet armed with atomic missiles back in time, from 1985 to 1918, to the days of aerial combat with the Lafayette Escadrille. The story is oblivious to the paradoxes of changing the past, and the hero never displays the slightest concern about how he rewrites history. I think it interesting that a recent "change-the-past" novel, *Lord Kelvin's Machine* (Blaylock) is being correctly promoted as fantasy and not as science fiction. Fiction for young readers has already begun to reflect an understanding of this point as well. For example, the time-traveling professor of history in *The Trolley to Yesterday* (Bellairs), who tries (and fails) to prevent a slaughter of innocents during the 1453 storming of Constantinople, shows he has learned that the past cannot be altered when he declares, "I was just trying to change the course of history, but it seems that there is some evil genie in the Universe that won't allow you to do things like that." A few years earlier, the novel *The Danger Quotient* (Johnson and Johnson) made a masterful presentation of the principle of self-consistency for young adults. And the children's novel *Voices After Midnight* (Peck) also gets the principle of self-consistency correct, with its story of time travelers to the past saving their own ancestors from premature deaths. Even the recent mainstream novel *Crossover* (Fubank), in which mind transference is the explanation for time travel, has its discussion of the paradoxes of changing the past right. Yet some writers of fiction still have a lot to learn. For example, two recent books for pre-teenage children connect

time travel with ghosts—*The Ghost Inside the Monitor* (Anderson) and *Something Upstairs* (Avi). The latter tale is particularly unfortunate, because it claims the past can be changed. Both of these short novels offer only the supernatural for explaining time travel. In the otherwise admirably written 1997 novel *Days of Cain* (Dunn) we again find the "time police." In the "time warp trio" series of children's books by Jon Scieszka, the mechanism behind time travel is a magical book entitled *The Book*. Those stories are directed at the 7-to-11 age group, but there are adults who also associate magic with time travel. For example, in the 1993 film *Army of Darkness,* chanting spells from the *Necronomicon* (made famous in H.P. Lorecraft's 1920s tales in the pulp horror magazine *Weird Tales*) is the way back to the past. The entire film is simply silly, and its "explanation" for time travel is no exception. And to the day he died, the great Isaac Asimov failed to understand the consistency issue. In his last robot story, Asimov tells of a robot sent two centuries into the future. At the start of "Robot Visions" the narrator says time travel to the past is impossible because the past is unchangeable and (of course) a time traveler would disturb history. This is simply a failure to distinguish the difference between changing and affecting the past—and a failure to see how the principle of self-consistency negates the issue of paradoxes. Then, when the robot returns from the future (and so backward time travel is *not* impossible!), he reports that his arrival had been expected, that history had recorded that he would appear. At the end we learn how the future knew this—it had read "Robot Visions"! So now Asimov *uses* the principle of self-consistency, with the narrator realizing that he *must* preserve his story so the future can read it. Not a very consistent story!

9. A.J. Accioly and G.E.A. Matsas, "Are There Causal Vacuum Solutions with the Symmetries of the Gödel Universe in Higher-Derivative Gravity?" *Physical Review D* 38 (August 15, 1988): 1083–1086.

10. In Volume 1 of *The Philosophy of Karl Popper,* ed. P.A. Schilpp, The Library of Living Philosophers (La Salle, IL: Open Court, 1974), p. 3.

11. The alternative point of view is expressed by Allen and Simon (1992). They conclude a discussion of time travel via cosmic strings (see Tech Note 10) with this assessment: "While there is still hope that one day a sufficiently clever design may make building a time machine possible, it is beginning to seem more and more improbable. Like the perpetual motion machines of the nineteenth century, the designs have an elegant simplicity (as well as enormous commercial potential), but it seems that Nature also may abhor them just as much." Of course, at one time it was thought that Nature abhorred a vacuum. Actually, Nature must love a vacuum, else why did she make so much of it? The big question, of course, is in which category Nature placed time travel: the one with perpetual motion or the one with vacuum.

What Time Is *Now*?

> If I can't make six clocks tick together, how in the world can I
> hope to make six nations tick together?
> —Charles the Fifth, Holy Roman Emperor,
> in a perhaps apocryphal story of his despair at
> achieving consensus ("political simultaneity")

The idea behind this quotation is to make explicit the first revolutionary insight into the nature of time (as opposed to mere metaphysical speculation) since the invention of the idea of time itself—the discovery that one clock's tick has nothing necessarily to do with another's tock; the discovery that one's present is not everybody else's present. The discarding of the intuitively "obvious" concept of universal time was the first (but not the last) of the repercussions of Einstein's ponderings on the nature of space and time. It was an idea that could have been conceived long before it actually was. It is interesting to note that some poets understood this at the intuitive level before most physicists. In an 1817 letter, for example, the British writer Charles Lamb wrote, "Your 'now' is not my 'now'; and again, your 'then' is not my 'then'; but my 'now' may be your 'then,' and vice versa. Whose head is competent to these things?" This last line still has meaning for students of time travel today.

The *present* for any given observer is that instant in time that separates her future from her past. To be just a little poetic perhaps, it is always the present for us, but as soon as the instant we call "now" arrives (from the future), it leaves us and recedes into the past. As Leonardo de Vinci wrote in one of his notebooks,[1] "The water you touch in a river is the last of that which has passed, and the first of that which is coming. Thus it is with time present." Or, finally, as one enthusiastic writer (whose name I cannot recall) said, "The present is the knife edge upon which the past and the future balance." We exist, or at least our conscious selves exist, only in the now, and we use it both to anticipate the future and to remember the past.

One early science fiction story, the 1938 "The Time Conqueror" (Eschbach), contains a long discussion (between a mad scientist and his victim, whose brain, in time-honored pulp magazine tradition, winds up in a vat!) on the nature of time. At one point the mad scientist delivers an interesting lecture on the present. After he gets the victim to declare that the past no longer exists and that the future does not yet exist, the mad scientist argues that such a position means "the present is the moment of transition of a phenomenon from one non-existence to another non-existence—since you say the future and past do not exist. But that short moment we term the present is after all fictional in character; it cannot be measured. We can never seize it. That which we did seize is always the past. So, according to your own conception of time . . . neither past, present nor future exists!" That argument is immediately rejected by the victim, who exclaims, "But that is absurd! You and I exist, and the world exists—and all in the present." To which responds the mad scientist, "Of course. And that proves the falsity of the popular conceptions of the divisions of time."

The *now* is a mathematical instant, the temporal equivalent of a dimensionless point. Physiologically, however, for living creatures it is an interval, one that for humans extends perhaps a few tens of milliseconds into the past; see Efron (NYAS). Sometimes it is even longer; as Oscar Wilde put it in *De Profundis,* "Suffering is a long moment." In his essay "Explorations in the World of Dreams" in the *New York Times Magazine* of July 10, 1927, H.G. Wells wrote of the now, stating, "Our mind can be considered as existing in the past and in the future, as extending, so to speak, both ways beyond what we consider to be the actual moment. I hope that does not strike the reader as too crazy a proposition. Most of us have given very little thought to what we mean by the actual moment. What do we mean by 'now'? How much time is it? Behind 'now' stretches the past, ahead is the future, but is it itself an infinitesimal instant?" For a fictional treatment of the possible result of this interval of now-ness extending even just a bit into the future, see the story "The Golden Man" (Dick). For tales of people stuck in the present, see "All the Time in the World" (Clarke) and "In Frozen Time" (Rucker).

Until Einstein it was universally believed that the present was a common experience, and so one could sensibly ask a question such as "I wonder what's happening right now on the largest inhabited planet (if any) in the Andromeda galaxy, two million light years away?" We subconsciously extrapolate our earthly belief that there is a planet-wide now (a belief made easy to accept by the proliferation of nearly "instantaneous" satellite communication links) to a belief that there is a cosmic now. When two people talk across a continent on the telephone, they seem to be sharing a common now. So why couldn't two creatures, each in a different galaxy, do the same?

No less a thinker than Isaac Newton codified this belief. In 1687 in the *Scholium* to the *Definitions* of his *Principia,* he wrote,[2] "Absolute rule and mathematical time, of itself, and from its own nature, flows equally without relation to anything external, and by another name is called *duration.*" We remember Newton's words because he was, of course, *Newton,* but as Newton himself said, he stood on the shoulders of giants, and the origins of his views about absolute time are no exception. Capek (1987) traces the so-called *absolutist* view of time backward from Newton, and he convincingly documents the cases of several analysts who held similar, earlier ideas. For example, Isaac Barrow, Newton's influential contemporary and immediate predecessor as occupant of the Lucasian chair at Cambridge, wrote in his *Lectiones Geometricae* (1670) that "whether things run or stand still, whether we sleep or wake, time flows in its even tenor." When Barrow wrote this, Newton was his student.

We find the same kinds of ideas even earlier in the French philosopher and scientist Pierre Gassendi's posthumous *Syntagma Philosophicum* (1658): "Even before there were things time flowed. . . . Even now, while they exist . . . time flows in the same tenor as it flowed before. . . . [I]f God would wish to recreate the Universe, time would flow in the interval between its destruction and recreation." The alternative view that time is inseparably related to changes in the configurations of matter, which is the *relational* view, has even more ancient roots. We find, for example, in the First Book of Lucretius' *On the Nature of Things,* "Time itself does not exist, but from things themselves there results a sense of what has already taken place, what is now going on, and what is to ensue. It must not be claimed that anyone can sense time by itself apart from the movement of things." And in his *Physics,* Aristotle declared that "time cannot be disconnected from change; for when we experience no changes in consciousness, or if we are not aware of them, no time seems to have passed. . . . [W]e are not aware of time when we do not distinguish any change."

A cosmic clock, eternal in its tick and perfectly uniform in its tock, and everywhere at once in its presence, was how Newton imagined God's absolute timepiece. For more than two centuries after Newton, this image was accepted by all as being beyond question. It was, after all, simply obvious and what more could you say? In 1905 Einstein showed that one definitely could say more; indeed, he showed that it just isn't true. Einstein demonstrated the "relativity of the present," the fact that two observers can watch the same physical process and yet disagree on the times when various events in that process occur. He thereby arrested forever the pendulum of Newton's clock—and perhaps equally astonishing is the fact that what Einstein did, Newton could have done. No new physical discoveries or deep mathematical theorems were required. Einstein arrived at his result by the simplest imaginable arguments, using mathematics no more advanced than arithmetic.[3]

Here is how he did it.

To begin, we have to ask ourselves (as Einstein did) what we mean when we say that two events are simultaneous. We mean, of course, that they happen at the same time, and, of course, this really begs the question. More precisely, we mean that we see the two events happen together in time. The word *see* is crucial. We see via light waves that travel from events to our eyes. However, it is clear that where we happen to position ourselves with respect to the two events will influence how long it takes the light waves from each event to reach our eyes. If the two events occur at the same location in space, there is no problem about which event is first. We can, however, imagine two events being simultaneous (such as the explosion of two widely separated sticks of dynamite connected by equal lengths of wire to a common electrical detonator box) but not appearing to be simultaneous because our eyes are closer to one explosion than to the other. We ordinarily do not notice these time delays in light propagation, because different places on Earth are not very far apart compared to the distances that can be traveled at the enormous speed of light in even very short time intervals.

We can eliminate the effects of these time delays due to the large but finite speed of light by arranging to observe two events from the position midway between them. Thus, so it would seem, the two delays are equal, and so if we see the two events happen at different times, then they really are not simultaneous—or so it would seem.

Now, imagine a train on a straight stretch of track. Part of the train is a boxcar with an experimenter in it. He is standing precisely in the middle of the boxcar, with a laser equipped with focusing lenses under his control via a switch. When he closes the switch, the laser will fire and two tightly focused light beams will be transmitted, one forward and the other rearward. On the front and rear walls of the boxcar are photocells that, when hit by a light beam, generate an electric current that triggers the illumination of an associated lamp. (Let us suppose that the forward lamp is red and that the rearward lamp is green.)

What could be more obviously true than the claim that if the experimenter closes the laser switch, then a very short time later he will see the red and green lamps glow simultaneously? After all, if the boxcar is of length L, then the laser is exactly $L/2$ distant from each photocell, and the two laser beams (each traveling at the speed of light—call it c) will take the same time to reach each photocell. Indeed, the red and green lamps will both turn on $L/2c$ seconds after the laser switch is closed. Because it takes the red and green light an additional $L/2c$ seconds to travel back to the center of the boxcar, the experimenter will see both lamps glow, at the same time, L/c seconds after he closes the switch.

Now imagine a second experimenter standing outside the boxcar on the edge of the track. Also imagine that the train is moving from left to right at a constant speed past this second experimenter. Matters are arranged so that at the instant the moving experimenter passes by his colleague on the ground, the laser switch is closed. Thus, at the start of the process that eventually results in the red and green lamps turning on, *both* experimenters are exactly midway between the two lamps. We have already decided what the experimenter inside the boxcar sees, but if the boxcar has transparent walls, what does the outside experimenter see?

He sees one laser beam shoot forward and another rearward, with the forward-traveling beam chasing after the red lamp's receding photocell and the rearward-traveling beam on a head-on collision course with the green lamp's approaching photocell. To the outside experimenter, the rearward-moving beam gets to its photocell first because it has to travel less distance than does the forward-moving beam, which has to catch up with its target photocell. On top of this, the rear photocell will be closer to the outside experimenter than the forward photocell will be, so the light from the green lamp has less distance to travel to arrive back to the outside experimenter's eyes than does the red light. The end result is that the outside experimenter sees the green lamp glow *before* he sees the red lamp—in stark disagreement with what the inside experimenter sees. The relative motion between the two observers has altered their perception of simultaneity.

I made no mention of motion in the first analysis for the simple reason that there is no way for the experimenter inside the boxcar to sense any such motion. (I am assuming, of course, that there are no windows in the boxcar and that the train wheels are silent.) We can sense motion only through *changes* in speed—that is, via accelerations or decelerations. This is, in fact, one of the fundamental postulates of dynamics, even before Einstein and relativity. Einstein's great insight was to extend the idea to all of physics, including, for example, electromagnetics and gravity. The laws of physics and the results of experiments are the same for all experimenters who are in relative, uniform motion with respect to each other. Physics on a merry-go-round is very different from our boxcar physics, yes, but that is because the merry-go-round is an *accelerated* system even if the rotation rate is constant. Thus, what the experimenter inside the boxcar sees is independent of whether the train is sitting motionless on the tracks or is hurtling along at 150 miles per hour. Similarly, pouring coffee on the Concorde at twice the speed of sound is indistinguishable (as long as there is no turbulence, which is another name for accelerated motion) from pouring it at home at the kitchen table. Indeed, in Galileo's writings (which, of course, pre-date even those of Newton), we find the example of dropping a stone from a ship's mast. (This experiment was actually performed by Galileo's contemporary, the Frenchman Pierre Gassendi.) Whether the ship is tied up in port or is under way at sea at a constant speed, the rock hits the deck at the same spot at the base of the mast. Uniform, unaccelerated motion has no effect on what sailor observers on the ship would see.

So revolutionary was this idea that Kurt Gödel declared (Herbert, 1987), "The very starting point of relativity theory consists in the discovery of a new and very astonishing property of time, namely the relativity of simultaneity." Einstein was not the first to suspect simultaneity was a central concept not nearly so obvious as it might appear. For example, the French philosopher Henri Bergson wrote in his 1888 doctoral dissertation (1910) that "the connecting link between . . . space and duration, is simultaneity, which might be defined as the intersection of time and space." But it was Einstein who showed the *relativity* of the concept. Not everybody shared this belief. The Irish mathematical physicist Alfred A. Robb (1873–1936) was so scandalized by it all that he wrote,[4] "From the first I felt that Einstein's standpoint and method of treatment were unsatisfactory. . . . In particular I felt strongly repelled by the idea that events could be simultaneous to one person and not simultaneous to another. . . . This seemed to destroy all sense of the reality of the external world and to leave the physical Universe no better than a dream, or rather, a nightmare." And a year later, in 1922, the venerable Oliver Lodge wrote,[5] "A theory which renders it uncertain whether the Fire of London preceded or succeeded the outburst of Nova Persei [a spectacular stellar explosion, first observed in 1901] . . . and whether a much-traveled man's death preceded his birth, should not be too positive when it leaves its own realm and enters the region of fact and reality. . . . There is room not only for Einstein. . . but also for Newton and Maxwell."

It is always hard to give up old ideas. We love these old ideas because they are so obvious, so intuitive, so *common-sense,* but, to echo Einstein himself,[6] "Common sense is that layer of prejudices laid down in the mind prior to the age of eighteen." Certainly when the president of the American Physical Society for 1911 gave his presidential address,[7] he showed that he—a professor of physics at Princeton—was still in the nineteenth century: "A description of phenomena in terms of four dimensions in space [he meant space*time*] would be unsatisfactory to me as an explanation, because by no stretch of my imagination can I make myself believe in the reality of a fourth dimension. The description of phenomena in terms of a time which is a function of the velocity of the body on which I reside will be, I fear, equally unsatisfactory to me, because, try I ever so hard, I cannot make myself realize that such a time is conceivable. . . . I do not believe that there is any man now living who can assert with truth that he can conceive a time which is a function of velocity or is willing to go to the stake for the conviction that his 'now' is another man's 'future' or still another man's 'past.' " What could Magie have meant by all this? He must have understood that *Einstein* was such a man. Einstein came to live in Princeton in 1933, and Magie lived until 1943, so almost surely the two men met. One wonders what they might have said to each other.[8]

Notes

1. Quoted from *The Artist by Himself: Self-Portraits from Youth to Old Age,* ed. J. Kinneir (New York: St. Martin's Press, 1980), p. 205.

2. Quoted from A. Koyre, *Newtonian Studies* (London: Champman Hall, 1965), p. 103.

3. Although this is pretty close to the truth, some might argue that perhaps I have slightly warped the facts to make a better story. Indeed, Einstein himself said that he was led to his ideas on special relativity by his boyhood ponderings about what a light beam would look like if he could race along with it. Knowing of Maxwell's description of light as electromagnetic waves (moving at the speed of light, of course), Einstein wondered how such waves would appear if he moved at the speed of light, too? Would they be frozen in shape? Could they then still be light? Newton, who thought of light as being composed of particles, just did not have such visual imagery to jump-start his curiosity, and of course there was no reason for Newton not to believe that light moved infinitely fast, which *would* result in a Universal now. Curiously, there are those who write on time, *these days,* who are still in Newton's time. In Wild (1954), for example, we have a philosopher who wrote (incorrectly), "Everything in the world seems to be engulfed in an irreversible flux of time which cannot be quickened or retarded, but flows everywhere at a constant rate."

4. Quoted from the Preface of Robb's book *The Absolute Relations of Time and Space* (Cambridge: Cambridge University Press, 1921). Robb was no fool. A Fellow of the Royal Society, he held three (!) doctorates and was esteemed by many British scientific luminaries. A man of independent means, he never held an official academic position, but he was allowed to reside at Cambridge and occasionally was even

invited to offer lectures on mathematics. When he died, his obituary in *Nature* was written by no less a follower of Einstein than the English scientific superstar Arthur Stanley Eddington.

5. "Relativity and the Aether," *Nature* 110 (September 30, 1922):446.

6. Quoted from R. Skinner, *Relativity for Scientists and Engineers* (New York: Dover, 1982), p. 27.

7. William Francis Magie, "The Primary Concepts of Physics," *Science* 35 (February 23, 1912):281–293, as well as in *Physical Review* 34 (February 1912):125–138.

8. Einstein was used to such reactions. One of his own close friends, for example, a talented physicist, wrote him a letter (dated November 24, 1919) in which he declared, "Einstein, my upset stomach hates your theory—it almost hates you yourself! How am I to provide for my students? What am I to answer to the philosophers?" See Martin J. Klein, *Paul Ehrenfest* (Amsterdam: North-Holland, 1970), p. 315.

Time Dilation via the Photon Clock

> Time as we know it is not universally absolute. The rate of its passage depends to a great extent upon the velocity of its observer with regard to some certain reference system. A moving clock will run slower with respect to a selected coordinate system than a stationary one.
>
> —a time traveler explains how his time machine works in the 1931 story "Via the Time Accelerator" (Bridge)

Imagine two horizontal, parallel mirrors, one positioned over the other and separated by distance d. The two mirrors are in the same frame of reference with an observer; that is, the observer is looking at two stationary mirrors. Between the two mirrors we further imagine that a particle of light, a photon, is bouncing endlessly back and forth, up and down, in relentless reflection. We define the time required for the photon to travel from one mirror to the other as a tick, and the return trip defines the tock. This simple system is called a *photon clock* or the Einstein–Langevin clock, after the French physicist Paul Langevin (1872–1946), and it has been part of physics for decades. The rate of timekeeping, the time interval separating consecutive ticks, is obviously

$$t' = 2\frac{d}{c}$$

where c is the speed of light.

Suppose we imagine now that the observer and the mirrors and the bouncing photon are moving at constant speed v, to the right across our line of sight. The mirror/photon system is then in a different frame of reference from ours, and we will not see the photon bouncing up and down vertically. Rather, we see the photon tracing out a triangular path, as shown in Figure TN2.1. One round trip of the photon evidently requires more time than before, because the distance in the stationary frame (our frame) is now greater than that in the moving observer's frame. In fact, if t is the time between consecutive ticks as seen by a stationary observer (us), then the total path length is

$$2\sqrt{d^2 + \left(\frac{vt}{2}\right)^2}$$

and so

$$t = \frac{2\sqrt{d^2 + (vt/2)^2}}{c}$$

This can be quickly manipulated algebraically and combined with the expression for t', the tick interval for an observer in the same frame as the clock, to give the tick interval for a moving clock as measured by a stationary observer (us):

$$t = \frac{t'}{\sqrt{1 - (v/c)^2}}$$

Note that this reduces to $t = t'$ when $v = 0$—that is, when the clock is stationary.

This is the famous Einstein time dilation formula,[1] which shows $t \geq t'$, and indeed $t = \infty$ when $v = c$. That is, the photon clock appears to run *slow* compared to clocks in our stationary frame of reference, and at the speed of light, time *stands still*. This is what one writer was referring to when he wrote that

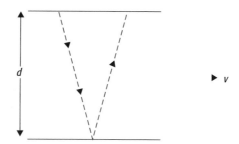

Figure TN2.1 The photon clock.

"according to the theory of relativity, if the observer is moving with the velocity of light, time remains unchanged. This must have been the case with the mad Hatter. With him it is always six o'clock, and always tea-time."[2] A curious anticipation of this association between light and timelessness can be found in a poem by the seventeenth-century English writer Henry Vaughan. The opening words to his "The World," which appeared in 1650 as part of *Silex Scintillans* ("Sparkling Flint"), are

> I saw Eternity the other night
> Like a great *Ring* of pure and endless light,
> All calm, as it was bright,
> And round beneath it, Time in hours, days, years
> Driven by the spheres
> Like a vast shadow moved, in which the world
> And all her train were hurled.

For $v > c$, the equation says that time becomes imaginary, and this is one reason for claiming that $v > c$ is not possible. Not everyone believes that this simple model is valid (though only crackpots dispute the conclusion); Schlegel (1980) offers a good discussion of the objections.

A similar modification in length (measured in the direction of motion) occurs when $v > 0$. Whereas an observer moving with an object will measure its length to be L', a stationary observer will "report" it contracted to the length

$$L = L' \sqrt{1 - \left(\frac{v}{c}\right)^2}$$

This effect is called the Lorentz–FitzGerald contraction. For "everyday" objects the contraction is extremely small. For example, for a low-altitude satellite 100 meters long, moving at 18,000 miles per hour (that is, at $v = 2.7 \times 10^{-5}c$), the contraction would be less than 4×10^{-6} cm.

I have put the word *report* in quotation marks for a reason. In 1959 the English mathematical physicist Roger Penrose made the astonishing discovery that a photograph of a sphere moving past an observer (who has a camera), at a speed that is an appreciable fraction of the speed of light, will still display a circular cross section, not the flattened ellipsoid that physicists had assumed since Einstein's 1905 paper. This directly contradicted Lorentz himself, who in 1922 had stated that the contraction could be photographed. A few months after Penrose's paper, the American physicist James Terrell generalized and extended the results and clearly stated what is happening. As he wrote,

> An observation of the shape of a fast-moving object involves simultaneous measurement of the position of a number of points on the object. If done by means of light, all the quanta

should leave the surface simultaneously, as determined in the
observer's system, but will arrive at the observer's position at
different times. . . . In such observations the data received
must be corrected for the finite velocity of light, using measured distances to various points of the moving object. In
seeing the object, on the other hand, or photographing it, all
the light quanta arrive simultaneously at the eye (or shutter),
having departed from the object at various earlier times.
Clearly this should make a difference between the contracted
shape which is in principle observable and the actual appearance of a fast-moving object.

Indeed, Terrell showed (as did Penrose) that the finite velocity of light is
just what is required to make the contraction invisible in a photograph. Terrell,
however, showed that there is an unexpected effect that *is* visible: The object
will appear *rotated* in the photograph. A little more than a year later, the American physicist C. W. Sherwin showed that a certain kind of radar system *could*
display the contraction on a monitor screen. It's too bad Einstein died just before these astonishing discoveries[3] were made about a fundamental misconception concerning the special theory, a misconception almost certainly shared
by Einstein himself.

In the early days of science fiction this concept was fascinating to readers,
but authors often got it wrong. For example, in "Faster Than Light" (Haggard), the 1930 story of a runaway spaceship falling into the Sun, we find
"When our racing [ship] was drawn from the Earth's gravity and fell at ever increasing speed toward the Sun it soon approached the speed of light. As we fell
faster and faster our length in the direction of the Sun progressed into nothingness. Then—it reached the speed of light—passed it. Now—mind you
this—when the [ship] attained the speed of light it was of a *minus* length." This
author managed to make four errors in three sentences! The same author
botched the Lorentz–FitzGerald contraction yet again in 1935 in "Relativity
to the Rescue" (Haggard) and added two more errors to his growing list. In
this story there is an episode of faster-than-light *radio* communication (which
by definition is *at* the speed of light) and a long lecture that actually *denies* special relativity's fundamental assertion that all inertial frames of reference are indistinguishable (see the next Tech Note). In another story of high-speed space
travel, the 1933 "A Race Through Time" (Wandrei), the author has the contraction working in the wrong *direction*—as the rocket ship moves faster and
faster, it gets longer and longer.

First prize for mangling the laws of physics, however, has to go to the 1936
story "Reverse Universe" (Schachner). In this tale a spaceship is on its way to

Alpha Centauri at near light-speed when the crew mutinies and puts the captain and first officer "overboard" in a space boat with six months' provisions. This happens at mid-voyage, about two light years from both home and destination, so matters look grim. The author tells us, several times, that things do look *very* bad. But are they? With a stated speed of 162,000 miles per second, the time dilation factor is 2.035, and because the space boat is traveling at $0.87c$ it will take a little more than thirteen months of *space boat time* to complete the journey. If the two men go on half-rations, then it seems they *could* survive. There is, of course, the problem of slowing down so as to arrive at Alpha Centauri at a reasonable speed, but that concern was ignored by Schachner. Indeed, his attention was directed to a much more dramatic conclusion. He had a faster-than-light planet (please don't ask how *this* was explained) collide from behind and carry the castaways on toward their destination! While they are on this planet, time runs backward (for what *really* happens at superluminal speeds, see Tech Note 7), and finally, in a repeat of Haggard's error, we are told that the Lorentz–FitzGerald contraction is negative for $v > c$. For the rather curious history of the Lorentz–FitzGerald contraction, dating from 1889, see Bork (1966) and Brush (1967). This effect has entered popular culture, in the form of the following well-known (at least among physicists) limerick

> There was a young man named Fisk
> Whose fencing was exceedingly brisk.
>   So fast was his action
>   The FitzGerald contraction
> Reduced his rapier to a disk.

Bawdier versions also exist, but I'm certainly not going to repeat them here.

The time-slowing (or size-shrinking) factor becomes pronounced only at values of v close to c; Table TN2.1 shows this effect. For example, the last entry shows that a clock traveling at 99.99% the speed of light will register the passage of one year while nearly seventy-one years pass on Earth. One science fiction writer got this dramatically wrong, even though he actually reproduced the Lorentz–FitzGerald equation in his 1950 story "To the Stars" (Hubbard). At one point he writes of the near light-speed rocket ship that stars in this tale, "If it [the ship's speed] was as slow as ninety-four percent [of the speed of light] . . . for every moment ticked by the clocks of the [ship] hundreds passed on Earth." In fact, the time dilation factor at that speed is "only" 2.93.

One possible objection to this entire Tech Note is that the analysis has been done for a *particular* clock. How do we know that another clock—one using wheels and pendulums, for example, instead of photons and mirrors—wouldn't

Table TN2.1. The Lorentz-FitzGerald time-slowing factor.

$\left(\dfrac{v}{c}\right)$	$\dfrac{1}{\sqrt{1 - \left(\dfrac{v}{c}\right)^2}}$
.1	1.005
.2	1.021
.5	1.155
.7	1.4
.9	2.294
.999	22.366
.9999	70.712

be affected differently by motion? The answer comes from relativity itself, which says that there is no way to detect uniform motion. If two clocks did behave differently, then this difference could be used as a motion detector. This is not possible according to relativity, however, and so *all* clocks, no matter what their internal mechanisms (including the complex *biological* clocks of our own bodies), must respond to uniform motion in precisely the same way.[4]

Notes

1. The direct observation of time dilation was reported by H. E. Ives and G. R. Stilwell, "An Experimental Study of the Rate of a Moving Atomic Clock," *Journal of the Optical Society of America* 28 (July 1938):215–226.

2. *Nature* 104 (February 12, 1920):627–630.

3. The papers of Penrose, Terrell, and Sherwin are, respectively, "The Apparent Shape of a Relativistically Moving Sphere," *Proceedings of the Cambridge Philosophical Society* 55 (January 1959):137–139; "Invisibility of the Lorentz Contraction," *Physical Review* 116 (November 15, 1959):1041–1045; and "Regarding the Observation of the Lorentz Contraction on a Pulsed Radar System," *American Journal of Physics* 29 (February 1961):67–69.

4. Resistance to this conclusion persisted for years. For example, see the letter "Relativity and Radio-activity" in *Nature* 104 (January 8, 1920):468. The author wondered whether a clock based on radioactive decay might not somehow beat the "conspiracy" of moving clocks running slow as compared to stationary ones.

3

The Lorentz Transformation

The aethereal gymnastics of the 19th century physicists amuse us. Will the 21st century physicist be similarly amused by our temporal gymnastics?
> —Indian physicist C.K. Raju, writing of the Michelson–Morley experiment.

If only he'd paid more attention to mathematics in school.
> —Albert Eustance Rossi, the first time traveler, in "Extempore" (Knight)

We begin by imagining two distinct frames of reference. One we take to be stationary, and the other as moving at a uniform speed v with respect to the first. The moving frame is said to be *boosted* with respect to the stationary frame. We can always orient these two coordinate systems so that the motion occurs along just one axis (the x direction, for instance, as shown in Figure TN3.1). I am using primed variables for the moving frame. Thus the two frames have coincident x axes and parallel y and z axes that are moving apart at constant speed v. Let us also imagine that there is a clock at the origin of each frame and that at the instant the origins match, we synchronize the clocks; that is, $t = t' = 0$ is the instant the two coordinate systems coincide.

We next imagine that there is an observer at the origin in each frame. At some arbitrary instant of time, each observer records the coordinates of the arbitrary point P in space, as measured in this system.

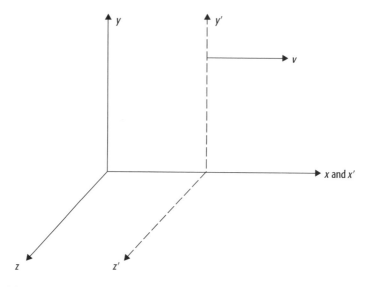

Figure TN3.1 Two spatial reference frames in relative motion.

The observers could, for example, agree to record the coordinates of P when each of their respective system clocks reads five seconds. It seems immediately obvious (as it was for Newton—see Tech Note 1) that $t = t'$; that is, time runs at the same rate in each frame and thus it makes sense to talk about "the same instant." (If you have already read Tech Note 2, you know this is *not true*, but temporarily forget that!) Thus at this "same instant" the stationary observer records (x, y, z) and the moving observer records (x', y', z'). What, we now wonder, are the relationships between the primed and unprimed coordinates? That is, what mathematical transformation converts from one frame to the other? The answer seems obvious:

$$y' = y$$
$$z' = z$$
$$x' = x - vt$$

This transformation, called the *Galilean transformation* after the Italian Galileo Galilei (1564–1642), satisfies the relativity principle, which says that uniform motion leaves the laws of physics unchanged. For example, in the stationary system the famous second law of Newton for a fixed mass m,

$$F = m\frac{d^2x}{dt^2}$$

becomes the identical-appearing form

$$F' = m\frac{d^2x'}{dt'^2}$$

More precisely, *all* the laws of mechanics are unchanged. Any frame of reference in which Newton's laws of mechanics hold true is said to be *inertial*. Given one inertial system, we can find infinitely many others simply by applying the Galilean transformation. When the mathematical laws of electrodynamics were discovered by Maxwell in the nineteenth century, however, it was found that the Galilean transformation did not leave Maxwell's equations unchanged; the transformed equations predict electromagnetic effects for the moving system that are not predicted to occur in the stationary system. This meant that there was theoretical support for the possibility that electromagnetic experiments might be devised to detect uniform motion, and this eventually led to the famous Michelson–Morley experiment of 1887.[1] This experiment, theoretically sensitive enough to detect the motion of Earth itself through space, failed to detect any such uniform motion. The conclusion is clear: The extra electromagnetic effects predicted by the Galilean transformation do not exist, and the transformation is wrong even though it works with the laws of mechanics. What is going on?

The answer is inspired, and again returns us to the cornerstone of relativity: the idea that the laws of physics, *all* the laws, should look the same to observers in uniform relative motion. That is, there is no special or preferred system of coordinates—*all* inertial systems are equivalent in all of physics. Evidence from an extremely broad variety of sensitive experiments had, by the end of the nineteenth century, convinced people that Maxwell's equations are correct. Thus, a new transformation was needed that leaves both the laws of mechanics *and* the laws of electrodynamics unchanged with motion. A single transformation that works on Maxwell's equations and on the mechanical laws would therefore mean that Newton's mechanical laws cannot be correct. This was a breathtaking conclusion, because Newton had been unchallenged for two centuries.

However, it turns out that Newton's laws were almost right. The only correction required is the idea that the mass of a moving body is not independent of speed but rather varies as

$$m = \frac{m_0}{\sqrt{1 - (v/c)^2}}$$

where m_0 is the rest mass when $v = 0$. The variation of mass with speed was experimentally observed in 1901 by the German Walter Kaufmann

(1871–1947). This effect shows that m is infinite at $v = c$ unless $m_0 = 0$ (as it is for a photon), which is one reason for the belief that accelerating a mass (such as a spaceship) up to the speed of light is impossible; it would require infinite energy. With this modification, the transformation that leaves all the laws unaltered in form by uniform motion is

$$y' = y$$
$$z' = z$$
$$x' = \frac{x - vt}{\sqrt{1 - (v/c)^2}}$$
$$t' = \frac{t - vx/c^2}{\sqrt{1 - (v/c)^2}}$$

These equations are called the *Lorentz transformation* after the Dutchman Hendrik Antoon Lorentz (1853–1928), who discovered them in 1904 by direct mathematical manipulation of Maxwell's equations [Lorentz, TPR]. (Some historians give the Irish physicist Joseph Larmor (1857–1942) priority for his study of the Maxwell equations, because his work was published in 1900.[2]) It was Einstein who in 1905 showed how to derive the transformation equations from a fundamental reexamination of space and time without concerning oneself about the details of specific laws.

By simple algebraic manipulation, the transform equations can be rewritten as

$$ct = \gamma ct' + \beta \gamma x'$$
$$x = \beta \gamma ct' + \gamma x'$$

where $\beta = v/c$ and $\gamma = 1/(1 - \beta^2)$ are both dimensionless constants. It is interesting to note that all the terms in both equations have the units of *space*, including the terms that involve time. In compact matrix form, the Lorentz transformation becomes

$$\begin{bmatrix} ct \\ x \end{bmatrix} = \begin{bmatrix} \gamma & \beta\gamma \\ \beta\gamma & \gamma \end{bmatrix} \begin{bmatrix} ct' \\ x' \end{bmatrix}$$

and the symmetrical 2×2 matrix is called the Lorentz boost matrix or simply the *boost*. Note that when $v = 0$ (zero boost), the boost matrix reduces to the identity matrix; that is, the two frames are one and the same with at most a constant shift in the location of the origin. Note, too, that $\beta = 0$ and $\gamma = 1$ for *any* v when c is infinite—that is, the boost matrix again reduces to the identity matrix and the Lorentz transform becomes the Galilean *if* c is infinite. But c is not infinite, and all of the implications of special relativity are the direct result of the *finite* speed of light.

The Lorentz transformation contains two results we have obtained in previous Tech Notes by using specialized arguments. For example, in Tech Note 1 we found that simultaneity is a relative concept in reference frames in relative motion. Therefore, now consider two events that occur specifically on the x-axis. They are simultaneous in the stationary system (at, say, time $t = T$) but are at different places (at, say, $x = X$ and $x = X + \Delta X$). Their occurrences in time for the moving observer are

$$t'_1 = \frac{T - vX/c^2}{\sqrt{1 - (v/c)^2}} \quad \text{and} \quad t'_2 = \frac{T - v(X + \Delta X)/c^2}{\sqrt{1 - (v/c)^2}}$$

For the moving observer, therefore, the two events are not simultaneous, being separated in time by

$$t'_1 - t'_2 = \frac{v\Delta X/c^2}{\sqrt{1 - (v/c)^2}}$$

Only if $\Delta X = 0$ (the two events occur at the same place) will $t'_1 = t'_2$. That is, only if $\Delta X = 0$ are simultaneous events in one frame also simultaneous in another frame in relative motion.

In Tech Note 2 we found that time runs slow in one frame as observed from another frame in relative motion. This result is easy to obtain formally by using the t' equation in the Lorentz transformation. Thus, if we differentiate with respect to t, we get

$$\frac{dt'}{dt} = \frac{1 - (v/c^2)dx/dt}{\sqrt{1 - (v/c)^2}}$$

But $\dfrac{dx}{dt} = v$, the speed of the moving frame as measured by the observer in the stationary frame. This gives

$$dt' = \sqrt{1 - \left(\frac{v}{c}\right)^2}\, dt$$

which is the same result we obtained by analyzing the photon clock.

The Lorentz transformation contains many other implications not explicitly stated in earlier Tech Notes. For example, I have often made special reference to the relativity principle, which says that uniform motion has no effect on the form of the physical laws—but how do we know who is moving and who is stationary? After all, a system moving to the right at speed v past a stationary system could just as well be viewed as a stationary system where the other system is moving to the *left* at speed $-v$.

In fact, it is possible to invert the Lorentz transformation and solve for the unprimed variables in terms of the primed ones, and this is exactly what we get—the Lorentz transformation back again, with v replaced with $-v$. That is, the Lorentz transformation is symmetrical, so two observers in different frames of reference *both* say that it is the other's clock that is running slow! This follows immediately, in fact, from the original transformation written in matrix form. That is, multiplying through by the inverse of the boost matrix, we get

$$\begin{bmatrix} ct' \\ x' \end{bmatrix} = \begin{bmatrix} \gamma & \beta\gamma \\ \beta\gamma & \gamma \end{bmatrix}^{-1} \begin{bmatrix} ct \\ x \end{bmatrix} = \begin{bmatrix} \gamma & -\beta\gamma \\ -\beta\gamma & \gamma \end{bmatrix} \begin{bmatrix} ct \\ x \end{bmatrix}$$

The only difference between the original boost matrix and its inverse (which is, of course, the new boost matrix for the new interpretation of which frame is moving) is a change in sign of β—that is, in the sign of v. The inverse transformation is

$$y = y'$$
$$z = z'$$
$$x = \frac{x' + vt'}{\sqrt{1 - (v/c)^2}}$$
$$t = \frac{t' + vx'/c^2}{\sqrt{1 - (v/c)^2}}$$

As a final example of what the Lorentz transformation tells us, consider the so-called addition-of-velocities problem. Suppose you are in a high-speed spaceship traveling away from Earth at speed v. Earth is the stationary system, and the spaceship is the moving system. (Assume that the x- and x'-axes are along the direction of motion.) Imagine next that you fire a gun toward the nose of your spaceship, the bullet exiting the muzzle at speed w. How fast is the bullet moving away from Earth? The common-sense Galilean transformation says $v + w$, but we now know that this transformation is wrong. What does the Lorentz transformation say? Inside the spaceship, the position of the bullet at time t' after the gun is fired (assuming that the gun is fired just as the rocket ship passes Earth) is

$$x' = wt'$$

From the inverse Lorentz transformation, the location of the bullet in Earth's frame is

$$x = \frac{x' + vt'}{\sqrt{1 - (v/c)^2}} = \frac{w + v}{\sqrt{1 - (v/c)^2}} t'$$

The transformation also tells us that

$$t = \frac{t' + vx'/c^2}{\sqrt{1 - (v/c)^2}} = \frac{1 + wv/c^2}{\sqrt{1 - (v/c)^2}} t'$$

The speed of the bullet in Earth's frame is, therefore,

$$\frac{x}{t} = \frac{w + v}{1 + wv/c^2}$$

Note that for a low-speed bullet ($w \ll c$) this result is close to $w + v$, but at high speeds it is very much different. Indeed, suppose we do not fire a gun at all, but instead replace it with a flashlight. Now, instead of a bullet, we shoot photons at $w = c$. The Galilean transformation would say that a stationary observer on Earth would see the photons moving at speed $v + c$, which is a superluminal speed. The Lorentz transformation says that the Earth observer would see a speed of

$$\frac{c + v}{1 + cv/c^2} = \frac{c^2(c + v)}{c^2 + cv} = \frac{c^2(c + v)}{c(c + v)} = c$$

That is, no matter what the speed of the moving observer is, he sees the light from his flashlight traveling at the same speed as does the stationary observer. This peculiar effect is unique to the speed of light ($w = c$). We have derived this effect as a consequence of the Lorentz transformation, but in fact Einstein actually did things in the reverse order. He began by postulating the invariance of the speed of light for all observers in uniform relative motion, combined this idea with the principle of relativity, which says that all physical laws look the same to these observers, and then derived the Lorentz transformation using no mathematics beyond algebra.

This result was actually first found by the great French theoretician Henri Poincaré (1854–1912) in June 1905, three months before the publication of Einstein's famous paper (TPR) on special relativity, which also derives the addition-of-velocities equation. It was Poincaré who gave the Lorentz transformation its name, and he was also the first to state specifically the principle of relativity as covering all the laws, electromagnetic as well as mechanical. In an address to the International Congress of Arts and Science in St. Louis, on September 24, 1904, Poincaré specifically stated[3] that in the mechanics of special relativity "no velocity could surpass that of light, any more than any temperature could fall below the zero absolute."

A mathematically elegant alternative derivation of the addition-of-velocities formula can be done by simply noticing that the condition of two successive boosts should be, itself, a boost. Thus, if we have a frame moving relative to a second frame (which itself is moving relative to a third frame), then the boost matrix of the first frame relative to the third frame is the product of the two individual boost matrices. That is,

$$\begin{bmatrix} \gamma_3 & \beta_3\gamma_3 \\ \beta_3\gamma_3 & \gamma_3 \end{bmatrix} = \begin{bmatrix} \gamma_2 & \beta_2\gamma_2 \\ \beta_2\gamma_2 & \gamma_2 \end{bmatrix} \begin{bmatrix} \gamma_1 & \beta_1\gamma_1 \\ \beta_1\gamma_1 & \gamma_1 \end{bmatrix}$$

From this it is easy to show that $\beta_3 = \dfrac{\beta_1 + \beta_2}{1 + \beta_1\beta_2}$. Substitution of $\beta_1 = \dfrac{w}{c}$ and $\beta_2 = \dfrac{v}{c}$ immediately gives the addition-of-velocities formula.[4]

A failure to understand the implications of the invariance of the speed of light resulted in two stupendous errors in "To the Stars" (Hubbard), a 1950 story of a near light-speed rocketship. At all times an officer stands watch on the bridge to be sure the ship doesn't accidently *reach* the speed of light. This is to be avoided (according to the author) because to reach the speed of light would cause the ship to "hang there forever, unmoving [in time] . . . locked, protected and condemned to eternity by zero time."[5] This transgression is so easy (according to the author, who was apparently unaware that it would require *infinite* energy) that occasionally the ship has to fire a "checkblast" from its forward rocket tubes to slow down! Equally absurd is the means by which the development of this fatal condition is detected: The nose of the ship mounts a forward-pointing light source, and the ship is getting too near the speed of light when it *overtakes* the photons emitted by the source! (The biggest puzzle of all, actually, is why the editor of *Astounding Science Fiction* let such a technically goofy story appear in a magazine recognized for its usual faithfulness to known science.)

A question of great interest to historians of science is what led Einstein to take the non-obvious invariance of the speed of light as a given. As best as anyone can reconstruct Einstein's early thoughts, it appears that the prediction by Maxwell's equations of a precise numerical value for the speed of light in empty space, independent of the speed of the observer's frame, may have provided the crucial stimulus. This was known, of course, to many others, but only Einstein had the nerve to use what Maxwell's equations said and pursue the prediction to its logical conclusion. Experimental verification of the invariance of the speed of light was finally achieved in 1932.[6]

As mentioned in Note 3 of Tech Note 1, the traditional account of Einstein's early start on relativity credits his curiosity about what a light beam would look like if one could race alongside it. This *is* pretty precocious for a teenager, even an Einstein. What could have put such a thought in his head? I am only speculating now (and I have not seen this elsewhere), but in 1895 the French astronomer Camille Flammarion published a novel that is entirely devoted to a discussion of the properties of light, including what might be seen if one could race along beside a light beam. Flammarion's ideas are mostly wrong, but could Einstein have got hold of a copy of this best-selling book? In 1895 Einstein was sixteen, the age at which he stated that he began to reflect on these matters.

Notes

1. An old but still authoritative paper is R. S. Shankland, "Michelson–Morley Experiment," *American Journal of Physics* 32 (January 1964):16–35. As Shankland points out, it was after reading a letter written by Maxwell in 1879 (the year of Maxwell's death) that Michelson first became interested in the problem that resulted in his performing the 1887 experiment and (at least to some extent) in his winning the 1907 Nobel Prize in physics. Maxwell's letter was reprinted in *Nature* 21 (January 29, 1880):314–315. See also "Clerk Maxwell and the Michelson Experiment," *Nature* 125 (April 12, 1930):566–567.

2. The case for Larmor is not strong. A better one might be made for Woldemar Voigt (1850–1919), a German physicist who in 1887 arrived at the Lorentz transformation while studying the Doppler effect. See M. N. Macrossan, "A Note on Relativity Before Einstein," *British Journal for the Philosophy of Science* 37 (June 1986): 232–234.

3. For Poincaré's address, see "The Principles of Mathematical Physics," *The Monist* 15 (January 1905):1–24.

4. The addition formula can be derived without knowledge of the Lorentz transformation; see N. D. Mermin, "Relativistic Addition of Velocities Directly from the Constancy of the Velocity of Light," *American Journal of Physics* 51 (December 1983):1130–1131.

5. This is not the proper interpretation of what relativity says happens to a space traveler when $v = c$. While still just seventeen years old, Isaac Asimov wrote a letter to the editor (December 1937) at *Astounding Stories* that has the correct interpretation: "The effect on time of increasing speeds is . . . well known. Relativity states that as speed approaches that of light, time slows up until at 186,000 miles a second, time (so to speak) stands still. This seems to refute statements found in so many astronomy books (and science fiction stories) that even at the speed of light it would take four years to reach the nearest star. No such thing! As time halts at the speed of light, a person traveling from Alpha Centauri to the solar system, or vice versa, would not be aware of any lapse of time. Only to us poor earth-bound mortals would four years seem to have passed. In that sense the speed of light is infinite (as was thought in ancient times). This, by the way, offers an entirely scientific (if impractical) means of travel into the future. Say that someone wants to see how the world would look a hundred years from now. His procedure would be as follows: getting into his spaceship, he would proceed to a spot fifty light-years away at the speed of light. The journey would, for him, be practically instantaneous (due to the curious behavior of time at the speed of light). But fifty years would have elapsed on earth. He then makes the return trip at the same speed. Another fifty years elapse on earth and he lands a hundred years after his time. With this system, however, it would be impossible to travel into the past, so I don't think it will ever be adopted." A more realistic version of the young Asimov's

suggestion is mathematically analyzed in Tech Note 6 (Asimov says nothing, for example, of the *important* "detail" of how to reverse the direction of travel of the spaceship for the return to Earth).

6. R. J. Kennedy and E. M. Thorndike, "Experimental Establishment of the Relativity of Time," *Physical Review* 42 (November 1, 1932):400–418.

Spacetime Diagrams, Light Cones, Metrics, and Invariant Intervals

There is one metaphor in the physicist's account of space-time which one would expect *anyone* to recognize as such, for metaphor is here strained far beyond the breaking point, i.e., when it is said that time is "at right angles to each of the other three dimensions." Can anyone really attach any meaning to this—except as a recipe for drawing diagrams?

—a philosopher's point of view (Mundle, 1967)

This is from the outset a study in descriptive metaphysics. In consequence, I shall have nothing to say about twice-differentiable Lorentzian manifolds, Minkowski diagrams, world-lines, timelike separations, space-time worms, or temporal parts. These are, to be sure, concepts with impeccable credentials, when not abused, but quite alien to [an everyday understanding of time].

—another philosopher's point of view (Rosenberg, 1972)

It is helpful in discussions about the spacetime of special relativity to use (despite what these two philosophers wrote) what are called *Minkowski spacetime* diagrams. These are plots of the spacetime coordinates of a particle; the resulting curve is called the *world line* of the particle. Such diagrams are four-dimensional—three space axes and one time axis—and thus hard to visualize (much less draw)! The convention is to make do, whenever possible, with a simplified spacetime that has just one space axis (horizontal) and one time axis (vertical). Thus, for a particle at rest in some observer's frame of reference, its spacetime diagram for that observer is a vertical world line. If the particle moves, its world line tilts away from the vertical, and for an accelerating particle the world line curves away from the vertical. Straight (uncurved) world lines represent unaccelerated particles—that is, particles experiencing no forces that are thus in free fall. I show in Figure TN4.1 the world lines for these various cases on the same axes. It is assumed in the figure that all three particles are at $x = x_0$ when $t = 0$.

Free-fall world lines are called spacetime *geodesics*. In ordinary use, a path that joins two points on a surface with the minimum length is called a geodesic of that surface. It is shown in Tech Note 5 that spacetime geodesics do indeed possess an extreme property, but rather than being a minimum, it is a *maximum* property. Spacetime diagrams can be misleading on this matter, so it is important to remember that such diagrams are not a perfect representation of all the properties of spacetime.

It has become customary to draw these diagrams with the speed of light as unity ($c = 1$). That is, a distance of 300,000 kilometers on the x-axis is represented by the same extension as is one second on the t-axis. This means that the world line of a photon is tilted away from the vertical time axis by 45°. Because photons can travel in both space directions in our simplified spacetime, and because the speed of light is the limiting speed (or so it is generally believed), we can represent the collection of all possible world lines as those paths that never tilt more than 45° away from the vertical, as in Figure TN4.2.

In Figure TN4.3 I have taken $x = 0$ at $t = 0$ for all of the possible world lines involving speeds below the speed of light. Let us agree to call this point in spacetime the Here-Now. Then, points in the upward region are in the future of Here-Now; similarly, all points in the downward region are in the past

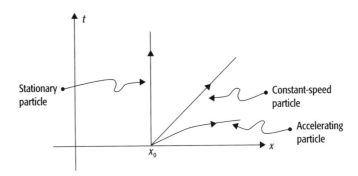

Figure TN4.1   World lines of three particles.

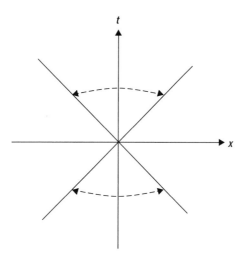

Figure TN4.2   World lines of photons.

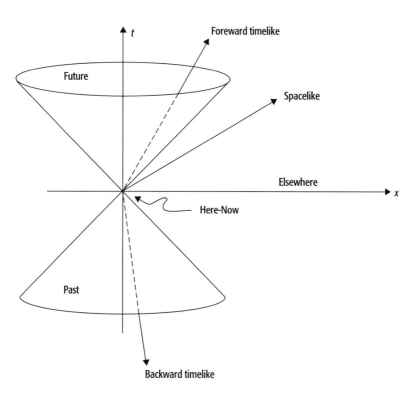

Figure TN4.3   A light cone with spacelike and timelike world lines.

of Here-Now. These regions are called *light cones* because if we included a second space dimension, say *y*, directed vertical to the page, then the upward and downward regions would be cones. We can draw a straight world line from Here-Now to any point in the Future cone with a tilt of less than 45° away from the vertical, which means that a particle could travel from Here-Now to that point at less than the speed of light. Similarly, a particle starting at any point in the Past cone could have reached Here-Now by traveling at less than the speed of light. Such world lines are called *timelike* because their projection on the time axis is greater than is their projection on the space axis, and they are the world lines of spacetime points that are at least potentially causally linked. That is, a cause at a spacetime point in the Past cone could have had an effect on the event at Here-Now even though its influence propagated at less than the speed of light. Also, a cause at Here-Now could potentially affect the event at any point in the Future cone.

Any points in the regions of spacetime outside the Future and Past cones *cannot* be reached from Here-Now except by world lines tilted more than 45° away from the vertical. Such world lines, which represent travel at a speed in excess of that of light, are called *spacelike* because their projection on the space axis is greater than is their projection on the time axis. It is impossible for these world lines to connect causally linked events, and collectively they form the Elsewhere of Here-Now. It is very important to realize that every point in spacetime has its own light cone. Thus, if B is in the Future cone of A, then A is in the Past cone of B.

This imagery of the light cone is often useful in making seemingly quite abstract ideas appear almost transparent. For example, consider the following ancient question: Can the future be predicted? Or, as Shakespeare had *Macbeth*'s Banquo ask (in Act I, scene 3),

> If you can look into the seeds of time,
> And say which grain will grow and which will not,
> Speak then to me . . .

More prosaically, can an observer predict his own future from perfect knowledge of his own past? The easy answer is "No, because quantum uncertainties prohibit perfect knowledge of even the present, much less the past." But suppose we ignore quantum mechanics and limit our question to a universe that obeys only classical physics (including special and general relativity). Surprisingly (perhaps), the answer is *still* no. Having perfect knowledge of your own past light cone doesn't include knowing the *entire* past, so if you attempt to predict your own future (say one minute from now), there can be influences in your current elsewhere that will arrive in the future (say fifty-nine seconds from now) about which you presently, by definition, *cannot* have any knowledge. And without that knowledge, you cannot predict. As Hogarth (1993) amusingly concludes a relativistic spacetime tutorial on this topic, "the prospect of predicting the future looks pretty bleak."

A spacetime diagram does not always have to have future-directed world lines. If a particle moves backward through time, assuming such a thing is possible, then the diagram can show this by having the world line double back on itself, as in Figure TN4.4. In this example the world line curves back and comes arbitrarily close to itself. This is the world line of a particle that visits itself in the past. Note that the world line does not actually touch or cross itself, because that would represent more than just a visit—it would represent a particle occupying the same spatial location, at the same time, as its earlier self. That would be catastrophic, and as we saw in Chapter Four, because it *did not* happen, it *cannot* happen. The arrows on the world line always point in the direction of the local future of the particle; if the particle is a human, for example, then memories are formed in the direction of the arrows. The time traveler at *B* has more memories than he does at *A* even though *A* and *B* are nearly identical locations in spacetime.

There is a problem here that the alert reader may have caught: It is impossible to draw such a doubled-back world line in such a way that *at all places* it

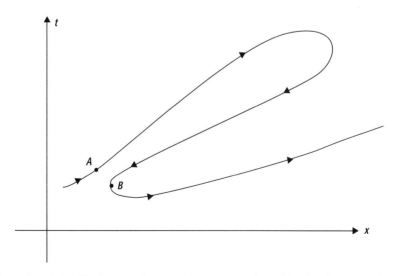

Figure TN4.4   World line of a particle traveling backward in time from *A* to *B*.

never tilts more than 45° from the vertical. That is, at least some portion of such a bent world line will have

$$\left|\frac{dx}{dt}\right| > 1$$

which represents superluminal motion. (I will return to this in more detail in Tech Note 7). One way to keep a bent-back world line always subluminal is to arrange for the light cones along the world line to be tilted relative to each other, as shown in Figure TN4.5, which is possible only in a *curved* spacetime. This is an illustration of how general relativity *locally* obeys special relativity's demand that nothing travels faster than light and yet *globally*, in curved spacetime, things are not so simple. In flat spacetimes light cones are always "aligned," but in curved spacetimes they (generally) are not, and from that can come time travel to the past.

This is the mechanism, for example, for backward time travel around a rotating Tipler cylinder, which we will be discuss in Tech Note 8. The mere presence of mass tips light cones, but the effect is unnoticeable in everyday life on Earth. A truly enormous mass density is required to tip nearby light cones over so that their Future noticeably opens up toward the massive body. If the massive body is set to rotating, then a further consequence of Einstein's general theory is that the local light cones are tilted additionally in the direction of rotation; that is, the Future cones in spacetime open up both toward the body

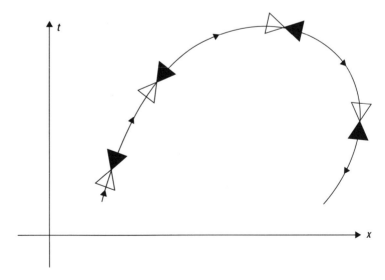

Figure TN4.5   Tilted light cones. The future halves are shaded.

and in the direction of the rotation. Light cone tipping is essential for time travel to the past, and Tech Note 8 has more on this point.

It should now be clear that the only way a world line can bend back on itself for a close-encounter visit is for both x and t to change. In other words, the world line of a particle that remains fixed in space and reverses just its time direction *runs into itself.* This is why the classic time machine of H.G. Wells could not possibly work. Any real-time machine must move in space as well as time, as does a Gödel rocket or the DeLorean time car in the *Back to the Future* films. The idea of warping world lines for time travel to the past entered science fiction quickly. For example, when the inventor of the time machine in the 1939 story "When the Future Dies" (Schachner) is asked about the principle underlying his gadget, he replies "An electro-magnetic warping of the spacetime continuum. The machine, if it works, will slide around the world-line of events and reappear at any specified time and place." Bent world lines appeared in the time travel "theory" (a so-called string theory!) attributed to the physicist-hero of television's "Quantum Leap." In that theory, human lives are like pieces of string: One end is birth, the other is death, and every day of life is some point on the string between the two ends. Tying the ends together gives a loop in time, and if you squeezed the loop together, "then each day of your life would touch another day." This poetic picture "explains" why that show's time travel was limited to within the traveler's own lifetime—but, alas, not a word on how to do it!

Using the Lorentz transformation equations from Tech Note 3, we can establish quite general relationships between events in the Future, Past, and Elsewhere regions of spacetime. For example, (1) All events in the Future/Past for the Here-Now observer are in the Future/Past for any other nearby, relatively moving observer; (2) Any event in Elsewhere can appear to be simultaneous with the Here-Now for some observer and not simultaneous for other observers; and (3) The temporal ordering (the relations of *before* and *after*) of causally related events is the same for all observers. This is not so for events that are not causally related; if two events have a spacelike separation, then two observers *can* disagree over the temporal ordering of the events. This is, in fact, the basis for the two-wormhole time machine and the cosmic string time machine discussed, respectively, in Tech Notes 9 and 10.

These statements are all easy to prove. Consider statement 1, for example. From Tech Note 3, we have (with $c = 1$, as before)

$$t' = \frac{t - vx}{\sqrt{1 - v^2}} \quad \text{and} \quad x' = \frac{x - vt}{\sqrt{1 - v^2}}$$

where t and x are the coordinates of some event A as measured by the observer in the stationary reference frame, and t' is the time measured by the observer in the reference frame moving at speed v. Thus

$$x'^2 - t'^2 = \frac{(x - vt)^2 - (t - vx)^2}{1 - v^2} = [\text{after a little algebra}]\, x^2 - t^2$$

For the stationary observer the criterion for an event to be in the Future cone is $t > |x|$—that is, $t^2 > x^2$ Thus $x^2 - t^2 < 0$ for all Future events. But the foregoing result then says $x'^2 - t'^2 < 0$, too, which is also the moving observer's criterion for the event being in his Future. The same sort of argument shows that the observers also agree on Past events.

Suppose next that two events A and B occur such that the stationary observer measures them to be $\Delta T = t_B - t_A$ apart in time. Then, we can establish statement 2 by writing

$$t'_A = \frac{t_A - vx_A}{\sqrt{1 - v^2}} \quad \text{and} \quad t'_B = \frac{(t_A + \Delta T) - vx_B}{\sqrt{1 - v^2}}$$

and so

$$\Delta T' = t'_B - t'_A = \frac{\Delta T + v(x_A - x_B)}{\sqrt{1 - v^2}}$$

From this we have $\Delta T' = 0$ (that is, simultaneity) for the two events for the special observer moving at the speed

$$v = \frac{\Delta T}{x_B - x_A}$$

and this speed is less than the speed of light for the condition $x_B - x_A > \Delta T$. This, of course, is the condition for B to be in the Elsewhere of A. In fact, we can even have $\Delta T' < 0$ (with $\Delta T > 0$) for $v < 1$ in this case of spacelike separation of A and B. That is, a stationary observer and a sublight-speed moving observer can disagree about the temporal ordering of events with spacelike separation.

Similarly, for event B to be in the causal Future of event A, we have the condition $x_B - x_A < \Delta T$. Then,

$$\Delta T' = \frac{\Delta T - v(x_B - x_A)}{\sqrt{1 - v^2}} > \frac{\Delta T - v\Delta T}{\sqrt{1 - v^2}} = \Delta T \frac{1 - v}{\sqrt{1 - v^2}}$$

Thus $\Delta T > 0$ says $\Delta T' > 0$ for $v < 1$, and this establishes statement 3.

If we were drawing diagrams with both axes representing space (a plot of y versus x, for example), we would normally define a *distance metric* for the diagram using our everyday ideas about distance. That is, we could say that if we make differential movements of dx and dy along the two coordinate axes, then the differential distance ds is given by

$$(ds)^2 = (dx)^2 + (dy)^2$$

This is, of course, just the Pythagorean theorem for the "Euclidean," or "as the crow flies," distance function. But it is not the only possible distance function. A distance function has several interesting properties, but the one we are particularly interested in here is its invariance with respect to the coordinate system. For example, if we draw a line segment on a flat sheet of paper, the distance between its ends does not depend on how we happen to select the x- and y-axes, a fact illustrated in Figure TN4.6. The coordinates for the endpoints A and B are obviously different in the two systems, but we still find that $(dx)^2 + (dy)^2 = (dx')^2 + (dy')^2$. We say that the Pythagorean distance function is *invariant*—that it is the same for all coordinate systems. Mathematicians have defined the general properties of distance as follows: If A and B are any two points, and if $d(A,B)$ is the distance between A and B, then (1) $d(A,B)$, $= d(B,A)$; (2) $d(A,B) = 0$ if and only if $A = B$; and (3) if C is any third point, then $d(A,B) \leq d(A,C) + d(C,B)$. The Pythagorean distance function possesses these three properties, but so do many other functions.[1]

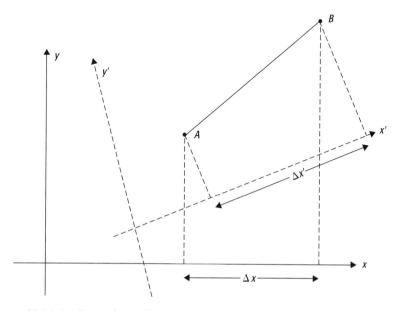

Figure TN4.6 Rotated coordinate systems.

We know that different observers, if in relative motion in the same space-time, will see different space and time coordinates for the same event. Thus it is natural to ask what the distance metric for flat spacetime is. Is there some metric that gives the same distance between two events for all observers? We might try to generalize in the obvious way from the Pythagorean theorem:

$$(ds)^2 = (dt)^2 + (dx)^2 + (dy)^2 + (dz)^2$$

where now all four dimensions are included. We would then ask ourselves whether it is true that

$$(ds)^2 = (ds')^2 = (dt')^2 + (dx')^2 + (dy')^2 + (dz')^2 \qquad ?$$

For our simple two-dimensional spacetime, this question reduces to asking whether

$$(dt)^2 + (dx)^2 = (dt')^2 + (dx')^2 \qquad ?$$

Using the Lorentz transformation equations from Tech Note 3, it is easy to discover that the answer is no. The "natural" generalization of Pythagorean distance for flat, two-dimensional spacetime fails when four dimensions are included. What do we do now? Recalling the first quotation that opened this Tech Note, we might wonder whether the problem could result from the fact that there is no fourth direction along which the time axis can point at right angles to the three space directions. At least there is no *real* direction—but perhaps there is an imaginary one. Accordingly, with $i = \sqrt{-1}$, let's try

$$(ds)^2 = (i\,dt)^2 + (dx)^2 + (dy)^2 + (dz)^2 = -(dt)^2 + (dx)^2 + (dy)^2 + (dz)^2$$

Using imaginary time, something that seems to be in the realm of science fiction, has resulted in a change in the sign of $(dt)^2$. This is a crucial change, however, because this new metric *is* invariant, as I'll now demonstrate. For a reason to be explained in Tech Note 5, I will use the negative of this metric (a choice that obviously has no impact on the invariance property). Thus

$$(ds)^2 = (dt)^2 - (dx)^2 - (dy)^2 - (dz)^2$$

For our simplified spacetime, with just one space dimension, this reduces to

$$(ds)^2 = (dt)^2 - (dx)^2$$

As before, the Lorentz transformation equations (with $c = 1$) are

$$x' = \frac{x - vt}{\sqrt{1 - v^2}} \quad \text{and} \quad t' = \frac{t - vx}{\sqrt{1 - v^2}}$$

If we calculate dx' and dt' from these equations, using

$$dx' = \frac{\partial x'}{\partial x} dx + \frac{\partial x'}{\partial t} dt$$

$$dt' = \frac{\partial t'}{\partial x} dx + \frac{\partial t'}{\partial t} dt$$

which are the fundamental relations for the total differential of a function of two variables, and insert the results into $(dt')^2 - (dx')^2$, we quickly discover the invariance of this quantity (a result we actually found earlier in a different way when we showed that observers in relative motion agree about what events are in the Future and what events are in the Past). Thus

$$(dt')^2 - (dx')^2 = (dt)^2 - (dx)^2$$

This quantity, on either side of the equality, is called the spacetime *interval* between the two events separated in flat spacetime either by dt, dx, dy, and dz or by dt', dx', dy', and dz'; that is, $(ds)^2 = (ds')^2$. The observers in the un-primed and primed systems see different individual space and time separations for two events,but they see the same interval. The single time and the three space coordinates are said to form a four-vector that is invariant under Lorentz transformation. There are other four-vectors that are also invariant under Lorentz transformation, such as the energy-momentum, the velocity, and the force four-vectors. All have invariants that are formed the same way, by taking the difference of the square of the time component and the sum of the squares of the space components.[2] The mixing of space and time appeared early in science fiction, but often in comically mangled form. In the 1930 tale "The Atom-Smasher" (Rousseau), for example, as the evil scientist uses his time machine to transport captives into the past, he tells them, "We've got a longish journey before us, ten thousand years more, multiplied by the fourth power of two thousand miles."

The intrusion of the square root of -1, in the time coordinate portion of the metric, seems a pretty clear indication that time *is* different from space. But still, the spatialization of time is nonetheless deeply embedded in Western culture. For example, I find it interesting that when writing her popular 1978 historical work drawing parallels between the fourteenth and twentieth centuries, Barbara Tuchman titled it *A Distant Mirror*, not *An Old Mirror*. She could be assured that her readers would understand the temporal nuances of *distant*. One science fiction writer had some particularly silly fun with the relationship between space and time in his novella *Out of Time* (Hogan). There, the world is invaded by *chronovores*, higher-dimensional bugs that eat time and *excrete* space.

Now, a little aside. In *general* relativity, the metric of any four-dimensional spacetime is of the symmetric quadratic Riemannian form

$$(ds)^2 = \sum_{i=1}^{4} \sum_{j=1}^{4} g_{ij}(dx_i)(dx_j), \quad g_{ij} = g_{ji}$$

where $x_1 = t, x_2 = x, x_3 = y$, and $x_4 = z$, and the $16 g$'s are all functions of these four variables. (Because of the symmetry condition, only 10 of the g's are independent.) In this notation, *flat* spacetime is characterized by $g_{11} = 1, g_{22} = g_{33} = g_{44} = -1, g_{ij} = 0$ for $i \neq j$. For a given spacetime, one can arbitrarily choose an infinity of coordinate systems. If just one of this infinity of systems is such that the $\pm 1, 0$ values for the g's occur, then that spacetime is *globally*— everywhere—flat. If no such coordinate system exists, then that spacetime is necessarily curved.

If this notation is extended to a fifth dimension by including an x_5, then there are an additional eight off-diagonal g's—that is, four new independent g's. These are just sufficient to describe the electromagnetic field, along with the gravitational field described by the other ten g's. This is what it means to say that five-dimensional spacetime "unifies" gravity and electromagnetism. There is a g_{55} term, too, which can either be set equal to 1 (like the other g_{ii}) or allowed to vary (which could model the slowly varying Newtonian gravitational constant suggested by Dirac in 1938). The g functions are the components of the so-called *metric tensor of the second rank* of that spacetime. The g's at each point in spacetime are related to the curvature of spacetime at that point, which in turn is dependent on the g's and on the energy density at that point. In fact, the equations for the g's, which are the Einstein gravitational field equations, are both nonlinear and coupled. That is, g_{ij} is in general a nonlinear function of g_{ik}, which accounts for the notorious difficulty in finding analytical solutions to the field equations except in certain highly special cases, such as around spinning spheres and rotating infinite cylinders.

As an example of this non-linearity, consider Roger Penrose's words on how gravitational collapse progresses in matter that is under great pressure, a condition we intuitively think should *resist* further collapse (1972): "In general relativity, pressure itself is a source of extra gravitational attraction. When the matter is sufficiently concentrated, the additional gravitational pull due to pressure more than overcomes the direct effect of the pressure itself. Greater pressure then *increases* the tendency to collapse." For a second example of non-linearity, the energy of a gravitational field produces, *itself*, a gravitational field, which of course has energy—and so on. This sort of thing is what is behind the theory's ability to explain totally the observed precession of

Mercury's orbit, an explanation that Newton's linear theory of gravity is not quite able to provide.

The sixteen g's are often written in the form of a 4×4 matrix. In fact, the metric tensor is of the *second* rank precisely because a matrix is a *two-dimensional* form; scalars and vectors, which have forms of zero and one dimension, are tensors of rank zero and one, respectively. The collection of the algebraic signs of the main diagonal terms (the g_{ii}) is called the *signature* of the metric tensor. The signature of flat (Minkowski) spacetime is thus written as $[+, -, -, -]$; in more general (curved) spacetimes, this same signature is called *Lorentzian*. By contrast, the signature of a four-dimensional Euclidean space is $[+, +, +, +]$. This signature is called *Riemannian*.

The Minkowski spacetime discussed in this Tech Note is said to be *flat* because for every straight-line geodesic there are infinitely many others parallel to it. This is a feature of a flat sheet of paper, but one must be careful not to carry the analogy too far. The geometry of our spacetime is not the Euclidean geometry of a flat sheet of paper, because the spacetime metric has both plus and minus signs. The geometry of flat, Minkowski spacetime is also called complex-Euclidean or pseudo-Euclidean; it *is* Euclidean geometry—with an imaginary coordinate axis for the time axis. Flat Minkowskian geometry, in turn, is a special case of curved Riemannian geometry. The curvature explains gravity, as developed in Einstein's general theory of relativity; our flat spacetime, by virtue of being uncurved, *has no gravity*. These features of the physics of spacetime had an enormous impact on one of the great minds of twentieth-century physics. Writing in 1972, Paul Dirac said of his first encounter with the metric of special relativity, while still an undergraduate,[3]

> Now, when I saw that minus sign [in $-(dt)^2$], it produced a tremendous effect on me. I immediately saw that here was something new. Perhaps I can explain the reason for this big effect from the fact that previously as a schoolboy I had been much interested in the relations of space and time. I had thought about them a great deal, and it had become apparent to me that time was very much like another dimension, and the possibility had occurred to me that perhaps there was some connection between space and time, and that we ought to consider them from a general four-dimensional point of view. However, at that time the only geometry that I knew was Euclidean geometry, and, if space and time were to be coupled in any way, they would have to be coupled with a plus sign here, and it was very easy to see that that would not work, and led to nonsense as soon as one tried to make any big change in one's time axis.

The idea of linking the curvature of a four-dimensional space to physical phenomena pre-dates Einstein. It can be seen, for example, in the work of the British mathematician William Kingdon Clifford (BST) in the 1870s—see Note 12 for Chapter Two. Indeed, one can find it mentioned even earlier, a full decade before Einstein's birth. In a footnote from Sylvester (1869) we read, "It is well known . . . we live in a flat or level space, our existence therein being assimilable to the life of the bookworm in a flat page: but what if the page should be undergoing a process of gradual bending into a curved form? Mr. W.K. Clifford has indulged in some remarkable speculations [supporting the idea that] our level space of three dimensions [is] in the act of undergoing in space of four dimensions . . . a distortion analogous to the rumpling of the page." Remember, this is in 1869! Clifford, in turn, found inspiration in the even earlier work of the German mathematician Bernhard Riemann (1826–1866).[4]

Typical of the contemporary reaction to such a radical idea was that of the great Clerk Maxwell (he knew Clifford through their mutual membership in the London Mathematical Society), who dismissed this part of Clifford's work as the speculations of a "space crumpler." Such ideas continued to be the subject of debate for many years after Clifford. For example, Kerszberg (1987) reports a 1917 exchange between the two British astronomers Sir James Jeans and Sir Arthur Stanley Eddington. When Jeans declared, "Einstein's crumpling up of his four-dimensional space may, for the present, be considered to be . . . fictitious," Eddington quickly replied that what is crumpled is "the ordinary space which we are persuaded to discard." Another advocate of curved space was the philosopher Charles Peirce, who in the 1890s argued that space is non-Euclidean and who even tried to determine the numerical value of the curvature of space from astronomical data (Dipert, 1977). Peirce also argued that space is four-dimensional (Dipert, 1978), so it would seem that he could indeed have thought of a curved four-dimensional space (though not of space-*time*) decades before Einstein. But Riemann was first.

Gauss (1777–1855) had studied the curvature of surfaces in three dimensions before Riemann, but it was Riemann who generalized curvature to *n* dimensions, most particularly to spaces of four dimensions. The prescient genius of Riemann is almost impossible to believe. In his 1854 lecture there is a throwaway line about spaces with a metric given by the positive fourth root of a quartic differential form. Such spaces are known today as *Finsler spaces* (after the German mathematician Paul Finsler, who developed Riemann's idea in his 1918 Göttingen doctoral dissertation "On Curves and Surfaces in General Spaces"), and their use in the theory of FTL speeds (see Tech Note 7) can be found in Shenglin (1992a,b, 1990, 1988.)

Geodesics in curved spacetimes play important special roles is general relativity. For example, a spacetime that contains singularities is *geodesically incomplete.* That is, there are points in such a spacetime beyond which the world lines of freely falling observers cannot be extended. As Davies (RTR) points out, general relativity predicts that spacetime can be stretched, bent, and twisted by various means (such as by a massive, rotating body), perhaps to the point where it "can tear, shrivel away, or curve so severely that it becomes cusp-like and stops. Any such regions would be considered as boundaries to spacetime. They could be in a spacelike direction . . . or timelike. . . ." That is, an incomplete spacelike geodesic terminates on a singularity in time ("an end to time"), and an incomplete timelike geodesic terminates on a singularity in space ("an end to space"). General relativity says that such bizarre happenings may be the normal state of affairs in the universe. The fact that we don't see such doings on Earth is what is exceptional! Because free-falling observers are not likely to have a spacelike world line, however, this sort of singularity may not really be very physical. As Earman (1996) put it, "Certainly the ghost of an observer whose world line is an incomplete timelike geodesic would have grounds for complaining that the spacetime is pathological."

There is also a third sort of singularity, the so-called *null incompleteness.* That is, if there are world lines for photons that cannot be followed arbitrarily far into the future or the past, then the spacetime has a null singularity. Of course, as Naber (1988) asks, why do we limit ourselves to the world lines of geodesics? As he writes, "accelerated observers seem to have as much right as free observers to object to their existence being abruptly curtailed. . . ." And, in fact, the idea of geodesic incompleteness as the signature of spacetime singularity is itself incomplete; see Geroch (1968) for an example of a spacetime that is timelike and null complete and yet still singular for at least one accelerating observer. Much more on this issue can be found in Clarke (1993) and Earman (1996).

Spacetime geometries are not easy concepts to understand, and the metrics of curved spacetimes are even more complicated than the metric of elementary flat spacetime. As one paper puts it,[5] "Experience has taught us that the space in which we live has a geometry that is three-dimensional and Euclidean. . . . We are very much at home with [that] geometry. . . . But the geometric properties of a Minkowskian space are so alien to us that we may well despair of visualizing them, and a Riemannian space . . . seems totally beyond comprehension."

An important feature of Riemannian geometry is that although it is generally not globally flat, it is always *locally* flat. Thus any sufficiently small region in a curved Riemannian spacetime can be approximated, with arbitrarily small error, by a flat pseudo-Euclidean Minkowskian spacetime. That is, at each point in Riemannian spacetime, there is some *particular* inertial frame of reference in which special relativity is all there is to spacetime physics *at that point* (see Tech Note 3). The particular inertial frame required is different, however, from point to point.

In a coordinate system different from rectangular, the flat Minkowskian metric can appear radically altered, but that is just an artifact of the mathematics and has no physical significance. For example, in spherical coordinates the Minkowskian metric becomes the equivalent

$$(ds)^2 = (dt)^2 - (dr)^2 - (rd\theta)^2 - (r\sin\theta d\phi)^2$$

where ϕ is the azmuthal angle and θ is the angle measured from the polar axis. A related metric occurs in the theory of spherically symmetric static (no time dependence) time machine wormholes (discussed in Tech Note 9), of the form

$$(ds)^2 = (e^{a(r)}dt)^2 - \frac{(dr)^2}{1 - \frac{b(r)}{r}} - (rd\theta)^2 - (r\sin\theta d\phi)^2$$

where $a(r)$ is the *redshift function* and $b(r)$ is the *shape factor.* These functions are arbitrary, subject only to the constraints that both $b(r)/r$ and $a(r)$ vanish as r goes to infinity (r is the radial distance from the throat of the wormhole mouth). Indeed, as r increases, this curved wormhole spacetime metric obviously reduces to that of flat Minkowski spacetime, so such a wormhole spacetime is said to be *asymptotically flat* [for wormhole spacetimes that are not asymptotically flat, see Narahara *et al.* (1994)].

An intuitive formulation of flatness can be visualized in terms of the parallel transport of a vector around a closed path. In a curved space the vector will experience a rotation, which will not occur in a flat space. Two examples of parallel transport in two-dimensional spaces are shown in Figure TN4.7. The spherical surface is curved because if you slide the vector from N to A to B and then back to N, all the while keeping it parallel to its previous orientation, then when it gets back to N the vector will point toward B, not toward A as it originally did. A similar trip on the cylindrical surface, however, will result in zero rotation of the vector. Thus a cylindrical surface is, despite superficial appearances, *not* curved.

Using the idea of the metric tensor discussed earlier makes possible a more formal demonstration that the surface of a sphere is *not* flat. On the surface of

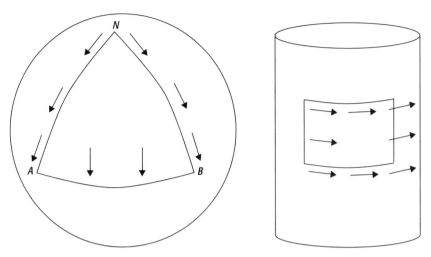

Figure TN4.7   The curvature of a space can be revealed by the process of parallel transport. For example, this figure shows the parallel transport of a vector around a closed path over two different two-dimensional surfaces. Transport over a curved—for example, spherical—surface results in a non-zero rotation of the vector, whereas transport over a flat—for example, cylindrical—surface produces zero rotation.

a sphere of radius a (on the surface $r = a$ everywhere, and so $dr = 0$), the measure of distance between two points is (in spherical coordinates)

$$(ds)^2 = a^2\sin^2\theta \, (d\phi)^2 + a^2(d\theta)^2$$

Writing $x_1 = \phi$ and $x_2 = \theta$ yields the more general form

$$(ds)^2 = g_{11} \, (dx_1)^2 + g_{12} \, (dx_1) \, (dx_2) + g_{21} \, (dx_2) \, (dx_1) + g_{22} \, (dx_2)^2$$

or, using the symmetry condition $g_{12} = g_{21}$,

$$(ds)^2 = g_{11} \, (dx_1)^2 + 2g_{12} \, (dx_1) \, (dx_2) + g_{22} \, (dx_2)^2$$

From this we immediately have $g_{11} = a^2\sin^2\theta$, $g_{22} = a^2$, and $g_{12} = g_{21} = 0$. Note carefully that this is a purely spatial problem, without time, and we are taking all of the metric coefficients as positive (unlike the case of the general *spacetime* metric).

Now, suppose we ask whether it is possible to find some new coordinate system (with variables x_1' and x_2') in which the invariant $(ds)^2$ is given by the flat Euclidean metric $(dx_1')^2 + (dx_2')^2$. In such a coordinate system (if it exists), we would have the "flatness" conditions of $g_{11}' = g_{22}' = 1$ and $g_{12}' = g_{21}' = 0$. With such a change of coordinates, each of our old ϕ, θ coordinates would generally be a function of both of the new coordinates—that is, $\phi =$

$\phi\ (x_1',x_2')$ and $\theta = \theta\ (x_1',x_2')$. Thus, writing the total differential of a function of two variables (just as we did earlier in this Tech Note) yields

$$d\phi = \frac{\partial\phi}{\partial x_1'}\,dx_1' + \frac{\partial\phi}{\partial x_2'}\,dx_2'$$

$$d\theta = \frac{\partial\theta}{\partial x_1'}\,dx_1' + \frac{\partial\theta}{\partial x_2'}\,dx_2'$$

Substituting these expressions into the original expression for $(ds)^2$ and collecting terms, we arrive at

$$(ds)^2 = a^2\left\{\left[\sin^2\theta\left(\frac{\partial\phi}{\partial x_1'}\right)^2 + \left(\frac{\partial\theta}{\partial x_1'}\right)^2\right](dx_1')^2\right.$$

$$\left. + \left[\sin^2\theta\left(\frac{\partial\phi}{\partial x_2'}\right)^2 + \left(\frac{\partial\theta}{\partial x_2'}\right)^2(dx_2')^2\right] + 2\left[\sin^2\theta\,\frac{\partial\phi}{\partial x_1'}\,\frac{\partial\phi}{\partial x_2'} + \frac{\partial\theta}{\partial x_1'}\,\frac{\partial\theta}{\partial x_2'}\right]\right\}$$

We can now immediately identify each of the g', and if we demand that they satisfy the "flatness" conditions, then we have the three statements

$$\sin^2\theta\left(\frac{\partial\phi}{\partial x_1'}\right)^2 + \left(\frac{\partial\theta}{\partial x_1'}\right)^2 = \frac{1}{a^2}$$

$$\sin^2\theta\left(\frac{\partial\phi}{\partial x_2'}\right)^2 + \left(\frac{\partial\theta}{\partial x_2'}\right)^2 = \frac{1}{a^2}$$

$$\sin^2\theta\,\frac{\partial\phi}{\partial x_1'}\,\frac{\partial\phi}{\partial x_2'} + \frac{\partial\theta}{\partial x_1'}\,\frac{\partial\theta}{\partial x_2'} = 0$$

The surface is to be globally flat (flat everywhere), so these three conditions must hold for all ϕ and θ, including the poles of the sphere (where $\sin\theta = 0$). At the poles the first two conditions reduce to

$$\left(\frac{\partial\theta}{\partial x_1'}\right)^2 = \frac{1}{a^2}$$

$$\left(\frac{\partial\theta}{\partial x_2'}\right)^2 = \frac{1}{a^2}$$

and the third reduces to the *incompatible* requirement that

$$\frac{\partial\theta}{\partial x_1'}\,\frac{\partial\theta}{\partial x_2'} = 0$$

Because of this inconsistency, we can conclude that there is in fact *no* primed coordinate system in which the g' coefficients are those of a flat metric.

Thus, unlike the surface of a cylinder, the surface of the sphere is *not* flat but curved.

One special case of particular interest will conclude this Tech Note. Imagine the following two events in spacetime: Event 1 is the emission of a photon at one point in spacetime, and event 2 is the absorption of that photon at some other point in spacetime. What is the interval between these two events? It might seem that we need to know more about the precise spatial and temporal coordinates of these two events, but in fact the interval is *always zero* for any two events connected by light. For our simple, flat Minkowski spacetime, this is easy to see. Just rewrite the metric as

$$\left(\frac{ds}{dt}\right)^2 = 1 - \left(\frac{dx}{dt}\right)^2$$

and then since $(dx/dt)^2 = 1$ (because the photon travels at the speed of light), we have $(ds)^2 = 0$. The world line of any photon is said to have a *null* interval. Indeed, null-interval world lines are always on the surface of light cones, whereas timelike world lines—in the interior of light cones, with $(dx/dt)^2 < 1$—have positive intervals; that is, $(ds)^2 > 0$. Spacelike world lines (in the exterior of light cones, in Elsewhere) have negative intervals [because $(dx/dt)^2 > 1$]; that is, $(ds)^2 < 0$. This is one of the significant differences between distances in space and intervals in spacetime—in space, the distance is always non-negative.

A strange implication of this is that in spacetime we can have the interval between A and B as zero and the interval between B and C as zero, but the interval between A and C may not be zero! To convince yourself of this, try plotting the Minkowski spacetime coordinates of A, B, and C as $(1, 3)$, $(2, 2)$ and $(1, 1)$, respectively. Null-interval world lines have an interesting interpretation as the result of this. As an early writer poetically put it,[6] "Any pair of points [in spacetime] which are separated by zero distance [our *interval*] are in *virtual contact*. In other words, I may say that my eye touches a star, not in the same sense as when I say that my hand touches a pen, but in an equally physical sense."

Notes

1. See, for example, E.F. Krause, *Taxicab Geometry* (New York: Dover, 1986), where the *city-block* distance function $ds = |dx| + |dy|$ is explored.

2. I will not pursue any of this here, but the interested reader will find a wonderfully readable exposition at the advanced undergraduate level in I.R. Kenyon, *General Relativity* (New York: Oxford University Press, 1990).

3. P.A.M. Dirac, "Recollections of an Exciting Era," in *History of Twentieth Century Physics*. Edited by C. Weiner. (New York: Academic Press, 1977).

4. See, for example, Clifford's translation of Riemann's 1854 lecture "On the Hypotheses Which Lie at the Bases of Geometry," *Nature* 8 (May 1, 1873):14–17, and continued in the next issue (May 8, 1873):36–37.

5. R.W. Brehme and W.E. Moore, "Gravitational and Two-Dimensional Curved Surfaces," *American Journal of Physics* 37 (July 1969):683–692.

6. G.N. Lewis, "Light Waves and Light Corpuscles," *Nature* 117 (February 13, 1926):236–238.

Proper Time, Curved World Lines, and the Twin Paradox

Before I realized the untenability of the special theory of relativity I was engaged in a vigorous controversy on the question whether relativity theory entailed the possibility of "asymmetrical ageing," e.g., of postponing the date of one's death, almost without limit, by high speed travel. It was, and still is, generally believed that relativity demands this possibility. . . . [I]f I am right, [this belief in] asymmetrical aging . . . shows that relativity theory is generally misunderstood.

— 1963 note to *Nature* by Herbert Dingle (former president of the Royal Astronomical Society), who rejected special relativity and engaged in a years-long effort to convince the world that there was a conspiracy to suppress him

What if Dingle is pulling the leg of the world? It is to me the most reasonable hypothesis to explain what is otherwise inexplicable to me. Knowing you as well as I do. . . . I cannot bring myself to believe that you are as stupid as you make yourself out to be. If my hypothesis is correct, I salute your sense of humor. No harm has been done. Printers have had good employment. My humiliation in having been taken in is swallowed up in my admiration at the way you have put the thing across.

— 1968 letter to Dingle, from the mathematical physicist J.L. Synge of the Dublin Institute of Advanced Studies

The spacetime interval introduced in Tech Note 4 has an important interpretation that will lead us to one of the more dazzling results of special relativity—time travel into the future. First, recall the flat spacetime metric

$$(ds)^2 = (dt)^2 - (dx)^2 - (dy)^2 - (dz)^2$$

in which the use of unprimed variables indicates that the measurements on the space and time coordinates of a moving particle are made with respect to a stationary observer's frame of reference. Now, suppose that the space and time coordinates of a moving particle are made with respect to the particle instead. Then, using primed variables of measurements made in this new frame of reference, we have $dx' = dy' = dz' = 0$ because the particle is always at the origin of it own coordinate system. Recalling the invariance of the spacetime interval for all observers, we conclude that

$$(ds')^2 = (ds)^2 = (dt')^2$$

That is, the spacetime interval between two events is the time lapse measured by a clock attached to a particle that moves from one event to the other. This time is called *proper time*, which gets its name from the idea that it belongs to, or is the *property* of, the moving particle. This is the technical reason why we took $(ds)^2 = (dt)^2 - (dx)^2$ rather than $(ds)^2 = (dx)^2 - (dt)^2$. The first choice avoids the somewhat awkward result of an imaginary proper time.

To establish one more preliminary result, I will next adopt what has come to be called the *clock hypothesis,* which states that an accelerated clock runs at the same instantaneous rate as an unaccelerated clock that is moving alongside at the instantaneously same speed. As shown in Tech Note 2, if the accelerated clock's instantaneous speed is v, then its rate of timekeeping (dt') is related to that of the "stationary" (unaccelerated) clock (dt) as

$$dt' = \sqrt{1 - \left(\frac{v}{c}\right)^2}\, dt$$

The clock hypothesis is generally assumed to be true. Einstein himself, in his 1905 paper, specifically took the rate of a clock's timekeeping to be velocity-dependent only. However, one can still find those who object. In this book I side with Einstein. When asked during a 1952 interview whether it is permissible to use special relativity in problems involving acceleration, Einstein replied, "Oh, yes, that is all right as long as gravity does not enter; in all other cases, special relativity is applicable. Although, perhaps the general relativity approach might be better, it is not necessary."[1] The clock hypothesis has long had experimental verification. For example, in the course of a 1960 experiment[2] performed for a different purpose, the timekeeping of accelerated atomic clocks was determined to be given precisely by the time dilation formula of special relativity, even when their direct mechanical acceleration reached levels greater than 66,000 gees. The accelerations generated in that experiment were produced with a rapidly spinning disk. And even more impressive are the time dilation results reported by Bailey *et al.* (1977), in which near light-speed charged particles orbited in a magnetic field. Excellent agreement with theory was observed even with accelerations well in excess of 10^{15} gees. One amusing (but dizzy) design for a time machine that would send its occupant into the future, then, would simply be a souped-up clothes dryer!

A more scholarly use of the idea of time dilation as a result of moving in a circle was presented by Pitowsky (1990). As Pitowsky writes,

> Suppose that M, a mathematician, is literally dying to know whether Fermat's conjecture is true or false. He takes a trip in a satellite which revolves around the earth. The satellite has an immense engine, which boosts it so hard that its instantaneous tangential velocity is $v(t) = c[1 - e^{-2t}]^{1/2}$, where t is the earth's time scale and c is the velocity of light. The engines are also oriented in such a way as to keep the satellite in a fixed orbit. If t' is the satellite's local time scale, then the time interval dt' is given by $\left[dt\sqrt{1 - (v/c)^2}\right] = e^{-t}dt$. Hence one second on the satellite's time scale corresponds to eternity on earth, since $\int_0^\infty e^{-t}dt = 1$. [For more on this last point, see Hogarth (1992).]
>
> While M peacefully cruises in orbit, his graduate students examine Fermat's conjecture one case after another, that is, they take quadruples of natural numbers (x,y,z,n), with $n \geq 3$, and check on a conventional computer whether $x^n + y^n = z^n$. . . . When they grow old, or become professors, they transmit the holy task to their own disciples, and so on. If a counterexample to Fermat's conjecture is ever encountered, a message is sent to the satellite. . . . If no message arrives, M disintegrates with a smile, knowing that Fermat was right after all.

For a reply to this, see Earman and Norton (1993), which explains why M "disintegrates." As Earman and Norton show, the *acceleration* of M becomes unbounded, so "M will be quickly crushed by g-forces." Rather than with a smile, Earman and Norton write, "The mathematician M disintegrates with a grimace. . . ." Of course, now that Fermat's conjecture has been settled (since 1995), this story would be better if the problem were changed, perhaps to

Riemann's conjecture on the location of the non-trivial complex zeros of the zeta-function (presently the outstanding, unsolved problem in mathematics).

The total elapsed time between two events A and B, as measured by the proper time of the accelerated clock making the journey, is

$$t' = \int dt' = \int_{t_A}^{t_B} \sqrt{1 - \left(\frac{v}{c}\right)^2}\, dt < t_B - t_A \ \text{ if } v \neq 0$$

where $t_B - t_A$ is the elapsed time between A and B as measured by the unaccelerated clock. This inequality results because for $v \neq 0$, the integrand is always less than 1.[3]

We know that in a spacetime diagram the world line of the unaccelerated clock is a straight line, whereas the world line of the accelerated clock is a curved line. Thus using Figure TN5.1 combined with the inequality $t' < t_B - t_A$ gives the following central result: The world line of *maximum* proper time is the one that *looks the shortest*—that is, the straight (or free-fall geodesic) world line. On the spacetime diagram the curved world line looks longer, but in fact any curved world line will have a smaller proper time than does the straight world line. This is a dramatic example of how Minkowskian *spacetime* geometry differs from Euclidean *space* geometry; in the latter geometry, there is no longest path between two points.

With these preliminary results established, you can now understand the famous paradox of the twins. Suppose we have twins Bob and Bill. Bill remains

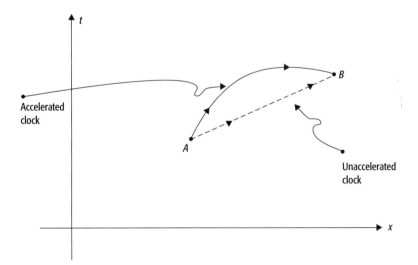

Figure TN5.1   World lines of two clocks.

on Earth, but Bob gets into a rocket ship and goes on a trip out into space, turns around by firing his engines, and comes home. The world lines of Bob and Bill are at first together, then they diverge, and then they come together again at the end of Bob's trip, as shown in Figure TN5.2. The details of Bob's trip are not important for a general statement of the paradox (although in Tech Note 6 the details for one possible trip are presented). All we need observe here is that Bill's world line from A to B is straight, whereas Bob's is curved. Bill's body (that is, his local clock) will therefore measure a greater proper time than will Bob's; that is, Bob will be younger than his stay-at-home twin! Equivalently, upon his return Bob will hear his Earthbound brother declare the date to be further in the future than Bob's trip lasted (according to Bob). Bob will conclude that he has traveled into the future. The difference in what each twin believes to be the date can, in fact, be truly astonishing.

At the beginning of a prose discussion of the twin paradox, University of Texas professor Alfred Schild (1960) states, "I have no doubt that if our technology should ever advance to the stage where large-scale twin effects become noticeable with our unaided senses, then [people] will have no difficulty in adjusting their concepts of time until the new phenomena seem quite natural." Well, perhaps. Oddly, this wonderful plot device seems not to have caught the attention of many science fiction writers. So far as I know, the twin paradox has been used explicitly, just as stated, by only one author: Robert Heinlein in his 1956 novel *Time for the Stars*. In a story set on a planet where time passes

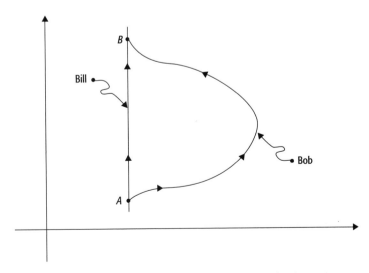

Figure TN5.2   World lines of unaccelerated and accelerated twins.

ever more slowly as one travels ever more upward in latitude, a story that offers no physical explanation for such time distortions, the bizarre human problems they would cause were explored in "Traveler's Rest" (Masson). For the twin paradox as a result of high-speed rocket travel, however, the physical explanation for such a time distortion is solid physics. And finally, Robert Forward uses biology to defeat the physics in his ironic "Twin Paradox." This story flips the asymmetric aging of the twins when, just after the traveling twin's departure, the secret of immortality is discovered. The treatment has to be administered no later than a certain age, however, and upon his return to earth, the traveling twin is just a bit too old for it to work. Hence he becomes the last person to die of old age.

This situation is called a paradox not because of the time travel aspect (there are no logical paradoxes with travel into the future) but rather because it seems to violate the very spirit of relativity. That is, from Bill's point of view, Bob at first travels away and then returns. But one might argue that from Bob's point of view, it is Bill who first recedes and then returns. Long after Einstein's 1905 publication of the special theory of relativity, this point still puzzled many. For example, in the 1923 Presidential Address to the Eastern Division of the American Philosophical Association, we read this very objection to the twin paradox. The conclusion there was that such a thing "could happen only in a universe in which all squares were round and the *principio contradiction* had been put to sleep."[4] So, back to the question of why it is *Bob* who is the younger. The classical physics answer is that the two points of view are actually not identical and that there is a definite asymmetry between Bill and Bob. After all, it is Bob who feels the acceleration from the rocket's engines—who feels forces—whereas Bill feels nothing unusual back on Earth. The more fundamental physics answer, however, is that Bob's world line in spacetime is curved, whereas Bill's is straight.

In an open flat spacetime, *curved* is indeed synonymous with *accelerated*, but this need not be so in a closed flat spacetime. In an open flat spacetime, the only way two world lines can diverge in the past and then meet again in the future is for at least one of them to curve, but in a closed but still flat spacetime, it is possible for two straight world lines to meet more than once. For example, in Figure TN5.3, our simplified two-dimensional spacetime is the surface of a cylinder (which Tech Note 4 showed is flat), rather than an infinite flat plane. The two world lines are both straight. (To visualize this, imagine cutting the cylinder open along the time dimension and then flattening it out.) Yet Bob's world line looks longer on the spacetime diagram, so Bob's proper time will still be less than Bill's when they again meet, even though now *neither* of them has experienced any acceleration.[5]

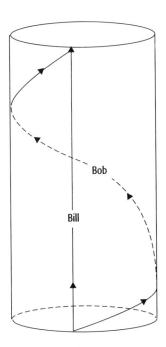

Figure TN5.3   Unaccelerated-twin paradox in a cylindrical spacetime.

The paradox of the twins has generated a vast literature; see Marder (1971) for an excellent discussion and extensive bibliography up to 1963, and North (1970). The twin paradox is hinted at in Einstein's 1905 paper, but it is in a 1911 address to the International Congress of Philosophy, in Bologna, by the French physicist Paul Langevin, that a space traveler (in a cannonball moving at a velocity close to that of light—an idea motivated by Langevin's reading of Jules Verne's 1872 *From the Earth to the Moon*!) is first introduced in this context. The *French* writer Pierre Boulle proudly mentioned Langevin in his astonishing time travel story "Time Out of Mind."

Fascination with the twin paradox continues to this day; see Prokhovnik (1989) and Cramer (1990b), for example. Not everybody accepts the results I have described here, however. Herbert Dingle, for one, strenuously objected to them. Dingle was not a crackpot; the late professor of history and philosophy of science at the University of London was, for most of his life, generally well regarded in scientific circles, but he went to his grave believing Einstein had got his sums wrong. The French philosopher Henri Bergson vigorously attacked Langevin's analysis of the clock paradox in a 1922 book, for which Dingle wrote a very interesting introduction (Bergson, 1965). Dingle's position is examined in Marder's book, and Dingle tries to make his case in the

1972 *Science at the Crossroads*[6] (from which the quotes that open this Tech Note come).

Notes

1. See R.S. Shankland, "Conversations with Albert Einstein," *American Journal of Physics* 31 (January 1963):47–57.

2. See H.J. Hay *et al.*, "Measurement of the Red Shift in an Accelerated System Using the Mössbauer Effect in Fe^{57}," *Physical Review Letters* 4 (February 15, 1960):165–166.

3. For a more mathematical presentation of this result, see Mary L. Boas, "The Clock Paradox," *Science* 130 (November 27, 1959):1471–1472.

4. W.P. Montague, "The Einstein Theory and a Possible Alternative," *The Philosophical Review* 33 (March 1924):143–170.

5. C.H. Brans and D.R. Stewart, "Unaccelerated-Returning-Twin Paradox in Flat Space-Time," *Physical Review D* 8 (September 15, 1973):1662–1666.

6. A interesting review of this book, and a summary of where Dingle went wrong, is offered by G.J. Whitrow, *British Journal for the Philosophy of Science* 26 (March 1975):358–362. An excellent review essay on the entire Dingle affair, with detailed analyses both of the twin paradox arguments used by Dingle and his critics and of what probably motivated Dingle to conduct a debate that destroyed his scientific reputation, is Mark Hogarth's "A Misunderstood Rebellion," *Studies in the History and Philosophy of Science* 26 (December 1993):741–790. Dingle is now gone from the scene, but his cause still has followers. For example, in the same year that Marder's book was published, Mendel Sachs, a professor of physics at the State University of New York at Buffalo, wrote a "Dingle-like" article entitled "A Resolution of the Clock Paradox," *Physics Today*, September 1971. That article provoked a reaction so strong that the editors devoted essentially the entire Letters column in the January 1972 issue to the replies, all of which rejected Sach's position.

6

A High-Speed Rocket Is a One-Way Time Machine to the Future

> A good many physicists believe that this paradox can only be resolved by the general theory of relativity. They find great comfort in this, because they don't know any general relativity and feel that they don't have to worry about this problem until they decide to learn general relativity. However, they are quite wrong. The twin effect . . . is one of the special theory of relativity.
>
> —Schild (1959)

> [The] equations of duo-quadrant lineations [have] been substantiated. . . . Our fourth-angle deviation from the six conceivable electronic dimensions did the trick all right. I went forward in Time.
>
> —a 1930s science fiction scientist babbles incoherent nonsense, not special relativity, about how to travel into the future in "He Who Masters Time" (Haggard)

The following analysis is essentially that of the German astronomer Sebastian von Hoerner, who published the basic ideas in *Science* in 1962 under the title "The General Limits of Space Travel." The following year it was reprinted in the classic anthology *Interstellar Communication*. Later, von Hoerner's somewhat muddy mathematics was cleaned up by Marder,[1] a British applied mathematician, who incorporated it into his brilliant, definitive book on the twin paradox of special relativity.

This analysis describes an accelerated rocket and assumes that the formulas of special relativity hold at every instant. This is a view that von Hoerner described as "generally assumed but not yet accepted."[2] To begin, our time traveler gets into his rocket ship at time $t = t' = 0$ (t is time measured on Earth, and t' is time measured on the rocket). The Earth and rocket clocks are synchronized at the instant of departure. We assume that the rocket trip is to be made in comfort, so let the rocket accelerate at a constant rate. In the numerical calculations I will, in fact, let this constant be one gee, which is equivalent to Earth's gravity, an acceleration we all live with through our entire lives. This is of practical importance, of course, because we do not want the experienced acceleration to be incompatible with the physical survival of the time traveler. All things considered, it is probably better to arrive in the future alive than to arrive dead from compression! The traveler travels this way for a time interval of T (as measured on Earth) and T' (as measured on the rocket). At that time the rocket is traveling at its maximum speed.

Then the traveler turns off the rearward engine and turns on a forward-mounted engine so as to experience a constant deceleration. Floor and ceiling interchange, but our traveler always weighs the same. If he does this for the same time interval (T in Earth time, T' in rocket time) as for the acceleration interval, then the rocket will be brought to rest with respect to Earth. At that time ($2T$ on Earth and $2T'$ on the rocket), the rocket is at its maximum distance from Earth. The time traveler then returns to Earth using the same acceleration/deceleration process. The time traveler thus arrives back home gently with a final speed of zero with respect to Earth (ignoring, of course, all the navigational problems due to the motion of Earth during the trip).

We must be careful, in this analysis, with how quantities are translated from one reference frame into another (that is, Earth's and the rocket's). In particular, what is meant by a *constant acceleration?* The acceleration of the rocket will not appear to be the same to observers on Earth and to the time traveler on the rocket. It is the time traveler who will *experience* the acceleration of the rocket, however, and we can physically arrange for this acceleration to be a constant if in the rocket we keep fixed the extension of a spring that has a mass attached to one end, that is free to move, and bolt the other end to a wall. From the point of view of an observer on Earth, of course, the acceleration continually decreases toward zero as the rocket's speed approaches the limiting speed of light.

The total duration of the trip has been $4T$ in Earth time and $4T'$ in rocket time. The claim is that $4T' < 4T$ (as is more generally shown in Tech Note 5), and in fact the inequality can be absolutely astonishing in its unbalance. To show this, I'll define the rocket speed in Earth's frame to be v. We can imagine that at any given instant there is a second reference frame moving at some fixed speed V (in Earth's frame) in a direction parallel to the rocket's motion. The location of the origin of this moving system is arbitrary; it is not necessarily at the instantaneous location of the rocket. The distance of the rocket from Earth is x, and the distance of the rocket from the origin of the second system is x'. The rocket's speed in the second system is v' and so, by the addition-of-velocities formula (see Tech Note 3),

$$v = \frac{v' + V}{1 + Vv'/c^2}$$

The times in the Earth system and the moving systems (t and t') are related via the Lorentz transformation as

$$t = \frac{1}{\beta}\left(t' + \frac{Vx'}{c^2}\right), \quad \beta = (1 - V^2/c^2)^{1/2}$$

Differentiating the first expression with respect to v', it is easy to show that

$$dv = \frac{\beta^2}{(1 + Vv'/c^2)^2}\, dv'$$

Using the chain rule for functions of multiple variables to find the total differential dt yields

$$dt = \frac{\partial t}{\partial t'}\, dt' + \frac{\partial t}{\partial x'}\, dx' = \frac{1}{\beta}\, dt'\left(1 + \frac{V}{c^2}\frac{dx'}{dt'}\right)$$

Because dx'/dt' is the velocity of the rocket in the second system (v'),

$$dt = \frac{1}{\beta}\, dt'\left(1 + \frac{V}{c^2}v'\right)$$

We can now write the rocket's acceleration as seen from Earth as the ratio of the dv and dt differentials:

$$\frac{dv}{dt} = \frac{\beta^3}{(1 + Vv'/c^2)^3}\frac{dv'}{dt'}$$

where dv'/dt' is the acceleration *according to the time traveler.* This establishes the fact that the accelerations in the two frames are not equal.

Now explicitly set $V = v$; that is, set V equal to the speed of the rocket at some (any) given instant. At this instant, then, the second system is at rest with respect to the rocket, and so $v' = 0$. The second system is called the comoving system of the rocket, and the so-called hypothesis of locality, which is generally invoked in relativity analyses, asserts the instantaneous equivalence of the accelerated time traveler with an observer in the co-moving system. Thus, with this choice for V, we have

$$\frac{dv}{dt} = \beta^3 \frac{dv'}{dt'}$$

The experienced acceleration, dv'/dt', is assumed to be a constant (call it a), so this is a particularly easy equation to solve. Thus, because $V = v$,

$$\frac{dv}{dt} = \beta^3 a = \left(1 - \frac{v^2}{c^2}\right)^{3/2} a$$

or

$$\frac{dv}{(1 - v^2/c^2)^{3/2}} = a \, dt$$

With $v = 0$ at $t = 0$, this integrates to give

$$v = \frac{at}{[1 + (at/c)^2]^{1/2}}$$

Inserting $t = T$ gives the maximum speed of the rocket:

$$\text{maximum rocket speed} = \frac{aT}{\left\{1 + \left(\frac{aT}{c}\right)^2\right\}^{1/2}}$$

Integrating once more to get the distance of the rocket from Earth ($x = 0$ at $t = 0$), we find that at time $t = T$, the rocket is at

$$x = \frac{c\{(a^2 T^2 + c^2)^{1/2} - c\}}{a}$$

and so the *maximum* distance of the rocket from Earth is *twice* this value, the distance at time $t = 2T$. To find the relationship between T and T', recall (see Tech Notes 2 and 3) that

$$dt' = \left(1 - \frac{v^2}{c^2}\right)^{1/2} dt$$

and integrate to get

$$T' = \int_0^T \left(1 - \frac{v^2}{c^2}\right)^{1/2} dt$$

Inserting the expression for v as a function of t yields

$$T' = \frac{c}{a} \sinh^{-1}\left(\frac{a}{c} T\right)$$

or, equivalently,

$$T = \frac{c}{a} \sinh\left(\frac{a}{c} T'\right)$$

Recall that the total round-trip time is $4T$ in Earth time and $4T'$ in rocket time. Table TN6.1 shows how $4T'$ compares to $4T$ for an experienced acceleration of a constant one gee. If humans can stand a two-gee acceleration for indefinite periods of time, then significantly further penetration into the future would be possible during a single lifetime, as seen in Table TN6.2. In response to von Hoerner's analysis, L.O. Pilgeram of the Arteriosclerosis Research Laboratory in Minneapolis wrote a letter[3] that objected to applying the laws of physics to biological systems. Incorrectly asserting that time dilation "has never been proved or disproved experimentally," Pilgeram further declared that "there is no known causal means by which greatly increased velocity could alter, without destroying the very biochemical basis of the life process, those metabolic changes which are responsible for the aging process." Von Hoerner gave a reasoned reply that can be found immediately after Pilgeram's letter.

Table TN6.1. Time travel to the future on a rocket ship accelerating at one Earth gravity.

EXPERIENCED ACCELERATION = ONE GEE			
4T' (rocket years)	4T (Earth years)	Maximum Distance (light-years)	Maximum Speed (× Light Speed)
1	1.01	0.065	0.252
2	2.09	0.26	0.475
5	6.5	1.85	0.86
7	11.5	4.1	0.95
10	25.5	10.9	0.9886
20	339	167	0.99993
30	4,478	2,237	0.9999996
40	59,223	29,610	

Table 6.2. Time travel to the future on a rocket ship accelerating at two Earth gravities.

EXPERIENCED ACCELERATION = TWO GEES			
4*T'* (rocket years)	4*T* (Earth years)	Maximum Distance (light-years)	*Maximum Speed* (× *Light Speed*)
1	1.04	0.13	0.475
2	2.4	0.56	0.775
5	12.7	5.47	0.9886
7	35.9	17	0.9985
10	169.3	83.7	0.99993
20	29,612	14,805	
30	5,180,000	2,589,279	
40	906 million (!)	453 million (!)	

If space travel is limited to sublight-speeds in normal space (see Tech Note 9 for an alternative), then it would seem that the only way to travel to the stars *within the lifetime of a human crew member* is to use the time dilation effect. But time travel to the future by this means would exact a terrible price. As the central character in the 1950 tale "To the Stars" (Hubbard) shouts, as he is shanghaied aboard a near light-speed rocket, " 'Let me go!' And there was real frenzy in him. . . . He knew all about the Lorentz–Einstein Relativity Equations. He knew what happened when a ship got to ninety-nine percent of the speed of light. . . . As mass approaches the speed of light, time approaches zero. It was his sentence . . . to forever."

Is it likely that such trips as these will someday be made by humans in and into the future? No.[4] Such trips result in speeds that are large fractions of, or even nearly equal to, the speed of light, as shown in the rightmost columns of Tables TN6.1 and TN6.2. To zip through space (which is not a perfect vacuum) at such speeds would result in a very high rate of collisions with stray hydrogen atoms (about one each cubic centimeter). The result of these energetic interactions would be the intense irradiation of the entire ship with a lethal dose of gamma rays long before the trip had even really begun. And, as von Hoerner showed in his paper, the energetics of such trips is mind-boggling.

An early (1933) science fiction story that used the time dilation effect from rapid motion to achieve time travel into the future is "A Race Through Time" (Wandrei). Initially set in 1950, this is a tale of two scientists, one evil and one good, who develop different methods for getting into the future. The evil one does it with a drug that slows the metabolic processes of the body, whereas the good one builds a high-speed, atomic-powered rocket in his home workshop! The evil scientist kidnaps the good one's girlfriend, injects her and himself with his drug, and seals the two of them in a crystal dome. He has arranged matters so that they will emerge from the dome in the year A.D. 1,000,000. Learning what has happened, the good scientist rushes to the dome, finds he can't break in, but sees an indicator dial pointing to "1,000,000." Returning to his rocket, the good scientist decides that he, too, will travel into the far future using time dilation rather than a drug. So off he goes, on a trip something like the one described in this Tech Note. The story ends with an ironic twist—the good scientist thought the "1,000,000" he saw through the dome meant one million years measured from 1950; he thus arrives back on Earth nearly two thousand years *after* his girlfriend and the evil scientist emerged from the dome. This story commits a terrible technical blunder, the author apparently believing that if a little speed gives a little time dilation, then *arbitrarily* more speed is even better. Thus on the outward leg of the rocket flight we read of the good scientist's "frightful speed—now thousands of light-years per Earth second." As Tech Note 7 discusses, if one could in fact break the light-speed barrier, then one would actually go *backward* in time, not forward.

Still, the romantic idea of a relativistic spaceship is a hard one to give up, and some *good* stories have been written using the concept; see, for example, Doris Pitkin Buck's "Story of a Curse," a tale about such a ship whose crew stays far too long away from home and about what they find has happened to Earth during their absence. Another poignant tale based on time dilation by high-speed space flight is "Semley's Necklace" (LeGuin).[5]

A possible use of time dilation occurs in the 1977 movie *Close Encounters of the Third Kind*. At the end of this film we see humans emerge from an alien spacecraft *decades* after they were first brought aboard, and yet they appear un-aged. If the craft had been traveling at near light speed for most of its absence from Earth, then time dilation is a possible explanation (another, of course, is that the aliens' guests were simply frozen by some technological process that is far beyond present capabilities).

Notes

1. The analysis done by von Hoerner is at best obscure. He does indeed arrive at correct analytical expressions, but even with correct formulas, many of his numerical evaluations are incorrect. My presentation follows Marder's, and my numbers agree with his. See L. Marder, *Time and the Space-Traveler* (London: George Allen & Unwin, Ltd., 1971).

2. Consider, for example, this remark by Nobel laureate Julian Schwinger about the twin paradox in his book *Einstein's Legacy* (New York: Scientific American Library, 1986): "The observer on the spaceship, however, is *not* in uniform, unaccelerated motion, because he must be accelerated to *reverse direction* after having reached the objective. The special theory of relativity does not apply to such an accelerated observer." In a footnote, Horwich (1975) makes the same assertion. However, this Tech Note takes the opposite view, which was put forth by C.W. Misner, K.S. Thorne, and J.A. Wheeler in their enormous (in every sense of the word) book *Gravitation* (San Francisco: W.H. Freeman, 1973). See, in particular, their Chapter 6.

3. *Science* 138 (December 7, 1962):1180.

4. In the mystical 1920 novel *Voyage to Arcturus* (Lindsay) we read of a spaceship that travels to Arcturus—the brightest star in the constellation Boötes—in just 19 hours of proper time. The technical details of the trip are not explained in the novel, so let's assume the acceleration/deceleration scenario developed in this Note: $2T' = 68,400$ seconds and $2x$ = distance to Arcturus = 36 light-years—the novel incorrectly says it is 100 light-years. (It is interesting to note that Robert Forward's "Twin Paradox," mentioned in Tech Note 5, uses slightly different details in that story's trip to Arcturus: accelerate at one gee to 99.2% of light-speed, coast at that speed, and then decelerate at one gee back down to zero. This strategy reduces the required fuel as compared to the strategy used in this Note and still gets Forward's hero to Arcturus almost as fast as a photon.) Using the equations derived in this Note for the variables T, T', and x, we can then solve for the required acceleration/deceleration. I get about 11,600 gees, which is of course *far* beyond anything the human body could survive. It is also easy to calculate that back on Earth, nearly 37 years pass during the 19 hours of ship time. That is only a little longer than light itself takes to make the trip; except for right at the beginning and the end of the journey, the ship is always moving almost at light-speed.

5. A similar theme can even be found in the song "39" by the rock-band Queen. The lyrics were probably interpreted by most listeners as simply telling of a long sea voyage to a new country, but it is clear that the song is really about a very long, high-speed rocket trip through space to find a new world. At one point, upon the return of the ship, we hear

> In the year of '39 came a ship in from the blue
> The volunteers came home that day
> And they bring good news of a world so newly born
> Though their hearts so heavily weigh
> For the earth is old and grey . . .
> :
> :
> For so many years have gone though I'm older but a year
> :
> :

Superluminal Speeds, Backward Time Travel, and Warp Drives, or Faster-Than-Light into the Past

We cannot fight the laws of nature.

Nature be damned! Feed more fuel into the tubes. We must break through the speed of light. . . . Give me a clear road and plenty of fuel and I'll build you up a speed of half a million miles a second. . . . What's there to stop it?

—words exchanged by the first officer and the captain of a starship on its way to Alpha Centauri in the 1936 tale "Reverse Universe" (Schachner); the captain we are told "had heard, of course, of the limiting velocity of light, but it meant nothing to him." Apparently not.

When we traversed five light-years of space in no appreciable time, we dropped back, also, through . . . time.

—from "Forgetfulness" (Stuart), originally published in 1937 by *Astounding Science Fiction* magazine

. . . [V]elocities greater than that of light . . . have no possibility of existence.

—Albert Einstein, 1905

So far I have limited the technical interpretation of relativity theory to speeds below the speed of light—that is, to the condition $v < c$, where v is the relative velocity of two reference frames. There was nothing, however, in the derivation of the Lorentz transformation equations that used this self-imposed constraint. What, in fact, happens for $v > c$? In a 1964 lecture to the British Association for the Advancement of Science, a philosopher (Robinson) posed this question and declared, "Faster-than-light travel remains a coherent, and possible concept, even though it is forbidden by relativity theory." The last half of this claim is wrong. That isn't to say we can have faster-than-light for free; there *is* a high price to pay: the price of causality violation (which probably accounts for Einstein's comment in the third opening quotation). If a material object goes faster than light (FTL), then the mathematics seems to say that it could physically travel into the past (as illustrated in 1978 in the first of the *Superman* films, when the caped hero induces time to reverse itself and run backward in order to *change* the past—specifically, to save Lois Lane from the death that has befallen her), and if a signal bearing information goes faster than light, then the mathematics seems to say that it could physically travel into the past. Such a signal might be, for example, a modulated beam of tachyons—if tachyons exist.

Several objections to the existence of tachyons have been raised.[1] The relativistic expressions for the energy and momentum of a particle with rest mass m_0 moving with speed v, for example, are, respectively,

$$E = \frac{m_0 c^2}{\sqrt{1 - (v/c)^2}} \quad \text{and} \quad p = \frac{m_0 v}{\sqrt{1 - (v/c)^2}}$$

For $v > c$, the radicals in these expressions become imaginary, whereas E and p must always be real-valued (because they can be observed and even measured as a result of the interactions the particle has with other matter). The energy and momentum can regain the property of being real-valued if we write $m_0 = i\mu$ (that is, $m_0^2 = -\mu^2$) for a tachyon, where μ is the real-valued (but unobservable) meta-mass (that is, $m_0^2 < 0$). This is a radical proposal, of course! It is $m_0^2 > 0$ that we are used to; as the Russian mathematician Yu I. Manin once wrote,[2] "What binds us to space-time is our [positive] rest mass, which prevents us from flying at the speed of light, when time stops and space loses meaning. In a world of light there are neither points nor moments of time; beings woven from light would live 'nowhere' and 'nowhen'; only poetry and mathematics are capable of speaking meaningfully about such things."

Well, perhaps so, but let's be brave here and go along with the tachyon idea. Then [see Parker (1969)],

$$E = \frac{\mu c^2}{\sqrt{(v/c)^2 - 1}} \quad \text{and} \quad p = \frac{\mu v}{\sqrt{(v/c)^2 - 1}}$$

An interesting consequence of this is that if tachyons lose energy, they speed up! This result, first stated by the German physicist Arnold Sommerfeld in 1904, means that if there is a mechanism for continuous energy loss (such as by Cerenkov radiation, either electromagnetic or gravitational), then tachyons will spontaneously accelerate without limit and enter what is called a transcendent state of infinite speed. Curiously, the foregoing equations show that such infinitely fast particles, though they possess zero energy, would carry a finite momentum of μc. University of Washington physicist John Cramer (1993) has described how one could use this property of tachyons (if they exist) to build a revolutionary new rocket propulsion system he calls the *tachyon drive*.

To see how backward time travel and faster-than-light issues are related, it is useful to establish a geometrical interpretation of the Lorentz transformation. As shown in Tech Note 3, if the x',t' system is moving with speed v in the x (or x') direction relative to the x,t system, then

$$x' = \frac{x - vt}{\sqrt{1 - (v/c)^2}} \quad \text{and} \quad t' = \frac{t - vx/c^2}{\sqrt{1 - (v/c)^2}}$$

These equations make sense for $v < c$, and in fact I will retain this condition for our two relatively moving frames of reference and will, in all that follows, use the symbol w to denote the speed of an FTL particle. Our human observers, however, will always be subluminal. As Svozil (1995) points out, "The assumption that *observers* move faster-than-light goes beyond superluminal signalling," and such an observer would have to be thought of as being built out of tachyons.

Now, recall what we mean by any line parallel to the x-axis; it is a line with a fixed time coordinate. Such a line is a *cosmic moment* line, with the equation $t = $ constant. Similarly, for the moving system we would write the equation of a cosmic moment line as $t' = $ constant, which, after the Lorentz transformation is applied, is equivalent to

$$t - \frac{vx}{c^2} = \text{constant}$$

In particular, the x'-axis (the $t' = 0$ cosmic moment line), which passes through the point $x = 0$ at $t = 0$, has the equation

$$t = \frac{vx}{c^2} = vx$$

with our usual convention of $c = 1$.

In a similar way, recall now what we mean by any line parallel to the t-axis; it is a line with a fixed space coordinate. Such a line is the world line of a stationary particle in the x,t frame, with the equation x = constant. Similarly, for the moving system we would write x' = constant as the equation of the world line of a particle stationary in that system. From the Lorentz transformation, this is equivalent to

$$x - vt = \text{constant}$$

In particular, the t'-axis (which is the $x' = 0$ world line of a particle stationary at the origin of the moving system) passes through the $x = 0, t = 0$ point, and it has the equation

$$x = vt$$

Thus, superimposed spacetime coordinate axes for the two frames look like those shown in Figure TN7.1. That is, the relative motion of the two frames results in a rotation of the spacetime axes; but it is a strange sort of rotation, with opposite senses for the space and time axes. The x'- and t'-axes rotate toward each other, to make equal angles α with the x- and t-axes, respectively, where

$$\alpha = \tan^{-1}(v)$$

If we limit the moving frame (the frame of a moving observer) to subluminal speeds $(0 \leq v < 1)$, then

$$0 \leq \alpha < 45°$$

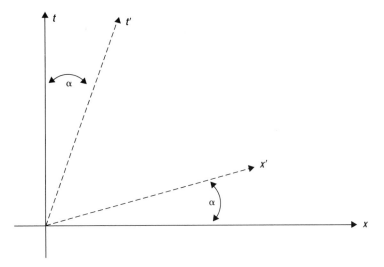

Figure TN7.1   Spacetime coordinate rotation by relative motion.

At the speed of light ($v = 1$) we have $\alpha = 45°$, and the x'- and t'-axes coincide—time and space have become indistinguishable.

It is important to realize that observers in either system would measure the same speed for a photon; that is, each would see the world line of a photon as a line with slope 1. This view of the world line of a photon is literally built into the Lorentz transformation because one of Einstein's fundamental postulates for special relativity is the invariance of the speed of light. The truth of this statement for the x,t system is obvious using the spacetime diagram. It is, perhaps, not so obvious with the x',t' system because of the non-traditional, non-perpendicular axes (as drawn on paper) in that system. In Figure TN7.2 the world line of a photon is shown in both systems. In the figure we emit the photon at $x' = 0, t' = 0$, and we later measure its coordinates at point A to be $x' = x'_A$ at time $t' = t'_A$. Note carefully how this is done. We draw lines from point A parallel to the x'- and t'-axes until they intersect the t'- and x'-axes, respectively. This is similar to the way we would get the spacetime coordinates of A in the more familiar x,t system, where we would draw lines parallel to the x- and t-axes.

It should be obvious now that x'_A and t'_A have the same extension, just as they do in the unprimed system, so

$$\frac{x'_A}{t'_A} = \text{the speed of light}$$

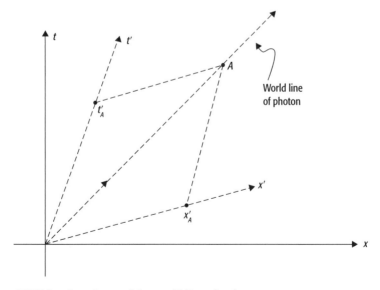

Figure TN7.2 Invariance of the world line of a photon.

The speed of light is the *only* invariant speed under Lorentz transformation. Indeed, the modern approach to special relativity emphasizes this invariance, rather than the idea of the speed of light being a limiting speed, as the central property of the speed of light—see Jones (1963).

This geometrical interpretation of the Lorentz transformation lets us quickly make another interesting (and, I think, not very obvious) observation: If a particle is faster than light in the x,t system, then there exists a subluminal x', t' frame for which a particle is *infinitely* fast! Figure TN7.3 shows the world line of an FTL particle in the x,t system. (It is, of course, below the world line of a photon; that is, it is spacelike.) Suppose the FTL particle has speed $w > c$ such that its world lines makes angle β with the x-axis. If we now pick v, the speed of the moving x', t' frame, to be such that $\alpha = \beta$, then the x'-axis will coincide with the world line of the particle, and the particle will appear to an observer in the x', t', frame to be *everywhere at once*—that is, to be infinitely fast. We have, then,

$$\beta = \tan^{-1}(v) = \tan^{-1}\left(\frac{1}{w}\right)$$

or $v = 1/w$, which seems to be dimensionally wrong. Recall, however, that with our convention of $c = 1$, the v in this result is a *normalized* speed. To return to the units of everyday use, we merely replace v with v/c and replace w with w/c; this transforms our result to

$$\frac{v}{c} = \frac{c}{w} \quad \text{or} \quad v = \frac{c^2}{w}$$

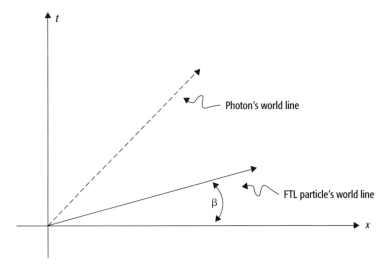

Figure TN7.3   World line of an FTL particle.

We can, of course, turn this result around. If an FTL particle moves with speed w in the x,t frame, then to an observer in the x', t' frame moving with subluminal speed v, the particle will appear to be infinitely fast if $w = c^2/v$. A particle with $w \geq c^2/v$ is said to be not just superluminal but *ultra*luminal.

If a particle has infinite velocity with $w = c^2/v$, then what happens if w is actually greater than c^2/v? The answer is easy to see from a spacetime diagram, as in Figure TN7.4, where the x'- and t'-axes have been extended back to negative values. In this figure I have labeled two arbitrary events A and B on the world line of an ultraluminal particle (which thus lies below the x'-axis) and have plotted their spacetime coordinates in both the x,t and x', t' frames. For the x,t frame we see that A is related to B by the relations $x_A < x_B$ and $t_A < t_B$; that is, the particle is moving forward in time from A to B and is moving in space along the increasing x-axis. In the x', t' system, however, A is related to B by the relations $x'_A < x'_B$ and $t'_B < t'_A$; that is, the time order of A and B is reversed for an observer in the x', t' frame. To this observer the particle appears to be traveling backward in time!

But this isn't quite the end of the story. Following the approach used in Bilaniuk and Sudarshan (1996b), we note that if the energy of the particle in the stationary system is E, then the energy in the moving system is[3]

$$E' = \frac{1}{\sqrt{1 - (v/c)^2}} \frac{\mu(c^2 - wv)}{\sqrt{(w/c)^2 - 1}}$$

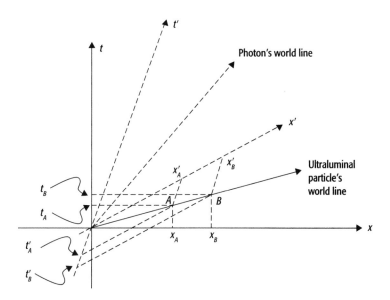

Figure TN7.4   World line of an ultraluminal particle.

Note that the sign of E' switches from positive to negative when w exceeds c^2/v, which is precisely the condition for the particle to go ultraluminal and begin to appear to be moving backward in time, as seen in the primed system. That is, negative energy moving backward in time in one system is positive energy moving forward in time in another. Historically this idea led to the reinterpretation principle (RP), which is discussed in Chapter Four. At one time the RP was claimed to be sufficient to avoid the issue of backward time travel and all the associated paradoxical implications. This is not the majority view today, however, and many physicists have been astonishingly inventive in constructing scenarios with causal paradoxes that the RP clearly fails to abolish. The literature on tachyon paradoxes, pro and con, is enormous. A long summary of all these paradoxes, together with a large bibliography and rebuttal analyses, is given by an advocate of tachyons in Recami (1986, 1987). For a strong rebuttal to the reinterpretation principle, see Savitt (1982).

Consider, for example, the situation described by Shoichi Yoshikawa of Princeton University, in one of the replies received by *Physics Today* to Bilaniuk and Sudarshan's 1969 article (see Note 1). Yoshikawa's example even *uses* the RP to arrive at the causal paradox! As shown in Figure TN7.5, Yoshikawa's example has an ultraluminal particle emitted by an observer P at A at time $t = 0$ in the stationary system. For P, the particle is moving forward in time and along the x-axis in the positive direction. A relatively moving observer (call him Q) receives the particle at B. Now, as we have already established, what Q

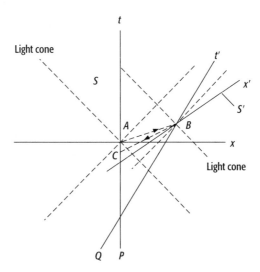

Figure TN7.5   An FTL causal paradox.

actually sees is a negative-energy particle absorbed at B, but because of the RP, he interprets it though it were a positive-energy particle emitted at B and traveling back down the x'-axis in the negative direction.

Now, let us say that at the instant Q observes this emission, he also emits a second tachyon that goes even faster down the negative x'-axis. (This tachyon is more ultraluminal than the one originally emitted at A.) This tachyon is then absorbed at C, an event observed by a past version of P, because C occurs earlier on the t-axis than does A! P sees the second tachyon as a negative-energy particle, of course, and (again because of the RP) he interprets it as the emission of a positive-energy particle. Thus, the emission of this tachyon at $t < 0$ has been caused by the emission of the original tachyon at $t = 0$. Thus we have backward causation *because* of the RP, with P seeing something happen at $t < 0$ because of something he *will* do at $t = 0$. Bilaniuk and Sudarshan admitted the strength of this example in undermining the RP and could give only a weak reply that involved "cosmological boundary conditions."

Of course, the problem now is the threat of a bilking paradox. Suppose that after P sees the $t < 0$ event, he then does *not* emit the original tachyon? In fact, Benford *et al.* (1970) use this puzzle as the concluding argument of their well-known "antitelephone" analysis of two people who both possess tachyonic transmitters that send messages one hour back in time: "Suppose A and B enter into the following agreement: A will send a message at three o'clock if and only if he does not receive one at one o'clock. B sends a message to reach A at one o'clock immediately on receiving one from A at two o'clock. Then the exchange of messages will take place if and only if it does not take place." In an analysis inspired by this, Gatlin (1980) makes a more extensive study of such backward-in-time agreements (or *contracts,* as Gatlin calls them), and she concludes with a provocative statement: "Although it will never be possible to design a communications system which transmits messages from the future [to the past] upon request, . . . messages from the future may sometimes reach [the past]." Perhaps a passive listening station for detecting messages from the future isn't so crazy after all!

Physicist/philosopher David Bohm earlier presented an essentially equivalent analysis.[4] As Bohm (who rejected the possibility of sending messages to the past) writes at the end, "In effect, S could communicate with his own past at M, and tell his past self what his future is going to be. But on learning this M could decide to change his actions, so that his future at S would be different from what his later self said it was going to be. For example, the past self could do something that would make it impossible for the future one to send the signal. Thus, there would arise a logical self-contradiction." The error in

Bohm's type of argument was discussed in Chapter Four (in "Communication with the Past").

Since Gregory Benford's novel *Timescape,* tachyons have often appeared in fiction—but not always faithfully. In *A Matter of Time* (Cook), for example, which at first has a tachyon generator sending messages into the past, we later find a character trying to build a tachyon communicator to transmit to the far future! For that, a note in a bottle is all that is needed. And this same story has a physicist utter the following muddled explanation of the RP: "A few years ago there was a flap over a hypothetical particle called a tachyon. At first it was supposed to move faster than light and have negative mass. Then it was supposed to have positive mass and a velocity below that of light, but was supposed to be moving backward in time." Oh, my.

The lure of superluminal travel is undeniable. Consider, for example, these words by the Russian physicist S. V. Krasnikov (1998a):

> Everybody knows that nothing can move faster than light. The regrettable consequences of this fact are also well known. Most of the interesting or promising candidates for colonization are so distant from us that the light barrier seems to make an insurmountable obstacle for any expedition. It is, for example, 200 pc [1 parsec is equal to about 3.2 light-years] from us to the Pole star, 500 pc to Deneb [the brightest star in the constellation Cygnus], and 10 kpc to the center of the Galaxy, not to mention other galaxies (hundreds of kiloparsecs). It makes no sense to send an expedition if we know that thousands of years will elapse before we receive its report. On the other hand, the prospects of being confined forever to the Solar system without any hope of visiting other civilizations or examining closely black holes, supergiants, and other marvels are so gloomy that it seems necessary to search for some way out.

So, to repeat the question asked at the end of Tech Note 6, will faster-than-light trips someday be made by humans in spaceships? Almost certainly not in our lifetimes, but *someday*—just maybe. This may seem an incredible response, inasmuch as I said in Tech Note 6 that it is most *unlikely* that humans would make very fast (but sublight) trips. The difference is that the FTL trips would be made in very *unordinary* spacetime. That is, continuing with Krasnikov's passage,

> The point . . . is that [whereas the light barrier exists in special relativity] in general relativity one can try to change the time necessary for some travel not only by varying one's speed [as in special relativity] but also . . . by changing the distance one is to cover.

To understand what Krasnikov is getting at, consider the theoretical analysis by Alcubierre (1994) on how actually to make a "Star Trek" kind of warp drive! He did this by demonstrating a spacetime metric that, by literally expanding and contracting spacetime, achieves space travel between any two points, no matter how far apart, in arbitrarily little elapsed time (for both the spaceship *and* the non-spaceship observers, there is no time dilation effect[5]).

Alcubierre observes that the idea of spacetime expanding is not original with him. Indeed, the concept is axiomatic in the Big Bang theory of the origin of the universe. Determining just *how* to achieve the spacetime warp, however, is far different from simply demonstrating that such a warp is consistent with the general theory of relativity. The 1996 movie *Star Trek: First Contact,* for example, is about the invention of the warp drive in the twenty-first century: The whole thing fits inside a discarded ICBM, which seems a *vast* underestimate for the machinery that would be required to control the energies needed for a real warp drive. One quite interesting feature of Alcubierre's warp drive is that the spaceship crew would experience no acceleration or tidal forces. This may explain—without having to invoke "inertial dampers"—why the *Enterprise* crew isn't flattened and/or torn apart when Mr. Sulu engages that ship's warp drive. The spaceship is surrounded by a "bubble" of warped spacetime that is swept along by the combined push-pull effect of the expanding spacetime behind the craft and the shrinking spacetime in the front. The ship, itself, however, resides in the *flat* spacetime interior of the bubble, which means the ship is always in *free fall.* The "only" problem is that Alcubierre's design for a starship warp engine requires "exotic matter"—stuff that violates all three of the usually assumed energy conditions of general relativity. (The energy conditions are defined in the Glossary and are briefly discussed in Tech Note 9.)

The weak, strong, and dominant energy conditions are all violated because the Alcubierre spacetime warp requires a *negative* energy density in the thin "skin" of the bubble—indeed, this requirement increases with the square of the speed of the warp. Negative energy density can be achieved on a microscopic scale and perhaps even on a macroscopic scale (through the use of what is called "exotic matter"); more is said about this in Tech Note 9, on wormholes. There are also two rather severe *operational* problems for the warp drive. First, any space matter encountered by the leading edges of the warp bubble (where spacetime is shrinking), such as interstellar dust, would certainly have its energy released as intense radiation. The ship, then, should carry plenty of shielding, which, curiously, would not be a problem because the energy density of the warp, itself, is *independent* of the mass in the interior, flat spacetime center of the bubble. In any case, the warp drive should obviously not be engaged anywhere near any sizeable chunk of matter, like a planet! Second, an even more severe problem was discovered by Krasnikov. In unpublished work described in Everett and Roman (1997), he showed that the ship at the center of the bubble is *not* causally connected to the edges of the bubble. That is, the ship's crew could not create a warp bubble on demand and, after it had been created, could not then control it on demand.

It is important to understand what this means and what it does *not* mean. The causality issue does *not* mean that Alcubierre warp bubbles are impossible to create (perhaps they are, but not because of a lack of causality). It only means that whatever action is required to change the spacetime metric to make a warp bubble *has to have already been done before the decision to use the bubble is made.* Thus a warp bubble wouldn't be of any use for a starship that needs to escape a sudden, unexpected threat. But, as Everett and Roman cautiously observe, the warp bubble might have a more mundane use. "Suppose space has been warped to create a bubble traveling from Earth to some distant star, e.g., Deneb, at superluminal speed. A spaceship, appropriately located with respect to the bubble trajectory, could then choose to enter the bubble, rather like a passenger catching a passing trolley car, and thus make the superluminal journey."

At the end of his paper Alcubierre speculated on the possibility of using his superluminal warp drive to build a time machine, but he didn't show how. This was done by Everett (1996), using, not surprisingly, an argument he called "reminiscent of the 'reinterpretation principle' . . . which played an important role in discussions of the physics of tachyons." A simple presentation on how to turn the Alcubierre warp drive into a time machine can be found in Parsons (1996).

In an attempt to avoid the Alcubierre warp bubble's causality problem, Krasnikov looked for a different, causal superluminal spacetime metric. This he succeeded in finding, as reported in Everett and Roman's paper and in Krasnikov (1998a), but rather than describing a bubble, Krasnikov's warp is in the shape of a *tube*. The interior of the tube is flat spacetime, just as in the case of the warp bubble, but unlike the case of the bubble there would be a causal link between a spaceship crew and the tube. Just as the warp bubble requires very thin walls of negative energy density (on the order of just a few thousand Planck lengths), so does the Krasnikov tube warp. Unlike the bubble warp, however, the tube warp stretches the entire length of any proposed trip, so the total negative energy in the warp is incredibly huge. For a tube a mere one meter long and one meter wide, the total negative energy is 10^{28} solar masses, and to create a tube from earth to just the *nearest* star would require 10^{44} solar masses of negative energy!

One curious feature of the Krasnikov warp is that the *outbound* leg of a round trip cannot be made in less time than that required by light. *But* on the return half of the journey, a traveler would find the spacetime metric so changed (because of mass-energy manipulations purposely made on the outbound leg) that she would move "backward in time." The net result in that the *round trip* could end arbitrarily soon after it started! As Everett and Roman cautiously conclude, the Krasnikov tube is a "very unlikely possibility." But it would make a *wonderful* science fiction gadget, don't you think?

Finally,[6] for a fascinating discussion of superluminal travel, and travel backward in time, in the *cylindrical* spacetime discussed in Tech Note 5, see Blau (1998).

Notes

1. Tachyons became a popular topic in the physics literature with the publication of Bilaniuk, Desphande, and Sudarshan (1962). After the appearance of that paper, Bilaniuk and Sudarshan published essentially the same ideas again (1969a) in somewhat expanded form in *Physics Today*. This second paper provoked a series of letters taking exception on a variety of interesting technical issues that are not addressed here but can be found (along with Bilaniuk and Sudarshan's replies) in the December 1969 issue of the magazine ["More About Tachyons," *Physics Today* (December 1969):47–52]. The "DeWitt Gambit" discussed in Chapter Four was first proposed in one of these letters, on pp. 49–50.

2. *Mathematics and Physics* (Boston: Birkhäuser, 1981), p. 84.

3. This expression for E' is the result of applying the Lorentz transformation to E. The interested reader can find the details of the procedure nicely worked out in A.P. French, *Special Relativity* (New York: W.W. Norton, 1968), pp. 208–210.

4. *The Special Theory of Relativity* (New York: W.A. Benjamin, 1965), pp. 155–160.

5. There is a similar absence of any time distortion in Larry Niven's story "At the Core," about a spacecraft that travels 60,000 light-years to the center of the Milky Way and back at a speed more than 420,000 times the speed of light. As Niven's hero says, "That's goddam fast." It doesn't make any scientific sense, either, and no explanation is given of why the ship's proper time remains in step with that of the folks back home.

6. And *really* finally, as this book goes to press, an electronic circuit that seems to implement the tachyonic bilking paradoxes described earlier in this Tech Note has recently been constructed. It is a feedback circuit for which the input is a signal with a peak value, and the output has a peak value as well; the interesting feature of the circuit is that the output peak occurs *before* the input peak. There is actually nothing paradoxical in this, but it is just a bit unusual. What does seem paradoxical, however, is that an amplitude detector subcircuit in the feedback path can be triggered by the output signal into disconnecting the input signal before it reaches *its* peak value. This "seems to open the way for a variant of the time travel paradox in which the traveller journeys to the past and kills his grandfather before his own father is

born," we read in Garrison *et al.*, "Superluminal Signals: Causal Loop Paradoxes Revisited," *Physics Letters A* 245 (August 10, 1998): 19–25. This electronic version of the grandfather paradox does indeed follow if one substitutes "input peak" for "grandfather" and "output peak" for "time traveller." But before you think this gadget is a time machine, be assured that its designers have also shown that, unlike the causally related grandfather and time traveller, the two peaks are not so related. You can read all about this very curious device in two papers by Morgan W. Mitchell and Raymond Y. Chiao: "Causality and Negative Group Delays in a Simple Bandpass Amplifier," *American Journal of Physics* 66 (January 1998): 14–19, and "Negative Group Delay and 'Fronts' in a Causal System: An Experiment With Very Low Frequency Bandpass Amplifiers," *Physics Letters A* 230 (June 16, 1997): 133–138.

TECH NOTE

8

Backward Time Travel According to Gödel and Tipler

A time machine? Nonsense. A bilgeful of crap. Physical, mathematical, logical impossibility. I proved it once, for a term paper in the philosophy of science.

—thoughts of a twentieth-century man
in *The Dancer from Atlantis* (Anderson), as he suspects
the truth after a malfunctioning time machine has
thrown him back into the far distant past

. . . within forty-eight hours we had invented, designed, and assembled a chronomobile. I won't weary you with the details, save to remark that it operated by transposing the seventh and eleventh dimensions in a hole in space, thus creating an inverse ether-vortex and standing the space-time continuum on its head.

—see De Camp (1981),
but this is probably *not* the way to build a time machine

In this Tech Note my intention is to demonstrate how you can visualize time travel to the past without the need of writing even a single line of mathematics. I will start with Gödel's example, which describes a rigid, non-expanding, uniformly rotating universe. Such rotating-universe models had been studied as early as 1924 by the Hungarian physicist Cornelius Lanczos, but it was Gödel who discovered their time travel property. The one result from general relativity that I will use here (without proof) is that the rotation of matter causes a distortion in spacetime that results in the tipping over of light cones, with the future half tilted in the direction of rotation. If you imagine a point in the universe about which the rotation takes place, then this tipping effect increases with the radial distance from that point.[1]

The fact that rotating masses tip light cones over in the direction of rotation was discovered very early in the history of general relativity (1918) by the German theoreticians Josef Lense and Hans Thirring. Originally called the Lense–Thirring effect, it now generally goes by the name of the "dragging of inertial frames" effect, but see Rindler (1994, 1997) and Bondi and Samuel (1997). See also Donald (1978), especially his Figure 9, for more on light cone tipping. Until recently the frame-dragging effect was strictly theoretical, but in late 1997 two separate groups claimed to have detected it in observational data from NASA's Rossi X-ray Timing Explorer satellite. An M.I.T. group claims to have seen frame dragging around spinning black holes, and another team at the Astronomical Observatory of Rome says it observed an effect around spinning neutron stars. These results are tentative, however, and confirmation will have to wait until the launch, in 2000, of NASA's Gravity Probe B satellite.

For Gödel spacetime, frame dragging and light cone tipping are crucial. At a certain critical distance, the future half of the light cone for a given point in spacetime will tilt into the past half of similarly tilted light cones for nearby points. This is illustrated in Figure TN8.1, which was taken from Herbert

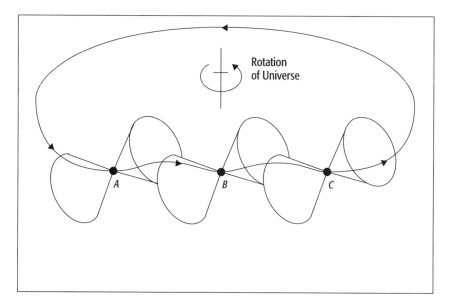

Figure TN8.1   Tilted light cones in a rotating universe.

(1988). A similar but somewhat more general picture can be found in Malament (1984). The figure shows a family of light cones in a rotating universe. Because light cones are tilted by a rotation-induced twist in spacetime, a traveler can move around a circular path on a trip into his local future and end up in his own global past without ever going faster than light. This kind of round trip, whose trajectory winds back into the past without ever becoming spacelike, is an example of a closed timelike curve. From this illustration it should be clear how a time traveler beginning at A can weave his way along a circular path that brings him back into the past half of the light cone at A. The path needs to have a radius at least as great as the critical value I mentioned. The traveler's world line is always inside the local light cone; that is, the world line of the time traveler is always timelike, and she never exceeds the speed of light. Because these timelike lines are present from the very beginning of the spacetime, Gödel's universe is called a *weak* time machine.

You can see how this works mathematically by taking the spacetime metric for Gödel's universe (with, as usual, the speed of light $c = 1$):

$$(ds)^2 = (dt)^2 - (dr)^2 - (dy)^2 + \sinh^2 r(\sinh^2 r - 1)(d\phi)^2$$
$$+ \sqrt{2} \sinh^2 r(d\phi)(dt)$$

where t, r, y, and ϕ are cylindrical coordinates in four-dimensional spacetime. Now imagine that our adventurer's world line is the helical curve $r =$ constant, $y = 0$, and $t = -\alpha\phi$. If, as usual, we take the time axis as vertical, then the time traveler's world line is a vertical helix in spacetime. For this curve, $dr = dy = 0$ and $dt = -\alpha d\phi$. This last differential means, in particular, that whatever the sign of the constant α, we can choose that one of the two senses of movement in the spatial ϕ dimension that gives $dt < 0$.

Continuing, we have

$$(ds)^2 = [\alpha^2 - 2\sqrt{2}\alpha \sinh^2 r + \sinh^2 r(\sinh^2 r - 1)](d\phi)^2$$

or, upon letting $u = \sinh r$, we have $(ds)^2 = [\alpha^2 - 2\sqrt{2}\alpha u^2 + u^2(u^2 - 1)](d\phi)^2$. Now, for $\alpha = 0$ we have $(ds)^2 = u^2(u^2 - 1)(d\phi)^2$, which is greater than 0 if $u > 1$. This condition holds if $\sinh(r) > 1$—that is, if $r =$ constant greater than $\ln (1 + \sqrt{2})$. In other words, for r sufficiently large (and now we know the critical value mentioned in the text), we have $(ds)^2 > 0$, which is the condition discussed in Tech Note 4 for a timelike interval. By continuity, then, we will continue to have $(ds)^2 > 0$ even with some small positive or negative value of α different from zero. Because ϕ is a periodic coordinate (we identify $\phi = 0$ with $\phi = 2\pi$), as the traveler moves on the curve, she returns repeatedly to the same spatial points, but her time coordinate is increasingly negative. That is, she is traveling into the past. Note, once again, that in Gödel's universe this property holds only for orbits with radii greater than a certain minimum. There are, however, other solutions of the Einstein field equations that have closed time-like curves at any radius, no matter how small; for example, see Reboucas (1979).

The Gödel universe is infinite in size and non-expanding, so it must rotate fast enough to counter the gravitational tendency to collapse. The more matter in the universe, the faster the minimum necessary rotation speed. If our universe were Gödelian, it would have to rotate once each 70 billion years, and the critical radius would be 16 billion light-years. The circular orbit in Figure TN8.1 would be at least 100 billion light-years around! To send a human on such a trip in a reasonable lapse of proper time would require a rocket ship that was moving at near light-speed, and this appears to be a fantastic requirement. An interesting interpretation of this "engineering" difficulty has been made by a philosopher in Sklar (1984). There, concerning time travel in general (not just in Gödel's universe), and after first arguing for the logical possibility of closed timelike curves, Sklar wrote,

> There is, of course, the *practical impossibility* [my emphasis] of traversing a closed timelike line . . . while this does not constitute an in principle rebuttal to the typical [grandfather paradox] objection to the possibility of closed timelike loops, it might be the ground for an argument to the following effect: "if there were such closed timelike loops, we would expect to observe many self-causing events. But we don't." Here the *practical impossibility* [my emphasis] of traversing the loop is meant only to explain the consequences of doing so wouldn't be commonly observed even if our world were causally pathological in this global sense. Plainly, as a refutation of the claim that closed causal loops are impossible since they *could* generate causal paradoxes, the mere invocation of practical impossibility of generating such a loop won't do.

A radio signal, of course, traveling at light-speed, *could* be used to send messages into the past—in a Gödelian universe. But our universe is not Gödelian, and we do not have any such natural paths available for time travel and/or time telegraphy. This conclusion would no doubt have greatly disappointed Gödel, who long after his 1949 papers retained a keen interest in any observational data that might support the thesis that our universe could be rotating—see Dawson (1989). Despite his admitted concern over the possibility of changing the past, or perhaps because of it, Gödel was utterly fascinated by the idea of time travel. Indeed, the author of Gödel's obituary notice in the Royal Society's *Biographical Memoirs* suggested that his dual interests in ghosts (!) and in time travel were somehow linked (are "ghosts" signals traveling on closed timelike loops?) It is also now well known that he was obsessed with his personal health and was terrified at the thought of death, and it has been speculated that Gödel's interest in time travel sprang from some idea that he had of its allowing a "reliving" of one's life.

Besides the rotational issue, there is yet another flaw in Gödel's time travel universe. Inherent in the metric for Gödel's spacetime is a non-zero value for the so-called cosmological constant, λ, a free parameter first introduced by Einstein himself in 1917 in the field equations (see Tech Note 10 for *why* he did this). In Gödel's universe, in fact, λ sets the local energy density at a constant, uniform value everywhere, a clearly unphysical situation if $\lambda \neq 0$. One can set λ to any value one wishes, and it shows up in Gödel's solution as determining the minimum radius for closed timelike curves; its square root appears in the denominator for the expression giving this minimum radius—see Deser and Jackiw (1992). In the observed universe, however, the cosmological constant[2] is very nearly zero. Perhaps, in fact, it is precisely zero because astronomical data, discussed in Hawking (1983), suggest that λ is probably no larger than 10^{-54} cm^{-2}. This says the minimum radius would be infinity, which is another way of saying there could be no naturally occurring closed timelike curves in our universe even if it were rotating.

Tipler's rotating, infinite cylinder is a mechanism for artificially producing the tipped-over light cone effect, thus *creating* closed timelike curves. For this reason, Tipler's cylinder is called a *strong* time machine. Figure TN8.2, taken from Tipler's 1976 dissertation, shows how his cylinder works. The cylinder is represented by the central vertical axis. Far away from the cylinder the light cones in spacetime are upright, but as we move inward they tip over, the future halves opening up into the direction of rotation. (Only the future halves of the light cones are shown.) This direction, which is the direction that far away from the cylinder measures *space,* near the cylinder measures *time* (just as in the Gödelian universe). That is, there has been a dimension reversal! This fantastic possibility has found its way into science fiction, as in Stephen Baxter's novel *The Time Ships,* wherein the Victorian narrator says, "If only one could *twist about* the Four Dimensions of Space and Time—transposing Length with Duration, say—then one could stroll through the corridors of History as easily as taking a cab in the West End!"

To travel back in time, therefore, all the time traveler need do is leave Earth and approach the cylinder until he is near enough to be in the tipped-over

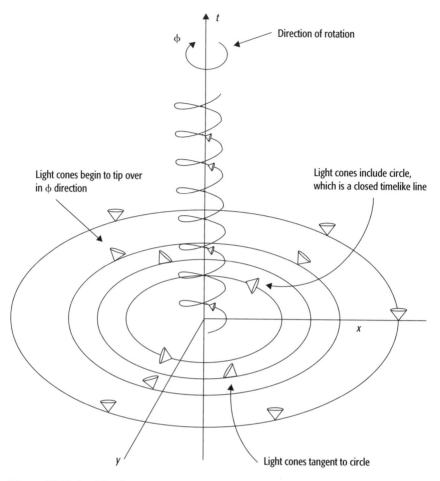

Figure TN8.2   The future halves of light cones point almost entirely in the +t direction far from rotating matter; they begin to tip over as the matter is approached. Note that there is a helical timelike path that moves locally into the *future* in the −t direction—that is, it goes into the past as seen by an observer far from the rotating matter. (The world lines of rotating matter are helixes in the +t direction.)

region of spacetime. Then he would follow a helical path around the cylinder and could spiral along the *negative* time direction as far back in time as desired—but no farther back than to the moment of the cylinder's creation. This motion is such that the time traveler is always moving into his local future, via the tipped-over light cones. Finally, he would withdraw from the cylinder and return to Earth—in the past. The time traveler had better be a good space navigator, of course, because Earth won't be where he left it!

NOTES

1. For a picture of this, see S.W. Hawking and G.F.R. Ellis, *The Large Scale Structure of Spacetime* (London: Cambridge University Press, 1973), p. 169.

2. For technical discussions of the history and meaning of the cosmological constant, see Novikov (1983) and the paper by Steven Weinberg, "The Cosmological Constant Problem," *Reviews of Modern Physics* 61 (January 1989):1–23.

Wormhole Time Machines

This fact reinforces the authors' feeling that [closed time loops] are not so nasty as people generally have assumed.
—from Friedman *et al.* (1990),
after a mathematical demonstration that time travel by wormhole does not conflict with the conservation of energy

Unless some really advanced beings have already made one of these things, we're not going back to visit the dinosaurs.
—Frank J. Tipler, commenting on the wormhole time machine in *Discover,* June 1989

Spacetime is friable. Wormholes riddle the fabric of spacetime on all scales. At the Planck length and below, wormholes arising from quantum uncertainty effects blur the clean Einsteinian lines of spacetime. And some of the wormholes expand to the human scale, and beyond—sometimes spontaneously, and sometimes at the instigation of intelligence.
—from Stephen Baxter's imaginative *Timelike Infinity,*
his 1993 novel of a far-future alien civilization able to control the energies of *constellations* of galaxies

Time loops are always accompanied by negative energy—disallowed as unphysical—or by violent objects such as blackholes and imploding universes. [Such objects] act as Nature's dragons, guarding time machines from fools who would rush in.
—A cautionary warning in "Time-Trippers Beware"
(*Scientific American,* February 1994)

The spacetime wormhole is presently the most promising of the approaches that have been advanced for building a time machine. Gödel rotated the entire universe in 1949, whereas Tipler reduced the problem in 1974 to "merely" spinning a cylinder of infinite length. In 1988 Kip Thorne scaled things down even more, this time to the other extreme. His idea calls for pulling a wormhole on the scale of the Planck length out of the topologically multiply connected quantum foam that spacetime is and then enlarging it somehow to human scale, all the while stabilizing it against self-collapse, and finally using the time dilation effect of special relativity to alter time at one mouth of the wormhole as compared to the other mouth. The rest of this Tech Note is devoted to explaining what all that means.

Wormholes themselves have been around in physics for decades, but they have always been thought to be so unstable as to exist only on paper in the mathematics of general relativity. In an analysis by Fuller and Wheeler (1962) published nearly forty years ago, for example, wormhole instability was shown to be so severe that not only would a human have no chance of getting through one, but also not even a single speedy photon could do so. Even at the speed of light, the photon could not zip through a wormhole before being trapped inside ("pinched off") in a region of infinite spacetime curvature. Wormholes would simply collapse too quickly after formation for even the so-called ultimate speed to save something inside. Indeed, the presence of mass-energy inside a wormhole actually *accelerates* its collapse. The dynamics of wormholes, it would seem, makes them simply untraversable. Stewart (1994) amusingly calls this the "catflap [door] effect": Moving a mass through a wormhole pulls the door shut "on your tail."

And anyway, how would one gain access to a wormhole in the first place? As suggested by Morris and Thorne (1988), one might perhaps imagine someday finding a rotating (Kerr–Newman) black hole that mathematically possesses, in its interior, so-called hyperspace tunnels to "other places"—either in our universe or in other universes (see Figure TN9.1). In the case of a wormhole connecting two places in the same universe, although the external distance between the places may be very large (megalight-years), it is conceivable that the distance through the wormhole itself could be very small. The time required to traverse the wormhole, as measured by the traveler's watch, might be essentially zero. This same paper, however, presents powerful arguments for why such Kerr–Newman wormholes probably do not exist and for why, if they do exist, they would be untraversable. But all is not lost. Morris and Thorne go on to state that there are other exact solutions to the Einstein field equations that describe hyperspace wormholes with none of the Kerr–Newman wormhole problems.

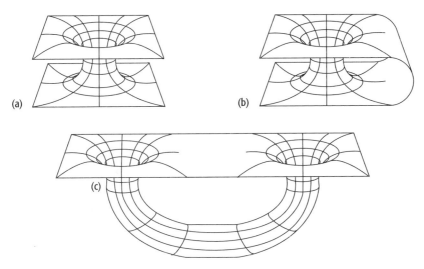

Figure TN9.1   These sketches are unavoidably misleading, being two-dimensional renditions of wormholes that connect two places in a three-dimensional space. Time machine wormholes, on the other hand, connect two places in a four-dimensional spacetime. In addition, the mouths of the wormholes will not appear as depressions into which the time traveler's rocket ship dives but rather as three-dimensional spheres (although for wormholes converted into time machines, spherical symmetry is replaced with axial symmetry, as discussed later in the text). The wormhole in part (a) connects two disjoint universes, whereas those in parts (b) and (c) are connections between two places in the same universe. As shown in these last two cases, the wormhole "handle" can be either long or short compared to the distance in external space between the wormhole mouths.

Such problems include the presence of a one-way event horizon, which precludes two-way travel (it seems reasonable to assume that time and space travelers might wish to return), and enormous gravitational gradients (tidal forces) that dismember (and worse) anything approaching and/or entering the hole. It is gravitational tidal forces that almost kill the narrator in Larry Niven's classic 1966 tale "Neutron Star." In fiction, of course, such problems are more often ignored. For example, the time machine in *A Bridge of Years* (Wilson) is in the form of a wormhole time tunnel that time travelers can simply *walk* through. In *Time Tunnel* (Leinster), too, characters simply walk back and forth between two openings in a wall that connect the same place in space separated in time by 160 years. And Henry Kuttner used a hole in spacetime that was a wormhole in everything but name for a time machine through which one could literally step from future to past and back again. That story, "Shock," originally published in 1943, also introduced the term *mugwump* for a time traveler who uses a time machine wormhole. To paraphrase Kuttner's time traveler as he transits the wormhole, his "mug" is in the past and his "wump" is in the future! On page 500 of Thorne (1994) there is a humorous illustration of Professor Thorne demonstrating wormhole mugwumping (without the time travel).

Hollywood quickly discovered the traversable time travel wormhole after the work of Morris and Thorne. In the perfectly awful 1992 film *Time Runner*, for example, the hero falls through one from 2022 to 1992. Once in the past, he changes history at will. He even watches his mother die in childbirth (his), and given that her death is a surprise to him, presumably that development is "new" as well. The television show "Sliders" (1995–) has a group of characters who find themselves in a new world each week, worlds they arrive at after "sliding" through what the opening scenes show as the tubelike interior of a wormhole.

Now a warning note on the sort of wormholes I am writing about in this Tech Note. There are, in fact, two kinds of wormholes in the physics literature today—see Sushkov (1991). The ones we are interested in, the ones suitable for time travel, are called *Lorentzian* wormholes, and they have a spacetime metric with Lorentzian signature $[+, -, -, -]$ (see Tech Note 4). Further, the wormhole is taken to be *static*—to have no time-varying behavior. At the end of this note I will have a few comments about recent work on non-static wormholes. Another type of wormhole occurs in the quantum theory of gravity [see Hawking (1978)]; it has a metric with Euclidean signature $[+, +, +, +]$. Such a wormhole is not suitable for time travel, because motion in a Euclidean-signature wormhole either involves imaginary momentum or takes place in imaginary proper time. Both situations are unphysical for a real time traveler transiting a wormhole. See Coleman (1988) and Visser (1995) for more discussion on Eculidean and Lorentzian signature wormholes and Pesic (1993) for more on Euclidean (hyper)spaces in general.

How, then, do we get our hands on one of these static, traversable wormholes? Morris, Thorne, and Yurtsever (1988) are honest—they don't know. Their best suggestion is that "one can imagine an advanced civilization pulling [such] a wormhole out of the quantum foam and enlarging it to classical size." "Quantum foam" refers to the idea that the topology of spacetime is always changing: The changes are topological fluctuations whose size is on the scale of the Planck length—Visser (1994a) uses the less technical term *Planck slop*. What does that mean? Like the ocean surface in-the-large, large-scale spacetime is simply connected. But just as one sees all sorts of transient structure as one looks at the water more closely (beginning with waves and then proceeding to the foam on the waves), spacetime too displays a fluctuating connectivity in-the-small. That is the "quantum foam."[1]

This isn't easy to visualize, but let's be brave and assume we *can* "pull" a wormhole out of the quantum foam. If so, then once it is inflated, the authors suggest, the wormhole could be stabilized against collapse by threading it with either matter or fields of stupendous negative (outward) tension—and by *stupendous* they mean STUPENDOUS. As Morris and Thorne (1988) show, if b_0 denotes the minimum radius of the wormhole (the size of the so-called *throat* of the wormhole), then the tension (radial pressure) at that location must be at least

$$\tau_0 = \frac{3.8}{b_0{}^2} \times 10^{36} \frac{\text{tons}}{\text{in}^2}$$

where b_0 is expressed in feet. The original equation for τ_0 was given in units of dynes/cm^2, with b_0 expressed in meters. I have converted these units to the more familiar units of tons per square inch (with b_0 in units of feet) for dramatic purposes, because the pressure (tension) unit of a dynes/cm^2 is so small that it can't be related to anything of everyday significance. For a wormhole with a throat radius of just several thousand feet, the value of τ_0 is of the same magnitude as the pressure at the center of the most massive neutron star. To stabilize a common sort of everyday wormhole, such as a subway tunnel, we can obtain the required tension/pressure by lining the tunnel with iron plates or concrete. But how, for a hyperspace wormhole, do we obtain "iron plates" that can achieve the required enormous tensions? As Morris, Thorne, and Yurtsever point out, such stuff could only be called "exotic."

One possible approach, in fact, does not use matter at all. If we make b_0 *very* large, then non-material fields will do the job—see Soen and Ori (1996) and Vollick (1997a,b) for more time machine speculation on this point. Indeed, suppose b_0 equals one light-year (a large wormhole by anybody's standards). Then τ_0 is "only" 4000 tons/in^2, and that is achievable by threading the wormhole throat with a magnetic field of "only" 2,700,000 gauss. To put this number in perspective, it is five million times greater than Earth's magnetic field. To generate such a field artificially is not impossible, but it is strictly a laboratory exercise at present (for example, experimental electromagnetic railguns used as hypervelocity, and armor-piercing antitank weapons incorporate transient magnetic fields in the megagauss range). To understand how the calculation of a magnetic field from a pressure requirement is accomplished note that pressure is dimensionally equivalent to field energy per unit volume, which in turn is given by a well-known result in electromagnetic theory.

As if all this weren't enough of a complication, Morris, Thorne, and Yurt-sever go on to show that there is another, even more curious problem. The geo-metrical requirement that the wormhole interior smoothly connect to the ex-ternal, asymptotically flat spacetime (see Tech Note 4) demands that the wormhole throat flare outward, as shown in Figure TN9.1. It turns out that this condition is mathematically equivalent to a requirement that τ_0 exceed the energy density of the throat material; and special relativity, in turn, says that for some timelike observers the energy density will then actually be *negative*. (The condition of τ_0 exceeding the energy density of the wormhole throat is, in fact, the technical definition of *exotic*. Pre-dating the work of Morris, Thorne, and Yurtsever on traversable wormholes is that of Balbinot (1985), which also uses the word *exotic* to describe the energy condition in the worm-hole throat.) Under everyday conditions, the exotic condition is never even re-motely approached. For example, the maximum tension necessary to pull a piece of steel apart (the so-called tensile strength, about 100,000 lb/in^2) is a trillion times less than the mass-energy density of steel. A negative energy den-sity can be interpreted as meaning that the exotic material keeps the wormhole open by exerting a *repulsive* gravitational force.

Such a repulsive force sounds like a property of *negative mass,* and although such a thing has never been observed (negative matter is *not* anti-matter, which *has* been observed and which does *not* repel "normal" matter), it was studied long ago, on paper, by the English cosmologist Hermann Bondi (1957). Bondi showed that negative mass would indeed have some truly bizarre prop-erties, but there is nothing in general relativity that forbids it. Wormholes, in fact, *may* offer a possible physical mechanism that displays negative mass that *may,* in turn, produce observable effects by which wormholes *may* be detected. I'll return to this point.

There is another interesting implication of a repulsive gravitational force. Just as Einstein's famous prediction (verified in 1919) from general relativity says that star light passing near the Sun's edge is bent inward by the Sun's at-tractive gravitational field (the so-called lensing effect, discussed at more length later in this Note), the repulsive, anti-gravity field in a wormhole will cause any light rays traveling through the wormhole to be bent outward. That is, a tight, narrow beam of radiation entering a wormhole will emerge *defo-cused*. This is crucial, in fact; as we'll soon see, a wormhole time machine could otherwise be destroyed by the light from the dimmest candle.

Now, in preparation for seeing how a static wormhole "works," we need to digress briefly. At one time it was almost a law of nature that no timelike ob-server should ever be able to measure a negative energy density at any point along that observer's world line (the so-called *weak energy condition* or WEC), but this is now known to be false. Indeed, the first hint that the possibility of a negative energy density is not crazy can be traced back as far as 1948, to a theoretical prediction by the Dutch physicist Hendrick Casimir. (The *strong* energy condition says gravity is always attractive, so static, traversable wormholes violate both the weak and the strong energy conditions.) As pointed out in Chapter 1, the Heisenberg uncertainty principle allows a temporary violation of conservation of energy to occur, the magnitude of the allowed violation increasing with decreasing time duration. Even in a vacuum, then, with particle/antiparticle creation and annihilation taking place, the *average* energy density being zero does not preclude fluctuations away from zero, becoming at times *negative*. What Casimir showed was that if one positioned two perfectly conductive plates parallel to each other, then the normal quantum fluctuations in this "vacuum sandwich" would be altered in such a way as to result in their mutual attraction—and this tiny effect was later actually observed.[2]

What does it mean to "alter the normal quantum fluctuations"? Consider the creation of a photon and its anti-particle, which is another photon. From the wave interpretation of particles, the parallel plates restrict the photons that appear in the vacuum layer to those that have wavelengths that "fit" because they have wavelengths that are submultiples of the plate separation (a perfectly conducting plate cannot support a non-zero tangential electric field). Photons with longer wavelengths than the plate separation cannot fit and thus do not appear. That is, the parallel plates have created boundary conditions that have quantized the electromagnetic field. The absence of these "longer-wavelength" photons lowers the average energy density between the plates, and because the average without the plates is zero, the altered average energy density must be negative. Indeed, the more the maximum allowed photon wavelength decreases with decreasing plate separation, the more negative the average energy density becomes in the enclosed Casimir vacuum.

The experimental detection of the Casimir effect was a remarkable event in physics. As mathematician Stephen A. Fulling wrote of it (in his 1989 book *Aspects of Quantum Field Theory in Curved Space-Time*), "No worker in the field of overlap of quantum theory and general relativity can fail to point this fact out in tones of awe and reverence." Robert Forward, an imaginative physicist who is an enthusiastic supporter of time travel, has described how the Casimir force might be used to extract energy literally *from a vacuum*. This is an idea as seemingly impossible as the plan of the professor in the 1937 story "The Time Contractor" (Binder) to squeeze energy out of *time*. As the professor asks his assistant, "But tell me, Bob, isn't that a ridiculous thought? To take time, something intangible, invisible, incomprehensible, and contract it—squeeze it together like a sponge?" Unlike the babble in Binder's story, however, Forward's idea is supported by solid mathematical physics.[3]

Epstein, Glaser, and Jaffe (1965) showed that an energy density that is everywhere and everywhen positive is not compatible with any quantum field theory that is local, as presumably quantum gravity will be. Requiring only that the *average* energy density over a complete null geodesic world line be non-negative is called the averaged weak energy condition (AWEC), and it has been shown in Yurtsever (1990a), Klinkhammer (1991), and Wald and Yurtsever (1991) to hold in a wide (but *not* exhaustively wide) range of spacetimes. The AWEC was introduced in Roman (1986). In the static, traversable wormholes studied by Thorne *et al.*, however, both the WEC and the AWEC *are* violated. Khatsymovsky (1994) showed that the required violation of the WEC in a wormhole throat can occur without the need of exotic Casimir forces, but simply via the normally occurring vacuum fluctuations of the electromagnetic field. Whether the AWEC is also violated by such fluctuations, as required if a static, traversable wormhole is to exist, remains an open question. Ford and Roman (1993) discuss what observers would "see" if they were in a region of negative energy, and Wheeler (1996) describes what one might see from inside a wormhole.

More recently, Ford and Roman (1997) have shown, on the basis of earlier work by Ford (1978), that although quantum field theory does not preclude negative energy densities, this does not mean it is possible to observe arbitrarily large negative densities for arbitrarily long times. In fact, they have established certain quantum inequalities (QI's), much like Heisenberg's, that place bounds on the magnitude and duration of the observable negative energy density. These QI's have the general form of

$$\hat{\rho}\, t_0^4 > -\, C$$

where C is a positive constant that depends on the nature of the particular quantum field being considered (they treat both the electromagnetic and the massive scalar field cases), t_0 is the time duration, and $\hat{\rho}$ is the integrated energy density along a finite section of a geodesic world line. This last condition is in contrast to the AWEC, which involves averaging over a *total* geodesic world line. The form of the QI shows that as t_0 increases, $\hat{\rho}$ must quickly decrease. For instance, if t_0 doubles, then $\hat{\rho}$ must decrease by a factor of *sixteen*.

The analysis of several different wormhole spacetimes, using their QI's, has led Ford and Roman to conclude that it "appears probable that nature will always prevent us from producing gross macroscopic effects with negative energy." Well, that's not too encouraging—and a paper three months later by Taylor, Hiscock and Anderson (1997) arrived at much the same conclusion, as had (by different means) an earlier paper by Flanagan and Wald (1996)—but these conclusions are for *static* traversable wormholes. Later, as I promised, I'll say a bit about how *dynamic* traversable wormholes might beat the QI problem. Finally, Woodward (1997) has a quite funny (as well as a compelling) argument that perhaps the QI's don't *really* prohibit time machines, at all! But Ford, Pfenning, and Roman (1998) continue to argue that the QI issue may well be central to why traversable wormholes will be hard to construct.

Returning to the historical development of wormhole time machines, Morris, Thorne, and Yurtsever (1988) use the Casimir effect to achieve the "exotic condition" *without* matter. They propose placing identical conducting, spherical plates that carry equal electric charges at each end of the wormhole. (Remember that the wormhole mouths are spherically symmetric.) The two identical charges repel each other, of course, but the charge value is adjusted so that the gravitational attraction of the plates precisely cancels the repulsion. The authors then calculate that the Casimir effect results in a negative energy density sufficient to provide the throat tension necessary to prevent wormhole collapse.

There are all sorts of weird aspects to this analysis, as one might suspect. For example, the authors assumed that the wormhole length is *very* small compared to its radius, such as 10^{-10} cm long and 200 million miles wide! The short length is required because it represents the separation of the wormhole plates, and the smaller the separation, the more negative the average energy density. (The functional dependence is as the inverse *fourth* power of the separation). Another problem is the balancing of the electrical repulsion and the gravitational attraction of the wormhole mouth plates. Such a balance is clearly an unstable one. Finally, because the two spherical plates completely fill the wormhole mouths, how would a traveler actually get through the wormhole? The "answer" is to drill a hole through the plates and hope that doesn't perturb the Casimir vacuum too much. Visser (1989a,b) shows how to build non-spherically symmetric wormholes that avoid this problem.

Now that I have sketched some of the difficulties static traversable wormholes have in simply existing, let's ignore all that and suppose we actually have one with both mouths in the same universe. Time machine wormholes, of course, connect to asymptotically flat regions in the *same* universe (as shown in Figure TN9.1b and c) and so spherical symmetry is too much to demand; each spherical mouth would ruin the asymptotic flatness condition for the other mouth. How can we imagine turning it into a time machine? Interestingly, although it is general relativity that gives us the wormhole, it is special relativity that adds the final touch of backward time travel. We begin by imagining that, somehow, one mouth of the wormhole can be moved with respect to the other mouth. For example, Friedman (1988) suggests using the gravitational attraction of a large asteroid to "drag" one end of the wormhole to induce a time-dilation effect.

Remembering that, as shown in Tech Note 2, a moving clock runs slow with respect to a stationary clock, we next suppose that we have two clocks A and B, one in each mouth of the wormhole. These two clocks, and other clocks in the flat spacetime outside the wormhole, are all initially running at the same rate and indicating the same time. This may strike you as paradoxical: I seem to be arguing for a universe-wide *now*, a concept that Tech Note 1 argues has no meaning. All I mean here, however, is that if we take into account the various light-speed transit times between any pairs of clocks and correct for these delays, then all the clocks agree. That is, an observer would be able to determine that the clocks were synchronized in the past. He would, of course, not be able to say they are all synchronized *now*—until sometime later in the future, when *now* would have become a past *then*. The issue of clock synchronization in a multiply connected spacetime such as in a wormhole spacetime, is treated in mathematical detail by Frolov (1993).

Stephen King's "The Jaunt" is a nice little horror story based on the idea that there is a distinction between time inside and time outside a space portal. In this tale—originally published in 1981 by *Twilight Zone* magazine, long before the interest in time travel wormholes developed—space travel through a wormholelike portal takes no time at all to the outside world, but the traveler's *subjective* time is much longer, so long in fact that all travelers enter the portal only while under gas-induced anesthesia. But then one traveler, a little boy whose curiosity gets the better of him, enters the portal *while holding his breath;* he is inside a *long* time, awake, and when he emerges . . .

Now, recalling the twin paradox (see Tech Note 5), let each mouth-clock play the role of one of the twins. Imagine that A and B are now separated because mouth B is placed on board a rocket ship. The rocket ship takes a long, high-speed trip out into space along the straight-line path joining A and B in external space, and then returns, just as described in Tech Note 6. We unload mouth B from the rocket ship and reposition it at its original location. What is the situation now? We can summarize matters as follows: (1) Clock A, in the non-moving mouth, remains in step with the local clocks in the space outside the mouth. (2) Clocks A and B, both inside the wormhole, have *not* moved with respect to each other because we are assuming a very short wormhole handle, as in part (b) of Figure NT9.1. We can arrange for the motion of mouth B to be such that the handle is *always* short, and so the distance between clocks A and B changes by an arbitrarily small amount. Thus clocks A and B remain in step with each other. (3) Clock B, because it has been moving with respect to its external space, arrives back at its starting position reading *behind* (that is, earlier than) the clocks in the space outside mouth B.

For the sake of argument, then, suppose the journey of B is such that there is a two-hour time-slip between clock B and its local, external clocks. Thus if clock B reads 9 A.M., the clocks outside of mouth B will read 11 A.M. But because clocks A and B are in step, clock A reads 9 A.M., as do the clocks outside of mouth A. That is, the wormhole connecting mouth A to mouth B is a connection between two parts of the same universe that are two hours apart in time. Now, suppose the journey from mouth A to mouth B can be made through external space in one hour. Then, one could leave mouth A at 10 A.M., rocket to mouth B by 11 A.M., enter mouth B, and travel back to mouth A via the wormhole to the starting point—where it is 9 A.M., one hour *before* the trip began! We could imagine repeating this cyclic process, going back one additional hour for each new loop through the wormhole. One clear restriction, however, is that we could not go back in time to *before* the creation of the wormhole time machine which is what Tipler meant by his quote at the start of this Tech Note. The wormhole works in the other direction, too. To see this, suppose that the space traveler leaves mouth B at 8 A.M. and rockets to mouth A, arriving at 9 A.M. Entering mouth A, he exits from mouth B (where he started) at 11 A.M., two hours in the future.

Not everybody is convinced by all this, of course. Visser (1990) raises the possibility that if one mouth of a wormhole is accelerated, then this will induce a time-dilation-canceling acceleration in the other mouth. Frolov and Novikov (1990) however, actually show how to convert a wormhole to a time machine *without* moving either mouth: Just place one mouth in an intense gravitational field (say, next to a neutron star) and this also leads to a time-dilation effect as shown in Tech Note 11. As Frolov and Novikov put it, almost any interaction with surrounding matter and gravity fields almost inevitably turns a wormhole into a time machine. Visser (1992), however, has more recently argued that it is not the method of achieving time dilation that will cause trouble with wormhole time machines, but rather the positioning of the two mouths. His analysis leads him to believe that the wormhole will inevitably be destroyed by space-time back-reaction effects before the two mouths can be brought close enough together to create a time machine—sufficiently close, that is, for the trip in external space between the mouths to be less than the time penetration into the past with each passage through the wormhole.

Another concern that has long bothered theoreticians is the possible instability of what is called the *Cauchy horizon,* the hyperspace surface in spacetime that separates the region where closed timelike loops can exist from the
region where they cannot exist. The name comes from the "Cauchy problem"—named after the nineteenth-century French mathematician Augustin
Cauchy—in the theory of partial differential equations. In this theory a
Cauchy initial-value problem is said to be well defined if the initial conditions
determine a unique solution and if a continuous variation in the initial condition gives a continuous variation in the solution.[4] In that part of spacetime
where closed timelike loops are not allowed, backward causation does not
occur (by definition), and the laws of physics (all expressed as differential
equations) satisfy the Cauchy condition. Outside of this *chronal* region, i.e.,
beyond the Cauchy horizon where physics is *dischronal,* however, the possibility of backward causation raises the possibility of violating the Cauchy condition, and in such a case the Cauchy horizon is also sometimes called the
chronology horizon.

The instability problem is caused by radiation that propagates in closed
timelike loops that thread through the wormhole on "straight lines." This radiation, as shown thirty years ago by Misner and Taub (1969), builds up unbounded energy density levels at the horizon and thus destroys the horizon. In
the particular wormhole time machine just described, for example, the horizon is thought to be unstable by some, but Morris, Thorne and Yurtsever
(1988) disagree. Those authors argue that the defocusing effect of the wormhole's repulsive gravity could possibly counter the disruptive energy buildup.

Novikov's later analysis (1989) examined another possible way to avoid
unbounded energy density on the Cauchy horizon. He imagined a wormhole
time machine that has a circular motion for mouth B; that is, mouth B orbits
around mouth A. The result is that the Cauchy horizon does seem to be stable,
because now there are no fixed, straight-line closed timelike loops threading
the wormhole from A to B. That is, B is a "moving target" and there is no
point on the Cauchy horizon where the energy density becomes unbounded.
Yet another artificial mechanism that might also disrupt destructive, circulating energy loops through a wormhole is a spherical reflection mirror being
placed between the two mouths of the wormhole, as proposed by Li (1994).
Li's mirror would divert all closed null geodesics representing circulating radiation (see Tech Note 5) that potentially thread through the wormhole. Those
potentially fatal geodesics would be scattered back into space, whereas a *pur-
poseful* traveler could navigate around the mirror and thus use the wormhole as
a time machine.

Cauchy horizon instability is central to Hawking's Chronology Protection Conjecture (1992). His analysis leads him to conclude that a physical entity—the stress energy tensor—becomes unphysical on the Cauchy horizon. That is, because of time-traveling quantum field fluctuations of the vacuum, that tensor diverges to infinity at the horizon. This results either in a failure of the horizon to form in the first place or, if it does form, in the creation of a singularity that "seals off" the horizon to any would-be time travelers attempting to gain access to the closed timelike loops beyond the horizon. A similar proposal concerning cosmic strings with spin (treated in Tech Note 10) that allows closed timelike curves, but seals them off from any possible use by embedding them in a "causal protecting capsule," has been presented by Novello and da Silva (1994).

Li, Xu, and Liu (1993) have argued that Hawking is mistaken in arguing that the divergence of the stress energy-momentum tensor on the Cauchy horizon will always forbid time travel. They have studied a complex-valued metric (and such a metric *is* allowed in the so-called "sum over all possible geometries, path integral" approach to the quantum theory of gravity) that has causal and non-causal spacetime regions separated not by a Cauchy horizon but rather by a region with complex geometry. Such a region, classically, would mean that the other two regions cannot be reached from one another. But by quantum mechanical tunneling an observer *could* travel between the two regions. In Li, Xu, and Liu's analyses, the stress-energy tensor is always physical and diverges nowhere.

Studies of the stress-energy divergence actually have a long history. The effect of an unphysical (infinite) gravitational and/or electromagnetic energy flux had been analyzed by Chandrasekhar and Hartle (1982) years before the wormhole time machine studies began. They studied the case of a potential traveler to "new worlds" who tries to cross the inner Cauchy horizon of an electrically charged, non-rotating black hole. An even earlier computer study by Simpson and Penrose (1973) concluded that for such a traveler, the attempt to cross that horizon "looks liable to prove a dangerous undertaking." The spacetime analogy of the stress-energy divergence analysis of Deutsch and Candelas (1979) was done by Hiscock and Konkowski (1982), who studied the stress-energy divergence at a Cauchy horizon in a non–time machine analysis.

And, in fact, it isn't at all clear that a theoretical divergence of the stress-energy *is* the signature of a failure of physics. One doesn't need anything as bizarre as a time machine for the stress-energy to diverge *on paper*. As long ago as 1979, for example, Deutsch and Candelas showed that such a divergence occurs in principle for the electromagnetic field near a perfectly smooth and conducting boundary. But it is simply the unphysical nature of a "perfectly conducting" boundary condition that causes the divergence, not the fact that the field actually exists near a conducting boundary. Similarly, other real-life considerations (quantum gravity) *may* keep the stress-energy physical everywhere in a time machine spacetime.

Well, what about paradoxes, you might ask? How do they come into play in a wormhole time machine? In an attempt to study the grandfather paradox, Echeverria, Klinkhammer, and Thorne (1991) studied self-interacting billiard balls traveling backward in time through wormholes. The authors use billiard balls (see Figures TN9.2 and TN9.3) rather than human time travelers for the same reason that Wheeler and Feynman used their pellet and shutter mechanism in their study of advanced electromagnetic waves (see Chapter Four) — to avoid any metaphysical questions about human free will. The central issue

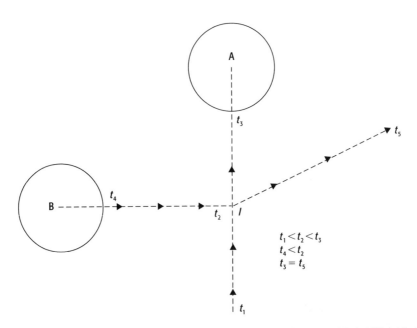

Figure TN9.2   The "grandfather paradox" in the billiard ball world. A billiard ball approaches mouth A of a time machine wormhole dead on center, and just before entering A, it passes without incident through point *I*. The ball then enters A and so exits mouth B *in the past,* just in time to pass through point *I* and hit *its younger self.* This impact knocks the younger ball away from A, so we have the familiar paradox of changing the past. That is, the impact did not occur when the ball "originally" passed through *I* on its way to A, and, of course, we also wonder how the ball manages to hit itself after leaving B if it doesn't enter A?

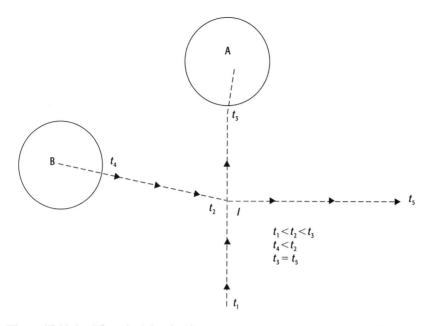

Figure TN9.3 The principle of self-consistency in the billiard ball world. Now the ball, in passing through point I on its dead-on path toward mouth A of the wormhole time machine, is suddenly hit a grazing blow by another ball that has just shot out of mouth B (and into the past) at an angle. The impact knocks the first ball slightly off its original trajectory, and it enters mouth A slightly off-center. Thus the ball emerges from B into the past slightly off-center and just in time to glance off itself at I—which explains *why* it emerged from B slightly off-center!

for these authors was to determine the multiplicity of trajectories for a single, self-interacting time-traveling ball, where the Cauchy condition for a well-defined trajectory is *unique* self-consistency. That is, for the trajectory to be well defined in the Cauchy sense, the authors expected a multiplicity of consistent trajectories *of one* for a self-interacting ball. A multiplicity of zero, of course, would be the physics declaring backward time travel to be nonsense—and the authors thought this a distinct possibility. Their actual results were startling. They found that under very general assumptions about the wormhole parameters, (1) there are no trajectories with zero multiplicity and (2) the multiplicity is also not one, but rather is *always* infinity! Thus the billiard ball form of the grandfather paradox *was* found not to be well defined, but not for the expected reason that there was no self-consistent solution. Instead it was because there were too many solutions. The authors credit physicist/science fiction writer Robert Forward, who uses these ideas in his novel *Timemaster,* for motivating part of their research.

This astonishing, completely unexpected result may be just what is needed to support a continued study of time machines; it may allow a definition of *well-defined* in the Cauchy sense and still permit an answer to the question of free will. The initial conditions of a time-traveling ball give rise to an infinity of self-consistent trajectories, each occurring in the same way that a random variable takes on different values with each new performance of the experiment that the random variable is defined on. And yet, there are still unique probability density functions for all sets of measurements that one might make anywhere along these trajectories. Thus the Cauchy problem is *stochastically* well defined; at the start of any trajectory, we do not know in detail what will happen except that whatever does happen will be self-consistent. In this probabilistic sense, then, wormhole time travel to the past *and* free will both make sense. Lossev and Novikov (1992), Mikheeva and Novikov (1993), and Mensky and Novikov (1996b) have continued the study of self-interacting billiard balls, with the additional complexity of non-elastic collisions and other forms of energy loss. This lets the authors study the thermodynamics of backward time travel. Most recently, Novikov and his colleagues have used the self-interacting billiard ball model to *deduce* self-consistency from the long-accepted principle of least action—see Carlini *et al.* (1995, 1996).

Is it reasonable to think static traversable wormholes can be made? As Visser (1989b) accurately sums it up, "The big open question, naturally, is whether exotic matter is in fact obtainable in the laboratory. The theoretical problems are daunting, and the technological problems seem completely beyond our reach." Woodward (1997) has repeated and extended Visser's words:

> [There are] two classes of exotic matter, the second being a subclass of the first: 1. plain old exotic matter (POEM)—matter which for some, but not all, observers with relative velocities $\leq c$ with respect to the matter see a negative mass-energy and 2. really exotic matter (REM)—matter which is negative for observers with zero relative velocity (that is, matter with negative proper mass-energy density). Contemplating POEM may fire the imagination, but REM is the stuff that dreams are made of. . . . The prospect of ever laying one's hands on some POEM, however, has seemed quite remote. Accumulating some REM . . . is widely deemed hopelessly beyond the pale. Indeed, inducing a wormhole where none existed before has appeared so formidable that the least implausible scenario for acquiring [one] has usually been taken to be enlargement—by unspecified means—of a preexisting wormhole in the putative Planck-scale quantum spacetime foam.

Morris, Thorne, and Yurtsever have carefully acknowledged these concerns of Visser and Woodward, writing (as we noted earlier) that it will take the skill of an "arbitrarily advanced civilization" to overcome them.

The curious phrase *arbitrarily advanced civilization* was given some specific definition years ago by astrophysicists interested in the topic of extraterrestrial life. Very roughly, Types I, II, and III civilizations are those, respectively, of a terrestrial technology that could control something like 10^{13} watts for inter-stellar radio broadcasts, a technology that could control the energy output of its parent star (10^{27} watts), and a technology that could control the energy output of its home galaxy (10^{38} watts). We are, today, clearly a long way short of being even a Type I civilization, and it would probably take *at least* a Type III to build a wormhole (and construct Li's mirror). Indeed, Stephen Baxter's novel *Timelike Infinity,* of beings who manipulate *constellations* of galaxies, seems to assume that a Type IV civilization will be required.

But perhaps there is no need for such a civilization: A paper by Hochberg and Kephart (1991) suggests that wormholes may form naturally at any time. Using a quantum mechanical process called vacuum squeezing—see Cramer (1992) for a non-technical exposition on this—such wormholes require none of the exotic matter that was the main concern of Morris, Thorne, and Yurt-sever. Indeed, vacuum squeezing automatically leads to negative energy densi-ties and to an unavoidable violation of the weak energy condition. And González-Diáz (1996) suggests that a spacetime with the topology of a torus (think donuts!) could, under certain conditions, form in a natural way from the collapse of rotating matter. Such "ringhole" spacetimes appear to be con-vertible to time machines with no violation of the AWEC.

Alternatively, the Big Bang version of the origin of the universe may not be entirely correct for the very earliest moments; other work has suggested that the universe underwent an initially transient period of enormous supercooling and extremely rapid growth—a process called *inflation*[5] that is driven by re-pulsive gravity. Such an inflationary universe would explain some of the still-puzzling aspects of the Big Bang cosmology, as discussed by Guth and Stein-hardt (NP). In particular, Moffat (1993) suggests that the so-called *symmetry breaking* associated with an inflationary universe might explain the thermody-namic (entropic) arrow of time and that arrow's connection with the cos-mological arrow of time. Roman (1993, 1994) has speculated that such an inflation could have enlarged primordial submicroscopic wormholes to mac-roscopic size. (This same process of symmetry breaking might also be the origin of the cosmic strings discussed in Tech Note 10). Although this is a one-time-only scenario for the creation of "big" wormholes—see Farhi and Guth

(1987)—it would allow for such entities to have existed somewhere in the universe since the beginning of time. And so, Tipler's words at the start of this Tech Note notwithstanding, perhaps the dinosaurs *are* accessible!

The latest approach to wormholes is to allow them to be *dynamic* (not static) structures in spacetime—that is, to allow one or more parameters to vary with time (perhaps the throat diameter can collapse). Then, according to Wang and Letelier (1995), it is possible to have a traversable wormhole *made of normal matter*—see also Anchordoqui *et al.* (1988)—that, even though it is collapsing, takes so long to do so that "a space adventurer will have enough time to pass through the throat of the wormhole from one asymptotically flat region to the other before the radius of the throat shrinks to . . . where the event horizon is developed." Such a dynamic wormhole, they claim, satisfies both the weak and the dominant energy conditions, but not the strong condition. Thus gravity is still repulsive in the throat, but they don't think that is a serious objection because such a condition is thought to have actually once occurred on a global scale during the inflationary stage of the Big Bang—see Note 5 again—although how that would help in the construction of a wormhole in the future is not clear to me. Kar (1994) also claims to have shown how Lorentzian wormholes can occur without violating the WEC but admits at the end that, for his wormholes, the "question of traversibility and possible models for time machines" have not been addressed and "will be discussed in a future article." Most recently, Schein and Aichelburg (1996) and Krasnikov (1998b) and Aichelburg and Schein (1998)—two papers published the same month in different journals—claim to describe wormholes that do not require exotic matter.

The daunting level of technology required to build a wormhole (with or without exotic matter) doesn't mean we couldn't search for existing wormholes—perhaps vast wormhole networks formed naturally at the Big Bang; as described in Gregory Benford's 1997 novel *Foundation's Fear*, wormholes are "leftovers from the Great Emergence [the Big Bang]." Or perhaps "advanced civilizations" have already constructed a vast, pan-galactic "subway" system of wormholes like the one described in Carl Sagan's 1985 novel *Contact* and illustrated in the 1997 film version. In fact, such searches are now actually in progress, although the original motivation was a bit less exotic than wormholes. Because of the flat rotational characteristics of spiral galaxies, including our own Milky Way, physicists have concluded that such galaxies are surrounded by halos of matter. The phrase *flat rotational characteristic* refers to the observed, unexpected orbital behavior of stars in the outer edges of galaxies, beyond the optically obvious luminous core where most of the luminous mass is located. To understand why I say "unexpected," recall how the planets orbit the Sun, which contains very nearly all of the mass of the solar system—the further out from the Sun, the slower the orbital speed of the planets. Indeed, if you simply set the gravitational force of the Sun on a planet at distance r equal to the centrifugal force on the planet, it immediately follows that the orbital speed, $v(r)$, decreases with r as $1/\sqrt{r}$. For most galaxies, however, the orbital speeds of their ever-more-distant stars *increase* with r, until one reaches stars that are distant by two or three times the radius of the luminous core. Then the orbital speeds become more or less constant; a plot of $v(r)$ versus r becomes "flat." This implies that there is unaccounted-for mass even farther out from the galactic core—that is, the dark-matter halo.

Because these halos are as yet undetected, they are said to consist of dark matter. The nature of this dark matter has been debated for years, and some argue that it must be very strange stuff (ordinary stars, gas, or dust would be easily detectable) collectively called weakly interacting massive particles, or WIMPs.[6] A massive neutrino, ordinarily thought of as having little or no mass, would be a WIMP, for example. Others have said that the dark matter need not be that esoteric and may perhaps simply be "Jupiters" (gas balls not quite large enough to initiate fusion and thereby become stars), neutron stars, or even black holes. This second class of possibilities is collectively called massive compact halo objects, or MACHOs. The most massive MACHOs are thought to be in the range of one solar mass, going down to about 10^{-4} solar mass.

How to detect WIMPs is unknown, but spotting MACHOs is easy (in principle). Their very massiveness is the key, because their attractive gravitational fields might be sufficient to bend the light of more distant stars in such a way as to *focus* the light sources and cause a temporary, observable intensification of brightness. That is, MACHOs might work as *gravitational microlenses*,[7] and the MACHO would give itself away in the characteristic, single peak of brightening and then fading of a distant light source, as the MACHO moved across the line of sight from Earth to the light source. That is the signature that the original (and still continuing) searches have been looking for, both in the Milky Way and in its two nearest galactic neighbors, the Large and Small Magellanic Clouds (which orbit the Milky Way at about 170,000 and 200,000 light-years, respectively). The searches for MACHOs are big searches. Millions of stars are watched over a period of years to detect any sudden, one-time light-spike (a MACHO of 10^{-2} solar mass would cause light-spike 10 days in duration). Such a spike would be totally unlike the light-spikes associated with either variable stars or exploding stars.

Now, what would happen if a *negative*-mass[8] wormhole mouth should cross the line of sight from Earth to a distant star? You might think that if a positive-mass MACHO focuses light, then a negative mass would defocus light, but this is not true. Surprisingly, a simple geometric analysis of such a line-of-sight crossing shows that what actually results is the formation of a conelike surface, with the wormhole mouth at the apex, such that the surface is a *caustic*—a surface of light *intensification* that can actually exceed that of a MACHO. Thus, as the line-of-sight crossing occurs, an observer of the star would first see the caustic surface (a light-spike),then the interior of the cone-like surface (no light), and then the other side of the caustic (a *second* light-spike). Calling *any* such negative mass in space a gravitationally negative anomalous compact halo object, or GNACHO, Cramer *et al.* (1995) observe that this light signature is very different from that of a MACHO, and yet both can be searched for in the same experiment. Indeed, one such MACHO search, a collaboration of Carnegie Observatory (U.S.A.) and the university observatories at Warsaw (Poland) and Princeton (U.S.A.) that is called Project OGLE (Optical Gravitational Lensing Experiment), actually detected such a double-peak light signature in 1994. It was later determined to be due not to a GNACHO, however, but rather to the lensing effect of an ordinary binary star. The search goes on.

Any such wormhole, if found, could, of course, be a very long way from Earth. It might be in another galaxy. So even if it exists, what good would the wormhole do anybody? In a remarkable analysis by Lossev and Novikov (1992), it is suggested such a wormhole might be very useful, even if its location is completely unknown! All that Lossev and Novikov assume is that the wormhole has existed for a "sufficiently long time" (what this means will be made explicit by the end of this Tech Note). With this assumption, they show how to make an information-creating time loop, the very sort of thing that David Deutsch argues against in Chapter Four (and that so many science fiction writers have used to good effect for decades, as also discussed in that chapter).

Lossev and Novikov begin their analysis by assuming that people have no knowledge of how to build a spacecraft that can make the interstellar voyage to the distant wormhole, even if they knew in which direction to go to reach the mouth that leads backward in time (mouth B). Instead, they build an automatic spacecraft construction plant that can follow any detailed sequence of instructions provided to it and then stockpile it with a supply of raw materials (energy, steel, plastic, computers, and so on). When the spacecraft construction is done (*how* that is done is explained in the next paragraph), the last step before launching the spacecraft toward mouth B will be to load its on-board computer with the following three pieces of information:

a. The detailed sequence of instructions used for the construction of the spacecraft
b. The direction from Earth to mouth B
c. The direction from mouth A (the wormhole exit mouth in the past) back to Earth

Thus people build the automatic plant, load it up with raw materials, *and then withdraw.* This last step is crucial, because it eliminates human free will from further consideration. That is, it removes any temptation to create a bilking paradox. So what happens next?

Lossev and Novikov suggest that what happens next is that a very old spacecraft suddenly appears in the sky and lands next to the automatic plant! In its on-board computer are items a, b, and c. Using item a, the automatic plant makes a new spacecraft, then it loads its new on-board computer with items a, b, and c from the very old spacecraft's computer, and then the new spacecraft is launched toward mouth B (using the information of item b). The very old spacecraft is given an honored place in a museum.

The new spacecraft arrives at the distant mouth B in the far future, by which time it is, of course, an old spacecraft (but not yet a very old one). It then plunges into mouth B and almost immediately emerges from mouth A, in the past. Indeed, it repeats this process as many times as required until it is in the far distant past, at a time even before it left Earth. It might seem that to do this, the spacecraft's computer memory needs a fourth piece of information, the direction from mouth A back to mouth B. But, in fact, items b and c are sufficient for the spacecraft to find its way from A to B. It is also now clear how long the wormhole must have been in existence. The spacecraft repeatedly uses the wormhole time machine until it is so far in the past that it can cruise back to Earth at normal speed (it knows the way back because of item c) and arrive as a very old spacecraft, just in time to be placed in the museum!

As Lossev and Novikov point out, this remarkable closed sequence of events has increased knowledge from what it was at the time just before the automatic plant was built. People now know both how to build an interstellar spacecraft and the locations of both mouths of the wormhole. They also now possess a very old, used spacecraft. It is curious to note that although the information in the spacecraft's computer memory has traveled on a closed time loop, the spacecraft itself has not. This is because the spacecraft left Earth when new but arrived back (before it left) as very old, whereupon it promptly entered a museum. There is therefore no question about the origin of the very old spacecraft, but where did the information of items a, b, and c come from? Lossev and Novikov say it came from the energy gained by the spacecraft as it interacted (will interact?) with the rest of the universe while on its journey. As Lossev and Novikov also point out, however, their analysis ignores all quantum mechanical considerations.

Not surprisingly, not everybody finds such a tale convincing. One unimpressed analyst is research professor of physics Matt Visser at Washington University in St. Louis, whose support for Hawking's Chronology protection Conjecture is discussed in the Epilogue. When I asked Professor Visser what he thinks of time travel, he replied (personal communication, August 24, 1992),

> If one insists on permitting time travel in one's theories of the Universe, then Novikov's "consistency conjecture" is probably the minimal modification of our current understanding of causality. I personally have a lot of trouble swallowing the "consistency conjecture" since it seems to me to be rather *ad hoc* [but recall my earlier comment in this Tech Note on Novikov's derivation of self-consistency from least action]. The "consistency conjecture" seems to be essentially equivalent to the statement: 'the Universe IS consistent, no matter what, since it MUST be consistent, come hell or high water.' I view this as begging the issue. . . . [O]nce one has opened Pandora's box by permitting time travel, I see no particular reason to believe that the only damage done to our notions of reality would be something as facile as the "consistency conjecture."

Visser repeated this, word for word, in Visser (1993) and later in his seminal 1995 book on Lorentzian wormholes. Visser may well be right in his pessimism about time travel, of course, but I do think demanding consistency in the universe is the *very heart* of physics. Without consistency, physicists could not even have understandable debates about their disagreements!

The one-wormhole, two-mouth time machines described so far were actually not the first kind of wormhole time machines mentioned in the literature. In their 1988 paper Morris and Thorne initially described a time machine in terms of *two* wormholes and added a *note in proof* at the end that they had just discovered how to build a time machine with one wormhole (the machine treated so far in this Tech Note). This reduction in required wormholes was believed to be a technical advance, of course, so the two-hole machine was put aside. But not for long. Soon thereafter, analyses began to hint at the possibility that one-hole machines might destroy themselves just at the instant their mouths were about to be threaded by closed timelike curves. As we noted earlier, however, wormholes have a defocusing property for electromagnetic radiation, so the initial concern that time-traveling photons would prove fatal for wormhole time machines was short-lived. But then it was found that the vacuum fluctuations of quantum fields are *not* so defocused—see Frolov (1991) for why they are not. Kim and Thorne (1991) showed that this failure to defocus time-traveling vacuum polarizations, as such fluctuations are called, would indeed result in an unphysical divergence of stress-energy on the Cauchy horizon of a one-hole time machine.

Kim and Thorne's analysis, however, also showed that the stress-energy divergence was extremely weak (it doesn't happen very fast). It was so weak, in fact, that they concluded that an actual infinity of the stress-energy would be precluded by the eventual intercession of quantum gravity. That is, the stress-energy would try to become unbounded as spacetime approached the formation of a time machine, but before it became so large as to destroy the time machine, quantum gravity would cut off the growth "in time" (so to speak!) to save the machine. Hence the resurrection of the two-hole time machine geometry. Perhaps *it* could avoid the self-destructive effect of time-traveling vacuum fluctuations. If a spacetime contains multiple wormholes, then it is called a *Roman* spacetime after Thomas Roman (at Central Connecticut State University), who was the originator of such time machine spacetimes. Each of these wormholes, individually, is *not* a time machine. Together, however, they form a time machine called a *Roman configuration* or, as Visser (1997b) calls it, a *Roman ring*. Here's how.

In Figure TN9.4 two pairs of wormhole mouths are labeled A, A' and B, B'. We imagine that the A, A' wormhole is stationary and that its two mouths are very far apart—so far apart, in fact, that if a traveler enters A and almost instantly (because the wormhole handle is very short in hyperspace) emerges from A', it will appear to an observer at rest with respect to the wormhole that the traveler has moved faster than light. That is, entering A' and exiting A' are events with a spacelike separation. Now, also imagine that the wormhole with mouths B, B' is moving past the first wormhole at speed v. To an observer in this second *moving* frame of reference, the spacelike separation of entering A' and exiting A' can result in the two events being temporally reversed (if v is sufficiently large but still less than the speed of light—see Tech Note 4 for the mathematics of this reversal).

Therefore, upon emerging from A' the traveler crosses normal space to the moving wormhole mouth B', enters the wormhole, then almost instantly emerges from mouth B, and finally travels again through normal space to

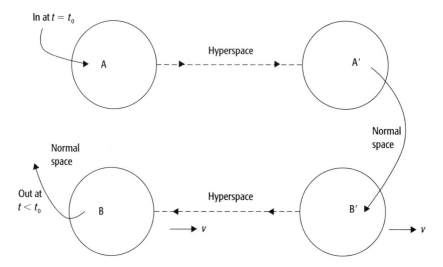

Figure TN9.4   A two-wormhole, Roman-configuration time machine.

mouth A. If the traveler can make the two trips in normal space in less time than the backward time shift achieved by the temporal reversal of entering A′ and exiting A′, then we have a time machine. Hawking (1992), however, disagreed, arguing that Kim and Thorne had made a crucial error in their calculations. According to Hawking—and Thorne has since stated (1994) that he finds Hawking's position hard to refute—the divergence of the stress-energy may indeed by cut off by quantum gravity, but not before the development of spacetime disturbances representing perhaps a hundred million times the energy levels associated with ordinary chemical binding energies. These would be sufficiently big disturbances to raise serious doubts about the physical survival of a one-wormhole time machine, even in the absence of a true stress-energy infinity.

But not everybody agrees with Hawking. Two nearly simultaneous analyses of this multiple-wormhole Roman-configuration time machine—Visser (1994a) and Lyutikov (1994)—both concluded that for suitable choices of sizes (the radii of the wormhole mouths, the wormhole lengths in normal space, the lateral offset of the two wormholes, and the relative speed of the wormholes), the stress-energy divergence *can* be limited by quantum gravity to an arbitrarily weak level. In particular, the saving effect of quantum gravity is enhanced with increased wormhole length in normal space (the separation of the mouths) and with decreased mouth radii. That is, the two-wormhole time machine is *not* necessarily destroyed by an unbounded stress-energy on the Cauchy horizon. On all this Visser and Lyutikov agreed, but on what to conclude from it all they quickly parted company.

Although he admitted that the quantum gravity cutoff of the stress-energy divergence does occur in a Roman configuration, Visser called the special sizing conditions that are required "bizarre" and asserted that the resulting time machine would be quite useless for a human traveler in any case. For example, he calculated that only if the mouths of the wormholes are separated in normal space by the radius of the universe, and only if the wormhole mouths have radii on the order of that of an atomic nucleus, will the cutoff be sufficient to allow the putative time machine to avoid destruction. When Visser reduced his wormholes from universe size to "merely" that of the distance between Sun and Earth, he concluded that it would require energy at the level of the Superconducting Supercollider accelerator to blast an information-bearing message through the narrow wormholes. And even then the "short" wormholes would provide a maximum penetration of eight minutes back into the past. As Visser put it, "This does not seem to be a workable recipe for studying tomorrow's *Wall Street Journal*." Indeed, he takes such calculations to be sufficiently convincing arguments against time machines that he concluded his analysis with the suggestion that quantum gravity should simply be taken as causal—that is, this should be regarded as an axiom rather than as something to be demonstrated. This seems to me, however, to repeat the mistake of Gödel, who also confused enormous engineering difficulties with fundamental violations of physics. Lyutikov, who mentions Visser's sizing calculations with the caveat that they are "far from obvious," takes a far less extreme stance. He concluded that although Visser's calculations, at face value, "make it very inconvenient for time travel [for humans]," nevertheless "the [principal] question of the possibility of transmitting information back in time through traversable wormholes would still remain."

Just a month after the analyses of Lyutikov and Visser appeared, Tanaka and Hiscock (1994) published an even more detailed study of the divergence of the stress-energy in time machine spacetimes, and their conclusion was most remarkable. Their work "suggests that quantum gravity may be unable to act as the protector of chronology," if the stress-energy divergence is supposed to be the fatal flaw in time machines. Later, Krasnikov (1996) wrote that a study of the quantum stability of wormhole time machines leads him to conclude that "there is actually *no* reason to expect that the energy density diverges at the Cauchy horizon in the general case," and even more emphatic is his statement "there is not a grain of evidence to suggest that the time machine must be unstable." This is remarkable because Hawking (1992), in his now-famous announcement of the chronology protection conjecture, wrote, "The philosophy of this paper is . . . to look for vacuum polarization [the divergence of the stress-energy] to enforce the chronology protection conjecture." He repeated this view in Hawking (1995).

As is becoming increasingly apparent, however, that may be the wrong place to look. As I write, Krasnikov's final words in his paper are the fairest I've seen: "It may well be that the vacuum fluctuations do make the time machine unstable, but nothing at present suggests this. All we have are a few simple examples. In some of them the energy density diverges at the horizon and in some does not. So, the time machine perhaps is stable and perhaps is not." Even Visser agrees. As he wrote at the end of his 1997b paper. "I view [the failure of the stress-energy divergence to inforce chronology protection] not as a vindication for time travel enthusiasts, but rather as an indication that resolving issues of chronology protection requires a fully developed theory of quantum gravity." And he was even more emphatic at the end of Visser (1997a): ". . . semi-classical quantum gravity is fundamentally incapable of answering the question of whether or not the universe is chronology protected. To really answer this question we will first need an acceptable theory of quantum gravity." It's hard to argue with any of those last sentences![9]

In fact, the case against Hawking's chronology protection conjecture may be even stronger, so strong that quantum gravity may not be necessary to refute it—or, at least, to refute the claimed protection mechanism of stress-energy divergence. In 1996 the Chinese physicist Li-Xin Li showed that under very general conditions, the presence of an absorber (opaque matter) in a spacetime that is on the verge of forming closed timelike curves can *smooth away* the disruptive effects of vacuum fluctuations; the stress-energy will then be bounded on the Cauchy horizon. That is, closed timelike curves *will form*. Li (1996) showed that under quite reasonable physical assumptions, the mass density of the absorber could actually be less than the observed mass density of the universe. This major result emboldened Li to the point where he proposed his *anti*-chronology protection conjecture: "there is no law of physics preventing the appearance of closed timelike curves."[10]

Hawking himself finally seems ready to be willing to concede on the issue of stress-energy divergence (if not chronology protection, itself), writing in Cassidy and Hawking (1998), "the fact that the energy-momentum tensor fails to diverge [in certain special cases of time machine spacetimes] shows that the back reaction does not enforce chronology protection." These two physicists have therefore looked for another mechanism for chronology protection and have shown, in certain interesting *special* situations (a scalar quantum field in a background spacetime that is forced to contain closed timelike lines), that the entropy diverges to negative infinity just as causality is about to be violated. That is, the density of available quantum states for the field approaches zero. This result hints at the possibility that *quantum mechanics* might prohibit time machines by the simple mechanism of providing a vanishing number of quantum states for any causality-violating field configurations. But, as Cassidy and Hawking state, "the crucial question to ask . . . is whether this result holds in the general case." Li and Gott (1998), however, have found a self-consistent, closed timelike solution to the field equations that is *not* forbidden by quantum effects.

As they say in another paper—Gott and Li (1998)—"New objections to spacetimes with CTC's can continue to surface, as old problems are put to rest, so it might seem that disproving the chronology protection conjecture would be a tall order. But, proving that there are no exceptions to the chronology protection conjecture, ever, would seem an equally daunting task. This is particularly true since we currently do not have either a theory of quantum gravity or a theory of everything."

Notes

1. The Chinese physicist Liao Liu published a paper (1993) claiming to show, for the first time, *how* this romantic imagery actually occurs in the context of general relativity, with the creation of Lorentzian wormholes taking place naturally as a result of vacuum fluctuations. However, see Anderson and DeWitt (1986) for a rebuttal to the idea of changes in the topology of spacetime. Also see Gibbons and Hawking (1992) and Chamblin, Gibbons, and Steif (1994).

2. A complete presentation of Casimir's analysis, with citations to the original literature, can be found in L.E. Ballentine, *Quantum Mechanics* (Englewood Cliffs, NJ: Prentice-Hall, 1990), pp. 399–403. For an historical, tutorial presentation, including Casimir's personal comments on how he was led to make his discovery, see P.W. Milonni and M.-L. Shih, "Casimir Forces," *Contemporary Physics* 33 (September/October 1992):313–322.

3. The theory of Forward's "vacuum-fluctuation battery" is given in his paper "Extracting Electrical Energy from the Vacuum by Cohesion of Charged Foliated Conductors," *Physical Review B* 30 (August 15, 1984):1700–1702. For a somewhat different view, see also the essay by Philip Yam, "Exploiting Zero-Point Energy," *Scientific American*, December 1997.

4. A classic work on the mathematics of Cauchy problems is J. Hadamard, *Lectures on Cauchy's Problem in Linear Partial Differential Equations* (New York: Dover, 1952). There is a curious bit of irony in this. In a section of his book, Hadamard uses spacetime to illustrate one possible four-dimensional space, and in passing he casually writes, "This conception was beautifully illustrated a good many years ago by the novelist Wells in his *Time Machine*." Hadamard wrote his book in 1923, and he would almost certainly have been astonished to have been informed that less than seventy years later, his work would play a central role in the non-fictional theory of time machines. A more modern tutorial on the Cauchy problem is Bruhat (GRAV).

5. During inflation the universe expanded at a rate far beyond human comprehension. It is currently estimated that during the first 10^{-35} second of the Big Bang, the universe doubled in each spatial dimension by a factor of two each 10^{-37} second; that is, there were about 100 such doublings. Thus there was an increase by a factor of $2^{100} = 10^{30}$ in each linear dimension of the universe—and the volume increased, of course, by the *cube* of that already enormous factor. See Alan H. Guth, *The Inflationary Universe* (Reading, MA: Addison-Wesley, 1997).

6. The invention of amusing acronyms is an occupational hazard for theoretical physicists. In Woodward (1997), for another example of this activity, you can read of TWISTs and TWITs: "traversable wormholes in spacetime" and "traversable wormholes in time." There is also mention of MUSH, a play on Hawking's chronology protection conjecture, which Hawking feels is required for "making the universe safe for historians." It is fun reading.

7. Gravitational lensing is an extension of the basic idea of gravity bending light (and, as discussed in Tech Note 10, it is the fundamental idea behind Gott's discovery of the cosmic string time machine). The experimental confirmation of this prediction from general relativity, during a 1919 solar eclipse, is what made Einstein famous. The first mention of *lensing* by a gravitational field was made by Oliver Lodge, in a letter to *Nature* ("Gravitation and Light," December 4, 1919, p. 354), just months after the eclipse data were analyzed. Many years later Einstein himself returned to light bending in a short note titled "Lens-Like Action of a Star by the Deviation of Light in the Gravitational Field" (*Science,* December 4, 1936, pp. 506–507). There he showed how, if star B passes through the line of sight (of an observer on Earth) to star A, then A's light will form a bright halo (now called an Einstein ring) around B. Larry Niven mentions this effect in his "Neutron Star." Einstein wrote that "Of course, there is no hope of observing this phenomenon" because of the limitations on "the resolving power of our instruments." He was wrong in this assessment, however, because he was thinking of Earth-based optical instruments—since the orbiting of the Hubble Space Telescope in 1990, the Einstein ring effect *has* been observed. Einstein also showed how *multiple* images of A could be seen from Earth because of B's gravitational field. Again he declared, "there is no great chance of observing this phenomenon," and again such multiple images *have* been observed via the HST. When the gravitational lensing effect is too weak to resolve the multiple images, the effect is called *microlensing*.

8. Although negative mass (like time travel) is not forbidden by general relativity itself, such a thing does imply (as does time travel) some consequences so bizarre that most physicists are unwilling to take the concept completely seriously (their reaction to time travel, too),—yet. For example, general relativity says that a negative mass will *repel* all other mass (positive *and* negative), whereas a positive mass will *attract* all other masses (positive *and* negative). Imagine, then, a negative mass attached to the nose of a positive-mass spaceship. The spaceship tries to move toward the negative mass, while the negative mass tries to move away from the spaceship. So off they both go into the sky like a cat chasing its tail! This so-called reactionless antigravity-drive, bizarre as it appears, does *not* violate either of the conservation laws of energy or momentum. Weird stuff, indeed, this negative mass. Something like negative mass appeared in fiction quite early, in the 1827 novel *A Voyage to the Moon* by "Joseph Atterley," a pseudonym for George Tucker, a professor of moral philosophy at the University of Virginia. (One of Tucker's students was Edgar Allan Poe, who almost surely was influenced by Tucker's book to write his own moon tale, the 1835 "The Unparalleled Adventure of One Hans Pfall.") The trip in Tucker's work was powered by a metal called *lunarium,* which repels Earth. This is *not* the same sort of stuff as Wells' "Cavorite," a metallic alloy that is "transparent" to gravity and that appears in his 1901 *The First Men In the Moon* (see also Note 2 for

the Epilogue). For a recent discussion of negative mass, see G. Cavalleri and E. Tonni, "Negative Masses, Even If Isolated, Imply Self-Acceleration, Hence a Catastrophic World," *Il Nuovo Cimento* 112B (July 1997):897–903. And James Glanz describes the skeptical reaction of most physicists to antigravity in "Astronomers See a Cosmic Antigravity Force at Work," *Science* 279 (February 27, 1998):1298–1299.

9. As this book goes to press, Visser has recently argued that the "no exotic matter required" wormhole analyses discussed earlier in this Tech Note are simply wrong even without invoking quantum gravity. He and his co-author David Hochberg also argue that dynamic wormholes actually have two throats (one for each direction of travel) that coalesce for the static case: see their paper "Null Energy Condition in Dynamic Wormholes," *Physical Review Letters* 81 (July 27, 1998): 746–749. Also, Edward Teo has shown that static wormholes can be made axially symmetric via rotation (which makes them traversable between regions of asymptotically flat spacetime in the *same universe*). And although such wormholes do indeed require exotic matter in their throats, Teo has shown that it is also possible to travel through them without ever encountering such matter. See his 1998 paper "Rotating Traversable Wormholes," *Physical Review D* 58:024014.

10. As a sort of last-ditch analysis to see if there might be *something* in semiclassical quantum gravity to refute Li's conjecture, the Russian physicist Sergey Sushkov has recently tried a novel approach. Noting that time-dependent processes (such as the formation of the chronology horizon of a wormhole evolving into a time machine) in the presence of quantum fields are always accompanied by the creation of particle pairs, he asked the following question: Can particle creation stop the horizon from forming in a spacetime that initially has no causal pathologies, that is, no timelike curves? He found, in fact, that the number of particles created is finite at all times, including the instant the horizon forms—an infinite particle count would, of course, be suspicious! He did not, however, calculate the back reaction of the created particles (i.e., their perturbation on the spacetime metric), so perhaps *that* would disrupt the horizon. In the face of the apparent ability of spacetime to survive even with time machines, however, I—and I suspect Sushkov, too—think that unlikely. See his 1998 paper "Particle Creation Near the Chronology Horizon," *Physical Review D* 58:044006.

"Solving" the Einstein Gravitational Field Equations, Unphysical Mass-Energy, and the Cosmic String Time Machine

It's an amazingly simple solution. It doesn't take much physics to understand it.

> —MIT astrophysicist Alan Guth, commenting in Travis (1992) on Gott's discovery of the cosmic string time machine

Louise, working out the spacetime geometry of a cosmic string is a hard problem in general relativity. But, given that geometry, all the rest of it is no more than Pythagoras' theorem. . . .

> —a character in Stephen Baxter's novel *Ring*, agreeing with Guth

The ten Einstein gravitational field equations are extraordinarily complicated; they are almost surely the most complicated equations in all of physics.[1] They are non-linear, coupled, partial-differential tensor equations, and *each one* of those adjectives implies great difficulty. It has required brilliant minds, over decades of inspired work, to find the relatively small number of exact solutions to the equations that are known. This immediately leads to an interesting question, one of direct importance in modern time machine studies: What does it mean to "solve" the field equations? It is useful to think of the equations schematically as follows:

$$\begin{matrix} \text{local geometry} \\ \text{of spacetime} \end{matrix} \quad \Leftarrow \quad \begin{matrix} \text{local density, momentum and stress} \\ \text{of the mass-energy of spacetime} \end{matrix}$$

where the direction of the arrowhead on the equality sign means that the "usual" practice is to assume the right-hand side (the so-called stress-energy tensor) as given and then attempt to calculate the left-hand side. If the attempted calculation can be done, then one has solved the field equations for the spacetime geometry that is associated with the assumed mass-energy distribution.

Suppose, however, that we reverse the direction of the arrowhead on the equality. That is, we assume a desired geometry. This is what Einstein did when he assumed the geometry of a *static* (non-expanding) universe and solved for the required mass-energy. What he found was just what had been observed up to that time—a multitude of "grains of matter" or what physicists call "dust," *plus* the infamous cosmological constant. The constant, with its repulsive gravity, was needed to counteract the ordinary gravitational attraction of the stars that tends to pull them together. The later discovery that the universe is not static but rather is expanding, of course rendered moot Einstein's original motivation for his solution.

Now let's go Einstein one better, and assume a geometry like a spacetime that contains closed timelike curves (a time machine, in other words) and then try to calculate the mass-energy distribution required by that spacetime. If that can be accomplished—and in fact the field equations themselves provide an algorithmic means of solution in this direction—then the physicist's work is done. The required mass-energy distribution requirements are put out to bid to "spacetime engineers," and the lowest bidder "simply" constructs that mass-energy distribution and so builds us our time machine! When the calculations are done, however, what has happened without fail, at least up to now, is that the resulting mass-energy distribution comes out with an 'unphysical' nature,[2] a technical way of saying our spacetime engineer wouldn't know *how* to assemble the required mass-energy distribution. In the case of a spacetime with a wormhole geometry, for example, the mass energy in the wormhole throats is negative, as discussed in Tech Note 9.

All is not lost for time travelers, however. Gott (1991) gives a way to achieve closed timelike loops without "unphysical" negative-mass wormholes, violated energy conditions, or topology changes. Gott gives exact solutions to Einstein's equations for cosmic strings that (1) unlike wormholes do not violate any of the energy conditions, (2) have no crushing singularities or event horizons, and (3) are not topologically multiply connected. Cosmic strings are fantastically thin (such as 10^{-28} cm in radius) filaments of pure energy that are speculated to have formed at Big Bang time—see Tech Note 9. According to current theories, they stretch the width of the universe and have an enormous linear mass-energy density on the order of 10^{22}g/cm. To generate the closed timelike paths, however, Gott requires either that two *fast-moving* (which means moving practically at the speed of light) parallel cosmic strings pass each other on a near-collision course or that there be a closed-loop, elliptical string that collapses in a slightly non-planar manner so that the opposite, nearly straight sides "just miss." The gravitational interaction of the two strings "warps" spacetime enough to produce closed timelike curves.

A hint at the possibility of violating causality with strings appeared before Gott's work, in Harari and Polychronakos (1988), but those authors didn't take time travel seriously. As they wrote, "We argue . . . that any realistic model [for a spinning string with angular momentum[3]] . . . will not have closed timelike curves." Gott, however, showed that as two strings pass each other, closed timelike loops encircle the strings. Gott, who appears to be far less rigid in his views toward time travel than are many of his fellow physicists, still held out an escape to those who pale at the very thought of time travel to the past. He suggested that one way to avoid string-induced time travel was to imagine that as the strings (or string-loop sides) pass, a black hole—following an analysis by Hawking (1990)—will form with an event horizon that seals off the time loops from any potential user. Another possible mechanism for escaping closed timelike curves might be thought to lie in the more realistic case of non-singular strings (strings with non-zero radius) and angular momentum "thick" strings that have "spin." Such analyses can be found in Soleng (1992), Jensen (1992), Jensen and Soleng (1992), and Novello and da Silva (1994). The time travel implications were found to remain intact, however.

Here's how the cosmic string time machine works. In 1985 it was discovered by Gott—and independently by Hiscock (1985)—that a cosmic string warps spacetime in a highly characteristic way,[4] as shown in Figure TN10.1. This stationary cosmic string is perpendicular to the xy-plane (the plane of the page) and passes through the page at the point $(0, d)$ on the y-axis. The warp produced by the string is as though a wedge of angle 2α (this angle is called the *deficit angle*[5]) were cut out of spacetime and the edges of the cut were then "glued" together; for example, points C and D are identified as identical. The

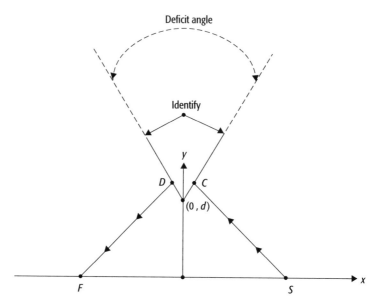

Figure TN10.1   The deficit angle in spacetime formed by a cosmic string.

reason for the name *deficit angle* is that at radius r from the string, a circular path around the string has the reduced length $(2\pi - 2\alpha)r$, not the usual $2\pi r$ (because although spacetime around the string is locally flat, it is "conical," as illustrated in Figure TN10.2.

Now consider the two points S and F on the axis at $(x_0, 0)$ and $(-x_0, 0)$. Suppose we want to send photons from S to F. In normal, "unwarped" space-time the direct path S to 0 to F has length $2x_0$. There is also another path, how-ever, S to C/D, to F, that loops out and around the cosmic string. Indeed, this second path is simply gravitational lensing (an observer at F would see *two* im-ages of S, as discussed in Tech Note 9), and this—*not* time travel!—is the issue that originally attracted Gott's attention to strings.[6] If the deficit angle were zero, then this alternative path would always be longer than $2x_0$ for any value of x_0. For the case of $2\alpha > 0$, however, if x_0 is large enough ($x_0 \gg d$), then it is possible for the "around the string and over the missing spacetime wedge" path to be shorter than the direct path.

The indirect path provides a way for a *subluminal* trip (say, by rocket) from S to F to beat a photon traveling on the direct path. That is, the two events of the rocket leaving S and the rocket arriving at F are spacelike separated. Thus it is possible (see Tech Note 4) to find a moving frame of reference in which those two events are *reversed in temporal order.* In that frame of reference, the

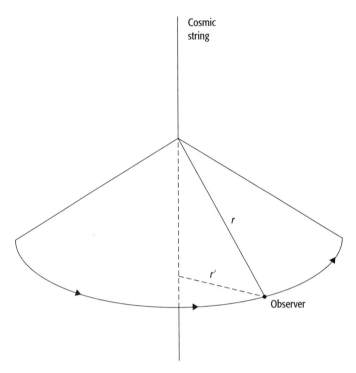

Figure TN10.2   The warped, conical spacetime around a cosmic string. An observer in this spacetime thinks she is distance r from the string, but a "meta-observer" sees she is actually distance r' from the string. Thus, if the observer follows a complete circular path around the string, she will travel a distance of $2\pi r' < 2\pi r$. The observer in the spacetime will interpret this result by saying that the 2π angle is really 2π—"a deficit."

cosmic string (which is stationary in the reference frame of S and in that of F) will move—at speed v, say—in the $+x$ direction, and in that frame of reference the rocket will arrive at F *before* it leaves S.[7] Then, to complete the construction of a closed timelike path, simply repeat the process as shown in Figure TN10.3. That is, have the rocket arriving at F turn around and fly back to S out-and-around and through the deficit angle warp in spacetime due to a *second* cosmic string on the negative y-axis and perpendicular to the xy-plane. This second string is moving at speed $-v$ (that is, opposite to the first string), so the rocket will arrive at S before it leaves F. But that means it arrives at S *before* it leaves S; that is, the rocket has traveled into the past. This entire process is precisely the same idea behind the two-wormhole, Roman-ring time machine described in Tech Note 9.

Now, instead of having two oppositely moving reference frames, one in which the top, stationary string at $(0, d)$ appears to be moving at $+v$ and

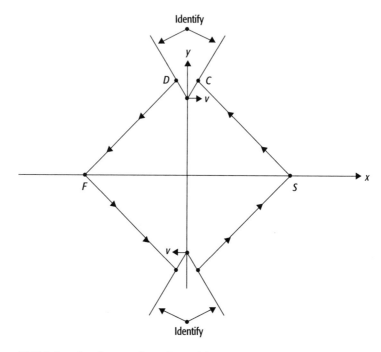

Figure TN10.3   Gott's spacetime, formed by joining two oppositely moving versions of the spacetime from Figure TN10.2.

another in which the bottom, stationary string appears to be moving at $-v$, we can imagine an observer in the stationary center-of-mass frame watching two strings that *are moving* at $+v$ and $-v$. This leaves the situation unchanged, so in the center-of-mass frame the rocket does travel into the past, arriving back at S before it leaves S. That is, the rocket has traveled all the way around a closed timelike world line. Note, too, that the geometric condition mentioned earlier of $x_0 \gg d$ immediately implies that for x_0 *not* sufficiently large, there isn't a closed timelike path from S to F and then back to S,—that is, there is a region in Gott's spacetime where such time travel journeys *cannot occur.*

Ori (1991a) pursued Gott's analysis in an attempt to see whether these "time travel paths" are *created* as the strings approach each other or, instead, the paths exist at other times as well. This important question gets to the idea of whether such time machine paths can be intentionally *created* by humans by a dynamical process (a *strong* time machine), or whether all such paths have existed since the formation of the universe (a *weak* time machine). Also, as Ori observed, this issue involves Hawking's so-called chronology protection con-

jecture, which asserts that the laws of physics will always prevent the creation
of a time machine. As discussed in Chapter One, one reason Hawking has repeatedly given for believing his conjecture is the apparent absence of time travelers from the future among us now (in their past). The only possible exception allowed by the conjecture is the creation of closed-timelike loops at the
moment the universe was created (at that moment there was *no past* for time
travelers to invade!). Ori proved that the closed timelike loops around Gott's
cosmic strings are *always* present. That is, a time machine is *not* created by the
near collision, so Hawking's conjecture is not refuted by Gott's spacetime.

One very curious issue is *where* the closed time loops are before the strings
pass one another. As briefly mentioned by Ori and later discussed in Deser *et
al.* (1992), Deser and Jackiw (1992), and Deser (1993), the time loops are initially at spatial infinity. As the strings pass each other, the time loops collapse
inward at the speed of light onto the interaction region of the strings—because of the twisted geometry of Gott's spacetime, the speed of light in it is
very fast, indeed! Hawking's conjecture assumes, by the way, that the Cauchy
horizon is compact—that is, all closed timelike loops are confined to a finite,
simply connected region in spacetime. This is another reason why Gott's
string time machine and the conjecture are not in conflict: As we noted earlier,
Gott's closed timelike loops are *not* so confined. Indeed, writing specifically
about Gott's analysis, Boulware (1992) states, "there is no singularity in the
stress energy [tensor on the Cauchy horizon] and, therefore, no mechanism
for chronology protection." This is still the source of some controversy, however, and Grant (1992) arrives at exactly the opposite conclusion! To the spatial-
infinity objection, Headrick and Gott (1994) made the following very strong
reply:

> [A problem] Deser *et al.* present with respect to the Gott
> spacetime is that it contains CTC's at spacelike infinity; this is
> supposed to be an unacceptable boundary condition. We
> wonder, however, how they know so much about boundary
> conditions at spacelike infinity. In our own Universe we do
> not know what spacelike infinity looks like (if it exists) since
> we have not seen it yet. We certainly have no way of knowing
> whether or not there are CTC's there. The working physicist
> is, of course, free to impose simple and convenient boundary
> conditions (e.g., asymptotic flatness) on a system in order to
> isolate and understand the processes occurring within it. *But
> boundary conditions are tools of physicists, and they should not be
> confused with laws of physics* [my emphasis]. There may be laws
> of nature that restrict the possible structure of spacelike in-
> finity, and even that prohibit CTC's there, but in the absence
> of evidence such laws should not be postulated *ad hoc*.

Still, as Ori (1993) observed, having closed time loops collapsing inward from infinity toward humans who might wish to use them is "a situation which has little to do with the creation of a time machine by a *human being* [my emphasis]." Cutler (1992) also shows that, unlike Gödel's spacetime, closed timelike loops are not present everywhere in Gott's spacetime; as mentioned earlier, there are regions in Gott's spacetime that can be visited only once and where time travel would therefore be impossible. Ori points out, however, that it is still an open question whether one can create closed timelike loops where none existed before by the process of *accelerating* two strings up to speed on a near-collision course. Gott himself raised this last point in his 1991 paper, when he wrote, "An advanced civilization could always in principle accelerate separate loops to high speed by towing them gravitationally with very massive rockets."

Two papers, Carroll *et al.* (1992, 1994), argue that Gott's "time machine" cannot be built in an open universe that has a timelike total momentum—that is, in a universe with subluminal net momentum. The authors claim to show this by demonstrating that Gott's cosmic strings, as a pair, have a total momentum that is tachyonic (even though each string, individually, is subluminal), which they call "an insurmountable obstacle to building a time machine." A similar conclusion is reached in Deser *et al.* (1992) and Kabat (1992). Headrick and Gott (1994) assert that these authors have simply made errors in their analyses and that the Gott cosmic string spacetime is *not* tachyonic.

The question of what happens in a closed universe is addressed in 't Hooft (1992), who shows that such a universe would collapse in a Big Crunch before closed timelike loops could form. These results were obtained under the assumption that all masses are non-spinning *point* particles in the toy spacetime of 2 + 1 gravity, a restriction generally taken to be unphysical. This restriction was removed in Menotti and Seminara (1993, 1994), and the conclusion that closed timelike loops would not be possible was extended to certain cases of non-moving (but rotating) *extended* masses.

The most damning objection to Gott's cosmic-string time machine comes, ironically, from Gott himself. The two-string spacetime, according to Li and Gott (1998), might actually be destabilized by the non-zero mass of any would-be time traveler. They suggest that this problem could perhaps be "solved" by assuming that the time traveler and her spaceship have a spherically symmetric mass distribution surrounded by a negative-mass shell to give a zero net mass (and thus a zero net gravitational field that would *not* destroy the closed timelike curves generated by the strings). But this, as they observe, negates the crucial advantage—no violation of the weak energy condition—that cosmic-string time machines have enjoyed over wormhole time machines.

Does a trip around a pair of cosmic strings present some problems *aside* from the sheer physics of the strings themselves? Well, I think "turning the rocket around at *F* and flying back to *S*" is a lot easier to write about than to do! The entire trip has to occur while the strings—moving at light speed—are in a position *to be* flown around. As one character says to another in Stephen Baxter's novel *Ring* "Louise, the strings are traveling just under the speed of light—within three decimal places of it, actually. [Our ship is] traveling at a little over half lightspeed. The turning curves, and the accelerations, are incredible. . . ." I should think so! And I do wonder who—or what—is actually controlling a *maneuvering* rocket traveling faster than $\frac{1}{2}c$?

Notes

1. An elementary, quite interesting discussion of the enormous computational complexity of the field equations is contained in the article by Richard Pavelle and Paul S. Wang. "MACSYMA from F to G," *Journal of Symbolic Computation* 1 (1985):69–100.

2. How does one know when something is "unphysical"? Intuition (also known as prejudice) is the first guide, but as Klinkhammer (1991) illustrates, it can be a poor guide: "At least twice in the past, physicists assumed without much proof that the stress-energy tensor was constrained in certain ways; they derived interesting constraints on spacetime curvature (gravity), and then, later, they discovered counterexamples to the stress-energy constraints and thereby lost their gravitational results." Klinkhammer then describes these two examples, one of which is the evolution of the energy conditions discussed in Tech Note 9.

3. The two strings in a Gott pair are not necessarily spinning, and no such assumption was made by Gott (1991). They don't even have to be parallel. If the strings have no spin, then it takes two strings to make a time machine. If spin is allowed, however, then just a single string will suffice for time travel—see Deser and Jackiw (1992).

4. Gott's paper appeared five months before Hiscock's, but it is evident that Hiscock's work was done before he became aware of Gott's. Both papers treat exact derivations of the deficit angle. The correct expression had actually been published four years earlier, but only as derived from a linearized form of the field equations—see Alexander Vilenkin, "Gravitational Field of Vacuum Domain Walls and Strings," *Physical Review D* 23 (February 15, 1981):852–857.

5. The deficit angle is equal to $8\pi\mu$ radians (in a system of units where G, Newton's gravitational constant, is 1) if the linear mass-energy density μ is expressed in units of Planck masses per Planck length. For example, $\mu = 1$ corresponds to 1.35×10^{28} g/cm (think of something on the order of the mass of Earth per inch of the string). For "more typical" values—say, a "mere" $\mu = 10^{22}$ g/cm, $2\alpha = 0.001°$.

6. Just as discussed in Tech Note 9 for wormholes, lensing may offer a way to detect strings; see the pair of papers by D.L. Ossipov, "Diffraction of Light by a Cosmic String," *JETP Letters* 62 (November 1995):765–771, and "Contribution of Strings to the Observed Variability of Extragalactic Sources of Radiation," *JETP Letters* 64 (September 1996):419–425.

7. For this to happen, however, v must be *very* close to the speed of light. Gott (1991) showed that with $v = \tanh(\theta)$, the condition for the rocket to arrive back at S before it leaves S is $\cosh(\theta)\sin(\alpha) > 1$, where α is one-half of the deficit angle. For $\mu = 10^{22}$ g/cm, this gives $v = 0.99999999995$ (times the speed of light).

11

Time and Gravity

If the clock were good enough [he told me], it could measure the effect of gravity on time—the red shift of the Earth's gravitational field. That gravity alters time seemed incredible to me. It still does.

> —Daniel Kleppner (M.I.T. professor of physics),
> remembering the words of his physics tutor when
> he was an undergraduate at Cambridge, England
> (from *Physics Today,* January 1994)

It is shown that the interaction of a wormhole with [a] gravitational field almost inevitably transforms it into a time machine.

> —Frolov and Novikov (1990)

Imagine a massive body in space. On the surface of this body a hot object emits electromagnetic radiation, and, though it isn't essential to the argument, imagine that the temperature of this object is sufficiently high that some of the radiation (emitted by the very atoms of the object) is in the visible-light portion of the spectrum. From elementary quantum theory we can also think of the object's atoms as emitting photons ("particles of light"), each of energy hf, where h is Planck's constant and f is the frequency in hertz (what used to be called cycles per second). The higher the temperature, the higher the photon energy and the higher the frequency. In the visible spectrum, f is on the order of 10^{15} Hz, a frequency one *billion* times higher than commercial broadcast AM radio frequencies.

The radiating atoms can be thought of as tiny clocks, with alternate half-cycles of radiation being ticks (the half-cycles in between are the tocks). The passage of time on the surface of the massive body can be measured by these atomic clocks in the hot, radiating object. To a *distant* observer, however, as she receives the photons from the hot object, this passage of surface time on the massive body will appear to occur at a reduced rate when compared with those photons emitted by her own identical hot object at the same temperature (the "local" clock). That is, the radiation that arrives at the distant observer, because it has traversed a gravitational field (a journey sometimes described as climbing out of a gravitational well), is shifted down in frequency toward the red end of the visible spectrum, and the effect is called either the gravitational red shift or the gravitational time dilation (or even the Einstein shift, because it was Einstein who first predicted the effect in 1907).

The gravitational red shift, in the "opposite" direction, is nicely described in Larry Niven's 1966 story "Neutron Star." A space traveler zooms down into a neutron star's intense gravity field (at half the speed of light!), passing within one mile of the star's surface. He observes, "All around me were blue-white stars. Imagine light falling into a savagely steep gravitational well. It won't accelerate. Light can't move faster than light. But it can gain in energy by increasing its frequency. The light was falling on me, harder and harder, as I dropped." To Niven's intrepid spaceman, the passage of time on those distance blue-white stars appears to be running *fast* compared to his local time. This shows that the effect could equally well be called the 'gravitational *blue* shift.' Note that gravitationally induced time alterations are *not* paradoxical in the same sense as is motion-induced time dilation. That is, for gravitational time dilation, observers located at each clock agree about which clock is running "slow" and which is "fast," unlike the motion-induced case where each of the relatively moving observers thinks it is the other observer's clock that is running slow. Science fictional discussions of both the red and the blue shift also appear in the novel *Out of Time* (Hogan).

In a qualitative way, one can think of a photon emitted by the hot object as something like a rock thrown upward. As the rock rises upward through the gravitational field, its *total* instantaneous energy is always constant, but this total, fixed energy is split between the kinetic and potential energies in an ever-changing way. That is, as the rock rises, its kinetic energy continually decreases (the rock slows down), whereas its potential energy continually increases. A photon is not a rock, however, and it certainly can't slow down as it rises up through a gravitational field (it must *always* move at the speed of light—the photon *is* light). The only way a photon can give up energy to balance the ever-increasing potential energy is to decrease its frequency. Hence the red shift as seen by a distant receiver of the photon.

We can understand this effect quantitatively by connecting three fundamental concepts. First, as mentioned at the start of this Tech Note, a photon of frequency f has energy $E = hf$. Second, from Einstein's famous mass-energy relation $E = mc^2$, we can associate with the photon an effective mass of

$$m = \frac{E}{c^2} = \frac{hf}{c^2}$$

Third, if a photon is in the gravitational field of a massive body, then via its effective mass, the photon can gain or lose (as would any other particle) potential energy.

To return to the physical situation at the start of this Tech Note, suppose the massive body is spherically symmetric, with radius R and mass M. Then, from Newton's theory of gravity,[1] we have the gravitational force F on a body of mass m, a distance $r > R$ from the center of the massive body, as

$$F = \frac{GMm}{r^2}$$

where G is the universal gravitational constant. Let W denote the potential energy of the mass m. We can calculate the differential increase in potential energy of the effective mass m of the photon as it moves from elevation r above the surface of the massive body to elevation $r + dr$:

$$dW = F \cdot dr = \frac{GMm}{(R + r)^2} dr$$

This differential increase in potential energy requires, because of conservation of energy, an equal differential *decrease* in the photon energy hf; that is,

$$h \cdot df = - \frac{GM(hf/c^2)}{(R + r)^2} dr$$

or

$$\frac{df}{f} = - \frac{GM}{(R + r)^2 c^2} dr$$

Integrating, with $f = f_0$ at $r = R$ (zero elevation) and $f = f_s$ at $r = R + s$ (an elevation of s), yields

$$\int_{f_0}^{f_s} \frac{df}{f} = - \frac{GM}{c^2} \int_{R}^{R+s} \frac{dr}{(R + r)^2}$$

or

$$\ln\left(\frac{f_s}{f_0}\right) = -\frac{GMs}{Rc^2(R+s)} = -\frac{GM}{R^2}\frac{s}{\left(1+\frac{s}{R}\right)c^2}$$

Thus

$$f_s = f_0\, e^{-\frac{GM}{R^2}\frac{s}{\left(1+\frac{s}{R}\right)c^2}}$$

We can put this in more convenient form, as a function of the surface gravity on the massive body, as follows. If a test mass m_T is on the surface, then we can set the inertial force it experiences equal to the gravitational force and write

$$m_T g = \frac{GMm_T}{R^2}$$

where g is the acceleration of gravity *at the surface*. That is,

$$g = \frac{GM}{R^2}$$

and so

$$f_s = f_0\, e^{-\frac{gs/c^2}{\left(1+\frac{s}{R}\right)}}$$

Because the exponential factor is less than 1 if $s > 0$, this result immediately tells us that an elevated observer, upon receiving the frequency down-shifted photon from the surface, would conclude that a clock on the surface of a massive body runs slower than her identical clock (elevated above the surface and so in a weaker gravity field).[2] That is, it would appear that the stronger gravity field present at the source of the emitted photon is "slowing time down." In the (usual) case of the exponent very small, we can approximate the exponential with just the first two terms of its power series expansion with negligible error:

$$f_s \approx f_0\left[1 - \frac{gs/c^2}{\left(1+\frac{s}{R}\right)}\right]$$

This is generally (but not always) a *very* small effect. It increases with increasing elevation, so let's consider first the case of $s >> R$, the "elevated clock at infinity" case. Then we have

$$\frac{f_s}{f_0} = 1 - \frac{gR}{c^2}$$

For the case of the surface clock on the surface of Earth, we use $g = 9.8$ m/sec^2, $R = 4000$ miles (6.44×10^6 m), and $c = 3 \times 10^8$ m/sec to give

$$\frac{f_s}{f_0} = 1 - 7 \times 10^{-10}$$

Thus if two identical clocks, one on the earth and one "at infinity," initially agree in time, then they will be just one second out of step after 45 years.

   This tiny frequency shift is even smaller if both clocks are close to the surface. In 1964, however, the minuscule difference in the rate of time was measured and found to agree with theory with less than 1% error. The experiment used photons traveling through a vertical distance of just 22.5 meters.[3] The experimental accuracy of this measurement was on the order of one part in a *million billion!* To see why, note that now we have $s < < R$ and so, with $s = 22.5$ meters,

$$\frac{f_s}{f_0} = 1 - \frac{gs}{c^2} = 1 - 2.45 \times 10^{-15}$$

The two identical clocks, if initially in agreement, will now require 13 *million* years to slip one second out of step.

   On Earth, then, the effect of gravity on time is seemingly nil. Over the entire age of Earth, for example, about four and a half billion years, the two clocks in the second calculation would be only about six minutes out of step. What if, however, we compare a clock on Earth with one on the *Sun?* That is, the formula for the "elevated clock at infinity" case will then let the Sun play the role of Earth (and so we should use the values for the Sun's radius and surface gravity), and the Earth-based clock will be the one "at infinity." This calculation assumes that Earth's gravity is negligible compared to the Sun's, and because the Sun's surface gravity is about 28 times greater, this seems a good approximation. Therefore, using $R = 6.95 \times 10^8$ m and $g = 272$ m/sec^2, we get

$$\frac{f_s}{f_0} = 1 - 2.1 \times 10^{-6}$$

Thus, over the age of Earth, the two clocks will be out of step by something like 9500 *years!* If, as in Gregory Benford's 1995 tale "A Worm in the Well," we someday discover a wormhole mouth in the Sun, and if we imagine that it has been there since Earth formed, then we will have a "time tunnel" through which one could travel 95 centuries into the past on each pass through the wormhole (less the time required to travel back to the entry mouth).

That last calculation is particularly stunning because the gravity of the Sun is, on the cosmic scale of things, pretty weak. In other regions of the universe, however, in the intense gravity fields of neutron stars and black holes, the effect of gravity on time can be far more dramatic. Gravitational time dilation is, in fact, the mechanism used by Frolov and Novikov (1990) to achieve a time shift between the two mouths of a wormhole time machine, as opposed to using time dilation by *moving* one of the mouths at high speed, as discussed in Tech Note 9.

Curiously, science fiction writers seem not to have used the gravity time dilation effect very much. Two well-known novels that do have high-gravity fields as a central feature are the 1954 classic *Mission of Gravity* by Hal Clement and the more recent (1980) *Dragon's Egg* by Robert L. Forward. The slowing of time by gravity, however, plays no role in either book. In Clement's novel the reasons are easy to understand; the gravity field on the giant planet Mesklin, though high by earthly standards (nearly 700 gees), would still produce only a tiny time dilation effect, and in 1954 these ideas were not in any case well known even inside the physics community. Forward's novel, on the other hand, is of intelligent life on a neutron star with a radius of 10 kilometers and a surface gravity of 67 *billion* gees. Putting these values into the result for an "elevated clock at infinity" gives

$$\frac{f_s}{f_0} = 1 - 0.07$$

A clock on Earth would run about 7% faster than one on the neutron star.[4] After just one year (as measured by the Earth clock), the neutron star clock would be more than 25 *days* behind the Earth clock. Forward actually has his intelligent beings living *faster* than earthlings, however. His explanation for this is interesting; life on a neutron star would be based not on chemical reactions (which are determined solely by weak interactions of electrons in the outer shells of atoms) but rather on much faster, *much* more energetic nuclear reactions. His beings live at the accelerated rate of more than one hundred years in an hour; that is, *biological* time runs a million times *faster* on the star than on Earth.

One story, however, *has* been written using the gravitational time dilation effect, and it is a beautiful, emotional story, indeed. Poul Anderson's 1969 tale "Kyrie" is about a starship's visit to a supernova, accompanied by a fantastic alien life-form—a ball of intelligent plasma that is telepathic. While the ship stands off at a distance of five hundred million kilometers, the plasma-being, named Lucifer, will approach much closer to the black hole at the center of the explosion and communicate its findings to a human telepath on the ship. A physicist in the crew is curious about one point:

> "I have wondered about an item. Presumably Lucifer will go quite near the supernova. Can you still maintain contact with him? The time dilation effect, will that not change the frequency of his thoughts too much?"

Lucifer, in fact, dies in the black hole even as he saves the ship from destruction, and the human telepath will hear (feel?) his death scream for the rest of her life. As the physicist explains to the captain, telepathy is instantaneous and has no limiting range (there is no known physical basis for believing any of this, but it is crucial for story effect):

> "Remember the time dilation. He fell from the sky and perished swiftly, yes. But in supernova time. Not the same as ours. To us, the final stellar collapse takes an infinite number of years. . . . He will always be with her."

A nice, illustrated discussion of how gravitational time dilation results in signals taking forever (even though they are emitted in a finite time) to travel from the event horizon of a black hole to a distant receiver is given by Thorne (1994).

Notes

1. This is, then, not a relativistically correct analysis. However, because Einstein's general theory reduces to Newton's theory in the case of low velocity and weak gravity, our results should be at least order of magnitude (or even better) if we stay away from extreme situations.

2. In *University of Chicago Graduate Problems in Physics: With solutions,* (ed. J.A. Cronin, D.F. Greenberg, and V.L. Telegdi (Reading, MA: Addison-Wesley, 1967), there is an interesting question with a frustating answer. In the mechanics section (p. 10, Question 22), we are asked, "If a watch is moved to a high altitude, does it run fast or slow?" From our discussion here, we would say (imperceptibly) fast. And indeed, in the answer section (p. 100) we learn that fast is the answer, but for an entirely different reason—the higher the altitude, the lower the air drag on the watch flywheel! There is not a word about gravity or about what the answer would be for a watch in a vacuum jar or for a watch *without* a flywheel. The editors tell an amusing wartime story about this problem that concerns two famous physicists, but I think the *real* lesson from the problem is how hard it can be for students to figure out what examination writers are thinking when they compose their questions.

3. R.V. Pound and J.L. Snider, "Effect of Gravity on Nuclear Resonance," *Physical Review Letters* 13 (November 2, 1964):539–540. An earlier (but less accurate) measurement was reported in 1960 by Pound and G.A. Rebka, Jr., "Apparent Weight of Photons," *Physical Review Letters* 4 (April 1, 1960):337–341. A later experiment involving a considerably greater distance, using a clock launched by rocket to an

elevation of 10^7 meters, was reported by Vessot *et. al.*, "Test of Relativistic Gravitation with a Space-Borne Hydrogen Maser," *Physical Review Letters* 45 (December 29, 1980):2081–2084. The history of the red shift, leading up to these experiments, is given by Klaus Hentschel, "Measurements of Gravitational Redshift Between 1959 and 1971," *Annals of Science* 53 (1996):269–295, and by Klaus Hentschel, *The Einstein Tower* (Stanford, CA: Stanford University Press, 1997). See also J. Dreitlein and J. Frazzini, "Aging in Gravitational Fields," *American Journal of Physics* 43 (July 1975):596–598.

4. The precise numerical value derived here for the difference in clock rates between Earth and the neutron star in Forward's novel shouldn't be accepted out of hand. Certainly, a gravity field of 67×10^9 gees (6.7×10^{11} m/sec^2) is not the "weak-field" approximation I made in deriving the formula for gravitational time dilation! Forgetting relativity and using Newton's theory, for example, we would calculate that a particle accelerating from rest in such a field would reach the speed of light in less than half a millisecond.

action: the integral over a **world line** of a quantity called the *Lagrangian*. When a massive particle is moving at non-relativistic speeds through a gravitational field, for example, the instantaneous value of the Lagrangian is the difference between the kinetic and potential energies of the particle. For other kinds of fields (such as the electromagnetic) and/or relativistic motion in any field, the Lagrangian is different. In any case, however, the actual world line of the particle is the one for which the integrated Lagrangian, i.e., the action, is minimized. See **least action.**

action at a distance: the direct interaction of two separated objects, without concern for the details of what (if anything) occurs in the region between the objects (see also **field**). Newton's theory of gravity is action at a distance, whereas Einstein's theory of gravity is a *field* theory.

advanced solution: the prediction by Maxwell's electromagnetic field equations of radio waves that travel into the past (see also **Dirac radio**).

anti-chronology protection conjecture: the assertion that there is no law of physics preventing the appearance of closed timelike curves (framed by Li) (see **chronology protection**).

anti-matter: quantum mechanical prediction (verified experimentally) that all fundamental particles of matter come in two versions (the "normal" version and the "anti-matter" version). The positron, for example, is the anti-particle version of the electron, differing only in the sign of its electric charge. The photon, on the other hand, is its own anti-particle. A **subluminal** anti-particle traveling forward in time can be thought of as its "normal" version traveling backward in time.

arbitrarily advanced civilization: a civilization with a technology sophisticated enough to construct a traversable **wormhole** in spacetime.

arrow of time: the statement that time appears to have a direction—that there is a difference between the past and the future. There are several different arrows: the psychological (we remember the past, we *anticipate* the future), the thermodynamic (organized systems evolve toward disorganization), the electromagnetic (radio waves propagate away from their sources), and the cosmological (the expansion of the universe is directed toward the future).

asymptotically flat: if the geometry of a curved spacetime is such that, as one moves ever farther from all matter and energy, the spacetime metric becomes that of flat Minkowski spacetime, then the curved spacetime is said to be *asymptotically flat.*

averaged null energy condition: the claim that the *averaged* value of the observed mass-energy density along the entirety of any **null geodesic** is not negative.

averaged weak energy condition: the claim that the averaged value of the observed mass-energy density along the entirety of any **timelike world line** is non-negative.

back reaction: the tendency of spacetime to resist the formation of **closed timelike lines** (see also **stress-energy divergence**).

Bell's theorem: an inequality that either holds or does not hold, depending on whether quantum mechanics is non-local or local, respectively.

Big Bang: the singular beginning of spacetime.

bilking paradox: what would happen if a causal loop were disrupted. For example, suppose a time traveler builds a time machine using plans he received years earlier from a mysterious stranger. He now realizes that the stranger was himself, using the time machine to travel back into the past to give his younger self the plans (see **causal loop**). A bilking paradox would be created if the time traveler built the time machine, verified that it works, and then decided *not* to visit his younger self to hand over the plans.

black hole: a region of spacetime where gravity is so strong that nothing can escape, including light. Black holes are formed when sufficiently massive stars burn out (see **white dwarf** and **neutron star**) and undergo *gravitational collapse.* A black hole of ten solar masses would have a radius of about twenty miles. Black holes might also have been created at the Big Bang, and if so, they theoretically could come in any mass and size; a black hole with the mass of the earth would have a diameter of less than half an inch.

block universe: a spacetime in which all world lines are completely determined, from beginning to end (a fatalistic universe). There is no free will in such a spacetime.

boost matrix: matrix formulation of the **Lorentz transformation.**

Cauchy horizon: a **Cauchy surface** at which **closed timelike lines** start to form.

Cauchy surface: a **spacelike** hypersurface that intersects exactly once every **timelike world line** that has no end point. Knowledge of the conditions on such a surface uniquely determines the spacetime at all other points. See also **global hyperbolicity.**

causal loop: a time loop containing an event caused by a *later* event that, itself, is caused by the earlier event (see the example in **bilking paradox**).

causality: the metaphysical claim that every event is caused by a *prior* event. Time travel to the past inherently violates causality.

chronal regions: those parts of spacetime that have no closed timelike curves.

chronology horizon: a (hyper)surface in spacetime that separates **chronal** and non-chronal regions. It is a special case of a **Cauchy horizon.**

chronology protection: the claim—as yet unproved—that time machines and time travel to the past are impossible because of the **back reaction** of spacetime that will lead to **stress-energy divergence.** Popularized among physicists as the *Hawking chronology protection conjecture* (1992), but it was actually stated earlier (1971) in somewhat different form by science fiction writer Larry Niven as "Niven's Law." Hawking has now admitted that stress-energy divergence is <u>not</u> sufficient to enforce his conjecture, and the Chinese physicist Li-Xin Li has denied the very truth of the conjecture with his **anti-chronology protection conjecture.**

chronon: science fiction name for **Planck time.**

closed timelike line (or curve): a **timelike world line** of finite length that has no ends, i.e., that forms a *closed loop* in spacetime. A region of spacetime containing closed timelike lines is said to be a **time machine.**

conservation law: physical quantities in interacting systems that remain unchanged are said to be conserved. Total energy and total momentum (linear and angular) are conserved quantities.

conservation of probability: the law that says the sum of the probabilities of a system evolving from any given initial state to all possible final states must be equal to unity. The law fails in the case of quantum fields in curved spacetime, and this failure is called a violation of **unitarity.**

cosmic string: hypothetical, threadlike spacetime structures with enormous mass-energy and density that may have been formed during the Big Bang. Cosmic strings may have been initially formed either as infinitely long or as closed loops, and it is the latter that are thought to be physically meaningful in the present-day universe. Cosmic strings do not violate the weak energy condition (as do wormholes), and they can theoretically create **closed timelike lines.**

cosmological constant: an extra term specifically added by Einstein to the general theory of relativity to keep that theory from predicting the expansion of the universe (which was later observationally confirmed). Einstein subsequently said that his failure to believe the general theory's original prediction of the expansion of the universe was the greatest mistake of his life. The constant (which is today believed to be almost zero, if not exactly zero) appears in Gödel's rotating time travel spacetime as a determining factor in the minimum radius of a closed timelike line.

determinism: the metaphysical belief that effects are uniquely determined by causes (which is *not* **fatalism**).

Dirac radio: science fiction gadget for sending information at infinite speed, which thus travels backward in time (see also **ultraluminal**).

dominant energy condition: the **weak energy condition plus** the claim that the observed energy flux is never **superluminal.**

electron: fundamental particle of mass (carrying one quantum of negative electric charge) that orbits the nuclei of atoms and that plays a central role in determining the chemical properties of the elements and their compounds.

elsewhen: the collection of spacetime events that cannot be reached from the here-now with a timelike world line.

entropy: a measure of the randomness of a system. Entropy plays the central role in the thermodynamic arrow of time.

ether: a substance once thought to fill all of space to allow radiation "something to propagate through" (as opposed to simply a vacuum). The special theory of relativity showed that the ether is an unnecessary concept because it has no observable effects (physicists argue that if something is impossible to detect, then it is meaningless to talk about it as being part of *science*).

event: a point in spacetime.

event horizon: the spacetime surface of a black hole or of a non-traversable wormhole, at which light can *just* escape to the outside universe. It is called a *horizon* because, by definition, an external observer can't see beyond it and into the interior of the hole. To see the inside of a hole you must enter the hole by crossing the horizon (but then you can't get out).

exotic matter: matter that violates one or both of the weak/strong energy conditions.

fatalism: the metaphysical belief that all events have been *pre*determined from the beginning of time.

field: the concept that if a physical law is **local,** then it is describable by differential equations that relate what is "happening" at every point in spacetime to what is "happening" at its closely located neighboring points. Electromagnetism is a field theory, for example, described by a set of partial differential equations called *Maxwell's equations.*

fourth dimension: either time or a *fourth* spatial dimension.

frame of reference: a spacetime coordinate system.

free will: the condition that prevails when we can choose to do what we do. There is no free will in a block universe.

future: the collection of spacetime events that can be reached from the here-now via timelike world lines directed toward a *later* time (for each individual, the future is what hasn't yet been experienced).

gamma ray: very high-energy, very high-frequency electromagnetic radiation. Gamma rays have frequencies on the order of *ten trillion* (10^{13}) times greater than those of AM broadcast radio waves.

general theory of relativity: Einstein's theory of curved spacetime, which explains gravity in terms of nothing but geometry. Its fundamental premise is that *all* the laws of physics should appear the same to all observers in *any* reference frame. It is believed the theory will fail when the local mass-energy density reaches a level of about 10^{94} grams/cm^3, a density so enormous (the density of water is just 1 gram/cm^3) that there is no known mechanism for achieving it anywhere in the universe except in another **Big Bang.** See also **Planck density.**

geodesic: the shortest path connecting two points in space. In spacetime, the world line of a particle in free-fall.

global: in the large.

global hyperbolicity: the mathematical form of a spacetime that obeys causality, i.e., contains no **closed timelike lines.**

Gödel universe: a spacetime that (unlike the one we live in) is rotating so fast that it automatically generates **closed timelike lines** and thus constitutes a *weak* time machine (see **time machine**). In such a universe, time travel to the past would be a natural phenomenon.

grandfather paradox: *the* classic time travel paradox, of a time traveler killing, while in the past and *before* the time traveler has been conceived, an ancestor directly linked to the future birth of the time traveler. A time traveler simply killing his own younger self is the most direct form of this type of paradox.

gravitational field equations: the set of coupled partial differential equations (non-linear, tensor) that *are* the general theory of relativity. These are considered to be the most complicated equations in all of mathematical physics. They show how the local curvature of spacetime depends on the local mass-energy of spacetime. The equations are independent of the **topology** of spacetime.

gravitational lensing: the ability of gravitational fields to bend and focus light.

graviton: the quantum particle of gravity.

Hawking radiation: the emission of particles (energy) by a black hole into the region *outside* its **event horizon,** which results in the eventual *evaporation* of the hole! This is a quantum mechanical effect.

here-now: the point or **event** in spacetime that separates the past, the future, and elsewhen.

hyperspace: any space of four or more dimensions (for example, four-dimensional spacetime is a hyperspace).

inertial frame: any frame of reference in which Newton's laws of mechanics are true (there are no *acceleration forces* in inertial frames, and so *rotating,* or "merry-go-round," frames are *not* inertial).

invariance: a quantity that remains the same in any frame of reference is an invariant. Two examples are the distance between any two points on a piece of paper (because it is independent of any particular coordinate system) and the speed of light (which is the same for all observers).

kaon: a particle that violates the **TCP theorem.**

Kerr–Newman black hole: a rotating black hole. It may or may not be electrically charged.

Klein–Gordon equation: differential equation for the wave propagation of particles with zero **spin,** such as **tachyons** and **kaons.** It is relativistically correct.

Krasnikov tube: a particular spacetime **metric** (or "warp") allowing **superluminal** travel, with the great difficulty of requiring *enormous* negative energy. Two Krasnikov tubes can be made into a time machine. Named after its Russian inventor.

least action: general principle in physics that asserts that the **world line** of a particle is the one that minimizes the **action.**

light cone: the lightlike surface in spacetime that, at each point in spacetime, separates the past from the future from elsewhen from the here-now.

lightlike: the world line of a photon (or of any other form of mass-energy traveling *at* the speed of light).

Li mirror: a perfectly reflecting, spherical surface that can be used to stabilize a **wormhole** against energy loops endlessly circulating through a wormhole time machine (thus creating unbounded energy levels that destroy the wormhole). Named after its Chinese inventor.

local: in the small.

Lorentz factor: the ubiquitous square-root expression that appears in many relativistic calculations, such as **time dilation,** length contraction, and the variation of mass with speed. For example, the mass m of a moving body is not independent of its speed v but rather varies as

$$ m = \frac{m_0}{\sqrt{1 - (v/c)^2}} , $$

where m_0 is the rest mass (that is, the mass when $v = 0$) and where c denotes the speed of light (186,200 miles per second). The denominator is the Lorentz factor.

Lorentz–FitzGerald contraction: the conclusion from special relativity that the appearance (to a stationary observer) of a moving object will be shortened in length along the direction of motion. Many years after Einstein's work, it was shown that the object will also appear to be *rotated.*

Lorentz transformation: equations from the special theory of relativity that describe how the space and time measurements of two relatively moving observers are related.

many-worlds interpretation: quantum mechanical view of splitting universes.

mass-energy: The famous $E = mc^2$, the equation behind atomic fission and nuclear fusion weapons.

metric: the *measure* of the separation between any two events in a spacetime.

mind travel: time travel of the consciousness *without* being accompanied by the body or any other physical artifact.

Minkowski spacetime: the flat spacetime of the special theory of relativity. In this spacetime there is no gravity, no spacetime curvature (hence it is *flat*), and no backward time travel.

neutron star: the end state of a star with one to three solar masses and a density of up to 10^{15} grams/cm^3.

non-Euclidean geometry: the geometry of spacetime, whether curved or flat. Spacetime is non-intuitive precisely because it is always hard to resist thinking in terms of high school Euclidean geometry, which is simply the *wrong* geometry.

null geodesic: the world line of a photon in spacetime.

observer: physicist's term for "somebody" equipped with recording instruments (such as a clock, a pencil and notepad, and the like).

parallel transport: a procedure for moving a vector around any closed curve in a space to determine whether that space is flat or curved.

parallel worlds: simultaneous existence of multiple (perhaps infinite) versions of reality in hyperspace.

past: the collection of spacetime events that can reach the here-now via time-like world lines directed from an earlier time (for each individual, the past is what has already been experienced).

photon: the quantum particle of electromagnetism. A photon of frequency f has energy hf, where h is **Planck's constant.**

Planck density: the density of mass-energy that distinguishes classical and quantum spacetimes; about 10^{94} grams/cm^3, equal to the **Planck mass** divided by the cube of the **Planck length.**

Planck length: the non-zero length in quantum theory (about 1.6×10^{-33} centimeter) below which quantum gravity effects will become important.

Planck mass: the fundamental mass in quantum theory (about 22×10^{-6} gram), but *not* the smallest non-zero mass in quantum theory.

Planck's constant: fundamental constant in quantum theory, h, associated with the discrete nature of quantum effects. (If h had the value zero, rather than its actual value of about 6.6×10^{-34} joule-second, then even the microworld would appear continuous).

Planck time: the time interval in quantum theory (about 5.3×10^{-44} second) below which quantum gravity effects become important. The time required to travel the **Planck length** at the speed of light.

positron: the electron's anti-particle (see **anti-matter**).

potential function: a function whose first derivatives make possible the construction of **fields.**

proper time: the timekeeping of an observer's clock.

pulps: the old science fiction magazines, through the 1940s, published on inexpensive, wood-pulp paper.

quantum coherence: the preservation over time of the state description of a quantum system.

quantum foam: see **topology.**

quantum gravity: the yet-to-be-discovered theory that unifies quantum field theory with the curved spacetime of general relativity.

quantum mechanics: the exact physics of the very small, i.e., of atoms and things smaller.

quantum theory: any theory in which physical quantities are not continuous but rather assume their values in discrete jumps (the size of the jump is the *quantum*).

recurrence paradox: the claim that if you wait long enough, then every system will return to every previous state infinitely often.

red shift: the *down* shift in the frequency of light received from all distant stars due to the Doppler effect induced by the expansion of the universe. The opposite effect is called a *blue shift*.

reinterpretation principle: asserts that negative mass-energy traveling forward in time is positive mass-energy traveling backward in time, and vice versa.

Reissner–Nordström black hole: a spherically symmetric, non-rotating electrically charged black hole.

reversibility paradox: the equations of physics contain no **arrow of time;** that is, they work equally well with time running forward or backward.

Roman ring: a time machine made of two or more traversable wormholes connected in a closed sequence.

scalar field: a field describable by a scalar **potential function,** such as the fields of a spin-0 particle such as the massive **kaon** and **tachyon.**

Schwarzschild black hole: a spherically symmetric, non-rotating black hole. It may or may not be electrically charged.

self-consistency: the assertion that the events on a **closed timelike line** must never be in contradiction. Generally attributed to the Russian physicist **Igor Novikov,** who with his colleagues has shown that it is not an independent assumption but rather an implication of the **principle of least action.**

sexual paradox: a special type of **causal loop,** where the connected events on a time loop are "coupled" (pun intended!) through reproductive sex. An example is a time traveler to the past who becomes her own ancestor.

singularity: either a region in spacetime where the curvature becomes infinite and the laws of physics fail or a point in spacetime beyond which world lines cannot be extended. Singularities of the first kind are called *curvature* or *crushing* singularities, and those of the second kind are called *incomplete* singularities. The **Big Bang** was a curvature singularity, as is the center of a black hole. In a **Schwarzschild black hole** the curvature singularity is a point, whereas in a **Kerr–Newman black hole** it is an extended region in the form of a ring.

spacelike: a world line on which propagating mass-energy would exceed the speed of light.

spacetime: the stuff out of which reality is built. Everything there is—the *universe*—is the total collection of events in spacetime. A *flat* spacetime has no gravity, whereas a curved spacetime is the *origin* of gravity.

special theory of relativity: Einstein's theory of *flat* spacetime, which assumes that gravity doesn't exist (gravity is the result of the geometry of *curved* spacetime). Its fundamental premise is that the laws of physics should appear the same to observers in different inertial frames.

spin: a measure of the intrinsic angular momentum of a particle, distinct from any orbital angular momentum it may also possess.

splitting universes: the idea that every decision causes reality to split into separate copies, identical in every respect except for each of the different possible results of the decision.

stargate: science fiction name for the mouth of a traversable **wormhole.**

stress-energy divergence: the unbounded growth of general relativity's measure of the density of mass-energy in spacetime.

strong energy condition: the claim that gravity is always (i.e., locally) attractive. **Wormholes** violate this condition.

subluminal: slower than light.

superluminal: faster than light.

tachyon: a particle (hypothetical, so far) that always travels faster than light, so its **world line** is always spacelike.

TCP theorem: the statement that a possible physical process transforms into another possible physical process under reversals of time, charge, and parity (the distinction between "left" and "right") See also **kaon.**

temporally orientable spacetime: any spacetime in which the direction of time at every point agrees with the direction of time at its local neighboring points.

tensor: mathematical generalization of the scalar and vector concepts. Einstein's gravitational field equations are tensor-differential equations (for example, the *metric* tensor contains information about the curvature of spacetime), whereas Newton's and Maxwell's equations are vastly simpler, scalar and vector-differential equations, respectively.

tensor field: a field describable by a tensor **potential function,** such as the field of a spin-2 particle like the massless **graviton.**

tidal force: force experienced by a non-point mass (one with spatial extension) in a non-uniform gravitational field. Such forces tend simultaneously to compress and stretch extended masses. Black holes and wormhole mouths can generate enormous tidal forces on extended masses as small as a human body. Interestingly, the *more* massive a black hole, the *less* severe its tidal forces at distances outside the **event horizon.** However, no matter what the black hole mass is, the tidal forces are infinite at the central curvature **singularity.**

time dilation: the altering of the rate of timekeeping by a clock, either by motion or by gravity.

timelike: a **world line** on which propagating mass-energy always travels more slowly than light.

time machine: (in the weak sense) a machine able to traverse **closed timelike world lines** inherent in a spacetime (e.g., a rocket in Gödel spacetime) but unable to create such world lines; (in the strong sense) a machine able to manipulate mass-energy in a finite or *compact* region of spacetime in such a way as to *create* closed timelike world lines.

time police: story characters in science fiction charged with the (unnecessary!) job of preventing time travelers from changing the past.

timewarp: science fiction name for a time machine.

Tipler cylinder: an infinitely long cylinder, made of super-dense matter, rotating so fast around its long axis that it warps spacetime enough to create closed timelike lines that encircle the cylinder. Can be used as a time machine to travel both into the future and into the past (but *not* to a time before the creation of the cylinder).

topology: the structure of a *space* (including *spacetime*) without regard to a metric. That is, topology is concerned only with how a space is connected together and not with how far apart points in the space are. Topologists consider stretching or compressing a space to be irrelevant, just as long as one doesn't *tear* it and so put holes in the space. The simplest topology is that of a *simply connected* space, in which if you construct any closed surface that lies totally in the space around any point in the space, then every other point inside the surface is also in the space. A space with a hole in it obviously fails this test and so is said to be *multiply connected*. A *quantum foam* spacetime has a multiply connected topology. The classical spacetime of general relativity is simply connected *until* the appearance of wormholes.

twin paradox: the conclusion from special relativity that a clock's rate of time-keeping slows with motion.

2+1 gravity: the Einstein gravitational field equations in a spacetime with one spatial dimension suppressed, leaving a spacetime with two spatial dimensions and one time dimension; also called a *toy spacetime*.

ultraluminal: motion sufficiently **superluminal** that mass-energy appears to travel backward in time (see also **Dirac radio**).

uncertainty principle: the statement in quantum mechanics that says certain pairs of quantities cannot simultaneously be measured with arbitrarily small error. Location and momentum are one such pair; energy and time are another.

unitarity: see **conservation of probability.**

vacuum fluctuation: the particle/anti-particle creation and annihilation processes allowed even in so-called empty space by the uncertainty principle of quantum mechanics.

vector field: a field describable by a vector **potential function,** such as the field of a spin-1 particle like the massless photon.

warp drive: science fiction name for the propulsion mechanism of a faster-than-light spaceship.

wave function: mathematical entity in quantum mechanics that measures the probability that a system is in any particular state.

weak energy condition: the claim that observed mass-energy density is always (i.e., locally) non-negative. Quantum mechanics predicts (and it has been experimentally confirmed) that there are exceptions.

white dwarf: a burnt-out star with less than 1.4 solar masses, of planetary size with a density up to 10^7 g/cm^3. The ultimate fate of our sun.

world line: the trajectory of mass-energy in spacetime.

wormhole: a spacetime structure (violating both energy conditions, weak and strong, if traversable) connecting two points of the same spacetime (or even two *different* spacetimes, i.e., two different universes), with a timelike path that requires less time to travel along than does a photon traveling between the two points *outside* of the wormhole. A wormhole is *traversable* if it has no event horizons, and such wormholes can apparently be made into a time machine, sometimes called a *time tunnel*—see **time machine (strong)**—by creating a time shift (using **time dilation**) between the two mouths of the wormhole *unless* quantum effects forbid time machines (still an open issue).

BIBLIOGRAPHY AND FURTHER READING

You will find it a very good practice always to verify your references, sir.

—advice given in 1847 to a young scholar by Martin Joseph Routh, president of Magdalen College, Oxford

Anthology Codes

AHT: *The Arbor House Treasury of Great Science Fiction Short Novels.* New York: Arbor House, 1980.

AM: *The Air of Mars.* Edited by M. Ginsburg. New York: Macmillan, 1976.

AO: *Alpha One.* Edited by R. Silverberg. New York: Ballantine, 1970.

ASF: *Australian Science Fiction.* Edited by V. Ikin. Chicago: Academy Chicago, 1984.

B51:   *The Best Science Fiction Stories: 1951.* Edited by E.F. Bleiler and T.E. Dikty. New York: Frederick Fell, 1951.

BFSF2:   *The Best from Fantasy and Science Fiction.* Vol. 2. Boston: Little, Brown, 1953.

BFSF3:   *The Best from Fantasy and Science Fiction.* Vol. 3. Garden City, NY: Doubleday, 1954.

BGA1:   *Before the Golden Age.* Edited by I. Asimov. Vol. 1. New York: Doubleday, 1974.

BGA2:   *Before the Golden Age.* Edited by I. Asimov. Vol. 2. New York: Doubleday, 1974.

BGA3:   *Before the Golden Age.* Edited by I. Asimov. Vol. 3. New York: Doubleday, 1974.

BIPT: *Basic Issues in the Philosophy of Time.* Edited by E. Freeman and W. Sellars. LaSalle, IL: Open Court, 1971.

BML: *The Best of Murray Leinster.* Edited by B. Davis. London: Corgi, 1976.

BO: *The Best from Orbit.* Edited by D. Knight. New York: Berkley, 1975.

BRW: *The Best from the Rest of the World.* Edited by D.A. Wollheim. Garden City, NY: Doubleday, 1976.

BSBSF: *The Bank Street Book of Science Fiction.* New York: Pocket, 1989.

BSF: *The Best of Science Fiction.* Edited by G. Conklin. New York: Crown, 1946.

BST: *Boston Studies in the Philosophy of Science.* Vol. 22. Boston: D. Reidel, 1976.

BT: *Beyond Time.* Edited by S. Ley. New York: Pocket, 1976.

BTS: *Beyond Time and Space.* Edited by A. Derleth. New York: Pellegrini and Cudahy, 1950.

BUFP: *Birth of the Universe and Fundamental Physics.* Edited by F. Occhionero. New York: Springer, 1995.

CA: *Coming Attractions.* Edited by M. Greenberg. New York: Gnome, 1957.

CBSF: *The Classic Book of Science Fiction.* Edited by G. Conklin. New York: Bonanza, 1982.

CJSF: *Criminal Justice Through Science Fiction.* Edited by J.D. Olander and M.H. Greenberg. New York: New Viewpoints, 1977.

CPP: *Cosmology, Physics and Philosophy,* by B. Gal-Or. New York: Springer-Verlag, 1981.

CPT: *Causality and Physical Theories.* Edited by W.B. Rolnick. New York: American Institute of Physics, 1974.

D: *Dinosaurs!* Edited by J. Dann and G. Dozois. New York: Ace, 1990.

DD: *Deals with the Devil.* Edited by B. Davenport. New York: Dodd, Mead, 1958.

DF: *Dinosaur Fantastic.* Edited by M. Resnik and M.H. Greenberg. New York: DAW, 1993.

DGR1: *Directions in General Relativity.* Vol. 1. Edited by B.L. Hu, M.P. Ryan, Jr., and C.V. Vishveshwara. Cambridge, England: Cambridge University Press, 1993.

DV: *Dangerous Visions.* Edited by H. Ellison. New York: Doubleday, 1967.

E: *Epoch.* Edited by R. Silverberg and R. Elwood. New York: Berkley, 1975.

ED: *The Expert Dreamers.* Edited by F. Pohl. Garden City, NY: Doubleday, 1962.

ET: *The Enigma of Time.* Edited by P.T. Landsberg. Bristol: Adam Hilger, 1982.

FCW: *The Fantastic Civil War.* Edited by F. McSherry, Jr. New York: Baen, 1991.

FM: *Fantasia Mathematica.* Edited by C. Fadiman. New York: Simon and Schuster, 1958.

FS: *Full Spectrum.* Edited by L. Aronica and S. McCarthy. New York: Bantam, 1988.

FSFS: *Famous Science-Fiction Stories.* Edited by R.J. Healy and J.F. McComas. New York: Random House, 1957.

FST: *The Far Side of Time.* Edited by R. Elwood. New York: Dodd, Mead, 1974.

FTI: *Futures to Infinity.* Edited by S. Moskowitz. New York: Pyramid, 1970.

FTL: *Faster Than Light.* Edited by J. Dann and G. Zebrowski. New York: Harper & Row, 1976.

GG: *Grumbles from the Grave.* Edited by V. Heinlein. New York: Del Rey, 1990.

GRAV: *Gravitation: An Introduction to Current Research.* Edited by L. Witter. New York: John Wiley, 1962.

GRSF7: *The Seventh Galaxy Reader of Science Fiction.* Edited by F. Pohl. Garden City, NY: Doubleday, 1964.

GRSF2: *The Second Galaxy Reader of Science Fiction.* Edited by H.L. Gold. New York: Crown, 1954.

GRSF1: *The Galaxy Reader of Science Fiction.* Edited by H.L. Gold. New York: Crown, 1952.

GSF: *Great Science Fiction Stories by the World's Great Scientists.* Edited by L. Asimov, M.H. Greenberg, and C.G. Waugh. New York: Donald I. Fine, 1985.

GSFS: *Great Science Fiction by Scientists.* Edited by G. Conklin. New York: Collier, 1962.

GSSF: *Great Stories of Science Fiction.* Edited by M. Leinster. New York: Random House, 1951.

HJ: *Hibbert Journal.* Vol. 37. 1938–1939.

HV: *Hitler Victorious.* Edited by G. Benford and M.H. Greenberg. New York: Berkley, 1987.

LDA: *Last Door to Aiya.* Edited by M. Ginsburg. New York: S.G. Phillips, 1968.

LME: *The Last Man on Earth.* Edited by I. Asimov, M. Greenberg, and C. Waugh. New York: Ballantine, 1982.

MBSFS: *My Best Science Fiction Story.* Edited by L. Margulies and O.J. Friend. New York: Merlin, 1949.

MDT: *Modern Developments in Thermodynamics.* Edited by B. Gal-Or. New York: John Wiley, 1974.

MI: *The Mirror of Infinity.* Edited by R. Silverberg. San Francisco: Canfield, 1970.

MIAM: *Maps in a Mirror: The Short Fiction of Orson Sott Card.* New York: Tor, 1990.

MM: *The Mathematical Magpie.* Edited by C. Fadiman. New York: Simon and Schuster, 1962.

MT: *Microcosmic Tales.* Edited by I. Asimov, M. Greenberg, and J. Olander. New York: Taplinger, 1980.

MTMW: *MATHENAUTS, Tales of Mathematical Wonder.* Edited by R. Rucker. New York: Arbor, 1987.

MWM: *Magic Without Magic: John Archibald Wheeler.* Edited by J.R. Klauder. San Francisco: W.H. Freeman, 1972.

NOT: *The Nature of Time.* Edited by R. Flood and M. Lockwood. Oxford, England: Basil Blackwell, 1986.

NP: *The New Physics.* Edited by P. Davies. New York: Cambridge University Press, 1989.

NT: *The Nature of Time.* Edited by T. Gold. Ithaca, NY: Cornell University Press, 1966.

NWF: *New Worlds of Fantasy.* Edited by T. Carr. New York: Ace, 1967.

NYAS: *Annals of the New York Academy of Science* 138. February 6, 1967.

OSF: *Omnibus of Science Fiction.* Edited by G. Conklin. New York: Crown, 1952.

OW: *The Other Worlds.* Edited by P. Stong. Garden City, NY: Garden City Publishing, 1941.

PSF: *Prize Science Fiction.* Edited by D.A. Wollheim. New York: McBride, 1953.

PWO: *The Penguin World Omnibus of Science Fiction.* Edited by B. Aldiss and S.J. Lundwall. Middlesex, England: Penguin, 1986.

QCST: *Quantum Concepts in Space and Time.* Edited by R. Penrose and C.J. Isham. New York: Oxford University Press, 1986.

RAG: *Relativity and Gravitation.* Edited by C.G. Kuper and A. Peres. New York: Gordon and Breach Science Publishers, 1971.

RRSF: *Pre-Revolutionary Russian Science Fiction.* Edited by L. Fetzer. Ann Arbor, MI: Ardis, 1982.

RTR: *Reduction, Time and Reality.* Edited by R. Healey. Cambridge, England: Cambridge University Press, 1981.

S1:   *Stellar 1.* Edited by J.-L. del Rey. New York: Ballantine, 1974.

SFAD: *Science-Fiction Adventures in Dimension.* Edited by G. Conklin. New York: Vanguard, 1953.

SFD: *The Science Fictional Dinosaur.* Edited by R. Silverberg, C. Waugh, and M. Greenberg. New York: Avon, 1982.

SFF: *Science Fiction of the 40's.* Edited by F. Pohl. New York: Avon, 1978.

SFS: *101 Science Fiction Stories.* Edited by M. Greenberg and C. Waugh. New York: Avenel, 1986.

SFSSS: *100 Great Science Fiction Short Short Stories.* Edited by I. Asimov, M. Greenberg, and J. Olander. New York: Avon, 1978.

SFT: *Science Fiction of the 30's.* Edited by D. Knight. New York: Avon, 1975.

SS: *Starships.* Edited by I. Asimov, M.H. Greenberg, and C.G. Waugh. New York: Ballantine, 1983.

SSFT: *50 Short Science Fiction Tales.* Edited by I. Asimov and G. Conklin. New York: Macmillan, 1976.

TC: *The Time Curve.* Edited by S. Moskowitz and R. Elwood. New York: Tower, 1968.

T3:   *The Time Travelers.* Edited by R. Silverberg and M.H. Greenberg. New York: Primus, 1985.

TIT: *Trips in Time.* Edited by R. Silverberg. New York: Thomas Nelson, 1977.

TOT: *Tales Out of Time.* Edited by B. Ireson. New York: Philomel, 1981.

TPR: *The Principle of Relativity.* Notes by A. Sommerfeld. New York: Dover, 1952.

TR: *Time Reversal—The Arthur Rich Memorial Symposium* (AIP Conference Proceedings 270). New York: American Institute of Physics, 1993.

TSF: *Treasury of Science Fiction.* Edited by G. Conklin. New York: Bonanza, 1980.

TT: *Time Travelers.* Edited by G. Dozois. New York: Ace, 1989.

TTF   *Time Travelers: Fiction in the Fourth Dimension.* Edited by P. Haining. New York: Barnes & Noble, 1998.

TTT: *The Traps of Time.* Edited by M. Moorcock. Middlesex, England: Penguin, 1970.

TW: *Time Wars.* Edited by C. Waugh and M.H. Greenberg. New York: Tor, 1986.

TZ: *The Twilight Zone Companion,* by M.S. Zicree. New York: Bantam, 1982.

U2:   *Universe 2.* Edited by T. Carr. New York: Ace, 1972.

VIT: *Voyagers in Time.* Edited by R. Silverberg. New York: Meredith Press, 1967.

VOT: *The Voices of Time.* Edited by J.T. Fraser. New York. George Braziller, 1966.

WMHB: *What Might Have Been*. Edited by G. Benford and M.H. Greenberg. New York: Bantam, 1989.

WT: *Words of Tomorrow*. Edited by A. Derleth. New York: Pellegrini & Cudahy, 1953.

Authors

Abbott, E.A. (1986). *Flatland*. New York: Penguin.

Abernathy, R. (OSF). "Heritage."

Abramenko, B. (1958). "On Dimensionality and Continuity of Physical Space and Time." *British Journal for the Philosophy of Science* 9 (August):89–109.

Abramowicz, M.A., and J.P. Lasota (1986). "On Traveling Round Without Feeling It and Uncurving Curves." *American Journal of Physics* 54 (October):936–939.

Adams, F.C., and G. Laughlin (1997). "A Dying Universe: The Long-Term Fate and Evolution of Astrophysical Objects." *Reviews of Modern Physics* 69 (April):337–372.

Adams, R.M. (1997). "Thisness and Time Travel." *Philosophia* 25 (April):407–415.

Aharonov, Y., J. Anandan, S. Popescu, and L. Vaidman (1990). "Superpositions of Time Evolutions of a Quantum System and a Quantum Time-Translation Machine." *Physical Review Letters* 64 (June 18):2965–2968.

Aharony, A., and Y. Ne'eman (1970). "Time-Reversal Symmetry Violation and the Oscillating Universe." *International Journal of Theoretical Physics* 3 (December):437–441.

Ahern, D.M. (1979). "Foreknowledge: Nelson Pike and Newcomb's Problem." *Religious Studies* 15 (December):475–490.

—— (1977). "Miracles and Physical Impossibility." *Canadian Journal of Philosophy* 7 (March):71–79.

Aichelburg, P.C., and R. Beig (1976). "Radiation Damping as an Initial Value Problem." *Annals of Physics* 98 (May):264–283.

Aichelburg, P.C., and F. Schein (1998). "Wormholes and Timetravel." *Acta Physica Polonica B* 29 (April):1025–1032.

Albert, D. (1986). "How to Take a Photograph of Another Everett World." *Annals of the New York Academy of Sciences* 480 (December 30):498–502.

Albrow, M.G. (1973). "CPT Conservation in the Oscillating Model of the Universe." *Nature Physical Science* 241 (January 15):56–57.

Alcubierre, M. (1994). "The Warp Drive: Hyperfast Travel Within General Relativity." *Classical and Quantum Gravity* 11 (May):L73–L77.

Aldiss, B.W. (1991). *Dracula Unbound*. New York: HarperCollins.

—— (1989). "The Failed Men." In *Man in His Time*. New York: Atheneum.

—— (SFD). "Poor Little Warrior!"

—— (1973). *Frankenstein Unbound*. New York: Random House.

—— (TTT). "Man in His Time."

—— (1967). *An Age*. London: Faber and Faber.

—— (1966). "T." In *First Flight*. Edited by D. Knight. New York: Lancer.

Alexander, L. (1963). *Time Cat*. New York: Holt, Rinehart, and Winston.

Alkon, P.K. (1987). *Origins of Futuristic Fiction*. Athens, GA: University of Georgia.

Allen, B., and J. Simon (1992). "Time Travel on a String." *Nature* 357 (May 7):19–21.

Allen, G. (BTS). "Pausodyne."

—— (1895). *British Barbarians*. London: John Lane.

Alvager, T., and M.N. Kreisler. (1968). Quest for Faster-Than-Light Particles." *Physical Review* 171 (July 25):1357–1361.

Amis, K. (1960). *New Maps of Hell*. New York: Harcourt.

Amis, M. (1991). *Time's Arrow*. New York: Crown.

Anchordoqui, L.A., *et al.* (1998). "Evolving Wormhole Geometries." *Physical Review D* 57 (January 15):829–833.

Anderson, A. (1920). "On the Advance of the Perihelion of a Planet, and the Path of a Ray of Light in the Gravitational Field of the Sun." *Philosophical Magazine* 39 (May):626–628.

Anderson, A., and B. DeWitt (1986). "Does the Topology of Space Fluctuate?" *Foundations of Physics* 16 (February):91–105.

Anderson, J.L. (1992). "Why We Use Retarded Potentials." *American Journal of Physics* 60 (May):465–467.

Anderson, K.J., and D. Beason (1991). *The Trinity Paradox.* New York: Bantam.

Anderson, M. (1937). *The Star-Wagon.* Washington, DC: Anderson House.

Anderson, M.J. (1990). *The Ghost Inside the Monitor.* New York: Knopf.

Anderson, P. (1991). "Earthman, Beware!" In *Alight in the Void.* New York: Tor.

—— (1988a). "The Man Who Came Early." In *The Great Science Fiction Stories.* Vol. 18. New York: DAW.

—— (1988b). "Delenda Est." In *The Great Science Fiction Stories.* Vol. 17. New York: DAW.

—— (SFS). "My Object All Sublime."

—— (SFD). "Wildcat."

—— (LME). "Flight to Forever."

—— (1981). "Time Patrol." In *The Guardians of Time.* New York: Pinnacle.

—— (1979). "Kyrie." In *The Road to Science Fiction.* Edited by J. Gunn. Vol. 3. New York: New American Library.

—— (1978). *The Avatar.* New York: Berkley.

—— (TIT). "The Long Remembering."

—— (FTL). "Dialogue."

—— (1973a). "The Little Monster." In *Science Fiction Adventure from WAY OUT.* Edited by R. Elwood. Racine, WI: Whitman.

—— (1973b). *There Will Be Time.* New York: Signet.

—— (1972). *The Dancer from Atlantis.* New York: Signet.

—— (1970). *Tau Zero.* Garden City, NY: Doubleday.

—— (VIT). "Time Heals."

—— (1966). *The Corridors of Time.* New York: Lancer.

Anderson, P., and G. Dickson (B51). "Trespass."

Anglin, W.S. (1981). "Backwards Causation." *Analysis* 41 (March):86–91.

Anonymous. (1951). "Missing One's Coach: An Anachronism." In *Far Boundaries.* New York: Pellegrini & Cudahy.

Anonymous (1856). "January First, A.D. 3000." In *Harper's New Monthly Magazine* 12 (January):145–157.

Anscombe, G.E.M. (1971). "The Reality of the Past." In *Philosophical Analysis.* Edited by M. Black. Freeport, NY: Books for Libraries Press.

—— (1956). "Aristotle and the Sea Battle." *Mind* 65 (January):1–15.

Anstey, F. (1891). *Tourmalin's Time Cheques.* New York: D. Appleton.

Anthony, P. and R. Fuentes (1994). *Dead Morn.* New York: Ace.

Apostolatos, T.A., and K.S. Thorne (1992). "Rotation Halts Cylindrical, Relativistic Gravitational Collapse." *Physical Review D* 46 (September 15):2435–2444.

Arntzenius, F. (1994). "Spacelike Connections." *British Journal for the Philosophy of Science* 45 (March):201–217.

—— (1990). "Causal Paradoxes in Special Relativity." *British Journal for the Philosophy of Science* 41 (June):223–243.

Arons, M.E., and E.C.G. Sudarshan (1968). "Lorentz Invariance, Local Field Theory, and Faster-Than-Light Particles." *Physical Review* 173 (September 25):1622–1628.

Arthur, R. (1959). "The Hero Equation." *Fantasy and Science Fiction,* June.

—— (1942). "Time Dredge." *Astounding Science Fiction,* June.

Arthur, R.T.W. (1988). "Continuous Creation, Continuous Time: A Refutation of the Alleged Discontinuity of Cartesian Time." *Journal of the History of Philosophy* 26 (July):349–375.

Ash, P. (SFD). "The Wings of a Bat."

Asimov, I. (TTF). "The Instability."

—— (1993). "Robot Visions." In *Robots in Time: Predator* by W.F. Wu. New York: Avon.

—— (1992). "Obituary." In *The Complete Stories*. Vol. 2. New York: Doubleday.

—— (1988). "The Last Question." In *The Great Science Fiction Stories*. Vol. 18. New York: DAW.

—— (SFS). "The Immortal Bard."

—— (1986a). "When It Comes to Time Travel, There's No Time Like the Present." *New York Times*, October 5, Sect. 2, pp. 1, 32.

—— (1986b). *The End of Eternity*. New York: Del Rey/Ballantine.

—— (T3). "The Ugly Little Boy."

—— (1984a). "Faster Than Light." *Asimov's Science Fiction Magazine*, November.

—— (1984b). "Time-Travel." *Asimov's Science Fiction Magazine*, April.

—— (1984c). "The Winds of Change." In *The Winds of Change*. New York: Del Rey/Ballantine.

—— (SFD). "A Statue for Father" and "Day of the Hunters."

—— (AHT). "The Dead Past."

—— (SFSSS). "A Loint of Paw."

—— (1978). "Fair Exchange?" *Asimov's Science Fiction Adventure Magazine*, Fall.

—— (1975). "Button, Button" and "Blank!" In *Buy Jupiter and Other Stories*. Garden City, NY: Doubleday.

—— (BGA3). "Big Game."

—— (1972). "The Red Queen's Race," "Time Pussy," and "The Endochronic Properties of Resublimated Thiotimoline." In *The Early Asimov*. Vol. 2. Greenwich, CT: Fawcett Crest.

—— (1968). "Impossible, That's All." In *Science, Numbers, and I*. New York: Doubleday.

—— (SFAD). "What If. . . ."

—— (DD). "The Brazen Locked Room."

Aspect, A., *et al.* (1982a). "Experimental Test of Bell's Inequalities Using Time-Varying Analyzers." *Physical Review Letters* 49 (December 20):1804–1807.

—— (1982b). "Experimental Realization of Einstein-Podolsky-Rosen-Bohm *Gedanken-experiment*: A New Violation of Bell's Inequalities." *Physical Review Letters* 49 (July 12):91–94.

—— (1981). "Experimental Tests of Realistic Local Theories via Bell's Theorem." *Physical Review Letters* 47 (17 August):460–467.

Augustynek, Z. (1976). "Past, Present and Future in Relativity." *Studia Logica* 35:45–53.

Avi (1988). *Something Upstairs*. New York: Orchard.

Bade, W. (SFAD). "Ambition."

Baierlein, R.F., D.H. Sharp, and J.A. Wheeler (1962). "Three-Dimensional Geometry as Carrier of Information About Time." *Physical Review* 126 (June 1):1864–1865.

Bailey, J. *et al.* (1977). "Measurements of Relativistic Time Dilatation for Positive and Negative Muons in a Circular Orbit." *Nature* 268 (July 28):301–305.

Bailey, J.O. (1972). *Pilgrims Through Space and Time*. Westport, CT: Greenwood Press.

Baker, A. (1970). "Time Reversal." In *Modern Physics and Antiphysics*. Reading, MA: Addison-Wesley.

Baker, G.L. (1986). "A Simple Model of Irreversibility." *American Journal of Physics* 54 (August):704–708.

Baker, J.R. (1972). "Omniscience and Divine Synchronization." *Process Studies* 2 (Fall):201–208.

Baker, L.R. (1975). "Temporal Becoming: The Argument from Physics." *Philosophical Forum* 6 (Spring):218–236.

Bakhnov, V. (AM). "Twelve Holidays."

Balbinot, R. (1985). "Crossing the Einstein-Rosen Bridge." *Lettere Al Nuovo Cimento* 43 (May 16):76–80.

Balderston, J.L. (1941). *Berkeley Square*. New York: Macmillan.

Baldner, S. (1989). "St. Bonaventure on the Temporal Beginning of the World." *New Scholasticism* 63 (Spring):206–228.

Ball, W.W.R. (1891). "A Hypothesis Relating to the Nature of the Ether and Gravity." *Messenger of Mathematics* 21:20–24.

Ballard, J.G. (TTF). "The Greatest Television Show on Earth."

—— (1978). "Time of Passage." In *Time of Passage*. New York: Taplinger.

—— (1971). "The Sound-Sweep" and "Chronopolis." In *Chronopolis*. New York: G.P. Putnam's Sons.

—— (TTT). "Mr F is Mr F."

—— (NWF). "The Lost Leonardo."

Band, W. (1988a). "Can Information Be Transferred Faster Than Light? II. The Relativistic Doppler Effect on Electromagnetic Wave Packets with Suboptic and Superoptic Group Velocities." *Foundations of Physics* 18 (June):625–638.

—— (1988b). "Can Information Be Transferred Faster Than Light? A *Gedanken* Device for Generating Electromagnetic Wave Packets with Superoptic Group Velocity." *Foundations of Physics* 18 (May):549–562.

Banerjee, A., and S. Banerji (1968). "Stationary Distributions of Dust and Electromagnetic Fields in General Relativity." *Journal of Physics A (Proceedings of the Physical Society)* 1:188–193.

Banks, R.A. (1961). "This Side Up." In *The Fifth Galaxy Reader*. Edited by H.L. Gold. Garden City, NY: Doubleday.

Barjavel, R. (1944). *Le Voyageur Imprudent*. Paris: Denoël.

Barr, R. (1989). "The Hour Glass." In *The Strong Arm*. New York: Frederick A. Stokes.

Baron, N. (1987). *Anatomy of Wonder: A Critical Guide to Science Fiction*. 3rd ed. New York: R.R. Bowker.

Barrow, J.D., and F.J. Tipler (1978). "Eternity Is Unstable." *Nature* 276 (November 30):453–459.

Bartlett, A.A. (1974). "A Simple Problem from the Real World That Can be Solved Through Time Reversal." *American Journal of Physics* 42 (May):416–417.

Bass, R.W., and L. Witten (1957). "Remark on Cosmological Models." *Reviews of Modern Physics* 29 (July):452–453.

Bates, H. (SFT). "Alas, All Thinking."

Baxter, S. (1995). *The Time Ships*. New York: Harper Prism.

—— (1994). *Ring*. New York: Harper Prism.

—— (1993). *Timelike Infinity*. New York: ROC.

Bayley, B.J. (1974). *The Fall of Chronopolis*. New York: DAW.

Bear, G. (MTMW). "Tangents."

Beason, D. (1990). "Ben Franklin's Laser." *Analog,* Mid-December.

Bechhoefer, J., and G. Chabrier (1993). "On the Fate of Stars in High Spatial Dimensions." *American Journal of Physics* 61 (May):460–462.

Beerbohm, M. (DD). "Enoch Soames."

Beichler, J.E. (1988). "Ether/Or: Hyperspace Models of the Ether in America." In *The Michelson Era in American Science 1870–1930*. Edited by S. Goldberg and R.H. Stuewer. New York: American Institute of Physics.

Bekenstein, J.D. (1993). "How Fast Does Information Leak Out from a Black Hole?" *Physical Review Letters* 70 (14 June):3680–3683.

Belinfante, F.J. (CPT). "Determinism, Time Arrow, and Prediction."

—— (1966). "Kruskal Space Without Wormholes." *Physics Letters* 20 (January 15):25–26.

Bell, E.T. (1934). *Before the Dawn*. Baltimore: Williams & Wilkins.

—— (1931). *The Time Stream*. Providence, RI: Buffalo Book Co.

Bell, J. (1979). "The Infinite Past Regained: A reply to Whitrow." *British Journal for the Philosophy of Science* 30 (June):161–165.

Bell, J.S. (1987). "Quantum Mechanics for Cosmologists," "On the Einstein-Podolsky-Rosen Paradox," and "Bertlmann's Socks and the Nature of Reality." In *Speakable and Unspeakable in Quantum Mechanics*. New York: Cambridge University Press.

Bellairs, J. (1989). *The Trolley to Yesterday*. New York: Dial.

Bellamy, E. (1983). *Looking Backward, 2000–1887*. New York: Bantam.

Benford, G. (1997). *Foundation's Fear*. New York: Harper Prism.

—— (1995). "A Worm in the Well." *Analog*, November.

—— (1994). "Not of an Age." In *Weird Tales from Shakespeare*. Edited by K. Kerr and M.H. Greenberg. New York: DAW.

—— (1993). "Time and *Timescape*." *Science-Fiction Studies* 20 (July):184–190.

—— (1991). "Down the River Road." In *After the King*. Edited by C. Tolkien and M. Greenberg. New York: Tor.

—— (HV). "Valhalla."

—— (1986). "Time Shards." In *In Alien Flesh*. New York: Tor.

—— (1980). *Timescape*. New York: Simon and Schuster.

—— (E). "Cambridge, 1:58 A.M."

—— (1970). "3:02 P.M., Oxford." *If*, September.

Benford, G.A., D.L. Book, and W.A. Newcomb (1970). "The Tachyonic Antitelephone." *Physical Review D* 2 (July 15):263–265.

Ben-Menahem, Y. (1993). "Struggling with Causality: Einstein's Case." *Science in Context* 6 (Spring):291–310.

Bennett, C.L. (1987a). "Precausal Quantum Mechanics." *Physical Review A* 36 (November 1):4139–4148.

—— (1987b). "Evidence for Microscopic Causality Violation." *Physical Review A* 35 (March 15):2409–2419.

—— (1987c). "Further Evidence for Causality Violation." *Physical Review A* 35 (March 15):2420–2428.

Bennett, J. (1971). "The Age and Size of the World." *Synthese* 23 (August):127–146.

Bennett, J.G., *et al.* (1949). "Unified Field Theory in a Curvature-Free Five-Dimensional Manifold." *Proceedings of the Royal Society of London* 198A (July):39–61.

Berenda, C.W. (1947). "The Determination of Past by Future Events: A Discussion of the Wheeler-Feynman Absorption-Radiation Theory." *Philosophy of Science* 14:13–19.

Berger, G. (1974). "Elementary Causal Structures in Newtonian and Minkowskian Space-Time." *Theoria* 40:191–201.

—— (1971). "Earman on Temporal Anisotropy." *Journal of Philosophy* 68 (11 March):132–137.

—— (1968). "The Conceptual Possibility of Time Travel." *British Journal for the Philosophy of Science* 19:152–155.

Berger, T. (1989). *Changing the Past*. Boston: Little, Brown.

Bergonzi, B. (1961). *The Early H.G. Wells: A study of the Scientific Romances*. Toronto: University of Toronto Press.

—— (1960). "*The Time Machine*: An Ironic Myth." *Critical Quarterly* 2 (Winter):293–305.

Bergson, H. (1965). *Duration and Simultaneity*. New York: Bobbs-Merrill.

—— (1910). *Time and Free Will*. New York: Macmillan.

Bernal, A.W. (1940). "Paul Revere and the Time Machine." *Amazing Stories*, March.

Bers, A., *et al.* (RAG). "The Impossibility of Free Tachyons."

Bester, A. (1986). "Hobson's Choice." *The Great Science Fiction Stories*. Vol. 14. New York: DAW.

—— (1983). "Of Time and Third Avenue." In *Magic for Sale*. Edited by A. Davidson. New York: Ace.

—— (1980). "The Push of a Finger." In *The Great Science Fiction Stories*. Vol. 4. New York: DAW.

—— (FTI). "The Probable Man."

—— (VIT). "The Men Who Murdered Mohammed."

Besterman, T. (1933). "Report of an Inquiry into Precognitive Dreams." *Proceedings of the Society for Psychical Research* 41:186–204.

Bierce, A. (1964). "An Occurrence at Owl Creek Bridge," "The Damned Thing," and "Mysterious Disappearances." In *Ghost and Horror Stories of Ambrose Bierce*. New York: Dover.

Biggle, L. Jr. (1967). "D.F.C." In *The Rule of the Door*. New York: Doubleday.

—— (1965). *The Fury Out of Time*. Garden City, NY: Doubleday.

Bilaniuk, O.M., and E.C.G. Sudarshan. (1969a) "Particles Beyond the Light Barrier." *Physics Today* 22 (May):43–51 [and the resulting discussion in *Physics Today* 22 (December):47–52].

—— (1969b). "Causality and Space-like Signals." *Nature* 223 (July 26):386–387.

Bilaniuk, O.M.P., V.K. Deshpande, and E.C.G. Sudarshan (1962). " 'Meta' Relativity." *American Journal of Physics* 30 (October):718–723.

Bilenkin, D. (1978). "The Uncertainty Principle," "Time Bank," and "The Inexorable Finger of Fate." In *The Uncertainty Principle*. New York: Macmillan.

Binder, E. (1953). "The Time Cylinder." *Science Fiction Plus,* March.

—— (1940). "The Time Cheaters." *Thrilling Wonder Stories,* March.

—— (1939). "The Man Who Saw Too Late." *Fantastic Adventures,* September.

—— (1938). "Eye of the Past." *Astounding Science Fiction,* March.

—— (1937). "The Time Contractor." *Astounding Stories,* December.

Birch, P. (1982). "Is the Universe Rotating?" *Nature* 298 (July 29):451–454.

Birrell, N.D., and P.C.W. Davies (1978). "On Falling Through a Black Hole Into Another Universe." *Nature* 272 (March 2):35–37.

Bishop, N.T. (1988). "Is Superluminal Travel a Theoretical Possibility? II." *Foundations of Physics* 18 (May):571–574.

—— (1984). "Is Superluminal Travel a Theoretical Possibility?" *Foundations of Physics* 14 (April):333–340.

Bisson, T. (1992). "Two Guys from the Future." *Omni,* August.

Bitov, A. (1989). "Pushkin's Photograph." In *The New Soviet Fiction*. Compiled by S. Zalygin. New York: Abbeville Press.

Bixby, J. (1954). "One Way Street." *Amazing Stories,* January.

Black, M. (1959a). "Linguistic Relativity: The Views of Benjamin Lee Whorf." *Philosophical Review* 68:228–238.

—— (1959b). "The 'Direction' of Time." *Analysis* 19 (January):54–63.

Blackford, R. (1985). "Physics and Fantasy: Scientific Mysticism, Kurt Vonnegut, and *Gravity's Rainbow*." *Journal of Popular Culture* 19 (Winter):35–44.

Blackwood, A. (1949). "Entrance and Exit" and "The Pikestaffe Case." In *Tales of the Uncanny and Supernatural*. London: Peter Nevill.

Blandford, R.D., C.F. McKee, and M.J. Rees (1977). "Super-luminal Expansion in Extragalactic Radio Sources." *Nature* 267 (May 19):211–216.

Blatt, J.M. (1956). "Time Reversal." *Scientific American,* August.

Blau, S.K. (1998). "Would a Topology Change Allow Ms. Bright to Travel Backward in Time?" *American Journal of Physics* 66 (March):179–185.

Blaylock, J.P. (1992). *Lord Kelvin's Machine*. New York: Ace.

Bleiler, E.F. (1990). *Science Fiction: The Early Years.* Kent, OH: Kent State University Press.

Blish, J. (1976). "Beep." In *Galactic Empires.* Edited by B.W. Aldiss. Vol. 2. New York: St. Martin's Press.

—— (MI). "Common Time."

—— (1956). "A Matter of Energy." In *The Best from Fantasy and Science Fiction.* Edited by A. Boucher. Garden City, NY: Doubleday.

—— (1942). "The Solar Comedy." *Future Fiction,* June.

—— (1941). "Weapon Out of Time." *Science Fiction Quarterly,* Spring.

Bloch, R. (1977). "The Past Master." In *The Best of Robert Bloch.* Edited by L. Del Rey. New York: Ballantine.

—— (DV). "A Toy for Juliette."

—— (GRSF7). "Crime Machine."

—— (1962). "The Hell-Bound Train." In *The Hugo Winners.* Edited by I. Asimov. Garden City, NY: Nelson Doubleday.

Bludman, S.A., and M.A. Ruderman: (1970). "Noncausality and Instability in Ultradense Matter." *Physical Review D* 1 (June 15):3243–3246.

—— (1968). "Possibility of the Speed of Sound Exceeding the Speed of Light in Ultradense Matter." *Physical Review* 170 (June 25):1176–1184.

Blumenthal, H.J., *et al.* (1988). *The Complete Time Traveler.* Berkeley, CA: Ten Speed Press.

Bohm, D. (BST). "Inadequacy of Laplacean Determinism and Irreversibility of Time."

Bolton, P. (1931). "The Time Hoaxers." *Amazing Stories,* August.

Bond, N. (1954). "Uncommon Castaway." In *No Time Like the Future.* New York: Avon.

—— (1946). "The Magic Staircase," "Johnny Cartwright's Camera," "The Einstein Inshoot," "Dr. Fuddle's Fingers," and "The Bacular Clock." In *Mr. Mergenthwirker's Lobblies and Other Fantastic Tales.* New York: Coward-McCann.

—— (1942). "Horsesense Hank in the Parallel Worlds." *Amazing Stories,* August.

—— (1941a). "The Geometrics of Johnny Day." *Astounding Science Fiction,* July.

—— (1941b). "The Fountain." *Unknown,* June.

—— (1940). "Parallel in Time." *Thrilling Wonder Stories,* June.

—— (1939a). "The Monster from Nowhere." *Fantastic Adventures,* July.

—— (1939b). "Lightship, Ho!" *Astounding Science Fiction,* July.

—— (1937). "Down the Dimensions." *Astounding Stories,* April.

Bondi, H. (1957). "Negative Mass in General Relativity." *Reviews of Modern Physics* 29 (July):423–428.

Bondi, H., and J. Samuel (1997). "The Lense–Thirring Effect and Mach's Principle." *Physics Letters A* 228 (April 7):122–126.

Bonner, W.B. (1987). "Arrow of Time for a Collapsing, Radiating Sphere." *Physics Letters A* 122 (June 22):305–308.

—— (1980). "The Rigidly Rotating Relativistic Dust Cylinder." *Journal of Physics A* 13 (June):2121–2132.

Borges, J.L. (1972). *Doctor Brodie's Report.* New York: E.P. Dutton.

—— (1970). "The Other Death." In *The Aleph and Other Stories 1933–1969.* New York: E.P. Dutton.

—— (1964). "The Garden of Forking Paths." In *Labyrinths.* New York: New Directions.

—— (1962). "The Secret Miracle." In *Ficciones.* New York: Grove.

Bork, A.M. (1966). "The 'Fitzgerald' Contraction." *Isis* 57:199–207.

—— (1964). "The Fourth Dimension in Nineteenth-Century Physics." *Isis* 55 (October):326–338.

Boucher, A. (1980a). "Barrier." In *The Great Science Fiction Stories.* Vol. 4. New York: DAW.

—— (1980b). "Snulbug." In *The Great Science Fiction Stories*. Vol. 3. New York: DAW.
—— (1953). "Elsewhen" and "The Other Inauguration." In *Far and Away*. New York: Ballantine.
—— (GSSF). "The Chronokinesis of Jonathan Hull."
Boulle, P. (1966). "Time Out of Mind." In *Time Out of Mind*. New York: Vanguard Press.
Boulware, D.G. (1992). "Quantum Field Theory in Spaces with Closed Timelike Curves." *Physical Review D* 46 (November 15):4421–4441.
Boussenard, L. (1898). *10,000 Years in a Block of Ice*. New York: F.T. Neely.
Bouwmeester, D., *et al.* (1997). "Experimental Quantum Teleportation." *Nature* 390 (December 11):575–579.
Bova, B. (1985). *As On a Darkling Plain*. New York: Tor.
Bradbury, R. (1988). "The Toynbee Convector." In *The Toynbee Convector*. New York: Knopf.
—— (TOT). "The Shape of Things."
—— (1981). "Forever and the Earth." In *Sinister, Strange and the Supernatural*. Edited by H. Hoke. New York: Elsevier/Nelson.
—— (1980). "A Sound of Thunder," "The Fox and the Forest," "Tomorrow's Child," and "A Scent of Sarsaparilla." In *The Stories of Ray Bradbury*. New York: Alfred A. Knopf.
—— (1971). "The Kilimanjaro Device." In *I Sing the Body Electric!* New York: Alfred A. Knopf.
—— (1966). "Time in Thy Flight." In *S Is for Space*. Garden City, NY: Doubleday.
—— (1950). *The Martian Chronicles*. New York: Doubleday.
Bradley, R.D. (1959). "Must the Future Be What It Is Going to Be? *Mind* 68 (April):193–208.
Brams, S.J. (1983). "Omniscience and Partial Omniscience." In *Superior Beings*. New York: Springer-Verlag.
Brennan, A. (1982). "Personal Identity and Personal Survival." *Analysis* 42 (January):44–50.
Brennert, A. (1990). *Time and Chance*. New York: Tor.
Bretnor, R. (1956). "The Past and Its Dead People." *Fantasy and Science Fiction*, September.
Breuer, M.J. (MM). "The Appendix and the Spectacles."
—— (GSFS). "The Gostak and the Doshes."
—— (FM). "The Captured Cross-Section."
—— (1932a). "The Finger of the Past." *Amazing Stories*, November.
—— (1932b). "The Einstein See-Saw." *Astounding Stories*, April.
—— (1930a). "The Time Valve." Wonder Stories, July.
—— (1930b). "The FitzGerald Contraction." *Science Wonder Stories*, January.
Bridge, F.J. (1931). "Via the Time Accelerator." *Amazing Stories*, January.
Brier, B. (1974). *Precognition and the Philosophy of Science: An Essay on Backward Causation*. New York: Humanities Press.
—— (1973). "Magicians, Alarm Clocks, and Backward Causation." *Southern Journal of Philosophy* 11:359–364 (and the reply by A. Flew, 11:365–366).
—— (1972). "An Atemporal View of Causality." *Journal of Critical Analysis* 4 (April):8–16.
Brill, D. (MWM). "Thoughts on Topology Change."
Broad, C.D. (1959). "A Reply to My Critics." In *The Philosophy of C.D. Broad*. Edited by P.A. Schilpp. New York: Tudor.
—— (1937). "The Philosophical Implications of Foreknowledge." *Aristotelian Society Supplement* 16:177–209.
—— (1935). "Mr. Dunne's Theory of Time in *An Experiment with Time*." *Philosophy* 10:168–185.

Broad, C.D., and H.H. Price (1937). "The Philosophical Implications of Precognition." *Aristotelian Society Supplement* 16:211–228.

Broido, M.M., and J.G. Taylor (1968). "Does Lorentz-Invariance Imply Causality?" *Physical Review* 174 (25 October):1606–1610.

Brotman, H. (1952). "Could Space Be Four-Dimensional?" *Mind* 61 (July):317–327.

Brown, B. (1992). "Defending Backwards Causation." *Canadian Journal of Philosophy* 22 (December):429–443.

Brown, F. (1986). "Hall of Mirrors." *The Great Science Fiction Stories.* Vol. 15. New York: DAW.

—— (MT). "Nightmare in Time."

—— (1977). "The Short Happy Lives of Eustace Weaver I, II, and III" and "The End." In *The Best of Fredric Brown*, New York: Ballantine.

—— (1958). "First Time Machine," "Blood," and "Experiment." In *Honeymoon in Hell.* New York: Bantam.

—— (1953). "Paradox Lost." In *Science Fiction Carnival.* Chicago: Shasta.

Brown, G. (1985). "Praying About the Past." *Philosophical Quarterly* 35 (January):83–86.

Bruhat, Y. (GRAV). "The Cauchy Problem."

Brumbaugh, R.S. (1980). "Time Passes: Platonic Variations." *Review of Metaphysics* 33 (June):711–726.

Brunner, J. (FST). "Lostling."

—— (1972). "Host Age." In *Entry to Elsewhen.* New York: DAW.

—— (1969). *Timescoop.* New York: Dell.

—— (1965). "Galactic Consumer Reports No. 1—Inexpensive Time Machines." *Galaxy Science Fiction,* December.

Brush, S.G. (1967). "Note on the History of the FitzGerald-Lorentz Contraction." *Isis* 58:230–232.

Bryant, E. (MT). "Paths."

Buchan, J. (1932). *The Gap in the Curtain.* London: Hodder and Stoughton.

Buck, D.P. (SS). "Story of a Curse."

Bulgarin, F. (RRSF). "Plausible Fantasies or a Journey in the 29th Century."

Bunge, M. (1958). "On Multi-dimensional Time." *British Journal for the Philosophy of Science* 9 (May):39.

Burger, D. (1983). *Sphereland.* New York: Harper & Row.

Burks, A.J. (1966). "When the Graves Were Opened." In *Black Medicine.* Sauk City, WI: Arkham House.

Burstein, M.A. (1998). "Cosmic Corkscrew." *Analog,* June.

Busby, F.M. (1987). "A Gun for Grandfather" and "Proof." In *Getting Home.* New York: Ace.

Butler, O.E. (1979). *Kindred.* Boston: Beacon Press.

Byram, G. (FCW). "The Chronicle of the 656th."

Cabot, J.Y. (1941). "Murder in the Past." *Amazing Stories,* March.

Cahn, S.M. (1967). *Fate, Logic, and Time.* New Haven, CT: Yale University Press.

Caldirola, P., and E. Recami (1981). "Causality and Tachyons in Relativity." In *Italian Studies in the Philosophy of Science.* Boston, MA: D. Reidel.

Calinon, A. (BST). "Geometrical Spaces."

Callahan, J.F. (1948). *Four Views of Time in Ancient Philosophy.* Cambridge, MA: Harvard University Press.

Calvani, M., *et al.* (1978). "Time Machine and Geodesic Motion in Kerr Metric." *General Relativity and Gravitation* 9 (February):155–163.

Campbell, J.W. Jr. (1948). "Twilight," "Night," and "Elimination." In *Who Goes There?* Chicago: Shasta.

Capek, M. (1987). "The Conflict Between the Absolutist and the Relational Theory of Time Before Newton." *Journal of the History of Ideas* 48 (October–December): 595–608.

—— (1983). "Time-Space Rather Than Space-Time." *Diogenes* 123 (Fall):30–49.

—— (BST). "The Inclusion of Becoming in the Physical World."

—— (1975). "Relativity and the Status of Becoming." *Foundations of Physics* 5 (December):607–617.

—— (VOT). "Time in Relativity Theory: Arguments for a Philosophy of Becoming."

—— (1965). "The Myth of Frozen Passage: The Status of Becoming in the Physical World." In *Boston Studies in the Philosophy of Science*. Vol. 2. New York: Humanities Press.

Card, O.S. (MIAM). "Closing the Timelid" and "Clap Hands and Sing."

Carlini, A., *et al.* (1996). "Time Machines and the Principle of Self-Consistency as a Consequence of the Principle of Stationary Action (II): The Cauchy Problem for a Self-Interacting Relativistic Particle." *International Journal of Modern Physics D* 5 (October):445–479.

—— (1995). "Time Machines: The Principle of Self-Consistency as a Consequence of the Principle of Minimal Action." *International Journal of Modern Physics D* 4 (October):557–580.

Carpentier, A. (1967). "Journey to the Seed." In *Latin American Writing Today*. Edited by J.M. Cohen. Baltimore, MD: Penguin.

Carr, J.D. (1951). *The Devil in Velvet*. New York: Harper & Brothers.

Carroll, J.A. (1959). "An Absolute Scale of Time." *Nature* 184 (July 25):260–261.

Carroll, S.M., *et al.* (1994). "Energy-Momentum Restrictions on the Creation of Gott Time Machines." *Physical Review D* 50 (November 15):6190–6206.

Carroll, S.M., *et al.* (1992). "An Obstacle to Building a Time Machine." *Physical Review Letters* 68 (January 20):263–266.

Carter, B. (1968). "Global Structure of the Kerr Family of Gravitational Fields." *Physical Review* 174 (October 25):1559–1571.

Carter, P. (1950). "Ounce of Prevention." *Fantasy and Science Fiction,* Summer.

Carter, P.A. (1977). *The Creation of Tomorrow*. New York: Columbia University Press.

Cartur, P. (SSFT). "The Mist."

Casella, R.C. (1969). "Time Reversal and the K° Meson Decays. II." *Physical Review Letters* 22 (March 17):554–556.

—— (1968). "Time Reversal and the K° Meson Decays. II." *Physical Review Letters* 21 (October 7):1128–1131.

Cassidy, M.J., and S.W. Hawking (1998). "Models for Chronology Selection." *Physical Review D* 57 (February 15):2372–2380.

Causey, J. (SSFT). "Teething Ring."

Chakrabarti, S.K., *et al.* (1983). "Timelike Curves of Limited Acceleration in General Relativity." *Journal of Mathematical Physics* 24 (March):597–598.

Chambers, R.W. (1895). "The Demoiselle d'Ys." In *The King in Yellow*. New York: F. Tennyson Neely.

Chamblin, A., G.W. Gibbons, and A.R. Steif (1994). "Kinks and Time Machines." *Physical Review D* 50 (August 15):2353–2355.

Champlin, C. (1989). *Back There Where the Past Was*. Syracuse, NY: Syracuse University Press.

Chandler, A.B. (ASF). "Kelly Country."

—— (1978). *The Way Back*. New York: DAW.

—— (SFAD). "Castaway."

—— (1948): "The Tides of Time." *Fantastic Adventures,* June.

Chandrasekhar, S. (NT). "Geodesics in Gödel's Universe."

Chandrasekhar, S., and J.B. Hartle (1982). "On Crossing the Cauchy Horizon of a Reissner–Nordström Black-Hole." *Proceedings of the Royal Society of London A* 384 (December 8):301–315.

Chandrasekhar, S., and J.P. Wright (1961). "The Geodesics in Gödel's Universe." *Proceedings of the National Academy of Sciences* 47 (March):341–347.

Chapman, T. (1982). "Time-Travel II." In *Time: A Philosophical Analysis.* Dordrecht, The Netherlands: D. Reidel.

Chari, C.T.K. (1960). "Time Reversal, Information Theory, and 'World-Geometry.'" *Journal of Philosophy* 60 (September 26):579–583.

—— (1957). "A Note on Multi-Dimensional Time." *British Journal for the Philosophy of Science* 8 (August):155–158.

—— (1949). "On Representations of Time as 'The Fourth Dimension' and Their Metaphysical Inadequacy." *Mind* 58 (April):218–221.

Charlton, N.J. (1978). "Some Properties of Time-Machines." *Journal of Physics A* 11 (November):2207–2211.

Cho, Y.M., and D.H. Park (1997). "Closed Time-like Curves and Weak Energy Condition." *Physics Letters B* 402 (June 5):18–24.

Christensen, F. (1993). *Space-Like Time.* Toronto: University of Toronto Press.

—— (1981). "Special Relativity and Space-like Time." *British Journal for the Philosophy of Science* 32 (March):37–53.

—— (1976). "The Source of the River of Time." *Ratio* 18 (December):131–144.

—— (1974). "McTaggart's Paradox and the Nature of Time." *Philosophical Quarterly* 24 (October):289–299.

Christensen, J.H., *et al.* (1964). "Evidence for the 2π Decay of the K°_2 Meson." *Physical Review Letters* 13 (July 27):138–140.

Christodoulou, D. (1984). "Violation of Cosmic Censorship in the Gravitational Collapse of a Dust Cloud." *Communications in Mathematical Physics* 93:171–195.

Chyba, C.F. (1985). "Kaluza–Klein Unified Field Theory and Apparent Four-Dimensional Space-Time." *American Journal of Physics* 53 (September):863–872.

Clarke, A.C. (1985). "About Time." In *Profiles of the Future.* New York: Warner.

—— (SSF). "Technical Error."

—— (FTL). "Possible, That's All!"

—— (1972). "Things That Can Never Be Done" and "Technology and the Future." In *Report on Planet Three and Other Speculations,* New York: Harper & Row.

—— (1959a). "Time's Arrow." In *Across the Sea of Stars.* New York: Harcourt.

—— (1959b). "The Wall of Darkness." In *The Other Side of the Sky.* New York: New American Library.

—— (PSF). "All the Time in the World."

Clarke, C.J.S. (1993). *The Analysis of Space-Time Singularities.* Cambridge, England: Cambridge University Press.

—— (1990). "Opening a Can of Wormholes." *Nature* 348 (November 22):287–288.

Clarke, C.J.S., and F. deFelice (1982). "Globally Non-Causal Space-Times." *Journal of Physics A* 15 (August):2415–2417.

Clarke, M. (ASF). "Human Repetends."

Clay, R.W., and P.C. Crouch (1974). "Possible Observation of Tachyons Associated with Extensive Air Showers." *Nature* 248 (March 1):28–30.

Clee, M. (1996). *Branch Point.* New York: Ace.

Cleugh, M.F. (1937). *Time and Its Importance in Modern Thought.* London: Methuen.

Click, J.H. (1929). "The Dimension Segregator." *Amazing Stories,* August.

Clifford, W.K. (BST). "On the Bending of Space" and "On the Space-Theory of Matter."

Clifton, M. (MM). "Star, Bright."

Clingerman, M. (SSFT). "Stair Trick."

—— (1958). "The Day of the Green Velvet Cloak." *Fantasy and Science Fiction,* July.

Cloukey, C. (1929). "Paradox." *Amazing Stories Quarterly,* Summer.

Coates, P. (1987). "Chris Marker and the Cinema as Time Machine." *Science-Fiction Studies* 14 (November):307–315.

Cobb, J.B. (1965). *A Christian Natural Theology.* Philadelphia: Westminster Press.

Coblentz, S.A. (1938). "Through the Time-Radio." *Marvel Science Stories,* August.

Cocke, W.J. (1967). "Statistical Time Symmetry and Two-Time Boundary Conditions in Physics and Cosmology." *Physical Review* 160 (August 25):1165–1170.

Coe, L. (1969). "The Nature of Time." *American Journal of Physics* 37 (August):810–815.

Cogswell, T.R. (DD). "Threesie" and "Impact with the Devil."

—— (GRSF2). "Minimum Sentence."

Cohen, J.M. (RAG). "The Rotating Einstein-Rosen Bridge."

Coleman, S. (1988). "Black Holes as Red Herrings: Topological Fluctuations and the Loss of Quantum Coherence." *Nuclear Physics B* 307 (October 3):867–882.

Collier, J. (1931). *No Traveller Returns.* London: White Owl Press.

Collins, L. (1959). "Triple-Time Try." *Amazing Stories,* October.

Compton, D.C. (1971). *Hot Wireless Sets, Aspirin Tablets, the Sandpaper Sides of Used Matchboxes, and Something That Might Have Been Castor Oil.* London: Michael Joseph.

Conrad, P. (1990). *Stonewords.* New York: Harper & Row.

Contento, W. (1978). *Index to Science Fiction Anthologies and Collections.* Boston, MA: G.K. Hall & Co.

Cook, G. (1985). *A Matter of Time.* New York: Ace.

Cook, M. (1982). "Tips for Time Travel." In *Philosophers Look at Science Fiction.* Edited by N.D. Smith. Chicago: Nelson-Hall.

Cooper, J.C. (1979). "Have Faster-Than-Light Particles Already Been Detected?" *Foundations of Physics* 9 (June):461–466.

Corben, H.C. (1976). "Thought Experiments at Superluminal Relative Velocities." *International Journal of Theoretical Physics* 15 (September):703–712.

Costello, M.J. (1990). *Time of the Fox.* New York: Penguin.

Cottle, T.J. (1976). "Fantasies of Temporal Recovery and Knowledge of the Future." In *Perceiving Time.* New York: John Wiley.

Cowper, R. (1977). "The Hertford Manuscript." In *The Best from Fantasy and Science Fiction.* Edited by E.L. Ferman. Vol. 22. Garden City, NY: Doubleday.

Cox, A.J. (1950). "Linguistics and Time." *Astounding Science Fiction,* August.

Craig, W.L. (1993). "The Caused Beginning of the Universe: A response to Quentin Smith." *British Journal for the Philosophy of Science* 44 (December):623–639.

—— (1990). " 'What Place, Then, for a Creator?': Hawking on God and Creation." *British Journal for the Philosophy of Science* 41 (December):473–491.

—— (1988). "Tachyons, Time Travel, and Divine Omniscience." *Journal of Philosophy* 85 (March):135–150.

—— (1986). "God, Creation and Mr. Davies." *British Journal for the Philosophy of Science* 37 (June):163–175.

—— (1985). "Was Thomas Aquinas a B-Theorist of Time?" *New Scholasticism* 59 (Autumn):475–483.

—— (1981). "The Finitude of the Past." *Aletheia* 2, 235–242.

—— (1980). "Julian Wolfe and Infinite Time." *International Journal for Philosophy of Religion* 11, 133–135.

—— (1979). "Whitrow and Popper on the Impossibility of an Infinite Past." *British Journal for the Philosophy of Science* 29 (June):165–170.

—— (1978). "God, Time, and Eternity." *Religious Studies* 14 (December):497–503.

Cramer, J.G. (1997). *Einstein's Bridge.* New York: Avon.

—— (1993). "The Tachyon Drive: Infinite Exhaust Velocity at Zero Energy Cost." *Analog,* October.

—— (1992). "Natural Wormholes: Squeezing the Vacuum." *Analog,* July.

—— (1991a). "Quantum Telephones to Other Universes, to Times Past." *Analog,* October.

—— (1991b). "Quantum Time Travel." *Analog,* April.

—— (1991c). *Twistor.* New York: Avon.

—— (1990a). "More about Wormholes—To the Stars in No Time." *Analog,* May.

—— (1990b). "The Twin Paradox Revisited." *Analog*, March.

—— (1989). "Wormholes and Time Machines." *Analog*, June.

—— (1988a). "Velocity Reversal and the Arrows of Time." *Foundations of Physics* 18 (December):1205–1212.

—— (1988b). "Paradoxes and FTL Communication." *Analog*, September.

—— (1985a). "Light in Reverse Gear." *Analog*, August.

—— (1985b). "The Other Forty Dimensions." *Analog*, April.

—— (1983). "The Arrow of Electromagnetic Time and the Generalized Absorber Theory." *Foundations of Physics* 13 (September):887–902.

—— (1980). "Generalized Absorber Theory and the Einstein-Podolsky-Rosen Paradox." *Physical Review D* 22 (July 15):362–376.

Cramer, J.G., *et al.* (1995). "Natural Wormholes as Gravitational Lenses." *Physical Review D* 51 (March 15):3117–3120.

Crawford, F.S. (1973). "Simple Demonstration of Time-Reversal Invariance in Classical Mechanics." *American Journal of Physics* 41 (April):574–577.

Cross, P. (1942). "Prisoner of Time." *Super Science Stories,* May.

Crowley, J. (1991). *Great Work of Time.* New York: Bantam.

Csonka, P.L. (1970). "Causality and Faster-Than-Light Particles." *Nuclear Physics B* 21 (August 15):436–444.

—— (1969). "Advanced Effects in Particle Physics." *Physical Review* 180 (April 25):1266–1281.

—— (1967). "Are Photon-Photon Scattering Experiments Feasible?" *Physics Letters B* 24 (June 12):625–628.

Cummings, R. (1946). *The Shadow Girl.* London: Gerald G. Swan.

—— (1941). "Bandits of Time." *Amazing Stories,* December.

—— (1929). *The Man Who Mastered Time.* Chicago: A.C. McClurg.

—— (1921). "The Time Professor." *Argosy-All-Story,* January 1.

Currie, G. (1992). "McTaggart at the Movies." *Philosophy* 67 (July):343–355.

Curry, T. (1931). "Hell's Dimension." *Astounding Stories,* April.

Cushing, J.T. (1990). *Theory Construction and Selection in Modern Physics: The S Matrix.* New York: Cambridge University Press.

Cutkosky, R.E. (1970). "Macroscopic Properties of Dense Noncausal Matter." *Physical Review D* 2 (October 15):1386–1389.

Cutler, C. (1992). "Global Structure in Gott's Two-String Spacetime." *Physical Review D* 45 (January 15):487–494.

Cvetic, M., *et al.* (1993). "Cauchy Horizons, Thermodynamics, and Closed Timelike Curves in Planar Supersymmetric Spaces." *Physical Review Letters* 70 (March 1):1191–1194.

Dainton, B.F. (1992). "Time and Division." *Ratio* 5 (December):102–128.

Dales, R.C. (1988). "Time and Eternity in the Thirteenth Century." *Journal of the History of Ideas* 49 (January–March):27–45.

Daley, J.B. (1957). "The Man Who Liked Lions." In *SF: '57 The Year's Greatest Science Fiction and Fantasy.* Edited by J. Merril. New York: Gnome.

Dalkey, K. (1991). "You Want It *When?*" In *2041.* Edited by J. Yolen. New York: Delacorte.

Daniels, D.R. (1935). "The Branches of Time." *Wonder Stories,* August.

Daniels, R.D., and G.M. Shore (1994). " 'Faster than Light' Photons and Charged Black Holes." *Nuclear Physics B* 425 (August 29):634–650.

Dann, J. (1997). *Timeshare.* New York: Ace.

Datta, A., D. Home, and A. Raychaudhuri (1987). "A Curious Gedanken Example of the Einstein-Podolsky-Rosen Paradox Using CP Nonconservation." *Physics Letters A* 123 (July 13):4–8.

Davidson, A. (1987). "Tachyonic Compactification." *Physical Review D* 35 (March 15):1811–1814.

Davidson, A., and D.A. Owen (1986). "Elementary Particles as Higher Dimensional Tachyons." *Physics Letters B* 177 (September 4):77–81.

Davies, P.C.W. (1984). "Inflation in the Universe and Time Asymmetry." *Nature* 312 (December 6):524–527.

—— (1983a). *God and the New Physics*. New York: Simon & Schuster.

—— (1983b). "Inflation and Time Asymmetry in the Universe." *Nature* 301 (February 3):398–400.

—— (ET). "Black Hole Thermodynamics and Time Asymmetry."

—— (1981). *The Edge of Infinity*. New York: Simon & Schuster.

—— (RTR). "Time and Reality."

—— (1977). *The Physics of Time Asymmetry*. Los Angeles: University of California Press.

—— (1972a). "Is the Universe Transparent or Opaque?" *Journal of Physics A* 5 (December):1722–1737.

—— (1972b). "Closed Time as an Explanation of the Black Body Background Radiation." *Nature Physical Science* 240 (November):3–5.

Davies, P.C.W., and J. Twamley (1993). "Time-Symmetric Cosmology and the Opacity of the Future Light Cone." *Classical and Quantum Gravity* 10 (May):931–945.

Dawson, J.H. (1989). "Kurt Gödel in Sharper Focus." In *Gödel's Theorem in Focus*. Edited by S.G. Shanker. London: Routledge.

Day, D.B. (1952). *Index to the Science-Fiction Magazines 1926–1950*. Portland, OR: Perri Press.

De, U.K. (1969). "Paths in Universes Having Closed Time-Like Lines." *Journal of Physics A*, 2:427–432.

de Beauregard, O. Costa (1987). *Time, the Physical Magnitude*. Boston: D. Reidel.

—— (1980). "A Burning Question: Einstein's Paradox of Correlations." *Diogenes* 110 (Summer):83–97.

—— (1977). "Two Lectures on the Direction of Time." *Synthese* 35 (June):129–154.

—— (1971). "No Paradox in the Theory of Time Anisotropy." *Stadium Generale* 24:10–18.

—— (NYAS). "Two Principles of the Science of Time."

De Brandt, D.H. (1992). *The Quicksilver Screen*. New York: Del Rey.

de Camp, L. Sprague (1993). "Miocene Romance," "The Satanic Illusion," and "The Big Splash." In *Rivers of Time*. New York: Baen.

—— (1981). "Some Curious Effects of Time Travel." In *Analog Readers' Choice*, New York: Dial.

—— (1979). "Balsamo's Mirror." In *The Purple Pterodactyls*. Huntington Woods, MI: Phantasia Press.

—— (1978). "Language for Time Travelers" and "A Gun for Dinosaur." In *The Best of L. Sprague de Camp*. New York: Ballantine.

—— (1972). "Aristotle and the Gun." In *Alpha Three*. Edited by R. Silverberg. New York: Ballantine.

—— (1970). "The Best-Laid Scheme." In *The Wheels of IF*. New York: Berkley.

—— (1957). "How to Talk Futurian." *Fantasy and Science Fiction*, October.

—— (1941). *Lest Darkness Fall*. New York: Henry Holt.

DeChancie, J., and D. Bischoff (1993). *Dr. Dimension*. New York: ROC.

Dee, R. (1954). "The Poundstone Paradox." *Fantasy and Science Fiction*, May.

Deeping, W. (1940). *The Man Who Went Back*. New York: Knopf.

de Felice, F. (BUFP). "Cosmic Time Machines."

deFord, M.A. (1960). "All in Good Time." *Fantasy and Science Fiction*, July.

—— (1958). "Timequake." *Fantasy and Science Fiction*, December.

—— (DD). "Time Trammel."

De La Cruze, V., and W. Israel (1967). "Gravitational Bounce." *Nuovo Cimento* 51A (October 1):744–760.

Delaire, J. (1904). *Around a Distant Star.* London: John Long.

del Rey, L. (SFF). "My Name Is Legion."

—— (TC). "Unto Him That Hath."

—— (VIT). ". . . And It Comes Out Here."

—— (1951a). "Fool's Errand." *Science Fiction Quarterly,* November.

—— (1951b). "Absolutely No Paradox." *Science Fiction Quarterly,* May.

de Maupassant, G. (1955). "The Horla." In *The Complete Short Stories of Guy de Maupassant.* Garden City, NY: Hanover House.

Denbigh, K.G. (1989). "The Many Faces of Irreversibility." *British Journal for the Philosophy of Science* 40 (December):501–518.

—— (1953). "Thermodynamics and the Subjective Sense of Time." *British Journal for the Philosophy of Science* 4 (November):183–191.

Denman, H.H. (1968). "Time-Translation Invariance for Certain Dissipative Classical Systems." *American Journal of Physics* 36 (June):516–519.

Denruyter, C. (1980). "Jocasta's Crime: A Science-Fiction Reply." *Analysis* 40 (March):71.

Dentinger, S. (1969). "The Future Is Ours." In *Crime Prevention in the 30th Century.* Edited by H.S. Santesson. New York: Walker.

Derleth, A. (1975). "An Eye for History" and "The Penfield Misadventure." In *Harrigan's File.* Sauk City, WI: Arkham House.

Deser, S. (1993). "Physical Obstacles to Time-Travel." *Classical and Quantum Gravity* 10 (Supplement):S67–S73.

Deser, S., *et al.* (1992). "Physical Cosmic Strings Do Not Generate Closed Timelike Curves." *Physical Review Letters* 68 (20 January):267–269.

Deser, S., and R. Jackiw (1992). "Time Travel?" *Comments on Nuclear and Particle Physics* 20 (September):337–354.

Deser, S., and A.R. Steif (DGRI). "No Time Machines from Lightlike Sources in $2+1$ Gravity."

Deutsch, A.J. (OSF). "A Subway Named Moebius."

Deutsch, D. (1991). "Quantum Mechanics Near Closed Timelike Lines." *Physical Review D* 44 (November 15):3197–3217.

—— (1985). "Quantum Theory, the Church–Turing Principle and the Universal Quantum Computer." *Proceedings of the Royal Society* 400A (July):97–117.

Deutsch, D., and P. Candelas (1979). "Boundary Effects in Quantum Field Theory." *Physical Review D* 20 (December 15):3063–3080.

Deutsch, D., and M. Lockwood (1994). "The Quantum Physics of Time Travel." *Scientific American,* March.

Deutch, J.M., and F.E. Low (1993). "Barrier Penetration and Superluminal Velocity." *Annals of Physics* 228 (November 15):184–202.

DeWitt, B.S. (1973). "Quantum Mechanics and Reality." In *The Many-Worlds Interpretation of Quantum Mechanics.* Edited by B.S. DeWitt and N. Graham. Princeton, NJ: Princeton University Press.

Dhar, J., and E.C.G. Sudarshan (1968). "Quantum Field Theory of Interacting Tachyons." *Physical Review* 174 (October 25):1808–1815.

Dick, P.K. (1993). *The World Jones Made.* New York: Vintage.

—— (1980). "The Golden Man" and "Meddler." In *The Golden Man.* New York: Berkley.

—— (1979). *Counter-Clock World,* Boston, MA: Gregg Press.

—— (1977). "Breakfast at Twilight," "Service Call," "Paycheck," and "A Little Something for Us Tempunauts." In *The Best of Philip K. Dick.* New York: Del Rey.

—— (1973). "Psi-Man." In *The Book of Philip K. Dick.* New York: DAW.

—— (1966). "Your Appointment Will Be Yesterday." *Amazing,* August.

—— (1954). "Jon's World." In *Time to Come.* Edited by A. Derleth. New York: Farrar, Straus and Young.

Dieks, D. (1988). "Special Relativity and the Flow of Time." *Philosophy of Science* 55 (September):456–460.

Diener, G. (1996). "Superluminal Group Velocities and Information Transfer." *Physics Letters A* 223 (December 16):327–331.

Dipert, R.R. (1978). "Pierce's Theory of the Dimensionality of Physical Space." *Journal of the History of Philosophy* 16 (January):61–70.

—— (1977). "Pierce's Theory of the Geometrical Structure of Physical Space." *Isis* 68 (September):404–413.

Dirac, P.A.M. (1949). "Forms of Relativistic Dynamics." *Reviews of Modern Physics* 21 (July):392–399.

—— (1938). "Classical Theory of Radiating Electrons." *Proceedings of the Royal Society A* 167 (August):148–168.

Disch, T.M. (1977). "Final Audit," "Genetic Coda," and "102 H-Bombs." In *The Early Science Fiction Stories of Thomas M. Disch.* Boston: Gregg Press.

Dixon, F.W. (1992). *Time Bomb.* New York: Pocket.

Dobbs. H.A.C. (1969). "The 'Present' in Physics." *British Journal for the Philosophy of Science* 19:317–324.

—— (1965). "Time and ESP." *Proceedings of the Society for Psychical Research* 54 (August):249–361.

—— (1956). "The Time of Physics and Psychology." *British Journal for the Philosophy of Science* 7 (August):156–160.

—— (1951a). "The Relation Between the Time of Psychology and the Time of Physics. Part II." *British Journal for the Philosophy of Science* 2 (November):177–192.

—— (1951b). "The Relation Between the Time of Psychology and the Time of Physics. Part I." *British Journal for the Philosophy of Science* 2 (August):122–137.

Donald, J.A. (1978). "Assumptions of the Singularity Theorems and the Rejuvenation of Universes." *Annals of Physics* 110 (February):251–273.

Donitz, H. (1928). "A Visitor from the Twentieth Century." *Amazing Stories,* May.

Dorling, J. (1970). "The Dimensionality of Time." *American Journal of Physics* 38 (April):539–540.

Dowe, P. (1997). "A Defense of Backwards-in-Time Causation Models in Quantum Mechanics." *Synthese* 112 (August):233–246.

Drake, D. (1993). "Time Safari" and "Boundary Layer." In *Tyrannosaur.* New York: Tor.

Dretske, F.I. (1962). "Moving Backward in Time." *Philosophical Review* 71 (January):94–98.

Driver, R.D. (1979). "Can the Future Influence the Present?" *Physical Review D* 19 (February 15):1098–1107.

Drummond, I.T., and S.J. Hathrell (1980). "QED Vacuum Polarization in a Background Gravitational Field and Its Effect on the Velocity of Photons." *Physical Review D* 22 (July 15):343–355.

Duclos, M. (1939). "Into Another Dimension." *Fantastic Adventures,* November.

du Maurier, D. (1969).*The House on the Strand.* Garden City, NY: Doubleday.

Dummett, M. (NOT). "Causal Loops."

—— (1969). "A Defense of McTaggart's Proof of the Unreality of Time." *Philosophical Review* 69 (October):497–504.

—— (1964). "Bringing About the Past." *Philosophical Review* 73 (July):338–359.

Dummett, M.E., and A. Flew (1954). "Can an Effect Precede Its Cause?" *Aristotelian Society Supplement* 28:27–62.

Dunn, J.R. (1997). *Davs of Cain.* New York: Avon.

Dunne, J.W. (1958). *An Experiment with Time.* London: Faber and Faber.

Dunsany, Lord (1948). "Lost." In *The Fourth Book of Jorkens.* Sauk City, WI: Arkham House.

—— (1929). *The Jest of Hahalaba*. In *Seven Modern Comedies*. New York: G.P. Putnam's Sons.

Dwyer, L. (1978). "Time Travel and Some Alleged Logical Asymmetries Between Past and Future." *Canadian Journal of Philosophy* 8 (March):15–38.

—— (1977). "How to Affect, But Not Change, the Past." *Southern Journal of Philosophy* 15:383–385.

—— (1975). "Time Travel and Changing the Past." *Philosophical Studies* 27 (May):341–350.

Dye, C. (1953). "Time Goes to Now." *Science Fiction Quarterly*, May.

Dyson, F.J. (1979). "Time Without End: Physics and Biology in an Open Universe." *Reviews of Modern Physics* 51 (July):447–460.

Earman, J. (1996). "Tolerance for Spacetime Singularities." *Foundations of Physics* 26 (May):623–640.

—— (1995a). *Bangs, Crunches, Whimpers, and Shrieks: Singularities and Acausalities in Relativistic Spacetimes*. New York: Oxford University Press.

—— (1995). "Outlawing Time Machines: Chronology Protection Theorems." *Erkenntnis* 42 (March):125–139.

—— (1976). "Causation: A Matter of Life and Death." *Journal of Philosophy* 73 (January 15):5–25.

—— (1974). "An Attempt to Add a Little Direction to 'The Problem of the Direction of Time' " *Philosophy of Science* 41 (March):15–47.

—— (1972). "Implications of Causal Propagation Outside the Null Cone." *Australasian Journal of Philosophy* 50 (December):222–237.

—— (1971). "Kant, Incongruous Counterparts, and the Nature of Space and Space-Time." *Ratio* 13 (June):1–18.

—— (1970a). "The Closed Universe." *Nous* 4 (September):261–269.

—— (1970b). "Space-Time or How to Solve Philosophical Problems and Dissolve Philosophical Muddles Without Really Trying." *Journal of Philosophy* 67 (May):259–276.

—— (1969). "The Anisotropy of Time." *Australasian Journal of Philosophy* 47 (December):273–295.

—— (1967a). "On Going Backward in Time." *Philosophy of Science* 34 (September):211–222.

—— (1967b). "Irreversibility and Temporal Asymmetry." *Journal of Philosophy* 64 (September):543–549.

Earman, J., and M. Friedman (1973). "The Meaning and Status of Newton's Laws of Inertia and the Nature of Gravitational Forces." *Philosophy of Science* 40 (September):329–359.

Earman, J., and J.D. Norton (1993). "Forever Is a Day: Supertasks in Pitowsky and Malament–Hogarth Spacetimes." *Philosophy of Science* 60 (March):22–42.

Echeverria, F. (1993). "Gravitational Collapse of an Infinite, Cylindrical Dust Shell." *Physical Review D* 47 (March 15):2271–2282.

Echeverria, F., G. Klinkhammer, and K.S. Thorne (1991). "Billiard Balls in Wormhole Spacetimes with Closed Timelike Curves: I. Classical Theory." *Physical Review D* 44 (August 15):1077–1099.

Eddington, A.S. (1935). "The End of the World." In *New Pathways in Science*. New York: Macmillan.

—— (1929). *The Nature of the Physical World*. New York: Macmillan.

Edmondson, G.C. (1965a). "The Misfit." In *Stranger Than You Think*. New York: Ace.

—— (1965b). *The Ship That Sailed the Time Stream*. New York: Ace.

Effinger, G.A. (WMHB). "Everything But Honor."

—— (1986). *The Bird of Time*. Garden City, NY: Doubleday.

—— (1985). *The Nick of Time*. Garden City, NY: Doubleday.

Efron, R. (NYAS). "The Duration of the Present."

Eggleston, G.C. (1875a). "The True Story of Bernard Poland's Prophecy." *American Homes* (June):80–84.

—— (1875b). "Who Is Russell?" *American Homes* (March):276–281.

Ehring, D. (1987). "Personal Identity and Time Travel." *Philosophical Studies* 52 (November):427–433.

Einstein, A. (1961). *Relativity: the Special and the General Theory.* New York: Crown.

—— (TPR). "On the Electrodynamics of Moving Bodies" and "The Foundation of the General Theory of Relativity."

—— (1949). "Reply to Criticisms." In *Albert Einstein,* Vol. 7 of *The Library of Living Philosophers.* Edited by P.A. Schilpp. Evanston, IL: Open Court, pp. 687–688.

—— (1939). "On a Stationary System with Spherical Symmetry Consisting of Many Gravitating Masses." *Annals of Mathematics* 40 (October):922–936.

—— (1935a). "The Particle Problem in the General Theory of Relativity." *Physical Review* 48 (July 1):73–77.

—— (1935b). "Can Quantum-Mechanical Description of Physical Reality Be Considered Complete?" *Physical Review* 47 (May 15):777–780.

—— (1931). "Knowledge of Past and Future in Quantum Mechanics." *Physical Review* 37 (March 15):780–781.

—— (1922). "La Théorie de la Relativité." *Bulletin de la Société Française de Philosophie* 17:91–113.

Eisenberg, L. (AO). "The Time of His Life."

Eisenstaedt, J. (1993). "Dark Bodies and Black Holes, Magic Circles and Montgolfiers: Light and Gravitation from Newton to Einstein." *Science in Context* 6 (Spring):83–106.

Eisentein, A., and P. Eisentein (1971). "The Trouble with the Past." In *New Dimensions 1.* Edited by R. Silverberg. Garden City, NY: Doubleday.

Ekert, A.K. (1991). "Quantum Cryptography Based on Bell's Theorem." *Physical Review Letters* 67 (August 5):661–663.

Eklund, G. (FST). "The Ambiguities of Yesterday."

—— (U2). "Stalking the Sun."

Elliot, R. (1982). "Going Nowhere Fast?" *Analysis* 42 (October):213–215.

—— (1981). "How to Travel Faster Than Light?" *Analysis* 41 (January):4–6.

Elliott, B. (FM). "The Last Magician."

Ellis, H.G. (1974). "Time, the Grand Illusion." *Foundations of Physics* 4 (June): 311–319.

—— (1973). "Ether Flow Through a Drainhole: A Particle Model in General Relativity." *Journal of Mathematical Physics* 14 (January):104–118.

Ellison, H. (1989). "Soldier." In *The Great Science Fiction Stories.* Vol. 19. New York: DAW.

—— (1970). "One Life, Furnished in Early Poverty." In *Orbit 8.* Edited by D. Knight, New York: Berkley.

—— (DV). "The Prowler in the City at the Edge of the World."

Ells, E. (1988). "Quentin Smith on Infinity and the Past." *Philosophy of Science* 55 (March):453–455.

England, G.A. (1905). "The Time Reflector." *The Monthly Story Magazine,* September.

Epstein, H., V. Glaser, and A. Jaffe (1965). "Nonpositivity of the Energy Density in Quantized Field Theories." *II Nuovo Cimento* 36 (April 1):1016–1022.

Ernst, P. (BSF). "The 32nd of May."

Eshbach, L.A. (1938). "Out of the Past." *Tales of Wonder,* Autumn.

—— (1932). "The Time Conqueror." *Wonder Stories,* July.

Everett, A.E. (1996). "Warp Drive and Causality." *Physical Review D* 53 (June 15):7365–7368.

Everett, A.E., and A.F. Antippa (CPT). "Tachyons, Causality, and Rotational Invariance."

Everett, A.E., and T.A. Roman (1997). "Superluminal Subway: The Krasnikov Tube." *Physical Review D* 56 (August 15):2100–2108.

Everett, H. III (1957). " 'Relative State' Formulation of Quantum Mechanics." *Reviews of Modern Physics* 29 (July):454–462.

Fair, D. (1979). "Causation and the Flow of Energy." *Erkenntnis* 14 (November):219–250.

Fallon, S.M. (1988). " 'To Act or Not': Milton's Conception of Divine Freedom." *Journal of the History of Ideas* 49 (July–September):425–449.

Farber, S.N. (1980). "Trans Dimensional Imports." *Isaac Asimov's Science Fiction Magazine,* August.

Farhi, E., and A.H. Guth (1987). "An Obstacle to Creating a Universe in the Laboratory." *Physics Letters B* 183 (January 8):149–155.

Farley, R.M. (1950). "The Man Who Met Himself," "I Killed Hitler," "Rescue into the Past," "The Immortality of Alan Whidden," "The Time-Wise Guy," "A Month a Minute," "The Invisible Bomber," and "Time for Sale." In *The Omnibus of Time.* Los Angeles: Fantasy Publishing.

—— (1939). "Visualizing Hyperspace," *Scientific American,* March.

Farmer, P.J. (1986). "Sail On! Sail On!" In *The Great Science Fiction Stories: Vol. 14.* New York: DAW.

—— (1977). "Sketches Among the Ruins of My Mind." In *Strangeness.* Edited by T.M. Disch and C. Naylor. New York: Charles Scribner's Sons.

—— (1975). *Time's Last Gift.* New York: Granada.

Farrell, J.W. (1949). "All Our Yesterdays." *Super Science Stories,* April.

Fast, H. (1959). "Of Time and Cats." *Fantasy and Science Fiction,* March.

Faye, J. (1989). *The Reality of the Future.* Denmark: Odense University Press.

—— (1987). "The Past Revisited." *Danish Yearbook of Philosophy.* Vol. 24, pp. 7–18. Copenhagen: Museum Tusculanum Press.

Fearn, J.R. (MBSFS). "Wanderer of Time."

Fehrenbach, T.R. (1963). "Remember the Alamo!" In *ANALOG 1.* Edited by J.W. Campbell. New York: Doubleday.

Feinberg, G. (1967). "Possibility of Faster-Than-Light Particles." *Physical Review* 159 (July 25):1089–1105.

Feinberg, G., D. Albert, and S. Levine (1992). "Knowledge of the Past and Future." *Journal of Philosophy* 89 (December):607–642.

Feynman, R.P. (1995). *Feynman Lectures on Gravitation.* Reading, MA: Addison-Wesley.

—— (1965). "The Distinction of Past and Future." In *The Character of Physical Law.* Cambridge, MA: M.I.T. Press.

—— (1963). "Relativity and the Philosophers." In *The Feynman Lectures on Physics.* Vol. 1. Reading, MA: Addison-Wesley.

—— (1949a). "The Theory of Positrons." *Physical Review* 76 (September 15):749–759.

—— (1949b). "Space-Time Approach to Quantum Electrodynamics." *Physical Review* 76 (September 15):769–789.

—— (1948). "A Relativistic Cut-Off for Classical Electrodynamics." *Physical Review* 74 (October 15):939–946.

Feynman, R.P. and A.R. Hibbs (1965). *The Path Integral Formulation of Quantum Mechanics.* New York: McGraw-Hill.

Findlay, J.N. (1978). "Time and Eternity." *Review of Metaphysics* 32 (September):3–14.

Findlay, J.N., and J.E. McGechie (1956). "Does It Make Sense to Suppose That All Events, Including Personal Experiences, Could Occur in Reverse?" *Analysis* 16 (June):121–123.

Finkelstein, D. (1958). "Past-Future Asymmetry of the Gravitational Field of a Point Particle." *Physical Review* 110 (May 15):965–967.

Finney, J. (FCW). "Quit Zoomin' Those Hands Through the Air."

―― (1986). "The Third Level," "I Love Galesburg in the Springtime," "Such Interesting Neighbors," "Of Missing Persons," "Where the Cluetts Are," "The Face in the Photo," "I'm Scared," "The Coin Collector," and "Second Chance." In *About Time*. New York: Simon and Schuster.

―― (TOT). "The Love Letter."

―― (1970). *Time and Again*. New York: Simon and Schuster.

―― (1966). "Double Take." In *The Playboy Book of Science Fiction and Fantasy*. Chicago: Playboy Press.

Fischer, J.M. (1984). "Power Over the Past." *Pacific Philosophical Quarterly* 65 (October):335–350.

Fisk, M. (1963). "Cause and Time in Physical Theory." *Review of Metaphysics* 16 (March):522–549.

Fitzgerald, F.S. (1944). "The Curious Case of Benjamin Button." In *Pause to Wonder*. New York: J. Messner.

Fitzgerald, P. (1985). "Four Kinds of Temporal Becoming." *Philosophical Topics* 13 (Fall):145–177.

―― (1974). "On Retrocausality." *Philosophia* 4 (October):513–551.

―― (1972). "Relativity Physics and the God of Process Philosophy." *Process Studies* 2 (Winter):251–276.

―― (1970). "Tachyons, Backwards Causation, and Freedom." In *Boston Studies in the Philosophy of Science*. Vol. 8, pp. 415–436.

―― (1969). "The Truth About Tomorrow's Sea Fight." *Journal of Philosophy* 66 (June 5):307–329.

Flagg, F. (1930a). "The Lizard-Men of Buh-lo." *Wonder Stories*, October.

―― (1930b). "An Adventure in Time." *Science Wonder Stories*, April.

―― (1927). "The Machine Man of Ardathia." *Amazing Stories*, November.

Flagg, F., and W. Wright (1947). "The Time Twister." *Thrilling Wonder Stories*, October.

Flammarion, C. (1897). *Lumen*. New York: Dodd, Mead and Co.

Flanagan, E.E., and R.M. Wald (1996). "Does Back Reaction Enforce the Averaged Null Energy Condition in Semiclassical Gravity?" *Physical Review D* 54 (November 15):6233–6283.

Fleisher, M.L. (1978). *The Encyclopedia of Comic Book Heroes*. New York: Warner.

Flynn, M. F. (1998). "The Forest of Time." In *Roads Not Taken*. Edited by G. Dozois and S. Schmidt. New York: Del Rey.

Folacci, A. (1992). "Averaged-Null-Energy Condition for Electromagnetism in Minkowski Spacetime." *Physical Review D* 46 (September 15):2726–2729.

Foote, B. (1991). *The Connecticut Yankee in the Twentieth Century: Travel to the Past in Science Fiction*. Westport, CT: Greenwood Press.

Ford, J.M. (FCW). "Slowly By, Lorena."

Ford, L.S. (1968). "Is Process Theism Compatible with Relativity Theory?" *Journal of Religion* 48 (April):124–135.

Ford, L.H. (1978). "Quantum Coherence Effects and the Second Law of Thermodynamics." *Proceedings of the Royal Society of London A* 364 (December 12):227–236.

Ford, L.H., M.J. Pfenning, and T.A. Roman (1998). "Quantum Inequalities and Singular Negative Energy Densities." *Physical Review D* 57 (April 15):4839–4846.

Ford, L.H., and T.A. Roman (1997). "Restrictions on Negative Energy Density in Flat Spacetime." *Physical Review D* 55 (February 15):2082–2089.

―― (1993). "Motion of Inertial Observers Through Negative Energy." *Physical Review D* 48 (July 15):776–782.

Forrest, P. (1985). "Backward Causation in Defense of Free Will." *Mind* 94 (April):210–217.

Fortenay, C.L. (1994). "An M-1 at Fort Donelson." *Analog*, July.

Forward, R. (1995). "Twin Paradox." In *Indistinguishable from Magic*. New York: Baen.
—— (1992). *Timemaster*, New York: Tor.
—— (1989). "Space Warps: A Review of One Form of Propulsionless Transport." *Journal of British Interplanetary Society* 42:533–542.
—— (1988). "Time Magic." In *Future Magic*. New York: Avon.
—— (1980). "How to Build a Time Machine." *Omni*, May.
Fox, D. (1952). "The Tiniest Time Traveler." *Astounding Science Fiction*, December.
Fox, R., C.G. Kuper, and S.G. Lipson (1970). "Faster-than-Light Group Velocities and Causality Violations." *Proceedings of the Royal Society of London* 316 A (May): 515–524.
—— (1969). "Do Faster-than-Light Group Velocities Imply Violation of Causality?" *Nature* 223 (August 9):597.
Frank, E. (1948). "Time and Eternity." *Review of Metaphysics* 2 (September):39–52.
Frank, P. (BST). "Is the Future Already Here?"
—— (1957). *Philosophy of Science*. Englewood Cliffs, NJ: Prentice-Hall.
Frankel, L. (1986). "Mutual Causation, Simultaneity, and Event Description." *Philosophical Studies* 49 (May):361–372.
Frankel, T. (1988). "Electric Currents in Multiply Connected Spaces." *International Journal of Theoretical Physics* 27 (August):995–999.
Franklin, H.B. (1966). *Future Perfect*. New York: Oxford University Press.
Frankowski, L.A. (1986). *The Cross-Time Engineer*. New York: Del Rey.
Franson, D. (MT). "Package Deal."
—— (1980). "One Time in Alexandria." *Analog*, June.
Fraser, J.T. (1978). *Time as Conflict*. Basel: Birkhauser Verlag.
—— (VOT). "Note Relating to a Paradox of the Temporal Order."
Frautschi, S. (1982). "Entropy in an Expanding Universe." *Science* 217 (August 13):593–599.
Freddoso, A.J. (1982). "Accidental Necessity and Power Over the Past." *Pacific Philosophical Quarterly* 63 (January):54–68.
Freedman, D.Z., and P. van Nieuweuhuizen (1985). "The Hidden Dimensions of Space-time." *Scientific American*, March.
Friedell, E. (1972). *The Return of the Time Machine*. New York: DAW.
Friedman, J.L., *et al.* (1992). "Failure of Unitarity for Interacting Fields on Spacetimes with Closed Timelike Curves." *Physical Review D* 46 (November 15):4456–4469.
Friedman, J.L., *et al.* (1990). "Cauchy Problem in Spacetimes with Closed Timelike Curves." *Physical Review D* 42 (September 15):1915–1930.
Friedman, J.L. (1988). "Back to the Future." *Nature* 336 (November 24):305–306.
Friedman, J.L., and M.S. Morris (1991). "The Cauchy Problem for the Scalar Wave Equation Is Well Defined on a Class of Spacetimes with Closed Time-Like Lines." *Physical Review Letters* 66 (January 28):401–404.
Friedman, M. (1983). *Foundations of Space-Time Theories*. Princeton, NJ: Princeton University Press.
Friend, B. (1982). "Time Travel as a Feminist Didactic in Works by Phyllis Eisenstein, Marlys Millhiser, and Octavia Butler." *Exploration* 23 (Spring):50–55.
Frolov, V.P. (1993). "Topology, Causality, and Chronology Protection." In *Proceedings of the First Iberian Meeting on Gravity*. Edited by M.C. Bento *et al*. Singapore: World Scientific.
—— (1991). "Vacuum Polarization in a Locally Static Multiply Connected Spacetime and a Time-Machine Problem." *Physical Review D* 43 (June 15):3878–3894.
Frolov, V.P., and I.D. Novikov (1990). "Physical Effects in Wormholes and Time Machines." *Physical Review D* 42 (August 15):1057–1065.
Frolov, V.P., *et al.* (1989). "Through a Black Hole Into a New Universe?" *Physics Letters B* 216 (January 12):272–276.
Fubank, J. (1992). *Crossover*. New York: Carroll & Graf.

Fuller, A.M. (1890). *A.D. 2000*. Chicago: Laird & Lee.

Fuller, R.W., and J.A. Wheeler (1962). "Causality and Multiply Connected Space-Time." *Physical Review* 128 (October 15):919–929.

Fulmer, G. (1983). "Cosmological Implications of Time Travel." In *The Intersection of Science Fiction and Philosophy*. Edited by R.E. Meyers. Westport, CT: Greenwood Press.

—— (1981). "Time Travel, Determinism, and Fatalism." *Philosophical Speculations in Science Fiction and Fantasy* 1 (Spring):41–48.

—— (1980). "Understanding Time Travel." *Southwestern Journal of Philosophy* 11 (Spring):151–156.

Gale, R. (1966). "McTaggart's Analysis of Time." *American Philosophical Quarterly* 3 (April):145–152.

—— (1965). "Why a Cause Cannot Be Later Than Its Effect." *Review of Metaphysics* 19:209–234.

—— (1964). "Is It Now Now?" *Mind* 73 (January):97–105.

—— (1963). "Some Metaphysical Statements About Time." *Journal of Philosophy* 60 (April 25):225–237.

Gallois, A. (1994). "Asymmetry in Attitudes and the Nature of Time." *Philosophical Studies* 76 (October):51–69.

Gal-Or, B. (MDT). "On a Paradox-Free Definition of Time and Irreversibility."

Gamow, G. (ED). "The Heart on the Other Side."

—— (1946). "Rotating Universe?" *Nature* 158 (October 19):549.

Gansovsky, S. (1984). "Vincent Van Gogh." In *Aliens, Travelers, and Other Strangers*. New York: Macmillan.

Gardner, M. (MTMW). "Left or Right?

—— (1982). *aha! Gotcha*. New York: W.H. Freeman.

—— (1980). "Mathematical Games." *Scientific American*, July.

—— (1979). "Mathematical Games." *Scientific American*, March.

—— (1974). "Mathematical Games." *Scientific American*, May.

—— (1969). "The Church of the Fourth Dimension." In *The Unexpected Hanging*. New York: Simon and Schuster.

—— (FM). "No-Sided Professor" and "The Island of Five Colors."

Garrett, R. (1994). "Time Fuze." In *The Ascent of Wonder: The Evolution of Hard SF*. Edited by D.G. Hartwell and K. Cramer. New York: Tor.

—— (TW). "Frost and Thunder."

Gaskin, R. (1997). "Peter Damian on Divine Power and the Contingency of the Past." *The British Journal for the History of Philosophy* 5 (September):229–247.

Gatlin, L.L. (1980). "Time-Reversed Information Transmission." *International Journal of Theoretical Physics* 19 (January):25–29.

Geach, P. (1969). *God and the Soul*. London: Routledge & Kegan Paul.

—— (1968). "Some Problems About Time." In *Studies in the Philosophy of Thought and Action*. Edited by P.F. Strawson. New York: Oxford.

Geroch, R. (1984). "The Everett Interpretation." *Nous* 18 (November):617–633.

—— (1968). "What Is a Singularity in General Relativity?" *Annals of Physics* 48 (July):526–540.

—— (1967). "Topology in General Relativity." *Journal of Mathematical Physics* 8 (April):782–786.

Gerrold, D. (1973). *The Man Who Folded Himself*. New York: Random House.

Ghirardi, G.C., A. Rimini, and T. Weber (1980). "A General Argument Against Superluminal Transmission Through the Quantum Mechanical Measurement Process." *Lettere Al Nuovo Cimento* 27 (March 8):293–298.

Gibbons, G.W., and S.W. Hawking (1992). "Kinks and Topology Change." *Physical Review Letters* 69 (September 21):1719–1721.

Gibbons, G.W., and R.A. Russell-Clark (1973). "Note on the Sato-Tomimatsu Solution of Einstein's Equations." *Physical Review Letters* 30 (February 26):398–399.

Gibson, W.M. (1969). *Mark Twain's Mysterious Stranger Manuscripts.* Los Angeles, CA: University of California.

Gillespie, A. (NWF). "The Evil Eye."

Gisin, N. (1990). "Weinberg's Non-linear Quantum Mechanics and Superluminal Communications." *Physics Letters A* 143 (January 1):1–2.

Giuliano, C.R. (1981). "Applications of Optical Phase Conjugation." *Physics Today* 34 (April):27–35.

Gödel, K. (BST). "Static Interpretation of Space-Time with Einstein's Comment on It."

—— (1952). "Rotating Universes in General Relativity Theory." *Proceedings of the International Congress of Mathematicians* 1:175–181.

—— (1949a). "A Remark About the Relationship Between Relativity Theory and Idealistic Philosophy." In *Albert Einstein: Philosopher-Scientist.* Vol. 7 of *The Library of Living Philosophers.* Edited by P.A. Schilpp. Evanston, IL: Open Court.

—— (1949b). "An Example of a New Type of Cosmological Solutions of Einstein's Field Equations of Gravitation." *Reviews of Modern Physics* 21 (July):447–450.

Godfrey-Smith, W. (1980). "Traveling in Time." *Analysis* 40 (March):72–73.

—— (1979). "Special Relativity and the Present." *Philosophical Studies* 36 (October):233–244.

Gold, H.L. (MT). "The Biography Project."

—— (1955). "The Old Die Rich." In *The Old Die Rich and Other Science Fiction Stories.* New York: Crown.

—— (SFAD). "Perfect Murder."

Gold, T. (MDT). "The World Map and the Apparent Flow of Time."

—— (1966). "Cosmic Processes and the Nature of Time." In *Mind and Cosmos.* Edited by R.G. Colodny. Pittsburgh, PA: University of Pittsburgh Press.

—— (1962). "The Arrow of Time." *American Journal of Physics* 30 (June):403–410.

Goldstone, C., and A. Davidson (1970). "Pebble in Time." *Fantasy and Science Fiction,* August.

Goldwirth, D.S., M.J. Perry, and T. Piran (1993). "The Breakdown of Quantum Mechanics in the Presence of Time Machines." *General Relativity and Gravitation* 25 (January):7–13.

Goldwirth, D.S., M.J. Perry, T. Piran, and K.S. Thorne (1994). "Quantum Propagator for a Nonrelativistic Particle in the Vicinity of a Time Machine." *Physical Review D* 49 (April 15):3951–3957.

Gonella, F. (1994). "Time Machine, Self-Consistency and the Foundation of Quantum Mechanics." *Foundations of Physics Letters* 7 (April):161–166.

González-Diáz, P.F. (1997). "Observable Effects from Spacetime Tunneling." *Physical Review D* 56 (November 15):6293–6297.

—— (1996). "Ringholes and Closed Timelike Curves." *Physical Review D* 54 (November 15):6122–6131.

Good, I.J. (1965). "Winding Space." In *The Scientist Speculates.* Edited by I.J. Good. New York: Capricorn.

Gor, G. (1969). "The Garden" and "The Minotaur." In *Russian Science Fiction.* Edited by R. Magidoff. New York: New York University Press.

Gordon, A. (1987). "*Back to the Future:* Oedipus as Time Traveller." *Science-Fiction Studies* 14 (November):372–385.

—— (1982). "Silverberg's Time Machine." *Extrapolation* 23 (Winter):345–361.

Gordon, R. (1955). *Utopia 239.* London: William Heinemann.

Gorovitz, S. (1964). "Leaving the Past Alone." *Philosophical Review* 73 (July):360–371.

Goswami, A. (1985). *The Cosmic Dancers: Exploring the Science of Science Fiction.* New York: McGraw-Hill, 1985.

Gott, J.R. (1991). "Closed Timelike Curves Produced by Pairs of Moving Cosmic Strings: Exact Solutions." *Physical Review Letters* 66 (March 4):1126–1129.

—— (1985). "Gravitational Lensing Effects of Vacuum Strings: Exact Solutions." *The Astrophysical Journal* 288 (January 15):422–427.

—— (1974). "A Time-Symmetric, Matter, Antimatter, Tachyon Cosmology." *Astrophysical Journal* 187 (January 1):1–3.

Gott, J.R., and M. Alpert (1984). "General Relativity in a (2 + 1)-Dimensional Space-Time." *General Relativity and Gravitation* 16 (March):243–247.

Gott, J.R., and L.-X. Li (1998). "Can the Universe Create Itself?" *Physical Review D* 58:023501.

Goulart, R. (1975). "Plumrose." In *Odd Job #101*. New York: Charles Scribner's Sons.

Gould, J.D. (1969). "Hypothetical History." *Economic History Review* 22:195–207.

Grant, J.D.E. (1993). "Cosmic Strings and Chronology Protection." *Physical Review D* 47 (March 15):2388–2394.

Graves, J.C., and J.E. Roper (1965). "Measuring Measuring Rods." *Philosophy of Science* 32 (January):39–56.

Grendon, E. (TSF). "The Figure."

Gribbin, J. (1990). "Don't Look Back." *Interzone,* October.

—— (1983). *Spacewarps.* New York: Delacorte.

—— (1979). *Time Warps.* New York: Delacorte.

Gribbin, J., and M. Rees (1989). *Cosmic Coincidences.* New York: Bantam.

Griffith, G. (1899). "The Conversion of the Professor." *Pearson's Magazine,* May.

Grigoriev, V. (LDA). "Vanya."

Grimwood, K. (1988). *Replay.* New York: Berkley.

—— (1976). *Breakthrough.* Garden City, NY: Doubleday.

Gross, C. (1985). "Twelfth-Century Concepts of Time: Three Reinterpretations of Augustine's Doctrine of Creation *Simul.*" *Journal of the History of Philosophy* 23 (July):325–338.

Gross, M. (SSFT). "The Good Provider."

Grünbaum, A. (1993). "Narlikar's 'Creation' of the Big Bang University Was a Mere Origination." *Philosophy of Science* 60 (December):638–646.

—— (1990). "Pseudo-creation of the Big Bang." *Nature* 344 (April 26):821–822.

—— (1989). "The Pseudo-Problem of Creation in Physical Cosmology." *Philosophy of Science* 56 (September):373–394.

—— (1976). "Is Preacceleration of Particles in Dirac's Electrodynamics a Case of Backward Causation? The Myth of Retrocausation in Classical Electrodynamics." *Philosophy of Science* 43 (June):165–201.

—— (1973). "Geometrodynamics and Ontology." *Journal of Philosophy* 70 (December 6):775–800.

—— (BIPT). "The Meaning of Time."

—— (1963). "Is There a 'Flow' of Time or Temporal Becoming?" In *Philosophical Problems of Space and Time.* New York: Knopf.

—— (1955). "Time and Entropy." *American Scientist* 43 (October).

Grünbaum, A., and A.I. Janis (1978). "Can the Effect Precede Its Cause in Classical Electrodynamics?" *American Journal of Physics* 46 (April):337–341.

—— (1977). "Is There Backward Causation in Classical Electrodynamics?" *Journal of Philosophy* 74 (August):475–482.

Gunawant, R., and B.S. Rajput (1985). "Tachyonic Dyons in Six-Dimensional Space." *Lettere Al Nuovo Cimento* 43 (July 1):219–222.

Gunn, J. (TTF). "The Reason Is with Us."

—— (1979). *The Road to Science Fiction.* Vol. 3. New York: New American Library.

—— (1975). *Alternate Worlds: The Illustrated History of Science Fiction.* Englewood Cliffs, NJ: A&W Visual Library.

Gunn, L.D. (1939). "The Time Twin." *Thrilling Wonder Stories,* August.

Guth, A.H., and P. Steinhardt (NP). "The Inflationary Universe."

Haggard, J.H. (1937). "He Who Masters Time." *Thrilling Wonder Stories*, February.

—— (1935). "Relativity to the Rescue." *Amazing Stories*, April.

—— (1930). "Faster Than Light." *Wonder Stories*, October.

Haining, P. (1987). *Doctor Who: The Time-Travellers' Guide*. London: W.H. Allen.

Haldane, J.B.S. (1928). "The Universe and Irreversibility." *Nature* 122 (November 24):808–809.

Haldeman, J. (1990). *The Hemingway Hoax*. New York: William Morrow.

—— (1985). "No Future in It." In *Dealing in Futures*. New York: Viking.

—— (1984). *The Forever War*. New York: Del Rey.

Hale, E.E. (1986). "Hands Off." In *Alternative Histories*. Edited by C.G. Waugh and M.H. Greenberg. New York: Garland Publishing.

Hall, C.F. (1938). "The Man Who Lived Backwards." *Tales of Wonder*, Summer.

Hamilton, E. (LME). "In the World's Dusk."

—— (1974). "The Man Who Evolved." In *Before the Golden Age*. Edited by I. Asimov. Vol. 1. Greenwich, CT: Fawcett.

—— (MBSFS). "The Inn Outside the World."

—— (1930). "The Man Who Saw the Future." *Amazing Stories*, October.

Harari, D., and A.P. Polychronakos (1988). "Gravitational Time Delay Due to a Spinning String." *Physical Review D* 38 (November 15):3320–3322.

Harness, C.L. (1994). "The Tetrahedron." *Analog*, January.

—— (FCW). "Quarks at Appomattox."

—— (1988). *Krono*. New York: Franklin Watts.

—— (TTT). "Time Trap."

—— (BFSF3). "Child by Chronos."

Harris, C.W. (1947). "The Fifth Dimension." In *Away From the Here and Now*. Philadelphia: Dorrance.

Harris, E.E. (1968). "Simultaneity and the Future." *British Journal for the Philosophy of Science* 19:254–256.

Harrison, B.K., *et al.* (1965). *Gravitational Theory and Gravitational Collapse*. Chicago: University of Chicago Press.

Harrison, H. (1983). *A Rebel in Time*. New York: Tor.

—— (1975). "The Secret of Stonehenge." In *The Ancient Mysteries Reader*. New York: Doubleday.

—— (1974). "The Ever-Branching Tree." In *School and Society Through Science Fiction*. Edited by J.D. Olander, M.H. Greenberg, and P. Warrick. Chicago: Rand McNally.

—— (1967). *The Technicolor Time Machine*. Garden City, NY: Doubleday.

Harrison, J. (1980). "Report on *Analysis* Problem No. 18." *Analysis* 40 (March):65–69.

—— (1979). "Jocasta's Crime." *Analysis* 39 (March):65.

—— (1971). "Dr. Who and the Philosophers, or Time Travel for Beginners." *Aristotelean Society Supplement* 45:1–24.

Harrison, K.L. (1940). "The Blonde, the Time Machine and Johnny Bell." *Thrilling Wonder Stories*, December.

Hartland-Swann, J. (1955). "The Concept of Time." *Philosophical Quarterly* 5 (January):1–20.

Hartshorne, C. (HJ). "The Reality of the Past, the Unreality of the Future."

Hawking, S.W. (1995). "Quantum Coherence and Closed Timelike Curves." *Physical Review D* 52 (November 15):5681–5686.

—— (1994). "The No Boundary Condition and the Arrow of Time." In *Physical Origins of Time Asymmetry*. Edited by J.J. Halliwell, J. Pérez-Mercader, and W.H. Zurek. Cambridge, England: Cambridge University Press.

—— (1992). "Chronology Protection Conjecture." *Physical Review D* 46 (July 15):603–611.

—— (1990). "Gravitational Radiation from Collapsing Cosmic String Loops." *Physics Letters B* 246 (August 23):36–38.

—— (1988). *A Brief History of Time.* New York: Bantam.

—— (1985). "Arrow of Time in Cosmology." *Physical Review D* 32 (November 15):2489–2495.

—— (1983). "The Cosmological Constant." *Philosophical Transactions of the Royal Society of London A* 310 (December 20):303–310.

—— (1982). "The Unpredictability of Quantum Gravity." *Communications in Mathematical Physics* 87 (December):395–415.

—— (1978). "Quantum Gravity and Path Integrals." *Physical Review D* 18 (September 15):1747–1753.

—— (1977). "The Quantum Mechanics of Black Holes." *Scientific American,* January.

—— (1976). "Breakdown of Predictability in Gravitational Collapse." *Physical Review D* 14 (November 15):2460–2473.

—— (1975). "Particle Creation by Black Holes." *Communications in Mathematical Physics* 43:199–220.

—— (1974). "Black Hole Explosions?" *Nature* 248 (March 1):30–31.

—— (1968). "The Existence of Cosmic Time Functions." *Proceedings of the Royal Society* 308A (December 11):433–435.

Hawking, S.W., and G.F.R. Ellis (1973). *The Large Scale Structure of Space-Time.* Cambridge, England: Cambridge University Press.

Hawking, S.W., R. Laflamme, and G.W. Lyons (1993). "Origin of Time Asymmetry." *Physical Review D* 47 (15 June):5342–5356.

Hawking, S.W., and R. Penrose (1970). "The Singularities of Gravitational Collapse and Cosmology." *Proceedings of the Royal Society* 314A (January 27):529–548.

Hawking, S.W., and J.M. Stewart (1993). "Naked and Thunderbolt Singularities in Black Hole Evaporation." *Nuclear Physics B* 400 (July):393–415.

Headrick, M.P., and J.R. Gott (1994). "(2 + 1)-Dimensional Spacetimes Containing Closed Timelike Curves." *Physical Review D* 50 (December 15):7244–7259.

Healey, R.A. (1984). "How Many Worlds?" *Nous* 18 (November):591–616.

Hedman, C.G. (1972). "On When There Must Be a Time-Difference Between Cause and Effect." *Philosophy of Science* 39 (December):507–511.

Heffern, R. (1977). *Time Travel: Myth or Reality?* New York: Pyramid.

Hegerfeldt, G.C. (1994). "Causality Problems for Fermi's Two-Atom System." *Physical Review Letters* 72 (January 31):596–599.

Heinlein, R.A. (1986). *The Door Into Summer.* New York: Del Rey.

—— (AHT). "By His Bootstraps."

—— (1980). *The Number of the Beast.* New York: Random House.

—— (1979). "Life-Line." In *The Great Science Fiction Stories.* Vol. 1. New York: DAW.

—— (MI). "All You Zombies—."

—— (1964). *Farnham's Freehold.* New York: G.P. Putnam's Sons.

—— (FM). "—And He Built a Crooked House."

—— (1955). *Tunnel in the Sky.* New York: Scribner's.

—— (1953). "Elsewhen." In *Assignment in Eternity.* Reading, PA: Fantasy Press.

Helliwell, T.M., and D.A. Konkowski (1983). "Causality Paradoxes and Nonparadoxes: Classical Superluminal Signals and Quantum Measurements." *American Journal of Physics* 51 (November):996–1003.

Henderson, L.D. (1983). *The Fourth Dimension and Non-Euclidean Geometry in Modern Art.* Princeton, NJ: Princeton University Press.

Hennelly, M.M., Jr. (1979). "*The Time Machine:* A Romance of 'The Human Heart.' " *Extrapolation* 20 (Summer):154–167.

Herbert, N. (1988). *Faster Than Light: Superluminal Loopholes in Physics.* New York: New American Library.

—— (1987). *Quantum Reality.* Garden City, NY: Anchor.

—— (1982). "FLASH—A Superluminal Communicator Based Upon a New Kind of Quantum Measurement." *Foundations of Physics* 12 (December):1171–1179.

Herbert, R.T. (1987). "The Relativity of Simultaneity." *Philosophy* 62 (October):455–471.

Herron, M.L., and D.T. Pegg (1974). "A Proposed Experiment on Absorber Theory." *Journal of Physics A* 7 (October):1965–1969.

Hesse, M.B. (1961). *Forces and Fields.* New York: Philosophical Library.

Hickey, H.B. (WT). "Like a Bird, Like a Fish."

Hilton-Young, W. (SSFT). "The Choice."

Hinckfuss, I. (1975). *The Existence of Space and Time.* London: Oxford University Press.

Hinton, C.H. (1980). *Speculations on the Fourth Dimension.* Edited by R. Rucker. New York: Dover.

Hiscock, W.A. (1985). "Exact Gravitational Field of a String." *Physical Review D* 31 (June 15):3288–3290.

Hiscock, W.A., and D.A. Konkowski (1982). "Quantum Vacuum Energy in Taub-NUT (Newman-Unti-Tamburino)-Type Cosmologies." *Physical Review* D 26 (September 15):1225–1230.

Hobana, I. (PWO). "Night Broadcast."

Hoch, E.D. (SFSSS). "The Last Paradox."

Hochberg, D., and T.W. Kephart (1991). "Lorentzian Wormholes from the Gravitationally Squeezed Vacuum." *Physics Letters B* 268 (October):377–383.

Hodgson, J.L. (1929). *The Time Journey of Dr. Barton.* Eggington, England: J. Hodgson.

Hodgson, W.H. (1908). *The House on the Borderland.* London: Holden and Hardingham.

Hoekzema, D.J. (1996). "Time Symmetry and Cosmic Age." *International Journal of Theoretical Physics* 35 (November):2391–2397.

Hogan, J.P. (1996). *Paths to Otherwhere.* New York: Baen.

—— (1993). *Out of Time.* New York: Bantam.

—— (1985). *The Proteus Operation.* New York: Bantam.

—— (1980). *Thrice Upon a Time.* New York: Ballantine.

Hogarth, J.E. (1962). "Cosmological Considerations of the Absorber Theory of Radiation." *Proceedings of the Royal Society* 267A (May 22):365–383.

Hogarth, M. (1993). "Predicting the Future in Relativistic Spacetimes." *Studies in the History and Philosophy of Science* 24 (December):721–739.

—— (1992): "Does General Relativity Allow an Observer to View an Eternity in a Finite Time?" *Foundations of Physics Letters* 5 (April):173–181.

Hollinger, V. (1987). "Deconstructing the Time Machine." *Science Fiction Studies* 14 (July):201–221.

Hollis, M. (1967a). "Times and Spaces." *Mind* 76 (October):524–536.

—— (1967b). "Box and Cox." *Philosophy* 42:75–78.

Holton, G. (1981). "Einstein's Search for the *Weltbild.*" *Proceedings of the American Philosophical Society* 125 (February):1–15.

—— (1965). "The Metaphor of Space-Time Events in Science." *Eranos-Jahrbuch* 34:33–78.

Horwich, P. (1987). *Asymmetries in Time.* Cambridge, MA: The M.I.T. Press.

—— (1975). "On Some Alleged Paradoxes of Time Travel." *Journal of Philosophy* 72 (August 14):432–444.

Horwitz, L.P., R.I. Arshansky, and A.C. Elitzur (1988). "On the Two Aspects of Time: The Distinction and Its Implications." *Foundations of Physics* 18 (December):1159–1193.

Hoyle, F. (1983). "Information from the Future" and "Loops in Time." In *The Intelligent Universe.* New York: Holt, Rinehart, and Winston.

—— (1975). "On the Origin of the Microwave Background." *Astrophysical Journal* 196 (March 15):661–670.

—— (1966). *October the First Is Too Late*. London: Heinemann.

Hoyle, F., and J.V. Narlikar (1974). *Action at a Distance in Physics and Cosmology*. San Francisco, CA: W.H. Freeman.

—— (1964). "Time Symmetric Electrodynamics and the Arrow of Time in Cosmology." *Proceedings of the Royal Society of London* 277A (January):1–23.

Hubbard, L.R. (FTI). "The Dangerous Dimension."

—— (1950). "To the Stars." *Astounding Science Fiction*, February and March.

Hudec, G. (PWO). "The Ring."

Huggett, W.J. (1960). "Losing One's Way in Time." *Philosophical Quarterly* 10 (July):264–267.

Hughes, R. (MM). "The Vanishing Man."

Hunter, G. (1951). "Journey." *Fantasy and Science Fiction*, February.

Hunter, L.R. (1991). "Tests of Time-Reversal Invariance in Atoms, Molecules, and the Neutron." *Science* 252 (April 5):73–79.

Hunter, N. (1933). "The Professor Invents a Machine." In *The Incredible Adventures of Professor Branestawm*. London: The Bodley Head.

Hunting, G. (1926). *The Vicarion*. Kansas City, MO: Unity School of Christianity.

Hurley, J. (1986). "The Time-Asymmetry Paradox." *American Journal of Physics* 54 (January):25–28.

Hutchinson, K. (1993). "Is Classical Mechanics Really Time-reversible and Deterministic?" *British Journal for the Philosophy of Science* 44 (June):307–323.

Infeld, L. (1941). "The Fourth Dimension and Relativity." In *Galois Lectures*. New York: Scripta Mathematica Library.

Inge, W.R. (1921). "Is the Time Series Reversible?" *Proceedings of the Aristotelian Society* 21:1–12.

Irwin, M. (BFSF2). "The Earlier Service."

Isham, C. (NP). "Quantum Gravity."

Israel, W. (1967). "Gravitational Collapse and Causality." *Physical Review* 153 (January 25):1388–1393.

Italiano, A. (1986). "How to Recover Causality in General Relativity." *Hadronic Journal* 9 (January):9–12.

Jackson, R. (1959). "Millennium." In *Anthology of Best Short Short Stories 7*. Edited by R. Oberfirst. New York: Frederick Fell.

Jakes, J. (1970). *Black in Time*. New York: Paperback Library.

Jakiel, S.J., and R.E. Levinthal (1980). "The Laws of Time Travel." *Extrapolation* 21 (Summer):130–138.

James, H. (1917). *The Sense of the Past*. New York: Charles Scribner's Sons.

Jameson, M. (GSSF). "Blind Alley."

—— (1941a). "Time Column." *Thrilling Wonder Stories*, December.

—— (1941b). "Dead End." *Thrilling Wonder Stories*, March.

—— (1940). "Murder in the Time World." *Amazing Stories*, August.

Janifer, L.M. (1973). "A Few Minutes." In *Ten Tomorrows*. Edited by R. Elwood. Greenwich, CT: Fawcett.

Jarry, A. (1965). "How to Construct a Time Machine." Translated from the French and reprinted in *Selected Works of Alfred Jarry*. Edited by R. Shattuck and S.W. Taylor. New York: Grove Press.

Jensen, B. (1992). "Notes on Spinning Strings." *Classical and Quantum Gravity* 9 (January):L7–L12.

Jensen, B., and H.H. Soleng (1992): "General-Relativistic Model of a Spinning Cosmic String." *Physical Review D* 45 (May 15):3528–3533.

Jeschke, W. (1982). *The Last Day of Creation*. New York: St. Martin's Press.

—— (BRW). "The King and the Dollmaker."

Jespersen, J., and J. Fitz-Randolph (1982). *From Sundials to Atomic Clocks: Understanding Time and Frequency*. New York: Dover.

Jeter, K.W. (1979). *Morlock Night*. New York: DAW.

Jobe, E.K. (1980). "Nature's Choice of Time." *Australasian Journal of Philosophy* 58 (December):347–359.

Johns, M. (1966). *Beyond Time.* New York: Arcadia House.

Johnson, A., and E. Johnson (1984). *The Danger Quotient.* New York: Harper & Row.

Jones, F.C. (1972). "Lorentz-Invariant Formulation of Cherenkov Radiation by Tachyons." *Physical Review D* 6 (November 15):2727–2735.

Jones, L. (BO). "The Time Machine."

—— (TTT). "The Great Clock."

Jones, R.F. (SFAD). "Pete Can Fix It."

—— (1950). "Sunday Is Three Thousand Years Away." *Thrilling Wonder Stories,* June.

Jones, R.T. (1963). "Conformal Coordinates Associated with Space-Like Motions." *Journal of the Franklin Institute* 275 (January):1–12.

Kabat, D.N. (1992). "Conditions for the Existence of Closed Timelike Curves in $2+1$ Gravity." *Physical Review D* 46 (September 15):2720–2722.

Kamoi, K., and S. Kamefuchi (1971). "Comments on Quantum Field Theory of Tachyons." *Progress of Theoretical Physics* 45 (May):1646–1661.

Kar, B.S. (1992). "The Principle of Locality and Quantum Field Theory on (Non Globally Hyperbolic) Curved Spacetimes." *Reviews in Mathematical Physics* Special Issue (December):167–195.

Kar, S. (1994). "Evolving Wormholes and the Weak Energy Condition." *Physical Review D* 49 (January 15):862–865.

Karageorge, M. (1965). "The Life of Your Time." *Analog,* September.

Kaufmann, W.J. (1977). *The Cosmic Frontiers of General Relativity.* New York: Little, Brown.

Kendig, J., Jr. (1930). "Fourth-Dimensional Penetrator." *Amazing Stories,* January.

Kent, K. (OW). "The Comedy of Eras."

—— (1939). "World's Pharaoh." *Thrilling Wonder Stories,* December.

Kerr, R.P. (1963). "Gravitational Field of a Spinning Mass as an Example of Algebraically Special Metrics." *Physical Review Letters* 11 (September 1):237–238.

Kerszberg, P. (1987). "The Relativity of Rotation in the Early Foundations of General Relativity." *Studies in History and Philosophy of Science* 18 (March):53–79.

Ketterer, D. (1984). *Science Fiction of Mark Twain.* Hamden, CT: Archon.

—— (1982). "Oedipus as Time Traveller." *Science Fiction Studies* 9 (November):340–341.

Khatsymovsky, V. (1994). "Can Wormholes Exist?" *Physics Letters B* 320 (January 13):234–240.

Kilworth, G. (1985). "Let's Go to Golgotha!" In *The Songbirds of Pain.* London: Victor Gollancz.

Kim, J. (1974). "Noncausal Connections." *Nous* 8 (March):41–52.

Kim, S.-W. (1992). "Particle Creation for Time Travel Through a Wormhole." *Physical Review D* 46 (September 15):2428–2434.

Kim, S.-W., and K.S. Thorne (1991). "Do Vacuum Fluctuations Prevent the Creation of Closed Timelike Curves?" *Physical Review D* 43 (June 15):3929–3947.

Kimberly, G. (FST). "Minna in the Night Sky."

King, S. (1985). "The Jaunt" and "Mrs. Todd's Shortcut." In *Skeleton Crew.* New York: G.P. Putnam's Sons.

Kirch, D. (1975). "Some Theoretical and Experimental Aspects of the Tachyon Problem." *International Journal of Theoretical Physics* 13 (June):153–173.

Kirkham, H.F. (1929). "The Time Oscillator." *Science Wonder Stories,* December.

Kirzhnitz, D.A., and V.L. Polyachenko (1964). "On the Possibility of Macroscopic Manifestations of Violation of Microscopic Causality." *Soviet Physics JETP* 19 (August):514–519.

Klass, P. (1974). "An Innocent in Time: Mark Twain in King Arthur's Court." *Extrapolation* 16 (December):17–32.

Klein, G. (BRW). "Party Line."

—— (1972). *The Day Before Tomorrow.* New York: DAW.

Klein, O. (1926). "The Atomicity of Electricity as a Quantum Theory Law." *Nature* 118 (October 9):516.

Klein, T.E.D. (MT). "Renaissance Man."

Kline, A.O. (1939). "Stolen Centuries." *Thrilling Wonder Stories,* June.

Klinkhammer, G. (1992). "Vacuum Polarization of Scalar and Spinor Fields Near Closed Null Geodesics." *Physical Review D* 46 (October 15):3388–3394.

—— (1991). "Averaged Energy Conditions for Free Scalar Fields in Flat Space-time." *Physical Review D* 43 (April 15):2542–2548.

Knight, D. (1991). "Azimuth 1,2,3. . . ," "The Time Exchange," and "The Man Who Went Back." In *One Side Laughing.* New York: St. Martin's.

—— (1990). "What Rough Beast." In *The Great Science Fiction Stories.* Vol. 21. New York: DAW.

—— (1987). "Anachron." In *The Great Science Fiction Stories.* Vol. 16. New York: DAW.

—— (1990). "I See You." In *The Best from Fantasy and Science Fiction.* Edited by E.L. Ferman. Vol. 23. New York: Doubleday.

—— (1961). "The Last Word," "Thing of Beauty," "Extempore," and "Time Enough." In *Far Out.* New York: Simon and Schuster.

—— (1956). "This Way to the Regress." *Galaxy Science Fiction,* August.

—— (OSF). "Catch That Martian."

—— (GRSF1). "Don't Live in the Past."

Knight, D.C. (1960). "The Amazing Mrs. Mimms." In *The Fantastic Universe Omnibus.* Edited by H.S. Santesson. Englewood Cliffs, NJ: Prentice-Hall.

Knight, N.L. (BSF). "Short-Circuited Probability."

—— (1940). "Bombardment in Reverse." *Astounding Science Fiction,* February.

Koestler, A. (1971). "The Boredom of Fantasy." In *Science Fiction: The Future.* Edited by D. Allen. New York: Harcourt.

Kolupayev, V. (AM). "A Ticket to Childhood."

Koontz, D. (1988). *Lightning.* New York: G.P. Putnam's Sons.

Kornbluth, C.M. (HV). "Two Dooms."

—— (1984). "The Little Black Bag." In *The Great Science Fiction Stories.* Vol. 12. New York: DAW.

—— (VIT). "Dominoes."

—— (CA). "Time Travel and the Law."

Korotkii, V.A., and Yu N. Obukhov (1991). "Kinematic Analysis of Cosmological Models with Rotation." *Soviet Physics JETP* 72 (January):11–15.

Kosso, P. (1988). "Spacetime Horizons and Unobservability." *Studies in History and Philosophy of Science* 19 (June):161–173.

Kotarbinski, T. (1968). "The Problem of the Existence of the Future." *Polish Review* 13:7–22.

Kowalczynski, J.K. (1984). "Critical Comments on the Discussion about Tachyonic Causal Paradoxes and on the Concept of Superluminal Reference Frame." *International Journal of Theoretical Physics* 23 (January):27–60 [and the reply by E. Recami (1987) 26 (September):913–919].

—— (1978). "Charged Tachyon in General Relativity: Can It Be Detected?" *Physics Letters A* 65 (March 20):269–272.

Kragh, H., and B. Carazza (1994). "From Time Atoms to Space-Time Quantization: The Idea of Discrete Time, ca 1925–1936." *Studies in History and Philosophy of Science* 25 (June):437–462.

Krasnikov, S.V. (1998a). "Hyperfast Travel in General Relativity." *Physical Review D* 57 (April 15):4760–4766.

—— (1998b). "A Singularity-Free WEC-Respecting Time Machine." *Classical and Quantum Gravity* 15 (April):997–1003.

—— (1997). "Causality Violation and Paradoxes." *Physical Review D* 55 (March 15):3427–3430.

—— (1996). "Quantum Stability of the Time Machine." *Physical Review D* 54 (December 15):7322–7327.

Kreisler, M.N. (1973). "Are There Faster-Than-Light Particles?" *American Scientist* 61 (March–April).

Kretzmann, N. (1966). "Omniscience and Immutability." *Journal of Philosophy* 63 (July 14):409–421.

Kriele, M. (1990a). "A Generalization of the Singularity Theorem of Hawking and Penrose to Space-Times with Causality Violations." *Proceedings of the Royal Society* 431A (December 8):451–464.

—— (1990b). "Causality Violations and Singularities." *General Relativity and Gravitation* 22 (June):619–623.

Kroes, P. (1985). "The Time Reversal Operator T*." In *Time: Its Structure and Role in Physical Theories*, pp. 124–129. Dordrecht, The Netherlands: D. Reidel.

—— (1984). "Objective Versus Mind-dependent Theories of Time Flow." *Synthese* 61 (December):423–446.

—— (1983). "Traces, Prediction, and Retrodiction." *Nature and System* 5 (September):131–146.

—— (1982). "Order and Irreversibility." *Nature and System* 4 (September):115–129.

Krueger, J. (1978). "Nietzschean Recurrence as a Cosmological Hypothesis." *Journal of the History of Philosophy* 16 (October):435–444.

Ksanda, C.F. (1946). "Forever Is Today." *Thrilling Wonder Stories,* Summer.

Kubilius, W. (1951). "Turn Backward, O Time." *Science Fiction Quarterly,* May.

Kummer, F.A., Jr. (1945). "Delvers in Destiny." *Thrilling Wonder Stories,* Spring.

Kuttner, H. (1953). "Shock." In *Ahead of Time.* New York: Ballantine.

Kuttner, H., and C.L. Moore (T3). "Vintage Season."

—— (1980). "The Twonky." In *The Great Science Fiction Stories.* Vol. 4. New York: DAW.

Lackey, D.P. (1974). "A New Disproof of the Compatibility of Foreknowledge and Free Choice." *Religious Studies* 10 (September):313–318.

Lafferty, R.A. (1991). "Rainbird." In *The Great Science Fiction Stories,* Vol. 23. New York: DAW.

—— (1971). "Through Other Eyes." In *Mind to Mind.* Edited by R. Silverberg. New York: Thomas Nelson.

—— (1970). "The Six Fingers of Time." In *Nine Hundred Grandmothers.* New York: Ace.

—— (AO). "Thus We Frustrate Charlemagne."

Lafleur, L.J. (1943). "Marvelous Voyages—H.G. Wells' *The Time Machine.*" *Popular Astronomy,* October.

—— (1940). "Time as a Fourth Dimension." *Journal of Philosophy* 37 (March 28):169–178.

Lake, K. (1992). "Precursory Singularities in Spherical Gravitational Collapse." *Physical Review Letters* 68 (May 25):3129–3132.

Lämmerzahl, C., and A. Macias (1993). "On the Dimensionality of Space-Time." *Journal of Mathematical Physics* 34 (October):4540–4553.

Landis, G.A. (1991). "Ripples in the Dirac Sea." In *Nebula Awards 25.* Edited by M. Bishop. New York: Harcourt.

—— (D). "Dinosaurs."

Lapedes, A.S., and K.C. Jacobs (1972). "Tachyons and Gravitational Cerenkov Radiation." *Nature Physical Science* 235 (January 3):6–7.

Laski, M. (1954). *The Victorian Chaise Longue.* Boston: Houghton Mifflin.

Laumer, K. (1991). "The Propitiation of Brullamagoo." In *Alien Minds.* New York: Baen.

—— (1971). *Dinosaur Beach.* New York: Charles Scribner's Sons.

—— (1964). *The Great Time Machine Hoax.* New York: Simon and Schuster.

Laurence, L. (1939). "History in Reverse." *Amazing Stories,* October.

Laverty, D. (1948). "No Winter, No Summer." *Thrilling Wonder Stories,* October.

Leftow, D. (1991). "Why Didn't God Create the Universe Sooner?" *Religious Studies* 27 (June):157–172.

LeGuin, U.K. (1975). "April in Paris" and "Semley's Necklace." In *The Wind's Twelve Quarters.* New York: Harper & Row.

Leiber, F. (1987). "The Man Who Never Grew Old." In *Great Science Fiction of the 20th Century.* Edited by R. Silverberg and M.H. Greenberg. New York: Avenel.

—— (SFS). "Try and Change the Past."

—— (1976). *The Big Time.* Boston, MA: Gregg Press.

—— (TC). "Nice Girl with 5 Husbands."

—— (1961). "Damnation Morning," and "The Oldest Soldier." In *The Mind Spider and Other Stories.* New York: Ace.

—— (1958). "Time in the Round." In *The Third Galaxy Reader.* Edited by H.L. Gold. Garden City, NY: Doubleday.

Leiby, D.A. (1987). "The Tooth That Gnaws: Reflections on Time Travel." In *Intersections: Fantasy and Science Fiction.* Edited by G.E. Slusser and E.S. Rabkin. Carbondale, IL: Southern Illinois University Press.

Leinster, M. (BML). "Time to Die," "Interference," and "Sam, This Is You."

—— (SFT). "The Fifth-Dimension Catapult."

—— (BGA2). "Sidewise in Time."

—— (1967). "The Runaway Skyscraper." *The Best of Amazing.* Edited by J. Ross. Garden City, NY: Doubleday.

—— (1964). *Time Tunnel.* New York: Pyramid Books.

—— (1960). "The *Corianis* Disaster." *Original Science Fiction Stories,* May.

—— (SFAD). "The Middle of the Week After Next."

—— (1953). "The Gadget Had a Ghost." In *Year's Best Science Fiction Novels.* Edited by E.F. Bleiler and T.E. Dikty. New York: Frederick Fell.

—— (GRSF1). "The Other Now."

—— (1950). "The Life-Work of Professor Muntz." In *The Best Science Fiction Stories.* Edited by E.F. Bleiler and T.E. Dikty. New York: Frederick Fell.

—— (1946). "Dead City." *Thrilling Wonder Stories,* Summer.

—— (OW). "The Fourth-Dimensional Demonstrator."

—— (1933). "The Fifth-Dimension Tube." *Astounding Stories,* January.

Lem, S. (1982). *Memoirs of a Space Traveler.* New York: Harcourt.

—— (1977). "Cosmology and Science Fiction." *Science-Fiction Studies* 4 (July):107–110.

—— (1976). "The Seventh Voyage of Ijon Tichy" and "The Twentieth Voyage of Ijon Tichy." In *The Star Diaries.* New York: Seabury Press.

—— (1974). "The Time-Travel Story and Related Matters of SF Structuring." *Science-Fiction Studies* 1 (Spring):143–154.

L'Engle, M. (1978). *A Swiftly Tilting Planet.* New York: Dell.

—— (1976). *A Wrinkle in Time.* New York: Dell.

Lenzen, V. (BST). "Geometrical Physics."

Le Poidevin, R. (1991). "Creation in a Closed Universe *Or,* Have Physicists Disproved the Existence of God?" *Religious Studies* 27 (March):39–48.

—— (1990). "Relationism and Temporal Topology: Physics or Metaphysics." *Philosophical Quarterly* 40 (October):419–432.

Leslie, J. (1976). "The Value of Time." *American Philosophical Quarterly* 13 (April):109–121.

Levin, M.R. (1980). "Swords' Points." *Analysis* 40 (March):69–70.

Lévy-Leblond, J.-M. (1990). "Did the Big Bang Begin?" *American Journal of Physics* 58 (February):156–169.

—— (1989). "The Unbegun Big Bang." *Nature* 342 (November 2):23.

Lewis, C.S. (1986). *The Grand Miracle*. New York: Ballantine.

—— (1978). *Miracles*. New York: Macmillan.

—— (1977). *The Dark Tower and Other Stories*. New York: Harcourt.

Lewis, D. (1976). "The Paradoxes of Time Travel." *American Philosophical Quarterly* 13 (April):145–152.

Lewis, G.N. (1930). "The Symmetry of Time in Physics." *Science* 71 (June 6):569–577.

—— (1926). "The Nature of Light." *Proceedings of the National Academy of Sciences* 12 (January 15):22–29.

Ley, W. (CA). "Geography for Time Travelers."

Li, L.-X. (1996). "Must Time Machines Be Unstable Against Vacuum Fluctuations?" *Classical and Quantum Gravity* 13 (September):2563–2568.

—— (1994). "New Light on Time Machines: Against the Chronology Protection Conjecture." *Physical Review D* 50 (November 15):R6037–R6040.

Li, L.-X., J.-M. Xu and L. Liu (1993). "Complex Geometry, Quantum Tunneling, and Time Machines." *Physical Review D* 48 (November 15):4735–4737.

Li, L.-X., and J. R. Gott (1998). "Self-Consistent Vacuum for Misner Space and the Chronology Protection Conjecture." *Physical Review Letters* 80 (April 6): 2980–2983.

Lieb, I.C. (1972). "Individuals and the Past." *Review of Metaphysics* 25 (June supplement):117–130.

Lightman, A. (1986). "Time Travel and Papa Joe's Pipe." In *Time Travel and Papa Joe's Pipe*. New York: Penguin.

Lindsay, D. (1971). *A Voyage to Arcturus*. London: Victor Gollancz.

Lindsay, R.B., and Margenau, M. (BST). "Time: Continuous or Discrete."

Linsky, L., and D.C. Williams (1954). "Professor Donald Williams on Aristotle" and "Professor Linsky on Aristotle." *Philosophical Review* 63 (April):250–255.

Lionel, R. (1964). *Time Echo*. New York: Arcadia House.

Liu, C. (1993). "The Arrow of Time in Quantum Gravity." *Philosophy of Science* 60 (December):619–637.

Liu, L. (1993). "Wormhole Created from Vacuum Fluctuation." *Physical Review D* 48 (December 15):R5463–R5464.

Locke, R.D. (PSF). "Demotion."

Lodge, O. (1921). "On the Supposed Weight and Ultimate Fate of Radiation." *Philosophical Magazine* 41 (April):549–557.

—— (1920). "The New World of Space and Time." *Living Age* 304 (January 24):240–244.

Löfgren, L. (1984). "Autology of Time." *International Journal of General Systems* 10:5–14.

Long, A.R. (1934). "Scandal in the 4th Dimension." *Astounding Stories*, February.

Long, F.B. (1970). *Monster From Out of Time*. New York: Popular Library.

—— (1953). "Throwback in Time." *Science Fiction Plus*, April.

—— (1948). "A Guest in the House." In *Strange Ports of Call*. Edited by A. Derleth. New York: Pellegrini and Cudahy.

—— (1947). "Collector's Item." *Astounding Science Fiction*, October.

—— (1937). "Temporary Warp." *Astounding Stories*, August.

Loomis, N. (CBSF). "The Long Dawn."

Lorentz, H.A. (TPR). "Electromagnetic Phenomena in a System Moving with Any Velocity Less Than that of Light."

Lorraine, L. (1930). "Into the 28th Century." *Science Wonder Quarterly*, Winter.

Lossev, A., and I.D. Novikov (1992). "The Jinn of the Time Machine: Nontrivial Self-Consistent Solutions." *Classical and Quantum Gravity* 9 (October):2309–2321.

Lovecraft, H.P. (1982). "The Shadow Out of Time" and "The Dreams in the Witch-House." In *The Best of H.P. Lovecraft*. New York: Ballantine.

—— (1970). "The Silver Key" and "Through the Gates of the Silver Key." In *The Dream—Quest of Unknown Kadath*. New York: Ballantine.

Lovecraft, H.P., and A. Derleth (1957). "The Ancestor" and "The Lamp of Alhazred." In *The Survivor and Others*. Sauk City, WI: Arkham House.

Lowenthal, D. (1985). *The Past Is a Foreign Country*. New York: Cambridge University Press.

Lundwall, S.J. (BRW). "Nobody Here But Us Shadows."

Lyutikov, M. (1994). "Vacuum Polarization at the Chronology Horizon of the Roman Spacetime." *Physical Review D* 49 (April 15):4041–4048.

MacBeath, M. (1993). "Time's Square." In *The Philosophy of Time*. Edited by R. Le Poidevin and M. MacBeath. New York: Oxford University Press.

—— (1986). "Clipping Time's Wings." *Mind* 95 (April):233–237.

—— (1983). "Communication and Time Reversal." *Synthese* 56 (July):27–46.

—— (1982). "Who Was Dr. Who's Father?" *Synthese* 51 (June):397–430.

MacCreigh, J. (1947). "A Hitch in Time." *Thrilling Wonder Stories*, June.

MacDonald, J.D. (SSFT). "Spectator Sport."

—— (1971). "Half-Past Eternity." In *The Human Equation*. Edited by W.F. Nolan. Los Angeles: Sherbourne Press.

—— (GRSF2). "A Game for Blondes."

Macey, S.L. (1980). *Clocks and the Cosmos: Time in Western Life and Thought*. Hamden, CT: Archon.

Macfadyen, A., Jr. (SFT). "The Time Decelerator."

Mackaye, H.S. (1904). *The Panchronicon*. New York: Charles Scribner's Sons.

Mackie, J.L. (1966). "The Direction of Causation." *Philosophical Review* 75:441–466.

Mackie, P. (1992). "Causing, Delaying, and Hastening: Do Rains Cause Fires?" *Mind* 101 (July):483–500.

MacKinnon, E. (1962). "Time and Contemporary Physics." *International Philosophical Quarterly* 2 (September):428–457.

Malament, D.B. (1987) "A Note About Closed Timelike Curves in Gödel Space-Time." *Journal of Mathematical Physics* 28 (October):2427–2430.

—— (1985). "Minimal Acceleration Requirements for 'Time Travel' in Gödel Space-Time." *Journal of Mathematical Physics* 26 (April):774–777.

—— (1984). " 'Time Travel' in the Gödel Universe." *Proceedings of the Philosophy of Science Association* 2:91–100.

Malotki, E. (1983). *Hopi Time*. New York: Mouton.

Malzberg, B.M. (1978). *Chorale*. Garden City, NY: Doubleday.

Mandino, O. (1980). *The Christ Commission*. New York: Lippincott and Crowell.

Manley, E.A., and W. Thode (1930). "The Time Annihilator." *Wonder Stories*, November.

Manning, H.P. (1960). *The Fourth Dimension Simply Explained*. New York: Dover.

Manning, L. (BGA2). "The Man Who Awoke."

Margenau, H. (1954). "Can Time Flow Backwards?" *Philosophy of Science* 21 (April):79–92.

Marlow, L. (1944). *The Devil in Crystal*. London: Faber and Faber.

Masson, D.I. (1968). "A Two-Timer." In *The Best SF Stories from "New Worlds."* Edited by M. Moorcock. New York: Berkley.

—— (VIT). "Traveler's Rest."

Martin, G.R.R. (SFSSS). "FTA."

Matheson, R. (1975). *Bid Time Return*. New York: Viking.

Matthews, G. (1979). "Time's Arrow and the Structure of Spacetime." *Philosophy of Science* 46 (March):82–97.

Mavrodes, G.I. (1984). "Is the Past Unpreventable?" *Faith and Philosophy* 1 (April):131–146.

Maxwell, N. (1993). "On Relativity Theory and the Openness of the Future." *Philosophy of Science* 60 (June):341–348.

—— (1988). "Are Probabilism and Special Relativity Incompatible?" *Philosophy of Science* 55 (December):640–645.

—— (1985). "Are Probabilism and Special Relativity Incompatible?" *Philosophy of Science* 52 (March):23–43.

Mayo, B. (1961). "Objects, Events, and Complementarity." *Philosophical Review* 70 (July):340–361.

—— (1955). "Professor Smart on Temporal Asymmetry." *Australasian Journal of Philosophy* 33:38–44.

McArthur, R.P., and M.P. Slattery (1974). "Peter Damian and Undoing the Past." *Philosophical Studies* 25:137–141.

McCall, S. (1976). "Objective Time Flow." *Philosophy of Science* 43 (September):337–362.

—— (1966). "Temporal Flux." *American Philosophical Quarterly* 3 (October):270–281.

McConnell, J. (1983). "Avoidance Situation." In *Starships.* Edited by I. Asimov, M.H. Greenberg, and C.G. Waugh. New York: Ballantine.

McDevitt, J. (FCW). "Time's Arrow.

—— (FS). "The Fort Moxie Branch."

McGivern, W.P. (1941a). "Sidetrack in Time." *Amazing Stories,* July.

—— (1941b). "Doorway of Vanishing Men." *Fantastic Adventures,* July.

McKenna, R. (BO). "The Secret Place."

McKeon, D.G.C., and G.N. Ord (1992). "Time Reversal in Stochastic Processes and the Dirac Equation." *Physical Review Letters* 69 (July 6):3–4.

McLaughlin, D. (1976). *Hawk Among the Sparrows.* New York: Charles Scribner's Sons.

McTaggart, J.E. (1908). "The Unreality of Time." *Mind* 17 (October):457–474.

Mead, G. (1950). "Time Is a Coffin." *Amazing Stories,* September.

Mehlberg, H. (BIPT). "Philosophical Aspects of Physical Time."

—— (1961). "Physical Laws and Time's Arrow." In *Current Issues in the Philosophy of Science.* Edited by H. Feigl and G. Maxwell. New York: Holt, Rinehart, and Winston.

Meiland, J.W. (1974). "A Two-Dimensional Passage Model of Time for Time Travel." *Philosophical Studies* 26 (November):153–173.

—— (1965). "Can the Past Change?" In *Scepticism and Historical Knowledge.* New York: Random House.

Mellor, D.H. (1987). "Fixed Past, Unfixed Future." In *Michael Dummett: Contributions to Philosophy.* Edited by B.M. Taylor. Dordrecht, The Netherlands: Martinus Nijhoff.

—— (1981). "Prediction, Time Travel and Backward Causation." In *Real Time.* New York: Cambridge University Press.

Mellor, F., and I. Moss (1990). "Stability of Black Holes in de Sitter Space," *Physical Review D* 41 (January 15):403–409.

Mendilow, A.A. (1952). *Time and the Novel.* New York: Peter Nevill.

Menotti, P., and D. Seminara (1994). "Stationary Solutions and Closed Time-Like Curves in $(2+1)$-Dimensional Gravity." *Nuclear Physics B* 419 (May 9):189–209.

—— (1993). "Closed Timelike Curves and the Energy Condition in $2+1$ Dimensional Gravity." *Physics Letters B* 301 (February 25):25–28.

Mensky, M.B., and I.D. Novikov (1996a). "Three-Dimensional Billiards with Time Machine." *International Journal of Modern Physics D* 5 (April):179–192.

—— (1996b). "Decoherence Caused by Topology in a Time-Machine Spacetime." *International Journal of Modern Physics D* 5 (February):1–27.

Meredith, D.W. (1949). "Next Friday Morning." *Astounding Science Fiction,* February.

Meredity, R.C. (1978). *Vestiges of Time.* Garden City, NY: Doubleday.

—— (1976). *Run, Come See Jerusalem!* New York: Ballantine.

Merrill, A.A. (1922). "The t of Physics." *Journal of Philosophy* 19 (April 27):238–241.

Meyerson, E. (1985). *The Relativistic Deduction, Boston Studies in the Philosophy of Science*. Vol. 83. Dordrecht, The Netherlands: D. Reidel.

Mikheeva, E.V., and I.D. Novikov (1993). "Inelastic Billiard Ball in a Spacetime with a Time Machine." *Physical Review D* 47 (February 15):1432–1436.

Miller, G. (1973). "The Inexactness of Time." *Foundations of Physics* 3 (September):389–398.

Miller, J.J. (1987). "Ouroboros." In *A Very Large Array*. Edited by M.M. Snodgrass. Albuquerque: University of New Mexico Press.

Miller, P.S. (FSFS). "As Never Was" and "The Sands of Time."

—— (1951). "Status Quondam." In *New Tales of Space and Time*. Edited by R.J. Healy. New York: Henry Holt and Co.

Miller, R.W., and G.B. Chamberlain (1982). "There Are Laws and Then There Are Laws." *Extrapolation* 23 (Fall):290–303.

Mines, S. (1951). "A Taxable Dimension." *Startling Stories,* May.

—— (1946). "Find the Sculptor." *Thrilling Wonder Stories,* Spring.

Mink, L.O. (1960). "Time, McTaggart and Pickwickian Language." *Philosophical Quarterly* 10 (July):252–263.

Minkowski, H. (BST). "The Union of Space and Time."

Mirrnan, R. (1975). "The Direction of Time." *Foundations of Physics* 5 (September):491–511.

—— (1973). "Comments on the Dimensionality of Time." *Foundations of Physics* 3 (September):321–333.

Misner, C.W. (1969a). "Quantum Cosmology." *Physical Review* 186 (October 25):1319–1327.

—— (1969b). "Absolute Zero of Time." *Physical Review* 186 (October 25):1328–1333.

Misner, C.W., and A.H. Taub (1969): "A Singularity-Free Empty Universe." *Soviet Physics JETP* 28 (January):122–133.

Misner, C.W., and J. Wheeler (1957). "Gravitation, Electromagnetism, Unquantized Charge, and Mass as Properties of Curved Empty Space." *Annals of Physics* 2 (December):525–603.

Mitchell, E.P. (1973). "The Clock That Went Backward." In *The Crystal Man*. New York: Doubleday.

—— (FM). "The Tachypomp: A Mathematical Demonstration."

Mitchell, M. (1936). *Traveller in Time*. New York: Sheed and Ward.

Moffat, J.W. (1993). "Quantum Gravity, the Origin of Time and Time's Arrow." *Foundations of Physics* 23 (March):411–437.

Moon, E. (1988). "Gravesite Revisited." *Analog,* mid-December.

Moorcock, M. (1969). *Behold the Man*. London: Allison & Busby.

Moore, C.L. (1975). "Tryst in Time." In *The Best of C.L. Moore*. New York: Ballantine.

Moore, W. (1953). *Bring the Jubilee*. New York: Farrar, Straus, and Young.

Moravec, H. (1991). "Time Travel and Computing." Unpublished.

Morpurgo, G., B.F. Touschek; and L.A. Radicati (1954). "On Time Reversal." *Nuovo Cimento* 12 (November):677–698.

Morris, J., and C. Morris (1984). *The Forty-Minute War*. New York: Baen.

Morris, M.S., and K.S. Thorne (1988). "Wormholes in Spacetime and Their Use for Interstellar Travel: A Tool for Teaching General Relativity." *American Journal of Physics* 56 (May):395–412.

Morris, M.S., K.S. Thorne, and U. Yurtsever (1988). "Wormholes, Time Machines, and the Weak Energy Condition." *Physical Review Letters* 61 (September 26):1446–1449.

Morris, R. (1990). *The Edges of Science*. Englewood Cliffs, NJ: Prentice-Hall.

—— (1984). *Time's Arrows*. New York: Simon and Schuster.

Morrison, P. (ET). "Time's Arrow and External Perturbations."

Morrison, W. (1943). "Forgotten Past." *Startling Stories,* January.

Moskowitz, S. (TC). "Death of a Dinosaur,"

Mundle, C.W.K. (1967). "The Space-Time World." *Mind* 76 (April):264–269.

—— (1964). "Does the Concept of Precognition Make Sense?" *International Journal of Parapsychology* 6:179–198.

—— (1954). "Mr. Dobbs' Two-Dimensional Theory of Time." *British Journal for the Philosophy of Science* 4 (February):331–337.

Murphy, M. (1951). "Time Tourist." *Fantasy and Science Fiction,* February.

Murray, L. (1989). "The Best Is Yet to Be." *Analog,* February.

Murphy, P.V.R. (1971). "Search for Tachyons in the Cosmic Radiation." *Lettere al Nuovo Cimento* 1(2) (March):908–912.

Naber, G.L. (1988). *Spacetime and Singularities.* Cambridge, England: Cambridge University Press.

Nahin, P.J. (1988). "Twisters." *Analog,* April.

—— (1985). "The Invitation." In *The Fourth Omni Book of Science Fiction.* Edited by E. Datlow. New York: Zebra.

—— (1979a). "Old Friends Across Time." *Analog,* May.

—— (1979b). "Newton's Gift." *Omni,* January.

Nambu, Y. (1950). "The Use of the Proper Time in Quantum Electrodynamics I." *Progress of Theoretical Physics* 5 (January–February):82–94.

Narahara, K., *et al.* (1994). "Traversable Wormhole in the Expanding Universe." *Physics Letters B* 336 (September 29):319–323.

Narlikar, J.V. (1992). "The Concepts of 'Beginning' and 'Creation' in Cosmology." *Philosophy of Science* 59 (September):361–371.

—— (1978). "Cosmic Tachyons: An Astrophysical Approach." *American Scientist* 66 (September–October).

—— (1965). "The Direction of Time." *British Journal for the Philosophy of Science* 15 (February):281–285.

—— (1962a). "Neutrinos and the Arrow of Time in Cosmology." *Proceedings of the Royal Society of London* 270 A (December):553–561.

—— (1962b). "Rotating Universes." In *Evidences for Gravitational Theories.* Edited by C. Moller. Proceedings of the International School of Physics. Course 20. New York: Academic Press.

Nathan, R. (1968). "Encounter in the Past." In *The Best from Fantasy and Science Fiction.* Edited by E.L. Ferman. Vol. 17. Garden City, NY: Doubleday.

Nearing, H., Jr. (MM). "The Hermeneutical Doughnut."

—— (BFSF3). "The Maladjusted Classroom."

—— (BFSF2). "The Hyperspherical Basketball."

Ne'eman, Y. (1970). "CP and CPT Symmetry Violations, Entropy and the Expanding Universe." *International Journal of Theoretical Physics* 3 (February):1–5.

Nelson, R.F. (1975). *Blake's Progress.* New York: Laser Books.

Nerlich, G. (1991). "How Euclidean Geometry Has Misled Metaphysics." *Journal for the Philosophy of Science* 88 (April):169–189.

—— (1982). "Special Relativity Is Not Based on Causality." *British Journal for the Philosophy of Science* 33 (December):361–388.

—— (1981). "Can Time Be Finite?" *Pacific Philosophical Quarterly* 62 (July):227–239.

—— (1979). "How to Make Things Have Happened." *Canadian Journal of Philosophy* 9 (March):1–22.

—— (1976). *The Shape of Space.* Cambridge, England: Cambridge University Press.

Neronov, A. Yu. (1998). "Quasiclassical Solutions of the Klein–Gordon Equation in a Space-Time with Closed Time-like Curves." *Soviety Physics JETP* 86 (January):1–5.

Neville, K. (1953). "Mission." *Fantasy and Science Fiction,* April.

New, C. (1992). "Time and Punishment." *Analysis* 52 (January):35–40.

Newcomb, S. (1898). "The Philosophy of Hyper-Space." *Science* 7 (January 7):1–7.
—— (1877). "Elementary Theorems Relating to the Geometry of a Space of Three Dimensions and of Uniform Positive Curvature in the Fourth Dimension." *Journal fur die Reine and Ungewandte Mathematik* 83:293–299.
Newton, R.G. (CPT). "Relativity and the Order of Cause and Effect in Time."
—— (1970). "Particles That Travel Faster Than Light?" *Science* 167 (March 20):1569–1574.
Newton-Smith, W.H. (1980). *The Structure of Time.* London: Routledge & Kegan Paul.
Ni, P. (1992). "Changing the Past." *Nous* 26 (September):349–359.
Nicholls, P. (1983). *The Science in Science Fiction.* New York: Knopf.
Nissim-Sabat, C. (1979). "On Grünbaum and Retrocausation in Classical Electrodynamics." *Philosophy of Science* 46 (March):118–135 [and the reply immediately following, 46 (March):136–160].
Niven, L. (WMHB). "The Return of William Proxmire."
—— (1979). "Wrong-Way Street" and "Rotating Cylinders and the Possibility of Global Causality Violation." In *Convergent Series.* New York: Ballantine.
—— (E). "ARM."
—— (S1). "Singularities Make Me Nervous."
—— (1973). "The Flight of the Horse," "Leviathan," "Bird in the Hand," "Death in a Cage," and "There Is a Wolf in My Time Machine." In *The Flight of the Horse.* New York: Del Rey/Ballantine.
—— (1971). "The Theory and Practice of Time Travel" and "All the Myriad Ways." In *All the Myriad Ways.* New York: Del Ray/Ballantine.
—— (1968). "At the Core" and "Neutron Star." In *Neutron Star.* New York: Del Rey.
Nolan, P. (1965). "The Time Jumpers." *Amazing Stories,* October.
Nolan, W.F. (1985). "Of Time and Kathy Benedict." In *Whispers V.* Edited by S.D. Schiff. Garden City, NY: Doubleday.
—— (SFSSS). "The Worlds of Monty Willson."
Nor, K.B.M. (1992). "A Topological Explanation for Three Properties of Time." *II Nuovo Cimento* 107B (January):65–70.
Norden, E. (1977). "The Primal Solution." *Fantasy and Science Fiction,* July.
North, J.D. (1970). "The Time Coordinate in Einstein's Restricted Theory of Relativity." *Stadium Generale* 23:203–223.
—— (1965). *The Measure of the Universe.* London: Oxford University Press.
Norton, A. (1970). *The Time Traders.* New York: Ace.
Norton, J. (1989). "Coordinates and Covariance: Einstein's View of Space-Time and the Modern View." *Foundations of Physics* 19 (October):1215–1263.
—— (1986). "The Quest for the One Way Velocity of Light." *British Journal for the Philosophy of Science* 37 (March):118–120.
Nourse, A.E. (SSFT). "Tiger by the Tail."
—— (1965). *The Universe Between.* New York: David McKay.
Novello, M., and M.C.M. da Silva (1994). "Cosmic Spinning String and Causal Protecting Capsules." *Physical Review D* 49 (January 15):825–830.
Novikov, I.D. (BUFP). "Black Holes, Wormholes, and Time Machines."
—— (1992). "Time Machine and Self-Consistent Evolutions in Problems with Self-Interaction." *Physical Review D* 45 (March 15):1989–1994.
—— (1989). "An Analysis of the Operation of a Time Machine." *Soviet Physics JETP* 68 (March):439–443.
—— (1983). *Evolution of the Universe.* Cambridge, England: Cambridge University Press.
—— (1970). "The Disturbances of the Metric When a Collapsing Sphere Passes Below the Schwarzschild Sphere." *Soviet Physics JETP* 30 (March):518–519.
—— (1966). "Change of Relativistic Collapse Into Anticollapse and Kinematics of a Charged Sphere." *JETP Letters* 3 (March 1):142–144.

Nusenoff, R.E. (1977). "Spatialized Time Again." *Philosophy* 52 (January):100–101.
—— (1976). "Two-Dimensional Time." *Philosophical Studies* 29 (May):337–341.
Oaklander, L.N. (1987). "McTaggart's Paradox and the Infinite Regress of Temporal Attributions: A Reply to Smith." *Southern Journal of Philosophy* 25 (Fall):425–431.
—— (1983). "McTaggart, Schlesinger, and the Two-Dimensional Time Hypothesis." *Philosophical Quarterly* 33 (October):391–397.
O'Brien, D.W. (1943). "Yesterday's Clock." *Fantastic Adventures,* February.
—— (1941a). "Beyond the Time Door." *Fantastic Advenures,* March.
—— (1941b). "The Man Who Lived Next Week." *Amazing Stories,* March.
O'Brien, F. (1976). *The Third Policeman.* New York: New American Library.
O'Brien, F.-J. (1988). "What Was It? A Mystery." In *The Supernatural Tales of Fitz-James O'Brien.* Edited by J.A. Salmonson. Vol. 1. New York: Doubleday.
Odle, E.V. (1923). *The Clockwork Man.* Garden City, NY: Doubleday, Page.
Odoevski, V.F. (RRSF). "The Year 4338."
O'Donnell, K.M. (1969). "We're Coming Through the Window." In *Final War and Other Fantasies.* New York: Ace.
Ohanian, H.C. (1976). *Gravitation and Spacetime.* New York: W.W. Norton.
Oliver, C. (GSF). "Transfusion."
—— (1955). "A Star Above It." In *Another Kind.* New York: Ballantine.
—— (1952). *Mists of Dawn.* New York: Holt, Rinehart, and Winston.
Olsen, B. (1951). "The Four-Dimensional Roller-Press." In *Every Boy's Book of Science Fiction.* Edited by D.A. Wollheim. New York: Frederick Fell.
—— (1934). "The Four Dimensional Auto-Parker." *Amazing Stories,* July.
—— (1933). "The Four Dimensional Escape." *Amazing Stories,* December.
—— (1928a). "Four Dimensional Transit." *Amazing Stories Quarterly,* Fall.
—— (1928b). "The Great Four Dimensional Robberies." *Amazing Stories,* May.
—— (1928c). "Four Dimensional Surgery." *Amazing Stories,* February.
Ori, A. (1993). "Must Time-Machine Construction Violate the Weak Energy Condition?" *Physical Review Letters* 71 (October 18):2517–2520.
—— (1991a). "Rapidly Moving Cosmic Strings and Chronology Protection." *Physical Review D* 44 (October 15):2214–2215.
—— (1991b). "Inner Structure of a Charged Black Hole: An Exact Mass-Inflation Solution." *Physical Review Letters* 67 (August 12):789–792.
Ori, A., and Y. Soen (1994). "Causality Violation and the Weak Energy Condition." *Physical Review D* 49 (April 15):3990–3997.
Otten, J. (1977). "Moving Back and Forth in Time." *Auslegung* 5:5–17.
Ouspensky, P.D. (1981). *Tertium Organum.* New York: Knopf.
—— (1948). *The Strange Life of Ivan Osokin.* London: Faber and Faber.
Overseth, O.E. (1969). "Experiments in Time Reversal." *Scientific American,* October.
Ozsvath, I., and E.L. Schucking (1969). "The Finite Rotating Universe." *Annals of Physics* 55 (October 15):166–204.
—— (1962). "Finite Rotating Universe." *Nature* 193 (March 24):1168–1169.
Padgett, A.G. (1989). "God and Time: Toward a New Doctrine of Divine Timeless Eternity." *Religious Studies* 25 (June):211–215.
Padgett, L. (1984). "Private Eye." In *The Great Science Fiction Stories.* Vol. 11. New York: DAW.
—— (1981). "Mimsy Were the Borogoves." *The Great Science Fiction Stories.* Vol. 5. New York: DAW.
—— (SFAD). "Endowment Policy."
—— (FSFS). "Time Locker."
—— (WT). "Line to Tomorrow."
—— (OSF). "What You Need."
—— (BTS). "When the Bough Breaks."

Page, D.N. (1993). "Information in Black Hole Radiation." *Physical Review Letters* 71 (December 6):3743–3746.

—— (1985). "Will Entropy Decrease If the Universe Recollapses?" *Physical Review D* 32 (November 15):2496–2499.

—— (1984). "Can Inflation Explain the Second Law of Thermodynamics?" *International Journal of Theoretical Physics* 23 (August):725–733.

—— (1983). "Inflation Does Not Explain Time Asymmetry." *Nature* 304 (July 7):39–41.

Page, D.N., and M.R. McKee (1981). "Matter Annihilation in the Late Universe." *Physical Review D* 24 (September 15):1458–1469.

Page, L. (1924). "Advanced Potentials and Their Application to Atomic Models." *Physical Review* 24 (September):296–305.

Paiva, F.M., M.J. Reboucas, and A.F.F. Teixeira (1987). "Time Travel in the Homogeneous Som-Raychaudhuri Universe." *Physics Letters A* 126 (December 28): 168–170.

Pal, G., and J. Morhaim (1981). *Time Machine II.* New York: Dell.

Palmer, R.A. (1938). "Matter Is Conserved." *Astounding Science Fiction,* April.

—— (1934). "The Time Tragedy." *Wonder Stories,* December.

—— (1930). "The Time Ray of Jandra." *Wonder Stories,* June.

Panov, V.F., and Y.G. Sbytov (1992). "Accounting for Birch's Observed Anisotropy of the Universe: Cosmological Rotation?" *Soviet Physics JETP* 74 (March):411–415.

Park, D. (1980). *The Image of Eternity.* Amherst, MA: The University of Massachusetts Press.

—— (1971). "The Myth of the Passage of Time." *Studium Generale* 24:19–30.

Parker, B. (1991). *Cosmic Time Travel.* New York: Plenum.

Parker, L. (1969). "Faster-Than-Light Inertial Frames and Tachyons." *Physical Review* 188 (December 25):2287–2292.

Parmentola, J.A., and D.D.H. Yee (1971). "Peculiar Properties of Tachyon Signals." *Physical Review D* 4 (September 15):1912–1915.

Parrinder, P. (1972). *H.G. Wells: The Critical Heritage.* London: Routledge & Kegan Paul.

Parsons, P. (1996). "A Warped View of Time Travel." *Science* (October 11):202–203.

Partridge, R.B. (1973). "Absorber Theory of Radiation and the Future of the Universe." *Nature* 244 (August 3):263–265.

Paul, T. (1983). "The Worm Ouroboros: Time Travel, Imagination, and Entropy." *Extrapolation* 24 (Fall):272–279.

Peacock, K.A. (1992). "A New Look at Simultaneity." In Vol. 1 of *PSA 1992.* East Lansing, MI: Philosophy of Science Association.

Pease, M.C. (1949). "Reversion." *Asounding Science Fiction,* December.

Peck, R. (1989). *Voices After Midnight,* New York: Delacorte.

Pei, M.A. (GSF). "The Bones of Charlemagne."

Penrose, R. (NOT). "Big Bangs, Black Holes, and 'Time's Arrow.' "

—— (1979). "Singularities and Time-Asymmetry." In *General Relativity: An Einstein Centenary Survey.* Edited by S.W. Hawking and W. Israel. Cambridge, England: Cambridge University Press.

—— (1978). "The Geometry of the Universe." In *Mathematics Today.* New York: Springer-Verlag.

—— (1972). "Black Holes and Gravitational Theory." *Nature* 236 (April 21):377–380.

Penrose, O., and I.C. Percival (1962). "The Direction of Time." *Proceedings of the Physical Society.* (London) 79 (March):605–616.

Peres, A. (1969). "Where Are Tachyons?" *Lettere Al Nuovo Cimento* 1:837–838.

Peres, A., and L.C. Schulman (1972). "Signals from the Future." *International Journal of Theoretical Physics* 6 (December):377–382.

Perry, G.P., and R.B. Mann (1992). "Traversible Wormholes in $(2+1)$ Dimensions." *General Relativity and Gravitation* 24 (March):305–321.

Pesic, P.D. (1993). "Euclidean Hyperspace and Its Physical Significance." *Il Nuovo Cimento* 108B (October):1145–1153.

Petkov, V. (1989). "Simultaneity, Conventionality and Existence." *British Journal for the Philosophy of Science* 40 (March):69–76.

Peyton, G. (CBSF). "The Ship That Turned Aside."

Pfarr, J. (1981). "Time Travel in Gödel's Space." *General Relativity and Gravitation* 13 (November):1073–1091.

Phillips, A.M. (FSFS). "Time-Travel Happens!"

Philmus, R.M. (1969). "*The Time Machine;* Or, the Fourth Dimension as Prophecy." *Publications of the Modern Language Association* 84 (May):530–535.

Pierce, J.R. (GSFS). "John Sze's Future."

—— (1954). "Mr. Kinkaid's Pasts." In *Beyond the Barriers of Space and Time*. Edited by J. Merril. New York: Random House.

—— (1943). "Unthinking Cap." *Astounding Science Fiction*, July.

—— (1942). "About Quarrels, About the Past." *Astounding Science Fiction*, July.

—— (1936). "Pre-Vision." *Astounding Stories*, March.

Piercy, M. (1976). *Woman on the Edge of Time*. New York: Knopf.

Pike, N. (1965). "Divine Omniscience and Voluntary Action." *Philosophical Review* 74 (January):27–46.

Pinkerton, J. (1979). "Backward Time Travel, Alternate Universes, and Edward Everett Hale." *Extrapolation* 20 (Summer):168–175.

Piper, H.B. (1983). "Flight from Tomorrow," "Time and Time Again," "Hunter Patrol," and "Crossroads of Destiny." In *The Worlds of H. Beam Piper*. Edited by J.F. Carr. New York: Ace.

—— (1981). *Paratime*. New York: Ace.

Pirani, F.A.E. (1970). "Noncausal Behavior of Classical Tachyons." *Physical Review* D 1 (June 15):3224–3225.

Pitkin, W.B. (1914). "Time and Pure Activity." *Journal of Philosophy, Psychology and Scientific Methods* 11 (August 27):521–526.

Pitowsky, I. (1990). "The Physical Church Thesis and Physical Computational Complexity." *Iyyun, A Jerusalem Philosophical Quarterly* 39 (January):81–99.

Plachta, D. (SFSSS). "The Man from When."

Plaga, R. (1997). "On a Possibility to Find Experimental Evidence for the Many-Worlds Interpretation of Quantum Mechanics." *Foundations of Physics* 27 (April):559–577.

Plantinga, A. (1986). "On Ockham's Way Out." *Faith and Philosophy* 3 (July):235–269.

Plass, G.N. (1961). "Classical Electrodynamic Equations of Motion with Radiative Reaction." *Reviews of Modern Physics* 33 (January):37–62.

Platt, C. (1974). "Time Machines for Domestic Use." *Harper's,* January.

Plimmer, D. (1941). "Man from the Wrong Time-Track." *Uncanny Stories*, April.

Podolny, R. (1970). "Invasion." In *The Ultimate Threshold*. Edited by M. Ginsburg. New York: Holt, Rinehart, and Winston.

Pohl, F. (SFSSS). "The Deadly Mission of Phineas Snodgrass."

—— (1975). "The Celebrated No-Hit Inning.": In *Run to Starlight*. Edited by M. Greenberg, J. Olander, and P. Warrick. New York: Delacorte.

—— (1956). "Target One," "Let the Ants Try," and "The Mapmakers." In *Alternating Currents*. New York: Ballantine.

Polchinski, J. (1991). "Weinberg's Nonlinear Quantum Mechanics and the Einstein-Podolsky-Rosen Paradox." *Physical Review Letters* 66 (January 28):397–400.

Poleshchuk, A. (LDA). "Homer's Secret."

Politzer, H.D. (1994). "Path Integrals, Density Matrices, and Information Flow with Closed, Timelike Curves." *Physical Review D* 49 (April 15):3981–3989.

—— (1992). "Simple Quantum Systems in Spacetimes with Closed, Timelike Curves." *Physical Review D* 46 (November 15):4470–4476.

Popper, K. (1978). "On the Possibility of an Infinite Past." *British Journal for the Philosophy of Science* 29 (March):47–48.

—— (1956). "The Arrow of Time." *Nature* 177 (March 17):538.

Porges, A. (SFS). "The Rescuer."

—— (CJSF). "Guilty as Charged."

Powers, T. (1983). *The Anubis Gates.* New York: Ace.

Poynting, J.H. (1920). "Overtaking the Rays of Light." In Poynting's *Collected Scientific Papers.* Cambridge, England: Cambridge University Press.

Preuss, P. (1981). *Re-Entry.* New York: Bantam.

Price, H. (1994). "A Neglected Route to Realism About Quantum Mechanics." *Mind* 103 (July):303–336.

—— (1991a). "The Asymmetry of Radiation: Reinterpreting the Wheeler–Feynman Argument." *Foundations of Physics* 21 (August):959–975.

—— (1991b). "Review Article." *British Journal for the Philosophy of Science* 42 (March):111–144.

—— (1984). "The Philosophy and Physics of Affecting the Past." *Synthese* 61 (December):299–323.

Priest, C. (1978). *The Space Machine.* New York: Popular Library.

Priestley, J.B. (1964). *Man and Time.* New York: Doubleday.

—— (1953). "Look After the Strange Girl," "The Statues," and "Mr. Strenberry's Tale." In *The Other Place.* New York: Harper and Brothers.

—— (1948). *Dangerous Corner.* In *The Plays of J.B. Priestley.* Vol. 1. London: Heinemann.

—— (1939). *I Have Been Here Before.* London: Samuel French Limited.

—— (1937). *Time and the Conways.* London: William Heinemann.

Prokhovnik, S.J. (1986). "The Twin Paradoxes of Special Relativity: Their Resolution and Implications." *Foundations of Physics* 19 (May):541–552.

Purcell, E. (1963). "Radioastronomy and Communication Through Space." In *Interstellar Communication.* Edited by A.G.W. Cameron. New York: W.A. Benjamin.

Purtill, R.L. (1974). "Foreknowledge and Fatalism." *Religious Studies* 10 (September):319–324.

—— (1973). "The Master Argument." *Apeiron* 7 (May):31–36.

Putnam, H. (1967). "Time and Physical Geometry." *Journal of Philosophy* 64 (April):240–247.

—— (1962). "It Ain't Necessarily So." *Journal of Philosophy* 59 (October 11):658–671.

Queenan, J. (1990). "Time Warp: Or, Investing in the Future Is a Bust." *Barron's* 70 (January 8):46.

Quinton, A. (1962). "Spaces and Times." *Philosophy* 37 (April):130–147.

Ramsaye, T. (1926). *A Million and One Nights.* Vol. 1. New York: Simon and Schuster.

Rao, B.S. (PWO). "Victims of Time."

Raphael, L. (1941). "The Man Who Saw Through Time." *Fantastic Adventures,* September.

Ray, C. (1991). "Time Travel." In *Time, Space and Philosophy.* New York: Routledge.

—— (1982). "Can We Travel Faster Than Light?" *Analysis* 42 (January):50–52.

Ray, R. (1934). "Today's Yesterday." *Wonder Stories,* January.

Reboucas, M.J. (1979). "A Rotating Universe with Violation of Causality." *Physics Letters* 70A (March 5):161–163.

Recami, E. (1987). "Tachyon Kinematics and Causality: A Systematic Thorough Analysis of the Tachyon Causal Paradoxes." *Foundations of Physics* 17 (March):239–296.

—— (1986). "Classical Tachyons and Possible Applications." *Rivista del Nuoro Cimento* 9 (6):1–178.

Redmount, I. (1990). "Wormholes, Time Travel and Quantum Gravity." *New Scientist,* 28 April.

Reeds, F.A. (1942). "Forever Is Not So Long." *Astounding Science Fiction,* May.

Rees, M.J. (1996). "Appearance of Relativistically Expanding Radio Sources." *Nature* 211 (July 30):468–470.

Reichenbach, H. (1958). *The Philosophy of Space and Time.* New York: Dover.

—— (1956). *The Direction of Time.* Los Angeles: University of California Press.

Reichert, M.Z. (1994). *The Unknown Soldier.* New York: DAW.

Reinganum, M.R. (1986). "Is Time Travel Impossible? A Financial Proof." *Journal of Portfolio Management* 13 (Fall):10–12.

Rementer, E.L. (1929). "The Time Deflector." *Amazing Stories,* December.

—— (1928). "The Space Bender." *Amazing Stories,* December.

Remnant, P. (1978). "Peter Damian: Could God Change the Past?" *Canadian Journal of Philosophy* 8 (June):259–268.

Reynolds, M. (1988). "Compounded Interest." *The Great Science Fiction Stories.* Vol. 18. New York: DAW.

—— (1986). "The Business, As Usual." In *The Great Science Fiction Stories.* Vol. 14. New York: DAW.

—— (1982). "Posted." In *Flying Saucers.* Edited by I. Asimov, M.H. Greenberg, and C.G. Waugh. New York: Ballantine.

—— (1959). "Unborn Tomorrow." *Astounding Science Fiction,* June.

—— (1953). "Advice from Tomorrow." *Science Fiction Quarterly,* August.

Reynolds, M., and F. Brown (GRSF1). "Dark Interlude."

Richards, J. (1984). "Deadtime." In *Universe* 14. New York: Doubleday.

Richards, J.T. (1965). "Minor Alteration." *Fantasy and Science Fiction,* December.

Richardson, C.A. (1944). *Happiness, Freedom and God.* London: George G. Harrap.

Rietdijk, C.W. (1987). "Retroactive Effects from Measurement." *Foundations of Physics* 17 (March):297–311.

—— (1981). "Another Proof That the Future Can Influence the Present." *Foundations of Physics* 11 (October):783–790.

—— (1978). "Proof of a Retroactive Influence." *Foundations of Physics* 8 (August):615–628.

—— (1976). "Special Relativity and Determinism." *Philosophy of Science* 43 (December):598–609.

—— (1966). "A Rigorous Proof of Determinism from the Special Theory of Relativity." *Philosophy of Science* 33 (December):341–344.

Rigaut, J. (1970). "Un brilliant sujet." In *Ecrits.* Paris: Gallimard.

Riggs, N.H. (HJ). "Immortality and the Fourth Dimension."

Riggs, P.J. (1997). "The Principal Paradox of Time Travel." *Ratio* 10 (April):48–64.

Rindler, W. (1997). "The Case Against Space Dragging." *Physics Letters A* 233 (August 18):25–29.

—— (1994). "The Lense–Thirring Effect Exposed as Anti-Machian." *Physics Letters A* 187 (April 18):236–238.

—— (1977). *Essential Relativity.* New York: Springer-Verlag.

Robertson, H.P. (BST). "Geometry as a Branch of Physics."

Robinson, G. (1964). "Hypertravel." *Listener* (December 17):976–977.

Rocklynne, R. (1980). "Time Wants a Skeleton." In *The Great Science Fiction Stories.* Vol. 3. New York: DAW.

—— (1940). "The Reflection That Lived." *Fantastic Adventures,* June.

Rogers, J.T. (1953). "Moment Without Time." In *The Best from Startling Stories.* Edited by S. Mines. New York: Henry Holt.

Rolnick, W.B. (1972). "Tachyons and the Arrow of Causality." *Physical Review D* 6 (October 15):2300–2301.

—— (1969). "Implications of Causality for Faster-Than-Light Matter." *Physical Review* 183 (July 25):1105–1108.

Roman, T.A. (1994). "The Inflating Wormhole: A MATHEMATICA Animation." *Computers in Physics* 8 (July–August):480–487.

—— (1993). "Inflating Lorentzian Wormholes." *Physical Review D* 47 (February 15):1370–1379.

—— (1986). "Quantum Stress-Energy Tensors and the Weak Energy Condition." *Physical Review D* 33 (June 15):3526–3533.

Rosborough, L.B. (1934). "Hastings—1066." *Amazing Stories,* June.

Rosen, R. (1985). *Anticipatory Systems.* New York: Pergamon Press.

Rosen, S.M. (1988). "A Neo-Intuitive Proposal for Kaluza–Klein Unification." *Foundations of Physics* 18 (November):1093–1139.

Rosen, S.P. (1968). "TCP Invariance and the Dimensionality of Space-Time." *Journal of Mathematical Physics* 9 (October):1593–1594.

Rosenbaum, J. (1957). "Now and Then." *Fantasy and Science Fiction,* November.

Rosenberg, J.F. (1972). "One Way of Understanding Time." *Philosophia* 2 (October):283–301.

Rosenberg, S. (1998). "Testing Causality on Spacetimes with Closed Timelike Curves." *Physical Review D* 57 (March 15):3365–3377.

Ross, M. (1950). *The Man Who Lived Backward.* New York: Farrar, Straus and Company.

Rothman, M.A. (1988). *A Physicist's Guide to Skepticism: Applying Laws of Physics to Faster-Than-Light Travel, Psychic Phenomena, Telepathy, Time Travel, UFO's and Other Pseudoscientific Claims.* New York: Prometheus.

—— (1960). "Things That Go Faster Than Light." *Scientific American,* July.

Rothman, T. (1987). "The Seven Arrows of Time." *Discover,* February.

—— (1985). "Grand Illusions: Further Conversations on the Edge of Spacetime." In *Frontiers of Modern Physics.* New York: Dover.

Rousseau, V. (1930). "The Atom Smasher." *Astounding Stories,* May.

—— (1917). *The Messiah of the Cylinder.* Chicago: A.C. McClurg.

Rucker, R. (1986). "In Frozen Time." In *Afterlives.* Edited by P. Sargent and I. Watson. New York: Vintage.

—— (1984). *The Fourth Dimension.* Boston: Houghton Mifflin.

—— (1977). *Geometry, Relativity, and the Fourth Dimension.* New York: Dover.

Rudavsky, T.M. (1988). "Creation, Time and Infinity in Gersonides." *Journal of the History of Philosophy* 26 (January):25–44.

—— (1983). "Divine Omniscience and Future Contingents in Gersonides." *Journal of the History of Philosophy* 21 (October):513–516.

Russell, E.F. (1980). "Mechanical Mice." In *The Great Science Fiction Stories.* Vol. 3. New York: DAW.

Russell, E.F., and L.T. Johnson (SFT). "Seeker of Tomorrow."

Russell, W.C. (1975). *The Frozen Pirate.* New York: Arno.

Ryan, J.B. (1940). "The Mosaic." *Astounding Science Fiction,* July.

Saari, O. (1941). "The Door." *Astounding Science Fiction,* November.

—— (1937). "The Time Bender." *Astounding Stories,* August.

Sachs, M. (1969). "Space, Time and Elementary Interactions in Relativity." *Physics Today* 22 (February):51–60.

Sachs, R.G. (1987). *The Physics of Time Reversal.* Chicago: University of Chicago Press.

—— (1972). "Time Reversal." *Science* 176 (May 12):587–597.

—— (1963). "Can the Direction of Flow of Time Be Determined?" *Science* 140 (June 21):1284–1290.

Sackville-West, V. (1932). "An Unborn Visitant." In *Thirty Clocks Strike the Hour and Other Stories*. Garden City NY: Doubleday, Doran.

Sadler, M. (1986). *Alistair's Time Machine*. Englewood Cliffs, NJ: Prentice-Hall.

Saint Augustine (1961). *Confessions*. New York: Penguin.

Sakharov, A.D. (1984). "Cosmological Transitions with Changes in the Signature of the Metric." *Soviet Physics JETP* 60 (August):214–218.

Salecker, H., and E.P. Wigner (1958). "Quantum Limitations of the Measurement of Space-Time Distances." *Physical Review* 109 (January 15):571–577.

Salmon, W.C. (1977). "The Philosophical Significance of the One-Way Speed of Light." *Nous* 11 (September):253–292.

Sauvy, A. (1990). "Time Reversal." In *One Step in the Clouds*. Edited by A. Salkeld and R. Smith. London: Daidem.

Savitt, S.F. (1994). "Is Classical Mechanics Time Reversal Invariant?" *British Journal for the Philosophy of Science* 45 (September):907–913.

—— (1982). "Tachyon Signals, Causal Paradoxes, and the Relativity of Simultaneity." *Proceedings of the 1982 Biennial Meeting of the Philosophy of Science Association*. Edited by P. Asquith and T. Nickles. East Lansing, MI. Vol. 1, pp. 277–292.

Sawyer, R.J. (1994). *End of an Era*. New York: Ace.

—— (DF). "Just Like Old Times."

Schachner, N. (BGA3). "Past, Present, and Future."

—— (1939). "When the Future Dies." *Astounding Science Fiction,* June.

—— (1937). "Lost in the Dimensions." *Astounding Stories,* November.

—— (1936). "Reverse Universe." *Astounding Stories,* June.

—— (1934). "The Time Imposter." *Astounding Stories,* March.

—— (1933). "Ancestral Voices." *Astounding Stories,* December.

—— (1932). "The Time Express." *Wonder Stories,* December.

Schachner, N., and A.L. Zagat (1930). "In 20,000 A.D. *Wonder Stories,* September.

Schaffer, S. (1979). "John Michell and Black Holes." *Journal for the History of Astronomy* 10 (February):42–43.

Schein, F., and P.C. Aichelburg (1996). "Traversable Wormholes in Geometries of Charged Shells." *Physical Review Letters* 77 (November 11):4130–4133.

Schein, F., P.C. Aichelburg, and W. Israel (1996). "String-Supported Wormhole Space-times Containing Closed Timelike Curves." *Physical Review D* 54 (September 15):3800–3805.

Schere, M. (1938). "Anachronistic Optics." *Astounding Stories,* February.

Schiff, L. (NT). "Nuclear Moments and Time Symmetry."

Schild, A. (1963). "Electromagnetic Two-Body Problem." *Physical Review* 131 (September 15):2762–2766.

—— (1960). "Time." *Texas Quarterly* 3 (Autumn):42–62.

—— (1959). "The Clock Paradox in Relativity Theory." *American Mathematical Monthly*. 66:1–18.

Schlegel, R. (1980). "The Light Clock: Error and Implications." *Foundations of Physics* 10 (April):345–351.

—— (VOT). "Time and Thermodynamics."

—— (1961). *Time and the Physical World*. East Lansing, MI: Michigan State University Press.

Schlesinger, G.N. (1985). "How to Navigate the River of Time." *Philosophical Quarterly* 35 (January):91–94.

—— (1982). "How Time Flies." *Mind* 91 (October):401–523.

Schmidt, H. (1978). "Can an Effect Precede Its Cause? A Model of a Noncausal World." *Foundations of Physics* 8 (June):463–480.

—— (1966). "Model of an Oscillating Cosmos Which Rejuvenates During Contraction." *Journal of Mathematical Physics* 7 (March):494–509.

Schmidt, J.H. (1998). "Newcomb's Paradox Realized with Backward Causation." *British Journal for the Philosophy of Science* 49 (March):67–87.

Schmidt, S. (1993). "Worthsayer." In *More Whatdunits*. Edited by M. Resnik. New York: DAW.

Schmidt, J., and R. Newman (1980). "Search for Advanced Fields in Electromagnetic Radiation (Abstract)." *Bulletin of the American Physical Society* 25 (April):581.

Schrödinger, E. (1950). "Irreversibility." *Proceedings of the Royal Irish Academy* 53A (August):189–195.

Schulman, L.S. (1977). "Illustration of Reversed Causality with Remarks on Experiment." *Journal of Statistical Physics* 16:217–231.

—— (1973). "Correlating Arrows of Time." *Physical Review D* 7 (May 15):2868–2874.

—— (1971a). "Tachyon Paradoxes." *American Journal of Physics* 39 (May):481–484.

—— (1971b). "Gravitational Shock Waves from Tachyons." *II Nuovo Cimento* 2B (March 11):38–44.

Schumacher, D.L. (1964). "The Direction of Time and the Equivalence of 'Expanding' and 'Contracting' World-Models." *Proceedings of the Cambridge Philosophical Society* 60:575–579.

Schuster, M.M. (1986). "Is the Flow of Time Subjective?" *Review of Metaphysics* 39 (June):695–714.

Schwebel, S.L. (1970). "Advanced and Retarded Solutions in Field Theory." *International Journal of Theoretical Physics* 3 (October):347–353.

Schweber, S.S. (1986). "Feynman and the Visualization of Space-Time Processes." *Reviews of Modern Physics* 58 (April):449–508.

Sciama, D.W. (1963). "Retarded Potentials and the Expansion of the Universe." *Proceedings of the Royal Society A* 273 (June 11):484–495.

—— (1958). "Determinism and the Cosmos." In *Determinism and Freedom in the Age of Modern Science*. Edited by S. Hook. New York: New York University Press.

Scieszka, J. (1991). *Knights of the Kitchen Table*. New York: Viking Penguin.

Scortia, T.N. (1981). "When You Hear the Tone." In *The Best of Thomas N. Scortia*. Garden City, NY: Doubleday.

Seabury, P. (1961). "The Histronaut." *Columbia University Forum* 4 (Summer):4–8.

Sears, F.W., and R.W. Brehme (1968). *Introduction to the Theory of Relativity*. Reading, MA: Addison-Wesley.

Sell, W. (SFAD). "Other Tracks."

Selleri, F. (1997). "The Relativity Principle and the Nature of Time." *Foundations of Physics* 27 (November):1527–1548.

Sellings, A. (1971). "The Last Time Around." In *World's Best Science Fiction*. Edited by D.A. Wollheim and T. Carr. New York: Ace.

Serling, R.E. (TZ). "Walking Distance," "Execution," "Back There," "The Odyssey of Flight 33," "A Hundred Yards Over the Rim," "Once Upon a Time," "No Time Like the Past," "A Kind of Stopwatch," "The 7th is Made Up of Phantoms," and "A Most Unusual Camera."

Shaara, M. (SFS). "Man of Distinction."

—— (1982). "Time Payment." In *Soldier Boy*. New York: Pocket Books.

Shallis, M. (1982). *On Time*. London: Burnett.

Shapiro, S. (1986). *A Time to Remember*. New York: Random House.

Shapiro, S.L., and S.A. Teukolsky (1991a). "Formation of Naked Singularities: The Violation of Cosmic Censorship." *Physical Review Letters* 66 (February 25):994–997.

—— (1991b). "Black Holes, Naked Singularities and Cosmic Censorship." *American Scientist*, July–August.

Sharkey, J. (1965). "The Trouble with Hyperspace." *Fantastic*, April.

Sharp, D.D. (1939). "Faster Than Light." *Marvel Science Stories*, February.

Shaw, B. (TW). "Skirmish on a Summer Morning."

—— (TOT): "Light of Other Days."

—— (1977). *Who Goes Here?* New York: Ace.

—— (1973). "What Time Do You Call This?" In *Tomorrow Lies in Ambush*. New York: Ace.

—— (U2). "Retroactive."

Shay, H.J. (1953). "The Ambassador from the 21st Century." *Startling Stories*, March.

Sheckley, R. (1990). "The World of Heart's Desire." In *The Great Science Fiction Stories*. Vol. 21. New York: DAW.

—— (MTMW). "Miss Mouse and the Fourth Dimension."

—— (TIT). "The King's Wishes."

—— (1960). "Double Indemnity." In *Notions: Unlimited*. New York: Bantam.

—— (1957). "The Deaths of Ben Baxter." *Galaxy Science Fiction*, July.

—— (1955). "A Thief in Time" and "Something for Nothing." In *Citizen in Space*. New York: Ballantine.

Sheldon, E. (1990). "Faster Than Light?" *Sky & Telescope* 79 (January):26–29.

Sheldon, W. (1950a). "A Bit of Forever." *Super Science Stories*, July.

—— (1950b). "The Time Cave." *Thrilling Wonder Stories*, April.

Shenglin, C. (1992a). "The Theory of Relativity and Super-Luminal Speeds IV: The Catastrophe of the Schwarzschild Field and Superluminal Expansion of Extragalactic Radio Sources." *Astrophysics and Space Science* 193 (July):123–140.

—— (1992b). "The Theory of Relativity and Super-Luminal Speeds III: The Catastrophe of the Space-time on the Finsler Metric." *Astrophysics and Space Science* 190 (April):303–315.

—— (1990). "The Theory of Relativity and Super-Luminal Speeds II: Theory of Relativity in the Finsler Space-time." *Astrophysics and Space Science* 174 (December):165–171.

—— (1988). "The Theory of Relativity and Super-Luminal Speeds I: Kinematical Part." *Astrophysics and Space Science* 145 (June):293–302.

Sherred, T.L. (1983). "E for Effort." In *The Great Science Fiction Stories*. Vol. 9. New York: DAW.

Shimony, A. (QCST). "Events and Processes in the Quantum World."

Shiner, L. (TT). "Twilight Time."

Shoemaker, S. (1969). "Time Without Change." *Journal of Philosophy* 66 (June 19):363–381.

Sigman, J. (1986). "Science and Parody in Kurt Vonnegut's *The Sirens of Titan*." *Mosaic* 19 (Winter):15–32.

Silverberg, R. (1991). "Hunters in the Forest." *Omni*, October.

—— (1990). *Letters from Atlantis*. New York: Antheneurn.

—— (1987). *Project Pendulum*. New York: Walker

—— (SFS). "The Assassin."

—— (1986). "In Entropy's Jaws," "Many Mansions," "Ms. Found in an Abandoned Time Machine," "When We Went to See the End of the World," "{Now+n, Now−n}," "Trips," and "What We Learned from This Morning's Newspaper." In *Beyond the Safe Zone*. New York: Warner.

—— (CJSF). "Hawksbill Station."

—— (TIT). "MUgwump 4."

—— (1970). *Vornan-19*. London: Coronet.

—— (1969). *Up the Line*. New York: Ballantine.

—— (1967). *The Time Hoppers*. Garden City, NY: Doubleday.

—— (VIT). "Absolutely Inflexible."

Simak, C.D. (1978). *Mastodonia*. New York: Ballantine.

—— (BGA1). "The World of the Red Sun."

—— (S1). "The Birch Clump Cylinder."

—— (1974a). "The Marathon Photograph." In *Thread of Time*. Edited by R. Silverberg. New York: Thomas Nelson.

—— (1974b). *Our Children's Children*. New York: G.P. Putnam's Sons.

—— (1969). "The Loot of Time." In *Alien Earth and Other Stories*. Edited by R. Elwood and S. Moskowitz. New York: Macfadden-Bartell.

—— (TC). "Over the River & Through the Woods."

—— (1962). "Project Mastodon." In *All the Traps of Earth*. Garden City, NY: Doubleday.

—— (1960). "Carbon Copy." In *The Worlds of Clifford Simak*. New York: Simon and Schuster.

—— (1951). *Time and Again*. New York: Simon and Schuster.

—— (1938). "Rule Eighteen." *Astounding Science Fiction*, July.

—— (1932). "Hellhounds of the Cosmos." *Astounding Stories*, June.

Simak, C.D., and C. Jacobi (1975). "The Street That Wasn't There." In *Creatures from Beyond*. Edited by T. Carr. New York: Thomas Nelson.

Simons, W. (FS). "Ghost Ship."

Simpson, G.G. (1996). *The Dechronization of Sam Magruder*. New York: St. Martin's Press.

Simpson, M., and R. Penrose (1973). "Internal Instability in a Reissner–Nordström Black Hole." *International Journal of Theoretical Physics* 7 (April):183–197.

Sinclair, M. (1946). "Where Their Fire Is Not Quenched." In *After the Darkness Falls*. Edited by B. Karloff. Cleveland, OH: World Publishing.

Skillen, A. (1965). "The Myth of Temporal Division." *Analysis* 26 (December):44–47.

Sklar, L. (1984). "Comments on Malament's" 'Time Travel' in the Gödel Universe." *Proceedings of the Philosophy of Science Association* 2:106–110.

—— (RTR). "Time, Reality and Relativity."

—— (1974). *Space, Time and Spacetime*. Los Angeles: University of California Press.

Sladek, J. (1981). "The Singular Visitor from Not-Yet" and "1937 A.D.!" In *The Best of John Sladek*. New York: Pocket.

Sleator, W. (1990). *Strange Attractors*. New York: E.P. Dutton.

—— (1986). *The Boy Who Reversed Himself*. New York: E.P. Dutton.

—— (1981). *The Green Futures of Tycho*. New York: E.P. Dutton.

Small, R. (1991). "Incommensurability and Recurrence: From Oresme to Simmel." *Journal of the History of Ideas* 52 (January–March):121–137.

—— (1986). "Tristram Shandy's Last Page." *British Journal for the Philosophy of Science* 37 (June):213–216.

Smalley, L.L., and J.P. Krisch (1994). "Energy Conditions in Non-Riemannian Spacetimes." *Physics Letters A* 196 (December 26):147–153.

Smart, J.J.C. (1981). "The Reality of the Future." *Philosophia* 10 (December):141–150.

—— (1972). "Space-Time and Individuals." In *Logic and Art*. Edited by R. Rudner and I. Scheffler. Indianapolis: Bobbs-Merrill.

—— (1967). "The Unity of Space-Time: Mathematics Versus Myth Making." *Australasian Journal of Philosophy* 25:214–217.

—— (1963). "Is Time Travel Possible?" *Journal of Philosophy* 60 (April):237–241.

—— (1958). "A Review of *The Direction of Time*." *Philosophical Quarterly* 8 (January):72–77.

—— (1955a). "Mr. Mayo on Temporal Asymmetry." *Australasian Journal of Philosophy* 33:124–127.

—— (1955b). "Spatialising Time." *Mind* 64:239–241.

—— (1954). "The Temporal Asymmetry of the World." *Analysis* 14 (March):79–83.

—— (1949). "The River of Time." *Mind* 58 (October):483–494.

Smith, C.A. (1970). "An Adventure in Futurity." In *Other Dimensions*. Sauk City, WI: Arkham House.

—— (1964). "Murder in the Fourth Dimension." In *Tales of Science and Sorcery*. Sauk City, WI: Arkham House.

—— (1932). "Flight Through Super-Time." *Wonder Stories,* August.

Smith, G., and R. Weingard (1990). "Quantum Cosmology and the Beginning of the Universe." *Philosophy of Science* 57 (December):663–667.

Smith, G.H. (MT). "Take Me to Your Leader."

Smith, G.O. (1948). "The Möbius Trail." *Thrilling Wonder Stories,* December.

Smith, J.S. (1958). "Encountering Things Past." *Library Journal* 83 (November 15):3265–3268.

Smith, J.W. (1985). "Time Travel and Backward Causation." *Cogito* 3:57–67.

Smith, N.J.J. (1997). "Bananas Enough for Time Travel?" *British Journal for the Philosophy of Science* 48 (September):363–389.

Smith, Q. (1994). "Did the Big Bang Have a Cause?" *British Journal for the Philosophy of Science* 45 (June):649–668.

—— (1991). "Atheism, Theism and Big Bang Cosmology." *Australasian Journal of Philosophy* 69 (March):48–66.

—— (1989). "A New Typology of Temporal and Atemporal Permanence." *Nous* 23 (June):307–330.

—— (1988). "The Uncaused Beginning of the Universe." *Philosophy of Science* 55 (March):39–57.

—— (1987). "Infinity and the Past." *Philosophy of Science* 54 (March):63–75.

—— (1986a). "The Infinite Regress of Temporal Attributions." *Southern Journal of Philosophy* 24 (Fall):383–396.

—— (1986b). "World Ensemble Explanations." *Pacific Philosophical Quarterly* 67 (January):73–86.

—— (1985a). "On the Beginning of Time." *Nous* 19 (December):579–584.

—— (1985b). "Kant and the Beginning of the World." *New Scholasticism* 59 (Summer):339–346.

—— (1985c). "The Mind-Independence of Temporal Becoming." *Philosophical Studies* 47 (January):109–119.

Snyder, H.S. (1947). "Quantized Space-Time." *Physical Review* 71 (January 1):38–41.

Soares, I.D. (1980). "Inhomogeneous Rotating Universes with Closed Timelike Geodesics of Matter." *Journal of Mathematical Physics* 21 (March):521–525.

Soen, Y., and A. Ori (1996). Improved Time-Machine Model." *Physical Review D* 54 (October 15):4858–4861.

Sokolowski, R. (1982). "Timing." *Review of Metaphysics* 35 (June):687–714.

Soleng, H.H. (1994). "Negative Energy Densities in Extended Sources Generating Closed Timelike Curves in General Relativity With and Without Torsion." *Physical Review D* 49 (January 15):1124–1125.

—— (1992). "A Spinning String." *General Relativity and Gravitation* 24 (January):111–117.

—— (1989). "Gödel-Type Cosmologies with Spin Density." *Physics Letters A* 137 (May 29):326–328.

Som, M.M., and A.K. Raychaudhuri (1968). "Cylindrically Symmetric Charged Dust Distributions in Rigid Rotation in General Relativity." *Proceedings of the Royal Society A* 304 (April):81–86.

Sorenson, R.A. (1987). "Time Travel, Parahistory and Hume." *Philosophy* 62 (April):227–236.

Sorkin, R.D. (1997). "Forks in the Road, on the Way to Quantum Gravity." *International Journal of Theoretical Physics* 36 (December):2759–2781.

South C. (1942). "The Time Mirror." *Amazing Stories,* December.

Spellman, L. (1982). "Causing Yesterday's Effects." *Canadian Journal of Philosophy* 12 (March):145–161.

Spinrad, N. (1980). "The Weed of Time." In *Alchemy & Academe.* Edited by A. McCaffrey. New York: Ballantine.

Sprague, C. (1951). "Time Track." *Startling Stories,* January.

Spruill, G. (1980). "The Janus Equation." In *Binary Star No. 4*. New York: Dell.

Stannard, F.R. (1966). "Symmetry of the Time Axis." *Nature* 211 (August 13):693–695.

Stapledon, O. (1953). *To the End of Time*. New York: Funk and Wagnalls.

Stapleton, D. (1942). "How Much to Thursday?" *Thrilling Wonder Stories,* December.

Stapp, H.P. (1994). "Theoretical Model of a Purported Empirical Violation of the Predictions of Quantum Theory." *Physical Review A* 50 (July):18–22.

Starobinsky, A.A. (1980). "A New Type of Isotropic Cosmological Models Without Singularity." *Physics Letters* 91B (24 March):99–102.

Statten, V. (1953). *Zero Hour*. London: Scion.

Stearns, I. (1950). "Time and the Timeless." *Review of Metaphysics* 4 (December):187–200.

Steif, A.R. (1994). "Multiparticle Solutions in 2+1 Gravity and Time Machines." *International Journal of Modern Physics D* 3 (March):277–280.

Stein, H. (1991). "On Relativity Theory and Openness of the Future." *Philosophy of Science* 58 (June):147–167.

—— (1984). "The Everett Interpretation of Quantum Mechanics: Many Worlds or None?" *Nous* 18 (November):635–652.

—— (1970). "On the Paradoxical Time-Structures of Gödel," *Philosophy of Science* 37 (December):589–601.

—— (1968). "On Einstein–Minkowski Space-Time." *Journal of Philosophy* 65 (January 11):5–23.

—— (1967). "Newtonian Space-Time." *Texas Quarterly* 10 (Autumn):174–200.

Stephenson, L.M. (1978). "Clarification of an Apparent Asymmetry in Electromagnetic Theory." *Foundations of Physics* 8 (December):921–926.

Stern, M.B. (1936). "Counterclockwise: Flux of Time in Literature." *Sewanee Review* 44 (July–September):338–365.

Stewart, I. (1994). "The Real Physics of Time Travel." *Analog,* January.

Stith, J.E. (DF). "One Giant Step."

Stone, L.F. (1935). "The Man with the Four Dimensional Eyes." *Wonder Stories,* August.

Strauss, E.S. (undated). *The MIT Science Fiction Society's Index to the S-F Magazines, 1951–1965*.

Stromberg, G. (1961). "Space, Time, and Eternity." *Journal of the Franklin Institute* 272 (August):134–144.

Stuart, D.A. (1972). "Forgetfulness." In *The Astounding-Analog Reader*. Vol. 1. Edited by H. Harrison and B. Aldiss. Garden City, NY: Doubleday.

Stump, E., and N. Kretzmann (1981). "Eternity." *Journal of Philosophy* 78 (August):429–458.

Sturgeon, T. (SFAD). "Yesterday Was Monday."

Sudarshan, E.C.G. (1969). "The Nature of Faster-Than-Light Particles and Their Interactions." *Arkiv for Fysik* 39:585–591.

Sudbery, T. (1997). "The Fastest Way from A to B." *Nature* 390 (December 11):551–552.

Sullivan, T. (D). "Dinosaur on a Bicycle."

—— (TT). "The Comedian."

Sushkov, S.V. (1991). "On the Existence of a Lorentz Wormhole." *Soviet Journal of Nuclear Physics* 53 (May):899–904.

Sutherland, R.I. (1983). "Bell's Theorem and Backwards-in-Time Causality." *International Journal of Theoretical Physics* 22 (April):377–384.

Svozil, K. (1995). "Time Paradoxes Reviewed." *Physics Letters A* 199 (April 3):323–326.

Swartz, N. (1991). *Beyond Experience: Metaphysical Theories and Philosophical Constraints*. London: University of Toronto Press.

—— (1973). "Is There an Ozma-Problem for Time?" *Analysis* 33 (January): 77–82.

Sweeney, L. (1974). "Bonaventure and Aquinas on the Divine Being as Infinite." *Southwestern Journal of Philosophy* 5 (Summer):71–91.

Swinburne, R.G. (1966). "Affecting the Past." *Philosophical Quarterly* 16 (October):341–347.

—— (1965a). "Conditions for Bitemporality." *Analysis* 26 (December):47–50.

—— (1965b). "Times." *Analysis* 25 (June):185–191.

Sycamore, H.M. (1959). "Success Story." *Fantasy and Science Fiction*, July.

Sylvester, J.J. (1869). "A Plea for the Mathematician." *Nature* 1 (30 December):237–239.

Szilard, L. (ED). "The Mark Gable Foundation."

Tanaka, S. (1960). "Theory of Matter with Super Light Velocity." *Progress of Theoretical Physics* 24 (July):171–200.

Tanaka, T., and W.A. Hiscock (1994). "Chronology Protection and Quantized Fields: Nonconformal and Massive Scalar Fields in Misner Space." *Physical Review D* 49 (May 15):5240–5245.

Taylor, B.E., and W.A. Hiscock (1997). "Stress-Energy of a Quantized Scalar Field in Static Wormhole Spacetimes." *Physical Review D* 55 (May 15):6116–6122.

Taylor, R. (1987). "Time and Life's Meaning." *Review of Metaphysics* 40 (June):675–686.

—— (1974). "Moving Forth and Back in Space and Time." In *Metaphysics*. 2nd ed. Englewood Cliffs, NJ: Prentice-Hall.

—— (1959). "Moving About in Time." *Philosophical Quarterly* 9 (October):289–301.

—— (1955). "Spatial and Temporal Analogies and the Concept of Identity." *Journal of Philosophy* 52 (October 27):599–612.

Teitelboim, C. (1970). "Splitting of the Maxwell Tensor: Radiation Reaction Without Advanced Fields." *Physical Review D* 1 (March 15):1572–1582.

Tenn, W. (1983). "Brooklyn Project." In *The Great Science Fiction Stories*. Vol. 10. New York: DAW.

—— (TSF). "Child's Play."

—— (1962). "Time Waits for Winthrop." In *Time Waits for Winthrop*. Edited by F. Pohl. Garden City, NY: Doubleday.

—— (1955). "Me, Myself, and I." In *Of All Possible Worlds*. New York: Ballantine.

Tertletskii, Y.P. (1968). *Paradoxes in the Theory of Relativity*. New York: Plenum.

Tevis, W. (1961). "The Other End of the Line." *Fantasy and Science Fiction*, November.

Theobald, D.W. (1976). "On the Recurrence of Things Past." *Mind* 85 (January):107–111.

Thom, R. (1975). "Time-Travel and Non-Fatal Suicide." *Philosophical Studies* 27 (March):211–216.

Thomas, T. (AO). "The Doctor."

Thompson, D. (BT). "Worlds Enough."

Thompson, R.G. (1941). "The Brontosaurus." *Stirring Science Fiction*, April.

Thompson, W.R. (1993). "The Plot to Save Hitler." *Analog*, September.

Thomson, J.J. (1965). "Time, Space, and Objects." *Mind* 74 (January):1–27.

't Hooft, G. (1992). "Causality in (2+1)-Dimensional Gravity." *Classical and Quantum Gravity* 9 (May):1335–1348.

Thorne, K.S. (1997). "Do the Laws of Physics Permit Wormholes for Interstellar Travel and Machines for Time Travel?" In *Carl Sagan's Universe*. Edited by Y. Terzian and E. Bilson. New York: Cambridge University Press.

—— (1994). *Black Holes & Time Warps*. New York: W.W. Norton.

—— (DGR1). "Misner Space as a Prototype for Almost Any Pathology."

—— (1992). "Closed Timelike Curves." In *General Relativity and Gravitation 1992*. Edited by R.J. Gleiser *et al.* Bristol, England: Institute of Physics.

—— (1991). "Do the Laws of Physics Permit Closed Timelike Curves?" *Annals of the New York Academy of Science* 631 (August 10):182–193.

—— (MWM). "Nonspherical Gravitational Collapse—A Short Review."

—— (1970). "Nonspherical Gravitational Collapse: Does It Produce Black Holes?" *Comments on Astrophysics and Space Physics* 2 (September–October):191–196.

—— (1965a). "Gravitational Collapse and the Death of a Star." *Science* 150 (December 24):1671–1679.

—— (1965b). "Energy of Infinitely Long, Cylindrically Symmetric Systems in General Relativity." *Physical Review* 138B (April 12):251–166.

Thouless, D.J. (1969). "Causality and Tachyons." *Nature* 224 (November 1):506.

Tilley, R.J. (1973). " 'Willie's Blues.' " In *The 1973 Annual World's Best SF.* Edited by D.A. Wollheim and A.W. Saha. New York: DAW.

Tipler, F.J. (1992). "The Ultimate Fate of Life in Universes Which Undergo Inflation." *Physics Letters B* 286 (July 23):36–43.

—— (1986). "Interpreting the Wave Function of the Universe." *Physics Reports* 137 (May):231–275.

—— (QCST). "The Many-Worlds Interpretation of Quantum Mechanics in Quantum Cosmology."

—— (1985). "Note on Cosmic Censorship." *General Relativity and Gravitation* 17:499–507.

—— (1980). "General Relativity and the Eternal Return." In *Essays in General Relativity.* Edited by F.J. Tipler. New York: Academic Press.

—— (1979). "General Relativity, Thermodynamics, and the Poincaré Cycle." *Nature* 280 (July 19):203–205.

—— (1978). "Energy Conditions and Spacetime Singularities." *Physical Review D* 17 (May 15):2521–2528.

—— (1977). "Singularities and Causality Violation." *Annals of Physics* 108 (September 22):1–36.

—— (1976a). "Causality Violation in Asymptotically Flat Space-Times." *Physical Review Letters* 37 (October 4):879–882.

—— (1976b). *Causality Violation in General Relativity.* University of Maryland Ph.D. dissertation. Unpublished. Available as Dissertation 76-29,018 from Xerox University Microfilms, Ann Arbor, MI.

—— (1975). "Direct-Action Electrodynamics and Magnetic Monopoles." *Il Nuovo Cimento* 28B (August 11):446–452.

—— (1974). "Rotating Cylinders and the Possibility of Global Causality Violation." *Physical Review D* 9 (April 15):2203–2206.

Tiptree, Jr., J. (D). "The Night-Blooming Saurian."

—— (1988). "Backward, Turn Backward." In *Crown of Stars.* New York: Tor.

—— (1978). "Houston, Houston, Do You Read?" In *Nebula Winners Twelve.* Edited by G.R. Dickson. New York: Harper & Row.

—— (1975). "The Man Who Walked Home." In *Looking Ahead.* Edited by D. Allen and L. Allen. New York: Harcourt.

Tolman, R.C. (1917). *The Theory of the Relativity of Motion.* Berkeley: University of California Press.

Tolman, R.C., and S. Smith (1926). "On the Nature of Light." *Proceedings of the National Academy of Sciences* 12 (May 15):343–347.

Trautman, A. (GRAV). "Conservation Laws in General Relativity."

Travis, J. (1992). "Could a Pair of Cosmic Strings Open a Route Into the Past?" *Science* 256 (April 10):179–180.

Tryon, E.P. (1973). "Is the Universe a Vacuum Fluctuation?" *Nature* 246 (December 14):396–397.

Tucker, B. (1952). "The Tourist Trade." In *Tomorrow, the Stars.* Edited by R.A. Heinlein. Garden City, NY: Doubleday.

Tucker, W. (1979). *The Year of the Quiet Sun.* Boston: Gregg.

—— (1958). *The Lincoln Hunters.* New York: Rinehart.

—— (1955). *Time Bomb.* New York: Reinhart.

Turtledove, H. (1992). *The Guns of the South.* New York: Ballantine.

Tute, R. (1940). "Space-Time: A Link Between Religion and Science." *Hibbert Journal* 38:261–270.

Updike, J. (1997). *Toward the End of Time.* New York: Knopf.

Upson, W.H. (MM). "Paul Bunyan Versus the Conveyor Belt."

—— (FM). "A. Botts and the Moebius Strip."

Uttley, A. (1964). *A Traveler in Time.* New York: Viking.

Vaidman, L. (1991). "A Quantum Time Machine." *Foundations of Physics* 21 (August):947–958.

Vance, J. (1973). "Rumfuddle." In *Three Trips in Time and Space.* Edited by R. Silverberg. New York: Hawthorn.

van Lorne, W. (BSF). "The Upper Level Road."

van Stockum, W.J. (1937). "The Gravitational Field of a Distribution of Particles Rotating About an Axis of Symmetry." *Proceedings of the Royal Society of Edinburgh* 57:135–154.

Van Vogt, A.E. (1990). "Secret Unattainable." In *The Fantastic World War II.* Edited by F. McSherry, Jr. New York: Baen.

—— (SS). "Far Centaurus."

—— (1980). "The Seesaw." In *The Great Science Fiction Stories.* Vol. 3. New York: DAW.

—— (1972). "The Timed Clock." In *The Book of Van Vogt.* New York: DAW.

—— (1960). "The Ghost." In *Zacherleys Midnight Snacks.* New York: Ballantine.

—— (1954). *Weapon Shops of Isher.* New York: Ace.

—— (OSF). "Recruiting Station."

Varley, J. (1983). *Millennium.* New York: Berkley.

Verne, J. (1996). *Paris in the Twentieth Century.* New York: Random House.

—— (CBSF). "In the Year 2889."

Verrill, A.H. (1927). "The Astounding Discoveries of Doctor Mentiroso." *Amazing Stories,* November.

Vidal, G. (1998). *The Smithsonian Institution.* New York: Random House.

—— (1992). *Live from Golgotha.* New York: Random House.

Vihvelin, K. (1996). "What Time Travelers Cannot Do." *Philosophical Studies* 81 (March):315–330.

Visser, M. (1997a). "The Reliability Horizon for Semi-Classical Quantum Gravity: Metric Fluctuations Are Often More Important Than Back-reaction." *Physics Letters B* 415 (December 4):8–14.

—— (1997b). "Traversable Wormholes: The Roman Ring." *Physical Review D* 55 (April 15):5212–5214.

—— (1995). *Lorentzian Wormholes:* New York: AIP Press.

—— (1994a). "Van Vleck Determinants: Traversable Wormhole Spacetimes." *Physical Review D* 49 (April 15):3963–3980.

—— (1994b). "Hawking's Chronology Protection Conjecture: Singularity Structure of the Quantum Stress-Energy Tensor." *Nuclear Physics B* 416 (April 4):895–906.

—— (1993). "From Wormhole to Time Machine: Comments on Hawking's Chronology Protection Conjecture." *Physical Review D* 47 (January 15):554–565.

—— (1990). "Wormholes, Baby Universes, and Causality." *Physical Review D* 41 (February 15):1116–1124.

—— (1989a). "Traversable Wormholes from Surgically Modified Schwarzschild Spacetimes." *Nuclear Physics B* 328 (December 11):203–212.

—— (1989b). "Traversable Wormholes: Some Simple Examples." *Physical Review D* 39 (May 15):3182–3184.

Voinovich, V. (1987). *Moscow 2042*. New York: Harcourt.

Vollick, D.N. (1997a). "How to Produce Exotic Matter Using Classical Fields." *Physical Review D* 56 (October 15):4720–4723.

—— (1997b). "Maintaining a Wormhole with a Scalar Field." *Physical Review D* 56 (October 15):4724–4728.

von Hoerner, S. (1963). "The General Limits of Space Travel." In *Interstellar Communication*. Edited by A.G.W. Cameron. New York: W.A. Benjamin.

Vonnegut, K., Jr. (1997). *Timequake*. New York: G.P. Putnam's Sons.

—— (1971). *Slaughterhouse-Five*. New York: Delacourte.

—— (1959). *The Sirens of Titan*. New York: Dell.

Wachhorst, W. (1984). "Time Travel Romance on Film." *Extrapolation* 25 (Winter):340–359.

Waelbroeck, H. (1991). "Do Universes with Parallel Cosmic Strings or Two-dimensional Wormholes have Closed Timelike Curves?" *General Relativity and Gravitation* 23 (February):219–233.

Wald, R.M. (1992). " 'Weak' Cosmic Censorship." In Vol. 2 of *PSA 1992*. East Lansing, MI: Philosophy of Science Association.

—— (1977). *Space, Time, and Gravity*. Chicago: University of Chicago Press.

—— (1974). "Gedanken Experiments to Destroy a Black Hole." *Annals of Physics* 83:548–556.

Wald, R. and U. Yurtsever (1991). "General Proof of the Averaged Null Energy Condition for a Massless Scalar Field in Two-Dimensional Curved Spacetime." *Physical Review D* 44 (July 15):403–416.

Waldrop, H. (1984). *Them Bones*. New York: Ace.

Walstad, A. (1980). "Time's Arrow in an Oscillating Universe." *Foundations of Physics* 10 (October):743–749.

Walton, H. (TSF). "Housing Shortage."

Wandrei, D. (1965). "The Man Who Never Lived" and "Infinity Zero." In *Strange Harvest*. Sauk City, WI: Arkham House.

—— (1933). "A Race Through Time." *Astounding Stories*, October.

Wang, A., and P.S. Letelier (1995). "Dynamical Wormholes and Energy Conditions." *Progress of Theoretical Progress* 94 (Letters) (July):137–142.

Ward, K. (1967). "The Unity of Space and Time." *Philosophy* 42:68–74.

Waterlow, S. (1974). "Backwards Causation and Continuing." *Mind* 83 (July): 372–387.

Watson, I. (1991). "In the Upper Cretaceous with the Summerfire Brigade" and "From the Annals of the Onomastic Society." In *Stalin's Teardrops*. London: Victor Gollancz.

—— (TT). "Ghost Lecturer."

—— (1989). *Chekhov's Journey*. New York: Carroll & Graf.

—— (1979). "The Very Slow Time Machine." In *The Best Science Fiction of the Year*. Edited by T. Carr. New York: Del Rey/Ballantine.

Watt-Evans, L. (1992). *Crosstime Traffic*. New York: Ballantine.

Watzlawick, P. (1976). *How Real Is Real?* New York: Random House.

Webb, C.W. (1977). "Could Space Be Time-Like?" *Journal of Philosophy* 73 (August):462–474.

—— (1960). "Could Time Flow? If So, How Fast?" *Journal of Philosophy* 57 (May 26):357–365.

Weeks, J.R. (1985). *The Shape of Space*. New York: Marcel Dekker.

Weinbaum, S.G. (1949). "The Worlds of If" and "The Circle of Zero." In *A Martian Odyssey and Others*. Reading, PA: Fantasy Press.

Weiner, A. (TT). "Klein's Machine."

—— (1983). "One More Time." In *Chrysalis 10*. Edited by R. Torgeson. Garden City, NY: Doubleday.

Weingard, R. (1979a). "Some Philosophical Aspects of Black Holes." *Synthese* 42 (September):191–219.

—— (1979b). "General Relativity and the Conceivability of Time Travel." *Philosophy of Science* 46 (June):328–332.

—— (1979c). "General Relativity and the Length of the Past." *British Journal for the Philosophy of Science* 29 (June):170–172.

—— (1977). "Space-Time and the Direction of Time." *Nous* 11 (May):119–131.

—— (1972a). "On Travelling Backward in Time." *Synthese*, 24:117–132.

—— (1972b). "Relativity and the Reality of Past and Future Events." *British Journal for the Philosophy of Science* 23:119–121.

Weir, S. (1988). "Closed Time and Causal Loops: A Defense Against Mellor." *Analysis* 48 (October):203–209.

Weisinger, M. (1944). "Thompson's Time Traveling Theory." *Amazing Stories*, March.

—— (1938). "Time on My Hands." *Thrilling Wonder Stories*, June.

Wellman, M.W. (1988). *Twice in Time*. New York: Baen.

—— (MT). "Who Else Could I Count On?"

—— (1949). ". . . backward, O Time!" *Thrilling Wonder Stories*, October.

—— (1939). "The Einstein Slugger." *Thrilling Wonder Stories*, December.

Wells, H.G. (1987). *The Time Machine*. In *The Definitive Time Machine*. Edited by H.M. Ceduld. Bloomington: Indiana University Press.

—— (1975). *H.G. Wells: Early Writings in Science and Science Fiction*. Edited by R.M. Philmus and D.Y. Hughes. Berkeley, CA: University of California Press.

—— (1966). "The Plattner Story," "The New Accelerator," "The Invisible Man," and "The Story of Davidson's Eyes." In *Best Science Fiction Stories of H.G. Wells*. New York: Dover.

—— (1924). *The Dream*. New York: Macmillan.

—— (1923). *Men Like Gods*. New York: Macmillan.

—— (1899). *When the Sleeper Wakes*. New York: Harper and Brothers.

—— (1895). *The Wonderful Visit*. New York: Macmillan.

West, W. (1963). *River of Time*. New York: Avalon Books.

Westall, R. (1978). *The Devil on the Road*. New York: Greenwillow.

Weyl, H. (1952). *Space-Time-Matter*. Translated from the 1921 fourth edition in German. New York: Dover.

Wheeler, J.A. (1981). "The Lesson of the Black Hole." *Proceedings of the American Philosophical Society* 125 (February):25–37.

—— (1979). "Frontiers of Time." In *Problems in the Foundations of Physics*. Edited by G.T. diFrancia. Proceedings of the International School of Physics. Course 72. New York: North-Holland.

—— (NT). "Three-Dimensional Geometry as a Carrier of Information About Time."

—— (1962a). *Geometrodynamics*, New York: Academic Press.

—— (1962b). "Curved Empty Space-Time as the Building Material of the Physical World: An Assessment." In *Logic, Methodology and Philosophy of Science*. Stanford, CA: Stanford University Press.

—— (1961). "Geometrodynamics and the Problem of Motion." *Reviews of Modern Physics* 33 (January):63–78.

Wheeler, J.A., and R.P. Feynman (1949). "Classical Electrodynamics in Terms of Direct Interparticle Action." *Reviews of Modern Physics* 21 (July):425–433.

—— (1945). "Interaction with the Absorber as the Mechanism of Radiation." *Reviews of Modern Physics* 17 (April–July):157–181.

Wheeler, J.C. (1996). "Of Wormholes, Time Machines, and Paradoxes," *Astronomy*, February.

Whiston, G.S. (1974). " 'Hyperspace' (The Cobordism Theory of Space-Time)." *International Journal of Theoretical Physics* 11 (December):285–288.

Whitaker, M.A.B. (1985). "On the Observability of 'Many Worlds.' " *Journal of Physics A* 18 (July 11):1831–1834.

White, C.B. (1927). "The Lost Continent." *Amazing Stories,* July.

Whitehead, A.N., *et al.* (BST). Discussions on "The Paradox of the Twins."

—— (BST). "The Inapplicability of the Concept of Instant on the Quantum Level."

Whitrow, G.J. (1988). *Time in History.* New York: Oxford University Press.

—— (1980). *The Natural Philosophy of Time.* Oxford, England: Oxford University Press.

—— (1978). "On the Impossibility of an Infinite Past." *British Journal for the Philosophy of Science* 29 (March):39–45.

—— (BST). " 'Becoming' and the Nature of Time."

—— (NYAS). "Reflections on the Natural Philosophy of Time."

Whorf, B.L. (1956). *Language, Thought, and Reality.* Cambridge, MA: The M.I.T. Press.

Wiener, N. (BST). "Spatio-Temporal Continuity, Quantum Theory and Music."

Wiggins, K.M. (1993). "Epic Heroes, Ethical Issues, and Time Paradoxes in *Quantum Leap.*" *Journal of Popular Film and Television* 21 (Fall):111–120.

Wigher, E.P. (1970). "On Hidden Variables and Quantum Mechanical Probabilities." *American Journal of Physics* 38 (August):1005–1009.

Wilcox, J.T. (1961). "A Question from Physics for Certain Theists." *Journal of Religion* 41 (October):293–300.

Wild, J. (1954). "The New Empiricism and Human Time." *Review of Metaphysics* 7 (June):537–557.

Wildsmith, B. (1980). *Professor Noah's Spaceship.* New York: Oxford University Press.

Wilhelm, K. (1992). "O Homo; O Femina; O Tempora." In *And the Angels Sing.* New York: St. Martin's Press.

—— (1975). "The Time Piece," In *The Infinity Box.* New York: Harper & Row.

Wilkerson, T.E. (1979). "More Time and Time Again." *Philosophy* 54 (January):110–112.

—— (1973). "Time and Time Again." *Philosophy* 48 (April):173–177.

Williams, C. (1963). *Many Dimensions.* London: Faber & Faber.

Williams, C.J.F. (1978). "True Tomorrow, Never True Today." *Philosophical Quarterly* 28 (October):285–299.

Williams, D.C. (1951a). "The Myth of Passage." *Journal of Philosophy* 48 (July):457–472.

—— (1951b). "The Sea Fight Tomorrow." In *Structure, Method and Meaning.* New York: The Liberal Arts Press.

Williams, J. (TC). "Terror Out of Time."

Williams, J., and R. Abrashkin (1963). *Danny Dunn, Time Traveler.* New York: Whittlesey House.

Williams, R.M. (TSF). "Flight of the Dawn Star."

—— (1942). "The Incredible Slingshot Bombs." *Amazing Stories,* May.

—— (1940). "The Tides of Time." *Thrilling Wonder Stories,* April.

Williamson, J. (1985). *"The Legion of Time."* New York: Bluejay-Books.

—— (1979). "Hindsight." In *The Great Science Fiction Stories.* Vol. 2. New York: DAW.

—— (1975). "The Meteor Girl" and "Through the Purple Cloud." In *The Early Williamson.* Garden City, NY: Doubleday.

—— (BGA1). "The Moon Era."

—— (1942). "Minus Sign." *Astounding Science Fiction,* November.

—— (1933a). "Terror Out of Time." *Astounding Stories,* December.

—— (1933b). "In the Scarlet Star." *Amazing Stories,* March.

Willis, C. (1997). *To Say Nothing of the Dog.* New York: Bantam.

—— (1992). *Doomsday Book.* New York: Bantam.

—— (1985). "Fire Watch." In *Fire Watch*. New York: Bluejay.

Wilson, R. (1942). "The Message." *Astounding Stories,* March.

Wilson, R.C. (1991). *A Bridge of Years*. New York: Doubleday.

Wilson, R.H. (SFT). "Out Around Rigel."

—— (1931). "A Flight Into Time." *Wonder Stories,* February.

Wimmel, H.K. (1972). "Tachyons and Cerenkov Radiation." *Nature Physical Science* 236, (April 3):79–80.

Windred, G. (1933–1934). "The History of Mathematical Time. II." *Isis* 20:192–219.

—— (1933). "The History of Mathematical Time. I." *Isis* 19:121–153.

Winterbotham, R.R. (1936). "The Fourth Dynasty." *Astounding Stories,* December.

Wolf, F.A. (1990). *Parallel Universes*. New York: Simon and Schuster.

Wolfe, G. (1974). "The Rubber Bend." In *Universe 5*. Edited by T. Carr. New York: Random House.

Wolfe, J. (1985). "The Impossibility of an Infinite Past: A Reply to Craig." *International Journal for Philosophy of Religion* 18:91.

—— (1971). "Infinite Regress and the Cosmological Argument." *International Journal for Philosophy of Religion* 2:246–249.

Wolterstorff, N. (1975). "God Everlasting." In *God and the Good*. Edited by C. Orlebeke and L. Smedes. Grand Rapids, MI: William B. Eerdmans.

Woodbridge, III, R.G. (1969). "Acoustic Recordings from Antiquity." *Proceedings of the IEEE* 57 (August):1465–1466.

Woodhouse, M.B. (1976). "The Reversibility of Absolute Time." *Philosophical Studies* 29 (June):465–468.

Woodward, J.F. (1997). "Twists of Fate: Can We Make Traversable Wormholes in Spacetime?" *Foundations of Physics Letters* 10 (April):153–181.

—— (1995). "Making the Universe Safe for Historians: Time Travel and the Laws of Physics." *Foundations of Physics Letters* 8 (February):1–39.

Woolfe, V. (1976). "A Sketch of the Past." In *Moments of Being*. Edited by J. Schulkind. Sussex, England: The University Press.

Worth, P. (1950). "Typewriter from the Future." *Amazing Stories,* February.

—— (1949a). "Window to the Future." *Amazing Stories,* May.

—— (1949b). "I Died Tomorrow." *Fantastic Adventures,* May.

Wright, P.B. (1979). "Immediate Interstellar Communications." *Speculations in Science and Technology* 2:211–213.

Wright, S.F. (1930). *The World Below*. New York: Longmans, Green and Co.

Wright, S.P. (1931). "The Man from 2071." *Astounding Stories,* June.

Wyndham, J. (T3). "Consider Her Ways."

—— (TC). "Operation Peep."

—— (1961). "Random Quest," "Odd," and "Stitch in Time." In *The Infinite Moment*. New York: Ballantine.

—— (1956). "The Chronoclasm" and "Opposite Number" In *The Seeds of Time*. London: Michael Joseph.

—— (1939). "Judson's Annihilator." *Amazing Stories,* October.

—— (1933). "Wanderers of Time." *Wonder Stories,* March.

Xin, Y. (1992). "Does Einstein's G.R. Theory Predict Gravitational Radiation of PSR 1913 + 16?" *Astrophysics and Space Science* 194 (August):159–163.

Yarov, R. (1968). "The Founding of Civilization." In *Russian Science Fiction*. Edited by R. Magidoff. New York: New York University Press.

Yndurain, F.J. (1991). "Disappearance of Matter Due to Causality and Probability Violations in Theories with Extra Timelike Dimensions." *Physics Letters B* 256 (February 28):15–16.

York, J.W. (1972). "Role of Conformal Three-Geometry in the Dynamics of Gravitation." *Physical Review Letters* 28 (April 17):1082–1085.

Young, R.A. (1935). "A Thief in Time." *Wonder Stories,* July.

Young, R.F. (SFD). "When Time Was New."

—— (1962). "The Dandelion Girl." In *7th Annual of the Year's Best SF.* Edited by J. Merril. New York: Simon and Schuster.

—— (B51). "Not To Be Opened—."

Yourgrau, P. (1991). *The Disappearance of Time.* New York: Cambridge University Press.

Yurtsever, U. (1990a). "Does Quantum Field Theory Enforce the Averaged Weak Energy Condition?" *Classical and Quantum Gravity* 7 (November):L251–L258.

—— (1990b). "Test Fields on Compact Space-Times." *Journal of Mathematical Physics* 31 (December):3064–3078.

Zacks, R. (1955). "Have Your Past Read, Mister?" *Startling Stories,* Winter.

Zahn, T. (1988). "Time Bomb." In *New Destinies* (Summer). Edited by J. Baen. New York: Baen Books.

Zebrowski, G. (BT). "The Cliometricon."

Zeh, H.D. (1992). *The Physical Basis of the Direction of Time.* 2nd ed. Berlin: Springer-Verlag.

Zeilicovici, D. (1989). "Temporal Becoming Minus the Moving-Now." *Nous* 23 (September):505–524.

—— (1986). "A (Dis)Solution of McTaggart's Paradox." *Ratio* 28 (December):175–195.

Zelazny, R. (1979). *Roadmarks.* New York: Ballantine.

—— (TIT). "Divine Madness."

Zemach, E.M. (1979). "Time and Self." *Analysis* 39 (June):143–147.

Zetterberg, J.P. (1979). "Letting the Past Be Brought About." *Southern Journal of Philosophy* 17:413–421.

Ziadat, A.A. (1994). "Early Reception of Einstein's Relativity in the Arab Periodical Press." *Annals of Science* 51 (January):17–35.

Zimmerman, E.J. (1962). "The Macroscopic Nature of Space-Time." *American Journal of Physics* 30 (January):97–105.

Zollner, J.C.F. (1878). "On Space of Four Dimensions." *Quarterly Journal of Science,* n.s. 8 (April):227–237.

Zwart, P.J. (1972). "The Flow of Time." *Synthese* 24 (15:2):133–158.

BIBLIOGRAPHIC ADIEU

> By the middle of the 21st century the volumes of *Physical Review* will be filling library shelves at a rate exceeding the speed of light. This will not violate the theory of relativity because no information will be transmitted.
>
> —joke told by technical librarians
> frustrated at the information explosion in science and mathematics

I make no claims for an exhaustive bibliography. The stories, books, and papers listed are the ones I found particularly useful, but if I have overlooked an item you feel should have been included, I would very much appreciate hearing about it. Indeed, there is one story that is missing from the bibliography precisely because I cannot recall either its author or its title. I hope some reader will be able to help me—all I can remember is having once read it. It is a "short-short" of a time traveler in disguise at Golgotha for the Crucifixion.

He has a camera, but of course he must hide it beneath his robe to avoid attracting attention. All goes well as he stands beneath the Cross at the Place of Bones, until he notices odd, clicking noises coming from all those standing near him. It is then he realizes the entire crowd is nothing *but* time travelers, from all through the ages, all with hidden cameras! This is a story—not to be confused with Gary Kilworth's horrifying "Let's Go to Golgotha!" in which the death of Christ has been reduced to a tourist attraction for time travelers—with much deeper meaning than just its surprise ending, and I would dearly love to be informed of where it can be found.

INDEX

The fictional titles that appear in this index are, of course, only a small fraction of the entries in the bibliography. The fictional works that are listed here are ones that I judged to be 'known' to many people, or are ones that I want to particularly direct to your attention. The selection of what to include is, naturally, subjective, and perhaps I've overlooked one or more works that you think should be in the index. For such oversights I apologize and can only hope that you enjoy reading the entire book from start to finish while you search for 'your' title(s)!